AIR POLLUTION XVI

WITPRESS

WIT Press publishes leading books in Science and Technology.
Visit our website for the current list of titles.
www.witpress.com

WITeLibrary

Home of the Transactions of the Wessex Institute.
Papers presented at Air Pollution XVI are archived in the WIT eLibrary in volume 116 of
WIT Transactions on Ecology and the Environment (ISSN 1743-3541).
The WIT eLibrary provides the international scientific community with immediate and
permanent access to individual papers presented at WIT conferences.
http://library.witpress.com

SIXTEENTH INTERNATIONAL CONFERENCE ON MODELLING,
MONITORING AND MANAGEMENT OF AIR POLLUTION

AIR POLLUTION XVI

CONFERENCE CHAIRMEN

C. A. Brebbia
Wessex Institute of Technology, UK

J.W.S. Longhurst
University of the West of England, UK

INTERNATIONAL SCIENTIFIC ADVISORY COMMITTEE

J. Baldasano
J.G. Bartzis
C. Borrego
F. Costabile
J.V. de Assuncao
D.M. Elsom
I. Hunova
J.G. Irwin
G. Passerini
F. Patania
L. Pignato
V. Popov
R. San Jose
K. Sawicka-Kapusta
F. Schwegler
E. Tiezzi
C. Trozzi

ORGANISED BY
Wessex Institute of Technology, UK

SPONSORED BY
WIT Transactions on Ecology and the Environment

WIT Transactions

Transactions Editor

Carlos Brebbia
Wessex Institute of Technology
Ashurst Lodge, Ashurst
Southampton SO40 7AA, UK
Email: carlos@wessex.ac.uk

Editorial Board

B Abersek University of Maribor, Slovenia
Y N Abousleiman University of Oklahoma, USA
P L Aguilar University of Extremadura, Spain
K S Al Jabri Sultan Qaboos University, Oman
E Alarcon Universidad Politecnica de Madrid, Spain
A Aldama IMTA, Mexico
C Alessandri Universita di Ferrara, Italy
D Almorza Gomar University of Cadiz, Spain
B Alzahabi Kettering University, USA
J A C Ambrosio IDMEC, Portugal
A M Amer Cairo University, Egypt
S A Anagnostopoulos University of Patras, Greece
M Andretta Montecatini, Italy
E Angelino A.R.P.A. Lombardia, Italy
H Antes Technische Universitat Braunschweig, Germany
M A Atherton South Bank University, UK
A G Atkins University of Reading, UK
D Aubry Ecole Centrale de Paris, France
H Azegami Toyohashi University of Technology, Japan
A F M Azevedo University of Porto, Portugal
J Baish Bucknell University, USA
J M Baldasano Universitat Politecnica de Catalunya, Spain
J G Bartzis Institute of Nuclear Technology, Greece
A Bejan Duke University, USA

M P Bekakos Democritus University of Thrace, Greece
G Belingardi Politecnico di Torino, Italy
R Belmans Katholieke Universiteit Leuven, Belgium
C D Bertram The University of New South Wales, Australia
D E Beskos University of Patras, Greece
S K Bhattacharyya Indian Institute of Technology, India
E Blums Latvian Academy of Sciences, Latvia
J Boarder Cartref Consulting Systems, UK
B Bobee Institut National de la Recherche Scientifique, Canada
H Boileau ESIGEC, France
J J Bommer Imperial College London, UK
M Bonnet Ecole Polytechnique, France
C A Borrego University of Aveiro, Portugal
A R Bretones University of Granada, Spain
J A Bryant University of Exeter, UK
F-G Buchholz Universitat Gesanthochschule Paderborn, Germany
M B Bush The University of Western Australia, Australia
F Butera Politecnico di Milano, Italy
J Byrne University of Portsmouth, UK
W Cantwell Liverpool University, UK
D J Cartwright Bucknell University, USA
P G Carydis National Technical University of Athens, Greece
J J Casares Long Universidad de Santiago de Compostela, Spain,
M A Celia Princeton University, USA
A Chakrabarti Indian Institute of Science, India

S K Chakrabarti Offshore Structure Analysis, USA
A H-D Cheng University of Mississippi, USA
J Chilton University of Lincoln, UK
C-L Chiu University of Pittsburgh, USA
H Choi Kangnung National University, Korea
A Cieslak Technical University of Lodz, Poland
S Clement Transport System Centre, Australia
M W Collins Brunel University, UK
J J Connor Massachusetts Institute of Technology, USA
M C Constantinou State University of New York at Buffalo, USA
D E Cormack University of Toronto, Canada
M Costantino Royal Bank of Scotland, UK
D F Cutler Royal Botanic Gardens, UK
W Czyczula Krakow University of Technology, Poland
M da Conceicao Cunha University of Coimbra, Portugal
A Davies University of Hertfordshire, UK
M Davis Temple University, USA
A B de Almeida Instituto Superior Tecnico, Portugal
E R de Arantes e Oliveira Instituto Superior Tecnico, Portugal
L De Biase University of Milan, Italy
R de Borst Delft University of Technology, Netherlands
G De Mey University of Ghent, Belgium
A De Montis Universita di Cagliari, Italy
A De Naeyer Universiteit Ghent, Belgium
W P De Wilde Vrije Universiteit Brussel, Belgium
L Debnath University of Texas-Pan American, USA
N J Dedios Mimbela Universidad de Cordoba, Spain
G Degrande Katholieke Universiteit Leuven, Belgium
S del Giudice University of Udine, Italy
G Deplano Universita di Cagliari, Italy
I Doltsinis University of Stuttgart, Germany
M Domaszewski Universite de Technologie de Belfort-Montbeliard, France
J Dominguez University of Seville, Spain
K Dorow Pacific Northwest National Laboratory, USA
W Dover University College London, UK
C Dowlen South Bank University, UK
J P du Plessis University of Stellenbosch, South Africa
R Duffell University of Hertfordshire, UK
A Ebel University of Cologne, Germany
E E Edoutos Democritus University of Thrace, Greece
G K Egan Monash University, Australia
K M Elawadly Alexandria University, Egypt
K-H Elmer Universitat Hannover, Germany
D Elms University of Canterbury, New Zealand
M E M El-Sayed Kettering University, USA
D M Elsom Oxford Brookes University, UK
A El-Zafrany Cranfield University, UK
F Erdogan Lehigh University, USA
F P Escrig University of Seville, Spain
D J Evans Nottingham Trent University, UK
J W Everett Rowan University, USA
M Faghri University of Rhode Island, USA
R A Falconer Cardiff University, UK
M N Fardis University of Patras, Greece
P Fedelinski Silesian Technical University, Poland
H J S Fernando Arizona State University, USA
S Finger Carnegie Mellon University, USA
J I Frankel University of Tennessee, USA
D M Fraser University of Cape Town, South Africa
M J Fritzler University of Calgary, Canada
U Gabbert Otto-von-Guericke Universitat Magdeburg, Germany
G Gambolati Universita di Padova, Italy
C J Gantes National Technical University of Athens, Greece
L Gaul Universitat Stuttgart, Germany
A Genco University of Palermo, Italy
N Georgantzis Universitat Jaume I, Spain
G S Gipson Oklahoma State University, USA
P Giudici Universita di Pavia, Italy
F Gomez Universidad Politecnica de Valencia, Spain

R Gomez Martin University of Granada, Spain
D Goulias University of Maryland, USA
K G Goulias Pennsylvania State University, USA
F Grandori Politecnico di Milano, Italy
W E Grant Texas A & M University, USA
S Grilli University of Rhode Island, USA
R H J Grimshaw, Loughborough University, UK
D Gross Technische Hochschule Darmstadt, Germany
R Grundmann Technische Universitat Dresden, Germany
A Gualtierotti IDHEAP, Switzerland
R C Gupta National University of Singapore, Singapore
J M Hale University of Newcastle, UK
K Hameyer Katholieke Universiteit Leuven, Belgium
C Hanke Danish Technical University, Denmark
K Hayami National Institute of Informatics, Japan
Y Hayashi Nagoya University, Japan
L Haydock Newage International Limited, UK
A H Hendrickx Free University of Brussels, Belgium
C Herman John Hopkins University, USA
S Heslop University of Bristol, UK
I Hideaki Nagoya University, Japan
D A Hills University of Oxford, UK
W F Huebner Southwest Research Institute, USA
J A C Humphrey Bucknell University, USA
M Y Hussaini Florida State University, USA
W Hutchinson Edith Cowan University, Australia
T H Hyde University of Nottingham, UK
M Iguchi Science University of Tokyo, Japan
D B Ingham University of Leeds, UK
L Int Panis VITO Expertisecentrum IMS, Belgium
N Ishikawa National Defence Academy, Japan
J Jaafar UiTm, Malaysia
W Jager Technical University of Dresden, Germany

Y Jaluria Rutgers University, USA
C M Jefferson University of the West of England, UK
P R Johnston Griffith University, Australia
D R H Jones University of Cambridge, UK
N Jones University of Liverpool, UK
D Kaliampakos National Technical University of Athens, Greece
N Kamiya Nagoya University, Japan
D L Karabalis University of Patras, Greece
M Karlsson Linkoping University, Sweden
T Katayama Doshisha University, Japan
K L Katsifarakis Aristotle University of Thessaloniki, Greece
J T Katsikadelis National Technical University of Athens, Greece
E Kausel Massachusetts Institute of Technology, USA
H Kawashima The University of Tokyo, Japan
B A Kazimee Washington State University, USA
S Kim University of Wisconsin-Madison, USA
D Kirkland Nicholas Grimshaw & Partners Ltd, UK
E Kita Nagoya University, Japan
A S Kobayashi University of Washington, USA
T Kobayashi University of Tokyo, Japan
D Koga Saga University, Japan
A Konrad University of Toronto, Canada
S Kotake University of Tokyo, Japan
A N Kounadis National Technical University of Athens, Greece
W B Kratzig Ruhr Universitat Bochum, Germany
T Krauthammer Penn State University, USA
C-H Lai University of Greenwich, UK
M Langseth Norwegian University of Science and Technology, Norway
B S Larsen Technical University of Denmark, Denmark
F Lattarulo, Politecnico di Bari, Italy
A Lebedev Moscow State University, Russia
L J Leon University of Montreal, Canada
D Lewis Mississippi State University, USA
S Ighobashi University of California Irvine, USA

K-C Lin University of New Brunswick, Canada
A A Liolios Democritus University of Thrace, Greece
S Lomov Katholieke Universiteit Leuven, Belgium
J W S Longhurst University of the West of England, UK
G Loo The University of Auckland, New Zealand
J Lourenco Universidade do Minho, Portugal
J E Luco University of California at San Diego, USA
H Lui State Seismological Bureau Harbin, China
C J Lumsden University of Toronto, Canada
L Lundqvist Division of Transport and Location Analysis, Sweden
T Lyons Murdoch University, Australia
Y-W Mai University of Sydney, Australia
M Majowiecki University of Bologna, Italy
D Malerba Università degli Studi di Bari, Italy
G Manara University of Pisa, Italy
B N Mandal Indian Statistical Institute, India
Ü Mander University of Tartu, Estonia
H A Mang Technische Universitat Wien, Austria,
G D, Manolis, Aristotle University of Thessaloniki, Greece
W J Mansur COPPE/UFRJ, Brazil
N Marchettini University of Siena, Italy
J D M Marsh Griffith University, Australia
J F Martin-Duque Universidad Complutense, Spain
T Matsui Nagoya University, Japan
G Mattrisch DaimlerChrysler AG, Germany
F M Mazzolani University of Naples "Federico II", Italy
K McManis University of New Orleans, USA
A C Mendes Universidade de Beira Interior, Portugal,
R A Meric Research Institute for Basic Sciences, Turkey
J Mikielewicz Polish Academy of Sciences, Poland

N Milic-Frayling Microsoft Research Ltd, UK
R A W Mines University of Liverpool, UK
C A Mitchell University of Sydney, Australia
K Miura Kajima Corporation, Japan
A Miyamoto Yamaguchi University, Japan
T Miyoshi Kobe University, Japan
G Molinari University of Genoa, Italy
T B Moodie University of Alberta, Canada
D B Murray Trinity College Dublin, Ireland
G Nakhaeizadeh DaimlerChrysler AG, Germany
M B Neace Mercer University, USA
D Necsulescu University of Ottawa, Canada
F Neumann University of Vienna, Austria
S-I Nishida Saga University, Japan
H Nisitani Kyushu Sangyo University, Japan
B Notaros University of Massachusetts, USA
P O'Donoghue University College Dublin, Ireland
R O O'Neill Oak Ridge National Laboratory, USA
M Ohkusu Kyushu University, Japan
G Oliveto Universitá di Catania, Italy
R Olsen Camp Dresser & McKee Inc., USA
E Oñate Universitat Politecnica de Catalunya, Spain
K Onishi Ibaraki University, Japan
P H Oosthuizen Queens University, Canada
E L Ortiz Imperial College London, UK
E Outa Waseda University, Japan
A S Papageorgiou Rensselaer Polytechnic Institute, USA
J Park Seoul National University, Korea
G Passerini Universita delle Marche, Italy
B C Patten, University of Georgia, USA
G Pelosi University of Florence, Italy
G G Penelis, Aristotle University of Thessaloniki, Greece
W Perrie Bedford Institute of Oceanography, Canada
R Pietrabissa Politecnico di Milano, Italy
H Pina Instituto Superior Tecnico, Portugal
M F Platzer Naval Postgraduate School, USA
D Poljak University of Split, Croatia

V **Popov** Wessex Institute of Technology, UK
H **Power** University of Nottingham, UK
D **Prandle** Proudman Oceanographic Laboratory, UK
M **Predeleanu** University Paris VI, France
M R I **Purvis** University of Portsmouth, UK
I S **Putra** Institute of Technology Bandung, Indonesia
Y A **Pykh** Russian Academy of Sciences, Russia
F **Rachidi** EMC Group, Switzerland
M **Rahman** Dalhousie University, Canada
K R **Rajagopal** Texas A & M University, USA
T **Rang** Tallinn Technical University, Estonia
J **Rao** Case Western Reserve University, USA
A M **Reinhorn** State University of New York at Buffalo, USA
A D **Rey** McGill University, Canada
D N **Riahi** University of Illinois at Urbana-Champaign, USA
B **Ribas** Spanish National Centre for Environmental Health, Spain
K **Richter** Graz University of Technology, Austria
S **Rinaldi** Politecnico di Milano, Italy
F **Robuste** Universitat Politecnica de Catalunya, Spain
J **Roddick** Flinders University, Australia
A C **Rodrigues** Universidade Nova de Lisboa, Portugal
F **Rodrigues** Poly Institute of Porto, Portugal
C W **Roeder** University of Washington, USA
J M **Roesset** Texas A & M University, USA
W **Roetzel** Universitaet der Bundeswehr Hamburg, Germany
V **Roje** University of Split, Croatia
R **Rosset** Laboratoire d'Aerologie, France
J L **Rubio** Centro de Investigaciones sobre Desertificacion, Spain
T J **Rudolphi** Iowa State University, USA
S **Russenchuck** Magnet Group, Switzerland
H **Ryssel** Fraunhofer Institut Integrierte Schaltungen, Germany
S G **Saad** American University in Cairo, Egypt
M **Saiidi** University of Nevada-Reno, USA
R **San Jose** Technical University of Madrid, Spain
F J **Sanchez-Sesma** Instituto Mexicano del Petroleo, Mexico
B **Sarler** Nova Gorica Polytechnic, Slovenia
S A **Savidis** Technische Universitat Berlin, Germany
A **Savini** Universita de Pavia, Italy
G **Schmid** Ruhr-Universitat Bochum, Germany
R **Schmidt** RWTH Aachen, Germany
B **Scholtes** Universitaet of Kassel, Germany
W **Schreiber** University of Alabama, USA
A P S **Selvadurai** McGill University, Canada
J J **Sendra** University of Seville, Spain
J J **Sharp** Memorial University of Newfoundland, Canada
Q **Shen** Massachusetts Institute of Technology, USA
X **Shixiong** Fudan University, China
G C **Sih** Lehigh University, USA
L C **Simoes** University of Coimbra, Portugal
A C **Singhal** Arizona State University, USA
P **Skerget** University of Maribor, Slovenia
J **Sladek** Slovak Academy of Sciences, Slovakia
V **Sladek** Slovak Academy of Sciences, Slovakia
A C M **Sousa** University of New Brunswick, Canada
H **Sozer** Illinois Institute of Technology, USA
D B **Spalding** CHAM, UK
P D **Spanos** Rice University, USA
T **Speck** Albert-Ludwigs-Universitaet Freiburg, Germany
C C **Spyrakos** National Technical University of Athens, Greece
I V **Stangeeva** St Petersburg University, Russia
J **Stasiek** Technical University of Gdansk, Poland
G E **Swaters** University of Alberta, Canada
S **Syngellakis** University of Southampton, UK
J **Szmyd** University of Mining and Metallurgy, Poland
S T **Tadano** Hokkaido University, Japan

H Takemiya Okayama University, Japan
I Takewaki Kyoto University, Japan
C-L Tan Carleton University, Canada
M Tanaka Shinshu University, Japan
E Taniguchi Kyoto University, Japan
S Tanimura Aichi University of Technology, Japan
J L Tassoulas University of Texas at Austin, USA
M A P Taylor University of South Australia, Australia
A Terranova Politecnico di Milano, Italy
E Tiezzi University of Siena, Italy
A G Tijhuis Technische Universiteit Eindhoven, Netherlands
T Tirabassi Institute FISBAT-CNR, Italy
S Tkachenko Otto-von-Guericke-University, Germany
N Tosaka Nihon University, Japan
T Tran-Cong University of Southern Queensland, Australia
R Tremblay Ecole Polytechnique, Canada
I Tsukrov University of New Hampshire, USA
R Turra CINECA Interuniversity Computing Centre, Italy
S G Tushinski Moscow State University, Russia
J-L Uso Universitat Jaume I, Spain
E Van den Bulck Katholieke Universiteit Leuven, Belgium
D Van den Poel Ghent University, Belgium
R van der Heijden Radboud University, Netherlands
R van Duin Delft University of Technology, Netherlands

P Vas University of Aberdeen, UK
W S Venturini University of Sao Paulo, Brazil
R Verhoeven Ghent University, Belgium
A Viguri Universitat Jaume I, Spain
Y Villacampa Esteve Universidad de Alicante, Spain
F F V Vincent University of Bath, UK
S Walker Imperial College, UK
G Walters University of Exeter, UK
B Weiss University of Vienna, Austria
H Westphal University of Magdeburg, Germany
J R Whiteman Brunel University, UK
Z-Y Yan Peking University, China
S Yanniotis Agricultural University of Athens, Greece
A Yeh University of Hong Kong, China
J Yoon Old Dominion University, USA
K Yoshizato Hiroshima University, Japan
T X Yu Hong Kong University of Science & Technology, Hong Kong
M Zador Technical University of Budapest, Hungary
K Zakrzewski Politechnika Lodzka, Poland
M Zamir University of Western Ontario, Canada
R Zarnic University of Ljubljana, Slovenia
G Zharkova Institute of Theoretical and Applied Mechanics, Russia
N Zhong Maebashi Institute of Technology, Japan
H G Zimmermann Siemens AG, Germany

AIR POLLUTION XVI

Editors

C.A. Brebbia
Wessex Institute of Technology, UK

J.W.S. Longhurst
University of the West of England, UK

WITPRESS Southampton, Boston

Editors

C.A. Brebbia
Wessex Institute of Technology, UK

J.W.S. Longhurst
University of the West of England, UK

Published by

WIT Press
Ashurst Lodge, Ashurst, Southampton, SO40 7AA, UK
Tel: 44 (0) 238 029 3223; Fax: 44 (0) 238 029 2853
E-Mail: witpress@witpress.com
http://www.witpress.com

For USA, Canada and Mexico

WIT Press
25 Bridge Street, Billerica, MA 01821, USA
Tel: 978 667 5841; Fax: 978 667 7582
E-Mail: infousa@witpress.com
http://www.witpress.com

British Library Cataloguing-in-Publication Data

A Catalogue record for this book is available
from the British Library

ISBN: 978-1-84564-127-6
ISSN: (print) 1746-448X
ISSN: (on-line) 1734-3541

The texts of the papers in this volume were set individually by the authors or under their supervision. Only minor corrections to the text may have been carried out by the publisher.

No responsibility is assumed by the Publisher, the Editors and Authors for any injury and/ or damage to persons or property as a matter of products liability, negligence or otherwise, or from any use or operation of any methods, products, instructions or ideas contained in the material herein. The Publisher does not necessarily endorse the ideas held, or views expressed by the Editors or Authors of the material contained in its publications.

© WIT Press 2008

Printed in Great Britain by Athenaeum Press Ltd.

All rights reserved. No part of this publication may be reproduced, stored in a retrieval system, or transmitted in any form or by any means, electronic, mechanical, photocopying, recording, or otherwise, without the prior written permission of the Publisher.

Preface

This volume contains the reviewed papers accepted for the Sixteenth International Conference on Modelling, Monitoring and Management of Air Pollution held in Skiathos, Greece in September 2008. This successful international meeting builds upon the prestigious outcomes of the 15 preceding conferences beginning with Monterrey, Mexico in 1993 and most recently in the Algarve, Portugal in 2007. These meetings have attracted outstanding contributions from leading researchers from around the world with the presented papers permanently stored in the WIT eLibrary as Transactions of the Wessex Institute (see http://library.witpress.com/)

Air pollution remains one of the most challenging problems facing the international community; it is widespread and growing in importance and has clear and known impacts on health and the environment. The human need for transport, manufactured goods and services brings with it often unintended, but none the less real, impacts on the atmospheric environment at scales from the local to the global. Whilst there are good examples of regulatory successes in minimising such impacts the rate of development of the global economy bring new pressures and the willingness of governments to regulate air pollution is often balanced by concerns over the economic impact of regulation.

Science is the key to identifying the nature and scale of air pollution impacts and is essential in the formulation of polices and regulatory decision-making. Continuous improvements in our knowledge of the fundamental science of air pollution and its application are necessary if we are to properly predict, assess and mitigate the air pollution implications of changes to the interlinked local, regional, national and international economic systems. Science must also be able to provide the evidence of improvements to air quality that result from implementation of the mitigation measure or the control regulation. The ability to assess and mitigate using the precautionary principle is a challenge that science must grasp and position itself to convince decision makers that uncertainty does not mean inertia. The outcomes of such activities must also be translatable into a suitable format to assist policy makers in reaching sustainable decisions and to build public acceptance and understanding of the nature and scale of the air pollution problem.

This volume brings together contributions from scientist around the world to discuss their recent work on various aspects of the air pollution phenomena. The conference series provides opportunities to foster scientific exchange, develop new collaborations amongst scientists and between scientists and policy makers or regulators, and provides a means for identifying new areas for scientific investigations.

The Editors of this book wish to thank the authors for their contributions and to acknowledge the assistance of the eminent members of the International Scientific Advisory Committee with the organisation of the conference and particularly with reviewing of the submitted papers.

WIT proceedings are regularly recorded and reported in abstracting services and databases including Crossref, Cambridge Scientific Abstracts, Scopus, Compendex, GeoBase, INSPEC, Thompson Index to Scientific and Technical proceedings, Scitech Book News and the Directory of Published Proceedings.

The Editors
Skiathos, 2008

Contents

Section 1: Air pollution modelling

Elevated PM10 and PM2.5 concentrations in Europe: a model experiment with MM5-CMAQ and WRF-CHEM
R. San José, J. L. Pérez, J. L. Morant & R. M. González 3

Analytic solutions of the diffusion-deposition equation for fluids heavier than atmospheric air
F. P. Barrera, T. Brugarino, V. Piazza & L. Pignato 13

Air quality forecasting in a large city
P. Perez 21

Meshing effects on CFD modelling of atmospheric flow over buildings situated on ground with high terrain
A. Karim, P. Nolan & A. Qubian 29

Enhanced evaluation of a Lagrangian-particle air pollution model based on a Šaleška region field data set
B. Grašič, P. Mlakar, M. Z. Božnar & G. Tinarelli 39

Modelling the multi year air quality time series in Edinburgh: an application of the Hierarchical Profiling Approach
H. Al-Madfai, D. G. Snelson & A. J. Geens 49

Definition of PM_{10} emission factors from traffic: use of tracers and definition of background concentration
E. Brizio & G. Genon 57

LNG dispersion over the sea
A. Fatsis, J. Statharas, A. Panoutsopoulou & N. Vlachakis 67

Modelling of air pollutants released from highway traffic in Hungary
Gy. Baranka 77

Dust barriers in open pit blasts. Multiphase Computational Fluid
Dynamics (CFD) simulations
J. T. Alvarez, I. D. Alvarez & S. T. Lougedo 85

Contribution of oil palm isoprene emissions to tropospheric ozone
levels in the Distrito Metropolitano de Quito (Ecuador)
R. Parra 95

Some reflections on the modelling of biogenic emissions of
monoterpenes in the boreal zone
K. M. Granström 105

An evaluation of SOA modelling in the Madrid metropolitan area
*M. G. Vivanco, I. Palomino, J. Plaza, B. Artíñano, M. Pujadas
& B. Bessagnet* 115

Comparison between ozone monitoring data and modelling data,
in Italy, from the perspective of health indicator assessments
*A. De Marco, A. Screpanti, S. Racalbuto, T. Pignatelli, G. Vialetto,
F. Monforti & G. Zanini* 125

Section 2: Air quality management

Dealing with air pollution in Europe
C. Trozzi, R. Vaccaro & C. Leonardi 137

The development and operation of the United Kingdom's air quality
management regime
*J. W. S. Longhurst, J. G. Irwin, T. J. Chatterton, E. T. Hayes,
N. S. Leksmono & J. K. Symons* 149

Integrating local air quality and carbon management at a regional and
local governance level: a case study of south west England
*S. T. Baldwin, M. Everard, E. T. Hayes, J. W. S. Longhurst
& J. R. Merefield* 159

Failures and successes in the implementation of an air quality
management program in Mexicali, Baja California, Mexico
M. Quintero-Nuñez & E. C. Nieblas-Ortiz 169

Emission management system in the Russian Federation: necessity
for reforming and future adaptation of the western experience
A. Y. Nedre, R. A. Shatilov & A. F. Gubanov 179

Outdoor air quality data analysis of Al-Mansoriah residential area
(Kuwait): air quality indices results
S. A. Al-Haider & S. M. Al-Salem 189

A modelling tool for PM10 exposure assessment:
an application example
E. Angelino, M. P. Costa, E. Peroni & C. Sala 197

Air pollution and management in the Niger Delta –
emerging issues
M. A. Fagbeja, T. J. Chatterton, J. W. S. Longhurst, J. O. Akinyede
& J. O. Adegoke 207

Air quality monitoring and management for the industrialized
Highveld region of South Africa
G. V. Mkhatshwa 217

Real time air quality forecasting systems for industrial plants and
urban areas by using the MM5-CMAQ-EMIMO
R. San José, J. L. Pérez, J. L. Morant & R. M. González 225

Section 3: Emission studies

Air quality in the vicinity of a governmental school in Kuwait
E. Al-Bassam, V. Popov & A. Khan 237

Emissions of nitrogen dioxide from modern diesel vehicles
G. A. Bishop & D. H. Stedman 247

Errors in model predictions of NO_x traffic emissions at road level –
impacts of input data quality
R. Smit 255

Air pollution from traffic, ships and industry in an Italian port
G. Fava & M. Letizia Ruello 271

Fugitive dust from agricultural land affecting air quality within the
Columbia Plateau, USA
B. S. Sharratt 281

Emission inventory for urban transport in the rush hour:
application to Seville
J. Racero, M. Cristina Martín, I. Eguía & F. Guerrero 291

Modeling carbon emissions from urban land conversion:
gamma distribution model
A. Svirejeva-Hopkins & H.-J. Schellnhuber 301

Reduction of CO_2 emissions by carbonation of
alkaline wastewater
M. Uibu, O. Velts, A. Trikkel & R. Kuusik 311

Section 4: Monitoring and measuring

Characterisation of inhalable atmospheric aerosols
N. A. Kgabi, J. J. Pienaar & M. Kulmala 323

Data handling of complex GC-MS signals to characterize
homologous series as organic source tracers in
atmospheric aerosols
M. C. Pietrogrande, M. Mercuriali & D. Bacco 335

Monitoring of trace organic air pollutants –
a developing country perspective
P. B. C. Forbes & E. R. Rohwer 345

NIST gas standards containing volatile organic compounds in support
of ambient air pollution measurements
G. C. Rhoderick 357

A comparison of EPA and EN requirements for nitrogen oxide
chemiluminescence analyzers
J. Barberá, M. Doval, E. González, A. Miñana & F. J. Marzal 367

A procedure for correcting readings in chemiluminescence nitrogen
oxide analyzers due to the effect of sample pressure
M. Doval, J. Barberá, E. González & F. J. Marzal 375

New measures of wind angular dispersion in three dimensions
P. S. Farrugia & A. Micallef 385

Variation of air pollution with related meteorological factors in
Tripoli (case study)
T. A. Sharif, A. K. El-Henshir & M. M. Treban 397

Section 5: Urban air management

Prediction of air pollution levels using neural networks:
influence of spatial variability
*G. Ibarra-Berastegi, A. Elias, A. Barona, J. Sáenz, A. Ezcurra
& J. Diaz de Argandona*..409

Environmental planning and management of air quality:
the case of Mexicali, Baja California, Mexico
E. Corona-Zambrano & R. Rojas-Caldelas...419

High-resolution air quality modelling and time scale analysis of
ozone and NOx in Osaka, Japan
K. L. Shrestha, A. Kondo, A. Kaga & Y. Inoue..429

Practical problems associated with assessing the impact of outdoor
smoking on outdoor air quality: an Edinburgh study
D. G. Snelson, A. J. Geens, H. Al-Madfai & D. Hillier......................................439

Role of leaf- and rhizosphere-associated bacteria in reducing air
pollution of industrial cities in Saudi Arabia
M. A. Khiyami...447

Section 6: Indoor air pollution

Indoor concentrations of VOCs and ozone in two cities of
Northern Europe during the summer period
*J. G. Bartzis, S. Michaelidou, D. Missia, E. Tolis, D. Saraga,
E. Demetriou-Georgiou, D. Kotzias & J. M. Barero-Moreno*..........................459

Comparative study of indoor-outdoor exposure against volatile
organic compounds in South and Middle America
O. Herbarth, A. Müller, L. Massolo & H. Tovalin...467

Sampling of respirable particle PM_{10} in the library at the
Metropolitana University, Campus Azcapotzalco, Mexico City
Y. I. Falcon, E. Martinez, A. Cuenca, C. Herrera & E. A. Zavala....................475

Impacts of ventilation: studies on "environmental tobacco smoke"
A. J. Geens, H. Al-Madfai & D. G. Snelson..483

PCB contamination in indoor buildings
*S. J. Hellman, O. Lindroos, T. Palukka, E. Priha, T. Rantio
& T. Tuhkanen*...491

Evaluation of Indoor Air treatment by two pilot-scale biofilters packed with compost and compost-based material
M. Ondarts, C. Hort, V. Platel & S. Sochard 499

Section 7: Aerosols and particles

The role of PM_{10} in air quality and exposure in urban areas
C. Borrego, M. Lopes, J. Valente, O. Tchepel, A. I. Miranda
& J. Ferreira 511

Spatial distribution of ultrafine particles at urban scale: the road-to-ambient stage
F. Costabile, B. Zani & I. Allegrini 521

Electromagnetic and informational pollution as a co-challenge to air pollution
A. A. Berezin & V. V. Gridin 533

Genotoxic and oxidative damage related to $PM_{2.5}$ chemical fraction
Sa. Bonetta, V. Gianotti, D. Scozia, Si. Bonetta, E. Carraro,
F. Gosetti, M. Oddone & M. C. Gennaro 543

Preliminary results of aerosol optical thickness from MIVIS data
C. Bassani, R. M. Cavalli & S. Pignatti 551

Section 8: Air pollution effects on ecosystems

Response of lichens to heavy metal and SO_2 pollution in Poland – an overview
K. Sawicka-Kapusta, M. Zakrzewska, G. Bydłoń & A. Sowińska 561

Forest ecosystem development after heavy deposition loads – case study Dübener Heide
C. Fürst, M. Abiy & F. Makeschin 571

Impact of biogenic volatile organic compound emissions on ozone formation in the Kinki region, Japan
A. Kondo, B. Hai, K. L. Shrestha, A. Kaga & Y. Inoue 585

Section 9: Policy studies

Potential contribution of local air quality management to environmental justice in England
I. Gegisian, M. Grey, J. W. S. Longhurst & J. G. Irwin 597

Are environmental health officers and transport planners in English local authorities working together to achieve air quality objectives?
A. O. Olowoporoku, E. T. Hayes, N. S. Leksmono, J. W. S. Longhurst & G. P. Parkhurst 607

A fuzzy MCDM framework for the environmental pollution potential of industries focusing on air pollution
R. K. Lad, R. A. Christian & A. W. Deshpande 617

A Model Municipal By-Law for regulating wood burning appliances
A. Germain, F. Granger & A. Gosselin 627

Author Index 637

Section 1
Air pollution modelling

Elevated PM10 and PM2.5 concentrations in Europe: a model experiment with MM5-CMAQ and WRF-CHEM

R. San José[1], J. L. Pérez[1], J. L. Morant[1] & R. M. González[2]
[1]*Environmental Software and Modelling Group,*
Computer Science School, Technical University of Madrid (UPM),
Campus de Montegancedo, Madrid, Spain
[2]*Department of Meteorology and Geophysics, Faculty of Physics,*
Complutense University of Madrid; Ciudad Universitaria, Madrid, Spain

Abstract

We have applied the MM5-CMAQ model to simulate the high concentrations in PM10 and PM2.5 during a winter episode (2003) in Central Europe. The selected period is January 15 – April 6 2003. Values of daily mean concentrations up to 75 $\mu g m^{-3}$ are found on average of several monitoring stations in Northern Germany. This model evaluation shows that there is an increasing underestimation of primary and secondary species with increasing observed PM10. The high PM levels were observed under stagnant weather conditions that are difficult to simulate. The MM5 is the PSU/NCAR non-hydrostatic meteorological model and CMAQ is the chemical dispersion model developed by EPA (US) used in this simulation with CBM-V. The TNO emission inventory was used to simulate the PM10 and PM2.5 concentrations with the MM5-CMAQ model. The results show a substantial underestimation of the elevated values in February and March 2003. An increase on the PM2.5 emissions (five times) produces the expected results and the correlation coefficient increases slightly. The WRF/CHEM model results show an excellent performance with correct emission database. The main difference between MM5-CMAQ simulations and WRF/CHEM is the MOSAIC particle models and the "classical" MADE/SORGAM particle model used in WRF/CHEM and CMAQ respectively. MOSAIC seems to make a better job than MADE particle model for this particular episode.

Keywords: emissions, PM10 and PM2.5, air quality models, air particles.

1 Introduction

Simulations of elevated PM10 and PM2.5 concentrations have been always underestimated by modern three dimensional air quality modelling tools. This fact has focused much more attention between researchers during last years. Three dimensional air quality models have been developed during the last 15–20 years and substantial progress has occurred in this research area. These models are composed by a meteorological driver and a chemical and transport module. Examples of meteorological drivers are: MM5 (PSU.NCAR, USA) [5], RSM (NOAA, USA), ECMWF (Redding, U.K.), HIRLAM (Finnish Meteorological Institute, Finland), WRF [15] and examples of dispersion and chemical transport modules are EURAD (University of Cologne, Germany) [13], EUROS (RIVM, The Netherlands) [7], EMEP Eulerian (DNMI, Oslo, Norway), MATCH (SMHI, Norrkoping, Sweden) [2], REM3 (Free University of Berlin, Germany) [14], CHIMERE (ISPL, Paris, France) [12], NILU-CTM (NILU, Kjeller, Norway) [3], LOTOS (TNO, Apeldoorm, The Netherlands) [8], DEM (NERI, Roskilde, Denmark) [4], OPANA model [9–11] based on MEMO and MM5 mesoscale meteorological models and with the chemistry on-line solved by [6], STOCHEM (UK Met. Office, Bracknell, U.K.) [1] and CMAQ (Community Multiscale Air Quality modelling system) [16], developed by EPA (USA). In USA, CAMx Environ Inc., STEM-III (University of Iowa) and CMAQ model are the most up-to-date air quality dispersion chemical models. In this application we have used the CMAQ model (EPA, U.S.) which is one of the most complete models and includes aerosol, cloud and aerosol chemistry.

In this contribution we present results from two simulations by two different models. The first air quality modelling systems is MM5-CMAQ which is a matured modelling system based on the MM5 mesoscale non-hydrostatic meteorological model and the dispersion and chemical transport module, CMAQ. The second tool is the WRF/CHEM [15] air quality modelling system, which is an on-line (one code, one system) tool to simulate air concentrations based on the WRF meteorological driver. In WRF/CHEM the chemistry transport and transformations are embedded into WRF as part of the code so that the interactions between many meteorological and climate variables and the chemistry if at hand and can be investigated. WRF/CHEM is developed by NOAA/NCAR (US) [15]. The advantage of on-line models is based on the capability to analyze all variables simultaneously and to account for all interactions (or at least, as much as possible) with a full modular approach.

2 PM10 and PM2.5 episode

During the period January 15 2003 to April 5 2003 in central Europe (mainly northern part of Germany), we observe three high peaks on PM10 and PM2.5 values in several monitoring stations located in the area of North-East of Germany. The daily averages of PM10 concentrations were close to 80 μgm^{-3} and higher than 70 μgm^{-3} for PM2.5 concentrations. These values are about 4–5 times higher than those registered as "normal" values. The first peak on PM10 and PM2.5 concentrations was developed after Feb. 1 until Feb. 15. During this period of time, Central Europe was under the influence of a high-pressure system coming from Russia through Poland and Souther Scandinavia. In Northern part of Germany, we found southeasterly winds and stable conditions with low winds. These meteorological conditions brought daily PM10 concentrations at about 40 $\mu g m^{-3}$. The second peak was characterized by a sharp gradient on PM10 concentrations after Feb. 15 and until March 7. These episode reached daily PM10 concentrations up to 70 $\mu g m^{-3}$. The meteorological conditions on March 2 (peak values) was characterized by a wind rotation composed by Southwesterly winds from Poland over the North of Germany and Northwesterly and Western winds in the Central part of Germany. Finally a third peak with values of about 65 $\mu g m^{-3}$ on March 27 starts on March 20, ending on April 5 2003 was having a similar structure and causes to the second one.

3 Emission data

In both models, we have applied the TNO emissions [17] as area and point sources with a geographical resolution of 0.125° latitude by 0.25° longitude and covering all Europe. The emission totals by SNAP activity sectors and countries agree with the baseline scenario for the Clean Air For Europe (CAFE) program [18]. This database gives the PM10 and PM2.5 emission for the primary particle emissions. We also took from CAFE the PM splitting sub-groups, height distribution and the breakdown of the annual emissions into hourly emissions. The PM2.5 fraction of the particle emissions was split into an unspecified fraction, elemental carbon (EC) and primary organic carbon (OC). The EC fraction of the PM2.5 emissions for the different SNAP sectors were taken from [19]. For the OC fraction, the method proposed by [20] is applied as follows: an average OC/EC emission ratio of two was used for all sectors, i.e. the OC fraction were set as twice the EC fractions, except if the sum of the two fractions exceed the unity. In this case ($f_{EC} > 0.33$), f_{OC} was set as: $f_{OC} = 1 - f_{EC}$. With this prepared input, the WRF/CHEM and CMAQ took the information as it is. The hourly emissions are derived using sector-dependent, monthly, daily and hourly emission factors as used in the EURODELTA (http://aqm.jrc.it/eurodelta/) exercise.

4 Observational data

Eighteen PM10 stations were selected for the comparison with the model results. Seventeen stations represent the rural background and one station represents the urban background in Berlin. All stations are located in flat or moderate hill terrain. Most of the stations are operated by the respective Federal State agencies. At four stations (Neuglobsow, Zingst, Westerland and Deuselbach, which are EMEP background stations run by the German Environmental Protection Agency, Umweltbundesamt), the observed concentrations of particulate sulphate, total nitrate ($HNO_3+NO_3^-$) and total ammonia ($NH_3+NH_4^+$) were available. Deuselbach, in the southwest of Germany, is located outside of the high PM10 concentration region. In addition, at the research station Melpitz [21] the concentrations of the components of secondary inorganic aerosols SO_4^{--}, NO_3^-, NH_4^+, as well as the concentrations of EC, OC and NH_3 were available.

The SO_2 and NO_2 concentrations at these five stations were also taken into account in the model comparison. PM2.5 observations were available at four stations: Melpitz, Waldhof, Deuselbach and Hannover. All PM10 and PM2.5 observations are based on gravimetric measurements, and the concentrations of the inorganic species in aerosol particles on ion chromatography. The chemical composition data at Melpitz result from the PM2.5 fraction, whereas the composition data from the other stations were analyzed from the PM10 particle concentrations. OC data were corrected by a factor of 1.4 to account for the non-C atoms in the particulate organic matter (OM) concentrations, which are currently not measured [22].

5 MM5-CMAQ and WRF-CHEM architectures and configurations

MM5 was set up with two domains: a mother domain with 60x60 grid cells with 90 km spatial resolution and 23 vertical layers and 61x61 grid cells with 30 km spatial resolution with 23 vertical layers. The central point is set at 50.0 N and 10.0 E. The model is run with Lambert Conformal Conical projection. The CMAQ domain is slightly smaller following the CMAQ architecture rules. We use reanalysis T62 (209 km) datasets as 6-hour boundary conditions for MM5 with 28 vertical sigma levels and nudging with meteorological observations for the mother domain. We run MM5 with two-way nesting capability. We use the Kain-Fritsch 2 cumulus parameterization scheme, the MRF PBL scheme, Schultz microphysics scheme and Noah land-surface model. In CMAQ we use clean boundary profiles for initial conditions, Yamartino advection scheme, ACM2 for vertical diffusion, EBI solver and the aqueous/cloud chemistry with CB05 chemical scheme. Since our mother domain includes significant areas outside of Europe (North of Africa), we have used EDGAR emission inventory with EMIMO 2.0 emission model approach to fill those grid cells with hourly emission data. The VOC emissions are treated by SPECIATE Version 4.0 (EPA, USA) and for the lumping of the chemical species, we have used the [24] procedure for 16 different groups. We use our BIOEMI scheme for biogenic emission modeling. The classical, Atkin, Accumulation and Coarse modes are used (MADE/SORGAM modal approach).

In WRF/CHEM simulation we have used only one domain with 30 km spatial resolution similar to the MM5. We have used the Lin et al. (1983) scheme for the microphysics, Yamartino scheme for the boundary layer parameterization and [23] for the biogenic emissions. The MOSAIC sectional approach is used with 4 modes for particle modeling.

6 Model results

The comparison between daily average values (averaged over all monitoring stations) of PM10 concentrations and modeled values has been performed with several statistical tools such as: Calculated mean/Observed mean; Calculated STD/Observed STD; bias; squared correlation coefficient (R2); RMSE/Observed mean (Root Mean Squared Error); percentage within +/- 50% and number of data

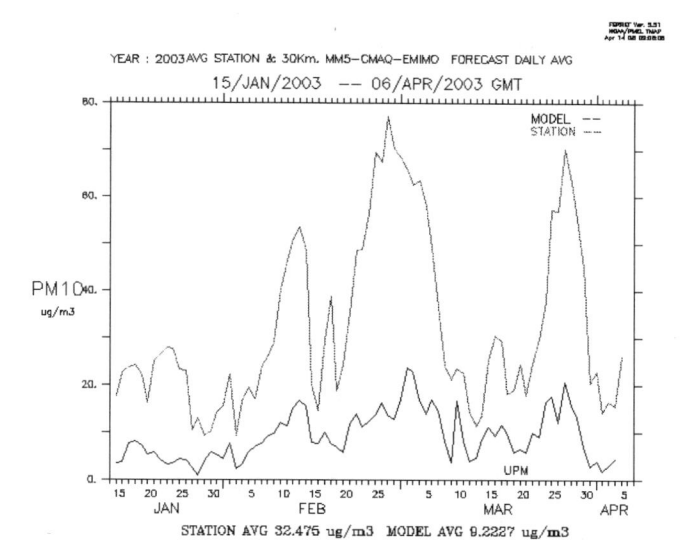

Figure 1: Comparison between daily average observed PM10 concentrations and model results produced by MM5-CMAQ. The model does not capture the magnitude of the PM10 peaks.

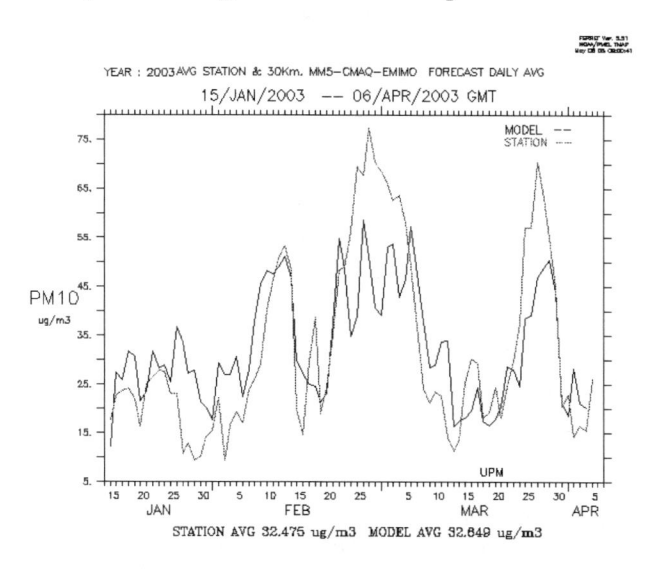

Figure 2: Comparison between daily average observed PM10 concentrations and model results produced by WRF/CHEM. The model captures quite well the magnitude of the PM10 peaks, particularly the first one.

sets. Figure 1 shows the comparison between PM10 observed averaged daily values and the modeled values by MM5-CMAQ. The results show that MM5-CMAQ underestimates about 4 times the observed peak values and particularly the

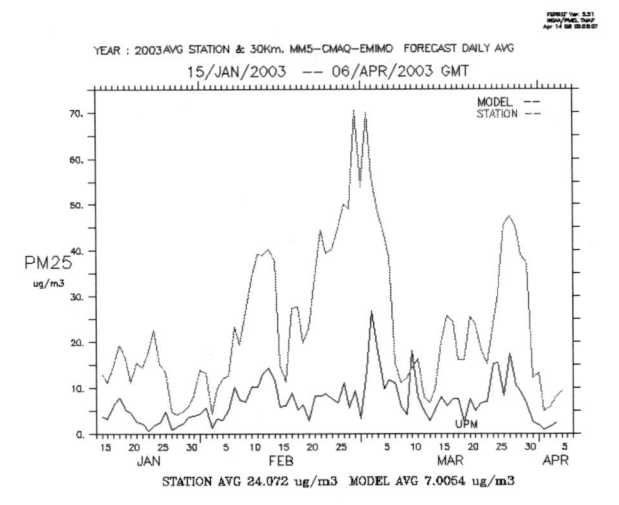

Figure 3: Comparison between daily average observed PM2.5 concentrations and model results produced by MM5-CMAQ. The model does not capture the magnitude of the PM2.5 peaks.

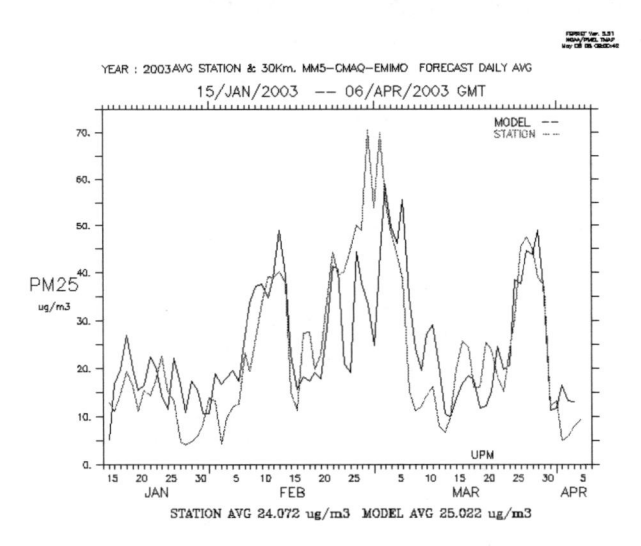

Figure 4: Comparison between daily average observed PM2.5 concentrations and model results produced by WRF/CHEM. The model captures quite well the magnitude of the PM10 peaks, particularly the last one.

highest one on March 2 2003. The R2 coefficient is 0.69. Figure 2 shows similar information but for the WRF/CHEM results. In this case WRF/CHEM captures quite well the magnitude of the peaks, particularly the first one. For the second and third peak, the model underestimates about 20% the peak values. The R2 coefficient is 0.61. In the case of PM2.5 Figures 3 and 4 show similar results to

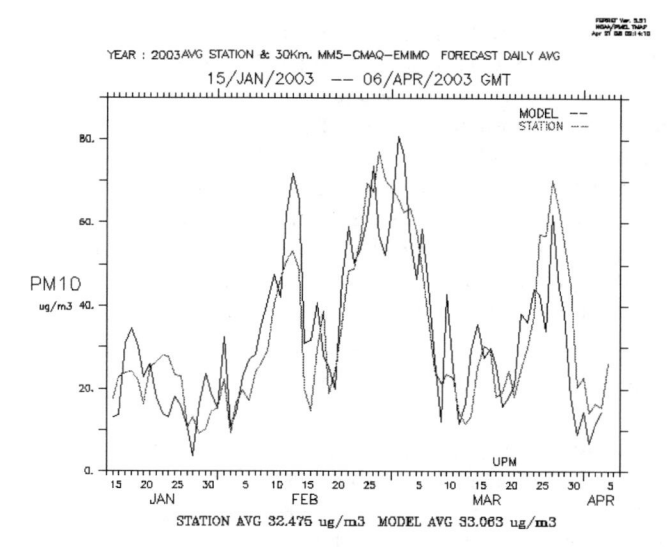

Figure 5: Comparison between daily average observed PM10 concentrations and model results produced by MM5-CMAQ with PM2.5 emissions multiplied by 5. The model captures quite well the magnitude of the PM10 peaks, particularly the second one.

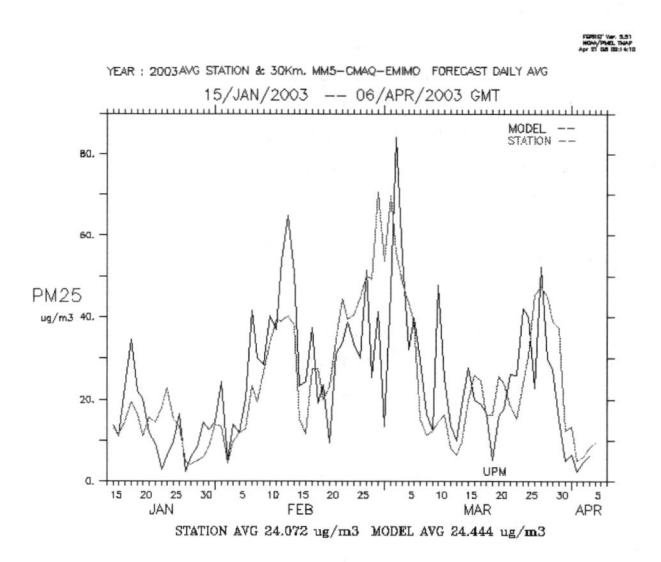

Figure 6: Comparison between daily average observed PM2.5 concentrations and model results produced by MM5-CMAQ with PM2.5 emissions multiplied by 5. The model captures quite well the magnitude of the PM10 peaks, particularly the third one.

figures 1 and 2. The R2 coefficients are 0.41 and 0.58. The squared correlation coefficient goes from 0.69 to 0.61 in the case of PM10 but increases substantially In the case of PM2.5, from 0.41 to 0.58. In WRF/CHEM both R2 coefficients (for PM10 and PM2.5) are quite close (0.58 and 0.61) but in the case of MM5-CMAQ, PM2.5 R2 coefficient is substantially lower than in the case of PM10.

We performed another full experiment with MM5-CMAQ. We multiply by 5 the PM2.5 emissions provided by TNO in the whole domain. The results are shown in Figures 5 and 6. The results are surprisingly good for both species. The R2 coefficient is 0.70 and 0.48 for PM10 and PM2.5 respectively. In both cases the correlation is improved and particularly for PM2.5 although just slightly. It is difficult to explain these results but it is a fact.

7 Conclusions

We have implemented and run two different models (MM5-CMAQ and WRF-CHEM) for the same episode over Northern part of Germany during the winter period of 2003 (Jan. 15-Apr. 5, 2003). WRF-CHEM made a better job than MM5-CMAQ, not only the patterns reproduce the peak values quite well but also the statistical parameters are good. The calculated mean values divided by thye observed mean value os exactly 1.0 for PM10 and WRF/CHEM on-line model. For the MM5-CMAQ this ratio is 0.28 and when we multiply the PM2.5 emissions by 5, the ratio is 1.02 which is also excellent. The bias values for WRF/CHEM, MM5-CMAQ and MM5-CMAQ (x5) are 0.09, -23.33 and 0.51 which are excellent values for WRF/CHEM and MM5-CMAQ (x5). No realistic explanation is found for the exercise related to multiply by 5 the PM2.5 emissions from TNO emission inventory. The main apparent reason why WRF/CHEM is doing much better job than normal MM5-CMAQ is the use of MOSAIC particle model based on sectional modal approach instead the "classical" approach based on MADE/SORGAM modal approach.

Acknowledgements

We would like to thank Dr. Peter Builtjes for the initial guidance and suggestion for this experiment and also COST 728 project (EU) where the inter-comparison experiment was proposed.

References

[1] Collins W.J., D.S. Stevenson, C.E. Johnson and R.G. Derwent, Tropospheric ozone in a global scale 3D Lagrangian model and its response to NOx emission controls, *J. Atmos. Chem.* **86** (1997), 223–274.
[2] Derwent R., and M. Jenkin, Hydrocarbons and the long-range transport of ozone and PAN across Europe, *Atmospheric Environment* **8** (1991), 1661–1678.

[3] Gardner R.K., K. Adams, T. Cook, F. Deidewig, S. Ernedal, R. Falk, E. Fleuti, E. Herms, C. Johnson, M. Lecht, D. Lee, M. Leech, D. Lister, B. Masse, M. Metcalfe, P. Newton, A. Schmidt, C Vandenberg. and R. van Drimmelen, The ANCAT/EC global inventory of NOx emissions from aircraft, *Atmospheric Environment* **31** (1997), 1751–1766.
[4] Gery M.W., G.Z. Whitten, J.P. Killus and M.C. Dodge, A photochemical kinetics mechanism for urban and regional scale computer modelling, *Journal of Geophysical Research* **94** (1989), D10, 12925–12956.
[5] Grell, G.A., J. Dudhia and D.R. Stauffer, A description of the Fifth-Generation Penn State/NCAR Mesoscale Model (MM5), NCAR/TN- 398+ STR. *NCAR Technical Note*, 1994.
[6] Jacobson M.Z. and R.P. Turco, SMVGEAR: A sparse-matrix, vectorized GEAR code for atmospheric models, *Atmospheric Environment* **28**(1994), 2, 273–284.
[7] Langner J., R. Bergstrom and K. Pleijel, European scale modeling of sulfur, oxidized nitrogen and photochemical oxidants. Model development and evaluation for the 1994 growing season, *SMHI report RMK No. 82*, Swedish Met. And Hydrol. Inst., SE-601 76 Norrkoping, Sweden, (1998).
[8] Roemer M., G. Boersen, P. Builtjes and P. Esser, *The Budget of Ozone and Precursors over Europe Calculated with the LOTOS Model*. TNO publication P96/004, Apeldoorn, The Netherlands, 1996.
[9] San José R., L. Rodriguez, J. Moreno, M. Palacios, M.A. Sanz and M. Delgado, Eulerian and photochemical modelling over Madrid area in a mesoscale context, *Air Pollution II, Vol I., Computer Simulation, Computational Mechanics Publications, Ed. Baldasano, Brebbia, Power and Zannetti.*, 1994, 209–217.
[10] San José R., J. Cortés, J. Moreno, J.F. Prieto and R.M. González, Ozone modelling over a large city by using a mesoscale Eulerian model: Madrid case study, *Development and Application of Computer Techniques to Environmental Studies, Computational Mechanics Publications, Ed. Zannetti and Brebbia*, 1996, 309–319.
[11] San José, R., J.F. Prieto, N. Castellanos and J.M. Arranz, Sensitivity study of dry deposition fluxes in ANA air quality model over Madrid mesoscale area, *Measurements and Modelling in Environmental Pollution, Ed. San José and Brebbia*, 1997, 119–130.
[12] Schmidt H., C. Derognat, R. Vautard and M. Beekmann, A comparison of simulated and observed ozone mixing ratios for the summer 1998 in Western Europe, *Atmospheric Environment* **35** (2001), 6277–6297.
[13] Stockwell W., F. Kirchner, M. Kuhn and S. Seefeld, A new mechanism for regional atmospheric chemistry modeling, *J. Geophys. Res.* **102** (1977), 25847–25879.
[14] Walcek C., Minor flux adjustment near mixing ration extremes for simplified yet highly accurate monotonic calculation of tracer advection, *J. Geophys. Res.* **105** (2000), 9335–9348.

[15] Janjic, Z. I., J. P. Gerrity, Jr. and S. Nickovic, 2001: An Alternative Approach to Nonhydrostatic Modeling. *Monthly Weather Review*, Vol. 129, 1164–1178
[16] Byun, D.W., J. Young, G. Gipson, J. Godowitch, F. Binkowsky, S. Roselle, B. Benjey, J. Pleim, J.K.S. Ching, J. Novak, C. Coats, T. Odman, A. Hanna, K. Alapaty, R. Mathur, J. McHenry, U. Shankar, S. Fine, A. Xiu, and C. Lang. 1998. *Description of the Models-3 Community Multiscale Air Quality (CMAQ) model.* Proceedings of the American Meteorological Society 78th Annual Meeting Phoenix, AZ, Jan. 11-16, 264–268.
[17] Visscherdijk, A. and H. Denier van der Gon, 2005. Gridded European anthropogenic emission data for NOx, SO2, NMVOC, NH3, CO, PM10, PM2.5 and CH4 for the year 2000. TNO-report B&O-AR, 2005/106.
[18] Amann, M., Bertok, I., Cofala, J., Gyarfas, F., Heyes, C., Klimon, Z., 2005. Baseline Scenarios for the Clean Air for Europe (CAFE) Programme. Final Report, International Institute for Applied Systems Analysis, Schlossplatz 1, A-2361 Laxenburg, Austria.
[19] Schaap, M., H. Denier van der Gon, A. Visschedijk, M. van Loon, H. ten Brink, F. Dentener, J. Putaud, B. Guillaume, C. Liousse, P. Builtjes, 2004a. Anthropogenic Black Carbon and Fine Aerosol Distribution over Europe, *J. Geophys. Res.*, **109**, D18207, doi:10.1029/2003JD004330.
[20] Beekmann, M., Kerschbaumer, A., Reimer, E., Stern, R., Möller, D., 2007. PM Measurement Campaign HOVERT in the Greater Berlin area: model evaluation with chemically specified observations for a one year period. *Atmos. Chem. Phys.* **7**, 55–68.
[21] Spindler, G., K. Mueller, E. Brueggemann, T. Gnauk, H. Herrmann, 2004. Long-term size-segregated characterization of PM10, PM2.5, and PM1 at the IfT research station Melpitz downwind of Leipzig (Germany) using high and low-volume filter samplers. *Atmospheric Environment* **38**, 5333–5347.
[22] Putaud, J., F. Raesa, R. Van Dingenen, E. Bruggemann, M. Facchini, S. Decesari, S. Fuzzi, R. Gehrig, C. Hueglin, P. Laj, G. Lorbeer, W. Maenhaut, N. Mihalopoulos, K. Mueller, X. Querol, S. Rodriguez, J. Schneider, G. Spindler, H. ten Brink, K. Torseth, A. Wiedensohler, 2004. A European aerosol phenomenology – 2: chemical characteristics of particulate matter at kerbside, urban, rural and background sites in Europe. *Atmospheric Environment* **38**, 2579–2595.
[23] Guenther et al., 1995 A. Guenther, C.N. Hewitt, D. Erickson, R. Fall, C. Geron, T. Graedel, P. Harley, L. Klinger, M. Lerdau, W.A. McKay, T. Pierce, B. Scholes, R. Steinbrecher, R. Tallamraju, J. Taylor and P. Zimmerman, A global model of natural volatile organic compound emissions, *Journal of Geophysical Research* **100** (1995), pp. 8873–8892.
[24] Carter, W. P. L. (2007): "Development of the SAPRC-07 Chemical Mechanism and Updated Ozone Reactivity Scales," Final report to the California Air Resources Board Contract No. 03-318. August. Available at http://www.cert.ucr.edu/~carter/SAPRC.

Analytic solutions of the diffusion-deposition equation for fluids heavier than atmospheric air

F. P. Barrera[1], T. Brugarino[2], V. Piazza[3] & L. Pignato[3]
[1]*Dip. di Ingegneria dei Trasporti,*
Università di Palermo Facoltà d'Ingegneria, 90128 Palermo, Italy
[2]*Dip. di Metodi e Modelli Matematici,*
Università di Palermo Facoltà d'Ingegneria, 90128 Palermo, Italy
[3]*Dip. di Ricerche Energetiche ed Ambientali*
Università di Palermo Facoltà d'Ingegneria, 90128 Palermo, Italy

Abstract

A steady-state bi-dimensional turbulent diffusion equation was studied to find the concentration distribution of a pollutant near the ground. We have considered the air pollutant emitted from an elevated point source in the lower atmosphere in adiabatic conditions. The wind velocity and diffusion coefficient are given by power laws. We have found analytical solutions using or the Lie Group Analysis or the Method of Separation of Variables. The classical diffusion equation has been modified introducing the falling term with non-zero deposition velocity.

Analytical solutions are essential to test numerical models for the great difficulty in validating with experiments.

Keywords: atmospheric pollution, diffusion equation, exact solutions.

1 Introduction

The classical form of the mean steady diffusion equation is valid for elementary particles of the fluid or when the foreign particles are of the same density as the fluid. If the density and dimensions are high enough to have terminal velocities v_s not negligible, the distribution of the particles will be affected in various ways [1–4].

A simple approximation is to consider that the particle sinks at a rate v_s and the ground acts as a permeable surface and retains all material passing through it.

Using a very simple model it is possible to examine various cases. It is possible to have exact solutions of the mean steady diffusion when the turbulent diffusivity k_z and the terminal velocity v_d depend somehow on the height [5–7].

2 Mathematical model

The sedimentation of the material may be allowed by introducing a convection term in the mean steady equation that becomes [8, 9]:

$$u(z)\frac{\partial c}{\partial x} = \frac{\partial}{\partial z}\left(k_z(z)\frac{\partial c}{\partial z} + v_s(z)c\right) \quad (1)$$

where $v_s(z)$ is the deposition velocity.

We assume that mean wind velocity $u(z)$, the eddy diffusivity in z-direction $k_z(z)$ and the deposition velocity $v_s(z)$ are:

$$u(z) = u_0 z^\alpha \quad (2)$$

$$k_z(z) = k_0 z^n \quad (3)$$

$$v_s(z) = v_0 z^q \quad (4)$$

3 Group analysis of the equation

Group analysis of the (1) is performed through the one-parameter Lie group of transformations:

$$\begin{cases} x^* = x + \epsilon X(x, z, c) + O(\epsilon^2) \\ z^* = z + \epsilon Z(x, z, c) + O(\epsilon^2) \\ c^* = c + \epsilon C(x, z, c) + O(\epsilon^2) \end{cases} \quad (5)$$

where X, Z, C are the infinitesimal generators of the transformations [10–12].

Equation (1) is invariant respect to the group (5) of transformations if c^* is the solution of eq. (1) in the star variables. In this case, the number of independent variables can be decreased.

A considerable difficulty lies in the amount of the auxiliary calculations involved. We performed the calculations of the generators of the transformations group on a P.C. using the *MATHEMATICA* package.

Since eq. (1) is linear, the infinitesimal generators of the group of invariance are of the form:

$$\begin{cases} X = X(x) \\ Z = Z(x, z) \\ C = A(x, z)c + B(x, z) \end{cases} \quad (6)$$

The function $B(x, z)$ must satisfy eq. (1) and, without compromising with the generality, can be assumed equal to zero.

If we normalize the parameters $u_0/k_0 \to v_0$ and the variable $k_0 x/v_0 \to x$, we have that X, Z and A must satisfy the following equations:

$$n z^{-1} Z - z^{-1} \alpha Z + X' - 2 Z_z = 0 \quad (7)$$

$$qv_0 z^{-2-n+q} Z + nqv_0 z^{-2-n+q} Z - q^2 v_0 z^{-2-n+q} Z - nz^{-1} A_z$$
$$- v_0 z^{-n+q} A_z - 2qv_0 z^{-1-n+q} Z_z - A_{zz} + z^{-n+\alpha} A_x = 0 \quad (8)$$

$$nz^{-2} Z + nv_0 z^{-1-n+q} Z - qv_0 z^{-1-n+q} Z - 2A_z - nz^{-1} Z_z$$
$$- v_0 z^{-n+q} Z_z + Z_{zz} - z^{-n+\alpha} Z_x = 0 \quad (9)$$

We show now some results.

4 Similarity solutions

Let us look at some similarity solutions.

4.1 α, n and q arbitrary ($n - \alpha \neq 2$)

In this case it possible to obtain from the eq.s (7–9) the generators of group of similarity:

$$\begin{cases} X = a_0 \\ Z = 0 \\ C = c_1 c \end{cases} \quad (10)$$

where a_0, and c_1 are arbitrary constants.
The characteristic equations are:

$$\frac{dx}{x_1} = \frac{dz}{0} = \frac{dc}{c_1 c}$$

The invariants are z and $ce^{-\frac{c_1}{x_1} x}$. If we assume $\frac{c_1}{x_1} = -\lambda^2$, the similarity solution, corresponding to the separation of variables, is

$$c = e^{-\lambda^2 x} Z(z)$$

where $Z(z)$ is solution of the following ordinary differential equation:

$$(qv_0 z^{1-n+q} + z^{2-n+\alpha} \lambda^2) Z(z) + z(n + v_0 z^{1-n+q}) Z'(z) + z^2 Z''(z) = 0$$

If $n = 2q - \alpha$, the solution is [13]

$$Z(z) = e^{\frac{z^{1-q+\alpha}(v_0 + \sqrt{v_0^2 - 4\lambda^2})}{-2+2q-2\alpha}}$$

$$\times \left[h_1 \Psi\left(\frac{2 - 3q + 2\alpha - \frac{qv_0}{\sqrt{v_0^2 - 4\lambda^2}}}{2(1 - q + \alpha)}, \frac{2 - 3q + 2\alpha}{1 - q + \alpha}; \frac{z^{1-q+\alpha} \sqrt{v_0^2 - 4\lambda^2}}{1 - q + \alpha} \right) \right.$$
$$\left. + h_2 L\left(\frac{2 - 3q + 2\alpha - \frac{qv_0}{\sqrt{v_0^2 - 4\lambda^2}}}{-2(1 - q + \alpha)}, \frac{1 - 2q + \alpha}{1 - q + \alpha}; \frac{z^{1-q+\alpha} \sqrt{v_0^2 - 4\lambda^2}}{1 - q + \alpha} \right) \right]$$

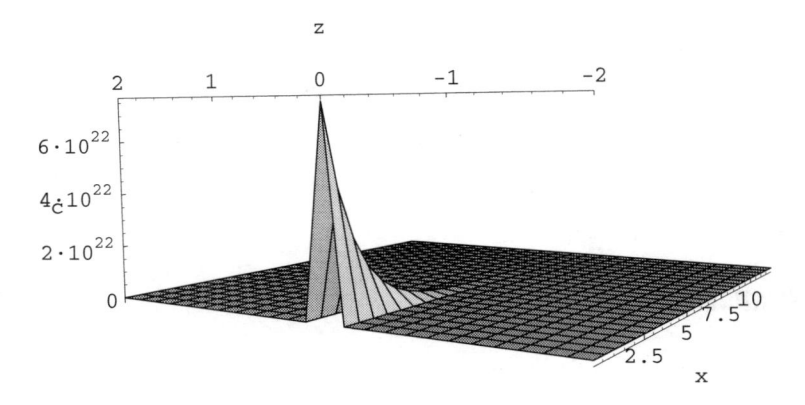

Figure 1: The $c(x, z)$, $v_0 = 20$, $q = 0.5$, $\lambda = 1$, $\alpha = 1.5$, $h_1 = 1$, $h_2 = 0$.

where h_1 and h_2 are arbitrary constants, $\Psi(-, -; \cdot)$ is the confluent hypergeometric function and $L(-, -; \cdot)$ is the generalized Laguerre polynomial. The concentration is

$$c = e^{-\lambda^2 x} e^{-\frac{(v_0 + \sqrt{v_0^2 - 4\lambda^2})}{1-q+\alpha} \frac{z^{1-q+\alpha}}{2}}$$

$$\times \left[h_1 \Psi\left(\frac{2 - 3q + 2\alpha - \frac{qv_0}{\sqrt{v_0^2 - 4\lambda^2}}}{2(1-q+\alpha)}, \frac{2 - 3q + 2\alpha}{1 - q + \alpha}; \frac{z^{1-q+\alpha}\sqrt{v_0^2 - 4\lambda^2}}{1 - q + \alpha} \right) \right.$$

$$\left. + h_2 L\left(\frac{2 - 3q + 2\alpha - \frac{qv_0}{\sqrt{v_0^2 - 4\lambda^2}}}{-2(1-q+\alpha)}, \frac{1 - 2q + \alpha}{1 - q + \alpha}; \frac{z^{1-q+\alpha}\sqrt{v_0^2 - 4\lambda^2}}{1 - q + \alpha} \right) \right]$$

4.2 $n = 1$, $q = 0$ and α arbitrary

We observe that in this case $k_0 = ku_*$ where k is the Von Karman constant and u_* is the friction velocity.

The generators of group of similarity are:

$$\begin{cases} X = a_0 + a_1 x + \frac{a_2}{2} x^2 \\ Z = \frac{a_1 + a_2 x}{1 + \alpha} z + c_0 z^{\frac{1-\alpha}{2}} \\ C = (b_0 + b_1 z^{1+\alpha} + b_1 x (1+\alpha)(1 + v_0 + \alpha)) c \end{cases} \quad (11)$$

where a_0, a_1, a_2, b_0, b_1 and c_0 are constants satisfying the conditions

$$c_0(1 + 2v_0 + \alpha) = 0, \quad a_2 + 2b_1(1+\alpha)^2 = 0$$

Now we consider the following subcases.

4.2.1 $c_0 = 0$, $a_2 = -2b_1(1+\alpha)^2$ and $\alpha \neq -1$

The characteristic equations are:

$$\frac{dx}{x} = (1+\alpha)\frac{dz}{z} = \frac{dc}{c}$$

The invariants are $\frac{c}{x}$ and $\xi = zx^{-\frac{1}{1+\alpha}}$; the concentration is

$$c = xf(\xi)$$

where $f(\xi)$ is solution of the following ordinary differential equation:

$$\xi^\alpha f(\xi) - \left(1 + v_0 + \frac{\xi^{1+\alpha}}{1+\alpha}\right)f'(\xi) - \xi f''(\xi) = 0$$

In this case we have:

$$f(\xi) = e^{-\frac{z^{1+\alpha}}{x(1+\alpha)^2}}\left[h_1\Psi\left(\frac{2+v_0+2\alpha}{1+\alpha}, 1+\frac{v_0}{1+\alpha}; \frac{z^{1+\alpha}}{x(1+\alpha)^2}\right)\right.$$
$$\left. + h_2 L\left(-\frac{2+v_0+2\alpha}{1+\alpha}, \frac{v_0}{1+\alpha}; \frac{z^{1+\alpha}}{x(1+\alpha)^2}\right)\right]$$

where h_1 and h_2 are arbitrary constants. The concentration is

$$c = xe^{-\frac{z^{1+\alpha}}{x(1+\alpha)^2}}\left[h_1\Psi\left(\frac{2+v_0+2\alpha}{1+\alpha}, 1+\frac{v_0}{1+\alpha}; \frac{z^{1+\alpha}}{x(1+\alpha)^2}\right)\right.$$
$$\left. + h_2 L\left(-\frac{2+v_0+2\alpha}{1+\alpha}, \frac{v_0}{1+\alpha}; \frac{z^{1+\alpha}}{x(1+\alpha)^2}\right)\right]$$

4.2.2 $(1 + 2v_0 + \alpha) = 0$, $a_0 = -1$, $b_1 = 0$ and $\alpha \neq -1$

The characteristic equations are:

$$\frac{dx}{-1} = \frac{dz}{z^{\frac{1-\alpha}{2}}} = \frac{dc}{c}$$

The invariants are ce^x and $\xi = x + \frac{2}{1+\alpha}z^{\frac{\alpha+1}{2}}$; the concentration is

$$c = e^{-x}f(\xi)$$

where $f(\xi)$ is solution of the following ordinary differential equation:

$$f(\xi) - f'(\xi) + f''(\xi) = 0$$

In this case the concentration is:

$$f(\xi) = e^{-\frac{x}{2}+\frac{z^{\frac{1+\alpha}{2}}}{1+\alpha}}\left[h_1\cos\sqrt{3}\left(\frac{x}{2}+\frac{z^{\frac{1+\alpha}{2}}}{1+\alpha}\right) + h_2\sin\sqrt{3}\left(\frac{x}{2}+\frac{z^{\frac{1+\alpha}{2}}}{1+\alpha}\right)\right]$$

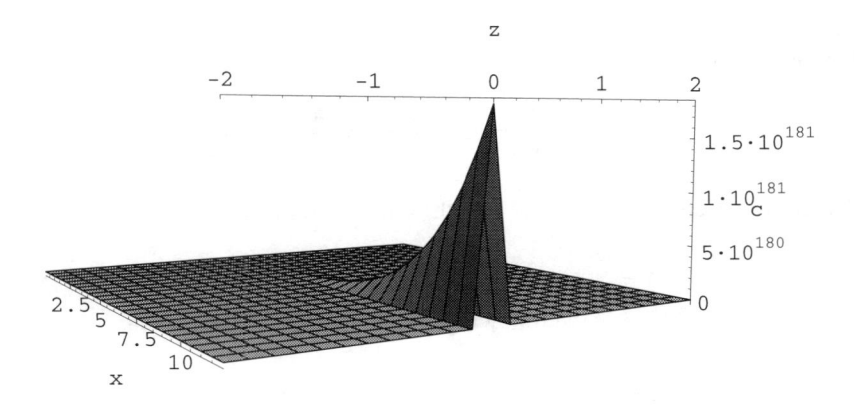

Figure 2: The $c(x, z)$, $v_0 = 11$, $\alpha = 1$, $h_1 = 1$, $h_2 = 0$.

4.3 $n = 2$, $q = 1$ and $\alpha = 0$

The generators of group of similarity are:

$$\begin{cases} X = a_0 + a_1 x + \dfrac{a_2}{2} x^2 \\ Z = b_0 z + \dfrac{1}{2}(a_1 + a_2 x) z \log z \\ C = \Big(c_2 - \dfrac{1}{8} x (2 a_1 (v_0 - 1)^2 + a_2 (2 + (v_0 - 1)^2 x)) \\ \quad - \dfrac{1}{8} \log z (2(v_0 + 1)(a_1 + a_2 x) + a_2 \log z) \Big) c \end{cases} \quad (12)$$

where a_0, a_1, a_2, b_0, and c_2 are arbitrary constants.

If we put: $a_0 = 0$, $a_1 = 1$, $a_2 = 0$, $b_0 = 0$, $c_2 = 0$, we have

$$\begin{cases} X = x \\ Z = \dfrac{1}{2} z \log z \\ C = -\dfrac{1}{4}((v_0 - 1)^2 x + (1 + v_0) \log z) c \end{cases} \quad (13)$$

The invariants are $c e^{\frac{1}{4}(v_0-1)^2 x} z^{\frac{1}{2}(1+v_0)}$ and $\xi = \dfrac{\log z}{\sqrt{x}}$; the concentration is

$$c = e^{-\frac{1}{4}(v_0-1)^2 x} z^{-\frac{1}{2}(1+v_0)} f(\xi)$$

where $f(\xi)$ is solution of the following ordinary differential equation:

$$f''(\xi) + \dfrac{\xi}{2} f'(\xi) = 0$$

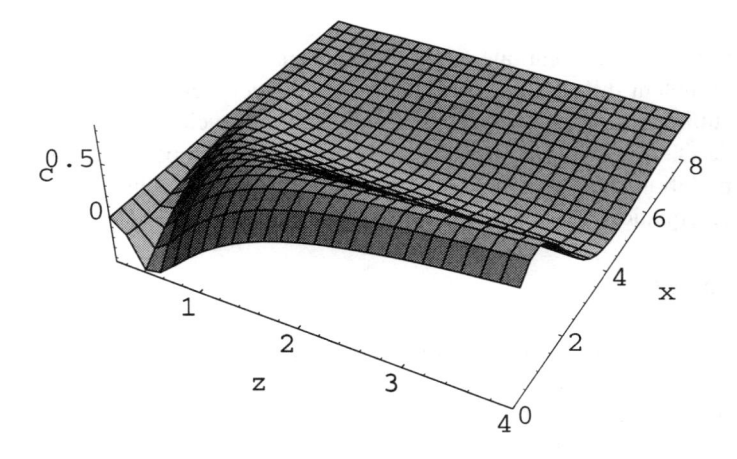

Figure 3: The $c(x, z)$, $\alpha = -3$, $h_1 = 1$, $h_2 = 1$.

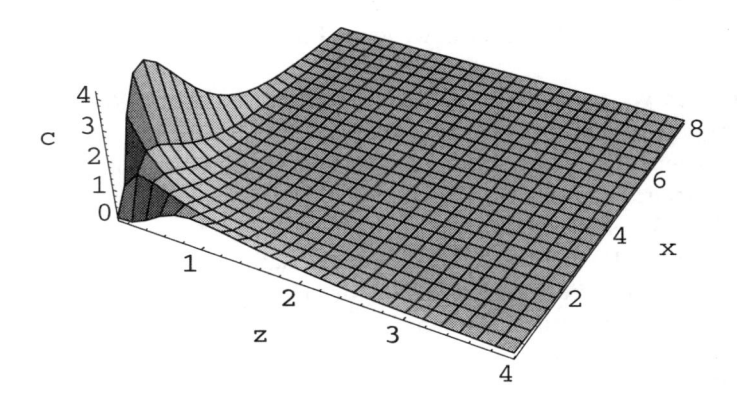

Figure 4: The $c(x, z)$, $v_0 = 3$, $h_1 = 1$, $h_2 = 1$.

the solution is

$$f(\xi) = h_1 + h_2 \operatorname{erf}\left(\frac{\xi}{2}\right)$$

The concentration is

$$c = e^{-\frac{1}{4}(v_0-1)^2 x} z^{-\frac{1}{2}(1+v_0)} \left(h_1 + h_2 \operatorname{erf}\left(\frac{1}{2}\frac{\log z}{\sqrt{x}}\right)\right)$$

5 Conclusion

We obtain analytical solutions, using Lie group analysis, for steady-state bi-dimensional turbulent diffusion equation with variable coefficients. The laws of wind speed, turbulent diffusion coefficients and terminal velocities are specified by power laws. The obtained solutions are more realistic respect to gaussian model in air pollution modeling. In future we intend, using our solutions, to solve the diffusion equation (1) for many boundary conditions.

References

[1] Arya S. P., *Air Pollution Meteorology and Dispersion* Oxford University Press (1999).
[2] Pasquill F. and Smith F. B., *Atmospheric Diffusion* Ellis Horwood (1983).
[3] Seinfeld J. H., *Atmospheric Chemistry and Physics of Air Pollution* Wiley (1986).
[4] Sutton O. G., *Micrometeorology* M.Graw Hill (1953).
[5] Godson W. L., The diffusion of particulate matter from an elevated source; Archiv für Meteor. Geophys. Bioklim., Ser. A., **10**, 305–327, (1958).
[6] Huang C. H., On solutions of the diffusion-deposition equation for point sources in turbulent shear flow; Journal of Applied Meteorology, 250–254, (1999).
[7] Rounds W., Solution of two dimensional diffusion equation; Trans. Amer. Geophys. Union, **36**, 395–405, (1955).
[8] Yamamoto G., Shimamuki A., Nishinomiya S., Diffusion of falling particles in diabatic atmospheres; Journal of the Meteorological Society of Japan, **48**, 417–424, (1970).
[9] Yeh G. T., Huang C. H., Three-dimensional air pollutant modeling in the lower atmosphere; Boundary Layer Meteorology, **9**, 381–390, (1975).
[10] Barrera P., Brugarino T., Group analysis and some exact solution for the thermal boundary layer; Advances in Fluid Mechanics VI, WIT Press, **52**, 327–337, (2006).
[11] Barrera P., Brugarino T., Some exact solutions of two coupled non linear diffusion-convection equations; Air Pollution XIII, Modelling, Monitoring and Management of Air Pollution, Cordova, (2005).
[12] Barrera P., Brugarino T., Pignato L., Solutions for a diffusion process in non-homogeneous media; Il Nuovo Cimento B., Vol. **116** B, 8, 951–958, (2001).
[13] Polyanin A. D., Zaitsev V. F. *Handbook of Exact Solutions for Ordinary Differential Equations* CRC Press (1995).

Air quality forecasting in a large city

P. Perez
Departamento de Fisica, Universidad de Santiago de Chile, Chile

Abstract

We describe the different air pollution statistical forecasting models that have been used in Santiago, Chile during the fall/winter period for the last ten years. Effort has been concentrated on particulate matter PM10 for which a standard of 150 $\mu g/m^3$ for the 24 h average is currently established. Inputs to the models are concentrations measured at several monitoring stations distributed throughout the city and meteorological information in the region. Outputs are the expected maxima concentrations for the following day at the site of the same monitoring stations. Forecast values using neural network models are compared with the results obtained with linear models and persistence. Recently, a clustering algorithm has appeared as a potentially useful tool to detect high concentration episodes in advance.
Keywords: particulate matter forecasting, neural networks, linear models.

1 Introduction

Air pollution has been a major concern in the metropolitan area of Santiago, the capital of Chile during the last 15 years. Together with Sao Paulo, Mexico City, and some Chinese cities it is considered as one of the most polluted in world. Several factors concur to create unfavorable conditions for air pollutant dispersion. The city is located in a valley that has an extension between 70 and 80 km in the north-south direction and approximately 40 km in the east-west direction. To the west we find the Andes Mountains and to the east a coastal range. Some elevations to the north and south trap the air and air pollutants in a region of poor air circulation, which is enhanced during fall and winter when strong thermal inversions prevent vertical dispersion. During this period of the year, the 24 hour moving average (24MA) of PM10 is used as an indicator of air quality.

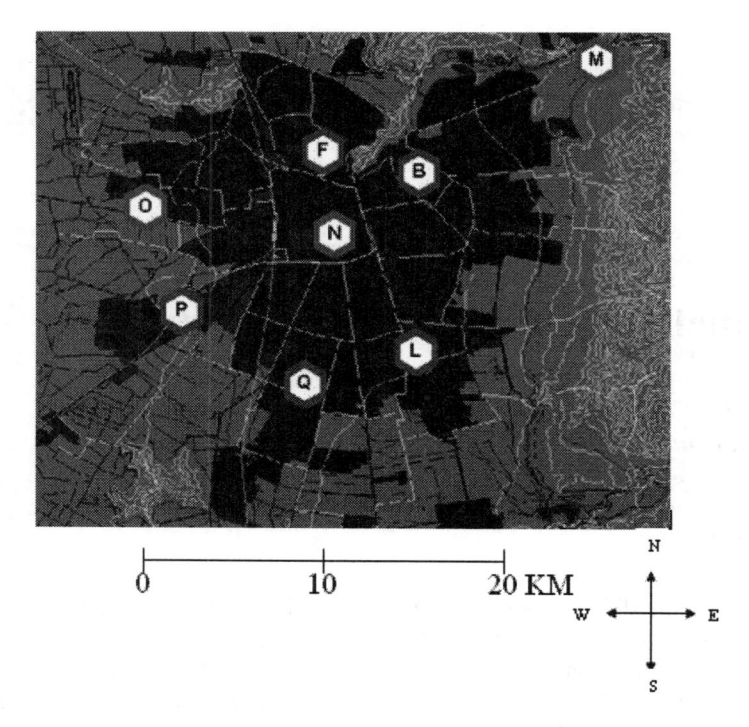

Figure 1: Location of air pollution monitoring stations in the city of Santiago, Chile. The black area represents the urban region. To the extreme right we see the Andes mountains, which spread in the north-south direction.

The main sources of PM10 in Santiago, with six million habitants, are vehicular traffic, industrial activity and heating. Although environmental policies in Santiago during recent years have implied a significant improvement in air quality, particle levels still are considerably high compared to international standards [1]. At present, the standard for the 24 hour average for PM10 in Santiago is 150 $\mu g/m^3$ and the standard for the one year average is 50 $\mu g/m^3$. Regulations that apply on episodes of high concentrations have to do with the definition of classes. According to the value of the maximum of the 24MA of PM10 (MO) the day is classified as class A (good) if MO < 195 $\mu g/m^3$, class B (bad) if 195 $\mu g/m^3 \le$ MO < 240 $\mu g/m^3$, class C (critical), if 240 $\mu g/m^3 \le$ MO $\le$ 330 $\mu g/m^3$ and class D if MO $\ge$ 330 $\mu g/m^3$. The condition of the city is given by the worst class among all official monitoring stations. On class B days, 40% of the motor vehicles without catalytic converters cannot circulate. On class C days 60% of the vehicles without catalytic converters and 20% of those with catalytic converters are not allowed to circulate. On class C days a number of industries identified as pollutant emitters are enforced to stop operation. On class D days 80% of vehicles without catalytic converters and 40% of those with catalytic converters are not allowed to circulate and more industries are required to stop. Fortunately, the last class D day in Santiago occurred in 2001. It appears very convenient to have a reliable PM10 forecasting model for the city, which can be used by the authorities in order to warn the population about adverse conditions and to implement palliative actions in advance when extreme conditions are foreseen. In recent years, atmospheric particulate matter forecasting models have been proposed as an aid for air quality management in different parts of the world. Perez et al. [2] have developed neural network, linear and persistence models in order to forecast hourly values of PM2.5 several hours in advance in Santiago, Chile. Three types of neural models, a linear model and a persistence model have been reported in order to forecast the daily averages of PM2.5 in El Paso (USA) and Ciudad Juarez (Mexico) [3]. The performance of multiple linear regressions and neural network models on the forecasting of PM10 in Athens was analyzed by Chaloulakou et al. [4]. Several types of neural network models and a linear model have been used for PM10 forecasting in Helsinki [5]. A multilayer perceptron with emphasis on a novel training algorithm has been used in order to forecast the 24 hour moving average of PM10 in Shanghai [6]. G. Corani has analyzed the performance of neural networks and a linear model locally trained to forecast the daily average of PM10 in Milan [7]. Multilayer neural networks have been used for PM10 forecasting in Santiago since 2002 [8,9]. The results of most of these studies show that neural network models are more accurate than linear models for atmospheric particulate matter concentrations forecasting. A hybrid clustering algorithm (HCA) has also been proposed for forecasting tasks, and it is claimed that it can outperform neural network models [11]. This approach was used for PM10 forecasting, and the results showed a 10% improvement over neural network models [12]. They also agree that sometimes, more important than the particular method, is the appropriate choice of input variables.

2 Forecasting models

The forecasting task may be represented by the implementation of a function of the form:

$$Y = F(x_1, ..., x_n, z_1, ..., z_m) \qquad (1)$$

where Y is a vector with components that are the maxima of tomorrow's 24MA at the site of the monitoring stations, $x_1, ..., x_n$ are past values of PM10 concentrations and $z_1 ..., z_m$ are measured and forecasted exogenous variables. Input variables may be selected by performing a correlation analysis with historical data.

In the late nineties, restrictions associated to classes B, C, D days in Santiago were applied on the basis of persistence. This means that if on a given day concentrations reached levels within class C, for example, on the following day restrictions associated to that class were applied. This action would make sense only if the episode lasted two or more days.

Since 2001, there is an official forecasting model, which consists of a set of linear equations, one for each monitoring station. The area where most of the times, the highest concentrations are observed is that covered by station O. The equation for this zone is:

$$Y_O = 39.4\, V_O + 0.33\, C_O + 2.06\, T_O + 0.21\, D_O - 21.7 \tag{2}$$

where:
Y_O: is the maximum of the 24 hour moving average of PM10 expected for the following day in $\mu g/m^3$.
V_O: forecasted atmospheric stability for the following day, which is a discrete variable ranging from 1 to 5.
C_O: 24 hour average of PM10 measured at 10:00 AM of present day in station O in $\mu g/m^3$.
T_O: temperature in °C of the 925 hPa level measured at 12 UTC of present day at a location 80 km west of Santiago.
D_O: change in the last 24 h for height of the 500 hPa level measured at 12 UTC of present day at a location 80 km west of Santiago (in meters).

These last two variables give important information about strength of thermal inversions expected in Santiago in the next hours.

The performance of the forecasting model (worst station) may be evaluated by building a contingency table, which for year 2004 is shown in table 1.

Table 1: Contingency table for official PM10 forecasting model between April 1 and August 16, 2004.

2004		FORECASTED					
		A	B	C	D	TOT	%O
O B S E R V E D	A	109	15	2	0	126	87
	B	1	6	2	0	9	67
	C	0	1	1	0	2	50
	D	0	0	0	0	0	x
	TOT	110	22	5	0	137	85
	%F	99	27	20	x		

In table 1, in columns A, B, C, D we see the number of days forecast to be in a given class against the class of the observed day, which appears in the corresponding arrow. The column %O displays the percentage of observed days by class that were forecast to be in that class. Arrow %F delivers the percentage of forecast days by class that were verified to occur. 100 - %F for each class corresponds to percentage of false forecasts. Numbers in the grey diagonal boxes are the successful forecasts by class. At the lower right corner, the overall rate of successful forecasts is registered. We observe that the performance for this year was reasonable for the identification of class B and class C days, but was poor for the large fraction of false positives on these two classes, which affects the model reliability.

Starting 2003, an alternative PM10 forecasting model for Santiago was presented with the idea to increase the reliability of the instrument on which city authorities base their decisions about restrictions. It was an artificial neural network model [9]. In this case, Equation (1) is a non linear function that can be schematically represented as a set of nodes connected by weights, in which an input layer contains the variables in parenthesis and the output layer contains the Y components. A hidden layer with a number of auxiliary variables was also included. The transfer function between layers was a sigmoid. Connection weights were calculated by an optimization algorithm that fitted historical data from the previous two years [12]. The inputs used in the neural model were: one hour averages of PM10 measured at 6 PM and 7 PM of the present day at each of five stations (those with highest concentrations in average), the observed difference between maximum and minimum temperature on the present day, the forecasted difference between maximum and minimum temperature on the next day and the forecasted value of an index called PMCA for the next day. This index is a discrete meteorological variable that ranges from 1 to 5 and it is a measure of atmospheric stability in the Santiago area.

Table 2 shows the 2004 contingency table for the neural network PM10 forecasting model.

Table 2: Contingency table for neural PM10 forecasting model between April 1 and August 16, 2004.

2004		FORECASTED					
		A	B	C	D	TOT	% O
OBSERVED	A	119	7	0	0	126	94
	B	2	5	2	0	9	56
	C	0	0	2	0	2	100
	D	0	0	0	0	0	X
	TOT	121	12	4	0	137	92
	% F	98	42	50	X		

We observe that the neural model is in overall more accurate than the official model (92% against 85%), it is better for identification of class C days (100% against 50%) and produces less false positives on class B and class C days. Due to change in the properties of emissions in the city, identification of high concentrations (especially class C days) has been poor with both the official model and neural model in the last two years. For year 2007, the contingency table for the neural model is shown in table 3.

This result seems poor considering that the population was exposed to high concentrations of particulate matter when a class C day was verified and no restrictions were applied. It is expected that when the restrictions associated to class C days are applied, they have the effect of lowering to some extent the concentrations. A way to correct the poor performance of the neural model on class C days identification is the proposal by Sfestos and Siriopoulos [10] and Vlachogiannis and Sfestos [11] of a clustering algorithm that may be applied for

Table 3:   Contingency table for the neural PM10 forecasting model between April 1 and September 15, 2007.

2007		FORECASTED					
		A	B	C	D	TOT	% O
O B S E R V E D	A	138	1	0	0	139	99
	B	14	7	0	0	21	33
	C	2	3	2	0	7	29
	D	0	0	0	0	0	-
	TOT	154	11	2	0	167	88
	% F	90	64	100	-		

Table 4:   Contingency table for the clustering PM10 forecasting model between April 1 and September 15, 2007.

2007		FORECASTED					
		A	B	C	D	TOT	% O
O B S E R V E D	A	120	14	1	0	135	89
	B	4	13	8	0	25	52
	C	0	1	6	0	7	86
	D	0	0	0	0	0	-
	TOT	124	28	15	0	167	83
	% F	97	46	40	-		

air quality forecasting. A natural adaptation of this clustering algorithm has been implemented in Santiago to solve our problem of class identification one day in advance. The algorithm works in the following manner:

For a period of three year training data, we have calculated the average values of the selected input variables within the respective classes (the same variables used in the neural model) A, B, C and D (four centroids). Within every class, we constructed linear or neural networks algorithms that reproduce the values of the output variables (the maxima of 24MA for the sites of the monitoring stations on the following day). Once we have the centroid patterns for each class, we can perform a test with the following year data, by assigning a given vector to the class with centroid to the least Euclidean distance from it. After class identification, we can calculate the numerical forecasted value by using the algorithm valid for that class. For an operational forecasting system, it would be desirable to generate the most accurate value for tomorrow's class and the expected numerical value of the maximum of the PM10 concentration. The implementation of the clustering algorithm described above with 2007 data produced table 4.

From this table we can verify that the clustering algorithm, having less overall accuracy compared with the neural model (83% against 88%), it has a significantly better performance in detecting class C days (86% against 29%). The false C forecasts would not be so critical considering that most of them were verified to be class B days, which also represent levels considered harmful for the people. A disadvantage of this clustering method is the discontinuity of the numerical forecasted value upon changing from one class to another.

3 Conclusion

With rather simple statistical models it is possible to generate relevant information regarding air quality for the population and authorities in a large city. We have presented several tools that have been used for air quality management in the city of Santiago, Chile and the choice of one of them over the others will depend on the goals we pursue with the forecasting. The models may be used, with the appropriate adaptations in other cities

Acknowledgement

We would like to thank Fondo Nacional de Ciencia y Tecnología (FONDECYT) for support through project 1070139.

References

[1] Koutrakis, P., Sax, S., Sarnat, J., Coull, B., Demokritou, P., Oyola, P., García, J., Gramsch, E., Analysis of PM_{10}, $PM_{2.5}$ and $PM_{2.5-10}$ Concentrations in Santiago, Chile, from 1989 to 2001 *J. Air Waste Manag Assoc* 55, 342–351 (2005).

[2] Perez, P., Trier, A., Reyes, J., Prediction of PM2.5 concentrations several hours in advance using neural networks in Santiago, Chile. *Atmospheric Environment* 34, 1189–1196 (2000).

[3] Ordieres, J. B., Vergara, E. P., Capuz, R. S., Salazar, R. E. Neural network prediction model for fine particulate matter (PM2.5) on the US-Mexico border in El Paso (Texas) and Ciudad Juarez (Chihuahua). *Environmental Modelling & Software* 20, 547–559 (2005).

[4] Chaloulakou, A., Grivas, G., Spyrellis, N. Neural Network and Multiple Regression Models for PM10 Prediction in Athens: A comparative Assessment. *J. Air Waste Manag Assoc* 53, 1183–1190 (2003).

[5] Kukkonen, J., Partanen, L., Karppinen, A., Ruuskanen, J., Junninen, H., Kolehmainen, M., Niska, H., Dorling, S., Chatterton, T., Foxall, R.,

Cawley, G., Extensive evaluation of neural network models for the prediction of NO2 and PM10 concentrations, compared with a deterministic modeling system and measurements in central Helsinki. *Atmospheric Environment* 37, 4539–4550 (2003).

[6] Jiang, D., Zhang, Y., Hu, X., Zeng, Y., Tan, J., Shao, D. Progress in developing an ANN model for air pollution index forecast. *Atmospheric Environment* 38, 7055–7064 (2004).

[7] Corani, G. Air quality prediction in Milan: feed-forward neural networks, pruned neural networks and lazy learning. *Ecological Modelling* 185, 513–529 (2005).

[8] Perez, P., Reyes, J. Prediction of maximum of 24-h average of PM10 concentrations 30 h in advance in Santiago, Chile. *Atmospheric Environment* 36, 4555–4561 (2002).

[9] Perez, P., Reyes, J. An integrated neural network model for PM10 forecasting. *Atmospheric Environment* 40, 2845–2851 (2006).

[10] Sfestos, A., Siriopoulos, C. Time series forecasting with a hybrid clustering scheme and pattern recognition. *IEEE Transactions on systems, man and cybernetics,* Part A 34, 399–405 (2004).

[11] Vlachogiannis, D., Sfestos, A., Time series forecasting of hourly PM10 values: model intercomparison and the development of localized linear approaches. *Air Pollution XIV*, edited by Longhurst, J. W. S. and Brebbia, C. A., WIT Press, 85–94 (2006).

[12] Rumelhart, D. E., Hinton, G. E., Williams, R. J., Learning Internal Representations by Error Propagation. Parallel Distributed Processing. The MIT Press, Cambridge, London, pp 318–364 (1986).

Meshing effects on CFD modelling of atmospheric flow over buildings situated on ground with high terrain

A. Karim, P. Nolan & A. Qubian
Chemical Engineering Department, London South Bank University, UK

Abstract

A Computational Fluid Dynamics (CFD) study was carried out on the wind environment and vehicle pollution dispersion from a newly built by-pass adjacent to the Whatman International site in Maidstone, UK.

The site buildings are sited on ground incorporating an area of high terrain, accordingly they were modelled using Geographical Information System (GIS). The site contains a substantial number of trees of differing species which were extensively surveyed and modeled using a simple 2D momentum sink dependent on the tree Leaf Area Density (LAD).

One of the important factors which has a significant effect on CFD results is the computational mesh. The purpose of this paper is to investigate the effects of using different mesh approaches for both easterly and southerly wind scenarios and the results are compared with that of field measurements taken at the site which include the CO concentration and wind velocity.

The CFD results showed that the hexahedral mesh delivers a higher level of agreement with field measurements than the tetrahedral dominant mesh and this is mainly due to the higher truncation error in the tetrahedral cell type.

It was also found that the tetrahedral dominant mesh can be significantly improved by applying different numerical solving techniques such as the Node-based gradient solver.

Keywords: CFD, pollution dispersion modelling, grid effect, urban environment.

1 Introduction

The Whatman International factory is sited in Maidstone, Kent / UK, the factory consists of a number of buildings which are sited on a ground characterized by terrain forming a small valley on either side of the river Medway, whilst a new dual carriageway (A229 bypass) is sited on the upper side of the terrain. The factory manufactures special types of filters which are highly sensitive to any type of air pollutants. The main objective of the study is to investigate the pollutant dispersion and wind environment from the A229 by-pass onto the sensitive manufacturing facility of this factory. Whatman International were seeking to obtain the concentration of certain pollutants at specific sensitive locations in their factory, therefore the pollutant concentration and velocity were measured at those points in the site.

The Computational Fluid Dynamics (CFD) approach was used to simulate the wind environment and to investigate the effects of CO dispersion produced by the traffic passing through the dual-carriageway on the sensitive manufacturing facility. The preprocessor GAMBIT of the commercial CFD code Fluent [4] was used to design both the geometry and mesh of this site and the Fluent solver is used to solve the Reynolds Averaged Navier Stokes (RANS) equations. However, the footprints and locations of the buildings and land terrain were obtained using GIS and those were used to build the 3D geometry for the CFD simulation, Fig. 1.

From the CFD analysis point of view, one of the main points to be considered is the meshing approach which has significant effects on the prediction and accuracy of the results [1], and this is particularly true of atmospheric flow problems with sometimes very complex geometries (buildings and terrain). Not only is this true of the mesh size but also in the mesh type used (structured / unstructured, hexahedral dominant / tetrahedral dominant etc).

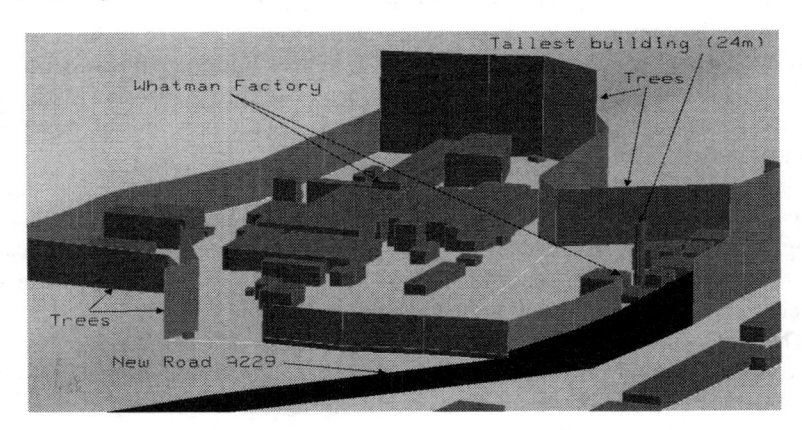

Figure 1: CFD model of the Whatman Intl site and surrounding area.

This paper presents the two main meshing approaches (hexahedral and tetrahedral). The CFD results are compared with that of the field measurements of velocity and CO concentrations. Accordingly, the mesh approaches and limitations are discussed briefly in the following section, and this is followed by a discussion of the modelling technique, and the paper finishes with the results of specific simulations and their findings.

2 Mesh generation

Generally speaking, the domain of a CFD analysis is usually divided into a number of cells (mesh) in which the RANS equations are solved. However, there are two main types of meshing approaches, structured and unstructured. The structured mesh approach generally ensures stability in convergence, however it is impractical for the real atmospheric flow cases [2,3]. On the other hand, the unstructured mesh approach would have higher numerical errors which significantly vary depending on the type of the cells (e.g hexahedral or tetrahedral) and the degrees of the cell quality.

It is worth discussing briefly the difference between the hexahedral cell shape and the tetrahedral cell shape with regards to the solution accuracy. The face value of the cells used in the Fluent software to calculate the various flow variables is dependent on the interpolation method used and that involves different schemes to calculate values from cell to cell.

For example the cell face value for any given variable ϕ_f can be determined using equation (1) (typical interpolation equation), and this is dependent on the cell centroid value ϕ_c, cell gradient $\nabla \phi$, and the geometrical features of the cell as defined in Fig. 2. Thus if we take two near wall adjacent cells of hexahedral and tetrahedral shape shown in Fig. 2 (here the flow is assumed to be parallel to the wall), with hexahedral cells the face is on a vertical angle whilst the tetrahedral cells are inclined and this gives rise to a non-zero gradient. This very well known phenomenon results in a higher truncation error.

$$\phi_f = \phi_c + \delta r (\nabla \phi)_c + \left(|\delta r|^2 \right) \tag{1}$$

However if the flow is not aligned to the cells (not parallel), the hexahedral mesh looses its "edge" on the tetrahedral mesh scheme. For atmospheric flow problems typically the flow can be easily and naturally aligned with the hexahedral cell arrangement.

Hexahedral face   **Tetrahedral face**

$[\delta r (\nabla \phi)_c] = 0$ $[\delta r (\nabla \phi)_c] \neq 0$

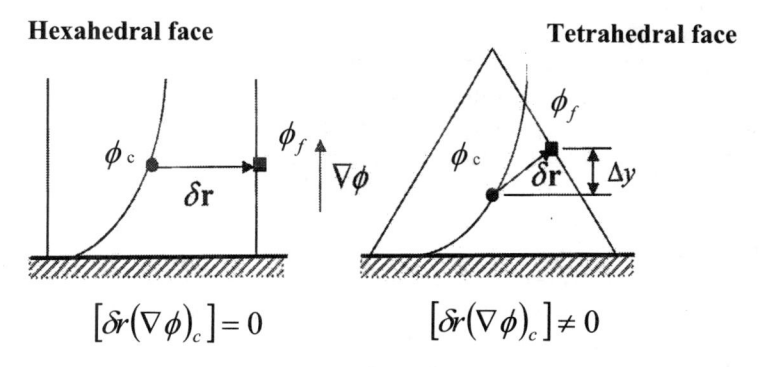

Figure 2: Diagram illustrating the effect of mesh shape on the simulation accuracy.

For the present CFD work, two mesh schemes were used; i) The Hexahedral Core (HC) type [4] which is a tetrahedral dominant mesh ii) The second mesh type used is the wholly hexahedral type, Hexahedral-Cooper (Cooper). The results of both approaches will be discussed in order to highlight their level of agreement with field measurements.

Due to the requirement of an orthogonal cell in the near wall area studies have shown improved simulations using prism shapes in the tetrahedral mesh [5]. However the present work has sought to compare between the tetrahedral mesh and hexahedral mesh primarily.

For the Hex-Cooper mesh, the maximum skewness was approximately 0.9 whilst the maximum aspect ratio was 200. The Cooper mesh contained 2.7 million cells in total.

The HC mesh, on the other hand, contained a maximum skewness of 0.95, however the aspect ratio is fairly low throughout the domain. The HC mesh was about 2.6 million cells.

3 Modelling technique

For the CFD modelling procedure, there are many important factors which should be carefully chosen for solving the RANS equations. Among them are the turbulence model, the domain boundary conditions and the calculation of the cell gradient values. Brief discussions on these issues are given below.

For the domain boundary condition, the side flow inlet is modelled using approximated profiles suggested by Richards and Hoxey [6]. The profiles would be calculated using meteorological data taken at the nearby East Malling station. The aforementioned profiles are commonly used in literature and seem to give suitable approximations to the flow variation, they include the wind speed, turbulent kinetic energy and dissipation [7]. All other domain boundaries are specified as symmetry whilst the outlet is set as an outflow boundary [8].

A line of substantially tall trees almost entirely surrounds the Whatman site, these were modelled as 2D momentum sinks, with turbulence generation inside the canopy ignored. This was based on data from the UK Forestry Commission on Leaf area Density (LAD).

For modelling turbulence the standard k-ε modified by Detering and Etling [9] is used, with standard wall functions. The vehicle pollutant from the road is released as an area emission with the turbulence generated by vehicles ignored.

For the cell gradient values; the Fluent code provides three methods for calculating the gradient values of each individual cell from the surrounding cells, Green-Gauss cell-based, Green-Gauss Node-based and Least squares cell-based, for more details see [4].

However, the CFD results could be significantly improved by using the Node-based method, this is particularly true for the arbitrary shape arrangement of the cells which are associated with the tetrahedral meshing scheme [4,10]. The Node-based method calculates the node weighted average for all surrounding cells which share the same node. This is especially critical for the tetrahedral mesh scheme because significantly more cells share any single node than the hexahedral arrangement, thus the numerical errors become naturally higher. However, the iteration time would be significantly increased, and this may be more severe (double or triple) when tackling more complex problems, e.g reactions.

Accordingly, in order to improve the result of the tetrahedral based mesh without seeking higher numerical schemes or more complex turbulence models the Node-based gradient method is utilized.

4 Results and discussions

In order to validate the CFD models against the results of the experimental work, the two dominant wind directions of both east and south were investigated. The CFD prediction involved the use of two meshing schemes, first the Cooper scheme which is an entirely hexahedral mesh, whilst, the second one, HC is a tetrahedral dominant mesh, and the main features of those schemes are presented below.

4.1 Wake flow behind the buildings

Generally speaking, for the dispersion of the CO pollutant, the wake flow behind the buildings represents one of the important fluid features which would participate in the changes of pollutant concentration locally and also in channeling and directing the flow movement downstream of the buildings.

For the Cooper meshing approach, the predictions of flow movement within the wake behind the buildings were predicted with some considerable clarity and accuracy in their sizes, flow separation and flow directions.

For the HC tetrahedral approach the wake formation is much less clear and the wake region is larger in size and this is typical of all the buildings in the domain. Another point worth mentioning here, is the flow reattachment point downwind of the buildings are further downstream in the HC mesh as compared to the Cooper scheme, and this is shown in Figs. 3 and 4 around a typical site building.

However, the differences in results can be explained as follows; for the HC meshing approach, despite a higher number of cells filling the wake region behind the building which in fact slightly exceeded that of the Cooper scheme, but due to the random structure arrangement of the tetrahedral cells, which mean that the cell centers are in different non-equidistant locations in relation to neighboring cells, the consequences of that, are the solver will interpolate values carrying a higher numerical error which is exponentially expanded downstream, Figs 5 and 6.

The hexahedral shape at the wall is known to give improved predictions due to the better resolution of turbulent energy normal to the wall [11]. Accordingly the HC mesh results showed that the flow near the ground terrain region is characterized by small fluctuations (bumps) in the velocity plot, whilst the Cooper mesh did not have these inconsistencies.

The features discussed above had a significant effect on the CO concentration as shown in Figs. 7 and 8. The CO concentration for the HC mesh is highly irregular with sporadic high concentration points.

Figure 3: Velocity contours in vertical plane (m/s)-East wind, (Cooper).

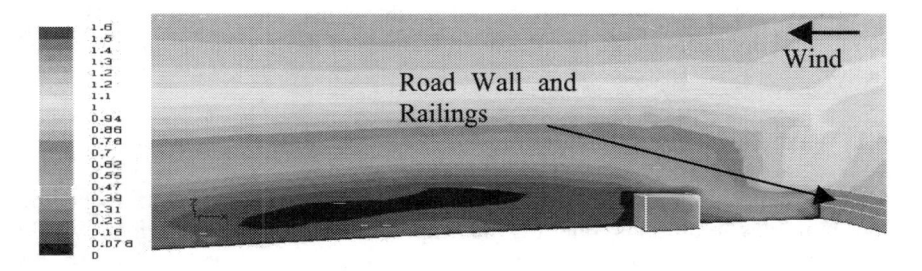

Figure 4: Velocity contours in vertical plane (m/s)-East wind, (HC).

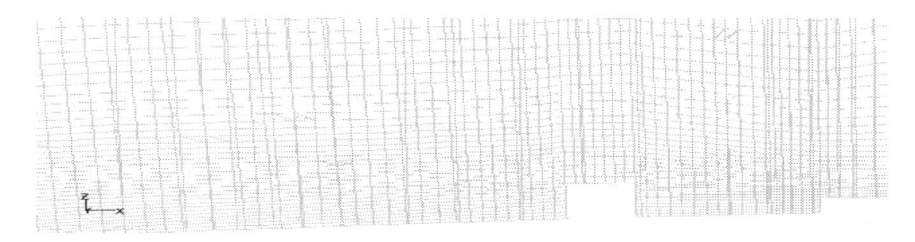

Figure 5: Mesh in vertical plane surrounding typical site building, (Cooper).

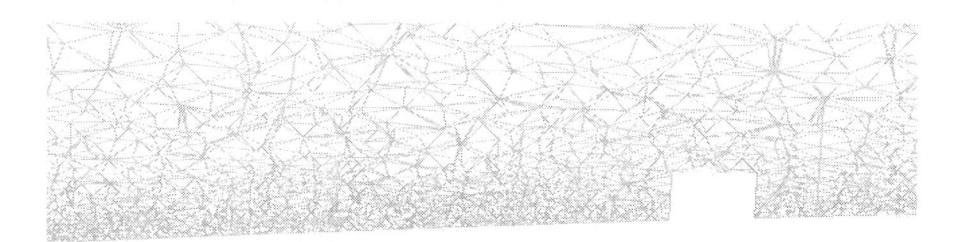

Figure 6: Mesh in vertical plane surrounding typical site building, (HC).

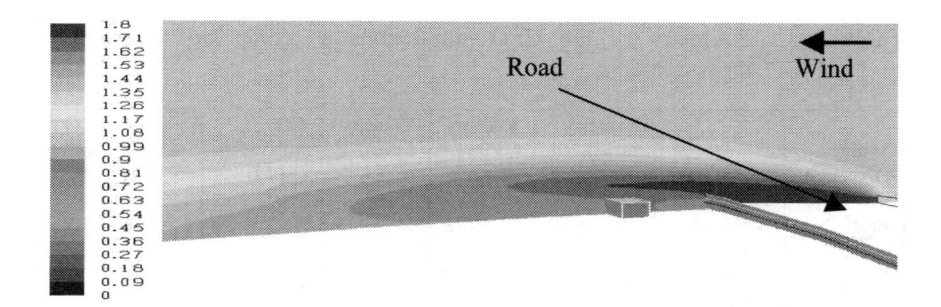

Figure 7: CO concentration contours in vertical plane (PPM)-East wind, (Cooper).

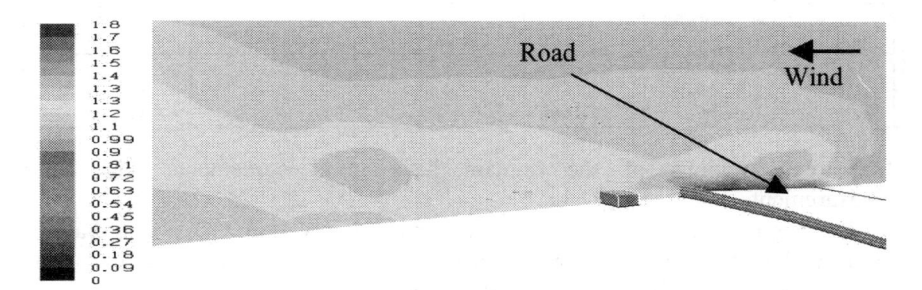

Figure 8: CO concentration contours in vertical plane (PPM)-East wind, (HC).

4.2 Wind velocity and Carbon Monoxide concentration

The measurements of CO concentration were carried out at a location 10 m away from the road, whilst the wind velocity measurements were carried out 60m away from the road. Generally speaking, the CFD predicted results are in good agreement with those measured, see Table 1.

For the CFD results, the Cooper mesh scheme produces better overall agreement with that of the measurements and the discrepancies of both CO concentration and wind velocity were 12.5% and 3% respectively for the east wind and 16% and 1% for the south wind.

Given the drastic differences between the meshing schemes in predicting wake size behind buildings and the boundary layer development, accordingly, in using the HC mesh scheme, the discrepancies of both CO prediction and wind velocity were 20% and 13% respectively for the east wind, whilst it is slightly higher for the southerly wind, 32% and 26% respectively.

Those high discrepancies which are usually associated with the tetrahedral meshing scheme would be significantly attributed to the numerical and truncation errors. As a result of that the distributions within the complex flow regions such as the wakes and boundary layer formation on the terrain would be significantly changed. However, the predicted velocity or CO obtained for the

Table 1: Summary of the CFD simulations as compared to the field measurements.

Wind Direction	Mesh Type	Velocity measured (m/s)	Velocity CFD (m/s)	Velocity Error (%)	CO measured (PPM)	CO CFD (PPM)	CO Error (%)
East	Hexahedral (Cooper)	0.70	0.72	3	1.69	1.48	12.5
East	Tetrahedral (HC)	0.70	0.60	13	1.69	2.03	20
East	Tetrahedral (HC-Node based)	0.70	0.73	4	1.69	1.98	17
South	Hexahedral (Cooper)	1.00	1.01	1	1.06	1.23	16
South	Tetrahedral (HC)	1.00	1.26	26	1.06	1.41	32
South	Tetrahedral (HC-Node based)	1.00	1.1	10	1.06	1.37	28

free-stream regions of the domain have good agreements with field measurements.

The highest discrepancies of both wind velocity and CO concentrations associated with the southern wind scenario can be related to the fact that the unsettled and unsteady flow phenomena such as the wakes, flow separation and recirculation become more significant within the reference points which are used in the measurements, accordingly, the predicted CFD results would have increased numerical and truncation errors.

On the other hand, for the HC tetrahedral mesh case, the CFD results were improved significantly by using the Node-based gradient method comparing to that of Cell-based method (default Fluent option), Table 1. The predicted CO and velocity at the measurement points become comparable to that obtained by the Cooper mesh scheme

5 Conclusion

Two CFD models for the Whatman international site in Kent, UK were designed and solved using fully hexahedral Cooper mesh and the HC tetrahedral dominant mesh with the same flow conditions for the easterly and southerly wind scenarios. Both models were solved with the same turbulence model and numerical schemes and with almost the same number of meshing cells.

Both schemes generally predicted the same overall flow characteristics; however the HC scheme compared to the Cooper scheme produced irregular plots of CO and velocity in the wake regions and along the ground which has a varying terrain, as well as differences in the prediction of the wake sizes behind buildings.

When compared to the field measurements, the HC mesh predicted lower localized velocities which led to over-prediction of the CO concentration. The HC mesh gave discrepancies of up to 16% in wind velocity and 25% in CO concentration, whilst the Cooper scheme produced discrepancies of 3% in the velocity and 16% in the CO concentration.

The results strongly suggested that the HC scheme, in non-complex flow regions of the domain (complex being areas of separation or recirculation regions) produced velocity and CO predictions which were quite reasonable. However the performance of the HC mesh is greatly influenced by the complex flow phenomenon which occurred for the south wind scenario and that produced much higher discrepancies, whilst the Cooper scheme was largely unaffected

Furthermore it was found that the HC mesh numerical errors could be greatly reduced by running the solver using the Node-based method for calculating the cells' gradient values instead of using the Cell-based method which is the typical method employed in most commercial codes.

Acknowledgements

The authors would like to thank Dr. Olga Grant of the East Malling Research Centre, and Dr. Rona Pitman and Chris Peachey of UK Forestry Commission.

References

[1] Cowan, I. R., Castro, I.P., Robins, A.G. Numerical considerations for simulations of flow and dispersion around buildings. *Journal of Wind Engineering and Industrial Aerodynamics* **67 & 68** pp. 535–545, 1997.

[2] Huber, A., Freeman, M., Spencer, R., Schwarz. W., Bell, B., Kuehlert, K. Development and applications of CFD simulations supporting urban air quality and homeland security, *AMS sixth symposium on the urban environment*, Atlanta, U.S, 2006.

[3] Kim, S., and Boysan, F. Application of CFD to environmental flows, *Journal of Wind Engineering and Industrial Aerodynamics*, **81** pp. 145–158, 1999.

[4] Fluent user guide. Fluent. Inc 2006

[5] Fothergill, C.E., Roberts, P.T., Packwood, A.R. Flow and dispersion around storage tanks. A comparison between numerical and wind tunnel simulations. *Wind and structures,* Vol. 5, No. **2-4** pp. 89–100, 2002

[6] Richards, P.J. and Hoxey, R. Appropriate boundary conditions for computational wind engineering models using the k-e model, *Journal of Wind Engineering and Industrial Aerodynamics*, **46, 47**, pp. 145–153, 1993.

[7] Hargreaves, D.M. and Wright, N. G. On the use of the k- ε model in commercial CFD software to model the neutral atmospheric boundary layer. *Journal of Wind Engineering and Industrial Aerodynamics*, doi: 10.1016/j.weia.2006.06.002, 2006

[8] Franke, J. et al Recommendations on the use of CFD in predicting pedestrian wind environment. COST Action C14 "Impact of wind and storms on city life and built environment" Working Group 2-CFD techniques, 2004

[9] Detering, H.W. and Etling, D. Application of the E-ε Turbulence Model to the Atmospheric Boundary Layer. *Boundary Layer Meteorology* **33** pp. 113–133, 1985
[10] Ferziger, J. H. and Peric, M. *Computational Methods for Fluid Dynamics*, Springer – Verlag: New York, 2002.
[11] Fothergill, C. E., Roberts, P.T., Packwood, A.R. Flow and dispersion around storage tanks. A comparison between numerical and wind tunnel simulations. *Wind and Structures*, Vol. 5, **2~4** pp. 89–100, 2002.

Enhanced evaluation of a Lagrangian-particle air pollution model based on a Šaleška region field data set

B. Grašič, P. Mlakar, M. Z. Božnar & G. Tinarelli
MEIS d.o.o., Mali Vrh pri Šmarju 78, 1293 Šmarje-Sap, Slovenia
ARIANET s.r.l., Via Gilino 9, 20128 Milano, Italy

Abstract

Lagrangian-particle air pollution model is required by Slovenian legislation for industrial air pollution control because it is the most efficient for small domains over complex topography. To determine its performance and efficiency for regulatory purpose a research is made. In this research a general purpose modelling system designed for local scale areas is used. The main goal of the research is to define an enhanced statistical analysis used to evaluate an air pollution model of this kind where an operational configuration of both input data and model parameters are used and a testing period with very complex dispersion conditions is used. This enhanced evaluation can help to better understand the general quality that a model can achieve in these conditions. It gives some idea on how to better evaluate and use some results that seem to be very negative simply looking to some statistical parameter.
Keywords: air pollution, model evaluation, Lagrangian particle model, field data set.

1 Introduction

In accordance with the European Council Directive of 28 June 1984 on combating air pollution from industrial plants (84/360/EEC) a Slovenian Government decree on the emission of substances into the atmosphere from stationary sources of pollution came into force in April 2007. The decree defines that the performance of the air pollution model used to reconstruct air pollution around stationary sources must meet the requirements of complex terrain as defined in paper by Grašič et al. [1].

The Lagrangian-particle air pollution model is the one that fully satisfies all these requirements [2] among the currently available air pollution models. The main purpose of the research is an enhanced evaluation of the Lagrangian-particle air pollution model that is used for regulatory purposes over complex terrain. We are looking at its performances in severe conditions and trying to better understand and interpret some results.

2 Enhanced evaluation method

Standard method to evaluate the quality and performance of an air pollution model is based on statistical analysis of measured and reconstructed data. Measured data is obtained from automatic air quality measuring stations located around source of air pollution. Reconstructed data is obtained from simulation results performed with selected air pollution model. Evaluation is performed by statistical analysis of data that is available for selected time interval T. For the selected time interval a set of data patterns must be prepared where each data pattern consist of a pair of measured and reconstructed concentration $\{Cm(t), Cr(t)\}$.

During the evaluation following three performance indices are determined:

- the correlation coefficient:

$$r = \frac{\frac{1}{T}\sum_{t=0}^{T}(Cm(t) - \hat{C}m)(Cr(t) - \hat{C}r)}{\sigma_{Cm}\sigma_{Cr}} \tag{1}$$

- the normalized mean square error:

$$NMSE = \frac{\frac{1}{T}\sum_{t=0}^{T}(Cm(t) - Cr(t))^2}{\hat{C}m \cdot \hat{C}r} \tag{2}$$

- and the fractional bias:

$$FB = 2\frac{\hat{C}m - \hat{C}r}{\hat{C}m + \hat{C}r} \tag{3}$$

where

$Cm(t)$ …measured concentration at time t,

$Cr(t)$ …reconstructed concentration at time t,

$\hat{C}$ …average concentration,

σ_C …concentration standard deviation,

T …interval length (number of concentrations).

To avoid effect of model's inaccuracy of position and time of reconstructed concentrations an enhanced evaluation method is used. It is based on presented statistical analysis where additional reconstructed ground level concentrations around measuring station are used in comparison procedure as presented on

Figure 1. In standard evaluation procedure only one reconstructed concentration is used from the cell where station is located $Cr(t) = Cr(t, m_s, n_s)$. In enhanced evaluation procedure a reconstructed concentration for comparison is selected from set of reconstructed concentrations $GCS(t)$ using best matching fuction BM. The function selects the reconstructed concentration that represents the best match according to measured concentration.

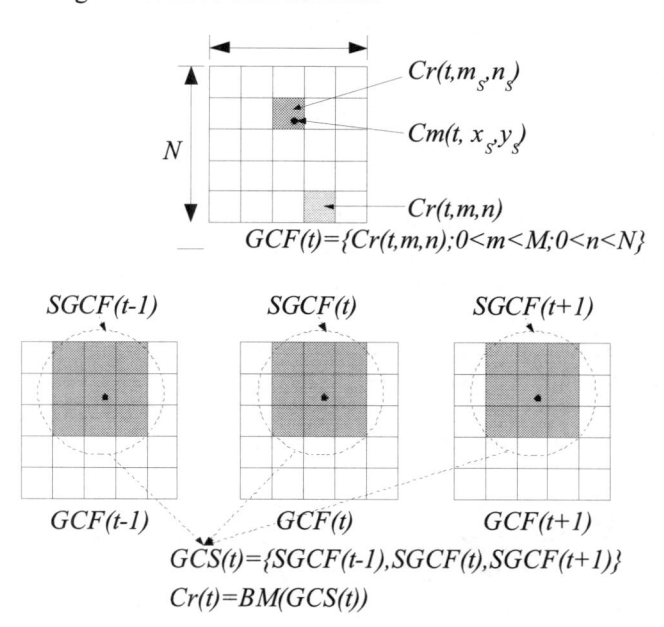

Figure 1: Determination of best matching reconstructed concentration within enhanced evaluation method where x_S, y_S is the position of the station, $Cm(t, x, y)$ is the measured concentration at position (x,y), m_S, n_S is the cell indexes at position of the station, $Cr(t, m, n)$ is the reconstructed concentration at cell (m,n), $GCF(t)$ is the reconstructed ground concentration field at time t, N is the number of cells in the east-west direction, M is the number of cells in the north-south direction and $BM()$ is the best matching function.

3 Šaleška region field data set

For enhanced evaluation of the performance of the selected Lagrangian particle-dispersion model, data from a measuring campaign performed from 15th of March to 5th of April 1991 across the Šaleška region is used. The measuring campaign was performed as the joint effort of three institutions: ENEL-CRAM and CISE from Milano in Italy and the Jozef Stefan Institute from Ljubljana in

Slovenia. The database from the measuring campaign was published and distributed on floppy disks in order to be available for further processing and research. The contents of the floppy disks are available as part of the final report [3].

In this measuring campaign the concentrations of SO_2 higher than 1 mg/m^3 were measured at surrounding stations. All of them were caused by the high emissions from the three stacks (100 m, 150 m and 230 m high) of a thermal power plant that did not have desulphurization plants installed at that time. Because all other local sources of emission can be neglected, the data obtained during that measuring campaign can thus be used as a tracer experiment. The database was constructed from different measurement sources, like the Environmental Informational System (EIS) of the Šoštanj TPP, one mobile Doppler SODAR, DIAL and an automatic mobile laboratory.

The Šaleška region field data set is selected for several reasons:
- it spreads over complex orography (basin surrounded by high hills and a semi-mountainous continuation of the Karavanke Alps) where almost all possible complex terrain conditions occur,
- the database of ambient measurements is available from the measuring campaign organized in spring 1991,
- and the emissions from the three stacks of thermal power plant Šoštanj represent the main air-pollution source in the region.

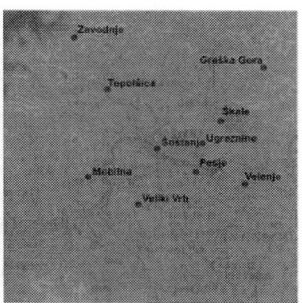

Figure 2: Šaleška region: complex topography of the valley and positions of automatic air quality and meteorological measuring stations.

4 Results and discussion

4.1 Air pollution model and simulation

The air pollution model is selected following the new Slovenian legislation: in complex terrain situation it is required to use a Lagrangian-particle air pollution model that is combined with corresponding meteorological pre-processor able to reconstruct a three dimensional diagnostic non-divergent wind field.

The Lagrangian-particle air pollution model termed *SPRAY* [4] is evaluated in this research. The exact description of selected air pollution model, parameters and options are given in the papers by Tinarelli et al. [4, 5]. Complex terrain situation is very common in Slovenia and almost all air pollution facilities are located at the bottom of basins, valleys or river canyons. The evaluation is needed because the Lagrangian-particle air pollution model evolved in the last ten years from research usage to usage for operational regulatory purposes [5, 6].

To describe particle velocity fluctuations and to generate ½ hour average ground concentration field at the same horizontal resolution as used by meteorological reconstructions a Thomson's 1987 scheme with Gaussian random forcing [7] has been used. Emission from three stacks of the Šoštanj thermal power plant have been considered where each source of emission was described by static and dynamic parameters as presented in Table 1. The plume rise undergone by hot stack plumes option has been simulated by means of the Anfossi's formulation [8].

Table 1:   List of emission parameters.

Stack	Position	Height	Diameter	Exit temperature	Exit velocity	Emission rate
Stack 1	46,373N 15,052E	100 m	6.50 m	155 ÷ 171°C	0.7 ÷ 2.9 m/s	0.01 ÷ 0.24 kg/s
Stack 2	46,372N 15,053E	150 m	6.34 m	155 ÷ 183°C	8.8 ÷ 12.3 m/s	0.87 ÷ 2.05 kg/s
Stack 3	46,371N 15,055E	230 m	6.20 m	172 ÷ 202°C	8.6 ÷ 12.7 m/s	0.53 ÷ 2.46 kg/s

The simulation has been performed for the full duration of the campaign that lasted from 15th of March until 5th of April 1991. The result of simulation is a set of reconstructed air pollution episodes. Each air pollution episode represents average air pollution situation in time interval of ½ hour described by a 3D concentration field, out of which only the 2D ground-level concentration field is relevant for the enhanced evaluation. 2D concentration field is described by a matrix of size 100x100 grid cells where each cell size is 150 m x 150 m. The full size of domain is 15 km x 15 km where the thermal power plant is located in the centre of domain as the source of emission.

4.2 Standard evaluation results

An evaluation is made at different directions around the thermal power plant where following four automatic air quality measuring stations are located: Graška Gora, Šoštanj, Veliki Vrh and Zavodnje. The measured and reconstructed SO_2 concentrations at locations are presented on Figures 3, 4, 5 and 6 where a rotation of the air pollution plume from power plant over the domain is evident.

In higher layers at height from 200 m to 250 m wind changed its direction for north-west to south-east at the beginning of simulated period. The SO_2 concentration on Graška Gora increased as a consequence of this change as presented in Figure 3. The figure also shows strong correlation of reconstructed SO_2 concentrations with the measured ones. Later the wind changed its direction toward the south. This change caused an increase of SO_2 concentrations at the Šoštanj and Veliki Vrh stations.

In Figure 4 the SO_2 concentrations at the Šoštanj station are presented. The comparison shows that the reconstructed values are underestimated. The first reconstructed peak at 11:30 is underestimated mainly due to positional inaccuracy. While the second reconstructed peak is underestimated mainly due to the short distance between the station and the power plant which is approximately 500 m. This short distance does not express strongly enough two effects: the stack tip down-wash effect and the effect of a low-wind speed directed towards the station in combination with convective turbulences.

The comparison of SO_2 concentrations at the Veliki Vrh station presented on Figure 5 shows that two peaks of air pollution were reconstructed lasting from 00:00 to 04:00 and from 06:00 to 12:00. The reconstructed peaks are again a consequence of model's inaccuracy in the position. In the real situations such peaks with sharp edge could be created just few hundred metres from the station without being measured.

SO_2 concentrations at the Zavodnje station are presented on Figure 6 where a similar process occurred. The phenomenon of inaccuracy in the position of the reconstructed peak is again very obvious. It caused the appearance of the first measured peak in the simulation that lasted from 00:00 to 04:00. The second measured peak at 13:30 was underestimated due to the phenomenon of air pollution accumulation [1] that was lost in the simulation due to the insufficient size of the area of interest.

Table 2 shows the results of statistical analysis resulting from a point-to-point comparison. Results appear to be quite dissatisfactory at a first view. Only the correlation at Graška Gora reaches a satisfactory value higher than 0.30.

4.3 Enhanced evaluation results

Enhanced evaluation has been performed to avoid both effects of model's inaccuracy of position and time. Measured ground level concentration at

Table 2: Standard statistical comparison.

Station	CORR	FB	NMSE
Graška Gora	0,34	1,60	40,42
Šoštanj	0,02	0,37	17,32
Veliki Vrh	0,13	0,09	8,70
Zavodnje	-0,004	0,10	38,35

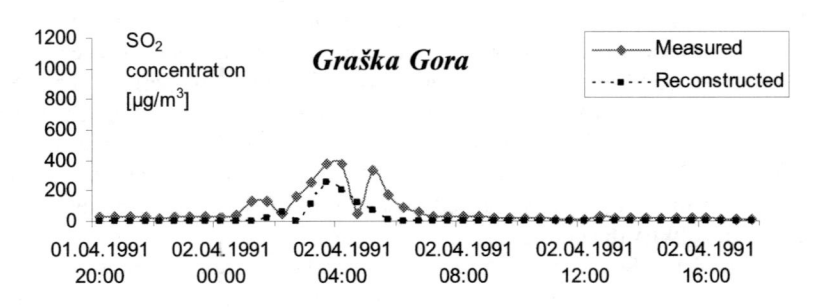

Figure 3: Standard comparison at Graška Gora.

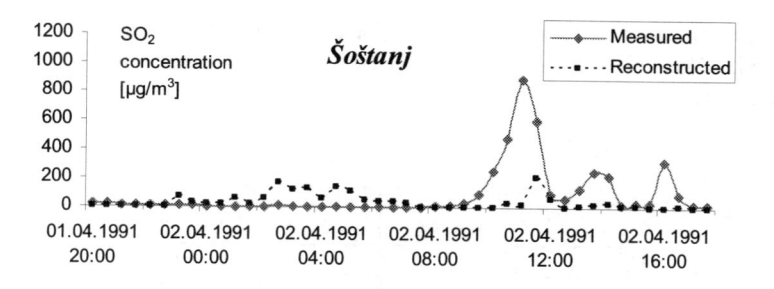

Figure 4: Standard comparison at Šoštanj.

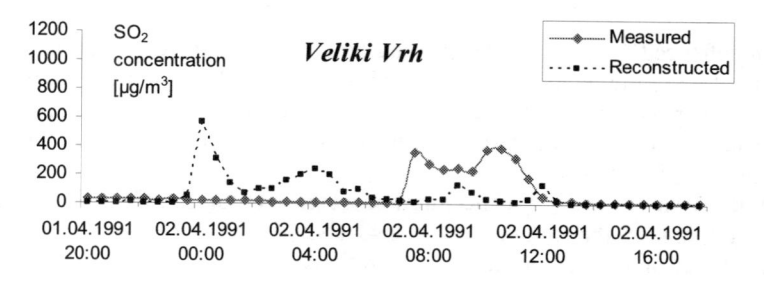

Figure 5: Standard comparison at Veliki vrh.

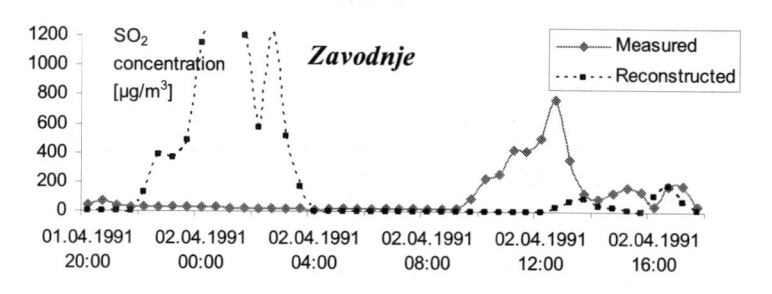

Figure 6: Standard comparison at Zavodnje.

locations of each four automatic station is compared to most similar reconstructed ground level concentration around the position of station in radius of 150 m and in time frame from -½ hour to +½ hour. The most similar reconstructed concentration for particular station is determined from the set of 27 combinations (3x3 cells around position of the station combined with 3 time shifts for time reconstructions of -½h, 0h, +½h).

The results of enhanced evaluation are presented on Table 3. All correlations between measured and reconstructed values are significantly improved. Values of fractional bias are generally increased because several of overestimated reconstructed concentrations were reduced which also resulted in decrease of mean value of reconstructed concentrations. Fractional bias values generally increased because several of overestimated reconstructed concentrations are reduced. This also resulted in decrease of mean value of reconstructed concentrations.

Table 3: Enhanced evaluation results.

Station	CORR	FB	NMSE
Graška Gora	0,69	1,14	10,38
Šoštanj	0,36	0,95	20,64
Veliki Vrh	0,74	0,61	3,26
Zavodnje	0,37	0,79	6,30

An enhanced evaluation is performed on the base of the comparison between measured and in time and space adjusted reconstructed SO_2 concentrations. It is made at the positions of four automatic air quality measuring stations to compare the results with standard evaluation. The results of this comparison are presented on Figures 7, 8, 9 and 10.

Figure 7 shows the SO_2 concentration increase at Graška Gora that occurs at the beginning of the simulation. It also shows well agreement between the reconstructed and measured SO_2 concentrations.

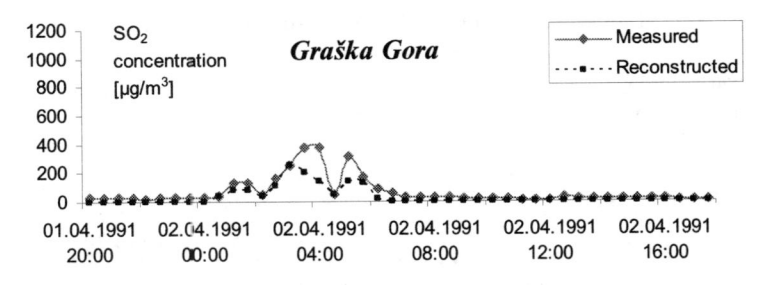

Figure 7: Enhanced comparison at Graška Gora.

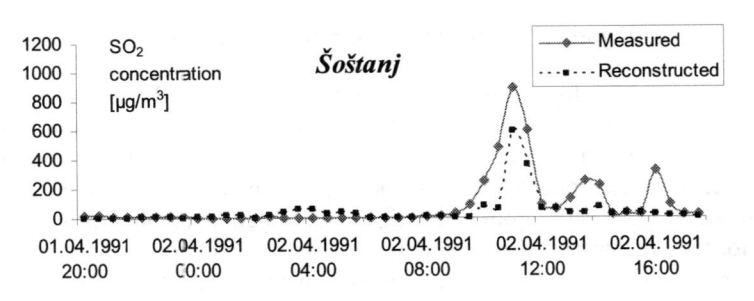

Figure 8: Enhanced comparison at Šoštanj.

In Figure 8 SO_2 concentrations at the Šoštanj station are depicted. It can be seen that the reconstructed values are now only slightly underestimated.

SO_2 concentrations at the Veliki Vrh station are depicted on Figure 9. Again the comparison shows two reconstructed peaks of air pollution that are very well correlated with measured ones.

At Zavodnje station comparison is presented on Figure 10. It is shown that the phenomenon of inaccuracy in the position of the reconstructed peak is strongly reduced. The lack of phenomenon of air pollution accumulation is still lacking.

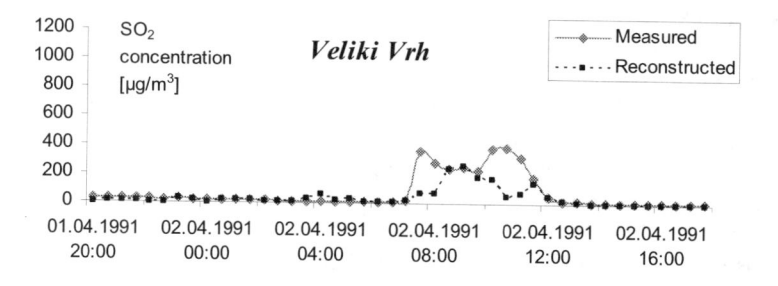

Figure 9:   Enhanced comparison at Veliki vrh.

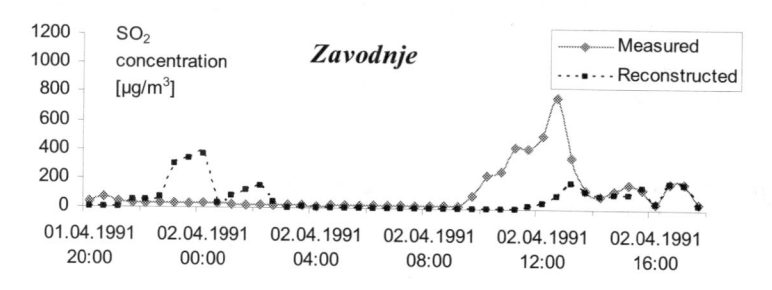

Figure 10:   Enhanced comparison at Zavodnje.

5   Conclusions

Enhanced evaluation of the general purpose Lagrangian-particle air pollution model designed for local scale complex terrain has been made. Evaluation has been performed to determine the performance and efficiency of an air pollution modelling technique that is required by Slovenian legislation for industrial air pollution control.

Enhanced statistical analysis has been defined and performed to minimize the model's effect of inaccuracy of position and time. Measured ground level concentration at certain station measured have been compared to most similar reconstructed ground level concentration around the position of station in radius of 150 m and in time frame from -1/2 hour to +1/2 hour.

Significant improvement has been shown by this analysis. It pointed out the sensitivity of the model's to input parameters. Air pollution model is especially sensitive to measured wind speed and direction which could be improved in future.

Acknowledgement

The authors gratefully acknowledge the contribution of the Ministry of Higher Education, Science and Technology of Republic of Slovenia, Grant No. 3211-05-000552.

References

[1] Grašič B., Božnar M. Z., Mlakar P., Re-evaluation of the Lagrangian particle modelling system on an experimental campaign in complex terrain, Il Nuovo Cimento C, Vol. 30, No. 6, pp. 19-, 2007
[2] Wilson J. D., Sawford B. L., Review of Lagrangian stochastic models for trajectories in the turbulent atmosphere. Boundary-Layer Meteorology Vol. 78, pp. 191-210, 1996
[3] Elisei G., Bistacchi S., Bocchiola G., Brusasca G., Marcacci P., Marzorati A., Morselli M. G., Tinarelli G., Catenacci G., CORIO V., DAINO G., ERA A., Finardi S., Foggi G., Negri A., Piazza G., Villa R., Lesjak M., BOŽNAR M., Mlakar P., Slavic F.: Experimental campaign for the environmental impact evaluation of Sostanj thermal power plant, Progress Report, ENEL S.p.A, CRAM-Servizio Ambiente, Milano, Italy, C.I.S.E. Tecnologie Innovative S.p.A, Milano, Italy, Institute Jozef Stefan, Ljubljana, Slovenia, 1991
[4] Brusasca G., Tinarelli G., Anfossi D., Particle model simulation of diffusion in low windspeed stable conditions", Atmospheric Environment Vol. 26, pp. 707–723, 1992
[5] Tinarelli G., Anfossi D., Bider M., Ferrero E., Trini Casteli S., A new high performance version of Lagrangian particle dispersion model SPRAY, some case studies., Air pollution modelling and its Applications XIII, S. E. Gryning and E. Batchvarova eds., Kluwer Academic / Plenum Press, New York, pp. 499–507, 2000
[6] Graff A., The new German regulatory model – a Lagrangian particle dispersion model., 8th International Conference on Harmonisation within Atmospheric Dispersion Modelling for Regulatory Purposes, October 14-17, Sofia, Bulgaria, pp. 153–158, 2002
[7] Thomson D.J., Criteria for the selection of stochastic models of particle trajectories in turbulent flows, Journal of Fluid Mechanics, Vol. 180, pp. 529–556, 1987
[8] Anfossi D., Ferrero E., Brusasca G., Marzorati A., Tinarelli G., A simple way of computing buoyant plume rise in Lagrangian stochastic dispersion models, Atmospheric Environment Vol. 27A, pp. 1443–1451, 1993

Modelling the multi year air quality time series in Edinburgh: an application of the Hierarchical Profiling Approach

H. Al-Madfai, D. G. Snelson & A. J. Geens
Faculty of Advanced Technology, University of Glamorgan, Wales, UK

Abstract

Modelling and forecasting of time series of concentrations of air pollutants is essential in monitoring air quality and assessing whether set targets will be achieved. While many established time series modelling approaches transform the data to stationarity a priori, the explicit modelling and presentation of the non-stationary components of the series in this application is essential to allow for further understanding of variability and hence more informed policies. The Hierarchical Profiling Approach (HPA) was used to model the multi-year daily air quality data gathered at St. Leonards in Edinburgh, UK spanning from 1st of March 2004 to 15th July 2007. The HPA is an avant-garde approach that explicitly models the non stationary component of time series data at different levels depending on the span of the component, so that within-year disturbances are at Level 1 and year-long variability such as seasonality is at Level 2, and so on. HPA decomposes the variability into deterministic, stochastic and noise and uses continuous models to describe the non-stationary components using the deterministic part of the model. The stationary stochastic component is then modelled using established approaches. The dataset modelled was the total daily concentrations of carbon monoxide. After modelling within-year events at Level 1, a harmonic regression with trend model was used to describe the weekly aggregates of the data at Level 2. This model was then sampled back in the daily domain and no evidence of a larger cyclical component was found. Wind-speed and a dummy intervention-variable indicating the implementation of the smoking ban were considered in a transfer function model. The concluding model included the present and one lagged observations of wind speed. The smoking ban variable was not significant.

Keywords: monitoring air quality, Hierarchical Profiling Approach, smoking.

1 Introduction

Air quality in our towns and cities are at their highest levels on busy streets, near to factories, and in inner-city areas. Poor air quality impacts on the young, the sick and elderly people's health and the environment [1]. The pollutant threshold/limit values for air quality are set out in the European Directives. The United Kingdom has National Air Quality Standards that defined levels that avoid significant risks to health.

Edinburgh is among the least-polluted of the European capitals but has a number of choke points where standing traffic causes pollutant levels to rise. Edinburgh has an Air Quality Management Area (certain areas of the city centre), which was declared on the 31st December 2000. The plan covered all the places where the annual average concentration of nitrogen dioxide is currently predicted not to attain the set target for 2005. Nitrogen dioxide is formed in a city from a build up of nitrogen oxides (NOx). Edinburgh studies have shown that 88 percent of nitrogen oxides come from road transport, with the remaining 12 percent coming from domestic heating and Edinburgh International Airport.

This investigation uses time series analysis to model and forecast air quality data, which was recorded at St. Leonards Air Quality monitoring station in the Southside of Edinburgh as a function of wind speed recorded at Edinburgh Gogarbank. It also attempts to evaluate the influence of the smoking ban introduced in Scotland on the 29th of March on the outdoor air quality recorded at that station. Specifically, this paper applies the Hierarchical Profiling Approach [2] to model the time series of CO levels recorded at the monitoring station. The HPA offers the advantage of explicitly quantifying and modelling the different components of a time series (e.g. trend and seasonality) and so provides an improved understanding of the underlying dynamics of the data.

The monitoring station is situated in a park adjacent to a medical centre car park. The nearest road is approximately 50 meters away which is a busy main road running into the city centre and out to the A7 South of Edinburgh. This Automatic point monitoring station produces high resolution measurements typically hourly or shorter period averages for particulates, oxides of nitrogen, sulphur dioxide carbon monoxide, benzene and 1,3-Butadiene. The air quality data used in this study is a United Kingdom wide monitoring network monitored by the Department for Environment, Food & Rural Affairs (defra). This data will be used for a larger study to measure background pollutant concentration levels for outdoor air quality where smokers are accommodated. Smoking causes pollutants to be released into the atmosphere. To measure the levels of the pollutants in the atmosphere from cigarettes background pollutant levels are required to be deducted to give a realistic concentration level.

2 Methodology

The Hierarchical Profiling Approach (HPA) is an event-driven time series modelling approach and can be seen as a generalisation of the Box-Jenkins intervention analysis. It has been developed in [3] and has been used in analysing difficult datasets in a number of disciplines including energy forecasting and crime modelling [4–7]. Assuming additivity, the HPA is based on decomposing the variability in time series into deterministic and stochastic and noise components. Analytically it model a time series, y_t, as

$$y_t = f(t) + Z_t, \dots \quad (1)$$

where $f(t)$, the deterministic component, is a collection of continuous functions built and fitted by the modeller and Z_t is the variable that holds the stochastic and noise components and can be modelled using a stochastic approach.

The deterministic function in Equation (1), $f(t)$ additively models the changes of the behaviour time series corresponding to identified and known events as well as the typical annual pattern of variability and the trend of the data. Hence, the first level of building $f(t)$ starts at the highest resolution of the data looking for changes in the time series corresponding to known events. The second level models the pattern at the next resolution level, and so on. For a daily dataset, for example, $f(t)$ at Level One models the within-year disturbances that are associated with salient exogenous events, at Level Two it models the weekly seasonality, at Level Three it models the annual seasonality of the data and so on. However, when estimating higher level profiles it may be necessary to aggregate the data to a lower resolution to smooth the volatility that is often observed at higher resolutions.

In theory, any deterministic function can be used in modelling the profiles and hence in building $f(t)$, so long as it is continuous and representative of the profile it aims to model. The parameters for the profile function can be estimated based on a number of criteria including least squares. Hence, the continuity of the profile functions allows the profiles to be estimated at a resolution different to the data's original and then be resampled at higher resolutions.

Assuming successful profiling of all levels, the profile-adjusted stochastic component, $Z_t = y_t - f(t)$, now likely to be weakly stationary, can then be modelled using established approaches such as ARIMA, State Space or Transfer Function models if explanatory variables are to be included in the analysis.

The HPA offers a number of advantages over other approaches in that it builds a catalogue of salient events and quantifies the corresponding changes in the time series corresponding to these events. This allows for an improved understanding of the underlying dynamics of the time series as well as the bulk forecasting of future values should the event occur again. The HPA models trend and seasonality explicitly thus allowing for further investigations of these components to be carried out. The HPA is also capable of dealing with time series with multiple seasonal components by modelling each seasonally component as an independent profile. And, the HPA can act as a powerful prewhitening technique to transform difficult datasets to stationarity.

3 Data

The raw dataset used in this research are the hourly measurements of carbon monoxide (CO) levels readings at St. Leonards Air Quality Monitoring Station in Edinburgh. These readings are made available in the public domain shortly after measurement on an Air Quality website [http://www.airquality.co.uk/archive/data_selector.php?u=7092d2de63c5e5fa319474c479f490d3]. The observations made available on the website are initially labelled as 'provisional' until they undergo a process of inspection and correction, if needed, to then be labelled as 'ratified'. This ratification process is in three stages and involves human intervention at its final stage. Consequently, it takes provisional observations up to 15 weeks to be 'ratified'. It is the assumption of this work that ratified data is representative and reliable.

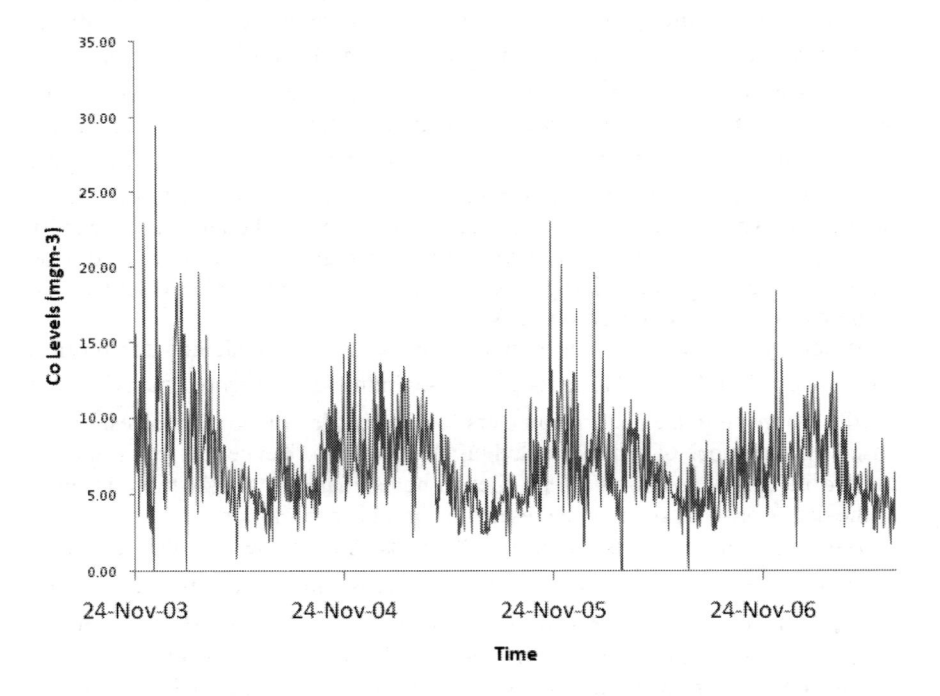

Figure 1: Daily CO levels reading at St. Leonard's Station. Annual seasonality and a number of exotic observations can be seen in the data.

The raw data is made available as hourly readings of CO Levels spanning from 24th November 2003 to 15th July 2007, a total of 31920 hourly observations, aggregated into the daily domain as the total CO readings per day. Table 1 shows a time plot of the daily CO time series in which a seasonal pattern and a number of exotic observations can be seen. The graph also shows that the first few weeks of data was erratic and seemingly ill behaved. This was confirmed by the data collection agency since these first few weeks were used to calibrate the equipment used in the measurements. Consequently, the data used in this work is the ratified daily aggregates of CO readings from 1st March 2004 to 15th July 2007 – a total of 1232 observations.

A number of exotic observations and time windows (identified subjectively) were noted in the data and great efforts have been put into trying to link these observations to events on the ground to allow for these events to be profiled. Unfortunately, to-date very little information was made available to this research about such events and information applied for under the Freedom of Information Act is still awaited leaving this application and work in progress.

In order to further the understanding of the underlying dynamics of the time series under investigation, two exogenous variables were included in the study. The maximum wind speed recorded on the day (in knots) and a binary intervention variable indicating the date of the smoking ban in Scotland (contained 0s up to the 26th of March 2006, the date of the smoking ban, and 1s afterwards) were included as explanatory variables in the study.

4 Analysis

The HPA was applied to create a profile for the normal behaviour of the data. To this end, exotic observations were subjectively identified at the dates 18th November 2005, 8th December 2005, 4th January 2006, 3rd February 2006, 17th & 18th December and 27th & 28th December 2006. In order to establish a reliable 'norm' for the data, these exotic observations were excluded from this stage of the analysis and replaced as missing values using the average of the available clean observations made at the same dates but different years.

The cleansed dataset was then aggregated to the weekly domain to reduce the volatility in the data and the harmonic regression with polynomial trend model given in Equation (2) was estimated:

$$f_2(t) = a + bt + ct^2 + \sum_{i=1}^{26}(a_i \sin(\alpha it) + b_i \cos(\alpha it)), \ldots \quad (2)$$

where $f_2(t)$ is the profile for the annual seasonality, t the time index variable, a, b and c are the trend parameters (to be estimated), $(a_i \sin(\alpha it) + b_i \cos(\alpha it))$ is harmonic i out of a total of 26, $\alpha = 2\pi/52$ is the harmonic angle, a_i and $b_i; i = 1,\ldots,26$ are the harmonic regression parameter (to be estimated).

Using the Levenberg-Marquardt procedure to fit the model in Equation (2) to the data, the following model was obtained:

$$f_2(t) = 47.975 - 0.023t + 12.206\sin(\alpha t) - 5.078\cos(\alpha t) + 2.68\sin(2\alpha t) + 3.11\sin(3\alpha t) + 1.759\sin(6\alpha t)$$

(3)

This profile was resampled in the daily domain (using $t/7$ as the time index) as shown in Figure 2. The deviations from this profile (i.e. $y_t - f_2(t)$) were inspected and no evidence for a further pattern in the data was observed. Hence, this concluded the application of the HPA with $f(t)$ set to be equal to the model in Equation (3).

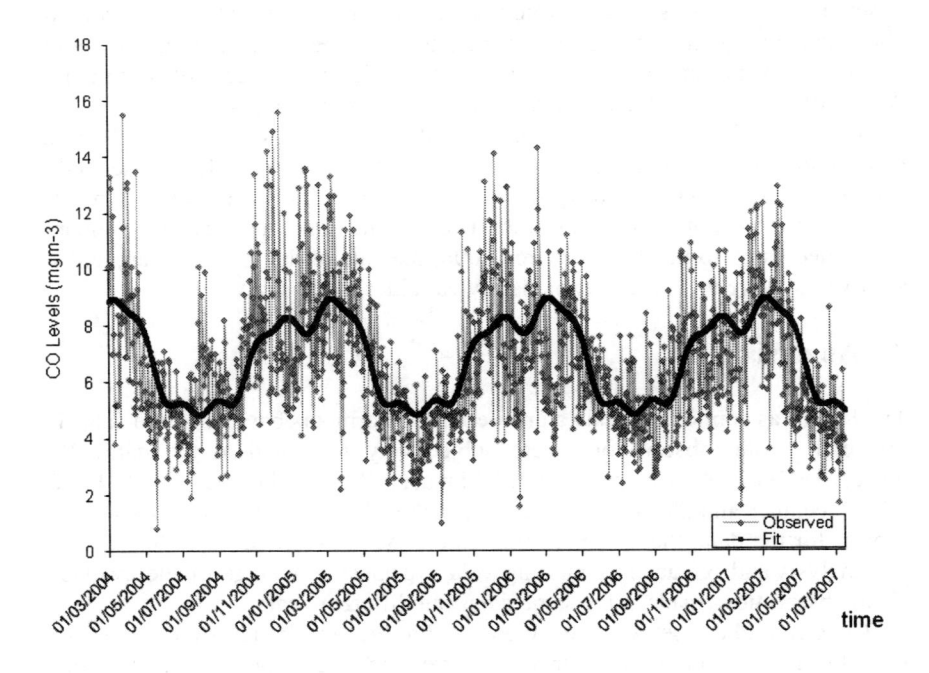

Figure 2: Daily total CO levels and HPA profile.

In the next step, the profile-adjusted stochastic component Z_t was calculated and modelled as a transfer function ARMA model using the Box-Jenkins approach with the maximum daily wind speed the smoking ban variables as inputs and the following model was obtained:

$$(1+0.65B^7)(1-B)Z_t = -.131(1-B)w_t$$
$$+ (1+0.71B+0.14B^2+0.089B^4)(1-0.74B^7)e_t \quad (4)$$

where B is the backshift operator so that $B^k Z_t = Z_{t-k}$, w_t is the maximum wind speed recorded at time t and e_t is white noise. The within sample forecasts for the original data $y_t(1)$ were then obtained by reintroducing the profile to the forecasts obtained from Equation (4) $Z_t(1)$ as:

$$y_t(1) = Z_t(1) + f(t), \dots \quad (5)$$

yielding a Residual Mean Square Error of 1.751 and Mean Absolute Error = 1.35, showing the observed and one step-ahead within sample forecasts for the total daily CO readings.

The time plot of the residuals e_t looked random with no evidence of a pattern remaining in the data and both the ACF and PACF of e_t show just one significant spike at lag 20. Figure 3 shows the observed CO readings and one step-ahead forecasts obtained from the model in Equation (4).

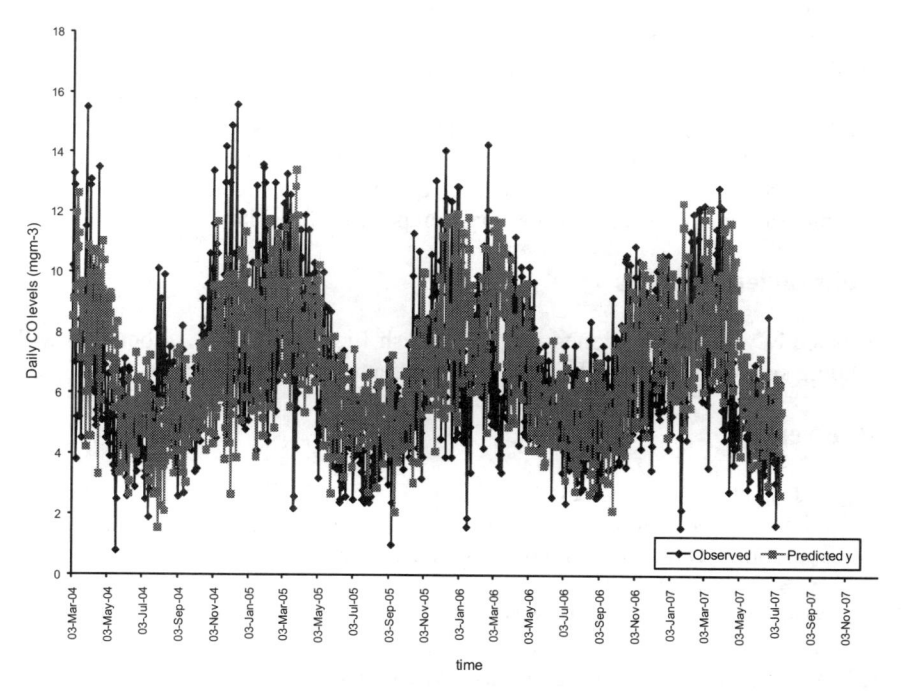

Figure 3:   Observed and within sample forecasts for CO levels.

5 Discussion

Using the HPA the different components that make up the variability in CO levels were quantified and explicitly modelled. Alongside the obvious annual seasonality in the data the HPA has identified a weak downwards trend in the data that is statistically significant. It was possible to prewhiten the data using the HPA and hence a successful transfer function model for the data was obtained. Therefore the HPA was a useful in modelling the data.

The Transfer Function model obtained for the data seems reliable since the model diagnostics all seem satisfactory and the significant spike observed at lag 20 in the ACF and PACF of the model's errors does not correspond to a logical or calendar cycle. This provided sufficient evidence to conclude that the CO levels recorded at this site are inversely related to the maximum wind speed recorded on the day. It is reasonable to speculate that this relationship relates to the role of the wind in the dispersion of airborne particles. However, there was insufficient evidence to conclude that the smoking ban had a significant effect on the levels of CO recorded at this station.

As in any statistical investigation, the reliability of the results of this study depends mainly on the reliability of the data used in the analysis. There was very little information available to this study for any Level one profiles to be constructed despite the efforts put in to identify any salient events that could be associated with the exotic observations identified in the data. In addition, there are only three seasonal cycles in the span of the data with only one complete cycle post the smoking ban. Therefore, while it is the authors' belief that the results presented in this work are as reliable as can be achieved given the available data, further analysis needs to be carried when more data becomes available. It is expected that the downwards trend that was just significant in this study (with 95% CI -0.046 to 0.000) will become stronger. In addition, having one or two more seasonality cycles in the data would most certainly yield a more reliable estimate of the annual seasonal component of the data.

Acknowledgements

This study was commissioned by the Scottish Licensed Trade Association with funding support from the UK Tobacco Manufacturers' Association.

References

[1] K-J. Chuang, C-C. Chan, T-C. Su, C-T. Lee and C-S. Tang, The effect of urban air pollution on inflammation, oxidative stress, coagulation, and autonomic dysfunction in young adults. *Am. J. Respir. Crit. Care Med.,* **176**, pp. 370–376, 2007.

[2] Al-Madfai, H., Ameen J., Ryley, A., Daily electricity demand forecasting: a Hierarchical Profiling Approach. *ETK/NTTS*, Crete, 2001.

[3] Al-Madfai, H., Weather corrected electricity demand forecasting. School Of Technology. 2002, University of Glamorgan: Pontypridd.

[4] Al-Madfai, H., Ivaha, C. & Ware A., The Dynamic Spatial Disaggregation Approach to Geo-Temporal Crime Forecasting. *The Ninth Crime Mapping Research Conference*, Pittsburgh, USA, US Department of Justice, 2007.

[5] Ivaha, C., Al-Madfai, H., Higgs, G., Ware A. & Corcoran J., The simple satial disaggregation approach to satio-temporal crime forecasting. *International Journal of Innovative Computing, Information and Control (IJICIC)*, **3**(3), pp. 509–523, 2007.

[6] AL-Madfai, H., Ivaha, C. & Ware, A., Hierarchical Profiling of daily crime time series data as a precursor to modelling. *International Symposium on Forecasting*, San Antonio, Texas, 2005.

[7] Al-Madfai, H., Ameen, J. & Ryley, A., The Hierarchical profiling approach to STLF of multi-year daily electricity demand in South Wales. *International symposium on forecasting*, Sydney, 2004.

Definition of PM_{10} emission factors from traffic: use of tracers and definition of background concentration

E. Brizio & G. Genon
Turin Polytechnic, Italy

Abstract

Within air quality management, one of the most important emission factors that should be known is PM_{10} (exhaust and non-exhaust flows) coming from traffic. The existing data and models produce very variable emission factors, also according to climate, sanding conditions, road material; thereby, a more general approach, based on the use of traffic tracer such as CO and NO_x, can be put into practice in order to have a reliable assessment of PM emissions. Within the tracer approach, the definition of the background concentration of pollutants is of prime importance but representative measurements could be not at disposal. The present work is an attempt to define background concentration by considering the average concentration measured in an urban area during the night (0-5 am), when traffic and industrial instantaneous contribution are negligible and the heating plants are switched off. The so called "night method" has been tested and validated by means of the OSPM model, an atmospheric dispersion model studied for street canyons, for CO and NO_x. In the case of CO, the results were surprisingly satisfactory and the method could be considered consistent, whereas NO_x turned out to be not reliable as a tracer because of the chemical reactions that occur in the troposphere.

Keywords: traffic, air quality, atmospheric modelling, background concentration, OSPM, CO, PM_{10}, NO_x, tropospheric chemistry.

1 Introduction

The air pollution situation of many European urban areas doesn't present indications of substantial improvement, in spite of the adoption of specific policies of limitation and emission reduction; actually, these actions, without other activities, like clear understanding of emissive and atmospheric phenomena influencing the result, are not able to lead the air quality back to desired standards. The air quality situation is even more critical in areas like Northern Italy, where the pollution levels (in particular PM_{10} and NO_2) are very high because of the low wind conditions of the Po Valley that don't help the dilution of the pollutants. In order to obtain some improvements for air quality, the regional decision makers are trying to define some intervention policies, such as the limitation of old vehicles, in particular diesel cars before EURO II and gasoline cars before EURO I. The present paper deals with PM emissions from traffic, considering exhaust and non-exhaust particles. The investigated area is the town of Cuneo, 50,000 inhabitants, placed in the South of Piedmont, N-W Italy. The mean wind speed in the area is quite low, around 1.4 m/s, with an high percentage of calm hours (< 1 m/s), almost 30%, and a typical bimodal behaviour around 40-60 degrees clockwise from the N.

Traffic flows for all the main street of the town, registered by magnetic counters, account for more than 308,000 vehicles per day.

2 PM emissions from traffic

PM emissions from traffic can be divided into three main groups (Ketzel et al. [1]):

- direct exhaust emissions, mainly fine fraction ($PM_{2.5}$), that can be calculated by means of different emission databases (i.e. COPERT, UBA, TNO, CORINAIR, UK-TLR);
- non-exhaust emissions deriving from brakes wear (PM_{10}-$PM_{2.5}$);
- non-exhaust emissions from road abrasion, tyre wear and road dust re-suspension that are found partly in the fine fraction ($PM_{2.5}$) and mostly in the coarse fraction (PM_{10}).

PM emissions are strongly influenced by external factors as road condition (wetness, salting, sanding, road material) and use of studded tyres.

Literature data reports several different model to define in particular non-exhaust emissions that can be very variable, as pointed out by Figure 1.

As one can easily understand, the provided database are quite variable and, most of all, they have been obtained in correspondence to precise conditions of weather and road characteristics that are strongly site specific. A more general and reliable approach could be the so called "tracer method", used within the Swedish Empirical Model [2] in order to obtain the total PM emission factor, including both direct emissions and emissions from the dust layer. The method can be written as follows, using for example CO (or NO_x) as tracer:

$$e_f^{PM} = e_f^{CO} \cdot \left(\frac{C_{PM}^{roadside} - C_{PM}^{background}}{C_{CO}^{roadside} - C_{CO}^{background}} \right)$$

where e_f^{CO} is the emission factor for CO (or NO_x), often more well known than the PM one.

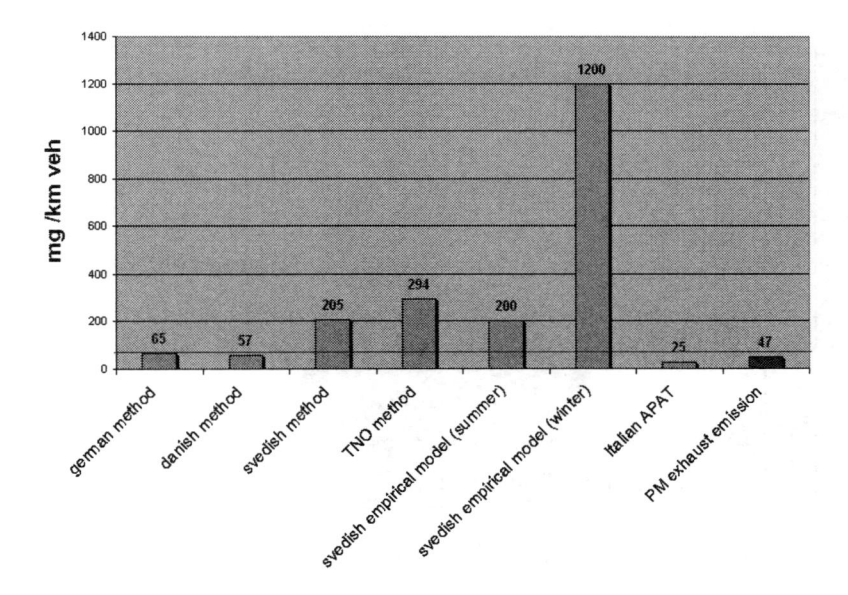

Figure 1: Non-exhaust PM$_{10}$ emission factors (fleet mix).

3 Background concentration and stagnation phenomena

As already described, the "tracer method" needs the definition of the background concentration of the involved pollutants, a very critical parameter. The background pollution level is usually measured in the countryside or at a rooftop within a urban context. In the analyzed area, all the monitoring stations are placed in urban areas and the measured values are almost the same. So we don't have any background monitoring station at disposal for our purposes. Moreover, on the basis of our experience, the background concentration that can be measured in the countryside is not the same as the one that can be measured in a urban environment; this can be observed, for example, when the traffic is totally stopped for sanitary reasons (the so called "no traffic Sundays") in the cities. This aspect is quite reasonable if one considers that the background concentration is also due to a stagnation effect of the pollutant emitted in the previous hours only partially dispersed by the wind and the atmospheric turbulence (mechanically and thermally induced); this way the background concentration is strongly dependent on the emission mixture and the dispersion capabilities of the area. For instance, the background concentration measurable in a street canyon would be mostly correlated to traffic as the main emission source and it would be probably higher than that measured in an outside area (or also in a urban background station placed on a rooftop, as indicated by Oemstedt et al. [2]) because of the low dispersion possibilities of a urban canyon if compared to a more open area. The described assumption can be better explained by Figure 2, taken from Hansson [3].

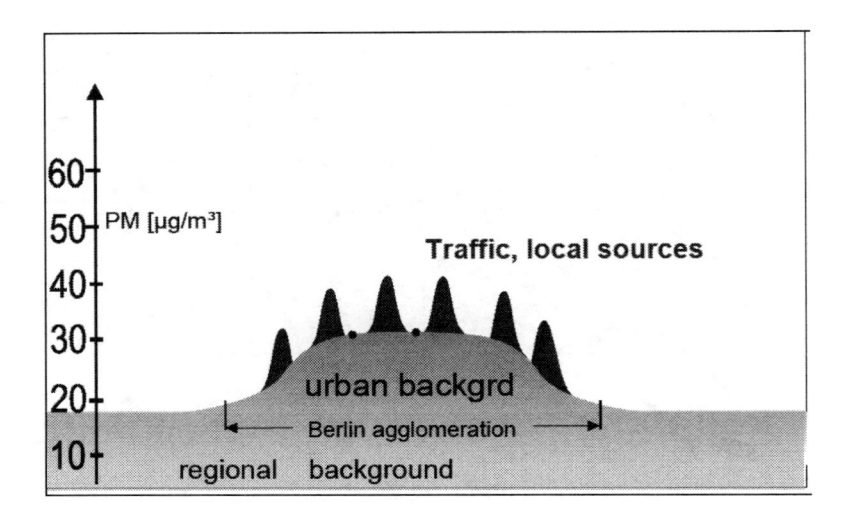

Figure 2:   Example of urban background concentration.

The difference between the background concentrations measured in a rural site, at a rooftop or at the bottom of a street canyon has been also observed in [4]. There, one can also notice that during the night the three measured levels are very close, so that this level seems to be a good approximation of the representative background concentration of the studied area.

Based on the reported arguments, we tried to define the background concentration for stable parameters, such as CO and also PM_{10}, by calculating the average concentration from 0:00 am to 5:00 am, when the traffic is low in small towns and the heating plants are not working; moreover, the instantaneous concentrations at the ground level due to industrial plants are very low, so that we assumed those contributions as negligible in this phase. The described way to define the background concentrations, called "night method", can be considered valid for pollutants such as CO or PM_{10}, that are quite stable in atmosphere (the lifetime is respectively in the order of months and weeks), but an attempt will be carried out also for NO_x.

Once the background concentration has been defined, the procedure can be validated by means of reliable atmospheric dispersion models that can predict the instantaneous effect of traffic (and the other sources) on air quality. This contribution can be added to the background concentration and compared to the measured daily concentration. In order to simulate the effects of traffic emissions on the urban air quality we used the Operational Street Pollution Model (N.E.R.I., Denmark [5]).

Figure 3 reports the comparison of measured and modelled CO daily concentrations at the monitoring station of Cuneo. As one can easily observe, the modelled values reproduce the measured one in a satisfactory way, the correlation coefficient is very high (r=0.977), so that the model and the approach can be considered reliable for our purposes.

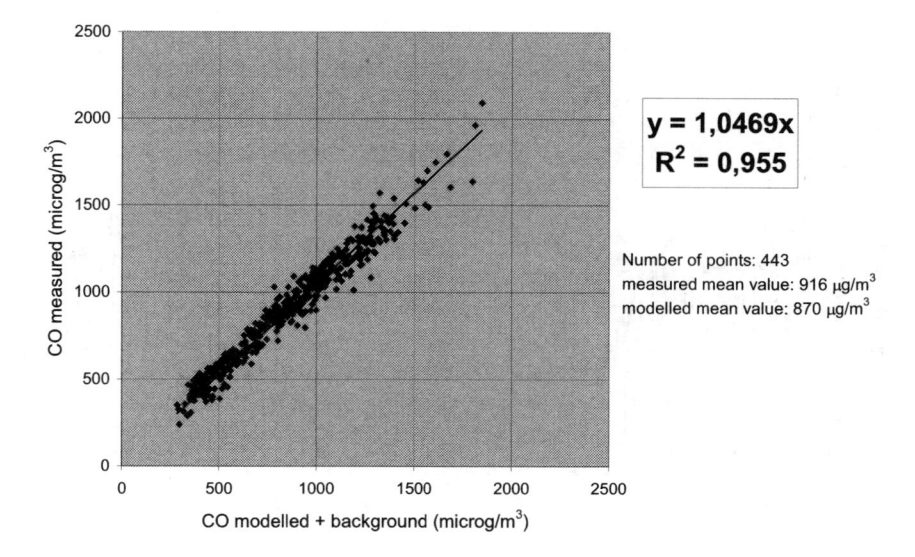

[Chart: CO measured (microg/m³) vs CO modelled + background (microg/m³); y = 1.0469x; R² = 0.955; Number of points: 443; measured mean value: 916 µg/m³; modelled mean value: 870 µg/m³]

Figure 3: Comparison of measured and modelled CO daily mean concentrations in Cuneo.

It should be remembered that, as far as the CO emissions deriving from the heating plants and the industrial activities are concerned, their effect on air quality can be considered around 5 µg/m³, as maximum daily concentration at the ground level; the reported levels is negligible if compared to the CO concentrations measured in the analyzed area (300-2000 µg/m³).

As a consequence, we may say that the instantaneous effect of sources, other than traffic, is very low on the air quality of the analyzed area and so it is acceptable to neglect them when applying the described method, as we assumed in the present paper. This reasoning can be extended to other pollutants as well, such as PM_{10} and NO_x, in the studied region.

4 The background concentration for unstable parameters: the case of NO_x in Southern Piedmont

As already pointed out in the previous chapters, the "night method" turned out to be a reliable tool in the case of stable parameters such as CO and PM_{10}. Nitrogen oxides (NO + NO_2) are considered and used as tracer of traffic emissions as the sum of NO (as NO_2) and NO_2 remain constant within the photolytic cycle.

The time scales characterizing these reactions are of the order of tens of seconds; in the case of residence time of pollutants in a street canyon comparable to the mentioned reaction time, the chemistry of NO_x could be restricted to the described cycle [6]. As a consequence, the "night method" applied to the sum of NO_2 and NO (calculated as NO_2) should be theoretically viable in order to obtain the background concentration.

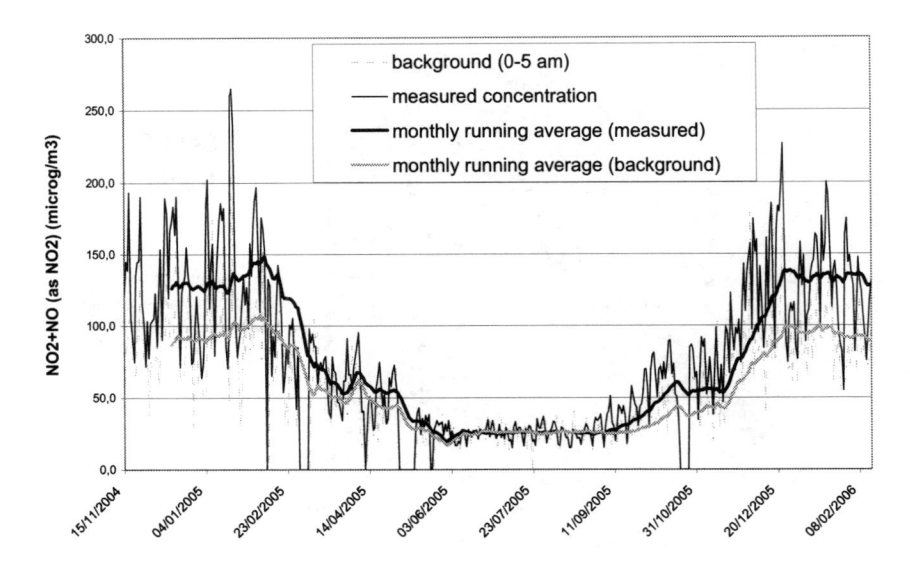

Figure 4: Measured and background daily concentrations for NO_x in Alba.

Thereby, we applied the same approach described in chapter 3 for carbon monoxide to NO_x for several urban contexts in Southern Piedmont (Cuneo, Alba, Bra, Asti). Here we found out an unexpected behavior of the NO_x background concentration calculated by means of the "night method", as pointed out by Figure 4. As a matter of fact, during the summer the night average concentration of NO_x is equivalent to the daily average concentration, and this can be observed for all the analyzed monitoring stations. The unreliability of the "night method" applied to unstable parameters is confirmed by the prediction of overall daily concentration (background + traffic instantaneous concentration calculated by OSPM at the monitoring station placed in Cuneo), reported in Figure 5. In this case, we won't find a satisfactory correlation, on the contrary a clear overestimation of the measured concentrations can be observed.

The behaviour of NO_x background concentration during the warm season seems to suggest that the sum of NO (as NO_2) and NO_2 is not constant within the residence time of these pollutants in the street canyon, on the contrary, other complex reactions involving a removal or a production of NO_x may occur, respectively during the daytime or at night. In this case, the presence of other chemical mechanism (in addition to the photolytic cycle of NO_x) with reaction times longer than seconds could be justified by large residence times of pollutants inside the street canyon, due to the stagnation phenomena within the Po basin.

The phenomenon has been analysed by comparing the daily trends of NO, NO_2 and O_3 during the year and the summer at some monitoring stations placed in the South of Piedmont (Brizio and Genon [7]). The study seems to suggest that, during the day, NO is consumed without being transformed in NO_2 and also NO_2 appears to be removed by chemical reactions different from photolytic cycle (that should maintain constant the sum of NO_x). Anyway, during the summer the trends of NO_x don't show the influence of daytime emissions, unlike the yearly behaviour, confirming the presence of a "sink".

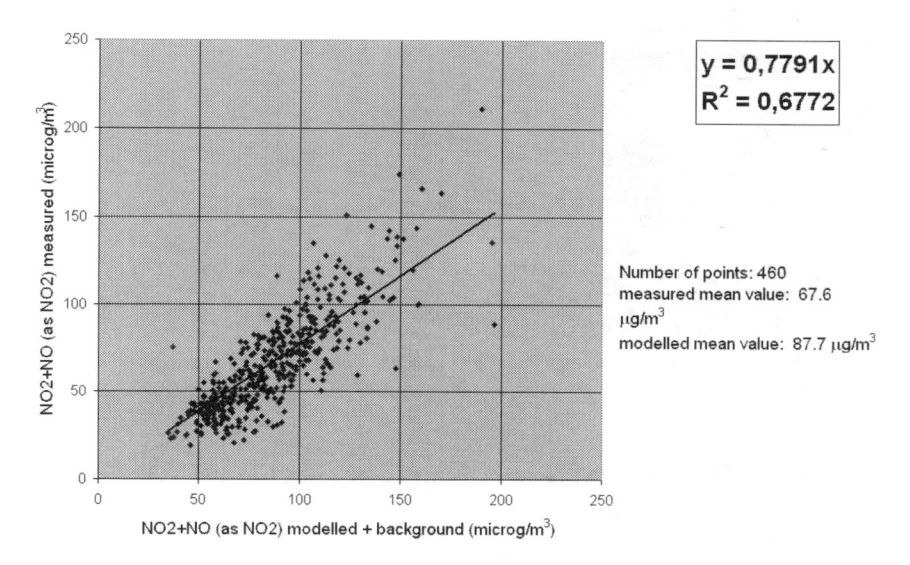

Figure 5: Comparison of measured and modeled NO_x daily mean concentrations in Cuneo.

5 Definition of PM_{10} emission factors by means of tracers

As described in chapter 2, pollutants emitted by traffic such as CO and NO_x can be used as tracer to define emission factors with an higher level of uncertainty, signally PM_{10}. Figure 6 points out PM_{10} emission factors (exhaust + non-exhaust) obtained in Cuneo by using CO and NO_x as tracer; as one can notice, the data are quite similar during the cold seasons, while during the summer the emission factors obtained through NOx tends to be strongly overestimated (see also Figure 7). As already described, NO_x is not as stable as the other pollutants, CO and PM_{10}, and this behaviour lead to a wrong definition of the reference NO_x concentrations to be run within the "tracer model". As a consequence, its use as a tracer is at least questionable.

On the contrary, CO turned out to be suitable for a reliable application of the "night method" for the background definition and the atmospheric dispersion modelling so that its use as a tracer seems to be the most adequate choice (Brizio et al. [8]). The calculated PM_{10} emission factor, as obvious, changes according to the season and the wetness of the atmospheric conditions. The emission factor calculated using CO as a tracer varies around a mean value of 257 mg/km/veh ± 164 mg/km/veh, with a maximum value of 1136 mg/km/veh; the reported data are much higher than the values referred by the "German method" and the "Danish method" while we can find a good agreement with the CEPMEIP-TNO

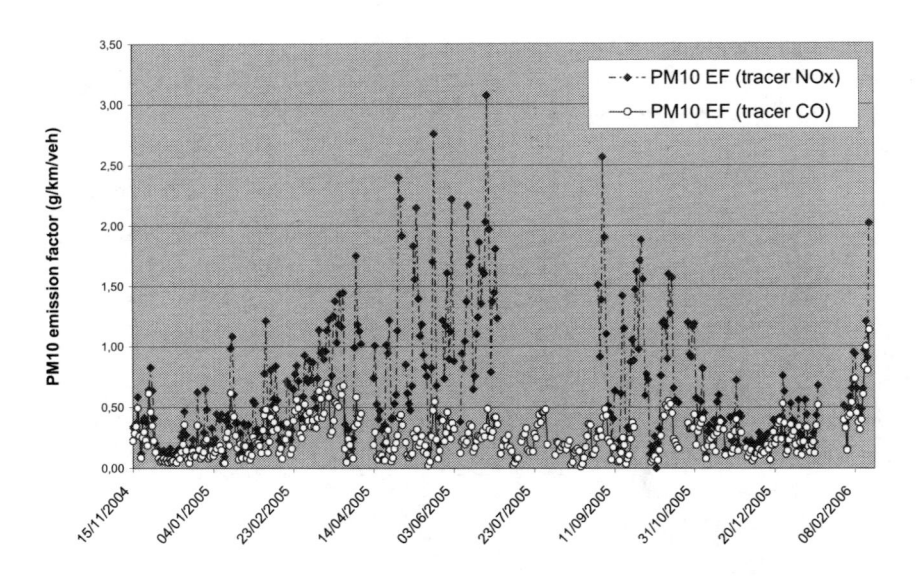

Figure 6: PM$_{10}$ emission factors calculated by using CO and NO$_x$ as tracers.

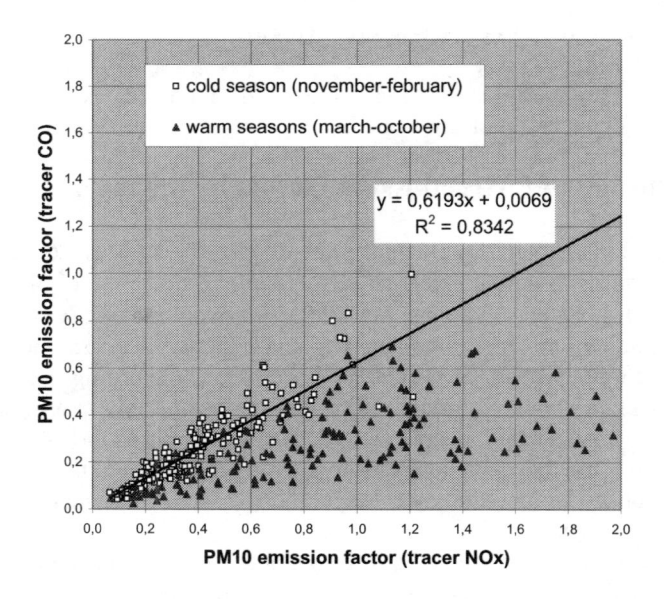

Figure 7: Comparison of calculated PM$_{10}$ emission factors.

suggested data and most of all with the Swedish values, in particular the described range 200-1200 mg/km/veh.

Based on the PM$_{10}$ emission factors obtained by means of the "tracer method" and the OSPM model, we calculated the PM$_{10}$ concentrations, as reported in Figure 8. Also in this case the correlation is very good (r=0.959), even though the model lightly underestimate the measured concentration. The mean deviation D, defined as follows:

$$D = \sum_{i=1}^{n} \frac{|Cm_i - Cc_i|}{n} \cdot \frac{1}{\overline{Cm}} \cdot 100$$

where Cm is the measured concentration and Cc is the calculated concentration, is less than 17%.

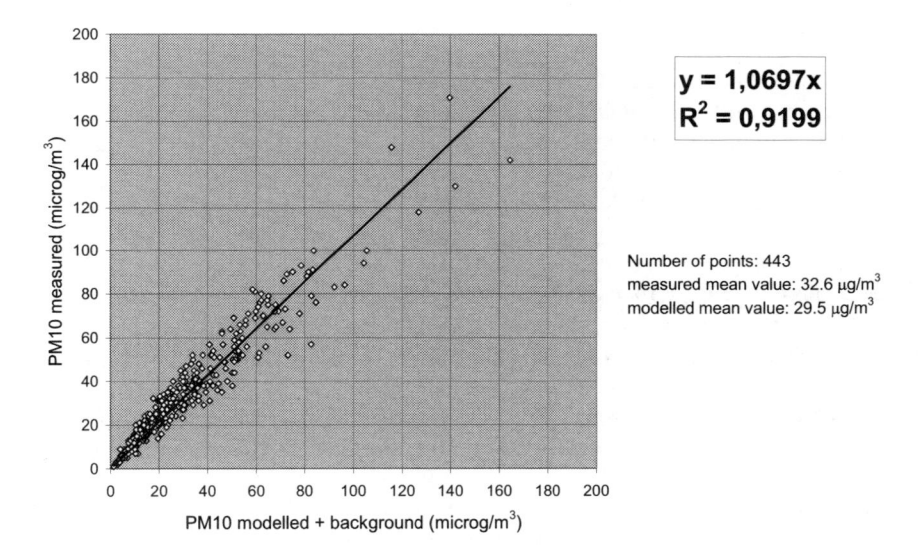

Figure 8: Comparison of measured and modelled PM_{10} daily mean concentrations.

6 Conclusions

PM_{10} emissions from traffic is a very important parameter when dealing with emission inventories, air quality management and policy making. Unfortunately, its definition is not straightforward as it depends on several site specific parameters so that an assessment based on tracers within the area of interest seems to be advisable. In this context, not all the pollutants emitted by vehicles are suitable for a reliable use of the "tracer method". In particular, NO_x is a very complex parameter and the prediction of its behaviour may be very complicated. Based on this conclusion, the application of the "night method" in order to define the NO_x background concentration could be not advisable, at least during the summer. More generally, in the analyzed area, the use of NO_x as a tracer for traffic could be questionable as well, because of the rapid chemical reactions that can occur also at time scales characterizing the dispersion of pollutants within a street canyon. On the contrary, the use of CO, a stable and easily predictable parameter, could lead to consistent results: in the studied area we obtained total

PM emission factors varying around a mean value of 257 mg/km/veh ± 164 mg/km/veh, with a maximum value of 1136 mg/km/veh, while the mean exhaust PM emission factor calculated by means of the Copert3 model is 47 mg/km/veh. The described methodology indicates that 80% of the total emitted PM_{10} originates from non-exhaust emissions and so it is evident that policies reducing the exhaust releases of the park or limiting diesel vehicles without particle traps can have a limited effect on the air quality.

References

[1] Ketzel, M., Omstedt, G., et al., *Estimation and validation of PM2.5/PM10 exhaust and non exhaust emission factors for street pollution modelling*, 5th International Conference on Urban Air Quality (UAQ 2005), Valencia, Spain, 29-31 March 2005.
[2] Omstedt, G., Bringfelt, B., Johansson, C., *A model for vehicle-induced non-tailpipe emissions of particles along Swedish roads*, Atmospheric Environment 39, 6088–6097, 2005.
[3] Hansson, H.C., Urban *aerosols, state of knowledge, on presence and effects*, http://akseli.tekes.fi/opencms/opencms/OhjelmaPortaali/ohjelmat/FINE/fi/Dokumenttiarkisto/Viestinta_ja_aktivointi/Seminaarit/Parnet_2005/HC_Hansson_Parnet.pdf.
[4] Johansson, C., Hansson, H.C., *PM10 and PM2.5 gradients through rural and urban areas in Sweden*, www.nilu.no/projects/CCC/tfmm/paris/se.doc, October 2006.
[5] Berkowicz, R., Olesen, H., Jensen, S., *Operational Street Pollution Model, User's Guide to Win OSPM*, National Environmental Research Institute, Denmark, 2003.
[6] Berkowicz, R., Hertel, O., larsen, S., Nielsen, M., Sorensen, N., *Modelling traffic pollution in street*, Ministry of Environment and Energy / National Environmental Research Institute, Denmark, 1997.
[7] Brizio, E., Genon, *Definition of background concentration of pollutants in a small urban area*, Process Safety and Environmental Protection (PSEP), Elsevier, submitted January 2008.
[8] Brizio, E., Genon, G., Borsarelli, S., *PM emissions in an urban context*, American Journal of Environmental Sciences 3 (3): 166–174, 2007, ISSN 1553-345X.

LNG dispersion over the sea

A. Fatsis[1], J. Statharas[2], A. Panoutsopoulou[3] & N. Vlachakis[1]
[1]Technological University of Chalkis,
Department of Mechanical Engineering, Psachna Evias, Greece
[2]Technological University of Chalkis,
Department of Aeronautical Engineering, Psachna Evias, Greece
[3]Hellenic Defence Systems, Athens, Greece

Abstract

A numerical study of heavy gas dispersion over sea from an LNG storage facility at Renythousa Island in Greece is examined for the various hazards imposed on the population of the surrounding areas. The study is done by using the code ADREA-HF solving the conservation equations of mass, momentum and energy for the mixture of the heavy gas and air, as well as the mass fraction of heavy gas. Distributions of methane, ethane and propane concentrations for various time instants after the gas release are obtained, identifying the riskiest zones using scenarios based on weather forecast data. Thus the contribution of this work is to show how the available time for the authorities to safely evacuate the nearest affected areas can be evaluated.

Keywords: heavy gas dispersion, Navier-Stokes equations, turbulence model, flammable plume zone.

1 Introduction

Large quantities of substances that are hazardous for the public and for the environment are produced or utilized in large quantities by most modern process industries that are often sited near populated areas. These substances being toxic, radioactive, flammable or explosive during an accident in the production process, transportation or storage may provoke a release of gas (or gas-liquid) clouds or plumes that are denser than air. The density difference of the formulated "heavy gas" introduces buoyant forces that can distort significantly the local wind pattern (Deaves [1]) and the other important atmospheric transport parameters such as the atmospheric stability and the turbulent structure and level.

The reason for studying dense gas dispersion is to analyze the hazards associated with the release as a function of space and time under specific conditions and accident scenarios, in order to predict how such hazardous material is dispersed, under the specific topographic and atmospheric conditions for the safety aspects of plant design and position, as well as for contingency plans for people in danger.

The ADREA–HF code was chosen to study the accident scenario at Revythousa Island in this manuscript due to its ability to describes best the effects of obstacles in dispersion of a dense gas (been successfully used in dispersion studies such as Liquid Ammonia Dispersion) as well as the practical aspect of the free access obtained by the "National Center Scientific Research" (NCSR) DEMOCRITOS Institute of Nuclear Technology and Radiation Protection in Greece.

2 Numerical modelling

The ADREA –HF code has been developed to calculate the dispersion of heavier than air gases on ground of irregular complexity. It assumes a mixture of two components: the heavier than air gas and the ambient air. The code solves the conservation equations of mass, momentum and energy for mixtures and the mass fraction of heavy gas. Furthermore the code includes a heat conduction equation for a thin layer of ground (20–50 cm). The following equations are solved for a particular geometry of region using a finite volume methodology. For closing the system of equations the ADREA-HF code uses turbulence closure scheme. In the calculation of this project one equation turbulence model has been used, because it was found that it best describes the effects of obstacles in dispersion of a dense gas.

In the code ADREA–HF, zero- equation, one- equation and two- equation turbulence models are available. The mathematical description of ADREA –HF model has been presented by Bartzis et al. [2].

The eddy viscosity is calculated as follows:

$$K_i^m = \begin{cases} C_\mu \cdot k^{1/2} \cdot l_t & \text{for } k > 0 \\ K_\infty & \text{for } k = 0 \end{cases} \quad (1)$$

where k is the turbulent kinetic energy, l_i $(i = x, y, z)$ is an effective length scale, $K_\infty = 0.5 \ m^2 / \sec$ and $C_\mu = 0.1887$ is an empirical constant.

For the space discretization of the equations, the finite volume technique by Patankar [3] was used in the ADREA-HF code. The computation domain is divided into a number of non-overlapping finite control volumes over which the differential conservation equations are integrated. The integral equations obtained can be expressed as a sum of four parts: the accumulation, the convection, the diffusion and the generation part.

For the time derivative in the accumulation term, a first order backward difference (implicit) scheme is applied. In the convection term, the "upwind difference" scheme is adopted. For the diffusion term, use of the "central differencing" is made. A pressure discretized equation is formed from the integrated mass conservation equation. A pressure Poisson equation is formed, utilizing the discretization equations for the velocity components.

3 The LNG Terminal at Revythousa Island

Revythousa. Island was chosen in 1988 to become the LNG receiving terminal from Algeria. This is located at the southern end of the natural gas system and about 30 km west of Athens, the main gas consumption center, figure 1. The islet, which is 500 meters from the nearest point on the mainland, is 600 meters in length, 250 meters in breadth, rising from the sea surface level to a height of 48 meters.

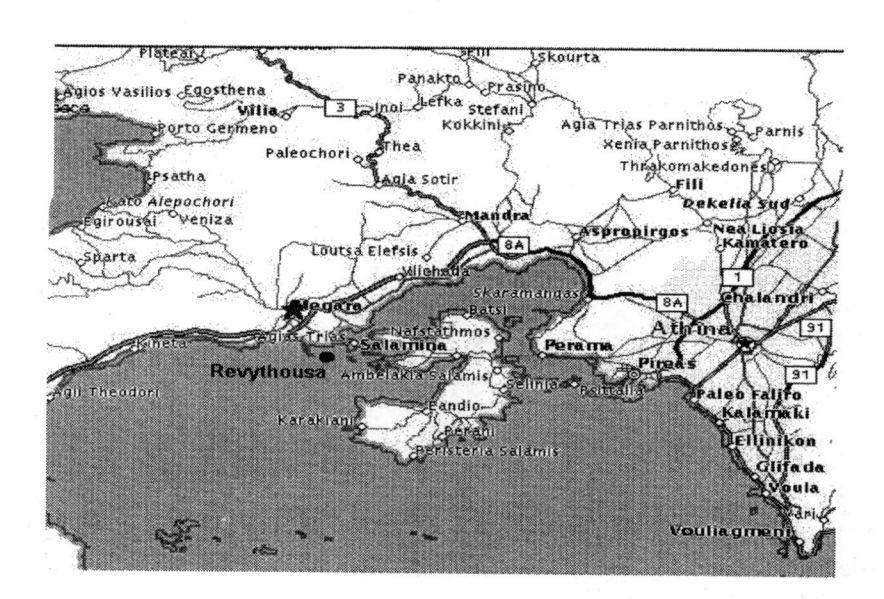

Figure 1:   Illustration of the region in which Revythousa island is located.

The terminal is designed for normal operation as well as peak send out for gas in the network as needed, from minimum send out to emergency peak send out of 830 m^3/h equivalent liquid LNG volume. Send out gas to the grid is in the range from 31 to 62 barg. Two identical full containment LNG storage tanks of total useful capacity 130000 m^3 have been constructed. The tanks have top connections only and internal piping for top and bottom loading. The operation of the terminal allows the ship unloading to one storage tank, while the other tank is on grid supply service (DEPA [4]).

4 Accident scenario – Computational domain – Topography

On the present release scenario considered, a great quantity of LNG leaks from a sudden break in the flank surface of a full cylindrical tank. The thermodynamic variations of LNG from storage conditions to ambient conditions can be considered as isentropic since these variations are very fast. The main quantity of Natural gas leaks in the ambient atmosphere at liquid phase.

It is assumed that the cylindrical tank is full and the whole quantity of LNG leaks to ambient air within 45 min through a parallelogram break. The average leakage mass of two phase mixture was calculated and was equal to 10709 kg/s. The exit temperature of the two phase mixture is assumed to be at the boiling point of Methane which is -160° C.

The three-dimensional domain consists of a parallelepiped which its two-dimensional projection is shown at figure 2 (dashed parallelogram).

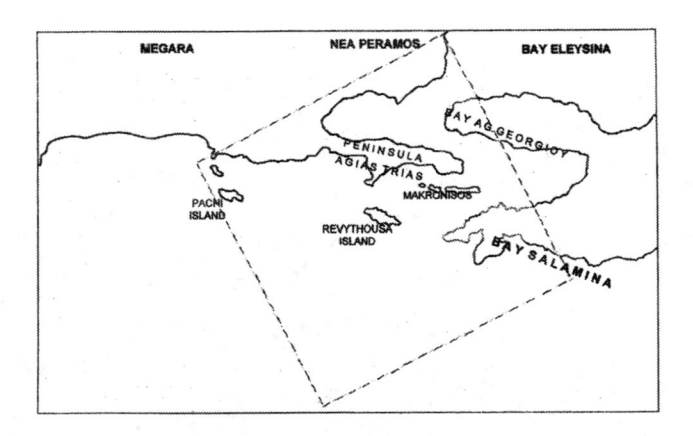

Figure 2: The Geographic region chosen as the calculation domain.

The dimensions of parallelepiped are the following: X-axis: 10km, Y-axis: 10 km, Z- axis: 1 km. The calculation domain is divided to series of smaller cells (right parallelepipeds) through out the complex 55(X-axis) 27 (Y-axis) 31(Z-axis) amounting 46035 cells. The three-dimensional grid is not uniform; it is more condense close to leakage source, which is located at the cell with coordinates i= 28, j=1, and k=2, and close to land. The smallest cell has length 40 m at X,Y directions and 2 m at Z direction. The computational calculations of the ADREA – HF code take in to consideration a thin layer of land. Geographic Military Service of Greece kindly provided the topography – input data at digital form.

5 The computational steps

In ADREA –HF, the complete simulation of the transient scenario consists of the following three stages:
- Calculation of the 1-D velocity profile

- Calculation of the 3-D steady state wind field
- Calculation of the 3-D cloud dispersion

In case of an unchanged wind field, only the last stage had to be repeated for the simulation of a new scenario.

5.1 Calculation of 1-D velocity profile

The velocity on the top plane was set to a specified input value in order to obtain a desired value at 10 m height. The ground temperature was set to 11°C. Vertical temperature profile was assumed to correspond to neutral stability conditions i.e.,

$$\frac{dT}{dz} = -\frac{g}{C_P} \qquad (2)$$

where g is the gravitational acceleration and C_P is the heat capacity (constant pressure).

5.2 Boundary conditions

The three-dimensional steady state wind field over the given geometry was calculated by solving the three momentum equations, the turbulent kinetic energy equation and the pressure equation.

For initialization the one-dimensional profiles calculated previously were applied at the three dimensional domain. The boundary conditions were specified as follows according to Statharas et al. [5]:
- On the ground the calculated parameters are assumed to have a specified value, except for temperature where a zero gradient was assumed.
- On the source surface a value plus a zero gradient was specified for all the parameters
- For the input domain plane, the boundary conditions are the same with the ones at the source surface, except for the vertical velocity component where a value was specified.
- For the other domain planes, it is assumed that the parameters have a zero gradient except for the vertical velocity component, which is specified in a value.

6 Results and discussion

The upper and lower flammability limits are particular for each component of Liquid Natural Gas. The riskiest zone takes place between lower and upper flammability limits. The worst instant time is, when the riskiest zone covers the greatest field of overland. Figure 3b shows the riskiest zone of methane at sea level at time t=500 s after the accident.

The methane concentration at sea level, at t=400 s is indicated by Figure 3a that with D2 wind condition and wind direction towards Megera town, the riskiest zone covers section of overland field as Peninsula Agias Trias, Bay

Salamina, and Makronisos Island. Furthermore, the width of flammable zone is significant at this instant time because of the dispersion of Methane dense cloud to atmosphere, which just begins.

The methane concentration at sea level, at t=500 s is indicated by figure 3b that with D2 wind condition and wind direction towards Megera town, the flammable zone continues to cover section of overland field as Peninsula Agias Trias, Bay Salamina, and Makronisos Island. According to the above graph, it can be seen that the flammable cloud field has slightly reduced due to continuative dilution by ambient air. Furthermore, it should be pointed out that the form and the extent of upper flammable limit are decreased and differ from the figure 3a.

The methane concentration at sea level at t=1000 s is significantly different with respect to previous situations. It is noted that with D2 wind condition and wind direction towards Megera town, the riskiest zone continues to cover section of overland region as Peninsula Agias Trias, Bay Salamina, and Makronisos Island. It is worth noting that the figure 3c shows the ring of flammable cloud significantly narrowed, while the upper flammability limit has been extended. This means that the lower flammability limit field is reduced and the upper flammability limit is increased. Consequently the flammable ring is being narrowed.

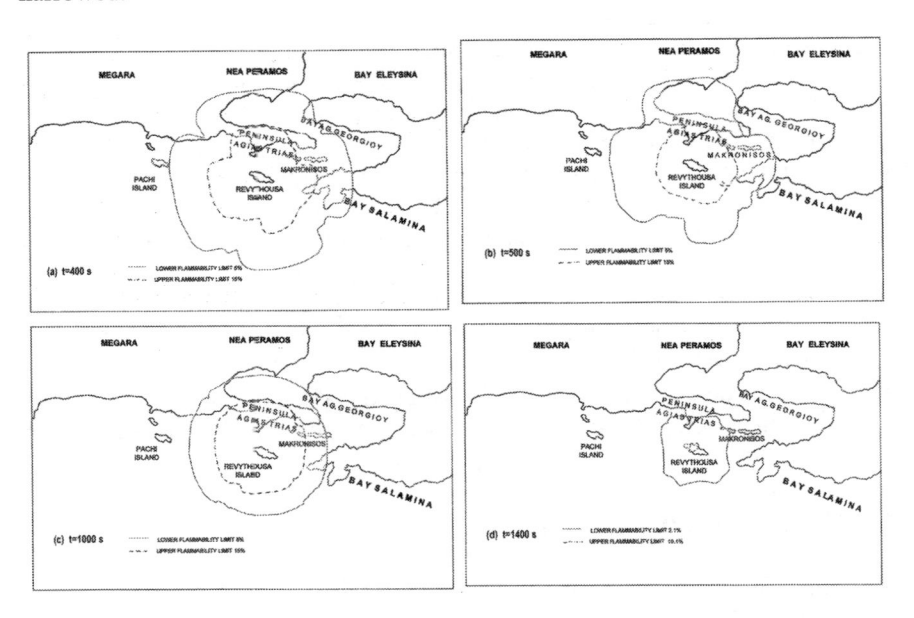

Figure 3: The methane distribution at different time instants after the release.

The shape of Methane dense cloud as it is illustrated at figure 3d of t=1400 s is similar to that of t=1000 s. It has a slight difference of upper flammability limit field, which is noted slightly reduced, but the flammable cloud continues to cover section of overland region as Peninsula Agias Trias, Bay Salamina, and

Makronisos Island. A 2-D view of the leakage plume at various time instants is shown in Figure 4.

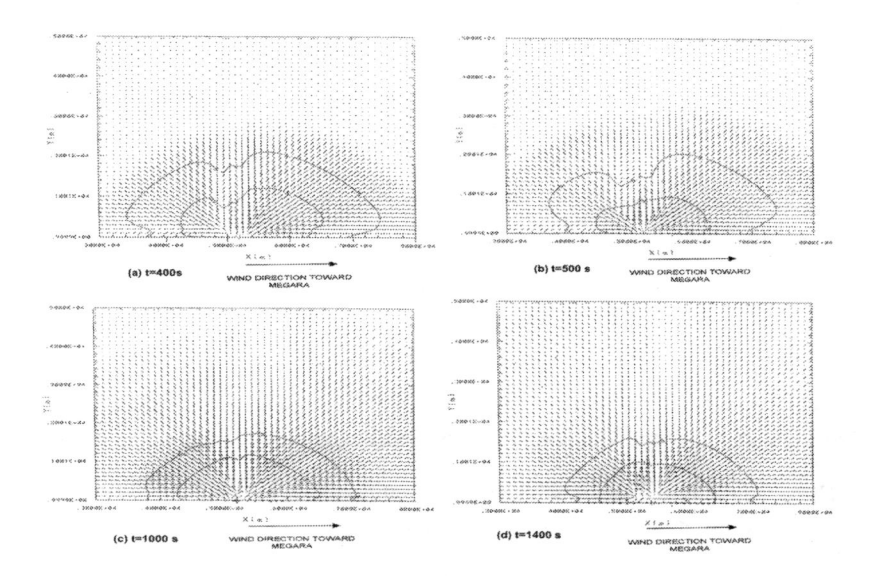

Figure 4: 2-D view of the methane leakage plume at various time instants.

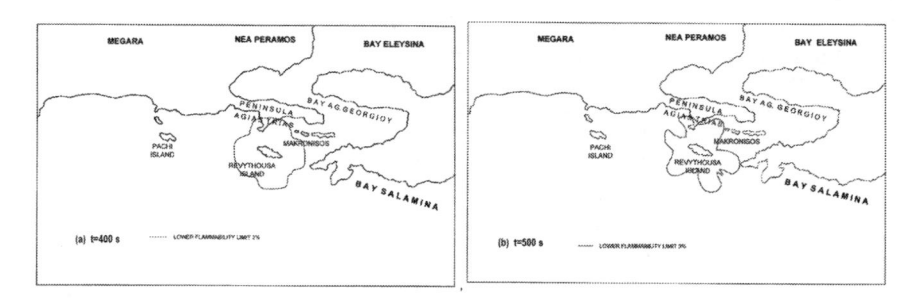

Figure 5: The ethane distribution at different time instants after the release.

The plume formed for ethane at sea level at t= 400 s with wind conditions D2 and with direction towards Megara town cover a section of overland region as Peninsula Agias Trias and Revythousa Island. Figure 5a has shown that the flammable cloud is in the form of umbrella and not in the formed ring. This happens because during the time, ethane has not reached at the upper flammability limits. The flammable cloud has not been diluted enough and the magnitude of flammable cloud field has not been stabilized yet.

Figure 6 shows a 2-D view of the leakage plume of ethane at various time instants.

The propane concentration at sea level and for instant time t=400 s, it is noted that with wind condition D2 and wind direction towards Megara, the riskiest flammable zone for propane covers a section of overland region such as

Peninsula Agias Trias, a part of Revythousa Island and a part of Makronisos Island. Figure 8a appears that the form of flammable zone is a very width ring. Also, it can be seen that the upper flammability limit is only limited on Revythousa island.

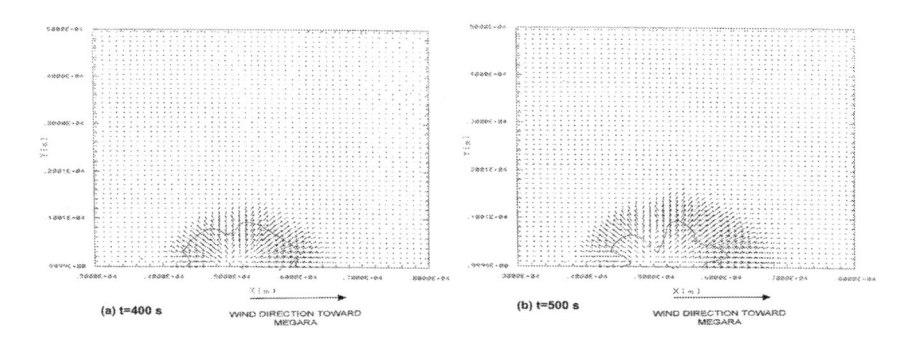

Figure 6: 2-D view of the methane leakage plume at various time instants.

The propane concentration at sea level and for instant time t=500 s, it is noted that with wind condition D2 and wind direction towards Megara, the flammable cloud continues to cover a section of overland region such as Peninsula Agias Trias, a part of Revythousa Island and a part of Makronisos Island. Figure 7b illustrates that the flammable field zone has been slightly increased due to continuative dilution by ambient air. While the shape and the extent of upper flammable limit are slightly different from the previous situation.

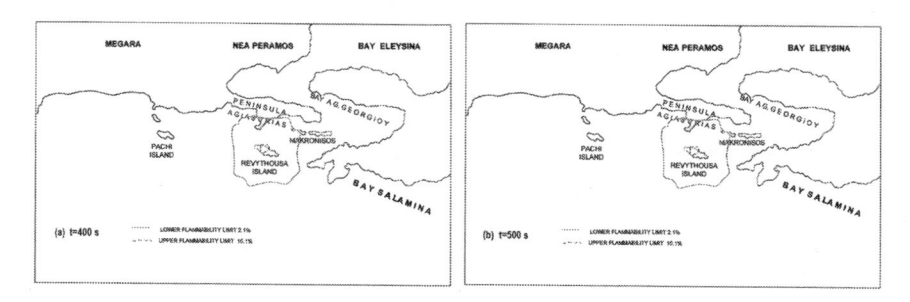

Figure 7: The propane distribution at different time instants after the release.

A 2-D view of the propane leakage plume at various time instants is shown in figure 8.

7 Conclusions

According to the above graphs, it can be seen that Megara region, which is the most populated at the surround area, is not in danger for a period of one hour and ten minutes, from the beginning of the leakage. It means that 45 minutes for

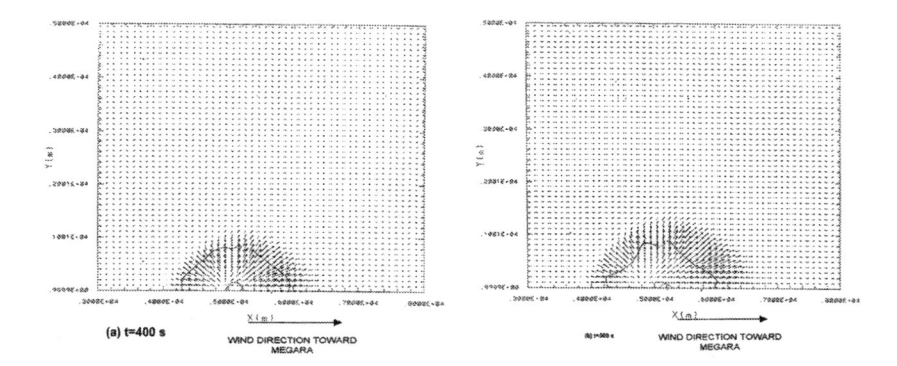

Figure 8: 2-D view of the propane leakage plume at various time instants.

complete discharge of the tank and 25 minutes approximately of modeling calculations, Megara town is out of risky zone. This is offered due to location of LNG terminal at Revythousa Island subsequently the flammable cloud covers mainly a section of overseas region.

However the regions such as Peninsula Agias Trias, Revythousa Island, Makronisos Island, Bay Salamina, are directly in danger due to flammable cloud of Methane, Ethane, Propane during one hour and ten minutes, in which is moved with coherence and not separately. This means that the cloud remains flammable out of flammable limits of each component.

Particularly the maximum risk occurs at time t= 400 s for all three components. At this time the maximum range in which each component reaches is: 2000 m for Methane, 1000 m for Ethane, and 800 m for Propane.

References

[1] Deaves D.M., Application of a turbulence flow model to heavy gas dispersion in complex situations, *Heavy Gas and Risk Assessment-II*, ed. Hartwing S, D. Reidel Publication Co., 1983.
[2] Bartzis J.G., Venetsanos A.G., Varvayanni M, Catsaros N. & Megaritou A., *ADREA-I: A Transient Three-Dimensional Transport Code for Complex Terrain and Other Applications,* Nuclear Technology, **94**, 1991, pp. 135–148.
[3] Patankar S.V., *Numerical Heat Transfer and Fluid Flow*, New York, Hemisphere Publishing Co, 1980.
[4] Public Gas Corporation of Greece S.A. (DEPA), *Plans for Natural Gas importation into Greece: The LNG terminal at Revythousa Island,* 1997.
[5] Statharas J.C., Bartzis J.C., Venetsanos A. & Wurtz J., *Prediction of Ammonia Releases using ADREA-HF Code,* Process Safety Progress, **12**, 1993, pp. 118–122.

Modelling of air pollutants released from highway traffic in Hungary

Gy. Baranka
Hungarian Meteorological Service, Hungary

Abstract

The aim of study is to determine the CO and NO_2 concentrations near Hungarian motorways. The expressway network in Hungary is comprised of 650 kilometres of motorway, 205 kilometres of clearways and highways as well as 188 kilometres of junction point sections. The CALINE4 dispersion model developed by the California Department of Transportation for predicting CO, NO_2 concentrations near roadways was adapted at Hungarian Meteorological Service. The dispersion equation is based on an analytic solution of the Gaussian diffusion equation for a finite line source. This model was combined with a "mixing zone" model segment. The "mixing zone" is defined as the region over the road characterized by uniform emissions and turbulence. The model treats traffic as an infinite line source divided into a series of elements located perpendicular to the wind direction. Vertical dispersion parameters take into consideration both thermal and mechanical turbulence caused by vehicles. A CALINE4 roadway link is assigned an equivalent line source strength based on the product of a fleet averaged vehicle emission factor (grams of pollutant per vehicle per mile traveled) and vehicle flow rate (vehicles per hour). The CALINE4 requires relatively minimal input from the user. Input data are traffic volume, emission factors, roadway geometry, wind speed and direction, ambient air temperature, mixing height, atmospheric stability class and coordinate of receptors. Using CALINE4 dispersion model 1-hour and 8-hour averages CO and NO_2 concentrations have been determined at the receptors within 500 meters of the Hungarian motorways and around the intersection links.

Keywords: emission of road traffic, turbulence, diffusion processes, CALINE4 model, spatial distribution.

1 Introduction

The aim of study is to investigate the possible impact of scenarios (including investment of shopping stores, ring roads and the variation of driving customs and the vehicle fleet mix) on air quality and to determine the number of exceedence of air quality limit values near motorways. Using dispersion models connected to the estimation of road emission for future years and to a weather forecast model the variation of air quality can be predicted. The ambient air quality effects of traffic emissions near Hungarian motorways were evaluated using the CALINE4 dispersion model.

2 Hungarian CO and NO$_x$ emissions from mobile sources

In Hungary the expressway network is comprised of 650 kilometres of motorway, 205 kilometres of clearways and highways (which are comprised of highways serving as expressways, highways with expressway features later converted to expressways, and highway sections with expressway features) as well as 188 kilometres of junction point sections, 52 kilometres of roads leading to rest areas, and 106 rest areas.

NOx emission from road traffic in Hungary
kt

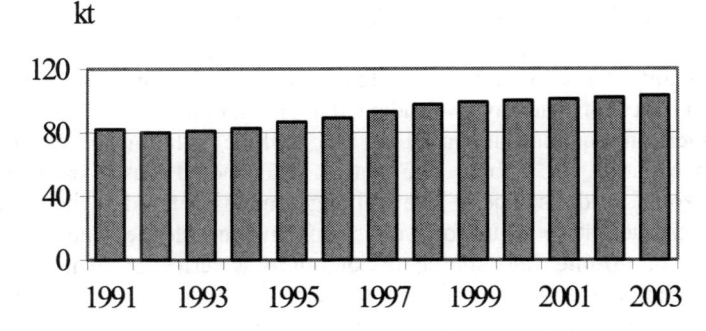

Figure 1:   NOx emission from road traffic calculated by Merétei [4].

The Hungarian CO and NOx emissions from mobile sources have been investigated by Merétei [4]. Emission of nitrogen oxides from road traffic was continuously increased, (fig. 1), while the emission of carbon monoxide from the same sources was decreased during the last twenty years (fig. 2). NOx emission from road traffic is growing (growing rate is 20%) due to the increasing of travelled kilometers by cars and duty. In the same time NOx emission of the European members was decreased by 20% due to the application of three steps catalizator in the new cars and the changes of the compound of the fuel. In general for EU-15 energy consumption for road traffic has been increased since 1990, but energy efficiency has been improved by 10% since 1990 (Ntziachristos et al [5]).

CO emission from road traffic in Hungary

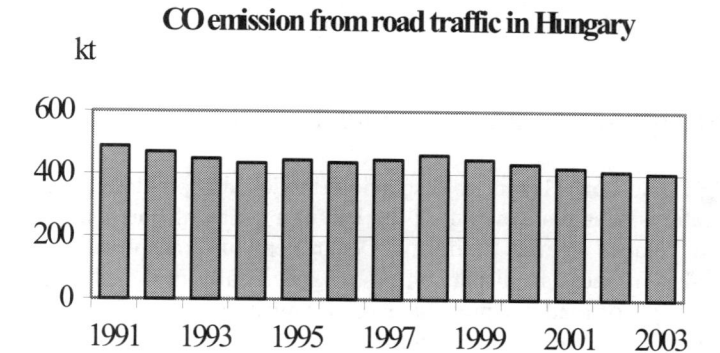

Figure 2: CO emission from road traffic calculated by Merétei [4].

Source emissions are a function of both the vehicle emissions factors and the vehicle activity (usually measured in vehicle kilometers travelled). The vehicle emission factors can differ greatly depending on the type of road, vehicle fleet mix, and traffic flow encountered. As a result, emissions factors were determined specifically for each roadside location to ensure that the appropriate input was used for the modelling of subsequent roadway emissions. These site-specific emissions factors were then combined with the vehicle kilometers travelled data to give a total emission rate for the road.

3 Diffusion processes in roadway modelling

The total diffusion is divided into primary diffusion affected by the traffic and the diffusion caused by ambient meteorological conditions. Contaminants outcome from tailpipe take part in primary process, on one hand the thermal turbulence caused by differential surface heating and by the buoyant exhausted plume creates eddy movements, on the other hand mechanical turbulence from wind flow and the traffic wake induce turbulence. Thermal influences interact with mechanical effects. There are interactions between the ambient wind speed, exhaust velocities of the tailpipe emissions and the traffic wake-induced turbulence. Primary turbulence is meaningful only in stagnant conditions (especially in case of cold-running) otherwise the atmospheric turbulence dominates the process. The plume is transported with the wind speed at the emission height. Behind a vehicle due to the thermal and mechanical turbulence a well-mixed zone can be occurred.

In general micro scale (characteristic lengths below 1 km) airflow is very complex, as it depends strongly on the detailed surface characteristics (i.e. form of buildings, their orientation with regard to the wind direction etc.). Although thermal effects may contribute to the generation of these flows, they are mainly determined by hydrodynamic effects (e.g. flow channelling, roughness effects), which have to be described well in an appropriate simulation model. In view of the complex nature of such effects, local scale dispersion phenomena (which are to a large extent associated with micro scale atmospheric processes) are mainly described with robust "simple" models in the case of practical applications, such as mobile source models.

4 Description of CALINE4 model

CALINE4 (Benson [1]) is a dispersion model for predicting air pollution concentrations near roadways. It is the last in a series of line source air quality models developed by the California Department of Transportation (Caltrans). CALINE4 is a Gaussian dispersion model specifically designed to evaluate air quality impacts of roadway projects. Each roadway link analysed in the model is treated as a sequence of short segments. Each segment of a roadway link is treated as separate emission source producing a plume of pollutants, which disperses downwind. Pollutant concentrations at any specific location are calculated using the total contribution from overlapping pollution plumes originating from the sequence of roadway segments. Given source strength, meteorology and site geometry, CALINE4 can predict carbon monoxide, nitrogen dioxide and suspended particle concentrations for receptors located within 500 meters of the roadway. It also has special options for modelling air quality near intersections, street canyons and parking facilities.

CALINE4 uses mixing zone concept. The model treats the region directly over the highway as a zone of uniform emissions and turbulence. This is designated as the mixing zone, and is defined as the region over the travelled way (traffic lanes not including shoulders) plus three meters on either side. The additional width accounts for the initial horizontal dispersion imparted to pollutants by the vehicle wake. Within the mixing zone, the mechanical turbulence created by moving vehicles and the thermal turbulence created by hot vehicle exhaust is assumed to be the dominant depressive mechanisms.

Concentrations of relatively inert pollutants such as carbon monoxide and suspended particle concentrations have been predicted, and Discrete Parcel Method chemical subrutin has been used to determine the nitrogen dioxide concentration in a Gaussian plume. Discrete Parcel Method uses the cycle NO-O_3-NO_2 solved analytically by Benson [1]. The processes associated with the expanding Gaussian plume and NOx-chemistry has been presented by Hanrahan [2]. The influence of plume dilution has been accounted according to the receptor oriented Discrete Parcel Method (Härkönen et al [3]), which is a modified version of the original method.

Historically, the CALINE series of models required relatively minimal input from the user. While CALINE4 uses more input parameters than its predecessors, it must still be considered an extremely easy model to implement.

5 Results

Using CALINE4 dispersion model hourly and 8-hour average concentrations can be obtained. The CO and NO_2 concentrations have been calculated at 100 m away from road axis for Hungarian motorway network. During worst-case meteorology NO_2 concentration is shown in fig. 3. Worst-case meteorology input is the combination of the worst wind speed, wind direction, stability class. The highest concentrations occur around Budapest. It can be explained by arterial roads structure in Hungary and by urban sprawling.

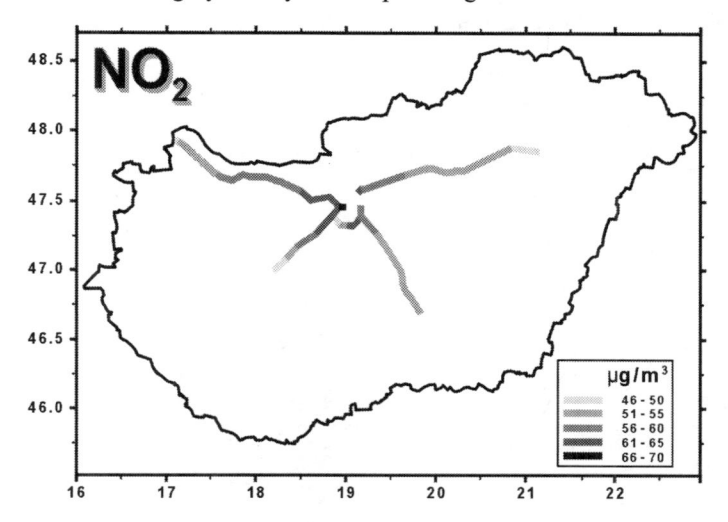

Figure 3: Worst-case 1-hour NO_2 level at 100 m away from road axis.

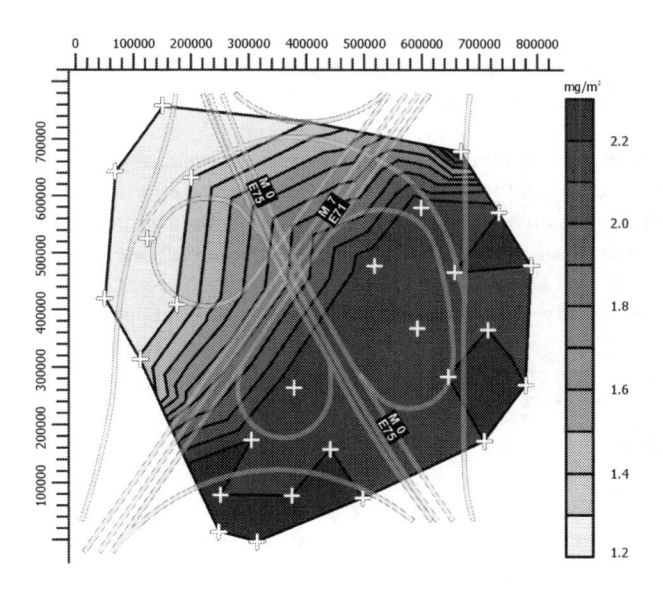

Figure 4: Hourly average of CO concentration around the intersection link No. M7 and M0.

CALINE4 model provides possibility to predict air pollutants near intersections. Spatial distributions of hourly average CO and NO_2 concentrations are shown around an intersection near Budapest in fig. 4 and fig. 5. The distance between receptors (indicated by cross in figures) is 100 m. Air pollutants are transported by the most frequent airflow to SW direction. At this intersection the traffic of motorway M7 is dominant.

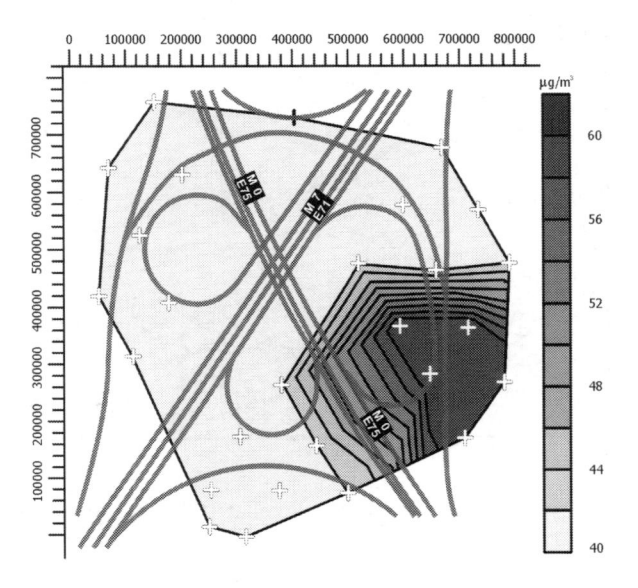

Figure 5: Hourly average of NO_2 concentration around the intersection link No. M7 and M0.

Using dispersion models connected to the estimation of road emission for future years and to weather prediction models the variation of air quality can be obtained on a given area. With the aid of this project we would like to develop an integrated system of meteorological and dispersion models to calculate and visualize forecasted air quality situation near crowded roads. During the further research the ability and the limitation of coupling of dispersion model (e.g. CALINE4) on mesoscale forecasting model (e.g. WRF) will be examined for air pollution and emergency forecasting purposes.

References

[1] Benson, P. E. CALINE4 a dispersion model for predicting air pollution concentrations near roadways. California Department of Transportation. Sacramento, CA. California Department of Transportation. 1989.
[2] Hanrahan, P.L. The plume volume molar ratio for determining NO2/NOx ratios in modelling – Part I: *Methodology. J. Air & Waste Manage. Assoc.* 49, pp. 1324–1331, 1999.
[3] Härkönen, J., Valkonen, E., Kukkonen, J., Rantakrans, E., Lahtinen, K., Karppinen, A. and Jalkanen, L. A model for the dispersion of pollution from

a road network. Finnish Meteorological Institute, *Publications of Air Quality 23*, Helsinki, 34 p. 1996.
[4] Merétei T. (editor) Determination of transport emission in Hungary (in Hungarian) *Institute for Transport Sciences Non-profit LTD*. Budapest 2006.
[5] Ntziachristos, L., Tourlou, P. M., Samaras Z., Geivanidis S. and Andrias A. National and central estimates for air emissions from road transport, *Technical report 74*, European Topic Centre on Air and Climate Change, 2002

Dust barriers in open pit blasts. Multiphase Computational Fluid Dynamics (CFD) simulations

J. T. Alvarez, I. D. Alvarez & S. T. Lougedo
*GIMOC, Mining Engineering and Civil Works Research Group,
Oviedo School of Mines, University of Oviedo, Spain*

Abstract

In the framework of the Research Project CTM2005-00187/TECNO, "Prediction models and prevention systems in the particle atmospheric contamination in an industrial environment" of the Spanish National R+D Plan of the Ministry of Education and Science, 2004-2007 period, there has been developed a CFD model to simulate the dispersion of the dust generated in blasts located in limestone quarries. The possible environmental impacts created by an open pit blast are ground vibrations, air shock waves, flying rocks and dust dispersion, being this last effect the less studied one. This is a complex phenomenon that was studied in previous successful researches through the use of several digital video recordings of blasts and the dust concentration field data measured by light scattering instruments, as well as the subsequent simulation of the dust clouds dispersion using Multiphase Computational Fluid Dynamics (CFD). CFD calculations where done using state of the art commercial software, Ansys CFX 10.0, through transitory models simultaneously using 7 different dust sizes with Lagrangian particle models crossing an Eulerian air continuous phase. This paper presents results and observations obtained through repeated simulations that virtually install physical barriers in the vicinity of the blast with the aim of diminishing the dust cloud dispersion ant the associated environmental impact. There have been tested several combinations of low and high barriers installed 100 or 200 m apart from the blasting area. Assuming full trapping of the dust in the solid barrier there are scenarios where dust retention in the model could be as high as 90%, which shows the potential use of barriers to avoid dust dispersion. Models also show the barriers shadow effect in the dust deposition area, useful for the definition of protection areas where sensible equipment or installation could be located.

Keywords: bench blasting, dust dispersion modelling, CFD, multiphase, discrete Lagrangian methods, dust barriers.

1 Introduction

Our research group is developing a project named CTM2005-00187/TECNO, "Prediction models and prevention systems in the particle atmospheric contamination in an industrial environment" granted by funds of the Spanish National R+D Plan of the Ministry of Education and Science, 2004-2007 period. Within the research objectives appears one relating to the determination of the amount of dust produced in a blast and its immersion in the atmosphere surrounding the quarry area by means of two main tools: measurement campaigns of the dust concentration using "Light scattering" dust sensors and computerized simulations through commercial CFD (Computational Fluid Dynamics) software. These are done through the combined use of Solidworks to generate the 3D models, ICEM CFD to adequately mesh the domain and Ansys CFX 10.0 in case of the calculation and analysis of the results.

There are dozens of methods to simulate the dust dispersion in open areas [1] and authors as Reed [2] have studied their application in dust generated in mining and civil works.

The authors have validated the use of CFD in order to simulate the dust movement both using CFD [3] and using more conventional dispersion models as ISC3 [4]. ISC3 allows the simulation of dust movement in large physical areas and time schedules but fails in the short range. CFD is the perfect method to simulate complicated geometrical configurations where wind and dust interact and then will be used to develop the simulations explained in this paper.

This paper will show simulations done where solid barriers 3 m and 20 m height are virtually installed in the vicinity of the quarry bench in order to evaluate their ability to trap the dust particles. Authors such as Cowherd and Grelinger [5] have made experiences in dust emitting roads surrounded by several species of trees and have observed diminishing in the dust levels due to their presence. The same phenomenon has to occur in a quarry bench configuration.

2 Previous studies and simulations

After detailed analysis of the topography of the quarry and the photographs and digital video recordings of the blast a 3D geometry model was developed using the parametric software Solidworks.

The simulated domain ranges 400x500x250 m including a 18 m height and 15° slope bench followed by a plain area where the measuring instrumentation was located, surrounded in the left side by a hill.

The blast was not only recorded in digital video but was also measured with dust concentration measurement equipment. The instrumentation sensed a peak level of 700 µg/m^3 in case of the dust sensor located at 100 meters from the blast and 200 µg/m^3 in case of the dust sensor located at 200 meters. The meteorological conditions were 24°C of temperature, 1015 mb of pressure, 2.5 m/s of sustained wind velocity, humidity of 55%, sunny day and null rain.

Following figure shows the model geometry including the virtual barriers domain and the dust injection points over the blasted bench.

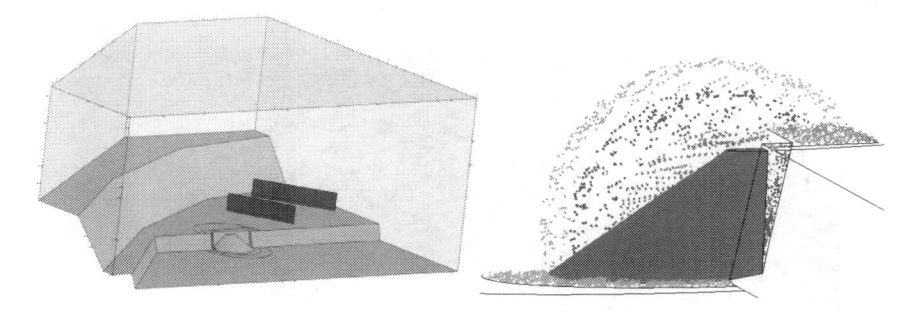

Figure 1:   Model geometry and dust injection points.

All the referred data was used to develop a multiphase CFD simulation, based in the numerical resolution of the differential equations that define the motion of a fluid.

The equations that govern the fluid flow are the Navier Stokes, eqn (1), which relates the velocity and pressure fields as well as density; the continuity equation, eqn (2), that express the mass conservation; and finally the energy equation, eqn (3), which relates also the temperature fields. These expressions create a system of differential equations that can only be solved, in the vast majority of cases, by numerical methods.

$$\rho \frac{D\overline{V}}{Dt} = -\overline{\nabla}p + \rho\overline{g} + \mu\nabla^2\overline{V} \qquad (1)$$

$$\frac{D\rho}{Dt} + \rho\overline{\nabla} \bullet \overline{V} = 0 \qquad (2)$$

$$\rho \frac{D\widetilde{u}}{Dt} = K\nabla^2 T - p\overline{\nabla} \bullet \overline{V} \qquad (3)$$

In turbulent flow there have also to be solved additional equations that allow the calculation of the velocity and pressure fields in all the domains, the so called "turbulence models". Taking into account just the RANS (Reynolds Averaged Navier Stokes) they vary from the more or less simple where just one equation is added to the calculations, as the Spalart-Allmaras model, to the two equations models, k-epsilon or even seven equations, Shear Stress Transport (SST) models. This paper will show calculations that use k-epsilon models after mesh and turbulence model dependence tests accomplished in simulations in similar configurations [6] and using related bibliography [7]. All the calculations are made using commercial code Ansys CFX 10.0 and its deep documentation is also revised [8] to select adequate parameters for the simulation.

The domain to be calculated is divided in finite volumes where the former equations will be solved by linear methods. The blowing air will be defined as a velocity profile following a classical logarithmic equation, as can be seen in documents from the U.S. Environmental Pollution Agency [9] as well as adequately oriented following the wind bearing measured in the field.

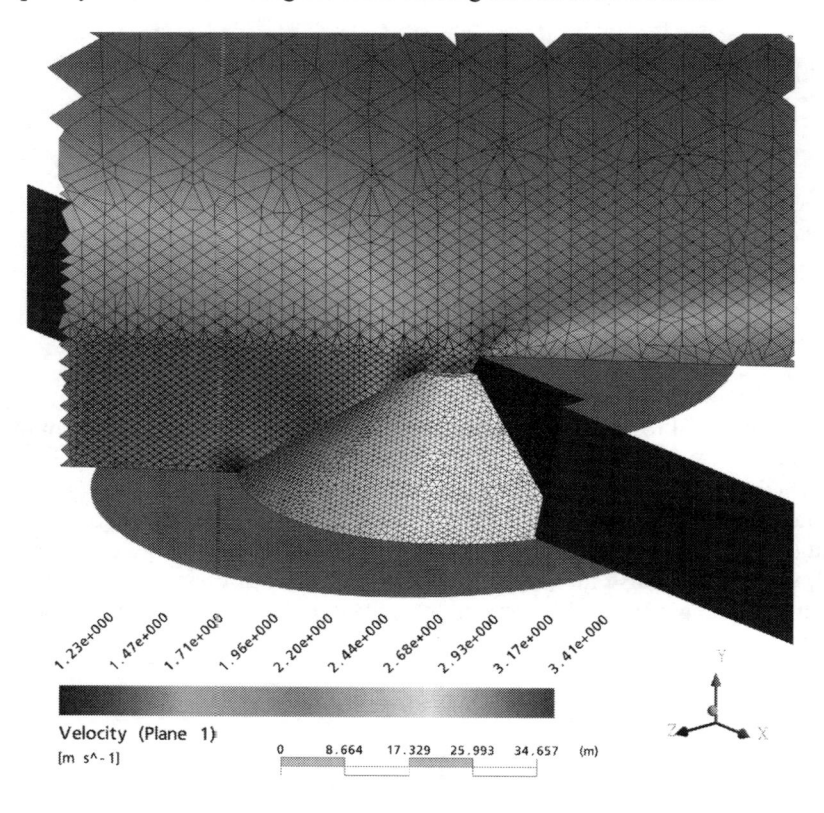

Figure 2:   Dust emission area and air velocity.

Figure 2 shows the area where the dust will be injected simulating the dust emission in the blast. The bench acts as a obstacle in the wind path and the pressure and velocity fields will be modified in its vicinity, thus modifying the dust dispersion mechanism. The flow in this area must be calculated in detail, so a high density mesh will be used near the blast generated cone. As can be seen in the figure mesh is rougher as height is higher, as the wind will not vary so much in that area.

The cut plane of figure 2 is coloured according to the air velocity. The logarithmic wind profile used in the inlet boundary condition, 2.5 m/s at 10 meters height, can be inferred at the left of the domain. In the bench the wind is accelerated by the bench presence to levels as high as 3.5 m/s.

Another important parameter that has to be defined is the particle size distribution of the material that forms the dust cloud. As is shown by Almeida et al.[10] and Jones et al. [11] the particulated material thrown to the air by a blast has two main sources. First, rocks pulverized by the several phenomena that take place in the blast (shock wave, high pressure gases or dynamic breaking mechanisms, etc.) and second the dusty products of the explosive chemical reaction. Both Almeida and Jones estimate the size distribution of the dust clouds in ranges from submicron sizes up to 50 microns. Particles over this size are also produced but have not been considered in the simulation as are quickly settled into ground by its own weight.

3 Installation of virtual barriers. Postprocessing

In open pit mines is quite common to find forests surrounding the installations. These trees constitute a barrier for the fugitive dust generated in the mine or quarry, thus avoiding its dispersion towards inhabited areas. Recent studies by Cowherd and Grelinger [5] about dust generation in unpaved roads determine that there are abnormally low levels of dust concentration dispersion in roads limited by trees. Cowherd studied the diminishing of the dust levels in presence of two vegetal species: low grass prairies and high oak trees forests.

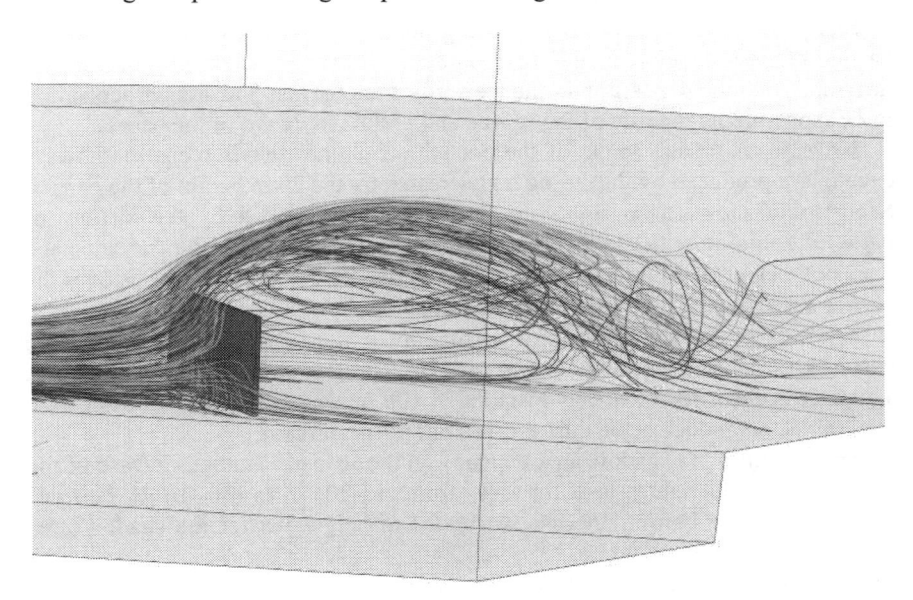

Figure 3: Particle paths (40 and 50 microns).

In this study barriers of two heights will be included. First one, 20 meters height will simulate "Eucaliptus Globulus" or "Eucaliptus Nitens", high speed growing trees species extremely extended in northern Spain. The second height, 3 meters, will simulate species as "Thuja Picata", "Prunas Lauroceasus" or "Viburnum tinus (laurustinus)", frequently used as wind and visual barriers both in industrial and home applications. Wide information about this and other species can be found in books like the Rushforth one [12] or Huxley [13].

In figure 3 there can be seen how the dust particles trajectories are affected by the presence of the barrier. Table 1 shows the characteristics of the different simulation scenarios.

Table 1: Simulation scenarios.

Distance to Dust source	Barrier height	SCENARIO
100 m	3 m	A
100 m	20 m	B
200 m	3 m	C
200 m	20 m	D

Figure 4 shows the deposition rate at ground level in all scenarios plus the no barrier scenario, which is the first one. In all barrier scenarios there is an interruption in the deposition in the leeward of the barrier. The higher deposition rates appear in the vicinity of the source and in the windward of the barrier.

If we focus on the shape of the deposition plumes the B scenario shows a deep curve produced by the strong eddy created by the 20 m height of the barrier. Scenario D shows deep right turn in the flow, produced by the airflow of particles colliding with the high barrier located at 200 meters. The curved shape does not mean that all dust will settle in that area. Part of the plume flies over the barrier and escapes from the simulated domain in the wind bearing.

In all cases there is a strong interruption of the deposition downwind the barrier which demonstrates the existence of a protection shadow behind them. This fact is clear in the barriers installed at 100 meters, while can not be seen in the 200 meters model as the dust escapes out of the domain.

The length of the shadow area depends on the scenario studied. In case of the barrier 3 meters height it lasts for areas from $(13-20) \cdot H$ to $(30-40) \cdot H$, being H the height of the barrier. In case of the 20 m height barrier the shadow area lengths distance around $4 \cdot H$.

On top of the qualitative studies referred above it would be extremely interesting to quantify which of the simulation scenarios would have the more dust retention ratio. It would be also interesting to know how much dust is retained in the barrier and how much dust settles down in the domain due to the flow modification in presence of the barrier.

The Ansys CFX Post processing module can filter the simulated particle by sizes, origins and destinies of each particle path. A total amount of 21.000 representative particles are injected in the domain at different rates depending on the size of the dust. Some of these particles will impact the barrier, others will settle in the ground and walls of the model and a third group will leave the

No barrier

Scenario A   Scenario B

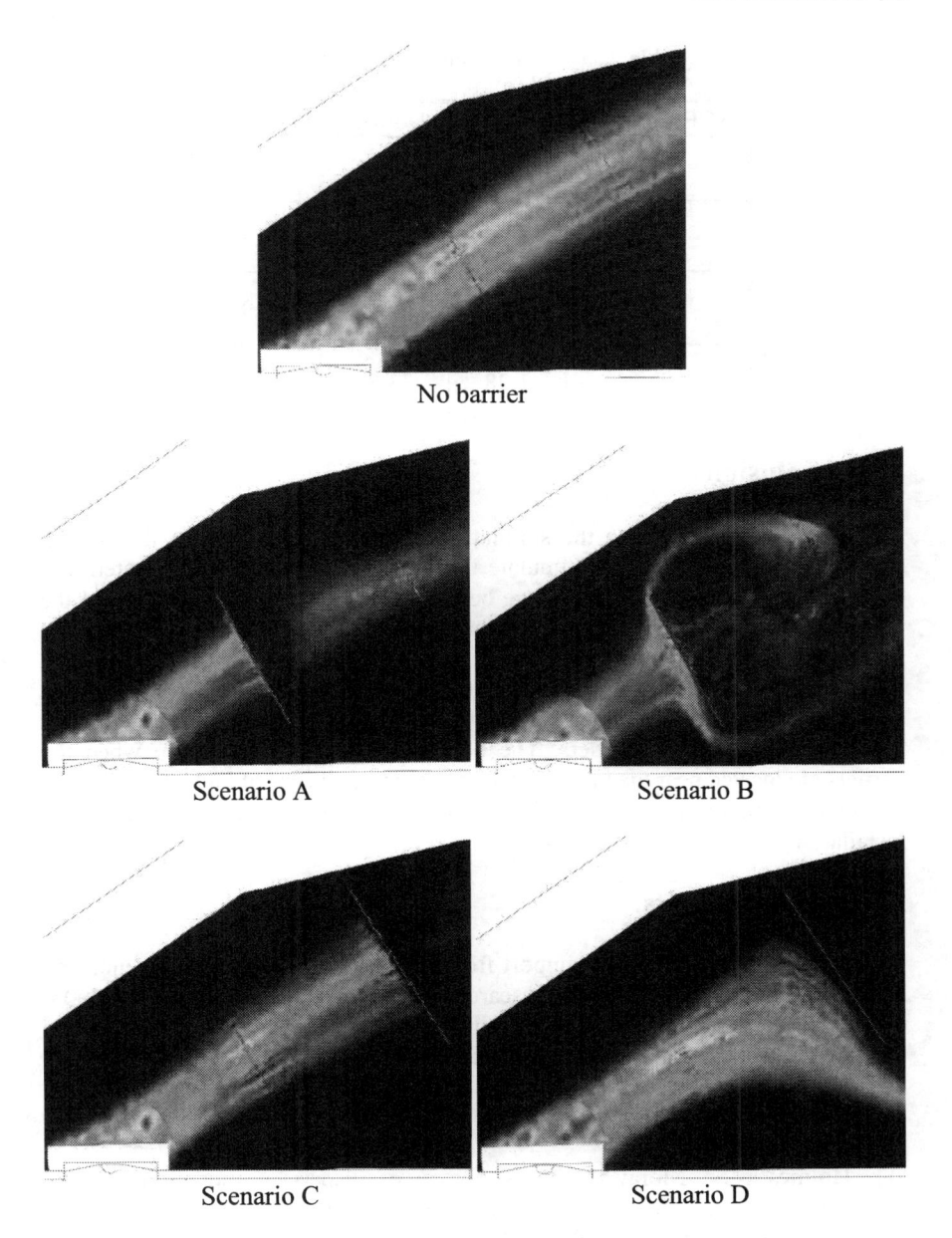

Scenario C   Scenario D

Figure 4:   Deposition plumes at 1m height.

domain. Table 2 shows the mass retention percentage of each of the three groups on each scenario. The barrier of scenario A is the one that traps more dust while scenario B is the one where less dust escapes out of the model, only a 3.3% of the injected particles. The dust trapped in the barriers is not so high but the effect of the barriers in the flow field makes the residence time of the particles in the domain bigger, thus creating more time to settle down.

Table 2: Mass retention percentage.

SCENARIOS	Barrier	Domain	Escapes
A	6.69	57.78	35.53
B	1.25	95.45	3.3
C	4.936	57.15	37.91
D	1.28	74.56	24.16

4 Conclusions

All the tools used to make the simulations, from the field data to the several software used can effectively simulate the dispersion of the dust generated in a quarry blast. In this paper there have been included obstacles in the path of the particles in order to evaluate their trapping and avoid its dispersion in the surrounding environment.

The barriers used are solid with heights of 3 m and 20 m located at 100 meters and at 200 meters from the blast. Simulations done quantify the dust emitted and the dust trapped in the barriers. The best trapping scenario is done when the barriers are located near the blast, with better trapping in case of high ones.

Future simulations will take into account porosity and holes in the barrier installations.

Acknowledgements

We want to acknowledge the support from the Spanish Ministry of Science and Education that granted these researches through the project CTM2005-00187/TECNO, "Prediction models and prevention systems in the particle atmospheric contamination in an industrial environment".

References

[1] Holmes N.S., Morawska L. A review of dispersion modelling and its application to the dispersion of particles: An overview of different dispersion models available. Atmospheric Environment 40 (2006) 5902–5928

[2] Reed,, W.R. Significant Dust Dispersion Models for Mining Operations. Information Circular 9478. National Institute for Occupational Safety and Health (NIOSH). September 2005

[3] J. Toraño et al. A CFD Lagrangian particle model to analyze the dust dispersion problem in quarries blasts. WIT Transactions on Engineering Sciences, Vol 56. ISSN 1743-3533. doi:10.2495/MPF070021. (2007)

[4] J. Toraño, R. Rodriguez, I. Diego and A. Pelegry, "Contamination by particulated material in blasts: analysis, application and adaptation of the existent calculation formulas and software". Environmental Health Risk III, pp. 209–219, (2004)
[5] Cowherd C., Grelinger, M.A.. Development of an Emission Reduction Term for Near-Source Dust Depletion. 15th International Emission Inventory Conference. New Orleans, USA, May 15-18, 2006.
[6] Toraño J., Rodríguez R. and Diego I. Surface velocity contour analysis in the airborne dust generation due to open storage piles. European Conference on Computational Fluid Dynamics. ECCOMAS CFD 2006, Delft The Netherlands, 2006
[7] Silvester S.A., Lowndes I.S. and Kingman S.W. The ventilation of an underground crushing plant. Mining Technology (Trans. Inst. Min. Metall. A), Vol. 113, 201–214. (2004)
[8] ANSYS CFX-Solver, Release 10.0: Theory; Particle Transport Theory: Lagrangian Tracking Implementation; page 173.
[9] Environmental Pollution Agency. AP-42, 13.2.5.1, Miscellaneous Sources. Pp2. 1998.
[10] Almeida, S.M. Eston, S.M. and De Assunçao, J.V. "Characterization of Suspended Particulate Material in Mining Areas in Sao Paulo, Brazil." I.T.. International Journal of surface Mining, Reclamation and Environment 2002, Vol. 16, no. 3, pp. 171–179
[11] Jones, T., Morgan, A. and Richards, R. "Primary blasting in a limestone quarry: physicochemical characterization of the dust clouds". Mineralogical Magazine, April 2003, Vol 67(2), pp. 153–162
[12] Rushforth, K. Trees of Britain and Europe. Harper Collins (1999).
[13] Huxley, A. New RHS Dictionary of Gardening. MacMillan (1992)

Contribution of oil palm isoprene emissions to tropospheric ozone levels in the Distrito Metropolitano de Quito (Ecuador)

R. Parra
Corporación para el Mejoramiento del Aire de Quito, Corpaire, Quito, Ecuador

Abstract

Volatile organic compound (VOC) emissions from vegetation are important in the *Distrito Metropolitano de Quito (DMQ)*. In particular, oil palm (*Elaeis guineensis*) isoprene emissions are significant near the DMQ. Oil palm cultivation is promoted because of the growing demand for biodiesel production. The contribution of isoprene from oil palm to tropospheric ozone (O_3) production was estimated by numerical simulation for the period 2–28 of September 2006, using estimates of emissions from the DMQ in the Eulerian chemical transport model WRF-Chem. On-road traffic, power facilities and other industries, vegetation, services stations and solvent use were considered as emission sources, under the following scenarios: 1) without oil palm, 2) with oil palm plantation in 2003; and, 3) with an expected future plantation. Two groups of emission rates were used: 1) a group of mid-range emission factors (up to 51 $\mu g\ g^{-1}\ h^{-1}$) and 2) a high isoprene emission factor (172.9 $\mu g\ g^{-1}\ h^{-1}$). Results indicate that increasing areas of oil palm plantations in the future could increase O_3 concentrations. Scenario 3) with the high emission factor, provides maximum hourly anomalies of 17.9 $\mu g\ m^{-3}$ over the urban zone. Scenario 3), with the mid-range emission factors, results in maximum hourly anomalies of 2.7 $\mu g\ m^{-3}$.

Keywords: Elaeis guineensis, WRF-Chem, numerical simulation, volatile organics compounds.

1 Introduction

Volatile organic compound (VOC) emissions from vegetation are important in the *Distrito Metropolitano de Quito* (***DMQ***). Isoprene (C_5H_8) is a reactive VOC whose emissions are indirectly related to photosynthesis. C_5H_8 is synthesized and emitted by different plant species when exposed to Photosynthetically Active Radiation (PAR). C_5H_8 emissions increase with temperature up to a maximum, and decrease again at higher temperatures (Guenther et al. [1]).

CORPAIRE [2] estimated that about 43% (21.7 kt a^{-1}) of total VOC emissions in the domain of the DMQ are produced by vegetation. About 57% of them (12.4 kt a^{-1}) are C_5H_8. About 56% of C_5H_8 is produced by oil palm (*Elaeis guineensis*) plantations. This species has been widely planted in Ecuador. According to statistics from the *Ministerio de Agricultura* [3], in 2005 oil palm coverage in Ecuador was about 207 000 ha. Oil palm was introduced near the Equator, replacing large stretches of native forest. Further expansion of oil palm plantations can be expected, owing to its high performance and low production costs in the region, and the growing demand for its oil for biodiesel production.

Ozone (O_3) can be produced when reactive VOCs mix with nitrogen oxides under incoming solar radiation. Typically O_3 concentrations are higher during midday and early afternoon hours, when both temperature and solar radiation are higher. Because of the magnitude of C_5H_8 emissions by oil palm and the expected increase of oil palm plantations, the contribution of oil palm isoprene emissions to tropospheric O_3 formation in the DMQ region was estimated.

2 Method

The numerical simulation of air quality was performed for the period 2–28 September 2006, using emissions provided by the *Sistema de Gestión del Inventario de Emisiones del DMQ (SIGIEQ)* (Mina and Parra [4]). At present, emissions from on-road traffic, power facilities and other industries, vegetation, service stations and by use of solvents are included within the SIGIEQ. Hourly gas-phase emissions were speciated according to the *Regional Acid Deposition Model (RADM)* chemical mechanism (Stockwell et al. [5]). They were used with the *Weather Research and Forecasting with Chemistry (WRF-Chem)* [6] model, to simulate air quality in the region of Quito.

Fig. 1 depicts the location of Quito, the master and the two nested domains used for simulation. The first two (d01: 27x27 km, d02: 9x9 km) are used only for meteorological simulations. For the second subdomain (d03: 3x3 km), the chemical transport option is also activated.

The DMQ is located near the Equator. It has a complex topography. Subdomain d03 has zones with height lower than 800 masl and higher than 4000. Quito has an average altitude of 2800 masl. The Guagua Pichincha volcano is allocated west of Quito (Fig. 2.) Subdomain d03 is a squared region (about 110 km x 110 km) with Quito allocated at its center. Fig. 2 also depicts the oil palm coverage in 2003, obtained from the SIGAGRO system, (*Ministerio de Agricultura y Ganadería del Ecuador* [7]). It covered about 8.7% of the Subdomain d03.

2.1 VOC emissions by vegetation

Isoprene, monoterpenes and other VOC emissions were estimated using the algorithm by Guenther et al. [8], considering temperature and solar radiation as

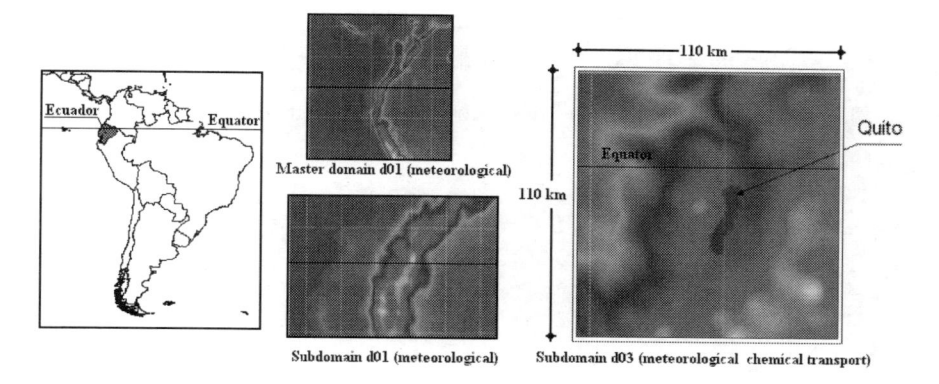

Figure 1: Location of Ecuador and domains for modeling. Subdomain d03 is a squared region (about 110 km by side) with Quito allocated at its center.

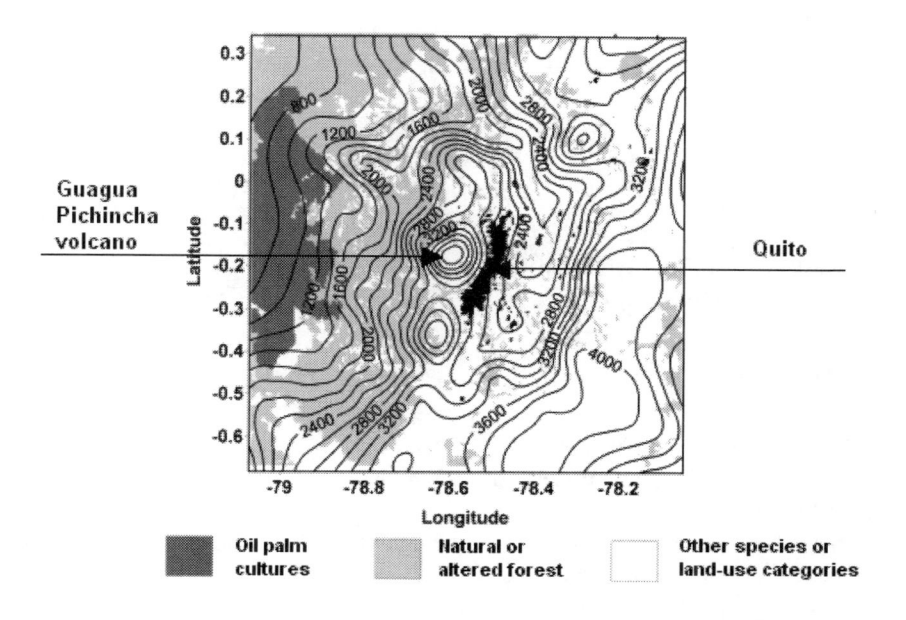

Figure 2: Location of Quito, topography (masl) and oil palm coverage in 2003.

physical drivers of emission factors. Hourly maps of these drivers were provided by the WRF-Chem model.

2.2 Isoprene emission factors for oil palm

Two groups of isoprene emission factors (EFs) were used:
  1) A group of mid-range EFs, up to 55 µg g^{-1} h^{-1} (at standard conditions, 30 °C of temperature and 1000 µmol m^{-2} s^{-1} of PAR), provided by Dr. Sue Owen

(CEH, Edinburgh, UK; personal communication). Fig. 3 depicts these EFs, which vary during daytime. Wilkinson et al. [9] found that they are under strong circadian control.

2) A high EF: 172.9 µg g^{-1} h^{-1} from Kesselmier and Staudt [10].

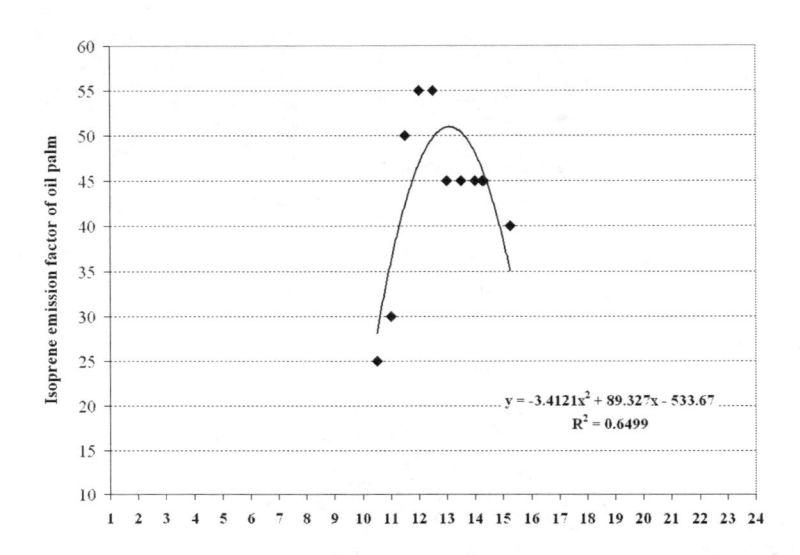

Figure 3: Isoprene emission factors (µg g^{-1} h^{-1}) of oil palm (Owen et al., in preparation).

3 Scenarios

Five scenarios were considered:

1) Without oil palm (**WOP**), replacing the 2003 oil palm coverage with natural forest

2) With oil palm according to the coverage of Fig. 2 (8.7% of the Subdomain d03) and the group of mid-range EFs (**OP1**)

3) With oil palm according to the coverage of Fig. 2 (8.7% of the Subdomain d03) and the high EF (**OP2**)

4) Projection of **more oil palm** (to 15.6% of the Subdomain d03) and the group of mid-range EFs (**MOP1**)

5) Projection of **more oil palm** (to 15.6% of the Subdomain d03) and the high EF (**MOP2**)

Fig. 4 shows the oil palm plantation for these scenarios. Oil palm in 2003 reached zones with average temperatures of 18 and 19 °C. Projected coverage for MOP1 and MOP2 scenarios was defined assuming that the increase of oil palm plantations will be produced over zones with average temperature equal or higher than of 19 °C, at the left upper zone of the Subdomain d03.

(a) Land-use map for WOP scenario.

(b) Land-use map for OP1 and OP2.

c) Land-use map for MOP1 and MOP2.

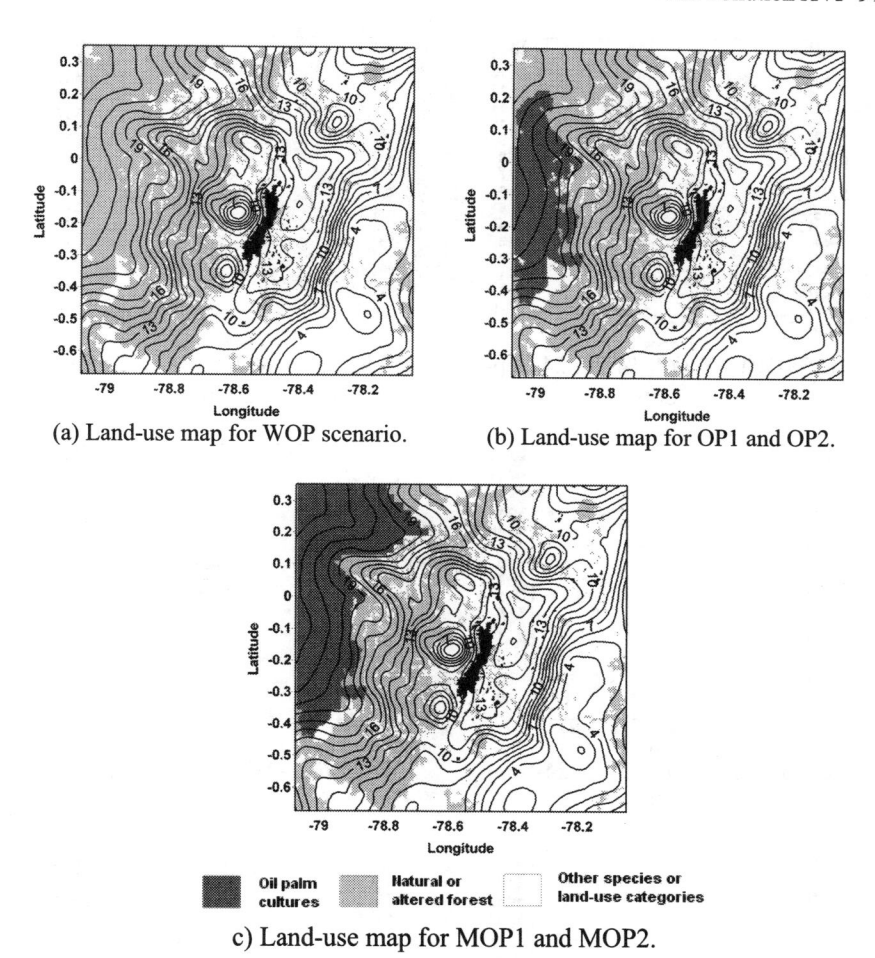

Figure 4: (a) Land-use map for the WOP scenario and average temperature curves (°C), (b) Land-use map for the year 2003 used for the OP1 and OP2 scenarios (oil palm coverage: 8.7% of the Subdomain d03), (c) Land-use map for the MOP1 and MOP2 scenarios (oil palm coverage: 15.6% of the Subdomain d03).

4 Results

Fig. 5 depicts the hourly O_3 anomalies between the MOP1, MOP2 and WOP scenarios, during 14:00 to 16:00 of 24-Sep-2006. They were the highest values found during the period of study. Over the urban region of Quito, anomalies reached 17.9 µg m^{-3} for the MOP2 scenario, and 2.7 µg m^{-3} for the MOP1 scenario.

(a) MOP1 - WOP. 14:00. 24-Sep-2006

(b) MOP2 - WOP. 14:00. 24-Sep-2006

(c) MOP1 - WOP 15:00. 24-Sep-2006

(d) MOP2 - WOP 15:00. 24-Sep-2006

(e) MOP1 - WOP 16:00. 24-Sep-2006

(f) MOP2 - WOP 16:00. 24-Sep-2006

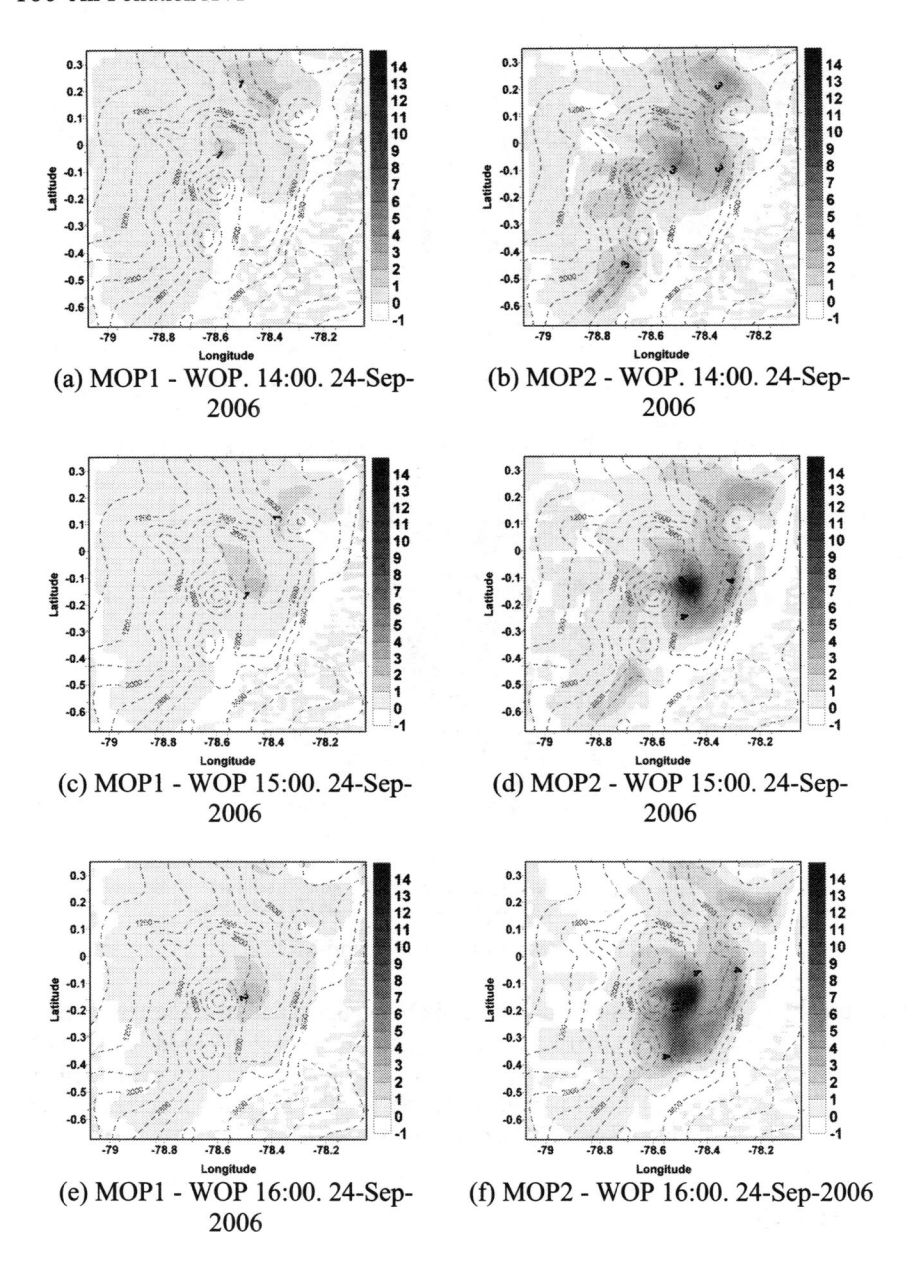

Figure 5: Hourly O_3 anomalies (μg m^{-3}) during 24 September 2006: (a) MOP1-WOP, 14:00, (b) MOP2-WOP, 14:00, (c) MOP1-WOP, 15:00, (d) MOP2-WOP, 15:00, (e) MOP2-WOP, 16:00, (f) MOP2-WOP, 16:00. They were the highest anomalies found during the period of study. Over Quito, anomalies reached 17.9 μg m^{-3} for the MOP2 scenario, and 2.7 μg m^{-3} for the MOP1 scenario.

For the period of study and over the urban region, the highest anomalies ranged from 5.5 to 17.9 µg m^{-3} for the MOP2 scenario, and from 1 to 2.7 µg m^{-3} for the MOP1 scenario. The highest anomalies ranged from 2.3 to 8.4 µg m^{-3} for the OP2 scenario; and from 0.4 to 1.4 µg m^{-3} for the OP1 scenario.

A network of 9 automatic air quality stations have been operating since July 2003 in the Quito region (Fig. 6). Data from Jipijapa (north of Quito; Fig. 6) were used to evaluate the model output.

Fig. 7 shows the hourly O_3 concentrations (µg m^{-3}), surface temperature (°C), and global solar radiation (W m^{-2}) during 23 to 28-Sep-2006, for the Jipijapa station (north of Quito, Fig. 6). Fig. 7 (a) shows clearly the highest increased O_3 concentrations during the afternoon of 24-Sep-2006. Other days the anomalies are lower.

Fig. 8 depicts the 8-hour O_3 average anomalies between the MOP1, MOP2 and WOP scenarios, during 11:00 to 18:00 of 24-Sep-2006. Ozone anomalies reached until 6.5 µg m^{-3} for the MOP2 scenario, and 1 µg m^{-3} for the MOP1 scenario.

During the period of study, hourly O_3 anomalies were higher at 15h00 and 16h00 for the OP1 and OP2 scenarios, but at 17:00 for the MOP1 and MOP2 scenarios.

Figure 6: Zoom in of Quito and allocation of the automatic air quality stations.

5 Conclusions

Oil palm plantations could increase the tropospheric O_3 in Quito. Potential range for maximum increase of hourly O_3 concentrations (2.7 to 17.9 µg m^{-3}) was established under a projected growth of oil palm plantation (to 15.6% of the Subdomain d03). Increased values of hourly O_3 concentrations for the 2003 coverage ranged from 1.4 to 8.4 µg m^{-3})

To reduce uncertainty, it is necessary to measure the isoprene emission factors over the region of study. The modeling domain should be resized, because there are more oil palm plantations to the west of the actual domain used for air quality modeling.

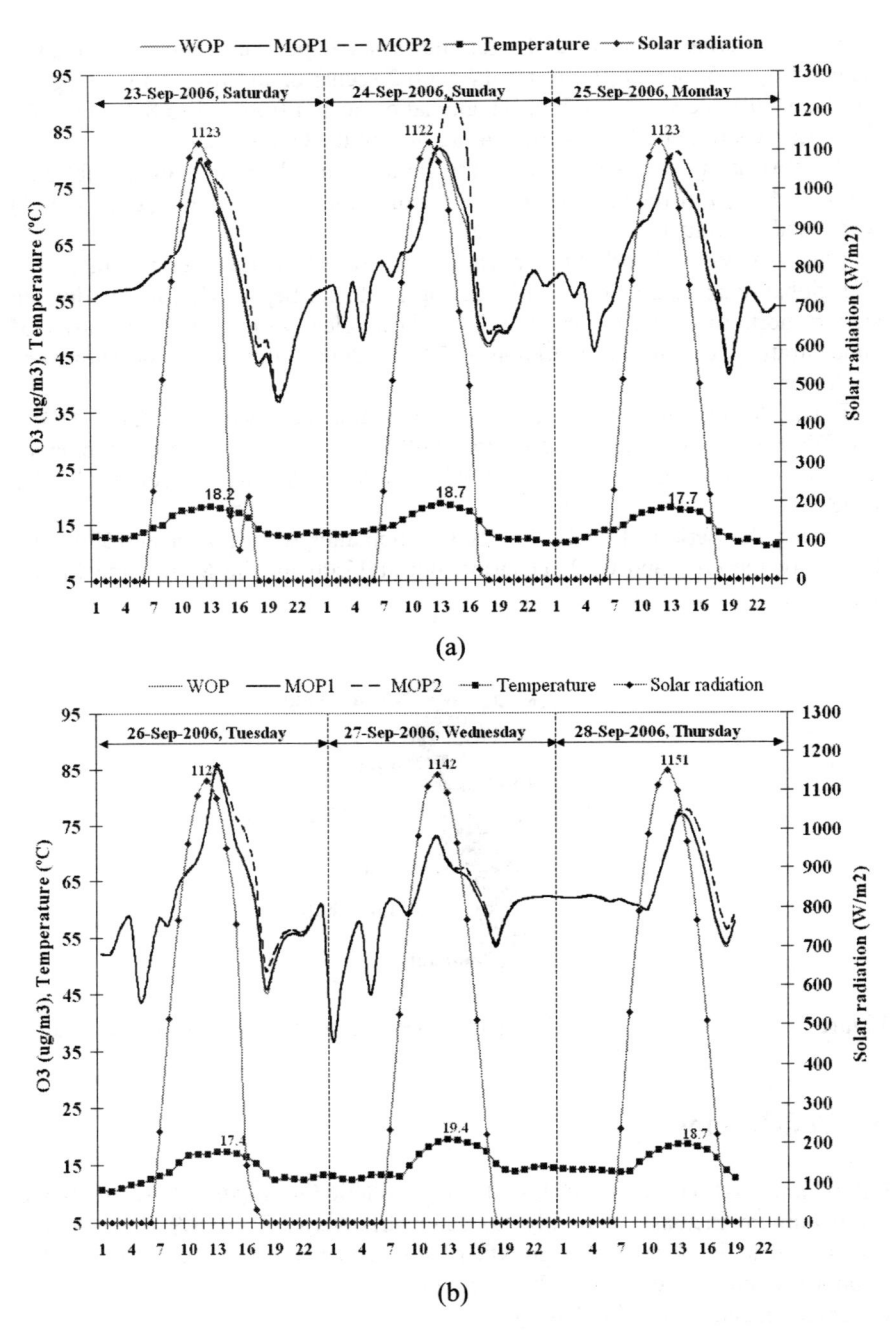

Figure 7: Hourly O_3 concentrations ($\mu g\ m^{-3}$), surface temperature (°C), and solar radiation ($W\ m^{-2}$) for the Jipijapa station: (a) 23 to 25, (b) 26 to 28 September 2006.

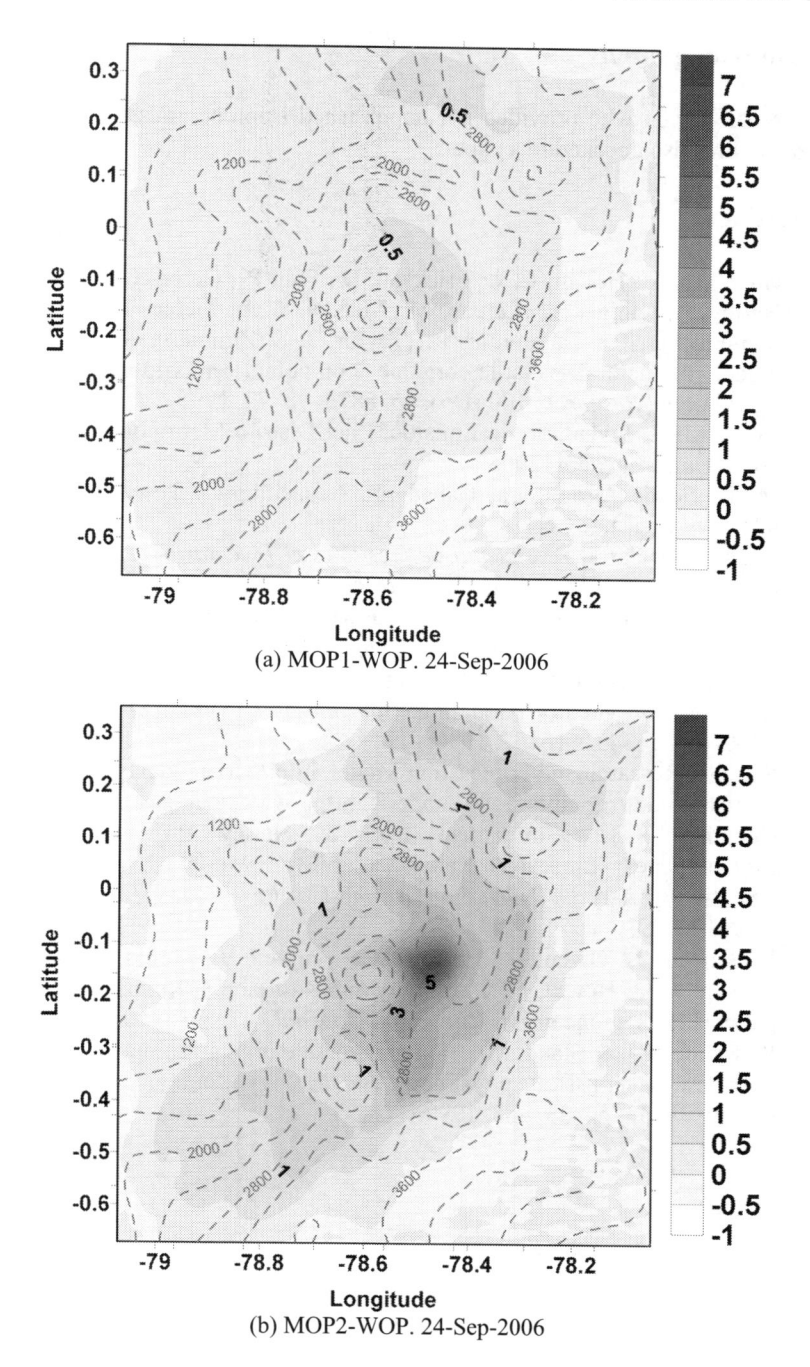

Figure 8: 8-hour O_3 average anomalies ($\mu g\ m^{-3}$) during 24 September 2006: (a) MOP1-WOP, (b) MOP2-WOP. Anomalies reached 6.5 $\mu g\ m^{-3}$ for the MOP2 scenario, and 1 $\mu g\ m^{-3}$ for the MOP1 scenario.

Acknowledgement

To Dr. Sue Owen who provided her unpublished isoprene emission factors for oil palm. She also checked this paper.

References

[1] Guenther, A.; Hewitt, C.N.; Erickson, D.; Fall, R.; Geron, C.; Graedel, T.; Harley, P.; Klinger, L.; Lerdau, M.; McKay, W.A.; Pierce, T.; Scholes, B.; Steinbrecher, R.; Tallamraju, R.; Taylor, J. & Zimmerman, P., A global model of natural volatile organic compound emissions. *Journal of Geophysical Research*, 100 (D5), pp. 8873 – 8892, 1995.

[2] CORPAIRE. Inventario de Emisiones del Distrito Metropolitano de Quito 2003. Quito-Ecuador, 2006.

[3] Ministerio de Agricultura, Ganadería, Acuacultura y Pesca del Ecuador, http://www.sica.gov.ec

[4] Mina, T. & Parra, R. *Sistema de Gestión del Inventario de Emisiones del Distrito Metropolitano de Quito (SIGIEQ)*, CORPAIRE, 1er Congreso Ecuatoriano Sobre la Gestión de la Calidad del Aire. Quito- Ecuador, 2006

[5] Stockwell, W., Middleton, P., Chang, J. & Tang, X., The Second Generation Regional Acid Deposition Model Chemical Mechanism for Regional Air Quality Modeling. *Journal of Geophysical Research*, 95 (D10), pp. 16343 – 16367, 1990.

[6] Weather Research and Forecasting Model, http://wrf-model.org

[7] Ministerio de Agricultura y Ganadería, http://www.mag.gov.ec/sigagro/

[8] Guenther A.B., Zimmerman P.R. & Harley P.C. Isoprene and Monoterpenes Emission Rate Variability: Model Evaluations and Sensitivity Analysis. *Journal of Geophysical Research*, 98(D7), pp. 12609 – 12617, 1993.

[9] Wilkinson, J.; Owen, S.; Possell, M.; Harwell, J.; Gould, P.; Hall, A.; Vickers, C. & Hewitt, N. Circadian control of isoprene emissions from oil palm (Elaeis guineensis). *The Plant Journal*, 47, pp. 960 – 968, 2006.

[10] Kesselmeier J. & Staudt M. Biogenic volatile organic compounds (voc): an overview on emission, physiology and ecology. Journal of Atmospheric Chemistry, 33, pp. 23 – 88, 1999.

Some reflections on the modelling of biogenic emissions of monoterpenes in the boreal zone

K. M. Granström
Department of Energy, Environmental and Building Technology, Karlstad University, Sweden

Abstract

Trees emit volatile organic compounds, mostly monoterpenes and isoprene. These biogenic substances are the dominant volatile organic compounds in air in forested regions. They contribute to the formation of thopospheric ozone and other photochemical oxidants if mixed with polluted air from urban areas. Increased ozone levels hamper photosynthesis and thus have a negative impact on the growth of forests and crops. Terpene flux estimations are needed for models of atmospheric chemistry and for carbon budgets. Several models of natural terpene emission have been constructed, both in a global scale and for various regions. Ideally, a model of natural terpene emissions should show the terpene flux at different times of day and year, at different weather conditions, and for different ecosystems. Its resolution should be sufficient to show short emission peaks. It should also be able to accommodate extreme events like pest outbreaks and serious storms, especially since those are expected to become more common due to global warming. An examination of the scientific literature on monoterpene content in trees and emission fluxes for the dominant boreal forest tree species shows that models aiming to predict terpene fluxes from natural sources over time should include the factors temperature and light intensity, and possibly also take into account the seasonal variation of terpene levels in trees. As wood tissue damage increases emissions, a base level of herbivory and insect predation should be estimated and included. When identification of high concentrations is important, models should have sufficient resolution to capture the emission peaks found, for example, at bud break. The temperature dependence is shown to vary sufficiently between different tree species to motivate using specific values for the ecosystems examined.

Keywords: biogenic, volatile, terpene, model, forest, boreal, spruce, pine, birch.

1 Introduction

Many plants emit biogenic volatile organic compounds (BVOC), mostly monoterpenes, isoprene and volatile carbonyl compounds. The boreal vegetation zone is heavily forested, mostly with pine and spruce. Birches dominate amongst the deciduous trees. In regions dominated by conifers or non-isoprene emitting deciduous tree species, monoterpenes may dominate BVOC emissions (Rinne *et al.* [1]). Estimates of BVOC emissions are important inputs for models of atmospheric chemistry (Geron *et al.* [2]). VOC from forests contribute to the formation of ozone and other photochemical oxidants if mixed with polluted air from urban areas. Increased ozone levels hamper photosynthesis and thus have a negative impact on forest and crops. The emissions of BVOC are also important when calculating carbon budgets. The aim of this paper is to elucidate which factors should be included in models of emissions of biogenic volatiles in the boreal zone. The species examined are norway spruce (lat. *Picea abies*), scots pine (lat. *Pinus silvestris*), and birches (lat. *Betula pendula, Betula pubescens*).

1.1 Emission models

Several models of BVOC emission flux have been constructed, both in a global scale and for various regions. According to a frequently used model created by Guenther *et al.* [3,4] the emission flux is given by the equation

$$F = D\varepsilon\gamma, \qquad (1)$$

where F is emission flux (μgC m^{-2} h^{-1}), D is foliar density (gdw m^{-2}), ε is an ecosystem dependent emission factor (μgC gdw^{-1} h^{-1}), and γ is an activity factor dependent on temperature calculated as

$$\gamma = \exp[\beta(T - T_s)], \qquad (2)$$

where β is an empirical coefficient (°C^{-1}), T is leaf temperature and T_s is a standard temperature of 30°C. Guenther *et al.* [3] cites estimates of β ranging from 0.057 to 0.144°C^{-1} with half of the values within 0.090±0.015, and suggest a best estimate of 0.09 °C^{-1} for all plants and monoterpenes. Seasonal and spatial changes in species composition and foliar density are further developed in Guenther [5].

In a model of global emissions by Guenther *et al.* [4], temporal variations are driven by monthly estimates of biomass and temperature and, for modelling of isoprene fluxes, by hourly light estimates. A model for terpene emissions from European boreal forest (Lindfors *et al.* [6]), based on the Guenther emission algorithms, takes into account the dependence of light intensity and the latitudinal variation of the biomass of conifers and the seasonal variability of the biomass of boreal deciduous trees.

2 Factors affecting emissions

There is a large natural variation in the extractive content of plants, and thus variations in the amount and type of emissions. There are also many external factors affecting biogenic terpene emissions, such as temperature, light intensity, the season (the plants' growth stadium), damage (herbivory, insects, fungi or touch), rain and air pollution.

2.1 Temperature

Monoterpenes are stored in special cells in the needles of conifers and leaves of deciduous trees and are emitted primarily through volatilisation. Emission rates are strongly dependent on the temperature of the leaf surface (Guenther *et al.* [3]). The infra-red radiation of a scots pine stand has a good correlation with monoterpene emissions from the canopy (Hakola *et al.* [7]).

2.2 Light

Marked diurnal variations with a maximum around noon have been observed for boreal coniferous forest in Sweden (Johansson and Janson [8]), Finland, (Spanke *et al.* [9]), and for a mixed Canadian deciduous forest (Fuentes *et al.* [10]). After normalisation to temperature, emission rates from scots pine and norway spruce still vary diurnally with a maximum at midday (Janson [11]), which implies an effect of light intensity. Likely, terpenes are not only released from storage pools but also released directly after synthesis. The terpene precursor isoprene is released immediately and its emission rate is strongly dependent on both temperature and light intensity. The storage pool emissions a percentage of total monoterpene emission potentials have been reported as 63% for scots pine and 64% for norway spruce in southern Germany (Steinbrecher *et al.* [12]).

2.3 Season (the plants' growth stadium)

The terpene concentration in wood is low during winter, increases rapidly in spring, reaches a maximum during summer, and drops off to the base level in autumn. The seasonal cycle in terpene concentration can be observed even in the absence of herbivory, probably due to increased attacks by herbivores in the evolutionary past (Lerdau *et al.* [13]). At bud break, the emission rates of monoterpenes to air increase sharply both for conifers (Lerdau *et al.* [14]) and for deciduous trees (Fuentes *et al.* [10]). Terpene emissions then decrease after the first few days or weeks after bud break (Hakola *et al.* [15]).

2.4 Damage

Monoterpene production is induced when the plant is attacked, for instance by herbivores or fungi or insects. Spruces rely on biosynthesis at the affected area when damage occurs, while pines have high level monoterpene resin transported via interconnected resin canals appearing at wound sites within seconds after an injury occurs (Lewinsohn *et al.* [16]). Calculated whole-canopy fluxes imply that a ponderosa pine forest with 10% damaged foliage will emit 2 times more monoterpenes, and 25% damage 3.6 times more than undamaged forest, while the same damage to a douglas-fir forest gives 1.6 times and 2.5 times more monoterpenes, respectively (Litvak et al. [17]). The increased emissions from the low level of grazing (10% damage) is sufficient to increase local tropospheric ozone production (Litvak et al. [17]). Debudding a scops pine branch increased its monoterpene emissions for about a week (Hakola et al. [18]). Touch is suspected to cause higher monoterpene emission rates (Guenther et al. [19]). This makes it very important to avoid rough handling of vegetation during enclosure experiments.

2.5 Rain

Emissions of monoterpenes increase temporarily during and after heavy rain (Janson [11]).

2.6 Air pollution

Considering that monoterpenes protect plants against ozone damage (which works in the absence of nitrogen oxides) monoterpene emissions ought to increase with increasing ozone concentration. Such a connection has been found for pine, while the results for spruce are ambiguous. Long-term ozone treatment of scots pine led to increased emissions of monoterpenes (Heiden et al. [20]). Traffic pollution increase monoterpene emission rate from norway spruce (Juttner [21]). When norway spruce in open-top chambers was exposed to air with or without added ozone, no significant difference in terpene emissions to air was found (Lindskog and Potter [22]).

3 Natural emissions of terpenes

3.1 Monoterpene flux estimates

Norway spruce has relatively stable emissions from the onset in late May until October (table 1). Norway spruce emit more terpenes than scots pine does (table 1). The E_{30}-value assumed in the model in Lindfors et al. [6], 1.5 µg gdw^{-1}, is a reasonable approximation for scots pine, but likely to lead to underestimations of emissions from norway spruce.

Both silver birch and downy birch emitted terpenes at bud break in May, after which the emission rate declined and was low while the leaves grew. When the leaves had reached their full size, the emission rate increased and remained high during the rest of the growing season (Hakola et al. [7,15]). In the model by Lindfors et al. [6], birch has an E_{30}-value of 0.64 in early summer and 5.6 in late summer.

3.2 Monoterpene ambient measurements

Ambient concentrations have been found to be larger at night (Petersson [25], Janson [26]). This despite that the terpene emissions are higher during the day (Rinne et al. [1]). The higher concentrations at night have been attributed to lower dispersion. In addition, ozone concentrations are higher in the daylight hours.

Table 1: Monoterpene fluxes (μg gdw^{-1} h^{-1}) for boreal trees; where ± is given it denotes 1 standard deviation; values are normalised using β=0.09.

Tree species	Month (Date)	E (μg gdw^{-1} h^{-1})	E$_{30}$ (μg gdw^{-1} h^{-1})	Location	Ref.
Norway spruce	May (5)	0.48	0.73	Sweden (60°N, 16°E)	J93
	May (29)	0.45	4.4	Sweden (60°N, 16°E)	J93
	June (2,4)	1.8	2.3±1.1	Sweden (60°N 17°E)	J99
	June (12)	0.20	2.1	Sweden (60°N, 16°E)	J93
	June (20)	3.8	5.2±0.5	Sweden (60°N 17°E)	J99
	July (8,9)	2.4	2.3±0.3	Sweden (60°N 17°E)	J99
	July (31)	0.12	4.3	Sweden (60°N, 16°E)	J93
	Aug (21)	0.57	3.4	Sweden (60°N, 16°E)	J93
	Sep (20)	2.0	8.3±6.1	Sweden (60°N 17°E)	J99
	Oct (5)	0.61	31	Sweden (60°N, 16°E)	J93
Scots pine	May (3)	1.3	1.0	Sweden (60°N, 16°E)	J93
	May (29,30)	0.8	0.67	Sweden (60°N, 16°E)	J93
	June (20)	0.22	2.6	Sweden (60°N, 16°E)	J93
	June (22)	0.46	1.3	Sweden (60°N, 16°E)	J93
	June (26)	0.38	3.7	Sweden (60°N, 16°E)	J93
	June (27)	0.32	1.1	Sweden (60°N, 16°E)	J93
	July (12)	0.28	2.7	Sweden (60°N, 16°E)	J93
	July (20)	0.28	3.4	Sweden (60°N, 16°E)	J93
	July	–	1.2	Finland (62°N, 30°E)	R00
	Aug (18)	2.0	1.1	Sweden (60°N 17°E)	J99
	Aug (19)	0.54	1.9	Sweden (60°N 17°E)	J99
	Aug (21)	0.1	0.77	Sweden (60°N, 16°E)	J93
	Aug (30)	0.074	1.3	Sweden (60°N, 16°E)	J93
	Aug (31)	0.11	0.86	Sweden (60°N, 16°E)	J93
	Sep (6)	0.34	2.7	Sweden (60°N 17°E)	J99
	Sep (7)	0.39	7.5	Sweden (60°N 17°E)	J99
	Oct (5)	0.58	45	Sweden (60°N, 16°E)	J93
Silver birch	May (24)	5.6	14	Finland (60°N 25°E)	H98
	June (12)	0.4	0.5	Finland (60°N 25°E)	H98
	June (25)	-0.2	-6	Finland (60°N 25°E)	H98
	Aug (14-22)	6-12	6-12	Finland (60°N 25°E)	H98
	Sep (3, 6)	0.05-1	1-5	Finland (60°N 25°E)	H98
Downy birch	Early summer		0.76	Finland (60°N 25°E)	H99
	Late summer		6.08	Finland (60°N 25°E)	H99

J93 Janson [11]; J99 Janson et al. [23]; R00 Rinne et al. [24]; H98 Hakola et al. [15]; H99 Hakola et al. [7].

The ambient monoterpene concentration varies over the year, with detectable concentrations beginning at bud break and a maximum at mid-summer. This was found by Janson [26] for scots pine, and for a deciduous forest by Fuentes et al. [10].

4 Implications for emission modelling

The monoterpene fluxes show considerable variations over time. Terpene emissions depend on temperature, but the diurnal variations are also an effect of variable light intensity. Bud break is an important emission period for both deciduous trees and conifers. To use monthly estimates of biomass and temperature would likely not reflect reality for zones with a variable climate, like the boreal zone where there is strong seasonal variability of temperature and light and also highly variable deciduous foliage density over the year. To use global averages for modelling of emissions in regions can not be expected to give accurate results, as the emission factors differ between regions.

Considering the effect of grazing, it is not surprising that forestry has a considerable impact on terpene emissions (Strömvall and Petersson [27]). This is outside the scope for models of natural BVOC emissions, but very important when the aim is to include both natural and anthropogenic emissions. For models of natural emissions, a base level of herbivory and pests could be included. This is probably not the case today, as branch enclosure experiments have been done on mostly visibly healthy branches, which does not reflect the emissions from damaged vegetation. While the healing of damages by oxidized resins is reasonably fast, the damage from grazing of young conifers is both considerable and frequent. If forest harvesting is included in emission models, the seasonal variation of terpene content in wood could be of interest.

The model by Lindfors et al. [6] includes both temperature and light intensity and the latitudinal variation of the biomass of conifers and the seasonal variability of the biomass of boreal deciduous trees. However, the monoterpene emission rates (ε in equation 1) are too schematic. While the model stretches from April to October, there are only two time periods for emission rates - early summer and late summer. This does not fully capture the seasonal variation, and the low E_{30}-value assigned to birch in early summer does not correspond to the high emissions during bud break found for Silver birch. Also, all conifers are assumed to have a normalised emission factor of 1.5 µg gdw^{-1} h^{-1} during both periods, which does not take into account the difference between Norway spruce and Scots pine.

The exponential increase in emissions with increased temperature in the commonly used equation

$$E = E_{30} \exp[\beta\ (T - 30°C)], \quad (3)$$

where E_{30} is the monoterpene emission normalised to 30°C (Guenther et al. [3]) and β and T are defined as in equation 2, is supported empirically but does not convey the full complexity of the temperature dependence of monoterpene emission of plants. For example, the increase of emissions with temperature is sometimes not observed during very hot and dry weather (Schade et al. [28]).

Several estimations of the coefficient β in equation 1 and 3 have been made for trees in the northern boreal region (table 2). A tentative latitudinal correlation can be seen. β-values have been reported as 0.24 for birch (Hakola et al. [15]). Although the value for birch is higher than those reported for spruce and pine, it is not sufficiently high to explain the high β-value found for a mixed forest by Rinne et al. [1]. The higher temperature dependence for deciduous trees is likely due to their high accumulation of monoterpenes in the cuticle, while conifer cuticles contains less monoterpenes (Schmid et al. [29]).

Table 2: Temperature dependence of the emission of monoterpenes from trees, as values of β in the Guenther algorithm $E=E_{30} \exp [\beta (T-30°C)]$.

Location	Norway spruce	Scots pine	Mixed forest*	Ref.
(60°N, 16°E)	0.07	0.07		Janson [11]
(60°N 17°E)	0.065-0.097	0.068-0.15		Janson and de Serves [30]
(62°N, 30°E)		0.15		Rinne et al. [24]
(68°N, 24°E)			0.96-1.7	Rinne et al. [1]

*Spruce 56%, birch 32%, Pine 10% as dry leaf mass.

Most emission rate values are reported as normalised to 30°C using $\beta=0.09°C^{-1}$, as found by Guenther et al. [3]. A terpene flux that is 2.7 µg gdw^{-1} h^{-1} if normalised to 30°C using $\beta=0.15°C^{-1}$ will be 1.2 µg gdw^{-1} h^{-1} with a β-value of $0.09°C^{-1}$ (Rinne et al. [24]). These values differ by a factor of 2. It is therefore important to use a locally correct value of β, and to recalculate reported values before they are used for modelling.

5 Concluding remarks

High quality data on emission factors is necessary for the modelling of canopy fluxes of terpenes. The terpene fluxes are needed in models of atmospheric chemistry and effects of photochemical oxidants, and also for carbon budgets. While biogenic terpenes are far from the only VOCs in air, they, together with isoprene, are dominant in forested regions. To be really useful a model should give the terpene and isoprene contribution to ozone formation at different times, weather, and regions. Its resolution should be sufficient to show short emission peaks. It should also work for extreme events like pest outbreaks and serious storms, especially since those are expected to become more common due to climate change.

A review of the scientific literature has shown that relevant factors for models of biogenic emissions from boreal forests include temperature, light intensity, seasonal factors, the effects of foreseeable damage to trees, and the effect of rainfall, whereas an effect of air pollution is more ambiguous. Also, models should use time intervals shorter than a month, especially at bud break. Furthermore, the temperature dependence varies sufficiently to motivate using specific values for the ecosystems examined.

It is clear that more research needs to be done on biogenic monoterpene emissions. Some areas I find to be of particular interest are:
- the sensitivity of northern boreal species to herbivory and pest outbreaks (the quantitative and qualitative differences between emissions from healthy trees and trees with various levels of damage);
- latitudinal and longitudinal variation in the boreal zone (are differences due to climate, soil, genetically different strands, or other factors?)

Experiments should include both measurements using branch enclosure and measurements in ambient air. Enclosure experiments are more precise, as they give information about emissions from a certain part of a certain plant. Measurements in ambient air are needed to test the models, and also to provide a solid empirical basis - including effects from herbivory and rain and other factors that may be excluded from detection in enclosure experiments. The results when measuring in ambient air are also affected by emissions from other vegetation and from the forest floor, which is not the case for enclosure experiments.

References

[1] Rinne J., Tuovinen J.P., Laurila T., Hakola H., Aurela M. & Hypen H. Measurements of hydrocarbon fluxes by a gradient method above a northern boreal forest. *Agricultural and Forest Meteorology,* 102:25–37, 2000.

[2] Geron C.D., Guenther A.B. & Pierce T.E. An Improved Model for Estimating Emissions of Volatile Organic-Compounds from Forests in the Eastern United-States. *Journal of Geophysical Research-Atmospheres,* 99:12773–12791, 1994.

[3] Guenther A.B., Zimmerman P.R., Harley P.C., Monson R.K. & Fall R. Isoprene and Monoterpene Emission Rate Variability - Model Evaluations and Sensitivity Analyses. *Journal of Geophysical Research-Atmospheres,* 98:12609–12617, 1993.

[4] Guenther A., Hewitt C.N., Erickson D., Fall R., Geron C., Graedel T., Harley P., Klinger L., Lerdau M., McKay W.A., Pierce T., Scholes B., Steinbrecher R., Tallamraju R., Taylor J. & Zimmerman P. A Global-Model of Natural Volatile Organic-Compound Emissions. *Journal of Geophysical Research-Atmospheres,* 100:8873–8892, 1995.

[5] Guenther A. Seasonal and spatial variations in natural volatile organic compound emissions. *Ecological applications,* 7:34–45, 1997.

[6] Lindfors V., Laurila T., Hakola H., Steinbrecher R. & Rinne J. Modeling speciated terpenoid emissions from the European boreal forest. *Atmospheric Environment,* 34:4983–4996, 2000.

[7] Hakola H., Rinne J. & Laurila T.: The VOC Emission Rates of Boreal Deciduous Trees. In *Biogenic VOC emissions and photochemistry in the*

boreal regions of Europe-BIOREP Air pollution research report. Lindfors V., Ed. Luxemburg, European Commission, 1999, p. 21–28
[8] Johansson C. & Janson R.W. Diurnal Cycle of O-3 and Monoterpenes in a Coniferous Forest - Importance of Atmospheric Stability, Surface Exchange, and Chemistry. *Journal of Geophysical Research-Atmospheres*, 98:5121–5133, 1993.
[9] Spanke J., Rannik U., Forkel R., Nigge W. & Hoffmann T. Emission fluxes and atmospheric degradation of monoterpenes above a boreal forest: field measurements and modelling. *Tellus Series B-Chemical and Physical Meteorology*, 53:406–422, 2001.
[10] Fuentes J.D., Wang D., Neumann H.H., Gillespie T.J., DenHartog G. & Dann T.F. Ambient biogenic hydrocarbons and isoprene emissions from a mixed deciduous forest. *Journal of Atmospheric Chemistry*, 25:67–95, 1996.
[11] Janson R.W. Monoterpene Emissions from Scots Pine and Norwegian Spruce. *Journal of Geophysical Research-Atmospheres*, 98:2839–2850, 1993.
[12] Steinbrecher R., Hauff K., Hakola H. & Rössler J. A revised parameterisation for emission modelling of isoprenoids for boreal plants. *Biogenic VOC emissions and photochemistry in the boreal regions of Europe-BIOREP* Laurila T. & Lindfors V., Eds., European Commission: Luxemburg, p. 29–43, 1999.
[13] Lerdau M., Litvak M. & Monson R. Plant-Chemical Defense - Monoterpenes and the Growth-Differentiation Balance Hypothesis. *Trends in Ecology & Evolution*, 9:58–61, 1994.
[14] Lerdau M., Matson P., Fall R. & Monson R. Ecological Controls over Monoterpene Emissions from Douglas-Fir (Pseudotsuga-Menziesii). *Ecology*, 76:2640–2647, 1995.
[15] Hakola H., Rinne J. & Laurila T. The hydrocarbon emission rates of tea-leafed willow (Salix phylicifolia), silver birch (Betula pendula) and European aspen (Populus tremula). *Atmospheric Environment*, 32:1825–1833, 1998.
[16] Lewinsohn E., Gijzen M. & Croteau R. Defense-Mechanisms of Conifers - Differences in Constitutive and Wound-Induced Monoterpene Biosynthesis among Species. *Plant Physiology*, 96:44–49, 1991.
[17] Litvak M.E., Madronich S. & Monson R.K. Herbivore-induced monoterpene emissions from coniferous forests: Potential impact on local tropospheric chemistry. *Ecological Applications*, 9:1147–1159, 1999.
[18] Hakola H., Tarvainen V., Back J., Ranta H., Bonn B., Rinne J. & Kulmala M. Seasonal variation of mono- and sesquiterpene emission rates of Scots pine. *Biogeosciences*, 3:93–101, 2006.
[19] Guenther A., Zimmerman P. & Wildermuth M. Natural Volatile Organic-Compound Emission Rate Estimates for United-States Woodland Landscapes. *Atmospheric Environment*, 28:1197–1210, 1994.
[20] Heiden A.C., Hoffmann T., Kahl J., Kley D., Klockow D., Langebartels C., Mehlhorn H., Sandermann H., Schraudner M., Schuh G. & Wildt J.

Emission of volatile organic compounds from ozone-exposed plants. *Ecological Applications,* 9:1160–1167, 1999.
[21] Juttner F. Changes of Monoterpene Concentrations in Needles of Pollution-Injured Picea-Abies Exhibiting Montane Yellowing. *Physiologia Plantarum,* 72:48–56, 1988.
[22] Lindskog A. & Potter A. Terpene Emission and Ozone Stress. *Chemosphere,* 30:1171–1181, 1995.
[23] Janson R.W., De Serves C. & Romero R. Emission of isoprene and carbonyl compounds from a boreal forest and wetland in Sweden. *Agricultural and Forest Meteorology,* 98-9:671–681, 1999.
[24] Rinne J., Hakola H., Laurila T. & Rannik U. Canopy scale monoterpene emissions of Pinus sylvestris dominated forests. *Atmospheric Environment,* 34:1099–1107, 2000.
[25] Petersson G. High Ambient Concentrations of Monoterpenes in a Scandinavian Pine Forest. *Atmospheric Environment,* 22:2617–2619, 1988.
[26] Janson R.W. Monoterpene Concentrations in and above a Forest of Scots Pine. *Journal of Atmospheric Chemistry,* 14:385–394, 1992.
[27] Strömvall A. & Petersson G. Conifer monoterpenes emitted to air by logging operations. *Scandinavian Journal Forest Research,* 6:253–258, 1991.
[28] Schade G.W., Goldstein A.H. & Lamanna M.S. Are monoterpene emissions influenced by humidity? *Geophysical Research Letters,* 26:2187–2190, 1999.
[29] Schmid C., Steinbrecher R. & Ziegler H. Partition-Coefficients of Plant Cuticles for Monoterpenes. *Trees-Structure and Function,* 6:32–36, 1992.
[30] Janson R. & de Serves C. Acetone and monoterpene emissions from the boreal forest in northern Europe. *Atmospheric Environment,* 35:4629–4637, 2001.

An evaluation of SOA modelling in the Madrid metropolitan area

M. G. Vivanco, I. Palomino, J. Plaza, B. Artíñano, M. Pujadas & B. Bessagnet
CIEMAT, Madrid, Spain

Abstract

The improvement of air quality represents an important challenge for our society. In urban areas, air quality standards are being exceeded. Like other tropospheric pollutants, aerosols are presently under regulation in the European Union. Control of aerosol concentration is an important objective, because high levels can affect human health. As an indirect effect, aerosols can alter earth's radiative budget by scattering or absorbing radiation, producing a change of ozone production. Because of both direct and indirect aerosol effects, it is important to know aerosol levels in the troposphere.

Models can be used as a tool for air quality management. Secondary organic aerosol (SOA) is presently one of the most important topics on air quality modelling. Many aspects of SOA modelling are still a challenge for the scientific community. Unfortunately, the quality of model results cannot be evaluated because SOA measurements are not available at air quality stations. The reason for that is that it is not possible to distinguish experimentally a primary organic aerosol from a secondary one.

In this paper, an air quality model was used to simulate hourly SOA concentrations during a 2003 summer period in the Madrid metropolitan area. A simple reaction scheme for SOA was used. Modelled SOA was compared against SOA estimations obtained from thermal OC and EC hourly measurements at an urban background site, using the OC/EC minimum ratio approach. Although a reasonable agreement is observed, higher-resolution simulations with higher-resolution emissions should be carried out in order to improve model predictions. Also, a more complex scheme of SOA formation should be tested to determine the origin of the discrepancies.

Keywords: secondary organic aerosols, air quality modelling, organic carbon, elemental carbon.

1 Introduction

In the urban atmosphere fine particulate matter includes primary and secondary organic and inorganic compounds. Secondary organic aerosol (SOA) can have a relevant presence, especially during smog episodes. Its formation involves the presence of oxidation products with a low pressure vapour in order to be able to partition in aerosol phase. Some research has been done over the last decade to describe the SOA formation from many reactive organic compounds [1–5].

It is difficult to represent atmospheric SOA formation in air quality models due to the complexity of the processes involved and the uncertainties affecting SOA precursor emissions. Some models use the lumped SOA yield based on [6]. According to the authors, oxidation reactions for each class of VOC are assumed to lead to a fixed fraction of SOA product.

The CHIMERE model has been extensively applied over the past year [7–11]. In Spain, evaluation of the model performance for O3 and NO2 has been shown in [12]. This study has indicated that O3 predictions are in a reasonably agreement to observations registered at rural sites. The capability to reproduce PM10 and PM2.5 has also been evaluated in [13].

As SOA cannot be directly recorded in the real atmosphere, techniques to estimate SOA formation from measurements of organic and elemental carbon are commonly used. Elemental carbon is defined as the material that will not thermally desorb from a filter sample and is generally attributed to graphitic, soot-like structures. It is assumed to be exclusively due to primary emission. Organic is defined as material that will thermally desorb from a filter sample and may be associated with either primary or secondary aerosol [14–16]. To estimate SOA values, a methodology based on the ratio of organic to elemental carbon in the primary aerosol is used. As this quantity is not accurately known, SOA estimates using this approach may be subject to inaccuracies. In spite of this limitation organic and elemental carbon measurements are frequently the only data about the carbonaceous component of atmospheric aerosol. That is the reason why this approach is commonly used.

2 OC-EC semicontinuous measurements

Measurements of carbonaceous aerosol concentration (EC/OC) were carried out using a thermal analyzer (Rupprecht and Patashnick model 5400) at a sub-urban site in Madrid (CIEMAT station).

The analyzer collects PM2.5 carbonaceous aerosol using an impactor with a cut-off diameter of 0.14 mm prior to its sequential oxidation with particle-free ambient air. The CO2 produced is analyzed and its concentration is related to carbon mass in both aerosol fractions (OC and EC). One-hour sampling and analysis cycle was set-up, being the combustion temperatures to obtain OC and EC concentrations 340°C and 750°C respectively. The EC concentration was corrected accounting for material loss due to the cut-off size of the impactor, from simultaneous 24h filter samples and EC and OC measurements using a thermal carbon analyzer (LECO) at the same split temperature.

3 Estimation of SOA from measurements

Based on the high temporal resolution of EC-OC measurements, a method to calculate primary OC/EC ratio from traffic has been applied in polluted days (EC hourly maximum > 4 µgm^{-3}). Background values have been considered to subtract from EC and OC values at the time of maximum EC slope during morning or evening rush hours. Usually the morning maximum slope in EC takes place before sunrise, thus this calculation avoids computing secondary OC [17]:

$$\left(\frac{OC}{EC}\right)_{primary} = \frac{OC_{\Delta EC_{max}} - OC_{background}}{EC_{\Delta EC_{max}} - EC_{background}}$$

The method to estimate SOA from OC and EC measurements in this suburban site was the subtraction of traffic primary OC from measured OC, i.e. the minimum ratio approach [18]):

$$SOA = OC_{measured} - (OC/EC)_{primary} \cdot EC)$$

The calculated OC/EC traffic primary ratio was 0.47 ± 0.26 (average ± std, n=59, 42 morning and 17 evening estimations from a 18-month period). Primary OC due to other minor sources is not considered in this method of SOA estimation, so the obtained value can be seen as an upper limit.

4 CHIMERE model SOA scheme

The V2006-par version of the CHIMERE model calculates the concentration of 44 gaseous species and both inorganic and organic aerosols of primary and secondary origin, including primary particulate matter, mineral dust, sulphate, nitrate, ammonium, secondary organic species and water. In this version a very simplified scheme for SOA formation is implemented in the chemical module MELCHIOR. Precursor volatile organic compounds able to form secondary aerosol species are high chain alkanes, aromatics and monoterpenes. Anthropogenic aerosol yields (ASOA) come from [1, 2, 6, 19]. ASOA in this scheme is originated by the reaction between alkanes with four or more carbon atoms and ortho-xylene with OH radical. For biogenic secondary organic aerosols (BSOA), aerosol yields for terpene oxidation are taken from [20]. Biogenic SOA are formed from alpha pinene reacting with OH radical and with ozone. The reactions included in MELCHIOR2 mechanism are presented in Table 1. ASOA and BSOA are partitioned between gas and aerosol phases. Mass transfer is not only driven by the gas phase diffusion but also by the thermodynamic equilibrium through a temperature dependent partition coefficient [21].

Table 1: Reactions in CHIMERE V2006par version for SOA formation.

Reaction		Products
$N\text{-}C_4H_{10} + OH$	$\rightarrow$ k1(T)	0.9 $CH_3COCH_2CH_3$ + 0.1 CH_3CHO + 0.1 CH3COO + 0.9 oRO_2 + 0.04 SOA
o-xylene + OH	$\rightarrow$ k2	MEMALD + MGLYOX + oRO2 + 0.2 SOA
α-pineno + OH	$\rightarrow$ k3(T)	0.8 CH_3CHO + 0.8 $CH_3COCH_2CH_3$ + obio + 0.07 SOA
α-pineno + O3	$\rightarrow$ k4(T)	1.27 CH3CHO +0.53 CH3COCH2CH3 + 0.14CO + 0.62ORO2 + 0.42HCHO + 0.85 OH + 0.1HO2 + 0.23SOA

MEMALD: unsaturated dicarbonyls, reacting like 4-oxo-2-pentenal; MGLYOX: methyl glyoxal; obio: operator representing peroxy radicals produced by C5H8 and APINEN + OH reaction; oRO2: operator representing peroxy radicals from OH attack to C2H5, NCHH10, C2H4, C3H6, OXYL, CH3COE, MEMALD, and MVK (methyl vinyl ketone).

$k1(T) = 1.36 \cdot 10^{-12} e^{0.0021\,T}$ $k2 = 1.37 \cdot 10^{-11}$ $k3(T) = 1.21 \cdot 10^{-11} e^{444/T}$
$k4(T) = 10^{-15} e^{-736/T}$

5 Model simulations

Simulations of photochemical compounds were carried out using the regional V200603par-rc1 version of the CHIMERE model. Modelling scheme to obtain SOA concentrations was the same as that described in [13]. First, the CHIMERE model was applied for a coarse domain of 0.5° of resolution at European scale, covering an area ranging from 10.5W to 22.5E and from 35N to 57.5N with 14 vertical sigma-pressure levels extending up to 500 hPa. A second domain was focused over the Iberian Peninsula (from 10.3W to 5.5E and from 35.5N to 44.5N), with a 0.2 degree resolution. A one-way nesting procedure was used; coarse-grid simulations forced the fine-grid ones at the boundaries without feedback.

The emissions for all the simulations were derived from the annual totals of the EMEP database for 2003 [22]. Original EMEP emissions were disaggregated taking into account the land use information, in order to get higher resolution emission data. The spatial emission distribution from the EMEP grid to the CHIMERE grid is performed using an intermediate fine grid at 1km resolution. This high-resolution land use inventory comes from the Global Land Cover Facility (GLCF) data set (http://change.gsfc.nasa.gov/create.html). For each SNAP activity sector, the total NMVOC emission is split into emissions of 227 real individual NMVOC according to the AEAT speciation [23], and real species emissions are aggregated into model species emissions. Biogenic emissions are computed according to the methodology described in [24], for alpha-pinene, NO and isoprene.

Boundary conditions for the coarse domain were provided from monthly 2003 climatology from LMDz-INCA model [25] for gases concentrations and from monthly 2004 GOCART model [26] for particulate species.

The MM5 model [27] was used to obtain meteorological input fields. The simulations were carried out also for two domains, with respective resolutions of 36 Km and 19 Km. The simulations were forced by the National Centres for Environmental Prediction model (GFS) analyses.

6 Results

In order to illustrate the capability of the CHIMERE model to simulate SOA levels, time series of both predicted and based on observations SOA concentrations are presented in Figure 1.

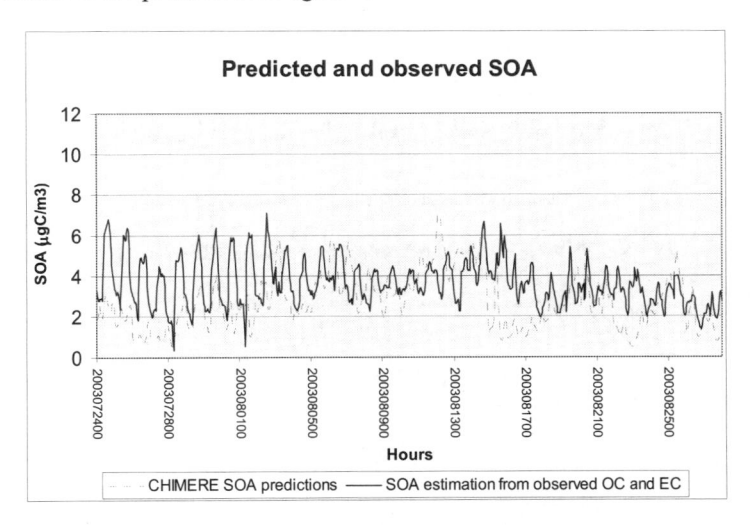

Figure 1: Time series of predicted and based on observations estimated SOA.

In general a good correlation between both SOA time series is found, although there are some discrepancies for some periods, such as August 3-5, August 11-13 and August 21-23. For the period 14-17, in spite of a good correlation a considerable underestimation is observed.

Figure 2 presents time series for other pollutants, such as O3, NO2, NOx and PM10 at "Barrio del Pilar" station, an urban traffic site close to CIEMAT station. Observed and MM5 predicted temperature, wind speed and wind direction are also included in this figure. Ozone temporal variability is quite well reproduced by the model. NO2 and NOx present a good correlation, although high observed concentrations are not reproduced by the CHIMERE model. As this monitoring site is located close to mobile sources, high levels are recorded. Figure 2 also indicates an important underestimation of PM10 model predictions. High resolution simulations and high-resolution emissions need to be used to reproduce those local effects.

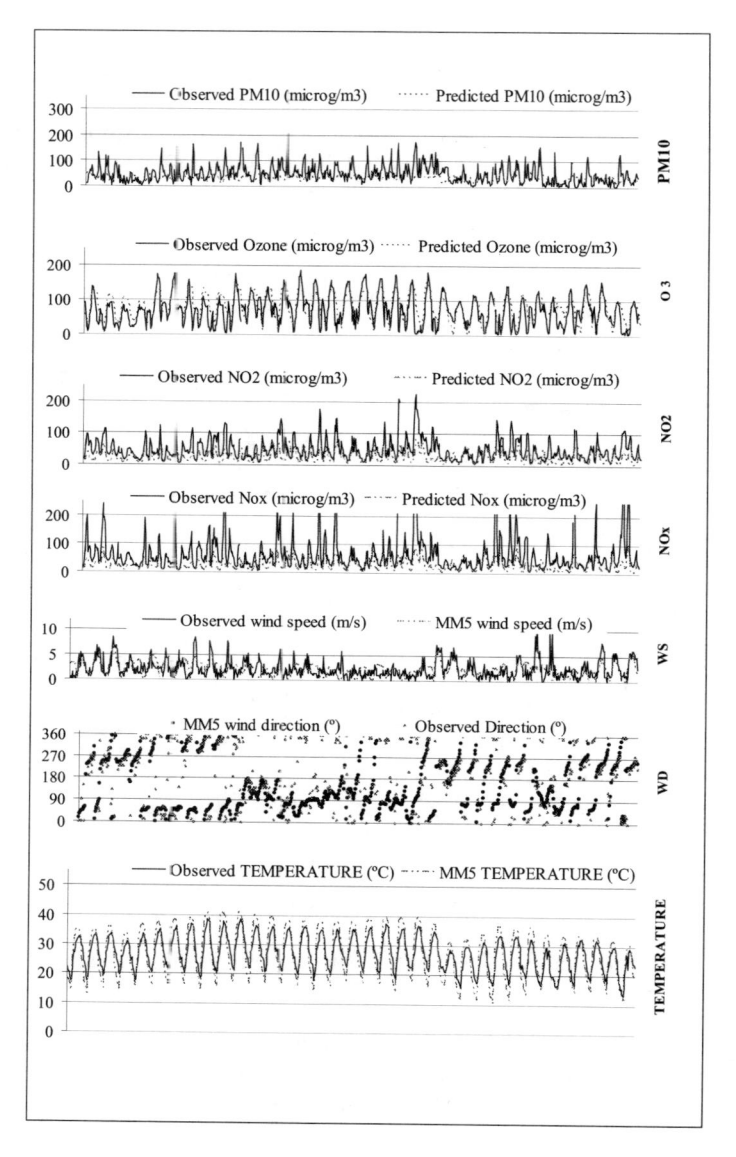

Figure 2: Time series presenting predictions and observations of some pollutant concentrations and some meteorology variables at a suburban site.

7 Conclusions

The present paper shows the results of a graphical evaluation of the SOA levels predicted by the CHIMERE photochemical model. Although, in general, model predictions present values similar to SOA levels estimated from EC and OC observations, some underestimation is found for some periods. These results have been obtained for 0.2° x 0.2° horizontal resolution. High-resolution simulations should be carried out in order to determine if the discrepancies are due to the coarse resolution. Also, high-resolution emissions should be used to better represent local effects. Presently, a 4x4km^2 emission inventory, developed at he Barcelona Supercomputing Centre (BSC) is being applied.

A new version of the CHIMERE model is now available. It incorporates more reactions involving SOA formation. This version is presently being evaluated in order to determine if an improvement of SOA model predictions is obtained.

Acknowledgement

The authors kindly acknowledge the support by the Spanish Environment Ministry.

References

[1] Odum, J. R.; T. Hoffmann, F. Bowman, D. Collins, R.C. Flagan, J.H. Seinfeld (1996). Gas/particle partitioningandsecondary organic aerosol yields. Environ. Sci. Technol., 30, 2580-2585.

[2] Odum J.R., Jungkamp T.W.P., R.J. Griffin, R.C. Flagan, J.H.Seinfeld (1997). The atmospheric aerosol formation potential of whole gasoline vapor. Science 274, 96-99.

[3] Zhang, S. H., M. Shaw, J.H. Seinfeld, and R.C. Flagan (1992), Photochemical aerosol formation from a- and b-Pinene, Geophy. Res., 97, 20717-20730.

[4] Griffin R.J., D. R. Cocker, J.H. Seinfeld, D. Dabdub, (1999). Estimate of global atmospheric organic aerosol from oxidation of biogenic hydrocarbon. Geophysical Research Letters, Vol. 26, No.17, 2721-2724.

[5] VanReken, T. M., J. P. Greenberg, P. C. Harley, A. B. Guenther, and J. N. Smith (2006). Direct measurement of particle formation and growth from the oxidation of biogenic emissions

[6] Grosjean, D. and J.H. Seinfeld (1989). Parametrization of the formation potential of secondary organic aerosol. Atmos. Environ., 23. 1733-1747.

[7] Vautard. R., B. Bessagnet, M. Chin and L. Menut (2005). On the contribution of natural Aeolian sources to particulate matter concentrations in Europe: Testing hypotheses with a modelling approach, Atmospheric Environment., 39, 3291-3303.

[8] Bessagnet, B., A. Hodzic, R. Vautard, M. Beekmann, L. Rouil, R. Rosset (2004). Aerosol modelling with CHIMERE –first evaluation at continental scal . Atmospheric Environment 38, 2803-2817.

[9] Derognat, C., M. Beekmann, Bäumle, D. Martin (2003). Effect of biogenic volatile organic compound emissions on tropospheric chemistry during the atmospheric pollution over the Paris area (ESQUIF) campaign in the Ile-de-France region. Journal of Geophysics Research 108 (D14), 4409-4423.

[10] Hodzic, A., R. Vautard, B. Bessagnet, M. Lattuati, and F. Moreto (2005). On the quality of long-term urban particulate matter simulation with the CHIMERE model. Atmospheric Environment, 39, 5851-5864.
[11] Schmidt, H., C. Derognat, R. Vautard, M. Beekmann (2001). A comparison of simulated and observed ozone mixing ratios for the summer of 1998 in western Europe. Atmospheric Environment 35, 6277-6297.
[12] Vivanco, M.G., I. Palomino, R. Vautard, B. Bessagnet, F. Martín, L. Menut, S. Jiménez (2008). Multi-year assessment of photochemical air quality simulation over SPAIN, Environmental Modelling & Software. In press.
[13] Vivanco, M.G., I. Palomino, F. Martín, M. Palacios (2007). Modelling Atmospheric Particles in the Iberian Peninsula and Balearic Islands. 11th International Conference on Hermonisation within Atmospheric Dispersion Modelling for Regulatory Purposes, Cambridge, UK. Volume 2, 48-52.
[14] Turpin, B. J., J.J. Huntzicker., S.M. Larson and G.M. Cass (1991) Los Angeles summer midday particular carbon: Primary and secondary aerosol, Environ. Sci. Technol., 25, 1788-1793.
[15] Turpin, B. J. and J.J. Huntzicker (1991) Secondary formation of organic aerosol in the Los Angeles basin: A descriptive analysis of organic and elemental carbon concentrations, Atmos. Environ., 25A, 207-213.
[16] McMurray, P. H., and X.Q. Zhang (1989) Size distributions of ambient organic and elemental carbon, Aerosol Science and Technology, 10, 430-437.
[17] Artíñano, B., J. Plaza, F.J. Gómez-Moreno, M. Pujadas, L. Núñez (2007), *Analysis of $PM_{2.5}$ OC-EC semicontinuous thermal measurements in Madrid*, 2nd ACCENT Symposium
[18] Castro, L.M., C.A. Pio, R.M. Harrison, D.J.T. Smith (1999), Carbonaceous aerosol in urban and rural European atmospheres: estimation of secondary organic carbon concentrations, Atmos. Environ. 33, 2771-2781
[19] Schell, B., Ackermann, I. J., Hass, H., Binkowski, F. S., and Ebel, A. (2001). Modeling the formation of secondary organic aerosol within a comprehensive air quality modeling system. J. Geophys. Res., 106:28275–28293.
[20] Pankow, J. F. (1994). An absorption model of gas/aerosol partition involved in the formation of secondary organic aerosol. Atmos. Environ., 28:189–193.
[21] Pankow, J.F., J.H. Seinfeld, W.E. Asher, and G.B. Erdakos (2001). Modeling the formation of secondary organic aerosol. 1. Application of theoretical principles to measurements obtained in the _-pinene, _-pinene, sabinene, _3-carene and cyclohexene/ozone systems. Environmental Sci. and Tech., 35:1164–1172.
[22] Vestreng, V., K. Breivik, M. Adams, A. Wagener, J. Goodwin, O. Rozovskkaya, J. M. Pacyna (2005). Inventory Review 2005, Emission Data reported to LRTAP Convention and NEC Directive, Initial review of HMs and POPs, Technical report MSC-W 1/2005, ISSN 0804-2446.

[23] Passant, N. R., 2002. Speciation of UK emissions of non-methane volatile organic compounds. AEAT/ENV/R/0545 Issue 1.
[24] Simpson, D., A. Guenther, C.N. Hewitt, and R. Steinbrecher (1995). Biogenic emissions in Europe 1. Estimates and uncertainties, Journal of Geophysical Research 100(D11), 22875-22890.
[25] Hauglustaine, D. A., F. Hourdin, L. Jourdain, M.-A Filiberti,., S. Walters., J.-F. Lamarque,., and E. A. Holland, (2004). Interactive chemistry in the Laboratoire de Météorologie Dynamique general circulation model: Description and background tropospheric chemistry evaluation, J. Geophys. Res., 109, doi:10.1029/2003JD003957.
[26] Chin, M., P. Ginoux, S. Kinne, B. N. Holben, B. N. Duncan, R. V. Martin, J. A. Logan, A. Higurashi, and T. Nakajima (2002). Tropospheric aerosol optical thickness fromt he GOCART model and comparisons with satellite and sunphotometer measurements, J. Atmos. Sci. 59, 461-483.
[27] Grell G.A., J. Dudhia, D.R. Stauffer (1995). A Description of the Fifth-Generation Penn State/NCAR Mesoscale Model (MM5).NCAR/TN-398 + STR. NCAR TECHNICAL NOTE.

Comparison between ozone monitoring data and modelling data, in Italy, from the perspective of health indicator assessments

A. De Marco, A. Screpanti, S. Racalbuto, T. Pignatelli, G. Vialetto, F. Monforti & G. Zanini
Italian Agency for New Technology, Energy and the Environment (ENEA), Italy

Abstract

The need for comparison between monitoring data and modelling data on ozone comes both from the qualitatively and quantitatively scarce outcome of the Italian ozone monitoring network and, at the same time, from the necessity for assessment and validation of the modelling methodology. Indeed, the distribution of the monitoring stations in Italy is not uniform and a dramatic lack of data is observed in all of the southern Italian areas. A number of different strategies can be applied to obtain a uniform distribution of data within the territory. The methodology of "spatialization" is described in the paper and applied to the health exposure indicator SOMO35 (developed by the WHO), pursuing the ultimate objective of identifying risk areas for the population. Such areas are then compared with similar areas from the analysis carried out by the Italian Integrated Assessment model RAINS Italy. The comparative analysis reported in this paper highlighted the differences, deepening the background rationale and ultimately increasing the robustness of the health risk analysis. Moreover, maps generated by the model could also be used to identify critical areas not covered by monitoring stations, so driving a more cost efficient allocation of expensive equipment.

Keywords: SOMO35, kriging, RAINS Italy, health risk area, tropospheric ozone.

1 Introduction

There is a more and more increasing interest in the analyses of the health effects caused by exposure of the population to ground level ozone. Scientific studies have demonstrated that exposure to ozone concentrations in the air has several significant adverse effects on human health, causing breathing problems, triggering asthma, reducing lung function and causing lung diseases (WHO [1]). In particular, the recent WHO update of the air quality guidelines to protect public health has concluded that these effects could occur at every ground level ozone concentration, and a possible threshold, if any, might be close to background levels and not determinable (TFH [2]).

A number of different strategies can be used to obtain a uniform distribution of ozone data, both measured and calculated, within the territory, to be used in the calculation of the health exposure indicator, pursuing the ultimate objective of identifying risk areas for the population. In this paper a health effect indicator, based upon a proper cumulated exposure indicator, the SOMO35, was selected for use following the suggestion given by the Joint UN/ECE-WHO Task Force on Health (TFH [2]). A possible approach to reaching the objective could be the "spatialization", the SOMO35 calculated from data measured through the ozone-monitoring network. Generally speaking, this approach requires a huge and uniform network of monitoring stations, which are very expensive to implement and indeed not necessary in those parts of the territory with a quite low density of population, where ozone does not represent a problem. Referring to the Italian context, the qualitatively and quantitatively scarce outcome of the Italian ozone-monitoring network makes this approach insufficient to achieve the objective. A different approach, complementary to the monitoring network, is the development of suitable models to reproduce the ozone concentration fields, with a sufficient degree of accuracy. The RAINS (Regional Air pollution INformation and Simulation) model, in its Italian version (RAINS-Italy), is an Integrated Assessment Model developed within joint research project ENEA-IIASA, (International Institute for Applied Systems Analysis), used along with a Eulerian Model for the dispersion of the pollutants and the chemistry of the atmosphere, is a suitable tool for estimating the ground level ozone concentration and its effects on human health. In this paper, the calculated values of SOMO35 are compared with the ozone-monitored value properly converted in SOMO35. Also, a preliminary assessment of the health impact from ozone, in terms of premature deaths, is shown.

2 Integrated Assessment Modelling System MINNI

The MINNI Modelling System is the result of a project led by ENEA and developed on behalf of the Italian Ministry for the Environment, Land and Sea, aimed at providing the policy with scientific support in the elaboration and assessment of air pollution policies, at international and national levels, by means of the more recent understanding of the atmospheric processes peculiarly characterizing the Italian territory. The two key components of MINNI (a full description can be found in Zanini et al, [3]), are the Atmospheric Modelling System (AMS-Italy), a multi-pollutant Eulerian model for dispersion of the pollutants and the chemistry of the atmosphere, and the integrated assessment model RAINS-Italy, developed in collaboration with IIASA, owner of the original RAINS_Europe model (fig. 1). AMS-Italy simulates the air pollution dynamics and the multiphase chemical transformations of the pollutants, providing concentrations and depositions on a hourly and yearly basis of SO_2, NO_x, NH_3, PM_{10}, $PM_{2.5}$ and O_3, with spatial resolutions from 20 km x 20 km to 4 km x 4 km. RAINS-Italy calculates emission scenarios for SO_2, NO_x, NH_3, PM_{10}, $PM_{2.5}$ and VOCs, as well as estimations of the costs associated with the implementation of the abatement technologies, and impact assessment on the environment and the human health. The RAINS-Italy model has a spatial resolution of 20 km x 20 km. The two components complement each other. AMS-Italy feeds RAINS-Italy with fine resolution Atmospheric Transfer Matrices (ATMs, source-receptor relationship coefficients) allowing RAINS-Italy to estimate average annual PM concentrations, SO_2 and NO_x depositions, for the purposes of acidification and eutrophication and health impact, in terms of Life Expectancy Reduction ($PM_{2.5}$) and Premature Deaths (ozone).

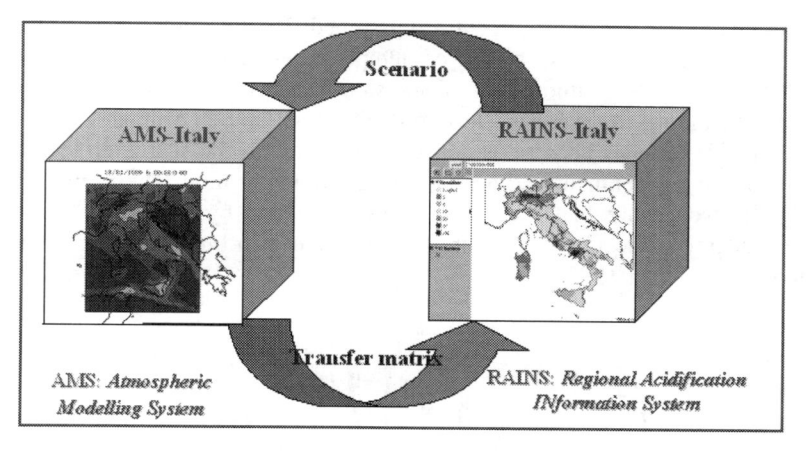

Figure 1:   Block scheme of the MINNI modelling system.

The core of the AMS-Italy is in the three-dimensional Eulerian chemical-transport model FARM (ARIANET [4]), which is derived from STEM (Carmichael et al. [5]). The code has been developed so that different chemical scheme can be implemented; within MINNI, the SPARC90 gas-phase scheme has been employed. As for PM, two approaches are available: the AERO0 simplified PM 'bulk' module, as implemented in the EMEP Eulerian unified model (EMEP [6]), considering the ammonia-nitric, acid-sulphuric, acid-water system, and the AERO3 Models-3/CMAQ module (Binkowski [7]), following the modal approach and also including secondary organic particulate. Atmospheric transfer matrices for RAINS-Italy have been calculated using the first approach, while the second one has been employed for specific studies. The reference meteorological year for AMS-Italy calculations is 1999.

RAINS-Italy has been used for the purposes of this paper to create maps of SOMO35 through the ATMs. RAINS-Italy (Vialetto et al. [8]) maintains the same features of the RAINS-Europe model (Amann et al. [9], IIASA web site [10]). The latest emission scenarios (D'Elia et al. [11]) elaborated with the EU NEC Directive review, have been used in this study.

3 Assessment of ozone concentrations by MINNI

The MINNI modelling system is suitable for reproducing the ozone concentration fields with a reasonable degree of accuracy. In figure 2, the comparison between the data monitored through the EMEP monitoring network stations and the data calculated with the AMS-Italy at the same locations for the year 1999 is shown. The results show a good agreement during the summer season, although the agreement seems to decrease along the rest of the year, where the model seems to overestimate the ozone concentrations. However, this result is not very surprising, for a number of reasons; the most important is that the model has been tailored in order to have the best performance at the higher temperatures, e.g. during the summer months, when the most adverse effects on human health and the environment are assumed. Indeed, the overestimation of the ozone concentrations, at lower temperatures, seems to be a general characteristic of the atmospheric dispersion models, as suggested from the analysis of the result of the recent City-Delta programme (Amman et al. [12]), although this fact has no serious consequences, due to the scarce effects of the ozone on the environment and health during the winter time. The results shown in fig. 2 also ensure that the ozone maps calculated by RAINS-Italy have a similar degree of accuracy, in reproducing ozone distribution, at the reference meteorological year 1999.

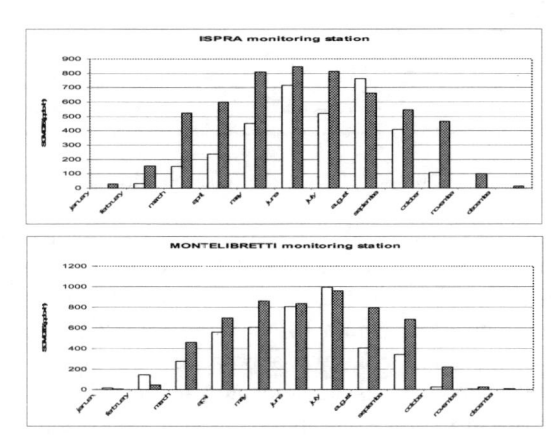

Figure 2:   Comparison between the data monitored through the EMEP monitoring network stations and the data calculated with the AMS-Italy, at the same locations, for the year 1999.

4 The Italian ozone monitoring station network

The Italian monitoring data come from regional and local databases (Regional Authority and Local Environment Protection Agencies, ARPA), based upon Air Quality monitoring stations and located both in urban and rural areas, although they are not uniformly distributed. In this paper, only rural or sub-urban stations are taken into account, because in the urban stations, measurements of tropospheric ozone are influenced by the presence of other pollutants, such as NO_x. Moreover, only those stations having at least 75% of yearly availability have been selected. Original data are recorded hourly. The monitoring network stations used in this work are shown in figure 3. SOMO35 is then calculated for the whole year, according to the following methodology:
- floating 8 hour mean value calculation;
- identification of the highest 8 hour mean value, over 24 hours;
- sum of the exceedances over 35 ppb/h, within the identified highest mean value.

Meteorological data come from UCEA (Central Office of Agrarian Ecology) database, which provides on-line monitoring data collected from stations, uniformly distributed over the national territory. In particular, average daily temperatures and average daytime humidity values have been used in the correlation for the ozone calculation (Hartkamp et al. [13]).

Figure 3: The monitoring network stations used for the calculation of the ozone concentrations.

With regard to the altitude, the Digital Elevation Model (DEM) used in this paper has a grid resolution of 500m x 500m. As is well known in literature, a relevant correlation exists between altitude and ozone concentrations

(Bronnimann et al. [14]), resulting in higher ozone concentrations in the mountain areas.

5 The spatialization methodology

With regard to the spatialization methodology, in this paper two spatialization methodologies have been used, as described below.

Kriging interpolation Kriging is a geo-statistical interpolation technique considering both the distance and the degree of variation between known data points, in estimating values in areas with no monitoring stations. A kriging estimate is a weighted linear combination of the known sample values around the point to be analysed. Applied properly, kriging allows deriving weights, which result in optimal and unbiased estimates. The construction of the semi-variogram of the data, used as weights nearby sample points during the interpolating, is also allowed by the kriging routine. The kriging routine also provides the users with a tool to evaluate the data trends. In the kriging methodology a proper variable called the semivariance is used to express the degree of relationship between points on a surface. The semivariance is simply defined as a half of the variance of the differences between all possible points spaced a constant distant apart. The semivariance at a distance d=0 is 0, because there are no differences between points that are compared to themselves. However, as a point is compared with other points at increasing distances, the semivariance increases. At some distance, the semivariance will become approximately equal to the variance of the whole surface itself. This is the maximum distance at which a point value on the surface is related to another point value. The calculation of semivariance between sample pairs at different distance combinations has been analyzed. The initial distance used is called Lag distance, which is increased by constant steps through the range. In the semivariogram graph, the variance between points is reported versus the distance at which the variance is calculated. The technique used in this paper is an ordinary kriging. This method assumes that the data set has a fixed variance but also a non-fixed mean value within the searching circle area. This method is highly reliable and is recommended for most of the data sets.

Cokriging interpolation. Cokriging has been traditionally used to minimize the variance of the error by taking into account the spatial correlation between the variables of interest and the secondary variables. The Cokriging is a moderately fast interpolator that can be exact or smoothed depending on the measurement error model. Cokriging uses multiple datasets and is very flexible, allowing investigating graphs of cross-correlation and auto-correlation. The flexibility of Cokriging requires a decision-process. Cokriging assumes the data are from a stationary stochastic process, while some methods assume normal-distributed data. In this paper, the daily mean temperature and the daily relative humidity (data derived from UCEA) have been used as a co-variable for the DEM.

6 Comparison between monitoring and calculated data

In figure 4 the maps of SOMO35 are shown, referred to the year 2000, obtained from the data measured in the Italian monitoring network stations (right side) by ordinary kriging and the data calculated with the RAINS-Italy model (left side). The map obtained with the RAINS-Italy model was processed by kriging to allow the comparison with the interpolation obtained by monitoring data.

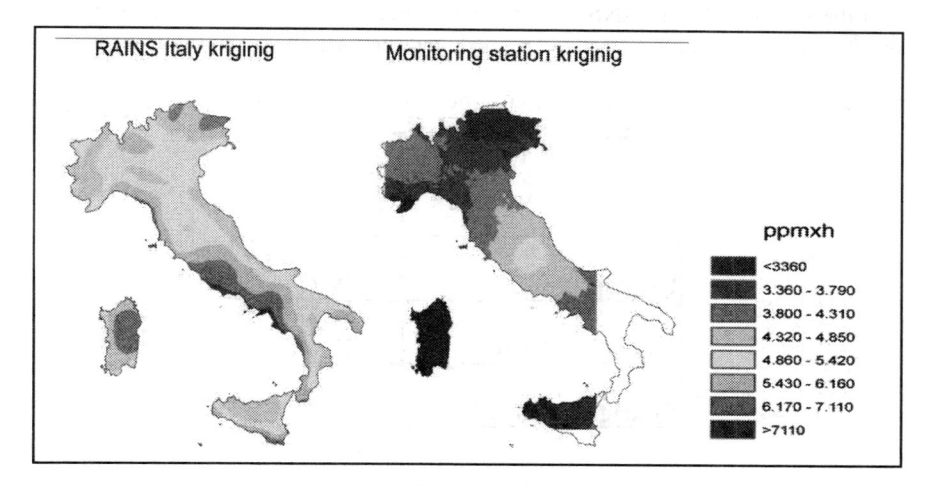

Figure 4: SOMO35 for the year 2000 in Italy calculated from the data measured in the monitoring network stations (right side) and with the RAINS-Italy model (left side).

By the comparison of the two different approaches it appears clear that the values are generally higher in the RAINS derived map in respect of the measured interpolated data. Moreover, there are some differences especially for the southern coastal zone in the RAINS map and for the central spot in the measured map. The differences are probably due to the starting data. In fact, the RAINS model, as previously described, is based on a pollutants atmospheric dispersion model, that covers all of the national territory and also the Mediterranean Sea, while in monitoring mapping the results take into account the distribution of the measurements stations that are obviously not present in the sea. The hot spot in the middle of Italy is related to high values of ozone recorded in the rural and mountain station of Leonessa.

As for other factors increasing the differences between the two maps, one is that the map calculated with the RAINS-Italy model refers to the meteorological year 1999, while the monitoring data obviously reflect the true meteorological condition of the year 2000. Moreover, the ozone directive, adopted by the Italian legislative decree, establishes a sampling efficiency threshold of 75% of available data to consider the dataset validated. This implies a possible underestimation of the real health effect for the measured maps.

All of these limitations are also found in the Cokriging interpolation of measured data (data not shown). The differences between the values estimated by the RAINS model and the interpolated measured data expressed as a percentage is reported in table 1 for all the different kinds of interpolation tested. As shown in the table, the use of a more complex interpolation technique is not expressed in a drastic improvement of the results. It has to be noted that the reduced sampling efficiency of the monitoring stations used could justify a lot of the discrepancies observed in the values.

Finally, using the RAINS-Italy model alone, preliminary maps of premature deaths due to ozone have been developed, and they are shown in figure 5.

Table 1: Mean percentage of distance between modelled and interpolated measured data.

	%
Kriging	29,23
Cokriging DEM	27,45
Cokriging meteo DEM	28,5

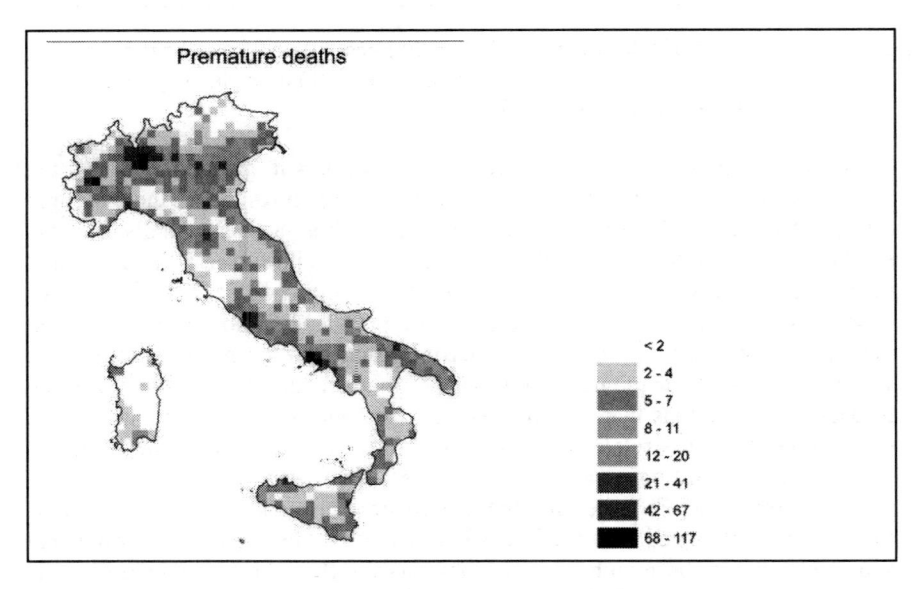

Figure 5: Preliminary map of premature deaths due to ozone concentrations, as calculated with the RAINS-Italy model (year 2000).

7 Conclusion

Considering the constraints and the differences in the two approaches, as discussed above, which result in indirectly comparable maps, some considerations can be drawn. In the North of Italy and in some regions in Central Italy, there is evidence of a strong potential health impact, which may result in considerable premature deaths. Due to the method of calculation applied, the measured map seems to underestimate the cumulative exposure to ozone and therefore the health impact. The lack of monitored data in a large part of the Italian territory, particularly in all the southern Italian areas, makes the modelling outcome the unique source of data in these areas. In light of the above discussion, the health impact analysis, carried out by the RAINS-Italy model, seems reasonably more reliable. From this point of view, the comparison has also highlighted the need for a significant improvement in the monitoring network, given the essential role played by the quality of the monitoring data in the model assessment and validation.

References

[1] WHO (2006). Use of the air quality guidelines in protecting public health: a global update. Fact sheet WHO/313, October 2006.
[2] TFH (2003). *Modelling and assessment of the health impact of the particulate matter and ozone.* EB.AIR/WG1/2003/11. United Nations Economic Commission for Europe, Task Force on Health, Geneve
[3] Zanini G., Pignatelli T., Monforti F., Vialetto G., Vitali L., Brusasca G., Calori G., Finardi S., Radice P., Silibello C. (2005); *The MINNI Project: An Integrated Assessment Modelling System for policy making.* In Zerger, A. and Argent, R.M. (eds) MODSIM 2005 International Congress on Modelling and Simulation. Modelling and Simulation Society of Australia and New Zealand, pp. 2005-2011. ISBN: 0-9758400-2-9. December 2005, http://www.mssanz.org.au/modsim05/papers/zanini.pdf
[4] ARIANET (2004) *FARM (Flexible Air quality Regional Model) Model formulation and user manual.* Arianet report R2004.04, Milano.
[5] Carmichael G.R., Uno I., Padnis M.J., Zhang Y. and Sunwoo Y., (1998) *Tropospheric ozone production and transport in the springtime in east Asia.* J. Geophys. Res., **103**, 10649–10671.
[6] EMEP, (2003) *Transboundary acidification, eutrophication and ground level ozone in Europe. Part I: Unified EMEP model description.* EMEP status Report 1/2003.
[7] Binkowski F. S., (1999). *The aerosol portion of Models-3 CMAQ.* In *Science Algorithms of the EPA Models-3 Community Multiscale Air Quality (CMAQ) Modeling System. Part II: Chapters 9–18.* D.W. Byun, and J.K.S. Ching (Eds.). EPA-600/R-99/030, National Exposure Research Laboratory, U.S. Environmental Protection Agency, Research Triangle Park, NC, 10-1-10-16.

[8] Vialetto, G., Contaldi, M., De Lauretis, R., Lelli, M., Mazzotta, V., Pignatelli, T. (2005); *Emission Scenarios of Air Pollutants in Italy using Integrated Assessment Models.* Pollution Atmosphérique, n.185, p.71, 2005.
[9] Amann, M., Cofala, J., Heyes, C., Klimont, Z. & Shopp, W. (1999). *The RAINS model: a tool for assessing regional emission control strategies in Europe*, Pollution Atmospherique, Numero speciale, December 1999.
[10] IIASA, www.iiasa.ac.at/~rains/index.html
[11] D'Elia, I., Contaldi, M.,, De Lauretis, R., Pignatelli, T., Vialetto, G. (2007); *Emission Scenarios of Air Pollutants in Italy (In Italian).* Submitted to Ingegneria Ambientale, 2007
[12] Amann, M., Cofala, J., Heyes, C., Klimont, Z. Mechler, R., Posch, M., Shopp, W. (2004). *The Regional Air Pollution Information and Simulation (RAINS) model: Review 2004.* IIASA Interim Reports, February 2004.
[13] Hartkamp AD, De Beurs K, Stein A, White JW: Interpolation Techniques for Climate Variable. NRG-GIS Series 99-01. Mexico, D.F.:CIMMYT, 1999
[14] Bronnimann S, Luterbacher J, Schmutz C, Wanner H): Variability of total ozone at Arosa, Switzerland, since 1931 related to atmospheric circulation indices. Geophys. Res. Let., 27:2213–2216, 2000

Section 2
Air quality management

Dealing with air pollution in Europe

C. Trozzi, R. Vaccaro & C. Leonardi
Techne Consulting srl, Via G. Ricci Curbastro, 34 00149 Roma, Italy

Abstract

Air pollution in Europe is managed and controlled through the implementation of International Conventions, dealing with global issues such as climate change, transboundary pollution and stratospheric ozone and ensuring compliance with European Legislation, regulating ambient air quality and emissions from stationary and mobile sources.

In this paper a complete description of the main provisions which represent the reference legislative framework for this sector is given.

Keywords: air pollution, legislation, European Union, LRTAP, UNFCC.

1 Introduction

Air pollution issues cover different geographical scales: from the local level, such as the sorroundings of isolated sources or urban locations to the global one, connected to all those phenomenon which invest wider areas. For this reason air quality management requests political strategies to be carried out and implemented at local, national and international level.

In Europe air pollution is managed and controlled through the implementation of International Conventions, dealing with global issues such as climate change, transboundary pollution and stratospheric ozone depletion and ensuring compliance with European Legislation, regulating ambient air quality and emissions from stationary and mobile sources.

In the following paragraphs a description of the main provisions which represent the reference legislative framework for this sector is given.

2 International conventions

2.1 United Nations Framework Convention on Climate Change

The United Nations Framework Convention on Climate Change (UNFCCC) came into force on March the 21st 1994 and represents a starting point to consider what can be done to reduce global warming and to cope with temperature increases [1]. Afterwards, governments agreed to an addition to the treaty, the Kyoto Protocol [2] that came into force on February the 16th 2005 and was approved by the European Community with Council Decision 2002/358/EC [3]. It shares the objective of the Convention and commits developed Countries to stabilize Green House Gas (GHG) emissions, with more powerful and legally binding measures.

The Protocol provides for three "flexibility mechanisms" that enable Annex I Parties to access cost-effective opportunities to reduce emissions, or to remove carbon from the atmosphere, in other Countries. The flexibility mechanisms are:

- the "Clean Development Mechanism" (CDM) provides for so called Annex I Parties to implement projects that reduce emissions in non-Annex I Parties, or absorb carbon through afforestation or reforestation activities, in return for certified emission reductions;
- the "Joint Implementation" (JI) provides for Annex I Parties to implement an emission-reducing project or a project that enhances removals by sinks in the territory of another Annex I Party;
- "Emissions trading" provides for Annex I Parties to acquire units from other Annex I Parties (Directive 2003/87/EC establishing a scheme for GHG emission allowance trading within the Community and amending Council Directive 96/61/EC).

The Intergovernmental Panel on Climate Change (IPCC) produced specific Guidelines for National GHG Inventories, used for estimating anthropogenic emissions by sources and removals by sinks of GHG and for calculating legally-binding targets during the first commitment period.

The last 2006 IPCC Guidelines [4] take into account also the related Good Practice reports. The new guidelines cover new sources and gases and update methods where technical and scientific knowledge have been improved. In particular, default values of the various parameters and emission factors required are supplied for all sectors. The guidance also integrates and improves earlier guidance on good practice in inventory compilation so that the final estimates are neither over- nor under-estimates as far as can be judged and uncertainties are reduced as far as possible.

Council Decision 93/389/EEC [5], as amended by Decision 99/296/EC [6], establishes a monitoring mechanism of Community CO_2 and other GHG emissions to help monitoring progress towards stabilisation of the total CO_2 emissions by 2000 at the 1990 level in the Community as a whole. European Union committed to a reduction of the emissions of the six Kyoto Protocol gases (carbon dioxide, methane, nitrous oxide, perfluorocarbons, hydrofluorocarbons and sulphur hexafluoride) by 8% in 2008-2012 from 1990 levels. Each year Member States shall report to the Commission on their anthropogenic emissions and removal by sinks for the previous calendar year and, on the most recent projected emissions for the period 2008-2012 and, as far as possible, for 2005.

2.2 Stratospheric ozone depletion

In 1985, scientific concerns about damage to the ozone layer prompted Governments to adopt the Vienna Convention on the Protection of the Ozone Layer [7], which established an international legal framework for action. Two years later, in 1987, legally binding commitments were adopted through the Montreal Protocol on Substances that Deplete the Ozone Layer [8], which required industrialized Countries to reduce their consumption of chemicals harming the ozone layer. Additional requirements have been added to the Montreal Protocol through amendments adopted in London (1990), Copenhagen (1992), Montreal (1997) and Beijing (1999).

As of November 2007, 191 Countries have ratified the Montreal Protocol, which sets out the time schedule to "freeze" and reduce consumption of ozone depleting substances (ODS). The Protocol requires all Parties to ban exports and imports of controlled substances to and from non-Parties.

Production and consumption of CFCs, halons and other ozone depleting chemicals have been phased out in industrialized Countries and a schedule is in place to eliminate the use of methyl bromide, a pesticide and agricultural fumigant. Developing Countries (Article 5 Parties) operate under different phase-out schedules, having been given a grace period before phase-out measures would apply to them, in recognition of their need for industrial development and their relatively small production and use of ODS. They have agreed to freeze most CFC consumption as of 1 July 1999 based on 1995-97 averages, to reduce this consumption by 50% by January the 1st 2005 and to fully eliminate these CFCs by January the 1st 2010. Other control measures apply to ODS such as halons, carbon tetrachloride and methyl chloroform. For methyl bromide, used primarily as a fumigant, developed Countries froze their consumption at 1995 levels and will eliminate all use by 2010, while developing Countries have committed to freeze consumption by 2002 based on average 1995-98 consumption levels.

2.3 Convention on Long-range Transboundary Air Pollution

The Convention on Long-range Transboundary Air Pollution [9] is one of the central means for protecting our environment.

In November 1979 the Convention, which now has 51 Parties, was signed by 34 Governments and the European Community. It was the first international legally binding instrument to deal with problems of air pollution on a broad regional basis. The Convention entered into force in 1983 and has been extended by eight specific protocols, among which the first one (on EMEP Programme) and the last three (on heavy metals, POPs and acidification, eutrophication and ground-level ozone) are still in force:

1984 Protocol on Long-term Financing of the Cooperative Programme for Monitoring and Evaluation of the Long-range Transmission of Air Pollutants in Europe (EMEP); 42 Parties. Entered into force on January the 28th 1988 [10] It is an instrument for international cost-sharing of a monitoring programme based on three elements: collection of emission data for SO_2, NO_x, VOCs and other air pollutants; measurement of air and precipitation quality; modelling of atmospheric dispersion.

1998 Protocol on Heavy Metals; 29 Parties. Entered into force on December the 29th 2003 [11]. It deals with three particularly harmful metals: cadmium, lead and mercury. Parties have to reduce their emissions below their levels in 1990 (or an alternative year between 1985 and 1995), cutting emissions from industrial sources, combustion processes and waste incineration. It lays down stringent limit values for emissions from stationary sources and suggests best available techniques (BAT). The Protocol requires Parties to phase out leaded petrol. It also introduces measures to lower heavy metal emissions from other products, and proposes the introduction of management measures for other mercury-containing products, such as electrical components, measuring devices fluorescent lamps, dental amalgam, pesticides and paint.

1998 Protocol on Persistent Organic Pollutants (POPs); 29 Parties. Entered into force on October the 23rd 2003 [12]. It deals with 16 substances, comprising eleven pesticides, two industrial chemicals and three by-products/contaminants. The ultimate objective is to eliminate any discharges, emissions and losses of POPs. The Protocol bans the production and use of some products outright (aldrin, chlordane, chlordecone, dieldrin, endrin, hexabromobiphenyl, mirex and toxaphene). Others are scheduled for elimination at a later stage (DDT, heptachlor, hexaclorobenzene, PCBs). Finally, the Protocol severely restricts the use of DDT, HCH (including lindane) and PCBs. The Protocol includes provisions for dealing with the wastes of products that will be banned. It also obliges Parties to reduce their emissions of dioxins, furans, PAHs and HCB below their levels in 1990 (or an alternative year between 1985 and 1995). For the incineration of municipal, hazardous and medical waste, it lays down specific limit values.

1999 Protocol to Abate Acidification, Eutrophication and Ground-level Ozone; 24 Parties. Entered into force on 17th May 2005 [13]. The Protocol sets emission ceilings for 2010 for four pollutants: sulphur, NO_x, VOCs and ammonia. Parties whose emissions have a more severe environmental or health impact and whose emissions are relatively cheap to reduce have to make the biggest cuts. Once the Protocol is fully implemented, Europe's sulphur emissions should be cut by at least 63%, its NO_x emissions by 41%, its VOC emissions by 40% and its ammonia emissions by 17% compared to 1990. The Protocol also sets tight limit values for specific emission sources and requires BAT to be used to keep emissions down. VOC emissions from such products as paints or aerosols will also have to be cut. Finally, farmers will have to take specific measures to control ammonia emissions. Guidance documents adopted together with the Protocol provide a wide range of abatement techniques and economic instruments for the reduction of emissions in the relevant sectors, including transport.

2.4 European Union legislation

2.4.1 Ambient air quality

The European Union legislation on ambient air quality and mainly the Framework Directive 96/62/CE [14] and the implementing provisions provided by the related four "Daughter" Directives [15–18], define the legislative basis for assessment and management of air quality in Member States.

The general aim is to define the basic principles of a common strategy to:
define and establish objectives for ambient air quality in order to avoid, prevent or reduce harmful effects on human health and the environment,
assess ambient air quality on the basis of common methods and criteria,
obtain adequate information on ambient air quality and ensure that it is made available to the public,
maintain ambient air quality where it is good and improve it in other cases.

According to legislation, air quality assessment can be done through measurements and any other techniques of estimation, including modelling; the *Framework Directive* defines that assessment "shall mean any method used to measure, calculate, predict or estimate the level of a pollutant…"

For this reason, Daughter Directives introduce assessments thresholds for each pollutant, that are to be used to establish where measurements are mandatory and where other techniques can be used, in particular they set:

- *Upper assessment thresholds*, below which a combination of measurements and modelling techniques may be used to assess ambient-air quality;
- *Lower assessment thresholds*, below which only modelling or objective-estimation techniques may be used.

When concentrations are over the upper assessment thresholds, measurements are mandatory and can be supplemented by other techniques.

Limit values, target values, long term objectives, alert and information thresholds are set in order to protect human health and the environment and specific provisions are established for air quality management through the implementation of plans to reduce concentrations of pollutants and obtain the best ambient air quality.

As one of the key measures outlined in the 2005 Thematic Strategy on air pollution, the new Air Quality Directive 2008/50/EC [19] has been recently adopted. It has the aim to simplify environmental legislation and in fact it merges in only one text the contents of the Framework Directive, the first three Daughter Directives and the Council Decision establishing a reciprocal exchange of information (EoI) and data from networks and individual stations measuring ambient air pollution within the Member States (Decision 97/101/EC, as amended by Decision 2001/752/EC) [20,21].

The new Directive doesn't change the existing air quality standards and the main contents of the previous legislative body but introduces some little changes about assessment criteria, natural contributions to pollutant levels and exemptions/derogations to the compliance to the limit values in particular conditions. Taking into account that it is now generally accepted that the fine fraction of particulate matter is the most dangerous for human health because it can lodge deeply in the lungs, really new provisions are those introduced about $PM_{2.5}$ monitoring and control. In this case a threshold below which no significant risks for human health appear has not yet been identified and for this reason $PM_{2.5}$ is to be controlled but not in the same way as other pollutants. In the new air quality Directive a double approach for this pollutant has been chosen.

The first one is aimed to reach a general reduction of concentrations in urban background locations, where the most part of population is likely to be exposed. This objective is pursued monitoring $PM_{2.5}$ in Urban Background locations sited in agglomerations and urban areas with more then 100.000 inhabitants; concentrations are assessed as an average value over 3 years. During the period 2010-2020 a reduction of the exposition has to be reached, as far as possible; starting from 2015, an exposure concentration obligation of 20 $\mu g/m^3$ is to be met and so the standard becomes mandatory.

The second approach is aimed to ensure a minimum degree of health protection everywhere, through establishing a target value (25 $\mu g/m^3$) by 2010 and a limit value (25 $\mu g/m^3$) by 2015.

2.4.2 Emissions

2.4.2.1 European Union Integrated Pollution Prevention and Control (IPPC) Directive Directive 2008/1/EC [22] is the codified version of the Directive 96/61/EC on Integrated Pollution Prevention and Control (IPPC) [23], that sets common rules on permitting for industrial installations; it merges in a single text all the previous amendments and introduces some linguistic changes and adaptations.

The Directive also provides that "an inventory of the principal emissions and sources responsible shall be published every three years by the Commission on the basis of the data supplied by the Member". The Commission Decision of 17 July 2000 on the implementation of a European Pollutant Emission Register (EPER) [24] requires Member States to report the Commission on emissions to air and water from all individual facilities with one or more activities for which "threshold values" are exceeded; both activities, pollutants and threshold values are specified in the Decision.

The Directive concern activities selected in the following sectors: energy industries, production and processing of metals, mineral industry, chemical industry, waste management, other activities (pulp; paper and board; fibers or textiles; tanning of hides and skins; slaughterhouses; animal and vegetable food products; milk; disposal or recycling of animal carcasses and animal waste; intensive rearing of poultry or pigs; surface treatment using organic solvents in particular for dressing, printing, coating, degreasing, waterproofing, sizing, painting, cleaning or impregnating; carbon (hard-burnt coal) or electro graphite by means of incineration or graphitization).

2.5 Other directives on stationary sources

- Directive 2001/80/EC on the limitation of emissions of certain pollutants into the air from Large Combustion Plants (LCP) [25], is aimed to reduce

emissions of acidifying pollutants, particles, and ozone precursors from LCPs, that are those whose rated thermal input is equal to or greater then 50 MW. The Directive establishes specific provisions for existing and new plants and provides for emission limit values for SO_2, NO_x and dust. For existing plants, Member States can also implement a National reduction plan to achieve overall reductions calculated using the emission limit values or a combined approach (limit values and plan).
- Directive 2000/76/EC on Waste Incineration [26], is aimed to prevent or reduce as far as possible negative effects on the environment (air, soil, surface water and groundwater) caused by the incineration and co-incineration of waste, through the application of operational conditions, technical requirements, and emission limit values for waste incineration and co-incineration plants.
- Directive 1999/13/EC, on the limitation of emissions of volatile organics compounds (VOC) due to the use of organic solvents in certain activities and installations [27], covers emissions of organic solvents from stationary, commercial and industrial sources by setting emission limits for such compounds and laying down operating conditions for industrial installations;
- Directive 2004/42/EC on the limitation of emissions of VOC due to the use of organic solvents in decorative paints and varnishes and vehicle refinishing products (Decopaint, amending Directive 1999/13/EC) [28].

2.5.1 Fuel quality

The subject is regulated by Council Directive 1999/32/EC relating to a reduction in the sulphur content of certain liquid fuels [29]; it is aimed to reduce the emissions of sulphur dioxide resulting from the combustion of certain petroleum-derived liquid fuels and thereby to reduce the harmful effects of such emissions on man and the environment. Such reductions shall be achieved by imposing limits on the sulphur content of such fuels as a condition for their use. The issue is also covered by Directive 2005/33/EC as regards the sulphur content of marine fuels [30].

European legislation also establishes some provisions about quality of fuels used for vehicles in order to reduce pollution coming from car emissions: a ban on the marketing of leaded petrol and an obligation to make sulphur-free fuels available within the Union.

The main Directives in this case is Directive 98/70/EC relating to the quality of petrol and diesel fuels [31] with the following amendments contained in Directive 2000/71/EC [32] and Directive 2003/17/EC [33].

2.5.2 Mobile sources

Further reductions on pollutants emissions are also obtained through regulations about Mobile sources. EU established emission standards for different kind of vehicles. In particular, for light-duty vehicles (cars and light vans) the emission standard currently in force is Euro 4, as defined by Directive 98/69/EC [34]. New Euro 5 and Euro 6 standards have already been agreed by Council and Parliament and formally adopted by the Council in 2007. The implementing legislation is currently under preparation. Euro 5 will enter into force in September 2009 while Euro 6 is scheduled to enter into force in January 2014.

For heavy-duty vehicles (trucks and buses), Directive 2005/55/EC [35] and Directive 2005/78/EC [36] define the emission standard currently in force, Euro IV, as well as the next stage (Euro V), which will enter into force in October 2008. In addition, a non-binding standard called Enhanced Environmentally-friendly Vehicle (EEV) is introduced; the Commission has also made a proposal for a new Euro VI stage in December 2007.

References

[1] United Nations, Framework Convention on Climate Change, 1992
[2] United Nations Framework Convention on Climate Change, Kyoto Protocol to the United Nations Framework Convention on Climate Change, Report of the Conference of the Parties on its third session, held at Kyoto from 1 to 11 December 1997, FCCC/CP/1997/7/Add., 18 March 1998
[3] Council Decision 2002/358/EC of 25 April 2002 concerning the approval, on behalf of the European Community, of the Kyoto Protocol to the United Nations Framework Convention on Climate Change and the joint fulfilment of commitments thereunder, *Off. Jour. Europ. Comm.* L 130, 15.5.2002, p. 1–3
[4] 2006 IPCC Guidelines for National Greenhouse Gas Inventories, Prepared by the National Greenhouse Gas Inventories Programme, Eggleston H.S., Buendia L., Miwa K., Ngara T. and Tanabe K. (eds), IGES, Japan.
[5] Council European Union Decision 93/389/EEC of 24 June 1993 for a monitoring mechanism of Community CO_2 and other greenhouse gas emissions, *Off. Jour. Europ. Comm.* L 167, 09.07.1993 p. 31
[6] Council European Union Decision 1999/296/EC of 26 April 1999 amending Decision 93/389/EEC for a monitoring mechanism of Community CO2 and other greenhouse gas emissions, *Off. Jour. Europ. Comm.* L 117, 05.05.1999 p. 35
[7] United Nations Environment Programme - Ozone Secretariat, The Vienna Convention for the Protection of the Ozone Layer, Secretariat for the Vienna Convention for the Protection of the Ozone Layer & The Montreal Protocol on Substances that Deplete the Ozone Layer, 2001
[8] United Nations Environment Programme - Ozone Secretariat, The Montreal Protocol on Substances that Deplete the Ozone Layer as adjusted and/or amended in London 1990, Copenhagen 1992, Vienna 1995, Montreal 1997, Beijing 1999, Secretariat for The Vienna Convention for the Protection of the Ozone Layer & The Montreal Protocol on Substances that Deplete the Ozone Layer, 2000
[9] Convention on long-range transboundary air pollution (CLRTAP), Geneva, 1979
[10] Protocol to the 1979 "Convention on long-range transboundary air pollution" on long-term financing of the cooperative programme for

monitoring and evaluation of the long-range transmission of air pollutants in Europe (EMEP), Geneva, 1984
[11] Protocol to the 1979 "Convention on long-range transboundary air pollution" on Heavy Metals, Aarhus, 1998
[12] Protocol to the 1979 "Convention on long-range transboundary air pollution", on Persistent Organic Pollutants (POPs), Aarhus, 1998
[13] Protocol to the 1979 "Convention on long-range transboundary air pollution" to abate acidification, eutrophication and ground-level ozone, Gothenburg, 1999
[14] Council European Union, Directive 1996/62/EC of 27 September 1996 on ambient air quality assessment and management, *Off. Jour. Europ. Comm.*, L 296, 21.11.1996, p. 55
[15] Council European Union, Directive 1999/30/EC of 22 April 1999 relating to limit values for sulphur dioxide, nitrogen dioxide and oxides of nitrogen, particulate matter and lead in ambient air, *Off. Jour. Europ. Comm.*, L 163, 29.6.1999, p.41
[16] Directive 2000/69/EC of the European Parliament and of the Council of 16 November 2000 relating to limit values for benzene and carbon monoxide in ambient air, *Off. Jour. Europ. Comm.*, L 313, 13.12.2000, p.12
[17] Directive 2002/3/EC of the European Parliament and of the Council of 12 February 2002 relating to ozone in ambient air, *Off. Jour. Europ. Comm.*, L 67, 09.03.2002, p. 14
[18] Directive 2004/107/EC of the European Parliament and of the Council of 15 December 2004 relating to arsenic, cadmium, mercury, nickel and polycyclic aromatic hydrocarbons in ambient air, *Off. Jour. Europ. Comm.*, L 23, 26.01.2005, p. 3-16
[19] Directive 2008/50/EC of the European Parliament and of the Council of 21 May 2008 on ambient air quality and cleaner air for Europe, *Off. Jour. Europ. Comm.*, L152, 11.06.2008, p. 1
[20] Council Decision 97/101/EC of 27 January 1997 establishing a reciprocal exchange of information and data from networks and individual stations measuring ambient air pollution within the Member States, *Off. Jour. Europ. Comm.*, L 35, 5.02.1997, p. 14–22
[21] Commission Decision 2001/752/EC of 17 October 2001 amending the Annexes to Council Decision 97/101/EC establishing a reciprocal exchange of information and data from networks and individual stations measuring ambient air pollution within the Member States, *Off. Jour. Europ. Comm.*, L 282, 26.10.2001, p. 69–76
[22] Directive 2008/1/EC of the European Parliament and of the Council of 15 January 2008 concerning Integrated Pollution Prevention and Control, *Off. Jour. Europ. Comm.*, L 24, 29.01.2008, p.8
[23] Council European Union, Directive 96/61/EC of 24 September 1996 concerning integrated pollution prevention and control, *Off. Jour. Europ. Comm.*, L 257, 10.10.1996, p.26
[24] Commission Decision 2000/479/EC of 17 July 2000 on the implementation of a European pollutant emission register (EPER) according to Article 15 of

Council Directive 96/61/EC concerning integrated pollution prevention and control (IPPC), *Off. Jour. Europ. Comm.*, L 192, 28.07.2000, p.36

[25] Directive 2001/80/EC of the European Parliament and of the Council of 23 October 2001 on the limitation of emissions of certain pollutants into the air from large combustion plants, *Off. Jour. Europ. Comm.*, L 39, 27.11.2001, p.1

[26] Directive 2000/76/EC of the European Parliament and of the Council of 4 December 2000 on the incineration of waste, *Off. Jour. Europ. Comm.* L 332, 28.12.2000 p. 91

[27] Council Directive 1999/13/EC of 11 March 1999 on the limitation of emissions of volatile organic compounds due to the use of organic solvents in certain activities and installations, *Off. Jour. Europ. Comm.* L 85, 29.03.1999 p. 1

[28] Directive 2004/42/CE of the European Parliament and of the Council of 21 April 2004 on the limitation of emissions of volatile organic compounds due to the use of organic solvents in certain paints and varnishes and vehicle refinishing products and amending Directive 1999/13/EC, *Off. Jour. Europ. Comm.* L 143, 30.4.2004, p. 87

[29] Council Directive 1999/32/EC of 26 April 1999 relating to a reduction in the sulphur content of certain liquid fuels and amending Directive 93/12/EEC, *Off. Jour. Europ. Comm.* L 121, 11.05.1999 p. 13

[30] Directive 2005/33/EC of the European Parliament and of the Council of 6 July 2005, amending Directive 1999/32/EC, *Off. Jour. Europ. Comm.* L 191, 22.07.2005 p. 59

[31] Directive 98/70/EC of the European Parliament and of the Council of 13 October 1998 relating to the quality of petrol and diesel fuels and amending Council Directive 93/12/EEC, *Off. Jour. Europ. Comm.* L 350, 28.12.1998 p. 58

[32] Commission Directive 2000/71/EC of 7 November 2000 to adapt the measuring methods as laid down in Annexes I, II, III and IV to Directive 98/70/EC of the European Parliament and of the Council to technical progress as foreseen in Article 10 of that Directive, *Off. Jour. Europ. Comm.* L 287, 14.11.2000 p. 46

[33] Directive 2003/17/EC of the European Parliament and of the Council of 3 March 2003 amending Directive 98/70/EC relating to the quality of petrol and diesel fuels, *Off. Jour. Europ. Comm.* L 76, 22.03.2003 p. 10

[34] Directive 98/69/EC of the European Parliament and of the Council of 13 October 1998 relating to measures to be taken against air pollution by emissions from motor vehicles and amending Council Directive 70/220/EEC, Off *Jour. Europ. Comm.* L 350, 28.12.1998 p. 1

[35] Directive 2005/55/EC of the European Parliament and of the Council of 28 September 2005 on the approximation of the laws of the Member States relating to the measures to be taken against the emission of gaseous and particulate pollutants from compression-ignition engines for use in vehicles, and the emission of gaseous pollutants from positive-ignition

engines fuelled with natural gas or liquefied petroleum gas for use in vehicles, *Off. Jour. Europ. Comm.* L 275, 20.10.2005 p. 1

[36] Commission Directive 2005/78/EC of 14 November 2005 implementing Directive 2005/55/EC of the European Parliament and of the Council on the approximation of the laws of the Member States relating to the measures to be taken against the emission of gaseous and particulate pollutants from compression-ignition engines for use in vehicles, and the emission of gaseous pollutants from positive ignition engines fuelled with natural gas or liquefied petroleum gas for use in vehicles and amending Annexes I, II, III, IV and VI thereto, *Off. Jour. Europ. Comm.* L 313, 29.11.2005 p. 1

The development and operation of the United Kingdom's air quality management regime

J. W. S. Longhurst, J. G. Irwin, T. J. Chatterton, E. T. Hayes,
N. S. Leksmono & J. K. Symons
Air Quality Management Resource Centre,
University of the West of England, Bristol, UK

Abstract

This paper considers the development and operation of the UK's air quality management regime since its introduction in the Environment Act, 1995. In the context of a changing and challenging air pollution climate, the development of an effects-based, risk management system for air quality regulation is evaluated. Key to the regime is the division of responsibility between central government and local government for both the diagnosis of areas of poor air quality and for the development of solutions to improve air quality.
Keywords: environment act, 1995, air quality management, United Kingdom.

1 Management in historical perspective

The United Kingdom (UK) has a long history of attempts to control air pollution. Initially attention was focused on the problem of smoke control in wintertime conditions as typified by the London Smog of 1952 [1,2] where air pollution was principally determined by patterns of coal consumption. The subsequent government appointed Beaver Committee [1] investigation into the causes, consequences and future control measures led to the introduction of the Clean Air Act, 1956. Regulatory attention was given to domestic sources primarily through this Act which required that domestic coal be burnt in approved appliances which were more efficient, and gave encouragement to the use of smokeless fuels and to other forms of fuel switching such as from coal to oil, gas or electricity. Following the introduction of the Act coal consumption declined in domestic and industrial markets throughout the 1960s and 1970s and the frequency and severity of coal smoke pollution within the UK's towns and cities progressively decreased. The primary control measure applied by local authority regulators was the designation of an area known as the 'smokeless zone' by issuing a smoke control order under the powers granted by the Act. In a smokeless zone, only specified fuels could be burnt, and grants were available to assist owners or occupiers to convert coal fires to allow them to use smokeless fuel.

The relatively slow timescale taken for these measures to be fully implemented should be noted; for example some large metropolitan areas did not complete their smoke control programmes until the early 1980s [3,4]. The effect of these control measures on smoke and SO_2 concentrations was pronounced, with significant and sustained reductions in concentrations being recorded across Great Britain as the measures were progressively applied. For example, in Manchester the annual mean concentrations of smoke fell by some 90 percent between 1959 and 1984. The bronchitic death rate – an indicator of the health effect of air pollution – reduced in line with the concentration of smoke [4].

Successful control of smoke and SO_2 in urban areas, regulation of lead in petrol and continuing downward pressure on industrial air pollution undoubtedly led to a prevailing view in the 1980s that urban air pollution was being successfully managed. In the 1980s and early 1990s the scale of the growth in road traffic and consequent emissions, allied with increasing public concern about possible health impacts, found the UK regulatory regime ill-prepared for the challenge of managing the impacts of vehicle related air pollution [5].

2 Developing an effects based air quality framework

In 1990, the UK Government set out its environmental strategy in the White Paper "This Common Inheritance" [5]. In the area of air quality, this recognised the growing concern over health effects and announced the establishment of an Expert Panel on Air Quality Standards (EPAQS). This was established with a remit to assess, on the basis of the best available epidemiological information, the public health impact of current UK air pollutant concentrations and to recommend a UK Standard for the protection of public health from named pollutants. An essential question in the risk assessment process undertaken by EPAQS was to identify the concentration and time interval over which adverse effects could occur, and to make recommendations to the Government taking account of uncertainties in the evidence base.

During the 1980s and early 1990s the UK had reasonably good air quality but was susceptible to occasional episodes of poor air quality which tended to occur with greater frequency and severity in heavily populated and industrialised areas. Typical examples of such episodes are the December 1991 London episode and the December 1994 episodes in Manchester, Birmingham and London [7]. Partly due to these episodes air quality began to receive increasing policy attention in the 1990s as public concern grew over the health impacts of air pollution [8].

The existing management framework and available policy tools were found to be inadequate for tackling these continued urban pollution episodes. Different.

emission sources were controlled by separate government departments and their associated agencies, resulting in a system of control that lacked co-ordination and integration particularly at the local scale. The Government response outlined in "This Common Inheritance" [6] was to explore mechanisms for a new approach which would make more effective use of air quality control mechanisms in an integrated and holistic way.

A new philosophy for air quality control was presented which built upon the existing technology-based controls by adding an effects-based, risk management approach founded on the new Air Quality Standards set by EPAQS. The new framework was introduced in the Environment Act 1995 (Part IV, Air Quality) which set out responsibilities for central and local government. Local control and management was placed at the heart of the response whilst maintaining, at a national level, a critical role in co-ordination and direction of local actions and the undertaking of such duties most effectively discharged at the level of the nation state. In Northern Ireland local authorities shadowed the process of LAQM until the Environment (Northern Ireland) Order, 2002 came into force and gave them equivalent responsibilities to their British counterparts.

3 Policy and practice in LAQM 1995-2005

The primary requirement of the 1995 Environment Act was the preparation of an Air Quality Strategy (AQS) by Government. The AQS was first published in 1997, and reviewed, updated and amended in 2000 and 2007 (with an Addendum in 2003). The AQS considered the historical legacy of air pollution, the contemporary nature of the air pollution challenge, and the adequacy of current controls, measures and priorities and laid out a new direction for the management of air quality in compliance with the requirements specified in the Environment Act (see [5,8–10]). Together, the Environment Act, 1995 and AQS provide a framework in which national and local actions are required to identify and remediate areas of poor air quality. The Act places a series of duties and responsibilities upon local authorities to review and assess local air quality against specific targets known as air quality objectives [9]. The original AQS was founded upon the principles of sound science, health effects based regulation, cost-effectiveness, proportionality, sustainability, precautionary approach and subsidiarity which to this day continue to inform its implementation.

The AQS sets out a series of air quality objectives for LAQM covering lead, CO, 1,3-butadiene, SO_2, NO_2, benzene and PM_{10}. These objectives were introduced as Regulations in 1997, revised in 2000 and amended 2002. A new AQS was published in 2007 [10]. This latest AQS does not remove any of the objectives set out previously but signals a change in approach to the management of particulate matter by introducing what is termed an 'exposure reduction' philosophy. This reflects opinion that there is no accepted threshold for health effects due to exposure to fine particles. This 'exposure reduction' approach has been adopted as a more efficient way to achieve future reductions in the health effects of air pollution by providing a driver to reduce concentrations across the UK rather than just in a small number of localised hotspots [10].

The objectives specified in Regulations represent the Government's judgement of achievable air quality by 2010 on the evidence of costs and benefits and technical feasibility. The objectives apply in areas where the public may be exposed for the averaging time of the relevant objective such as building facades, public open spaces, pavements and gardens. Objectives are not exceeded if there is no relevant public exposure and local authorities undertaking assessments of air quality are under a duty to identify if public exposure exists in any area where exceedence of an air quality objective is identified [8]. Local authorities are then required to declare areas where relevant exposure and exceedence of the objectives exist as Air Quality Management Areas (AQMAs). AQMAs are similar in concept to the idea introduced in the Clean Air Act, 1956 of a smokeless zone, although in practice they are more complex entities as they can relate to a wider number of pollutants. Following declaration of an AQMA a local authority is required to develop an Action Plan to pursue the achievement of the air quality objectives detailing both the measures to be taken and the time-scale for their implementation.

The Environment Act places a duty on Government to support local authorities through the provision of guidance and other initiatives. This has included the development of high quality national monitoring networks, the creation of 1km resolution emissions inventories, the provision of training for local authority personnel, the development of Technical and Policy Guidance to assist local authorities in their LAQM duties, and the provision of additional financial assistance to help authorities purchase monitoring equipment and other technical resources. An important support element has been the development of web, telephone and email help desks and other support materials [8] to assist, often non-specialist, air quality officers in carrying out their duties.

The AQS is dynamic and subject to rolling review to reflect developments in European legislation, technological and scientific advances, improved air pollution modelling techniques and an increasingly better understanding of the socio-economic issues involved in managing air quality and implementing effective Action Plans [8]. Similarly, the Air Quality Regulations, which give statutory weight to the air quality objectives, are reviewed periodically to reflect new developments in knowledge and understanding of specific pollutants and their impact on human health.

In parallel with UK developments, the EU has continued its development of strategic air quality management including the Air Quality Framework Directive (96/62/EC) and subsequent daughter directives. As new Daughter Directives emerge these are transposed into UK legislation and, subject to the scale and complexity of the management challenge implied by the Directive they may be incorporated in to the requirements of LAQM. Local authorities have the responsibility to work towards securing the objectives through the process of review and assessment and through the development of action plans. Under the Act Local authorities carry out reviews and assessments of air quality to define the contemporary and future state of local air quality in their area. This work is considered by Government to be a continuous process and local authorities are encouraged to consider it as such, although there are distinct phases of work known as Rounds. During each Round, the Government exercises its power as a Statutory Consultee to appraise the work of local authorities and acts to assure itself that the statutory duties of the local authorities are being undertaken in an appropriate manner.

Local authorities in Great Britain began the process of Review and Assessment in 1998. The first round of the process concluded in 2001 and resulted in some 129 local authorities declaring one or more AQMAs. A second round of Review and Assessment began in 2003, and a third round commenced in 2006. In the interval between Rounds local authorities are required to issue an annual Progress Report, the purpose of which is to report new monitoring data, describe any new developments that might affect air quality and to maintain the momentum of air quality management in the local authority.

At the end of Round 2, some 192 local authorities had declared AQMAs and by the summer of 2008 more than 200 UK local authorities had one or more AQMAs. These AQMAs have principally been declared for NO_2, with a significant number of PM_{10} and SO_2 declarations. An AQMA represents the conclusion of a technical assessment of air quality carried out in accordance with central government guidance against the air quality objectives. In declaring an AQMA a local authority will have satisfied itself, relevant stakeholders and Government that a risk of exceeding one or more objectives by the date the objective is to be achieved has been demonstrated in an area in which public exposure for a relevant period is or will be present. Details of local authorities with AQMAs may be seen at http://www.airquality.co.uk/archive/laqm/laqm.php. Figure 1 provides a time series of AQMA declarations in the UK.

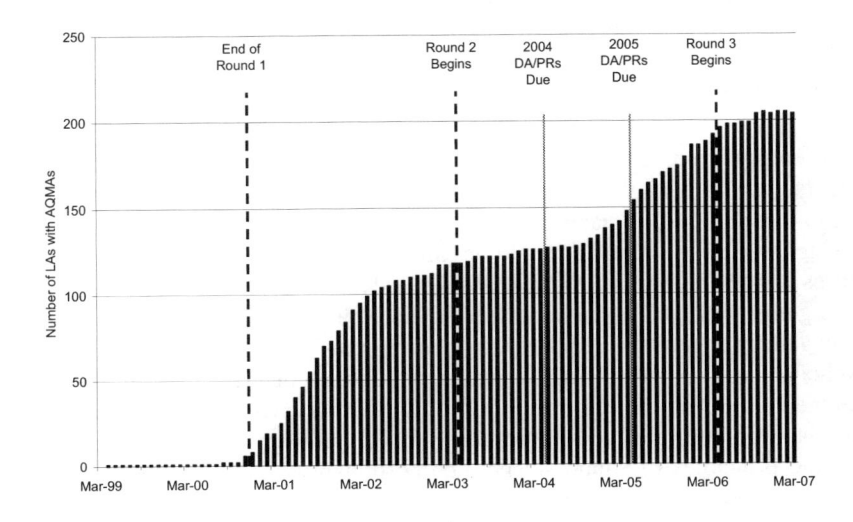

Figure 1: Number of local authorities with air quality management areas.

Following the declaration of an AQMA, a local authority must undertake a Further Assessment of air quality as required by Section 84 of the Environment Act. This is designed to confirm the appropriateness of the original decision, define the boundaries of the declared area and provide information on the emission sources contributing to the exceedence in order to support the preparation of an Action Plan. Following the declaration of an AQMA, the local authority is required to develop an Action Plan under section 84 of the Act. Although the Act does not specify a timescale for this Policy Guidance states that this is expected within 12 to 18 months of the designation of the AQMA [11]. In practice, the development of plans has proceeded at a slower rate than the technical Review and Assessment work. Even where a plan is being implemented there are limited examples of success to report. By the summer of 2007, some 122 local authorities had produced an Action Plan and details of the current situation may be seen at http://www.airquality.co.uk/archive/laqm/actionplan.php. Given the complexity of the contemporary air quality challenge with political, economic, public opinion and technical barriers to overcome this lack of significant progress is perhaps not surprising and it could be seen as reflecting a reluctance of society to adapt behaviour in order to achieve environmental goals. Changes in policy set out in the UK White Paper on Local Government [12] have meant that English local authorities with AQMAs associated with emissions from transport sources have been able to incorporate their Action Plans into their Local Transport Plan (LTP) process, a separate Government requirement operating to a different reporting timescale to that of LAQM. Despite these resulting administrative issues, the integration of LAQM and LTP offers the prospect of a more holistic approach within a local authority for the remediation of poor air quality. Importantly, local authorities that integrate their Action Plans into the LTP are eligible to receive funding under the LTP settlement grant to support the combined Action Plan.

4 Concluding remarks

The journey from an exclusively source-control approach to a complex but integrated, risk management effects-based process of air quality management has been a lengthy one. It has resulted in an integration of European, UK and local policies, strategies, regulation and action in order to improve air quality in the places where it matters most. Having evolved a comprehensive air quality management system it is important to note that the dynamic nature of the management challenge requires that the system continues to be periodically evaluated and renewed. This happens through regular reviews of the AQS, routinely assessing the challenges posed by poor air quality and determining the most appropriate management response to the problems identified using inter alia the principles of sound science, proportionality and risk assessment. This calibrated response enables actions to be appropriately targeted at the local level when evidence suggests that this is both effective and efficient at delivering improvements in air quality "hot spots", whilst reserving for national application and action those elements of policy and management best discharged at the level of the nation state. This is in line with the principle of subsidiarity.

The LAQM process has been refined over a series of rounds of activity dating back to the late 1990s and is now a routine activity for local authorities who have become capable and competent in discharging their air quality responsibilities. The outcome of their efforts is an efficient process for the identification of areas of poor air quality and the quantification of the temporal and spatial domain of these areas. If the judgement of effectiveness is taken as solely an improvement in air quality resulting from the application of the process then the system might be considered to be ineffective. However, this would not be a fair assessment. The development and implementation of Air Quality Action Plans has not as yet had the anticipated effect on local air quality. This outcome is not surprising if one considers the nature of the countervailing forces of growth in road traffic, commerce and industry that often lie wholly or in part outside of local authority control.

The LAQM process is administered by Defra and the Devolved Administrations through local authority Environmental Health or other Environmental Service departments. In practice, most of the sources of air pollution are related to the remits of the transport, land-use planning or economic development areas of local government and there is little direct ability for the review and assessment process to have a strong effect on these policy areas [9]. In the preparation of Action Plans the Environmental Health department will undertake negotiations with transport, land-use planning or economic development areas of local government in order to align, as best as is possible, the air quality resolution within existing policies and processes of governance. Where the local authority has control over the sources of pollution and has been able to establish effective internal and/or external co-ordination and communication systems between different functions, then significant air quality improvements can be seen at the local level even if they do not always succeed in reducing pollution levels to below the objective concentrations.

The current UK approach to the management of air quality, particularly at the local level, can be seen as a strong example of a public health orientated environmental management programme, setting out a risk-based framework, leading to targeted, proportionate and cost-effective actions focussed on a single area of the environment. There are clear health-based standards, indicating 'acceptable' and 'unacceptable' levels of air pollution. National policy objectives have been set based on these standards but taking into account technical and economic issues that might affect their attainment. This provides clear target dates for their achievement within a framework allowing assessment of problems at both local and national levels.

In the 11 years of LAQM since the first air quality strategy was published in 1997 there has been a significant enhancement in the ability of decision makers to take account of air quality in routine decision making. The quality of information available to decision makers has improved as the LAQM process has developed appropriate methods for local, repetitive, comprehensive, and quality assured review and assessment procedures. These reviews and assessments are able to draw upon high quality emissions and monitoring data via a number of specially commissioned web resources and support structures guiding the LAQM process. Local authorities have begun the difficult transition from simply defining the state of the local atmosphere towards securing AQAP goals. This represents a transition from procedural compliance with the process of LAQM towards achievement of improved air quality outcomes. They are doing this through new means of internal communication and co-operation and external consultation.

Whilst some of the processes and procedures used need refinement and enhancement to become fully effective, overall the position does provide encouragement that the continuing challenge posed by poor air quality will continue to be managed efficiently and effectively. The flexible, responsive system, which has created a strong flow of information between national and local government, in both directions, is at the heart of the difference between contemporary air quality management and traditional strategies of pollution control.

The Air Quality Management regime will evolve in the period covered by the new AQS [10]. Particularly important, will be the introduction of exposure reduction as a means to increase public health protection. This new policy initiative will itself bring new challenges for implementation but offers the promise of maximising public health gains at spatial scales greater than local areas of non-compliance with numerical objectives. However, the overall regime must continue to sit within the framework of sustainable development, and must take better account of social equity and environmental justice concerns. A critical issue for the further development of air quality management policy and practice will be to ensure appropriate integration with policies and strategies for the management of greenhouse gases.

References

[1] Brimblecombe, P. The Big Smoke. A history of air pollution in London since medieval time, Routledge, 1987.
[2] National Society for Clean Air. The Clean Air Revolution 1952 -2052. National Society for Clean Air. Brighton, 2002.
[3] Longhurst, J.W.S., Lindley, S.J., Conlan, D.E., Rayfield, D.J. & Hewison, T. Air quality in historical perspective. Urban Air Pollution III. eds Power, H. & Moussiopoulos, N. CMP: Southampton, pp.69 – 109, 1996.
[4] Manchester Area Council for Clean Air and Noise Control, 25 Year Review. MACCANC, Manchester, 1982.
[5] Longhurst, J.W.S., Lindley, S.J., Watson, A.F.R., Conlan, D.E. The introduction of local air quality management in the United Kingdom. A review and theoretical framework. Atmospheric Environment 30, 3975-3985, 1996
[6] HM Government. This Common Inheritance, CM 1200. Her Majesty's Stationery Office: London, 1990

[7] Elsom, D.M., Smog. More than a London Problem. The Clean Air Revolution 1952 -2052. National Society for Clean Air: Brighton, pp 13 – 18, 2002.
[8] Longhurst, J.W.S., Beattie, C.I., Chatterton, T.J., Hayes, E.T., Leksmono, N.S. & Woodfield, N.K., Local Air Quality Management as a risk management process: assessing, managing and remediating the risk of exceeding an air quality objective in Great Britain. *Environment International* 32, 934-947, 2006
[9] Beattie, C.I., Longhurst, J.W.S., Woodfield, N.K. Air quality management: evolution of policy and practice in the UK as exemplified by the experience of English local government. Atmospheric Environment 35, 1479-1490, 2001.
[10] Department for Environment, Food and Rural Affairs. The Air Quality Strategy for England, Scotland, Wales and Northern Ireland, Volume 1. Defra: London, 2007.
[11] Department for Environment, Food and Rural Affairs Local Air Quality Management Policy Guidance LAQM.PG(03), Defra: London, 2003.

Integrating local air quality and carbon management at a regional and local governance level: a case study of south west England

S. T. Baldwin[1], M. Everard[2], E. T. Hayes[1], J. W. S. Longhurst[1] & J. R. Merefield[3]
[1]*Faculty of Environment and Technology, University of the West of England, UK*
[2]*Environment Agency, UK*
[3]*Department of Life Long Learning, University of Exeter, UK*

Abstract

This paper examines the relationship between the policy process for local authority management of air quality and local government initiatives and strategies for carbon mitigation. It seeks to explore the policy and process linkages between the sources of carbon emissions and air quality pollutants in order to assess the potential benefits and/or limitations of an integrated approach for their co-management at a local and regional governance level. Local authorities, as environmental regulators, have a significant role in the UK's attempts to tackle the problems associated with climate change. This paper describes the extent to which non-statutory management of carbon emissions is undertaken at a local governance level in south west England and examines the extent to which carbon emissions and local air quality management are integrated and co-managed at local and regional governance levels. Results are presented from a questionnaire survey of local authorities in the south west and selected others from England conducted in 2007 and presents interim conclusions.
Keywords: Local Air Quality Management (LAQM), carbon management, local government, local authority.

1 Introduction

To date, the UK has endeavoured to fulfil its international obligations on mitigation of carbon emissions mainly through policies driven and implemented at a national level. Climate change is a global concern; however, it is at a local level where many of the mitigation measures can be implemented. Central government is increasingly recognising the contribution that local government can make towards delivering the UK's carbon reduction targets, such as those outlined in the Kyoto Protocol. Hilary Benn (Secretary of State for Environment, Food and Rural Affaires), speaking at the annual conference of the Local Government Association 2007, asserted this shift in attitude: *"Tackling climate change is the greatest challenge of our generation. Local government is not just a partner in this fight. You are one of the leaders of this fight"*.

Local Authorities are uniquely placed to provide vision and leadership to their local communities, and their wide range of responsibilities and stakeholder contacts means that they must be critical to delivering the UK's Climate Change Programme [1]. This makes integration of carbon management into local environmental management policy a desirable objective [2]. While it is important to build carbon management policies into the full spectrum of local authority duties and responsibilities, integrating carbon management into specific environmental policy frameworks could prove particularly beneficial. The potential benefits of integrated air quality and carbon management policies have been widely discussed and accepted but despite this, the two areas continue to be managed largely in isolation.

1.1 Benefits of integrated policies

The commonality of anthropogenic carbon and traditional air pollution sources (i.e. combustion of fossil fuels and agricultural practices) means that integrated management of emissions contributing to climate change and air quality could deliver considerable ancillary benefits, both fiscally and in terms of human/environmental health. The economic benefits of co-management techniques can be seen as two fold: efficient utilisation of available resources; and ancillary benefits for regional air pollution ensuing reductions in carbon emissions (where driven by demand reduction methods for energy, goods, services etc.). The Stern Review of the Economics of Climate Change addressed the economic benefits of co-management stating, *"Policies to meet air pollution and climate change goals are not always comparable. But if government wishes to meet both objectives together, there can be considerable cost savings compared to pursuing them separately"* [3]. Moreover, several studies have indicated that a considerable shared investment in climate change policies in Europe could be partially recovered by resultant lower costs in air pollution control of between $20 - 30\%$ [4–6].

Through the Local Air Quality Management (LAQM) process, local authorities are required to periodically review and assess traditional local air pollutants, predominantly arising from transport, industry and domestic sources.

These are similar sources to those of carbon emissions at a local level [2]. On the basis of this observation, it is hypothesised that integrating climate change strategies into aspects of the LAQM process will contribute substantially to local authority-driven reductions in carbon emissions.

1.2 Local authority climate change initiatives in the UK

A comprehensive and statutory framework for local government management of carbon emissions has yet to emerge from the policy rhetoric. Despite this there are numerous voluntary declarations, initiatives, and guidance documents available to local authorities in the UK to enable locally-driven carbon reduction.

The most widespread of these initiatives is the Nottingham Declaration on Climate Change. The first of its kind in the UK, the Nottingham Declaration, is a local authority initiative signed by 329 local government bodies to date (approximately 70%) which acts as a public statement for local governments to take action to tackle climate change issues at a local level. Local authorities signing the declaration commit to three broad aims; acknowledging that climate change is occurring, welcoming and engaging with the government targets and committing to working at a local level on carbon management [7].

The popular uptake of this initiative has established it as the first step for local governments that want to display their commitment to action. However there is little evidence to suggest that the declaration has resulted in any tangible reductions of carbon emissions at a local level. This may be due to the permissive nature of the declaration which, as a non-statutory initiative, presents no mandatory targets or performance indicators and lacks any sanctions if a council fails to deliver on a particular aspect of the declaration.

The International Council for Local Environmental Initiatives' Cities for Climate Protection (CCP) [8] campaign shares a common objective with the Nottingham Declaration on Climate Change: local government action on climate change and carbon abatement. However, only 49 English local government bodies have joined the CCP to date (approximately 10%), the majority (43) of which are also Nottingham Declaration signatories. The CCP campaign commits participants to undertaking five milestones: conducting a baseline emissions inventory; adopting a local reduction target; developing a local action plan; implementation of emission reduction policies; and monitoring progress of measures to reduce greenhouse gas emissions. The comparatively small participation in CCP may simply be due to a lack of awareness about the campaign. However, it is also probable that this underlines unwillingness among local authorities to commit to a more resource demanding programme.

1.3 Local Air Quality Management (LAQM) and carbon abatement

The current system for air quality management in the UK was legislated through the Environment Act 1995 [9]. This required the publication of the first National Air Quality Strategy, which introduced statutory requirements for local authorities in relation to the assessment and control of air quality. Local authorities are required to periodically review air quality in their area for specific pollutants and to assess current and projected future levels of air quality [9]. These reviews are assessed against a number of national Air Quality Objectives (AQOs). Where an area exceeds, or is likely to exceed these AQOs by a stated date, local authorities are required to designate an Air Quality Management Area (AQMA) and develop an Air Quality Action Plan (AQAP) outlining measures they will take to work towards remediating the problem [10]. The application of this process requires many of the same methods, skills and collaborative networks that would be required for an effective carbon management framework at the local level; the production of a robust emissions inventory, embedding carbon in the local and regional Air Quality Strategies, collaborative networks between key stakeholder influencing atmospheric emissions, and joint carbon/air quality action plans. This paper will investigate the opportunities and barriers for local authorities to co-manage carbon emissions at a local level through existing air quality management processes.

2 Methodology

The results presented in this paper are taken from a questionnaire survey conducted in 2007 of local government bodies in south west England, and a reference set of English authorities. The survey (a component of an ongoing longitudinal study) was designed to investigate the extent to which traditional air pollutants and carbon emissions are being co-managed at a local governance level in the south west region of England. In England, local government functions are delivered either entirely through a unitary authority, or are split over two tiers: district authorities, and County Councils. In order to investigate the convergence of what are at present considered two separate and parallel policy areas, it is recognised that they are currently managed by three key stakeholders within the various structures of local government:
- Environmental Health Officers (EHOs) responsible for LAQM in a district or unitary council
- Officers dealing with carbon management in a district or unitary council
- Officers dealing with carbon management in a County Council

A selection of local authorities from outside the south west were invited to participate as 'reference' authorities based on their proactive engagement with a combination of voluntary local authority climate change initiatives. The initiatives used as the selection criteria were the Nottingham Declaration on Climate Change [7], the Carbon Trust's Local Authority Carbon Management Program [11], Cities for Climate Protection [8], Sustainable Energy Beacon Councils [12] and Delivering Clean Air Beacon Councils [13]. All local governments participating in these schemes were entered into a database in Microsoft Excel and were selected using a pivot table based on the engagement with at least 3 of the target initiatives. Questionnaires consisted predominantly of closed questions to allow statistical analysis. Where opinions were required, Likert Scale questions were used. All data obtained was analysed using SPSS for Windows.

3 Research findings

Results are presented from 20 South West local authority EHOs and 11 officers with responsibility for carbon management, referred to as Climate Change Officers. 12 local authority EHOs and 10 Climate Change Officers from the reference survey group are also included. Results are presented for four thematic areas identified as import indicators of policy convergence: communication, action plans, emissions inventories and strategies.

3.1 Communication

Due to the multiplicity of sources for both air pollution and carbon emission at a local level, the success of control mechanisms relies on inter-profession communication between a numbers of stakeholders. One of the most successful ways of achieving participation between internal departments in local authorities and between external stakeholders is through the formation of multi-disciplinary groups, often referred to as steering groups [14]. These steering groups could include local authority officers across different departments (and in some circumstances neighbouring authorities) and often requires the support of outside bodies, businesses and local community groups. The percentage of EHOs (dealing with air quality) that are involved in an internal and/or external steering group for inter-professional communication on LAQM is shown in Table 1. Data are also provided for the responding local authority Climate Change Officers. In order for local authorities to maximise their impact on local carbon emissions beyond emissions arising from their own operations (vehicle fleet, procurement policy, estate, street lighting etc), it is necessary for steering groups to involve external stakeholders from the wider community. The importance of this has clearly been recognised by the 'reference' authorities of whom 70% have established such a group. However, while 83% of south west Climate Change Officer respondents reported establishing an internal steering group for climate change issues; just 27% had established an external steering group. By contrast, the importance of collaborating with external stakeholders, due to the multiplicity of sources of atmospheric emissions at a local level, became apparent soon after the implementation of the LAQM process. Indeed, 65% of south west respondents have established such a group, containing many of the relevant stakeholders to local carbon emissions.

Because of the integrated nature of air quality management, the creation of a strong collaborative working relationship between EHOs responsible for LAQM and other stakeholders is imperative to the success of the process. Carbon management at a local level will also necessitate collaboration between key stakeholders with influence or control over climate-active emissions. This will involve many of the same factions as those with influence on air pollution (e.g. local transport professionals, land-use planners, economic development officers, sustainability officers and business/commerce as well as EHOs) [15]. Thus, joint steering groups could have considerable benefits in terms of resource efficiency due to many of the networks and steering groups, established by local authorities

Table 1: Percentage of local authority respondents involved in internal/external steering groups for AQ/carbon management and percentage of relevant LA functions in attendance.

	EHO		Climate Change Officer	
	South West	English	South West	English
Internal steering group for LAQM/ Carbon management	10%	54%	83%	100%
External steering group for LAQM/ Carbon management	65%	38%	27%	70%

to assist them in their air quality duties, being intrinsically representative of those with influence on carbon emissions at a local level.

3.2 Air Quality Action Plans

Air Quality Action Plans (AQAPs) are the mechanisms by which local authorities identify the measures they will implement to work toward meeting the AQOs. However, actions taken to improve local air quality can result in synergistic or trade-off outcomes for greenhouse gas emissions and vice versa. Therefore, it is important that AQAPs are developed with non-air quality impacts such as the potential effect on GHGs emissions taken into account. Table 2 describes the percentage of south west local authority respondents that considered non air quality issues in the development of their action plan and the level of priority assigned to carbon mitigation measures. Of the respondents that have an active AQAP, only 27% of south west respondents considered issues other than air quality in their action plan, compared to 57% of the corresponding reference authorities. However, the mean score assigned to the level of priority carbon mitigation is given is low for both survey groups suggesting that in AQAPs, adequate attention is not been given to the impact measures may have on local carbon emissions.

Table 2: Percentage of local authority respondents that have an AQAP and the level of priority assigned to non air quality issues.

	EHO	
	South West	English
Current Air Quality Action Plan	55%	54%
AQAP considering non AQ issues	27%	57%
Priority of carbon mitigation in AQAP (mean)	1.82	1.43
Priority of climate change adaptation in AQAP (mean)	1.82	2.43

(scale: 1-6, 1= low 6=high)

There is a significant opportunity for a combined air quality and carbon management action plan where synergistic outcomes are given priority over those actions that may result in a trade-off situation, i.e. where an action taken to reduce traditional air pollutants result in an increase of carbon emissions or vice versa.

3.3 Emissions inventories

Emissions inventories are well established in the LAQM process and are important for establishing baseline data, however, if used effectively, local authorities could utilise their existing skills to produce combined air pollution and carbon emission inventory for the purpose of integrated planning, actions planning and strategies [16]. Table 3, shows the percentage of EHO and Climate Change Officer respondents that have produced an emissions inventory for LAQM pollutants and CO_2 emissions respectively for their local authority operations and their administrative area. The table also describes the percentage of those local authority respondents that have produced a combined emissions inventory for LAQM pollutants and CO_2 emissions. The importance of an emission inventory for carbon emissions is highlighted by the contrast between south west respondents (36%) and the 'reference' authority respondents (80%) who have produced an inventory for CO_2 emissions arising from their own operations. In order for local authorities to maximise the impact they have on locally-driven carbon emissions, it would also require them to tackle emission from sources outside of their direct influence. The results show that a comparatively small percentage of respondents have produced an emissions inventory for their administrative area for both LAQM pollutants (26% and 46% of south west and 'reference' authority respondents respectively) and carbon emissions (27% and 40% of south west and 'reference' authority respondents respectively).

Table 3: Percentage of local authority respondents that have conducted an emissions inventory of LAQM pollutants/CO_2 for their own operations and/or their administrative area.

	EHO		Climate Change Officer	
	South West	English	South West	English
Conducted an emission inventory of LAs CO_2	-	-	36%	80%
Conducted an emission inventory of LAQM pollutants/CO_2 in administrative area	26%	46%	27%	40%
CO_2 emission considered in LAQM emission inventory	40%	33%	-	-

The technical and non-technical skills displayed by local authorities producing emissions inventories for LAQM purposes would therefore lend themselves to the production of emission inventories for a basket of atmospheric emissions encompassing both 'traditional' air pollutants and carbon emissions. Utilising the existing skill set in this area could therefore be an effective and efficient use of local authority resources in managing carbon emissions, reducing the need for capacity-building associated with implementing new auditing and management processes.

3.4 Local air quality and climate change strategies (LAQS and CCS)

When undertaking an LAQS a steering group is usually established representing some of the key stakeholders with responsibility or control over emissions to the atmosphere. Currently, the majority of CCS at a local level are being developed separately from their air quality counterpart. A combined LAQS and CCS could bring all relevant stakeholders together and ensure the two areas are prioritised effectively at a local and regional level. The percentage of south west local authority respondents that have produced a LAQS and CCS and the level of priority that carbon management is given in LAQS is show in Table 4. Despite the benefits of a LAQS, only 37% of south west local authority respondents have produced one. However, more that half (55%) of the south west respondents have produced a CCS. This is probably accounted for by the fact the LAQM review and assessment process is prescriptive and cyclical in nature, thus, reducing the need for a voluntary LAQS. A prescriptive and uniform process for the management of carbon emissions at a local level does not yet exist, making CCSs a useful strategic tool. It could also be argued that climate change has become a more prevalent issue than air quality in public opinion, leading to CCS being used as a public relations tool.

Table 4: Percentage of local authority respondents that have produced a LAQS and/or CCS and the level of priority given to carbon in LAQS.

	EHO		Climate Change Officer	
	South West	English	South West	English
Produced a LAQS/CCS	37%	39%	55%	60%
Usefulness of LAQS for LAQM	2.59	3.80	-	-
Priority of carbon mitigation in LAQS	1.86	1.80	-	-
Priority of carbon adaptation in LAQS	2.29	1.80	-	-

(scale: 1-6, 1= low 6=high)

3.5 Additional observations

It is apparent from the results presented that local authorities face a number of barriers to achieving an integrated system for the co-management of air quality and carbon emissions. The LAQM process is now well established, with all local authority respondents employing an officer with primary responsibility for air quality. Conversely, only 20% of south west local authority respondents employ an office with primary responsibility for carbon management, compared to 78% of the 'reference' authorities from outside the south west. This seemingly large disparity between the two groups is due to the lack of ability, or willingness, to allocate scarce resources to an issue that is not yet a statutory responsibility. Those respondents that, as yet, had not produced a CCS were asked to rank a compendium of reasons why they had chosen not do so. The highest scoring reason given was 'not a statutory requirement' followed by 'other issues having higher priority' and 'lack of time'. This suggests that a strong statutory framework is required to drive management of carbon emissions at a local level and necessitate the allocation of time and resources. Integration of carbon emissions into the LAQM framework could prove an efficient use of these resources. When the LAQM process was introduced, many local authorities express concern that they did not have the technical skills or capacity to fulfil their statutory duties. In response to this Defra and the Devolved Administrations (Scotland, Wales and Northern Ireland) established a range of LAQM guidance documents and helpdesks to assist local authorities in their needs. In general local authorities seem happy with the level of guidance they now receive, with the south west respondents rating the LAQM guidance documents as 'very useful'. However, the same group strongly agreed with the statement 'There is a need for more prescriptive guidance in the policy and technical guidance documents for carbon management within LAQM', expressing the need for more comprehensive guidance incorporating carbon emissions.

4 Conclusion

The absence of a statutory requirement for carbon management was consistently reported by south west local authorities as being the main reason for inaction. This statutory requirement seems critical to facilitate the necessary allocation of time and resource. It is apparent from the results presented that despite the increased attention carbon management receives at an international and national level, it has yet to be effectively and comprehensively embedded at a local government level. While many local authorities now recognise the influence they can have on carbon emissions in their respective areas, and are proactively seeking to engage with initiatives for carbon management, there is only limited integration between the LAQM process and local government's management of carbon emissions. Reducing emissions of carbon and minimising the likely impact of future developments and lifestyles needs to come from a range of local functions such land use planning, building control, transport planning and waste management. However, the LAQM process is unique as the only local function that links these areas together. LAQM is a process designed to review, assess and manage certain atmospheric emissions. Therefore it provided many of the technical skill required to review and assess locally derived carbon emissions and already engages many of the relevant stakeholders.

References

[1] Department for Environment, Food and Rural Affairs, *Climate Change: the UK Programme*. The Stationary Office: London, 2006.
[2] Hayes, E.T., Chatterton, T.J., Leksmono, N.S. and Longhurst, J.W.S., Integrating climate change management into the local air quality management process at a local and regional governance level. *Proc. of the 14th Int. Conf. On Modelling, Monitoring and Management of Air Pollution*, eds. J.W.S. Longhurst & C.A. Brebbia, WIT Press: Southampton, pp. 439–446, 2006.
[3] Stern, N., *The Economics of Climate Change*, Cambridge University Press: Cambridge, UK, 2007.

[4] Alternative Policy Study for Europe and Central Asia; United Nations Environment Programme, Global Environment Outlook 2000, Online. http://www-cger.nies.go.jp/geo2000/english/text/0229.htm
[5] RIVM, EFTEC, NTUA and ILASA, *European Environmental Priorities: An Integrated Economic and Environmental Assessment.* National Institute of Public Health and the Environment: Bilthoven, 2001.
[6] Van Harmelen, T., Bakker, J., de Vries, B., van Vuuren, D.P., den Elzen, M.G.J., Mayerhofen, P., An analysis of the costs and benefits of joint policies to mitigate climate change and regional air pollution in Europe. *Environmental Science and Policy* **5 (4)**, pp. 349–365, 2002.
[7] Nottingham Declaration on Climate Change; Energy Savings Trust, Online. http://www.energysavingtrust.org.uk/housingbuildings/localauthorities/NottinghamDeclaration
[8] Cities for Climate Protection; International Council for Local Environmental Initiatives, Online. http://www.iclei.org/index.php?id=800
[9] HM Government, Environment Act 1995. The Stationary Office: London, 1995.
[10] Longhurst, J.W.S., Beattie, C.I., Chatterton, T.J., Hayes, E.T., Leksmono, N.S., and Woodfield, N.K., Local air quality management as a risk management process: assessing, managing and remediating the risk of exceeding an air quality objective in Great Britain. *Environment International* **32**, pp. 934–947, 2006.
[11] Local Authority Carbon Management Programme; Carbon Trust, Online. http://www.carbontrust.co.uk/carbon/ public sector/la/
[12] Sustainable Energy Beacon Councils; Improvement and Development Agency, Online. http://www.beacons.idea.gov.uk/
[13] Delivering Clean Air Beacon Councils; Improvement and Development Agency. http://www.beacons.idea.gov.uk/
[14] Beattie, C.I. & Longhurst, J.W.S., Joint Working within Local Government: air quality as a case study. *Local Environment* **5**, pp. 401–414, 2002.
[15] Beattie, C.I., Longhurst, J.W.S., and Woodfield, N.K., Air Quality Action Plans: early indicators of urban local authority practice in England. *Environmental Science & Policy* **5**, pp. 463–470, 2002.
[16] Hayes, E.T., Leksmono, N.S., Chatterton, T.J., Symons, J.K., Baldwin, S.T., and Longhurst, J.W.S., Co-management of carbon dioxide and local air quality pollutants: identifying the 'win-win' actions. *Proc. Of the 14th IUAPPA World Congress,* Brisbane, Australia, 2007.

Failures and successes in the implementation of an air quality management program in Mexicali, Baja California, Mexico

M. Quintero-Nuñez[1] & E. C. Nieblas-Ortiz[2]
[1]*Instituto de Ingeniería, Universidad Autónoma de Baja California, México*
[2]*Secretaría de Protección al Ambiente, Gobierno del Estado de Baja California, México*

Abstract

The Program to Improve the Air quality of Mexicali 2000-2005 (PROAIR) was released in the year 2000. The implementation of the program was based on 5 strategies and their specific actions that if applied according to the program would have helped to reduce the emissions of the different contaminants that exceeded the acceptable values of air quality. The PROAIR followed national guidelines to evaluate, what at the time were considered the most polluted cities in Mexico, such us Mexico City, Guadalajara and Jalisco among others. The program's strategies and their actions were directed to the following components and work areas: industry, commerce and services, vehicles, urban management and transport, ecological recovery and research, and international agreements. Mexicali, being located at the California-Baja California border has received many environmental benefits derived from border protocols established between Mexico and the USA, such as the Border 2012 Program, still running. Seven years after the implementation of PROAIR in Mexicali, very little progress has been observed in the actions outlined in the Programs' strategies. The analysis of the actual situation of the Program is presented in this document along with recommendations on how to improve it to benefit community health.

Keywords: air quality management program, Mexicali, Baja California, Mexico, airshed, Border 2012.

1 Introduction

1.1 Brief description of the Valley of Mexicali

The Valley of Mexicali, Baja California, with 878,000 inhabitants, is located in a strategic place at the border of Mexico with the USA, that emerges as agricultural land at the start of the twentieth century (Fig. 1). The valley is characterized by very hot weather during the summer months and an air pollution problem caused primarily by suspended particles arising from the desert environment [1], a large agricultural sector, vehicular activity, and unpaved streets in Mexicali. The location of the Valley of Mexicali and its neighbor Imperial Valley, CA, USA, where the cities of Calexico, El Centro and Brawley are located, give cause to consider the region as a unique international atmospheric airshed, which, combined with the increasing production activity, makes it more important, not only at the border but at a national level.

Due to the development and evolution of the region, the air quality has deteriorated in recent years. Actually, Imperial Valley does not comply with the North American air quality standards for PM10, and in Mexicali the values exceed the Mexican official norms for PM10, carbon monoxide and ozone [2].

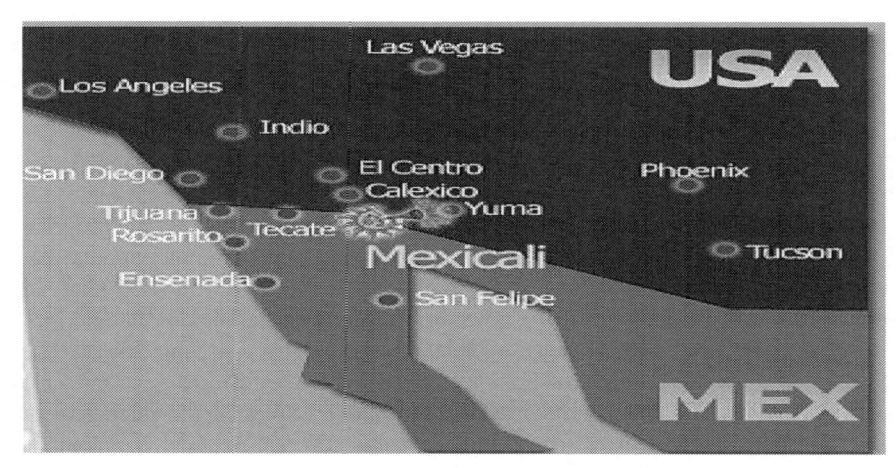

Figure 1: The city of Mexicali is located in the northwest of Mexico.

The Program to Improve the Air Quality of Mexicali 2000-2005 represents [3] a joint effort of society, local economic sector and the three levels of government to design and implement a set of actions which have as a final goal to control the sources of pollutants that degrade the air quality of the city.

The proliferation of a great number of industrial, commercial and service activities, as much as an accelerated motorization, have caused a degradation of the air quality of Mexicali, especially due to the poor conditions of the public transport and private automobiles and in particular, due to the importation of preowned vehicles that generally failed to pass the smog check in the USA [2]. Additionally, the situation gets exacerbated due to the emissions of particles and dust from the urban and agricultural clandestine burns [4] and emissions from paved and unpaved streets [5].

2 Description of the program to improve the air quality of Mexicali 2000-2005 (PROAIR)

The "PROAIR" had as a general objective to protect the health of the population, reducing the concentration of pollutants in the atmosphere, by the application of coordinated actions that assist in controlling the emissions generated by industry, commerce, services, transport and soil [3].
   -Objectives and general strategies of the program
   An analysis of the air quality was performed at the integration of the program. A detailed emissions inventory was prepared and it was divided into the following sectors: industrial, commerce, and services, and in a different category, motor vehicles. An integral diagnosis of the pollution problem was realized, with the actual air pollutants data obtained from known production sources.

2.1 Strategies and actions

The "PROAIR" contains five strategies and each one of them group a different number of specific actions that once applied according to the program, would reduce the emissions of the different pollutants that exceed the acceptable values of the air quality. The strategies are oriented to the following components and areas of work.

- Industry, commerce and services
- Motor vehicles
- Urban and transport management
- Ecological recovery
- Research and international agreements

With a total of 27 actions focused on diverse sectors and with responsibilities well identified and defined for each sector. The number of actions per sector were: industry, commerce and services (7 actions); motor vehicles (5 actions); urban management and transport (9 actions); ecological recovery (2 actions); research and international agreements (4 actions).

2.2 Work that was proposed in the PROAIR (27 actions)

Seven years after the initial proposals to improve the quality of the air in Mexicali few actions have been taken. The following table describes each one of them and their progress.

Table 1: Actions, responsibilities and realization of the program.

Industry, commerce and services

Actions	Responsible	Realization
1. Reduce the emissions of the most polluting industry through the installation of control equipment and bench marking processes	CANACINTRA, Industrialist	Yes
2. Implement a program to recover vapours in storage terminals and gasoline service stations	PEMEX, Owners, SEMARNAT	No
3. Strengthen the inspection and vigilance of industrial, commercial and service establishments	SEMARNAT, St. Gov., Mun. Gov.	Yes
4. Convene with the industry the implantation of a program of reduction of VOCs	Association of Maquiladoras, SEMARNAT	In progress
5. Realize on behalf of the CFE an environmental impact assessment of air emissions produced by Cerro Prieto geothermoelectrical power plant and, if necessary, a program of actions to reduce them in a maximum of one year	CFE	No
6. Realize environmental audits and autoregulation actions in the industrial sectors	St. Gov, Mun. Gov., CANACINTRA and Assoc. of Maquiladoras	In progress
7. Integrate a registry of emissions and pollutants (RETC) for Mexicali	SEMARNAT, St. Gov, Mun. Gov., CANACINTRA and Association of Maquiladoras	In progress

Motor vehicles

Actions	Responsible	Realization
8. PEMEX will evaluate the possibility of supplying oxygenated gasoline and low pressure vapor (PVR)	Pemex	Yes
9. Design a model, concense and application of a smog check program by the municipality, SCT	Mun. Gov, SCT	In progress
10. Condition the importation of second hand vehicles to the certification of smog check of the original country	SHCP, SECOFI	No

Table 1: (continued).

Actions	Responsible	Realization
11. Promote the utilization of LP gas and natural gas in public transportation	Private drivers	Partially
12. Design and implement a program to stop vehicles polluting ostentatiously	Mpal. Gov, SCT	Yes

Urban and transport management

Actions	Responsible	Realization
13. Application of soil stabilizers for the control of PM10 emissions in streets, unpaved areas and intense urban traffic	SEMARNAT, State Gov. Mpal. Gov.	Partially
14. Intensify a program to pave streets and roads	St. Gov., Mpal. Gov	Yes
15. Participation of PEMEX in a preferential way in fuels and asphalt prices to be utilized in paving infrastructure	PEMEX	Partially
16. Convene the transfer, operation and maintenance of the air quality monitoring net	Mpal. Gov., St. Gov.	Yes
17. Develop an integral study and execution of improvement of public transport	Mpal. Gov.	In progress
18. Promote a program of social participation and environmental education	Mpal. Gov.	Partially
19. Consolidate a program of epidemiological vigilance associated with pollution, as much as implanting corrective and preventive measures	SSA	No
20. Develop a program of fiscal stimuli for people, institutions and organisms to promote prevention and pollution control programs	SHCP	No
21. Integrate the Ecology Municipal Commission as a part of the COPLADEMM to follow up the Program	Mpal. Gov.	No

Table 1: (continued).

Ecological Recovery

Actions	Responsible	Realization
22. Study and establish emission factors and control options	SEMARNAT, SAGAR, St. Gov.	No
23. Design a reforestation program and preservation of grove of trees areas	St. Gov. Mpal Gov., SEMARNAT, Sedena	In progress

Research and International Agreements

Actions	Responsible	Realization
24. Review and update periodically the emissions inventory and the PROAIR, SEMARNAT	St. Gov., Mpal.Gov.	Partially
25. Establish agreements with Higher Education Institutions to realize studies in relation to pollution	St. Gov., Municipal Gov., SEMARNAT	Yes
27. Establish agreements with international Institutions to perform training activities and studies in relation to air pollution	St. Gov. Mpal. Gov. SEMARNAT	Yes
28. Reinforce the actions of the Border XXI Program and subsequent binational programs	SEMARNAT, St. Gov., Mpal Gov., CANACINTRA, SECOFI, CFE, SAGAR, SCT, SHCP, PEMEX, SSA.	Yes

CANACINTRA – National Chamber of the Transformation Industry
CFE – Federal Electricity Commission
Maquiladora – In bond industry
Mpal. Gov. – Municipal Government
SAGAR – Secretary of Agriculture and Social Development
St. Gov.– State Government
SECOFI – Secretary of Commerce and Industrial Promotion
SCT – Secretary of Communications and Transport
SHCP – Secretary of the Exchequer
SEMARNAT – Secretary of Environment, Natural Resources and Fishery
SEDENA – National Defense Secretary
PEMEX – Petróleos Mexicanos
SSA – Secretary of Health and Welfare
RETC – Registry of emissions and pollutants.

3 What has partially or totally worked in Mexicali in relation to the "ProAir"

- Incorporation of the air quality monitoring net of Mexicali under the management of the State of Baja California.
  On June 24, 2004, the agreement of collaboration between SEMARNAT and US Environmental Protection Agency (US EPA) was signed in Tijuana, Baja California, for the transfer of the air quality monitoring net of Mexicali, Tijuana, Tecate and Rosarito for the year 2006 from USEPA which managed and financed the monitoring of the net since 1997. In June 2007 the state government took over the management of the monitoring net.
- Public Information on the state of the air quality in real time
  The data collected by the monitoring stations on air quality is published in a site on the web: www.airebajacalifornia.gob.mx The information shown is the air quality index (IMECAS for its acronym in Spanish) of all the reference pollutants where recommendations are given to protect the health of the community in case of unhealthy air. The website of the air quality information on the neighboring city of Imperial Valley is www.imperialvalleyair.org for anyone interested in comparing the air quality between the two valleys.
- Development of an integral study and improvement of the public transport.
  Up to the present time the municipality of Mexicali has improved its transportation after several studies have been carried out such as: the Integral Study of the Transit and Urban Transport in 2002; the Determination of New Buses Routes in Mexicali in 2004; and the Update of the Transit and Public Transportation System of Mexicali in 2007. These types of projects resulted in the acquisition of 200 new buses equipped with air conditioned systems that although the fare increased it gave a different feature to the urban transportation in the city, and to and improved the air quality.
- Integral Paving Program and Air Quality (PIPCA for its acronym in Spanish)
  At the state level, the PIPCA is being developed in the five municipalities of the state including Mexicali. This project is in charge of the Urban Board of Mexicali, which is part of the Secretary of Infrastructure and Urban Development of Baja California (SIDUE for its acronym in Spanish). It was calculated a 14.50% reduction of PM10 after paving [6].

4 What has not worked of PROAIR as relevant issues

- Realize an environmental impact assessment of the Compañia Federal de Electricidad (CFE) on the air emissions produced by Cerro Prieto geothermoelectrical power plant.
  Cerro Prieto Geothermoelectric power plant with a capacity of 720 MW is a major producer of H_2S and CO_2 as the main pollutants. It is located 30 miles to the south of Mexicali and there has not been any attempt to control the

pollutants, either by reinjecting the non condensable gases o installing an abatement process to stop polluting the neighboring settlements [7].

- Program to recover vapours in storage terminals and gasoline service stations

  This is an important issue taking into account the number of gasoline filling station in Mexicali and the size of the vehicular fleet. As it is known the transport urban system is very inefficent in Mexicali, forcing people to buy old vehicles to fulfill their transportation needs. By establishing this program the amount of VOCs will be severely reduce.

- Develop a program of fiscal stimuli for people, institutions and organisms to promote prevention and pollution control programs

  An environmental tax was approved by the Baja California State Congress which was published by the State Official Newspaper (POE) on December 22nd 2002. This tax was created to tax emissions of CO to the atmosphere that are generated by fixed sources that used fossils fuels in their industrial processes [8]. Although the tax was approved in December of 2002 and put into effect in January 2003, it was accorded an exemption up to April 2003 due to an inconformity by the owners of the plant. The measure was derogated since it was vetoed by the state Governor based on the fact that the main CO polluters was the old vehicular fleet of Mexicali and not so much the industry.

  An attempt was made to establish environmental taxes in Baja California at a municipal level in particular after two thermoelectric power plants of combined cycle, the Thermoelectric of Mexicali and Rosita Power Plants were built up in the outskirts of the city. The tax was approved by the municipal Council based on the pollutants that would be generated by these two power plants but later cancelled by the Mayor of the city of Mexicali based on the facts that the tax was approved without taken into account the parties involved.

- Importation of pre-owned vehicles

  The Mexican American border has been characterized by the importation of second hand cars into Mexico at lower prices than the national vehicles since the *"free zone"* was created in Northern Mexico. Cars aged 5 years and older are allowed to be imported to Mexico, without fulfilling the smog check verification imposed on American cars in USA. The main fleet in the city is integrated by this type of cars (58%) which constitute the main contributors for CO and O_3 pollution in the city.

- Program of epidemiological vigilance associated to pollution

  Although air pollution has a direct effect on health not a single program of epidemiological vigilance associated to pollution has been implanted by any level of government based on lack of economical resources. Therefore, no corrective and preventive measures have been considered so far. Mexicali is classified as the city with the highest rate of allergic and asthma cases in Mexico.

5 Conclusions

-It is recommended to appoint a committee to follow up the air program from the start. The permanent follow up of the advancement in the development of the PROAIR will allow to evaluate its efficiency and to orient its course in a dynamic way.
-The solution to the problem of atmospheric pollution in this airshed would be possible if it involves the people that live and work in the region and the adequate coordination of the authorities in the application of the necessary measures.
-Strengthen the citizens' conscience on the importance of their role for protecting the environment and achieving a bigger participation, will be necessary.
-Instrument mechanisms that promote the participation of the private sector through economic incentives including cross border investment.
-Binational programs such as Border XXI, and Border 2012 [9] established between USA and Mexico are extremely important to have a cleaner air.
-Out of the 5 strategies considered in the air program some actions are urgently needed to be fulfilled:

- Industry, commerce and services: recovery of vapor in storage terminals and gasoline service stations; environmental impact assessment at Cerro Prieto geothermal power plant; implantation of a program to reduce VOCs in industry.
- Motor vehicles: condition the importation of pre-owned vehicles to the certification of smog check of the original country.
- Urban and transport management: consolidate a program of epidemiological vigilance associated to pollution, as much as implanting corrective and preventive measures; develop a program of fiscal stimuli for people, institutions and organisms to promote prevention and pollution control programs.

References

[1] Quintero N. M. & Vega R. A. Estudio comparativo de las tendencias de la calidad del aire en la ciudad de Mexicali (1997-2004) (Capitulo 1). *Contaminación y Medio Ambiente en Baja California* por Quintero N. M. ed. UABC y PORRUA. pp 9–42, 2006
[2] Quintero, N.M. & Sweedler, A., Air quality evaluation in the Mexicali and Imperial Valleys as an element for an outreach program, *Imperial-Mexicali Valleys: Development and Environment of the U.S.-Mexican Border Region*, ed. K. Collins, P. Ganster, C. Mason, E. Sánchez. & M. Quintero N., Institute for Regional Studies of the Californias and SDSU Press, pp 263–280, 2004
[3] INE, SEMARNAT, Gobierno del Estado de Baja California, Municipio de Mexicali, El Programa para Mejora del Aire de Mexicali 2000–2005, 2000

[4] Moncada A. A.M. & Quintero N. M., Quema Agrícola en los Valles de Mexicali e Imperial, *Ciencia y Desarrollo*, Conacyt, Dic., 31(190). pp 6–11, 2005
[5] Meza, T.L.M & Quintero N.M., Metodología para el cálculo de emisiones de particulas atmosféricas PM_{10} y $PM_{2.5}$: caso de estudio calles pavimentadas y no pavimentadas de la ciudad de Mexicali, Coloquio de Posgrado Maestria y Doctorado, Noviembre, Mexicali, B.C. pp 330–339. 2007
[6] UABC. Estudio para la Evaluación del Impacto del Programa Integral de Mejoramiento de Calidad de Vida. Universidad Autónoma de Baja California y la Junta de Urbanización del Estado de Baja California. Nov. Mexicali, Baja California, México. 2004.
[7] Gallegos O. R., Quintero, N.M & Garcia C.R.,. H_2S dispersion model at Cerro Prieto geothermoelectric power plant, World Geothermal Congress, May 28-June 10, Japan, 2000
[8] Avilés G.,. Eximen del Impuesto Ambiental en el 2003. La Crónica. Mexicali, B.C. Miércoles 16 de abril, pag 14/A. 2003
[9] SEMARNAT, U.S. EPA. Border 2012: US-Mexico Environmental Program, 2003

Emission management system in the Russian Federation: necessity for reforming and future adaptation of the western experience

A. Y. Nedre, R. A. Shatilov & A. F. Gubanov
Federal State Unitary Enterprise "Scientific Research Institute for Atmospheric Air Protection", Russian Federation, Russia

Abstract

Currently, emission management is carried out in the Russian Federation on the basis of hygienic standards which is, in turn, based on the calculations of emission dispersion from source. The aim of this standardization is bringing into compliance with air quality standards for the human health maximal allowable concentration (MAC) of pollutants in the atmospheric air for human health. This article gives a brief description of this standardization method. This approach of standardization has been used in Russia since 1986, but, unfortunately, it seems in vain. There is continuous growth of pollutant emissions in Russia from stationary and mobile sources and, therefore, there is an increase in the number of cities where the air quality standards are exceeded. This article covers the historical data concerning the emission growth and related increase of the number of cities where air quality standards are exceeded. Moreover, this standardization method does not engage the introduction of the best available techniques (BATs) neither does it include evaluation and regulation of the impact on vegetation. The western approach which is based on the usage of BATs and critical loads is the most perspective from our point of view. However, this process must be coordinated with the economic capabilities of the Russian business community because the new requirements would be a significant additional financial burden. This paper contains information on the experience gained by the individual industries which began to proceed with the determination of the BATs. Finally, the concept of the modernization of the emission management system in Russia, for the years 2008-2010 is also described in brief.

Keywords: emission management, Russian Federation, BAT, MAC MPE, sanitary standard, atmospheric air quality standards.

1 The trends of atmospheric pollutant emissions and atmospheric pollution status in the cities of the Russian Federation

Observations over atmospheric pollution are being carried out in 251 cities of the Russian Federation. The regular observations are being carried out on the 629 stationary sites.

The results of these observations show that the level of atmospheric pollution in the cities is still significantly high. The level of atmospheric pollution in 69% of all cities is extremely high whilst a low level of atmospheric pollution is observed only in 17% of all cities [1].

On the whole, 38% of the urban population lives within the territories where the observations over the atmospheric pollution are not carried out while 55% (58,2 million of people) – within the cities with high and extremely high level of atmospheric pollution. In the year 2006 65 million people lived in the 206 cities where the average annual concentrations of one or several substances exceeded the MAC. During the reporting period the average annual suspended materials concentrations exceeded the MAC in 64 cities, benzo(a)pyrene concentrations – in 160 cities, nitrogen dioxide – in 102 cities, formaldehyde concentrations – in 125 cities.

The priority list consists of 10 cities with the aluminium industry and ferrous metallurgy, 7 cities with the chemical and petrochemical enterprises, oil production and distribution enterprises, as well as a lot of cities with the enterprises of the fuel and energy sector.

The high level of atmospheric pollution in almost all of these cities is associated with high benzo(a)pyrene and formaldehyde concentrations, in 26 cities – with high nitrogen dioxide concentrations, in 14 cities – with high particulate matter concentrations, in 11 cities – with high phenols concentration.

In the year 2006 the maximum concentrations of the pollutant substances in the atmospheric air exceeded 10 MAC level in 26 cities with the total population equal to 14,7 million people [1].

After the USSR collapse and denationalization of the economics the constant reduction of emissions generated by stationary sources in the Russian Federation was occurring during the years 1990-1998 [2]. The increase of emission amount was detected only after the economic default which occurred in the year 1998. The increase of road transport emissions had started a little bit earlier – since 1997; thereby, the contribution of these emissions into total emissions rises permanently because of growing motor-vehicle pool (mainly due to the individual owners (private cars)) (See Table 1).

Table 2 includes the list of the cities, where the contribution of the road transport into the total emissions exceeded 80% (in the year 2006).

Table 1: The dynamics of total pollutant emissions in ambient air in the Russian Federation during the years 1990 – 2005.

Total emissions	1990	1995	1997	1998	2000	2005
Stationary sources, mln. of tons	34,7	22,2	19,8	18,9	19,4	20,3
Automotive transport, mln. of tons	20,9	14,5	14,4	15,3	15,7	17,4
Total from stationary sources and automotive transport, mln. of tons	55,6	36,7	34,2	34,2	35,1	37,7
St. sources/ Automotive transp. (%)	37,6	39,5	42,1	44,7	44,7	46,2

Table 2: The list of the cities with dominant emission contribution from road transport into the total urban emissions (80% and more) subject to the amount of road transport emissions equal to, at least, 50 000 tons per year.

City	% of road transport contribution	Road transport emissions (thousand of tons per year)
Sochi	96,8	92,65
Vladikavkaz	94,5	89,15
Moscow	92,5	1233,0
Yekaterinburg	92,2	201,2
Krasnodar	92,0	113,1
Rostov-on-Don	91,1	119,8
Saint-Petersburg	90,5	500,9
Voronezh	89,1	102,4
Tyumen	85,6	71,1
Kaliningrad	83,7	57,7

2 The basic reasons for the necessity of the air protection activity reforming

The basic reasons which ground the necessity for reforming of the air protection activity in the Russian Federation could be described as follows:
  - the application of the existing principles of atmospheric air quality management which was developed in the 1970s of the last century resulted in the exceedance of the atmospheric air quality standards in more than 200 cities. Notably, this number changed slightly during the years 2000-2005 in spite of the intensive development of the system of sanitary regulation and ecological monitoring;

- the existing system concerning atmospheric air quality management is based on the bulky sanitary calculations of the dispersion. It prevents the switching on the advanced clean technologies;
- the necessity for the integration of the Russian Federation into the international economical and political systems requires the approximation of the national air protection legislation and the current legislative systems of the economically developed countries and their international associations which are based on other principles.

3 Regulatory norms are the basis for the emission management in the Russian Federation

The most important trend of the air protection activity in the Russian Federation is the norm setting for emissions of the pollutant substances into the atmosphere since it defines the main target of the air protection activity – emission reduction value which is necessary for the attainment of the appropriate atmospheric air quality standards.

3.1 Types of regulation

There are three types of regulation in the Russian air protection legislation:
- sanitary regulation for the attainment of air quality standards allowable for public health;
- ecological regulation for the attainment of air quality standards allowable for the elements of the environment;
- technical regulation for the attainment of the specific discharges per the production unit with respect to the best available technology.

The combined and/or the selective application of these three regulation methods allows solving nearly any task in the field of environment protection against the pollution.

3.2 Sanitary regulation

For many years, the existing Russian regulation system concerning the emissions of the pollutant (harmful) substances into the atmospheric air is based upon the accounting of the sanitary standards related to the atmospheric air quality which regulate the allowable concentration levels of the harmful substances in the atmospheric air for the residential areas.

Sanitary standards have been developed for more than 2000 harmful substances and they allow calculation of the adverse effects on the atmospheric air caused by a great number of harmful substances. In order to calculate pollutant ventilation in the atmospheric air we use the general calculation method OND-86 which is based upon the solution of equation concerning atmospheric diffusion. The calculation results concerning the pollution ventilation in the atmospheric air caused by different sources, as well as background air pollution are used for the development of the maximum permissible emissions and temporary agreed emissions for the enterprises.

During the application of this method the following problems have been revealed:
- the inconsistency of its application in sparsely populated regions where the residential area is located far from the emission source (for example, gas-transmission station in tundra), that could be used for the justifying of any standard by means of dispersion calculations for the hundred of kilometres;
- the inconsistency of its application in populated regions where the monitoring of atmospheric pollution shows the stable absence of any MPE exceedance since in that case the actual emissions automatically turn into standards;
- the absence of the accounting for road transport emissions and mutual influence of the neighbouring enterprises.

3.3 Ecological regulation

The trend of air protection activity could be fully reflected by ecological regulation which supposed the adoption of such emission limitations which allow the compliance of the all maximum permissible (for the objects of ecological interest) atmospheric air quality standards. There are two world widespread standards of this kind:
- the maximum permissible concentrations of the pollutant substances for the vegetation and wildlife objects (in the Russian Federation such MPC are adopted for the several species of vegetation for the region of the Lake Baikal and "Jasnaya Polyana" zone in the Tula region);
- the critical loads in terms of maximum allowable depositions per unit area. (The given standards are adopted for the European part of the Russian Federation within the framework of the Convention on Long-Range Transboundary Air Pollution – the Russian Federation is a part of this Convention.)

4 Trends of the technical regulation development in the Russian Federation

The most favourable tool to introduce the advanced clean technologies in the developing economy is the technical regulation. It is supposed that this regulation type will be implemented in the Russian Federation in several stages. The first stage includes the determination of the specific emissions per production unit of the enterprise which belong to one sector but use different technologies. The second stage includes the determination (by means of the update values) of the best available techniques while the appropriate specific emissions turn into the technical standards. The third stage includes the determination of the attainment dates (within the discussions with the business community). The final stage includes the adoption of these standards by means of the special legislative act.

The existing Russian law "on the atmospheric air protection" requires granting the emission permit of the pollutant substances is the subject to the technical emission standards. The abovementioned law also provides the definition of the technical emission standards.

Our institute made its own contribution and has developed the Programme for the implementation of the technical standards in the Russian Federation.

4.1 Programme for the implementation of the technical standards in the Russian Federation

As we see it, the switch on of the implementation of the technical emission standards for ambient air should be performed in two stages:

1st stage:
- development of the national methodological and information centre on the implementation of the technical regulation;
- analysis of the actual situation + the gathering of the appropriate data;
- adoption of the specific emission standards for the existing enterprises (sectoral specific emission standards).

In order to implement this stage it is necessary to develop the following national documents:
- the method for development of the specific emission standards for the existing enterprises;
- the order of consideration and agreement of the specific emission standards for the existing enterprises;
- the list of objects (inventories) which should be regulated with the help of the technical emission standards;
- the programme on the development of the federal sectoral technical emission standards;
- the order of the accounting of the specific emission standards during the adoption of the standards (limits) concerning the emissions of the pollutant substances into the atmosphere;
- the order of the accounting of the specific emission standards during the adoption of the standards (limits) concerning the emissions of the pollutant substances into the atmosphere for the projected enterprises.

2nd stage:
– development of the federal technical emission standards and its implementation within the state regulation of ambient air pollutant emissions.

In order to implement this stage it is necessary to develop the following national documents:
- the order of the development and adoption of the technical emission standards;
- the order of the development and exploitation of the data related to the best available techniques during the development and adoption of the technical emission standards subject to the ecological and economical criteria;
- the guidance on the accounting of the technical emission standards during the adoption of maximum permissible emissions and temporary agreed emissions (emission limits) of the pollutant substances into the atmospheric air.

4.2 The list of objects which should be regulated with the help of the technical emission standards

Moreover, our institute prepared in 2007 "The list of objects (inventories) which should be regulated with the help of the technical emission standards". This list includes those types of economic activities (production, technological processes) which made the biggest contribution into the emissions of the pollutant substances into the atmosphere of the Russian Federation and become a reason for the high pollution level in the cities.
This list includes:
Mining operations
- Oil extraction enterprises
- Crude oil and associated gas extraction

Manufacturing activity
Metallurgical production
- Cast-iron and steel production
- Nonferrous metal production from the ores and concentrates
- Aluminium production
- Copper production
- Production of the other nonferrous metals

Electricity, steam and hot water generation
- Electricity generation by thermal power plants
- Steam and hot water generation (thermal energy) by thermal power plants
- Steam and hot water generation (thermal energy) by other power plants and industrial isolated generating plants
- Steam and hot water generation (thermal energy)

Petrochemicals production
Chemical production
- Synthetic rubber production
- Rubber article production
- Rubber tire, casing and innermost tire production
- Production of the non vulcanized caoutchouc and products made out of non vulcanized caoutchouc

4.3 The scope of the technical emission standards

We plan to use the technical emission standards subject to the BAT within the framework of Russian emission regulation system as follows:
- in the populated areas where the exceedance of the maximum one-time permissible concentrations (developed for the protection of public health) is registered, it is necessary to adopt the maximum allowable hygienic standards (not higher than technical emission standards) derived from the summary dissemination calculations,
- in the populated areas where the exceedance of the maximum one-time permissible concentrations (developed for the protection of public health) is not registered, it is necessary to adopt the maximum allowable emission standards on the basis of the technical emission standards,

- for the enterprises which affect the environmental compartments, it is necessary to adopt the maximum allowable ecological emission standards (not higher than technical emission standards) on the basis of the dispersion calculations,
- during the decision-making process concerning the allocation of the new enterprise, it is necessary to access its contribution into the total pollution on the basis of the dispersion calculations provided that its emissions will not exceed the technical emission standards,
- during the development of programmes concerning the economic advancement of the territories, it is necessary to access the contribution of the new enterprises into the depositions of the pollutant substances in order to check the compliance with the critical loads provided that emissions of these enterprises will not exceed the technical emission standards.

4.4 Russian experience concerning the adaptation of the technical regulation

Currently, there is only one rule "on the requirements concerning the emissions of the harmful (pollutant) substances generated by the vehicles put into circulation at the territory of Russian Federation" (RF Government regulation from the 12th October of 2005, №609). It includes direct references to the non-mandatory documents (UNECE rules) – analogues of the international standards. This decision was supported by all concerned parties, and uppermost, by the motor industry which is against the double standards. The phase-in of the international standards in Russian Federation (subject to the economic development and material and technical base) is stipulated up to the year 2014.

This situation became possible not only because of growing foreign-made car fleet but also because of the allocation of automotive production to the territory of the Russian Federation. These enterprises (technologies) entailed the introduction of the western standards concerning the emissions generated by the automotive vehicles.

The situation in other economic sectors is different. Thus, we believe it is necessary to adopt the technical emission standards for the most of the Russian producing units on the basis of the inventories concerning the existing technologies (in the country). Several major Russian companies have already begun to assess the ecological compatibility of the existing technologies and equipment in comparison with the international analogues. As a part of this work the specific emission values and sectoral standards related to the specific emissions have been adopted.

In particular, the work on the adoption of the specific emissions for the transpolar branch of the JSC "MMC Norilsk Nickel" must be completed in the nearest future; the work on the adoption of the specific emission values for the JSC "MMC Kolskaya" has been started recently – the list of regulated substances has been already adopted while the standard values for the specific producing units are being developed at the moment; the same work has been conducted since 2007 in the UC RUSAL company (the development of the sectoral method for determination of the technical emission standards for the UC RUSAL enterprises). It is planned to conduct the work (to be agreed with the

Scientific Institute for Atmospheric Air Protection) on the adoption of the specific emission standards for the basic equipment which is applied in the heat power industry. There are also some interesting best practices concerning the paper-pulp and glass industries.

It is planned to adopt, on the basis of the obtained results, the sectoral specific emission standards and, subsequently, the federal technical emission standards.

References

[1] State report about the state and protection of the environment in the Russian Federation in 2006, Part 1 Atmospheric air, Moscow, pp. 9–10, 2007.
[2] Annual report on pollutant emissions in atmospheric air of cities and regions of the Russian Federation in 2006, St. Petersburg, pp. 83–85, 2007.

Outdoor air quality data analysis of Al-Mansoriah residential area (Kuwait): air quality indices results

S. A. Al-Haider[1] & S. M. Al-Salem[2]
[1] Applied Sciences Department, College of Technological Studies, Public Authority for Applied Education and Training, PO Box: 14561, Al-Faiha 70700, State of Kuwait
[2] Petrochemical Processes Program Element, Petroleum Research and Studies Center, Kuwait Institute for Scientific Research, PO Box: 24885, Safat 13109, State of Kuwait

Abstract

Environmental awareness is of growing concern in the state of Kuwait, especially after the recognition of Kuwait Environment Public Authority (KUEPA) as a separate entity with legal power back in 2001. In this communication, the outdoor air quality data collected over the period of five years (2000-2004) were analyzed for Al-Mansoriah residential area in Al-Asemah "Capital" Governorate (Kuwait). Data points were in two different time spans; five minute original data points and hourly mean averages. Air Quality Indices (AQI) based on the following pollutants: SO_2, CO, O_3 and NO_2 were calculated. Based on CO and SO_2, the AQI was determined "good" ranging between an annual mean of 0.06-0.15. Based on ground level ozone and nitrogen dioxide AQI resulted in the "moderate" category for the period of study.
Keywords: KUEPA, SO_2, CO, O_3, NO_2, AQI.

1 Introduction

Many Gulf Council Countries (GCC) suffer from air pollution health effects especially when it comes to respiratory system chronic diseases and cancers associated with such airborne pollutants [1]. Kuwait is no exception, being a petroleum industry oriented country. Many pollution sources are linked with the downstream/upstream industry in the state but yet still little action is taken by the concerned parties. One way of monitoring such pollution levels is what has become a standard approach of Air Quality Indices (AQI) calculation. Each Air Quality Index is a standardized indicator of the air quality in a given location. It measures mainly ground-level ozone and particulates (except the pollen count), but may also include sulphur dioxide, and nitrogen dioxide. Various agencies around the world measure such indices, though definitions may vary between places. In the US, EPA calculates the AQI for five major air pollutants regulated by the Clean Air Act: ground-level ozone, particle pollution (also known as particulate matter), carbon monoxide, sulfur dioxide, and nitrogen dioxide. For each of these pollutants, EPA has established national air quality standards to protect public health. Ground-level ozone and airborne particles are the two pollutants that pose the greatest threat to human health in US.

Many scientists have devoted their work towards air pollution monitoring and standardizing rules and regulations governing cities around the world. This could be determined by literature available regarding the matter. An air quality monitoring methodology was presented by Landulfo et al. [2], by employing an elastic backscattering Lidar, sunphotometer data, air quality indexing and meteorological data in the city of São Paulo, Brazil, a typical Urban Area. This procedure was made aiming to gather information from different optical atmospheric techniques and add this information to the air quality data provided regularly by the environmental agencies in the city. The parameters obtained by the Lidar system, such as planetary boundary layer height, aerosol optical thickness and aerosol extinction and backscattering aerosol coefficients are correlated with air quality indexes/reports provided by state environmental control agencies in order to extend the database information concerning pollution assessment and abate policies.

In India [3], the measured 24 h average criteria pollutants such as sulfur dioxide, oxides of nitrogen, respirable suspended particulate matter and suspended particulate matter for the period from 1997 to 2005 at three air quality monitoring stations were used for the development of AQIs. The results indicated that the air pollution at all the three air quality monitoring stations can be characterized as 'good' and 'moderate' for SO_2 and NO_x concentrations for all days from 1997 to 2004. The Pollution Standards Index (PSI) was initially established in response to a dramatic increase in the number of people suffering respiratory irritation due to the deteriorating air quality. The PSI was subsequently revised and implemented by the USEPA in 1999, and became known as the (AQI) that includes data relating to particle suspension, $PM_{2.5}$, and a selective options of either 8-hour or 1-hour ozone concentration during increased O_3 periods. This was discussed in the publication of Cheng et al. [4]. An aggregate AQI based on the combined effects of five criteria pollutants (CO, SO_2, NO_2, O_3 and PM_{10}) taking into account the European standards was developed previously [5]. An evaluation was carried out for each monitoring station and for the whole area of Athens, Greece. A comparison was made with a modified version of Environmental Protection Agency/USA (USEPA) maximum value AQI model adjusted for European conditions. Hourly data of air pollutants from 4 monitoring stations, available during 1983–1999, were analyzed for the development of the proposed index.

The objective of this communication is to investigate the ambient air quality of Al-Mansoriah residential area using standard AQI calculations with respect to four airborne pollutants. These pollutants are: sulfur dioxide (SO_2), carbon monoxide (CO), ground level ozone (O_3) and nitrogen dioxide (NO_2). No previous attempts (to the authors knowledge) have been initiated or published regarding this matter in Kuwait.

2 Investigated area description

Al-Mansoriah residential area hosts Kuwaiti residents of mid/high class and could be considered a posh area. Fig.1 shows the area from satellite imagery, indicating main locations including Al-Arabi sports club, Cairo St. and receptor point (i.e. area's polyclinic).

The receptor point of the area was the polyclinic of Al-Mansoriah situated near the area's Co-Op, which associated with restaurant and human related pollution emissions. These pollutants include n-CH_4 (resulting from restaurants), NOx and CO (emissions of automobile and other burning sources) and VOCs from the gas and gasoline dispensing station. On the other hand, Al-Arabi sports

Figure 1: Al-Mansoriah residential area adapted from Google earth imagery. Main area locations are indicated.

club is considered one of the largest sports and family functions facility in Kuwait, in terms of children entertainment lounges and other function spaces. All of these sections in the club are always adjacent to a food court or lounges which also emit gases through vents and other outlets.

Cairo St. is situated to the north side of the receptor point. It is one of the busiest streets of Kuwait especially when it comes to school time linking three main residential areas together in the governorate, i.e. Al-Mansoriah, Al-Qadisiah and Al-Daeiah. Automobile vehicles operated with gasoline are the main source of pollution associated with it. Rush hours are usually between 7:30 am to 9:00 am (weekdays); and 8:00 pm to 9:30 pm (weekends).

3 Data collection and methods

The data used in this study were secured form the Kuwait Environment Public Authority (KUEPA), more precisely Al-Mansoriah monitoring station covering the period from Jan 1st 2000 – Dec 31st 2004. The station is operated with a number of air sampling devices and analyzers with a tolerance of 1%. Air probe was approximately 15 m above sea level. All data were stored and manipulated with *EnviDas* data acquisition software which did store up to three months worth of data points. Pollutants collected by the station included the following: CH_4 (ppm), n-MHC (ppm), CO (ppm), CO_2 (ppm), NO (ppb), NO_2 (ppb), NOx (ppb), VOCs (ppb), mp-Xylenes (ppb), NH_3 (ppb), H_2S (ppb) and O_3 (ppb). Metrological conditions were collected via a fixed weathering station recording the following: wind speed (ms^{-1}) and direction (^{0}), relative humidity (%) and ambient temperature (^0C).

Data points were treated and filtered before performing any analysis. Filtration procedure was performed as indicated by [6-11]. NOx, NO_2 and NO points exceeding 200 ppb were deleted from the spreadsheets, in order to eliminate any automobile point source effect on data collection. CH_4 levels below 1.3 ppm were also deleted to avoid ion presence and instrumentation chocking points. Calibration and span check points were also deleted.

The AQIs were calculated using standards USEPA methodology. The purpose of the AQI is to help you understand what local air quality means to health. To make it easier to understand, the AQI is divided into six categories (Table 1).

Table 1:   Air Quality Index (AQI) category in accordance with the USEPA.

AQI Value Range	AQI Conditions	Audience Color
0 to 50	Good	Green
51 to 100	Moderate	Yellow
101 to 150	Unhealthy for Sensitive Groups	Orange
151 to 200		Red
201 to 300	Unhealthy	Purple
301 to 500	Very Unhealthy Hazardous	Maroon

The AQI is calculated every hour for each air quality parameter using the formulas indicated below (Table 2). The highest number calculated for a specific hour is used as the AQI for that hour. Four pollutants were chosen to be studied which were: SO_2, CO, NO_2 and O_3.

Table 2: Air Quality Index (AQI) Calculation formulas in accordance to the USEPA.

Parameter Name	Concentration	Units	Formula
Carbon Monoxide	If <= 13	ppm	AQI = 1.92 x Concentration
	If > 13	ppm	AQI = (1.47 x Concentration) + 5.88
O_3	If <= 0.05	ppm	AQI = 500 x Concentration
	If > .05 <= 0.08	ppm	AQI = (833 x Concentration) - 16.67
	If > 0.08	ppm	AQI = (714 x Concentration) - 7.14
SO_2	All	ppm	AQI = 147.06 x Concentration
Nitrogen Dioxide	If <= 0.21	ppm	AQI = 238.09 x Concentration
	If > 0.21	ppm	AQI = (156.24 x Concentration) + 17.19
$PM_{2.5}$	If <= 30	ugm^{-3}	AQI = 0.8333 x Concentration
	If > 30	ug/m^{-3}	AQI = (0.5 x Concentration) + 10

4 Results and discussion

Based on the standard USEPA methodology the AQIs of the pollutants SO_2, CO, NO_2 and O_3, were calculated. Table 3 shows the results obtained by the running calculations of AQI for the five years period based on yearly averages.

In terms of SO_2 and CO, good air quality resulted from the calculations based on the standards calculations and annual mean. Both pollutants recorded a minimum in the last two years of study (2003 and 2004). Maximum SO_2 AQI calculated was in the year 2002 recording an AQI of 71.77. In terms of NO_2, moderate air quality was characterized for Al-Mansoriah with a maximum value for the first year of study of 160.46. Moderate air quality is also characterized for O_3 in the period of study.

Fig.2 shows the scatter left by the points of average AQI vs. years of study. A clear increase in the CO AQI, when comparing the 2000 (AQI=6.8) and 2004 (AQI=14.2) values. AQI of CO has increased by more than 100% in the period of study. In terms of ground level ozone (O_3), there was a general decrease in the annual average trend. Unlike the case of SO_2, where there was a general decrease but a shooting off point does exist in the middle range. NO_2 had also a general decrease in the trend. AQI in 2000 was equal to 37.46 while in 2004 it was 31.8.

Table 3: Air Quality Index (AQI) calculated for Al-Mansoriah residential area in the time of the study (2000-2004).

Year	Pollutant	Avg. AQI	Min. AQI	Max. AQI	Category
2000	SO_2	1.84	0.15	47.06	Good
	CO	6.80	0.06	26.53	Good
	NO_2	37.46	0.24	160.46	Moderate
	O_3	37.72	0.50	391.99	Moderate
2001	SO_2	2.73	0.15	37.94	Good
	CO	1.93	0.06	17.53	Good
	NO_2	31.40	0.24	133.75	Moderate
	O_3	38.26	1.0	84.24	Moderate
2002	SO_2	3.97	0.147	71.77	Good
	CO	18.52	0.50	25.0	Good
	NO_2	37.45	0.23	168.58	Moderate
	O_3	36.56	0.50	80.65	Moderate
2003	SO_2	4.9	0.1	68.2	Good
	CO	14.85	0.1	27.6	Good
	NO_2	34.6	0.2	105.9	Moderate
	O_3	33.31	0.5	65	Moderate
2004	SO_2	3.2	0.1	52.2	Good
	CO	14.2	0.1	27.6	Good
	NO_2	31.8	0.5	93.6	Moderate
	O_3	33.10	0.5	79.3	Moderate

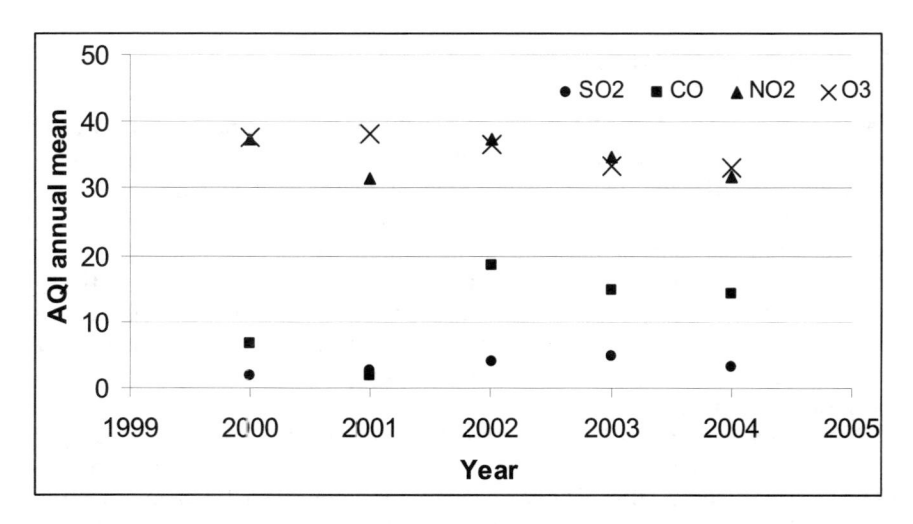

Figure 2: Average Air Quality Index for the studied pollutants in the period of study.

5 Conclusion

A general decrease was witnessed in the AQI values along the period of study. The four pollutants studied (*i.e.* SO_2, CO, NO_2 and O_3) resulted in a "good" and "moderate" ambient air quality. This leads to the understanding that Al-Mansoriah residential area exhibited a moderate air quality during the five years of investigation. In terms of SO_2 and CO, both pollutants recorded a minimum in the last two years of study (2003 and 2004). Maximum SO_2 AQI calculated was in the year 2002 recording an AQI of 71.77. For NO_2, moderate air quality was characterized for Al-Mansoriah with a maximum value for the first year of study of 160.46. Moderate air quality is also characterized for O_3 in the period of study. A clear increase in the CO AQI, when comparing the 2000 (AQI=6.8) and 2004 (AQI=14.2) values. AQI of CO has increased by more than 100% in the period of study. KUEPA has no records of any AQI running investigation in their current plan. More strict regulations should be applied for monitored area in the state of Kuwait.

Acknowledgements

Both authors would like to thank the KUEPA for providing the dataset used for this study. Gratitude and thanks must be given to everyone who supported this work, especially Dr. A.R. Khan (EUD/KISR) for his magnificent encouragement.

References

[1] Al-Salem, S.M. and Bouhamrah, W.S., Ambient concentrations of benzene and other VOCs at typical industrial sites in Kuwait and their cancer risk assessment, *Res J Chem Environ*, **10(3)**, pp. 42–46, 2006.
[2] Landulfo, E., Matos, C.A., Torres, A.S., Sawamura, P. and Uehara, S.T., Air quality assessment using a multi-instrument approach and air quality indexing in an urban area, *Atm Res*, **85(1)**, pp. 98–111, 2007.
[3] Shiva Nagendra, S.M., Venugopal, K. and Jones, S.L., Assessment of air quality near traffic intersections in Bangalore city using air quality indices, *Trans Res Part D: Transport and Environment*, **12(3)**, pp. 167–176, 2007.
[4] Cheng, W.L., Chen, Y.S., Zhang, J., Lyons, T.J., Pai, J.L., Chang, S.H., Comparison of the Revised Air Quality Index with the PSI and AQI indices, *Sci Total Environ*, **382(2-3)**, pp. 191–198, 2007.
[5] Kyrkilis, G., Chaloulakou, A. and Kassomenos, P.A., Development of an aggregate Air Quality Index for an urban Mediterranean agglomeration: Relation to potential health effects, *Environ Int*, **33(5)**, pp. 670–676, 2007.
[6] Al-Salem, S. and Al-Fadhlee, A., Ambient levels of primary and secondary pollutants in a residential area: population risk and hazard index calculation over a three years study period, *Am J Environ Sci*, **3(4)**, pp. 244–228, 2007.
[7] Al-Salem, S.M. and Al-Haddad, A.A., Pollutants monitoring and source determining: Effect of oil refineries on a residential area, In: *Proc 2nd Int*

Con on Scientific Computing to Computational Engineering, Edited By: Prof. Demos T. Tsahalis, Set No: 960-530-080-X, Athens, Greece, $5^{th} - 8^{th}$ July, 2006.
[8] Al-Salem, S.M. and Khan, A.R., Methane dispersion modeling and source determination around urban areas in Kuwait, In: Proc *1st Int Con & Exhib Green Industry*, Manama, Bahrain, $20^{th} - 22^{nd}$ November, 2006.
[9] Al-Salem, S.M., Al-Haddad, A.A. and Khan, A.R., Primary pollutants monitoring and modeling using chemical mass balance (CMB) around Fahaheel residential area, *Am J Environ Sci*, **4(1)**, pp. 13–21, 2008.
[10] Khan, A. and Al-Salem, S., Primary and secondary pollutants monitoring around an urban area in the state of Kuwait: a three years study, *Res J Chem Environ*, **11(3)**, pp. 77–81, 2007.
[11] Khan, A. and Al-Salem, S., Seasonal variation effect on airborne pollutants in an urban area of the state of Kuwait, *J Environ Res Dev*, **1(3)**, pp. 215–218, 2007.

A modelling tool for PM10 exposure assessment: an application example

E. Angelino, M. P. Costa, E. Peroni & C. Sala
Environmental Protection Agency of Lombardy Region, Italy

Abstract

A tool has been developed with the aim of calculating population exposure to PM10 concentrations. It integrates two main modules: a linker to the outputs of a chemistry and transport model (outdoor ambient concentrations) and a time-variable exposure module. The leading concept behind exposure modelling is that time-weighted average exposure can be considered as the sum of the partial exposures determined by both the concentration and the time spent in each so called micro-environment. A run of a CTM (Chemical and Transport Model) over the whole year of 2006 provided the PM10 hourly outdoor concentration field. In the absence of specifically measured data, some input data, such as infiltration parameters and indoor sources concentrations, have been chosen from relevant literature data. This paper reports the main input data used and the results obtained during an application to the domain under study, located in the North of Italy. The results confirm that the population habits and indoor sources contributions are the most critical parameters when assessing exposure.

Keywords: exposure modelling, particulate modelling, air quality management.

1 Introduction

A modelling system for exposure assessment must be able to integrate numerous and various pieces of information such as land use, emissions, meteorology, dispersion and chemical reactions of pollutants, population activity patterns and distribution. In recent years a growing integration of all this information, generally obtained by employing different and separate models in the past, into more sophisticated and comprehensive modelling systems has occurred. In the past, US modelling groups have supported the most comprehensive efforts in this area (Seigneur [14], Burke *et al* [3]), but recent European efforts have also produced interesting experiences in many parts of Europe. The FUMAPEX project [6] gives a wide overview of exposure modelling assessment applications, characterised by different approaches and computational tools, input data sources, spatial and temporal resolution in relation to their final use and output dissemination media.

2 Methodology

A methodology for determination of population exposure using modelling techniques and air pollution observations has been developed in the framework of a research project carried out by ARPA Lombardia and supported by the APAT (Italian National Environmental Protection Agency). Exposure computation is based on the algorithm of eqn. (1):

$$E = \frac{1}{t_{avg}} \sum_{i=1}^{n} t_i \times C_i = \sum_{i=1}^{n} f_i \times C_i, \qquad (1)$$

where E is a time-weighted average exposure level across the visited microenvironments (i) calculated as the sum of partial exposures in each one of them. The partial exposures are calculated by multiplying the microenvironment concentration by the fraction of time (f) spent in there. This approach assumes that a person's time-integrated exposure is the combination of the concentrations of a specific set of micro-environments, concentrations that are considered to be constant and homogeneous. Concentration in the indoor and in-vehicle microenvironments are calculated by separating ambient and indoor generated PM levels by the following equation as in Hänninen et al [8]:

$$C_{ig} = C_i - C_{ai}, \qquad (2)$$

where C_i (µg/m^3) is the total indoor concentration, C_{ai} (µg/m^3) the contribution from ambient outdoor concentration that has infiltrated indoors and C_{ig} (µg/m^3) the concentrations caused by indoor sources. On the basis of the steady-state mass balance equation, assuming uniform mixing within the building and steady state conditions, i.e. that the penetration efficiency P (unitless), the air exchange rate a (h^{-1}) and the indoor decay rate K (h^{-1}) do not change with time, then concentration fractions in eqn (2) can be expressed as:

$$C_{ig} = \frac{Q}{V(a+k)}, \qquad C_{ai} = \frac{P \cdot a}{a+k} C_a, \qquad (3)$$

where Q represents the source strength (µg h^{-1}), and V the interior volume of the building (m^3). By defining the indoor-outdoor ratio of the ambient pollutant concentration (C_{ai}/C_a) as an infiltration factor (F_{inf}), it is possible to express the indoor concentration as:

$$C_i = C_{ig} + C_{ai} = C_{ig} + F_{inf} C_a. \qquad (4)$$

Combining (1) and (4), exposure can be computed as:

$$\overline{E} = \frac{1}{t_{avg}}\sum_{i=1}^{n} t_i \times C_i = \sum_{i=1}^{n} f_i \times (F_{inf}C_a + C_{ig}). \tag{5}$$

The outdoor ambient concentrations can be estimated from monitoring data or simulated by an air quality model. F_{inf} and C_{ig} can be derived using regression of concurrent indoor and outdoor PM measurements. To date several experimental studies have been performed aiming to quantify how much outdoor sources contributed to the indoor concentrations and to characterise the air exchange rate, in relation to the building insulating materials, home or vehicle ventilation habits, seasonal factors etc. Attention has been paid also on estimation of indoor PM sources (cigarette smoking, wood burning, house work such as cooking, dusting, carpet cleaning, spraying such as using paints, cleaners and other consumer products in spray). Among the most relevant ones, the following studies can be mentioned: the Harvard six cities study (Dockery and Spengler [5]), the New York State ERDA (Koutrakis et al [11]), the Particle Total Exposure Assessment Methodology Study (PTEAM) conducted in California (Ozkaynak et al [13]), the characterisation of indoor particle sources in Boston (Abt et al [1]), the European EXPOLIS (Hänninen et al [8]), the Relation among Indoor, Outdoor and Personal Air (RIOPA) (Weisel et al [18]), the NERL residential ultrafine, fine and coarse PM study (Wallace et al [17]). Several papers give brief descriptions of them and summarise the main results (Wallace [16], Hänninen et al [8], Georgopoulos et al [7], Monn [12], Jantunen et al [10]).

3 The modelling tool

The modelling tool, represented in figure 1, is composed of two main sub-systems. The air quality modelling system is based on the FARM Eulerian chemical transport model (Flexible Air quality Regional Model, Silibello et al. [15]); the other components of the system are two pre-processors used to reconstruct flows and turbulence parameters, an emission pre-processor and a module preparing the initial and boundary concentrations from a large scale model results and observational data. In this study, the FARM model has been applied with the SAPRC-90 gas-phase chemical mechanism and the CMAQ aero3 modal aerosol module. The exposure modelling subsystem is based on equation (4) and has the following characteristics: the population is divided into several groups with specific activity patterns; the study area can be divided into several zones (including one or grouped municipalities).

A GIS module allocates population to the air quality field grid cells; the population dynamics follow the movement of a population group both through different microenvironments and different zones as a function of time. Infiltration factors can be specified both for zones and for each transit/indoor micro-environment. For each of them concentrations are calculated on the basis of both outdoor concentrations and contributions from selected indoor sources.

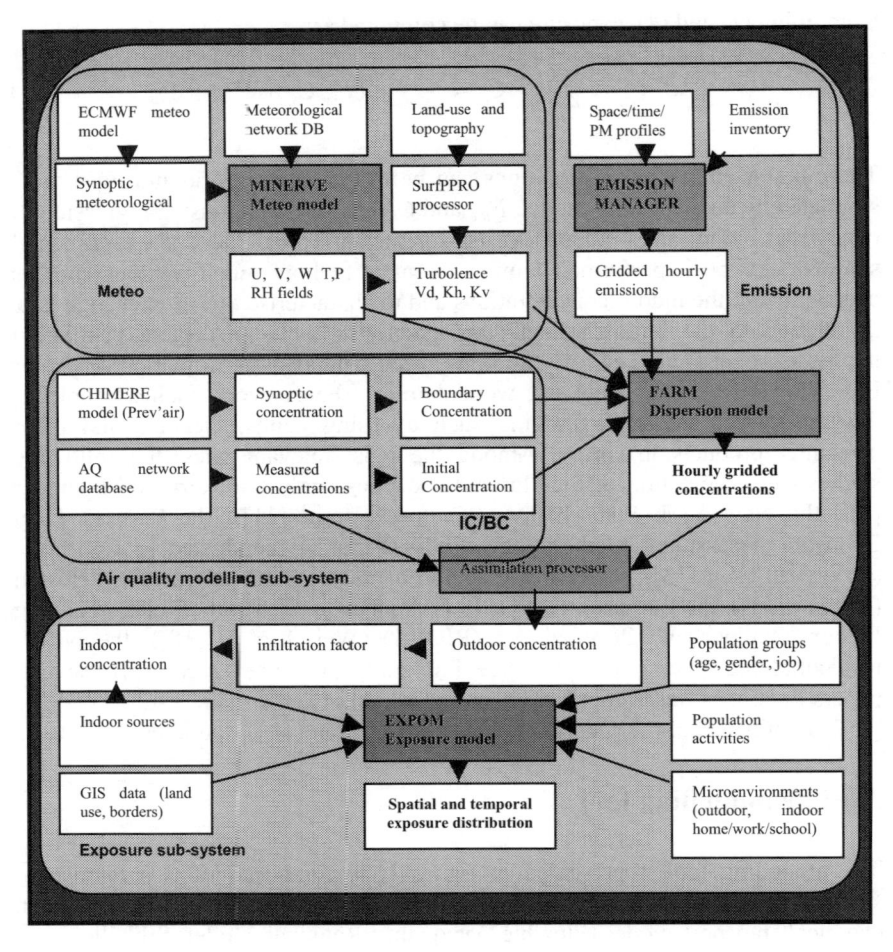

Figure 1: Modelling system used in the study.

4 Input data

To compute the outdoor concentration field an hourly run of the air quality modelling system has been done over the whole 2006 year (Silibello et al [15]). The system has been applied to a 4 Km cell-size 244 x 236 km^2 domain, including the whole Lombardy region, in the Po-Valley Basin in Northern Italy. The hourly meteorological fields were obtained from the meteorological ARPA Lombardia network and from ECMWF synoptic model fields; the boundary and initial conditions were provided by the Prev'air system (www.prevair.org) based on the CHIMERE model; the emission inputs were derived from the regional INEMAR 2003 emission inventory. Different assimilation procedures have been tested and used to guarantee consistency with observations. In this application the tool was run for six different population groups: "class 1": infants (0-3 years), "class 2": schoolchildren (3-14 years), "class 3": youths (15-24 years), "class 4": younger adults (25-34 years), "class 5": adults (35-64 years), "class 6": elderly people (over 64 years). Four microenvironments have been considered: "1" indoor home, "2" indoor work/school/other, "3" transit, "4" outdoor. In the absence of local measured data, concentrations for each indoor microenvironment have been derived as a function of ambient concentrations plus a contribution due to indoor sources. The infiltration parameters and indoor sources concentrations were chosen from range values shown by the previously mentioned literature. The infiltration factors selected for the application were: 0.8 for indoor micro-environments, 0.9 for transit, 1 for outdoor. Each micro-environment and each population group were associated to two sets of time profiles, inferred from data collected during a survey by the Italian National Statistical Institute ISTAT [9] and from the EXPOLIS project over the city of Milan. In both cases, the only available data is the total average amount of time spent in a specific activity or environment, without further specification on which hours of the day are dedicated to each. As a consequence, the profiles were arbitrarily set up making reasonable assumptions on which hours to assign to each microenvironment, and then making sure that the total number of hours matched the available data. In Figure 1 and Figure 2 the profiles are represented by a set of rectangles with varying height: a height of 1 indicates that in the corresponding hour of the day the population class resides in microenvironment "indoor home", 2 indicates "indoor work/school/other", 3 "transit", 4 "outdoor".

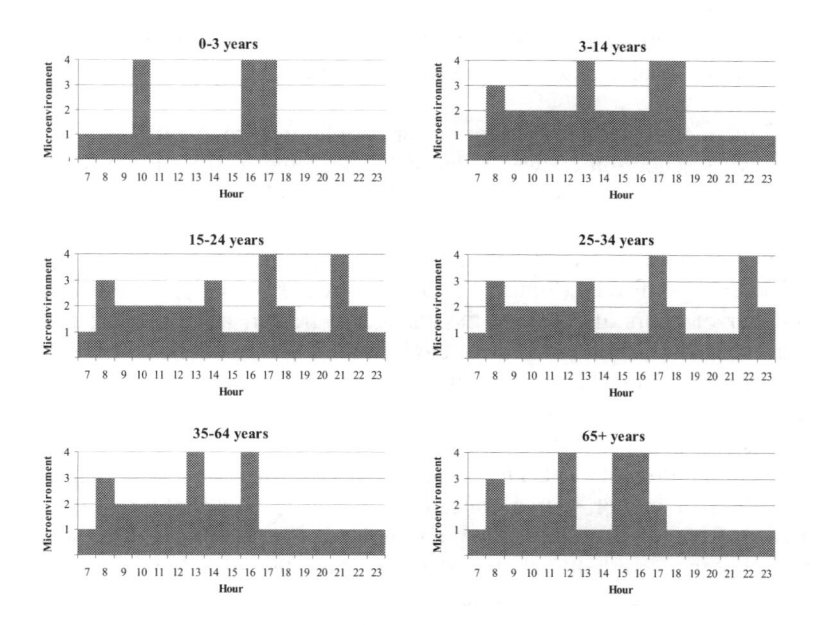

Figure 2:   Time profiles from ISTAT data.

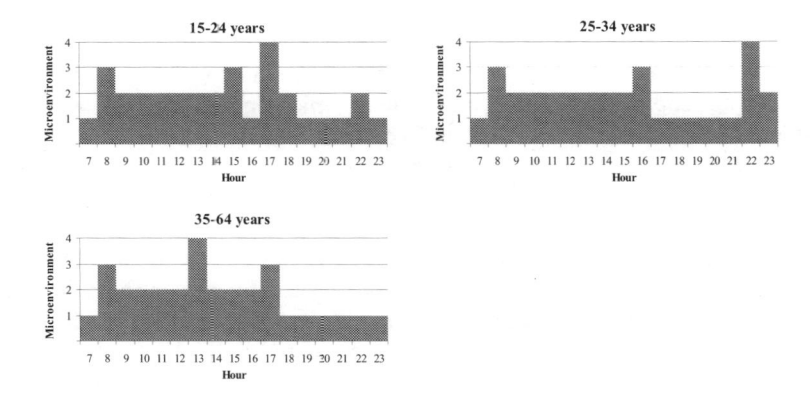

Figure 3: Time profiles from EXPOLIS data, only for the classes that differ from ISTAT.

5 Results

The tool was first fed with the ISTAT time profiles, obtaining a base case result to use as a reference for the following tests. By modifying the profiles according to the EXPOLIS data, the resulting overall exposure reduces. This is reasonable considering that the amount of time spent outdoors is higher according to ISTAT, and that in the absence of indoor sources, as supposed in this case, this will lead to higher levels of average exposure. The two tests were then repeated under different assumptions:
- two indoor sources were introduced: a 50 µg/m^3 source between 7 and 8 pm for cooking, and a 20 µg/m^3 source between 7 and 8 am for personal care.
- the infiltration factor for microenvironment 2 (indoor work/school/other) was reduced to the value of 0.5, in the hypothesis that homes are less insulated than workplaces, schools and public places.

The reduction pattern of the EXPOLIS-driven exposure values comparing to the ISTAT ones is confirmed (Figure 4), with the greatest variations occurring when the number of hours affected by the change is greater, as in the last test. The exposure distribution pattern is quite close to the PM10 modelled outdoor yearly average concentration (Figure 5); this follows from the assumption that has been made, i.e. the same population habits have been applied to the whole domain and no internal dynamics – shifts occurring from one concentration area to another, as when commuting – have been considered. In order to set up more detailed test cases, values of indoor source concentration due to household activities have been taken from Wallace [16]. They are probably not well related to the Italian lifestyle, but they can help one to understand the effect of indoor sources on the total exposure. A total of 12 model runs divided into three groups have been performed. For each group, only one kind of source was separately activated in the first three tests, while on the fourth all of them were taken into account in calculating the exposure (Table 1).

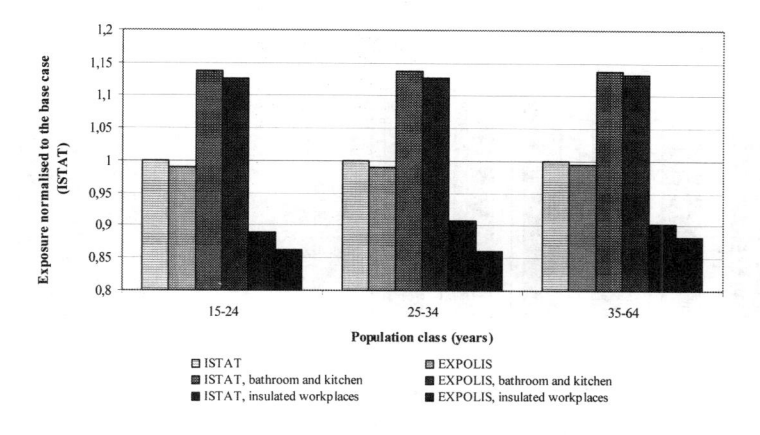

Figure 4: Population class average exposure, normalised to the base case.

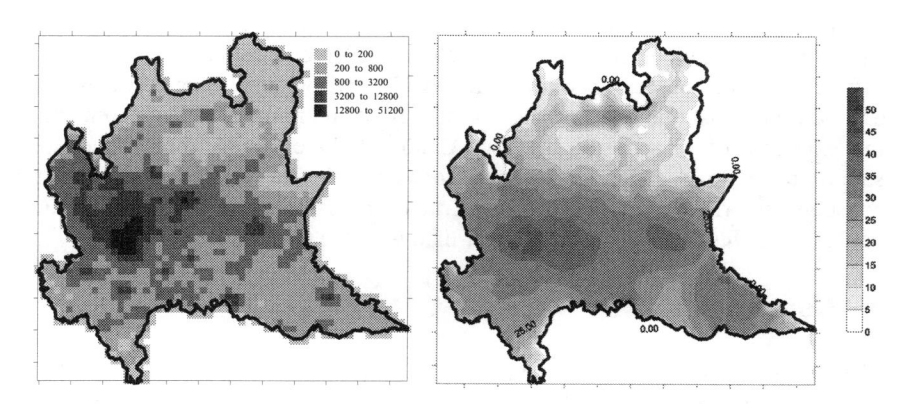

Figure 5: Population distribution and PM10 average exposure values over the study domain for class 3 under base case conditions.

Table 1: Indoor source tests.

Indoor source	TEST A	TEST B	*TEST C*
Cooking	50 µg/m³ 7 – 8 pm for all classes	Same as test A, 50% concentration reduction	*50 µg/m³ 7-8 am and 7-8 pm for all classes*
Personal care	20 µg/m³ 7 – 8 am for all classes	Same as test A, 50% concentration reduction	*20 µg/m³ 7-8 am and 9-10 pm for all classes*
Household cleaning	30 µg/m³ in the morning working hours for classes 4 and 5	Same as test A, 50% concentration reduction	*30 µg/m³ 8-9 am for all classes*

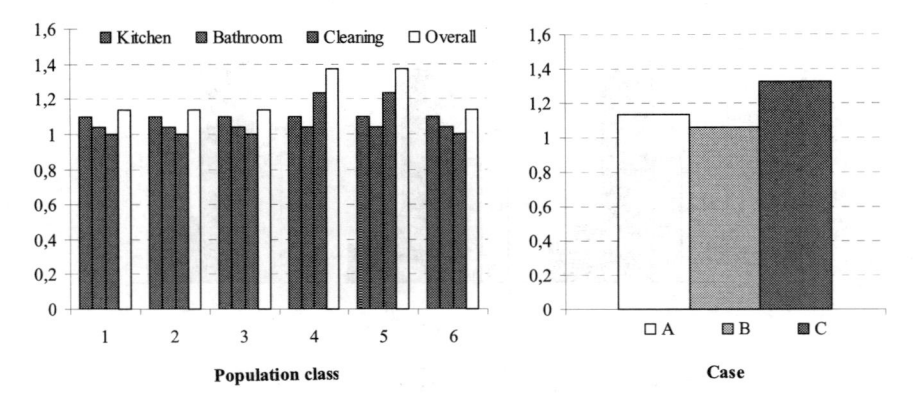

Figure 6: Household activities indoor sources: exposure normalised to the base case. Left: case A runs. Right: cases A, B, C for class 6, with all sources activated.

Figure 6, on the left, shows that the exposure values (normalised to the base case, where no indoor sources were included) correctly reflect the presence of different indoor sources. Cooking ("Kitchen") and personal care ("Bathroom") activities take just one hour, causing only a slight and homogeneous increment in the exposure of all classes. In case A, household cleaning hours have been supposed to span a longer interval for class 4 and 5, and the exposure consequently rises more for the concerned classes, even if the concentration value due to this indoor source is lower than that due to other activities.

6 Conclusions

The proposed results give insight into indoor source contribution to PM10 exposure. The developed tool proved itself sensible to the variations introduced into the input data. The need for more detailed information on time profiles and indoor sources is evident, as these factors control the dependency of exposure on outdoor measured or modelled concentration. It must be said that the required information on people habits and home characteristics is expected to vary depending on the country/region where the tool is going to be applied; for this reason a great effort is needed in order to obtain quality data from local measuring campaigns and population surveys.

Acknowledgments

This work has been supported by the APAT (National Environmental Protection Agency). The authors would thank Tiziana Forte and Luciana Sinisi from APAT and Angelo Giudici from ARPA Lombardia for their useful suggestions, and Pierantonio Gusmini for his precious support to the project.

References

[1]   Abt, E., Suh, H.H., Catalano, P. & Koutrakis, P., Relative contribution of outdoor and indoor particle sources to indoor concentrations. *Environmental Science & Technology*, **34 (17)**, pp. 3579–3587, 2000.

[2]   Brunekreef, B., Janssen N. A. H., de Hartog J. J., Oldenwening M., Meliefste K., Hoek G., Lanki T., Timonen K. L., Vallius M., Pekkanen J. & Van Grieken R., Personal, Indoor, and Outdoor Exposures to PM2.5 and its components for groups of cardiovascular patients in Amsterdam and Helsinki. HEI Health Effects Institute Research Report, **127**, 2005.

[3]   Burke, J. M., A population exposure model for particulate matter: case study results for PM2.5 in Philadelphia, PA, 2001, Online http://epa.gov/heasd/pm/pdf/exposure-model-for-pm.pdf

[4]   Carmichael, G. R., Uno I., Phandis M. J., Zhang Y. & Sunwoo, Y., (1998) Tropospheric ozone production and transport in the springtime in east Asia. *Journal of Geophysical Research J. Geophys. Res.*, **103**, pp. 10649–10671, 1998.

[5]   Dockery, D.W. & Spengler, J.D., Indoor–outdoor relationships of respirable sulfates and particles. *Atmospheric Environment* 1, **15**, pp. 335–343, 1981.

[6]   FUMAPEX Integrated Systems for Forecasting Urban Meteorology, Air Pollution and Population Exposure – Refined and validated population exposure models, work package 7: Population exposure modelling, Deliverable D7.1 2003 Online http://fumapex.dmi.dk/.

[7]   Georgopoulos, P.G., Jayjock, E., Sun, Q., Isukapalli, S.S., Kevrekidis, P.G., Lazaridis, M. & Drossinos, Y., Mechanisms controlling the outdoor/indoor relationships of fine particle levels and characteristics Final Technical Report CERM: 2002-3 Prepared under United States Environmental Protection Agency (USEPA) STAR Grant R826768, 2002, Online http://ccl.rutgers.edu/reports/cerm

[8]   Hänninen, O. O., Lebret E., Ilacqua V., Katsouyanni K., Kunzli N., Sram R.J & Jantunen M., Infiltration of ambient $PM_{2.5}$ and level of indoor generated non-ETS PM2.5 in residence of four European cities *Atmospheric Environment*, **38**, pp.6411–6423, 2004.

[9]   ISTAT Indagine multiscopo sulle famiglie "Aspetti della vita quotidiana". Anno 2005, Online www.istat.it/dati/catalogo/20070406_01/.

[10]   Jantunen, M.J., Katsouyanni, K., Knoppel, H., Kunzli, N., Lebret, E., Maroni, M., Saarela, K., Sram, R. & Zmirou, D., EXPOLIS Air Pollution Exposure in European Cities: the EXPOLIS study – Final Report, 2003, Online www.kt.fi/expolis.

[11]   Koutrakis, P., Briggs, S. L. K. & Leaderer, B.P., Source apportionment of indoor aerosols in Suffolk and Onondaga Counties, New York. *Environmental Science and Technology*, **26**, 521–527, 1992.

[12]   Monn, C., Exposure assessment of air pollutants: a review on spatial heterogeneity and indoor/outdoor/personal exposure to suspended

particulate matter, nitrogen dioxide and ozone, *Atmospheric Environment*, **35**, pp. 1–32, 2001.
[13] Ozkaynak, H., Xue, J., Spenler, J., Wallace, L., Pellizzari, E. & Jenkins, P., Personal exposure to airborne particles and metals: results from the particle TEAM study in Riverside, California. *Journal of Exposure Analysis and Environmental Epidemiology*, **1**, pp. 57–78, 1996.
[14] Seigneur C., Review of mathematical models for health risk assessment: VI. Population exposure, *Environmental Software*, **9**, pp. 133-145, 1994.
[15] Silibello, C., Calori, G., Brusasca, G., Giudici, A., Angelino, E., Fossati, G., Peroni, E. & Buganza, E., Modelling of PM10 concentrations over Milano urban area: validation and sensitivity analysis of two aerosol modules, *Environmental Modelling & Software*, **23 (3)**, pp. 333–343, 2008.
[16] Wallace, L., Indoor particles: a review. *Journal of Air and Waste Management Association*, **46**, pp.98–126, 1996.
[17] Wallace, L., Williams, R., Rea, A. & Croghan, C., Continuous weeklong measurement of personal exposure and indoor concentrations of fine particles for 37 health-impaired North Carolina resident for up to four seasons, *Atmospheric Environment*, **40**, pp. 399–414, 2006.
[18] Weisel, C.P., Zhang, J, Turpin, B.J., Morandi, M.T., Colome, S., Stock, T.H. & Spektor, D.M., Research Report: Relationships of Indoor, Outdoor, and Personal Air (RIOPA). HEI Research Report **130**, NUATRC Research Report **7**, 2005, Online www.sph.uth.tmc.edu.

Air pollution and management in the Niger Delta – emerging issues

M. A. Fagbeja[1,2], T. J. Chatterton[1], J. W. S. Longhurst[1], J. O. Akinyede[2] & J. O. Adegoke[3]
[1]Air Quality Management Resource Centre, University of the West of England, Bristol, UK
[2]National Space Research and Development Agency (NASRDA), Abuja, Nigeria
[3]University of Missouri - Kansas City (UMKC), Missouri, USA

Abstract

This paper considers the various sources of air pollution in the Niger Delta and reviews some of the possible management strategies available to the authorities. The identified sources include burning of fossil fuels for transportation and industrial power generation, use of fuel wood and kerosene for domestic cooking and lighting, and gas flaring and are a function of urbanisation and industrialisation in an oil rich environment. The level of pollution due to gas flaring is considered. Gas flaring in the region is identified as one of the main sources of both CO_2 and CO, and it is expected to be a major contributor to NOx, and NMVOC concentrations in the Niger Delta. Challenges confronting air quality and carbon management in Nigeria are highlighted. Finally, an introduction to the National Space Research and Development Agency (NASRDA) funded research on air quality and carbon management in the Niger Delta is presented.
Keywords: air pollution, Niger Delta, gas flaring, remote sensing, GIS.

1 Introduction

Pollution is defined by the European Union 1996 Council Directive on Integrated Pollution Prevention and Control (IPPC) as "the direct or indirect introduction as a result of human activity, of substances, vibrations, heat or noise into the air, water or land which may be harmful to human health or the quality of the environment, result in damage to material property, or impair or interfere with amenities and other legitimate uses of the environment" [1]. Inorganic and organic air pollutants cause negative health and environmental effects such as respiratory ailments, premature deaths. Air pollution-related deaths worldwide are estimated to be up to 2 million per annum [2]. Other environmental consequences of air pollution include acidification of soil and water and loss of plant and animal life.

Air quality assessment studies in Nigeria have focused mainly on urban centres where industrial processes, domestic activities and traffic congestion constitute major sources of air pollution [3–5]. Most of these studies are independent as there are no systematic measurements of air quality by the government [3]. Measurements from selected dumpsites, industrial estates and heavy traffic stations in Lagos revealed that average concentrations of carbon-monoxide (CO) in heavy traffic stations was 49.32ppm, while at industrial estates was 36.75ppm and at dumpsites, 10.76ppm. Sulphur dioxide (SO_2) averages were 0.166ppm at the traffic stations, and 0.670ppm levels were detected at both industrial and dumpsites. The NO_X concentrations were 0.220ppm at the dumpsites and 0.333ppm at both industrial and traffic stations [3]. WHO standards for CO, SO_2 and NOx are 5ppm for 8-hour average, 0.45ppm for 24-hour average and 0.25ppm for 24-hour average respectively [6]. These results clearly suggest that there are air quality problems arising from transportation, industrial and domestic activities and waste disposal in Nigeria. Another identified source of significant air pollution in Nigeria is gas flaring [7], which occurs mostly in the rural communities of the Niger Delta region of Nigeria, where majority of the oil and gas exploration takes place. However, there is little currently known about the impact of gas flaring due to a paucity of emissions measurements in the region. Considering the existence of gas flaring in the region since the 1950s when oil and gas exploration commenced in the region, it is expected that many studies should have been carried out on the extent of pollution from the activity. However, this is not the case. Shell Petroleum Development Corporation (SPDC) in its 2006 Annual report [8] stated that the oil industry submitted a proposal to the oil industry regulatory body, the Department of Petroleum Resources (DPR) for air quality assessment in the Niger Delta only in 2006. This effort to assess air quality in the Niger Delta came after five decades of oil exploration, and gas flaring, in the region.

The Niger Delta is located in the southern part of Nigeria. The region suffers from human and environmental issues of both national and international concern, in terms of the environmental pollution; impoverishment of the local people despite the wealth being generated from the region; security of human lives; property and infrastructure due to militancy; community agitations and youth unrest. This paper gives a brief description of the Niger Delta environment. It also highlights the major anthropogenic activities resulting in air pollution in the Niger Delta which include transportation, burning of fossil fuels for industrial and domestic use and waste disposal. Whilst the focus of the paper is on air pollution related to emissions from oil production, with specific emphasis on gas flaring, the challenges of air quality management in Nigeria as a whole are considered.

2 The Niger Delta

The Niger Delta is formed as a result of the splitting up of the 4,100km-long Niger River [9] into estuaries through which it flows into the Atlantic Ocean at the Gulf of Guinea. There have been various definitions of the size of the Niger Delta based on hydrological, ecological and political boundaries [8, [10–13]. These definitions give an estimated range of the area covered by the Niger Delta from between 19,100km^2 and 30,000km^2. For this paper, the Niger Delta is defined by the area covered by oil producing states in Nigeria [10]. As such, the Niger Delta comprises of nine states out of the thirty-six states of the Nigerian Federation, namely Abia, Akwa-Ibom, Bayela, Cross-River, Delta, Edo, Imo, Ondo and Rivers. In view of this, the Niger Delta covers an estimated area of 110, 445.98km^2 (figure 1). The Niger delta is a coastal lowland that consists of ecological zones of both vast fresh water swamps and mangrove forests, covering estimated areas of 11,700km^2 and 6,000km^2 respectively [11]. The region is rich in crude petroleum and natural gas reserves, one which makes it the twelfth richest petroleum province in the world [14]. Nigeria's oil and gas reserves are estimated to be 35.9 billion barrels and 185 trillion cubic feet respectively, making them the largest oil and gas reserves in Sub-Saharan Africa [15].

Figure 1: Map showing the approximate area defined as Niger Delta [10, 12] Source: Map derived from NASRDA data.

3 Air pollution in the Niger Delta

Air pollution in the Niger Delta has been on the increase since the commencement of oil and gas exploration in the region. The underlying factors responsible for air pollution in the Niger Delta are industrialisation, a general high rate of urbanisation that is unaccompanied with commensurate economic growth in Africa [16]; and poverty. The high rate of urbanisation with the attendant low growth is responsible for the level of urban poverty and inadequacies of infrastructure or amenities such as power and clean fuel for domestic use in African urban centres including those in the Niger Delta. The United Nations Human Development Report for 2007-2008 indicated that 92.4% of Nigerians lived below the income poverty line of $2 a day [17]. In the Niger Delta region, 66.1% of the population self-classify themselves as poor [18].

3.1 Sources of air pollution in the Niger Delta

Due to lack of infrastructure for providing clean energy to homes and poverty, majority of both the rural and urban populace in the Niger Delta depend on the use of fuel wood and kerosene for domestic cooking and lighting. This results in indoor air pollution and also contributes to ambient air pollution [19]. NOx and CO from these sources result in an estimated 79,000 deaths a year in Nigeria [20]. The Nigerian National Bureau of Statistics (NBS) identifies 70.0% of Nigerians rely on firewood for cooking. 26.6% depend on the use of kerosene or oil, while 0.52% and 1.11% use electricity and gas respectively [21].

Burning of fossil fuels for transportation in the inland urban and rural centres of the Niger Delta is another major source of air pollution in the region. The majority of Nigerians can not afford new vehicles due to prohibitive costs and, as a result, depend on vehicles which are older than the majority of cars used in Europe. Such vehicles can emit five times more hydrocarbons and carbon monoxide and four times more nitrogen oxides than vehicles commonly found in developed countries [22].

Due to the presence of the oil industry in the Niger Delta, as well as the region's proximity to the ocean, many industries are located in the main urban centres of Port Harcourt, Warri and Aba. In addition, there is oil and gas related activities in many rural, riverine areas where most of the oil fields are located. Nigeria achieved a 46% electrification rate between 2000 and 2005, with an estimated 71.1 million people out of the estimated total 2005 population of 141.4 million without access to electricity [17]. Electricity supply in Nigeria has dropped even lower in recent times due to poor management and vandalism of existing power generation, transmission and distribution infrastructure. Consequently, most industries in the Niger Delta depend on the use of diesel-fuelled generators. Some rural communities also depend solely on use of an industrial generator for electricity supply, due to non-connection to the national grid. For example, in Burutu, a riverine community in Delta State, the local government administrator provided an industrial generator to supply electricity to the community between 6pm and 6am daily. The majority of the community dwellers could not afford to buy personal generators.

The contributions of household and industrial emissions of particulates, NOx and non-methane volatile carbon (NMVOC) have been assessed in selected cities and state of the Niger Delta by Ajao & Anurigwo (2002) [4]. The results show that Port Harcourt city accounts for a high concentration of particulates, NOx and NMVOC. The emissions from Port Harcourt city are greater than the total emissions from an entire state (Delta State) (Table 2). This is an indication of the relatively high level of urbanisation and industrialisation of the city. The projected population of Port Harcourt for 2008 is 1,899,372 [23]. Together with

Table 1: Total air emissions from industry and households in Port Harcourt, Delta State and Calabar. Source: [4].

Location/Source	Particulates (tonnes yr^{-1})	NOx (tonnes yr^{-1})	NMVOC (tonnes yr^{-1})
Port Harcourt - Trans Amadi Industrial Area, including NAFCON and refinery, exclusive of solvents. - Industries and households	10,496 Not available	779 Not available	292 3,775
Delta State - Industries in Delta - Solvents, small industries and households in Warri	760.84 1,535.5	384.41 Not available	1,047.7 Not available
Calabar - Industries - Solvents, small industries and households	35.43 1,593.3	20.83 Not available	29.93 Not available

Warri, Port Harcourt serves as the operational base for most of the multinational and local oil companies in Nigeria.

3.2 Pollution from gas flaring in the Niger Delta

Since June 1956 when commercial oil exploration started at Oloibiri in Bayelsa State, the gas associated with the oil production has been flared. In 1970, 99% of the gas produced in Nigeria was flared (figure 2). This reduced to an estimated 72% in 1997, even though the volume steadily increased between 1970 and 1996 [24]. Due to efforts by the Nigerian government at utilising the associated gas through the development of liquefied natural gas (LNG) plant in Bonny, supply to industries and electricity generation, the volume of gas flared in Nigeria reduced to approximately 23.0 million cubic meters (812.24 billion cubic feet (BCF)), representing 39% of total gas produced in 2004/2005 [18].

In Nigeria, gas flaring contributed an estimated 12.7 million metric tonnes of carbon in 2004 through CO_2 emissions. This represents 41% of the estimated total CO_2 emissions in Nigeria [25]. In 1995, the total CO emission from Nigeria was 21.42Tg/yr CO, with gas flaring in the Niger Delta being the third main contributor after combustion of bio-fuels and agriculture. Gas flaring contributed estimated 2.49Tg/yr CO representing 12% of total CO emissions in Nigeria in 1995 [26]. Other pollutants from gas flaring include sulphur dioxide, non-methane volatile organic carbons (NMVOC), nitrogen oxides and methane. The quantities of emissions of these pollutants from gas flaring in Nigeria are yet to be ascertained due to unavailability of data. It is, however, expected that gas flaring in the Niger Delta will be the major single contributor of the emissions of these pollutants into the Nigerian atmosphere.

4 Challenges of air pollution management in Nigeria

The Nigerian government identifies emissions of CO_2 and other green house gases (GHGs) that contribute to climate change as an environmental problem [27]. However, the report did not include air quality as part of the environmental concerns for the Nigerian government. This could explain the unavailability of consistent emissions inventory for the country. Although there has been some independent research into air quality assessment in parts of Nigeria [5], more needs to be done by the Nigerian regulatory bodies to systematically enforce regulations aimed at improving air quality in Nigeria. The Federal Environmental Protection Agency (FEPA) was established under the Amended decree No. 59 of 1992 in the Laws of the Federation of Nigeria to undertake the following among other things [28]:

- prepare a comprehensive national policy for the protection of the environment and conservation of natural resources, including procedures for environmental impact assessment for all development projects;
- prepare, in accordance with the National Policy on the Environment, periodic master plans for the development of environmental sciences and technology and advise the Federal Military Government on the financial requirements for the implementation of such plans;
- promote co-operation in environmental science and conservation technology with similar bodies in other countries and with international bodies connected with the protection of the environment and the conservation of natural resources;
- co-operate with Federal and State Ministries, Local Governments, statutory bodies and research agencies on matters and facilities relating to the protection of the environment and the conservation of natural resources.

Consequently, FEPA established the National Air Quality Standards (NAQS) in Nigeria in 1991 (table 3) [29].

Some of the practices employed by FEPA and the Nigerian Federal Ministry of Environment (FMEnv) are the imposition of emission taxes, and the enforcement of emission abatement control mechanisms by industries. For example, the Federal Government of Nigeria recently increased gas flaring tax on oil companies in the Niger Delta. The new tax regime is $3.50 (about £1.75) on every 1,000 standard cubic feet of gas flared, representing an estimated 4,000% increase from the previous charge of about $0.08 (about £0.04), which had been in effect since the early 1960s [30]. In 1996, FEPA closed down an iron and steel company in Lagos for failing to implement measures to reduce emissions from its furnace [31]. However, enforcement has been one of the key issues confronting air pollution control in Nigeria. Many industries contributing to air pollution, such as the oil refineries and National Fertiliser Company

(NAFCON) are being run by the government, although in recent times, efforts are being made to privatise them, with the Government still holding some percentage of the shares. These companies often escape sanctions from government regulatory agencies and as a result, there is limited consideration of pollution control measures in these industries.

Table 2: National Air Quality Standards (NAQS) in Nigeria. Source: [29].

Pollutants	Ambient Limits	Limit from stationary sources (For 24 hrs)
Particulates	250µg/m³ (Daily average of daily values 1 hour)	0.15-0. mg/ m3
Sulphur Dioxides (SO₂)	**0.01ppm** (26µg/m³) 0.1 ppm (260 µg/m³) (Daily average of hourly values 1 hour)	0.05 – 0.5mg/ m³
Non-methane hydrocarbon	**160µg/m³** (Daily average of 3-hourly values)	2.0 – 5.0mg/ m³
Carbon monoxide	**10ppm** (11.4 mg/m³) 20 ppm (22.8µg/m³) (Daily average of hourly values 8-hours)	1.0 – 5.0mg/ m³
Nitrogen Oxides (NO2)	**0.04ppm - 0.06 ppm** (75.0–113 mg/m³) Daily average of 1-hourly values (range)	0.004 – 0.1mg/ m³
Photochemical Oxidant	0.06ppm (Hourly values)	5133.0ppm

NAFCON produces SO_2 as a by-product. The company allows sulphur wastes to be emitted as SO_2 instead of converting the waste to a more useful sulphuric acid, since the company escapes payment of charges or taxes on the emissions [32].

Another major limitation to air pollution and carbon management in Nigeria is the lack of data on concentrations of the major air pollutants and greenhouse gases (GHGs) in Nigeria. From the World Data Centre for Greenhouse Gases (WDCGG) website, Nigeria has no monitoring stations that contribute to data on concentrations of GHGs (CO_2, CH_4, CFCs, N_2O and surface O_3) and other air pollutants (CO, NOx, SO_2, and VOCs). The WDCGG collects data under the World Meteorological Organisation's (WMO) Global Atmospheric Watch (GAW) and other programmes [33]. This clearly suggests that the regulatory ministry and agencies in Nigeria – the Federal Ministry of Environment (FMEnv), Federal Environmental Protection Agency (FEPA) and the Nigerian Meteorological Agency (NIMET) – do not measure concentrations of air pollutants and GHGs on a systematic and consistent basis that would enable a nationwide assessment of the air quality situation in the country upon which proper legislation and efforts will be based at achieving clean air in Nigeria. There may be independent (e.g. oil industry) and/or research-based (universities and research institutions) measurements, but these are not readily available to the general public. This limitation necessitates and justifies the consideration of the use of satellite data in conjunction with independent in-situ measurements of air quality and carbon parameters for this research in order to apply remote sensing and GIS techniques to air quality and carbon management, a case study of gas flaring in the Niger Delta. The research is being funded by the Nigerian National Space Research and Development Agency (NASRDA).

5 Conclusions

Air pollution in the Niger Delta region is an issue that requires attention by the Nigerian authorities and operators of the oil and gas industry because it has been occurring for a long time without proper control and management. The Nigerian authorities need to fully explore the potentials for gas utilisation in Nigeria. Efforts should be intensified to ensure that hitherto flared gas is used to provide adequate power generation for the nation. Gas pipelines should be laid across the major industrial areas of Nigeria to supply gas for the running of industries. This will reduce industrial dependence on burning of liquid and solid fossil fuels for energy and ensure the use of cleaner fuels for running industrial operations in Nigeria. In addition, there is need for the development of a robust monitoring and management system, which ensure that high quality information on the extent and impact of air pollution can be used as the basis for legislation to curtail the pollution and develop a mechanism that will enhance clean air when gas flaring ends in Nigeria. The Nigerian government intends to phase out gas flaring in the Niger Delta in the near future. In view of this, the Nigerian National Space Research and Development Agency (NASRDA) is funding the research titled, *"Applying remote sensing and GIS techniques to air quality and carbon management, a case study of gas flaring in the Niger Delta."* The research, which commenced in January 2008 at the University of the West of England, aims to integrate in-situ measurements of ambient concentrations and emissions with satellite remote sensing data to assess air quality emissions and CO_2 concentrations resulting from gas flaring in the Niger Delta. The available satellite technology resources at NASRDA will combine with the European expertise in air quality studies available at UWE to proffer solutions to air pollution and air quality management in the Niger Delta.

References

[1] IPPC, Council Directive 96/61/EC of 24 September 1996 concerning integrated pollution prevention and control. Official Journal L 257, 10/10/1996 P. 0026 – 0040. 1996. Online: http://eur-lex.europa.eu/LexUriServ/LexUriServ.do?uri=CELEX:31996L0061:EN:HTML.
[2] World Health Organisation, WHO, Reducing risks, promoting healthy life. The World Health Report 2002. Geneva. 2002. Online. http://www.who.int/whr/2002/en/whr02_en.pdf.
[3] Taiwo, O., The case of Lagos – air quality improvement project. Presentation for the Lagos Metropolitan Area Transport Authority. 2005 Online. www.cleanairnet.org/ssa/1414/articles-69320_Taiwo.pdf.

[4] Ajao, E.A. & Anurigwo, S., Land-based sources of pollution in the Niger Delta, Nigeria. Ambio, **31**, pp. 442–445. 2002.
[5] Baumbach, G., Vogt, U., Hein, K.R.G., Oluwole, A.F., Ogunsola, O.J., Olaniyi, H.B., and Akeredolu, F.A., Air pollution in a large tropical city with high traffic density – results of measurements in Lagos, Nigeria. The Science of the Total Environment, **169,** pp. 25–31. 1995.
[6] World Bank, Air Quality Standards. 1995 Online. http://www.worldbank.org/html/fpd/em/power/standards/airqstd.stm#who
[7] Isuwa, S., Nigeria: transportation, major cause of air pollution. Report of the Director General of National Environmental Standards and Regulations Enforcement Agency (NESREA) at a seminar on WHO Air Quality Guidelines. 2008. Online: www.allafrica.com/stories/200804040160.html.
[8] Shell Petroleum Development Corporation, SPDC, People and the environment. Shell Nigeria Annual Report 2006. Shell Visual Media Services, London. 2006.
[9] FAO, Irrigation potentials in Africa: a basin approach. FAO Land and Water Bulletin – 4. FAO Land and Water Development Division, Rome. 1997.
[10] United Nations Development Programme (UNDP), Niger delta human development report. UNDP, Abuja, Nigeria 2006.
[11] Moffat, D. & Linden, O., Perception and realities: assessing priorities for sustainable development in the Niger River delta. Ambio, **24**, pp. 527–538. 1995
[12] ERML, Environmental and socio-economics characteristics of the Niger Delta. Niger Delta Environmental Survey Report. 1997
[13] Abam, T.S.K., Regional hydrological research perspectives in the Niger Delta. Hydrological Sciences, 46, pp. 13–25. 2001
[14] Klett, T.R., Ahlbrandt, T.S., Schmoker, J.W., and Dolton, J.L., Ranking of the world's oil and gas provinces by known petroleum volumes. U.S. Geological Survey Open-file Report-97-463. 1997.
[15] Library of Congress, Federal Research Division, Country Profile: Nigeria. Published June 2006, pp. 12–13, 21–22. 2006.
[16] Hicks, F.J., Enhancing the productivity of urban Africa. International Conference on Research Community for the Habitat Agenda Forum of Researchers on Human Settlements. Geneva., July 6-8, 1998. Online. http://wbln0018.worldbank.org:80/External/Urban/UrbanDev.nsf/Urban+Development/06FCBF7D26772E6885256B18007BA532?OpenDocument.
[17] United Nations Development Programme (UNDP), UNDP human development report 2007/2008 – Fighting climate change: human solidarity in a divided world. UNDP, New York. 2007. Online. http://hdr.undp.org/media/hdr_20072008_en_complete_pdf.
[18] National Bureau of Statistics, (NBS), The Nigerian statistical fact sheets on economic and social development. National Bureau of Statistics, Abuja. 2006.

[19] Anozie, A.N., Bakare, A.R., Sonibare, A.J., and Oyebisi, T.O., Evaluation of cooking energy cost, efficiency, impact on air pollution and policy in Nigeria. Energy, **32**, 1283–1290. 2007.
[20] World Health Organisation (WHO), Country profiles on environmental burden of disease. Geneva, 2007. Online: http://www.who.int/quantifying_ehimpacts/countryprofilesafro.pdf
[21] National Bureau of Statistics (NBS), Poverty profile for Nigeria. Abuja. 2005.
[22] United Nations Environment Programme (UNEP), Global environmental outlook 2000. Nairobi. (2000). Online: http://www.unep.org/Geo/geo2000.
[23] World Gazetteer, Nigeria: largest cities and towns and statistics of their population. 2008. Online: http://www.world-gazetteer.com
[24] Ukoli, M.K., Environmental factors in the management of the oil and gas industry in Nigeria. 2001. Online: http://www.cenbank.org/OUT/PUBLICATIONS/OCCASIONALPAPERS/RD/2001/OWE-01-2.PDF
[25] Marland, G., Boden, T. and Andres, R. J., *Global, regional, and national CO_2 emissions. Trends: A Compendium of Data on Global Change.* Carbon Dioxide Information Analysis Centre, Oak Ridge National Laboratory, U.S.A 2007. Online. http://cdiac.ornl.gov/ftp/ndp030/nation.1751_2004.ems
[26] Global Emissions Inventory Activity, GEIA, Source grouping in country summary tables. 2004. Online: http://www.mnp.nl/geia/data/Carbon_Monoxide/Documentation.jsp
[27] Adeyinka, M.A., Bankole, P.O., and Olaye, S., Environmental Statistics: situation in Federal Republic of Nigeria. Country report presented at the Workshop on Environmental Statistics, Dakar, Senegal, from 28th February – 4th March 2005.
[28] Federal Republic of Nigeria, Constitution of the Federal Republic of Nigeria 1999. Online: http://www.nigeria-law.org/Federal%20Environmental%20Protection%20Agency%20(Amendment)%20Decree%20No.%2059%201992.htm
[29] Federal Environmental Protection Agency (FEPA), National interim Guidelines and Standards for Industrial effluents, Gaseous emissions and Hazardous wastes. Environmental Pollution Control Handbook. Lagos, FEPA. 1991.
[30] Ibiyemi, S., FG approves N410/scf gas-flaring fee. Financial Standard Newspaper, 13 May 2008. Financial Standard, Lagos. 2008. Online. http://www.financialstandardnews.com/
[31] Izeze, E.M., FEPA to inspect CISCO. *The Guardian Newspapers*, 22nd April 1996. Lagos. 1996
[32] Uchegbu, S.N., Environmental management and protection. Precision Printer and Publishing, Enugu. 1998.
[33] World Meteorological Organisation, WMO, WMO Global Atmospheric Watch: World Data Centre for Greenhouse Gases (WDCGG). 2001. Online: http://gaw.kishou.go.jp/wdcgg/

Air quality monitoring and management for the industrialized Highveld region of South Africa

G. V. Mkhatshwa
Eskom Research and Innovation Department, South Africa

Abstract

The rapid economic growth in developing countries has led to increases in energy consumption patterns and industrial activity leading to air quality degradation. Due to the rapid industrial expansion in South Africa, there has been a great demand for increased electrical generation. About 89% of electricity generated in the Southern African Development Community (SADC) region is from coal, mostly produced in South Africa. Eskom generates 95% of the electricity used in South Africa and has undertaken extensive air quality monitoring on a regional basis since the 1970s. The aim of the monitoring is to assist Eskom in the assessment of environmental impacts related to its activities and to facilitate appropriate control strategies. The South African Highveld is impacted by emissions from motor vehicles, heavy and light industry, power stations, coal mining and numerous large townships and informal settlements. This paper presents air quality monitoring and management trends adopted by Eskom for the past 20 years over the Highveld region of South Africa and how these have changed with time. An overview of air quality monitoring methods is presented and ambient air quality trends over a minimum of a 10-year period at specific monitoring stations are discussed. Generally ambient concentrations of pollutants such as sulphur dioxide (SO_2), nitrogen oxides (NOx) and fine particulate matter (FPM) have been decreasing with time in the Highveld region. This can be attributed mainly to control strategies adopted by Eskom and to the electrification of households.
Keywords: Eskom, electricity, power stations, Highveld, air quality, monitoring, Elandsfontein, Palmer, Leandra, pollutants.

1 Introduction

Since the 1970s Eskom has undertaken extensive air quality monitoring on a regional basis. The aim of this monitoring is to assist Eskom in the assessment of environmental impacts related to its activities and to facilitate appropriate control strategies (Rorich and Galpin [1]).

Although the highest levels of air pollution at ground level are still found in informal settlements in South Africa, it is the contribution that industry makes to the already polluted air that is a contested issue. Industrial and power station emissions influence a much greater area, however, and the detrimental effects of the pollutants on ecosystems have yet to be properly assessed. Industrial activities in the Mpumalanga Highveld introduce a wide range of anthropogenic primary pollutants into the atmosphere. During coal combustion processes carbon dioxide, carbon monoxide, hydrocarbons, nitrogen oxide and nitrogen dioxide, sulphur dioxide and soot are released into the atmosphere (Eskom [2]). Other major source categories of anthropogenic gases and aerosols over the Highveld are metallurgical processes, biomass burning, mining activity, smouldering dumps and domestic combustion.

The economy of the Highveld region (figure 1) is based primarily on mining, agriculture, forestry and tourism and about 44% of all commercial plantations in South Africa are in this region. The economics of locating power stations near coalmines has dictated that 78% of Eskom's electricity energy comes from the Highveld region (Eskom [3]). Atmospheric pollution on the South African Highveld is perceived as a concern because of the combination of heavy industry, motor vehicle emissions, domestic fuel combustion and climatic features that prevail in the region. The frequent occurrence of surface inversions (80-90% of days in winter months) permits the accumulation of pollutants near ground level (van Tienhoven [4]). Although industrial stacks and those of power stations in particular, are generally able to emit gaseous and particulate pollutants over the boundary layer, looping and fumigation of plumes may occur under convective conditions.

2 Air quality monitoring trends in the Highveld

The most heavily industrialized area in South Africa is the Central Mpumalanga (Highveld), which is impacted by emissions from motor vehicles, heavy and light industry, power stations, coal mining and numerous large townships and squatter settlements. Eskom has, for the past two decades, undertaken extensive ambient air quality monitoring in this region and has data sets dating as far back as the early 1980s.

Ambient air quality monitoring is conducted by the air quality team in Eskom's Research and Development Department (ERID). Routine station visits are done on a two weekly basis. During the visits, data that was stored on the data logger during the previous two weeks of monitoring is downloaded onto a laptop computer and returned to Eskom, where it is transferred onto a central

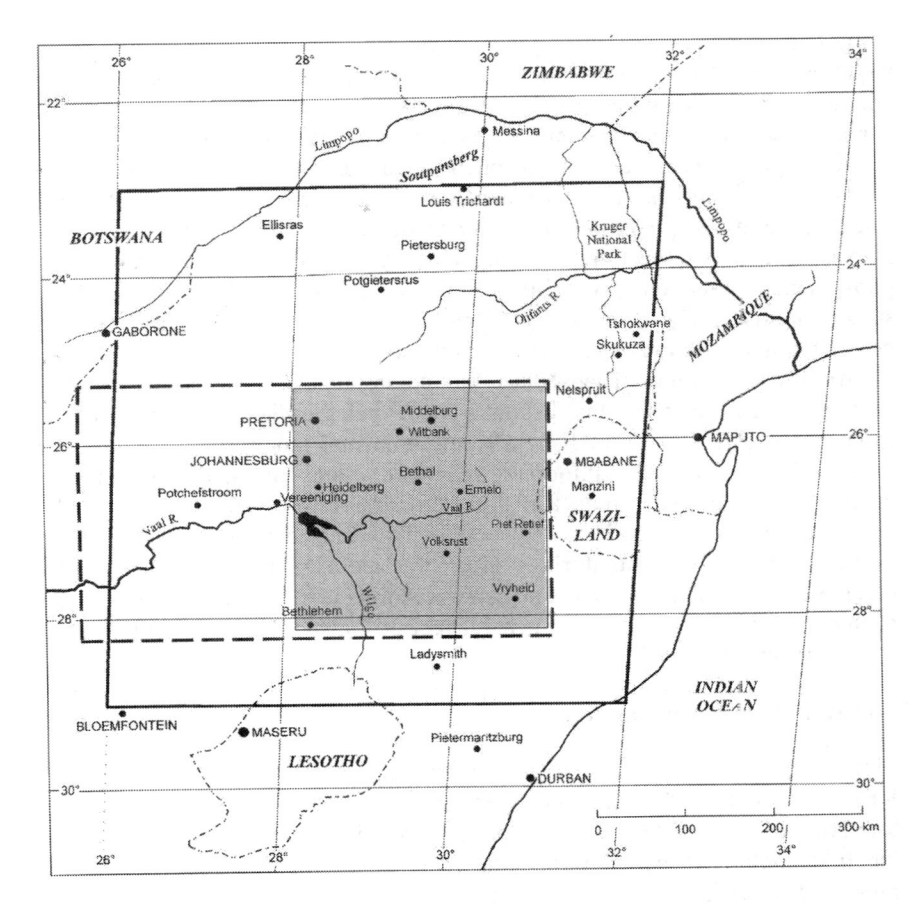

Figure 1: Highveld region (shaded area) as defined in Held *et al.* [5].

computer for processing, editing and verification. The processing, editing and verification process is carried out by expert assessment using the site logbook and site notes. After verification, copies of both the raw data and verified data are archived for report generation and future reference.

Analyzer zero and span checks and dynamic calibration audits are carried out on each of the analyzers at routine intervals and the discrepancies logged, these calibration checks can then be used to assist in data verification and validation at Eskom. This is possible as the analyzers are corrected to the audit concentrations after the initial checks if required. The air quality monitoring station internal temperature and mains voltage are continuously monitored and logged. These parameters are not monitored for compliance but for data verification purposes only.

Eskom recognized the importance of credible data and assisted local authorities in establishing accreditation of laboratories and monitoring networks in South Africa through the South African National Accreditation System (SANAS). Eskom uses the SANAS guideline figure of 80% per parameter monitored as a standard for representative data capture. This describes the required completeness of the data set for the reporting of averages, which is defined as 80%. This figure is based on standard arithmetic calculations. The completeness calculations for data sets exclude zero and span data and times where service and/or maintenance is being conducted on the instruments in question.

Data availability is a management definition related to system reliability. The availability target is not set in terms of data quality criteria and has no associated quality objectives. A target of 90% availability has been set for performance evaluation. Availability is reported per station as a measure of total station uptime. The measure is an average of the uptime for all variables monitored in the station but excludes quality control check data (Eskom [6]).

An Environmental metadatabase for environmental research datasets has been established to ensure that Eskom's data is being utilized to its full potential by presenting the history of Eskom's environmental monitoring activities. The metadatabase provides information about data available within the environmental sciences group and gives descriptive information on the quality and characteristics of air quality data. This metadatabase gives detailed information on the monitoring sites; the pollutants and meteorological parameters measured at each site; site performance; annual percentage data capture; collection methods and general feel of logistics relating to the running of each monitoring site.

2.1 Monitoring progression

Monitoring of Eskom's ambient air quality began in 1978 using 2 Mobile units for monitoring SO_2, the data was stored on magnetic tape; in 1979 Eskom set up 9 remote stations for SO_2 using strip chart recorders and in the very same year the first conference paper introducing monitoring work done by Eskom was presented. In 1983, 22 remote stations for SO_2 using PC logging facilities were set up and from 1984 onwards selective introduction of NOx nephelometers and ozone monitors has been done at almost all Eskom monitoring sites (Turner [7]). In 1989 the monitoring network was rationalised and reduced to 5 research stations with additional pollution parameters being monitored.

2.2 Ambient air quality trends

Long term trends of SO_2, NOx and FPM (PM10) have been investigated at 3 sites in the Highveld; Elandsfontein (considered to be of prime importance to regional impact studies and pollution trend analysis); Palmer (a rural site aimed at monitoring and assessing the impacts of plumes from power stations on forest plantations in the region); and Leandra (set up to determine corresponding reduction in air pollution as a result of the electrification of a township).

2.2.1 Elandsfontein

Elandsfontein monitoring station was commissioned in May 1985. Due to its central location in the Highveld and the large amount of historical data, this site is considered to be of prime importance to regional impact studies and pollution trend analysis. Long-term trend analyses of sulphur dioxide (figure 2) indicate a slight increase in concentration levels over the past twenty years covering the south west to northern wind sectors. Concentration levels of nitrogen oxides levels recorded at Elandsfontein have been decreasing with time. The reduction in NOx concentrations is possibly due to the mothballing of older power stations in central Mpumalanga. Fine particulate matter levels recorded at Elandsfontein indicate increasing levels of particulates due to increases in domestic combustion largely due to low cost housing developments and increased squatter areas in the region.

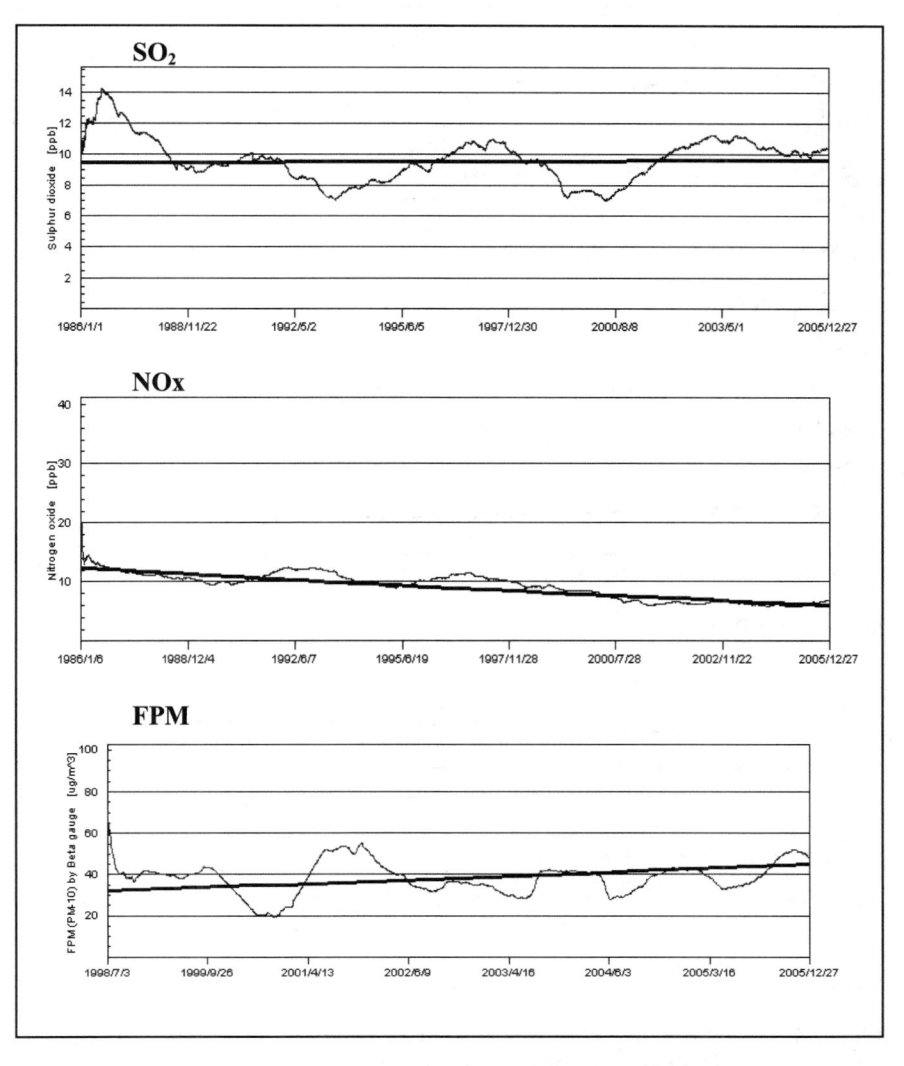

Figure 2: Elandsfontein long term trends from all wind sectors (1986–2005).

2.2.2 Palmer

Air quality monitoring at Palmer monitoring station began in 1989 for 15 years. This site was set up specifically to monitor the impact of plumes from central Mpumalanga on the eastern Escarpment forest plantations. Concentrations of SO_2 and NOx levels have been decreasing with time (figure 3) and increases of FPM were recorded at Palmer as was seen at Elandsfontein. FPM at Palmer only began in 1997; 8 years after monitoring of other pollutants began.

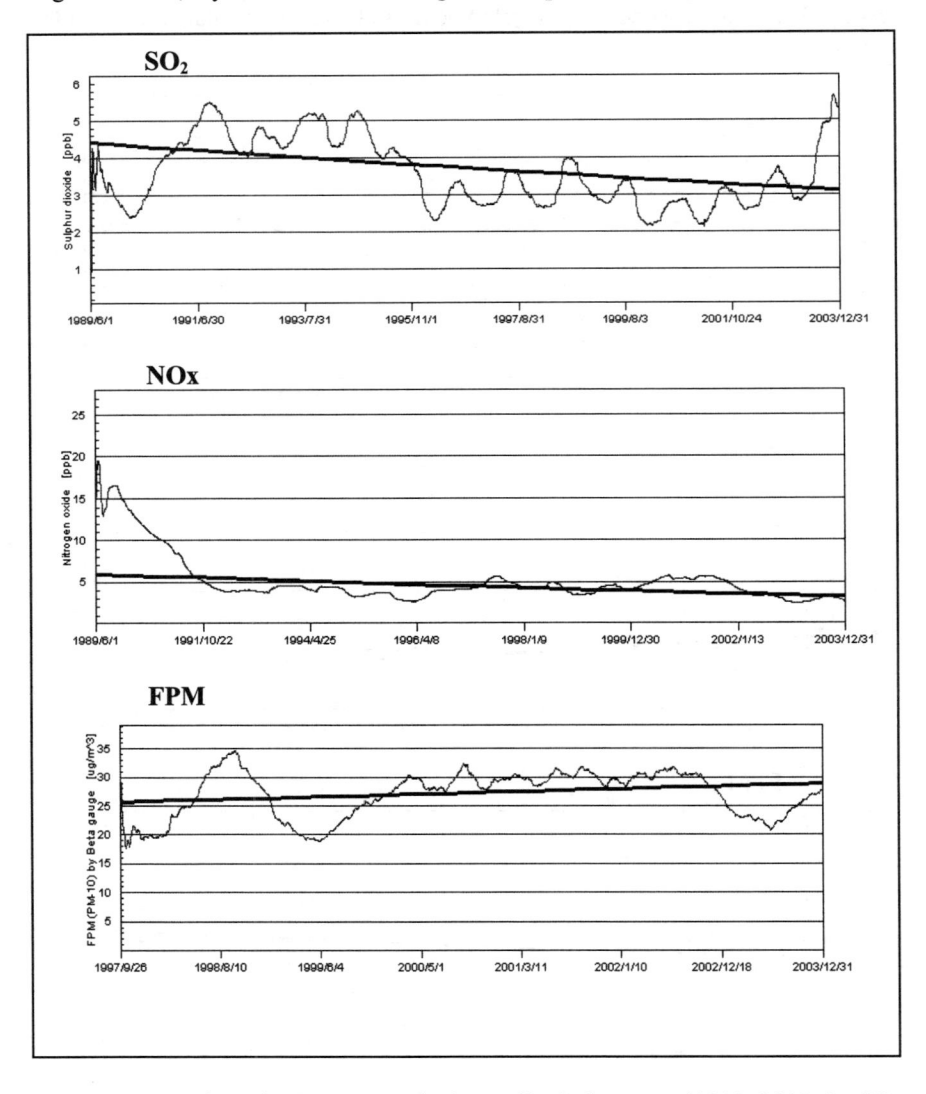

Figure 3: Palmer long term trends from all wind sectors (1989–2003 for SO_2 and NOx) and 1997–2003 for FPM.

2.2.3 Leandra

Monitoring at Leandra has been ongoing since 1995. This site was located downwind of a township to monitor changes in pollution levels with the introduction of electricity to the area. Electrification of households in Leandra and possibly the use of more efficient methods of domestic fuel usage, like the "basa njengo magogo" initiative in neighbouring townships, have resulted in decreased ambient SO2 and FPM concentrations in this area (figure 4).

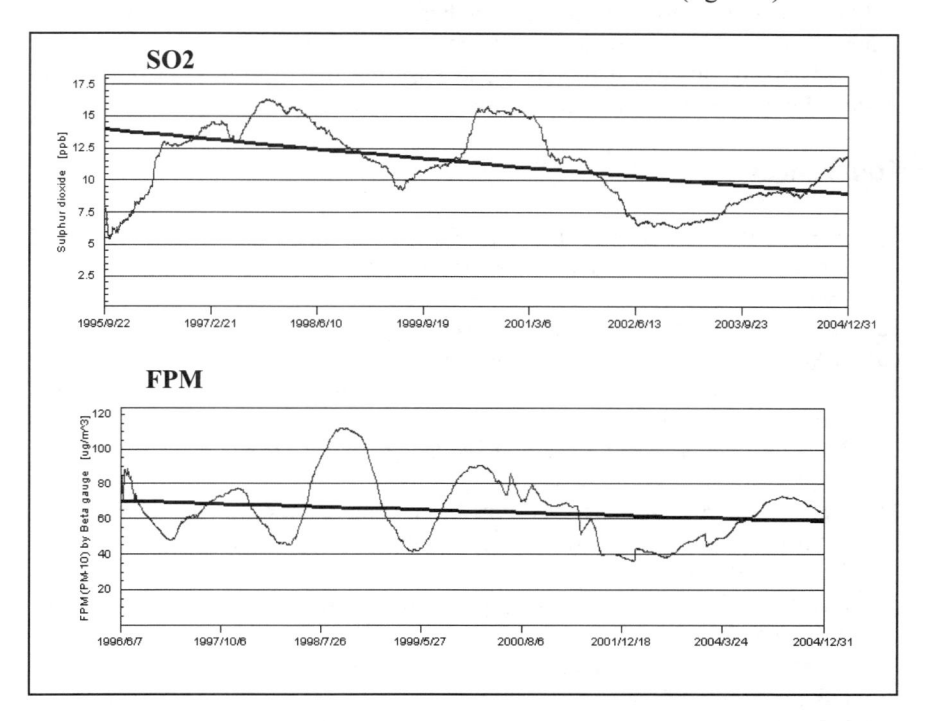

Figure 4: Leandra long term trends from all wind sectors (1996–2004).

3 Air pollution control strategies

Eskom's approach to emission control technology was previously centred on particulate emission management. In 1988 Eskom implemented a major emission reduction program aimed at retrofitting appropriate technologies to reduce emissions at existing plant. This reduced total emissions by 88% whilst in the same period generation of electricity increased by 48%. Eskom's operating units have been retrofitted with Pulse Jet Fabric Filters, and 66% of the remaining Electrostatic Precipitators have had their performance improved by flue gas conditioning. Low NOx boilers have been introduced on all Eskom power stations.

Research has strengthened and complimented this program and also kept cognizant with emerging technologies. Several research projects have and continue to focus on the reduction of gaseous emissions using technologies that are best suited to the local environment and to resource availability. It is planned that any future expansion will require the reduction of sulphur dioxide, nitrogen oxides in addition to particulate removal. Another principle is that only proven and reliable technologies will be utilised and in many cases this has required up front research, testing and modelling in order to demonstrate the effectiveness of the technology (Hansen [8]).

Eskom's current air quality strategy is based on 3 principles; ensuring that Eskom's emissions to atmosphere, whether from power station stacks or from other sources, are managed to acceptable levels; ensuring that Eskom complies with all legally binding regulations and ensuring that Eskom's operations are sustainable in the short, medium and long term (Turner [9]).

References

[1] Rorich, R.P. & Galpin, J.S., Air Quality in the Mpumalanga Highveld region, South Africa. *South African Journal of Science*, 94, pp 109–114, 1998
[2] Eskom power stations, www.eskom.co.za.
[3] Eskom 2002 Environmental report, Johannesburg, www.eskom.co.za.
[4] van Tienhoven, A. and Fey, M.V., Air pollution impacts on local soil properties near a power station in South Africa. *Proceedings of the 11th World Clean Air and Environment Congress*, Durban, 13–18 September 1999
[5] Held, G., Schiefinger, H. & Snyman, G., Recirculation of pollutants in the atmosphere of the South African Highveld, *South African Journal of Science*, 90, pp 91–97, 1994
[6] Eskom Annual ambient air quality data report. Johannesburg, South Africa, www.eskom.co.za.
[7] Turner, C.R., Personal communication, 4 April 2007, Corporate Consultant, Eskom, South Africa.
[8] Hansen, R.S., Personal communication, 29 March 2007, Independent consultant, Air pollution control technologies, South Africa.
[9] Turner, C.R., air quality significant achievements. *Proceedings of the 1st ERID air quality gap analysis workshop*, Midrand, 4–6 April 2007.

Real time air quality forecasting systems for industrial plants and urban areas by using the MM5-CMAQ-EMIMO

R. San José[1], J. L. Pérez[1], J. L. Morant[1] & R. M. González[2]
[1]*Environmental Software and Modelling Group,*
Computer Science School, Technical University of Madrid (UPM),
Campus de Montegancedo, Boadilla del Monte 28660 Madrid, Spain
[2]*Department of Meteorology and Geophysics, Faculty of Physics,*
Complutense University of Madrid; Ciudad Universitaria,
28040 Madrid, Spain

Abstract

During the last five years, we have been working intensively with the new generation of air quality modeling systems such as MM5-CMAQ (PSU/NCAR/EPA, US) and using our own emission model, EMIMO (UPM, 2006), the first version of which was developed several years ago and is in the process of continuous adaptation. Recently we have incorporated an adapted version of the CFD model MIMO (University of Karlsruhe, 2000) which includes a sophisticated cellular traffic model, CAMO (UPM, 2006), to model street level pollution with the numerical Eulerian models approach with a few meters of spatial resolution. The integrated system formed by an adapted version of MIMO and CAMO is called MICROSYS. The mesoscale air quality model MM5-CMAQ-EMIMO produces air concentrations in real-time and forecasting mode for urban and regional areas with 1 km spatial resolution. The MICROSYS model is implemented to simulate the air concentrations in one 1 km grid cell with 5–10 m spatial resolution and up to 200–300 m in height (over the maximum building heights in the 1 km grid cell). The MICROSYS model is run in diagnostic mode and uses the boundary and initial conditions from MM5-CMAQ-EMIMO. The system MM5-CMAQ-EMIMO-MICROSYS produces reliable air quality forecasts for urban areas with street level detail over the Internet. In this contribution we will show the special example applied to Las Palmas (Canary Islands, Spain). Additionally, the MM5-CMAQ-EMIMO has been used to find the air quality impact of several industrial plants such as combined cycle power and cement plants. The system uses sophisticated cluster technology to take advantage of distributed and shared memory machines in order to perform the parallel runs in an efficient and optimal way since the process should operate on a daily basis. We will also show the methodology and results of these applications in two industrial complexes in the surrounding area of Madrid City.

1 Introduction

There has been an increased interest in managing and controlling the air quality in regional and urban areas during the last five years. The interest in understanding and knowing the air quality over large domains and with high temporal and spatial resolution is growing rapidly together with the interest in knowing the relative quantitative and qualitative impacts on a specific area (grid) and time. The advances in air quality modeling have been substantial in the past decade. The third generation of air quality modeling systems use the so-called "one-atmosphere" approach which means that the dispersion and chemical transformation of the different pollutants emitted to the atmosphere is treated in an integrated and unique way. The first approaches – first and second generation – are related to the treatment of point emissions insolated from the surrounding atmosphere. The third generation of air quality modeling systems is representing the reality that occurs in the transport and transformation of chemical pollutants in a more realistic way. The further development of these models will move to integrate the effects of water and other ecosystems and feedback effects on the atmosphere and vice versa. The increase in computer power generated in the last 20 years has contributed substantially to this parallel advance in knowledge and efficiency.

The information technology progress has played an essential role in this spectacular advance in air quality modeling systems. Since the computer power required to run the complex FORTRAN codes which are developed to incorporate the complexity of the atmospheric dynamics is phenomenal, the technology involved to carry out complex air quality impact assessments and furthermore the real-time and forecasting application is quite important.

The cluster approaches open new scenarios for many applications and particularly in the atmospheric dynamics simulations. The atmospheric models have also reached highly sophisticated levels that include the simulation of the aerosol processes and cloud and aqueous chemistry. These models include sophisticated land use information and deposition/emission models [10]. The atmospheric models traditionally include two important modules: a) meteorological modeling and b) transport/chemistry modules. These two modules work in a full complementary mode, so that the meteorological module provides full 4D datasets (3D wind components, temperature and specific humidity) to the transport/chemistry modules. CPU time is mainly used for transport/chemistry (75%). This modeling system requires important initial and boundary data sets to simulate properly specific time periods and spatial domains, such as landuse data, digital elevation model data, global meteorological data sets, vertical chemical profiles and emission inventory data sets. In this experiment we have used AVN (NCEP/NOAA, USA) global meteorological information as input for the MM5 meteorological model. The emission inventory for the proper spatial domain and for the specific period of time (at high spatial and temporal resolution) is possibly the most delicate input data for the sophisticated meteorological/transport/chemistry models. The accuracy of emission data is much lower than the accuracy of the numerical methods used for solving the partial differential equation systems (Navier–Stokes equations) for meteorological models [5] and the ordinary differential equation system for the chemistry module [9–11]. Typical uncertainty associated to emission data is 25–50%. However, in our application it is more important to see the relative impact of the industrial emissions in the mesoscale domain – where the tested industrial plant is located – than to quantify and qualify the absolute pollutant concentrations in the atmosphere due to the emission uncertainty.

The emission inventory is a model that provides in time and space the amount of a pollutant emitted to the atmosphere. In our case we should quantify the emissions due to traffic, domestic sources, industrial and tertiary sector and also the biogenic emissions in the three model domains with 9 km, 3 km and 1 km spatial resolution. The mathematical procedures to create an emission inventory are essentially two: a) Top-down and b) Bottom-up. In reality a nice combination of both approaches offers the best results. Because of the high non-linearity of the atmospheric system, due to the characteristics of the turbulent atmospheric flow, the only possibility to establish the impact of the part of the emissions (due to traffic or one specific industrial plant, for example) in air concentrations, is to run the system several times, each time with a different emission scenario.

Examples of "state-of-the-art" meteorological models are: MM5 (PSU/NCAR, USA), RSM (NOAA, USA), ECMWF (Redding, U.K.), HIRLAM (Finnish Meteorological Institute, Finland), etc. Examples of "state-of-the-art" of transport/chemistry models – also called "third generation of air quality modeling systems" – are: EURAD (University of Cologne, Germany), [13], EUROS (RIVM, The Netherlands), (Langner et al. [7]), EMEP Eulerian (DNMI, Oslo, Norway), MATCH (SMHI, Norrkoping, Sweden), [2], REM3 (Free University of Berlin, Germany), [14], CHIMERE (ISPL, Paris, France), [12], NILU-CTM (NILU, Kjeller, Norway), [3], LOTOS (TNO, Apeldoorm, The Netherlands), [8], DEM (NERI, Roskilde, Denmark), [4], STOCHEM (UK Met. Office, Bracknell, U.K.), [1]. In USA, CAMx Environ Inc., STEM-III (University of Iowa) and CMAQ model are the most up-to-date air quality dispersion chemical models. In this application we have used the CMAQ model (EPA, U.S.) which is one of the most complete models and includes aerosol, cloud and aerosol chemistry.

The OPANA system is an air quality modelling system developed in the 90's by the Environmental Software and Modelling Group of the Computer Science School of the Technical University of Madrid (ESMG-FI-UPM). The system is used in forecasting and historical modes. The system includes different state-of-the-art mesoscale meteorological and chemical transport models such as MM5 and/or CMAQ. In the 90's the meteorological module was the MEMO model (University of Karslruhe (Germany), 1994) and the chemistry was included on-line in the meteorological model by using the CBM-IV scheme [6,7]. Further versions of the model included MM5 and CMAQ as part of the meteorological module and chemistry transport module. The system includes and emission model, EMIMO, which has also different version which includes biogenic and anthropogenic emissions with different updated version of the global or

European emission database. Actually, OPANA V4 includes a sophisticated CFD code (based on the MIMO model (University of Karslruhe (Germany), 1996)), which runs in diagnostic mode over a 1 km x 1 km model domain over highly dense populated areas in cities (Madrid, Las Palmas de Gran Canaria, etc.). The CFD code is called MICRFOSYS and has a resolution 1-10 m receiving traffic emission data from a sophisticated cellular automata model (CAMO) – also developed by the ESMG-FI-UPM. This CFD code receives the initial and boundary conditions from the mesoscale part of the model (typically MM5-CMAQ-EMIMO). OPANA V3 is used in several forecasting applications for urban and industrial sites but it does not include the CFD part. The CMAQ model is implemented in a consistent and balanced way with the MM5 model [4]. The CMAQ model is fixed "into" the MM5 model with the same grid resolution (MM5 grid cells are used at the boundaries for CMAQ boundary conditions). The system can be implemented in any domain over the world. As an example a domain architecture is shown in Figure 1. MM5 is linked to CMAQ by using the MCIP module which is providing the physical variables for running the dispersion/chemical module (CMAQ) such as boundary layer height, turbulent fluxes (momentum, latent and sensible heat), boundary layer turbulent stratification (Monin-Obukhov length), friction velocity, scale temperature, etc. We have run the modeling system (MM5-CMAQ) with USGS 1 km landuse data and GTOPO 30'' for the Digital Elevation Model (DEM). The system uses EMIMO model (EMIssion MOdel) to produce every hour and every 1 km grid cell the emissions of total VOC's (including biogenic), SO_2, NO_x and CO. This model uses global emission data from EMEP/CORINAIR European emission inventory (50 km spatial resolution) and EDGAR global emission inventory (RIVM, The Netherlands). In addition the EMIMO model uses data from DCW (Digital Chart of the World) and USGS land-use data from AVHRR/NOAA 1 km satellite information. The EMIMO model includes a biogenic module (BIOEMI) developed also in our laboratory based on the algorithms for natural NO_x, monoterpene and isoprene emissions in function of LAI (leaf Area Index) and PAR (photosintetic active radiation).

2 Results

Different applications have been carried out over different domains and emission sources. We show different applications: (1) Different simulations to know the air quality impact of a combined cycle power plant in Madrid area; (2) Similar application than in Madrid (Spain) area but in Andalusia (Spain); (3) A study of the impact of the emissions of an incinerator in the Basque Country (Spain) and (4) A real-time air quality forecasting system for Las Palmas de Gran Canaria (Canary Islands, Spain).

2.1 Combined cycle power plant in Madrid domain

In this section we show results for an application over Madrid domain designed for a specific study of the impact of a future power plant construction for different years. Several studies of this type have already been conducted at different areas in the Iberian Peninsula for different industrial type plants as mentioned above. In Figure 1 we showed the scheme designed for the study in the Madrid domain. Similar architecture has been used for different areas. Figure 2 shows the comparison between observed and modelled data in the Alcalá monitoring station in Madrid Community (Spain) for the year 2005. We observe that the comparison is excellent, particularly if we consider that the maximum concentration in this particular period of time (July, 9–15, 2005) is quite important. Values near 200 µgm^{-3} are found for 12th of July 2005. The modelled data and observed data are particularly satisfactory. On 14th of July, during the night time the model reproduces values higher than observed.

Figure 3 shows the impact at 15:00 GMT on July 25, 2006 for ozone concentrations. The map is produced by making the differences between ON and OFF scenarios, so that the only difference between both scenarios is the air concentrations produced as consequence of the emissions of the power plant. The nonlineal process in the atmosphere produces increases (4%) and decrease (28,5%) in the air concentrations. The changes are important hour per hour, so that, a real-time air pollution system is considered essential in order to control the important changes due to the atmospheric dynamics and chemical process.

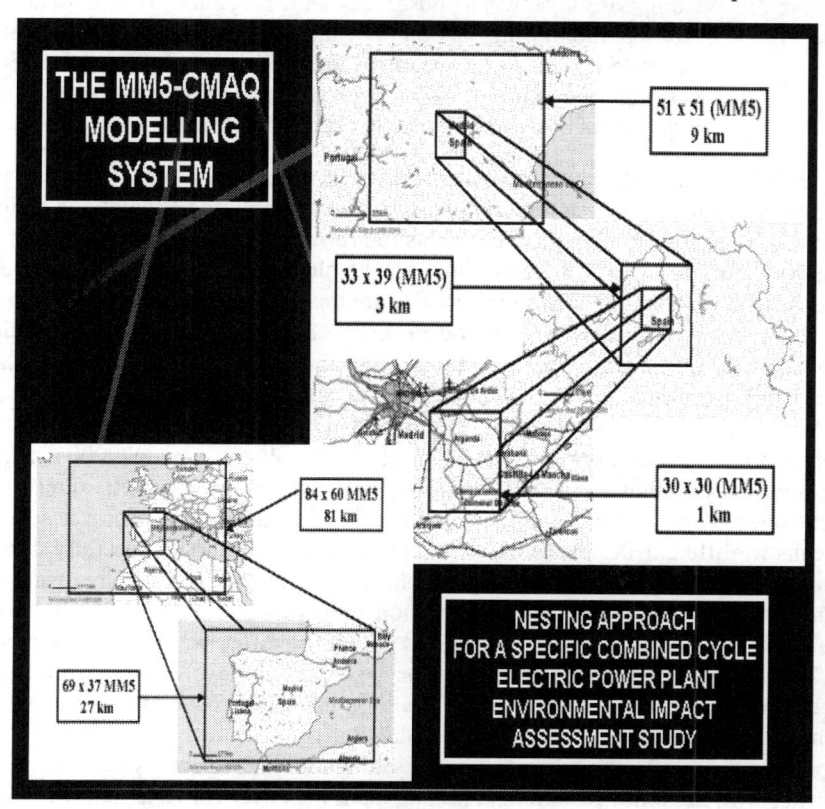

Figure 1: MM%_CMAQ-EMIMO modelling architecture for use in forecasting and historical modes.

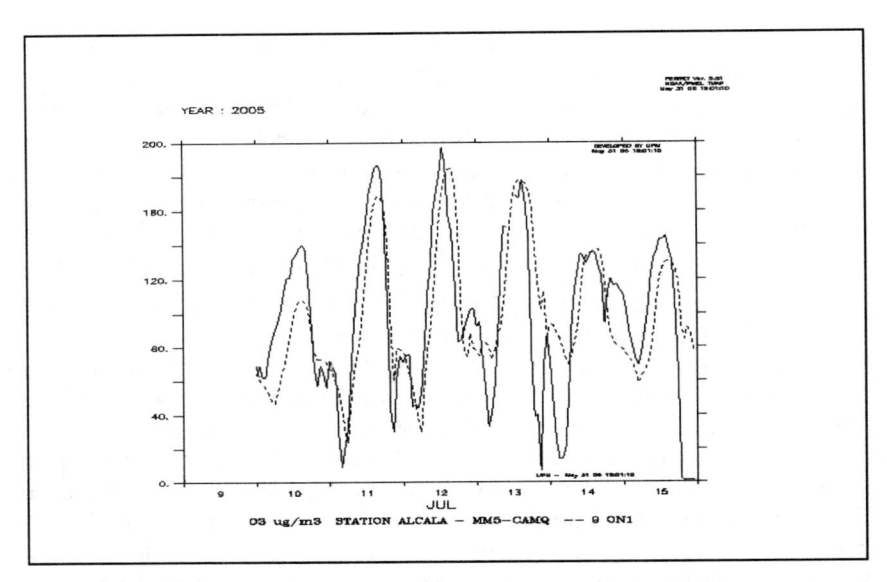

Figure 2: Comparison between observed and modelled ozone data for Alacalá monitoring station in Madrid Community area (located about 35 km east of the Madrid metropolitan area). The comparison between both data sets is particularly good taking into account the high values observed during such a period of time.

These results show an excellent agreement between observations and modeling results in the calibration phase (before running the simulations adding the emissions from the planned industrial power plant). This agreement is essential for the reliability of the final results although the differences between the concentrations in ON and OFF modes are the most important relative results in these types of studies. We should underline that the amount of information obtained for a typical air quality impact study of an industrial and power plant for 120 hours periods along 12 month a year and for five criteria pollutants, 3 different nesting levels (9 km, 3 km and 1 km) produces an amount of information (every hour analysis) of about 5 Gbytes and 400000 images (examples are shown in this contribution). The whole system should be controlled by the corresponding scripts running in automatic mode over several weeks in different PC platforms. In real-time mode we should carefully design our architecture (generally over a cluster platform) and assure that the simulations of ON, OFF and all emission reduction scenarios (X%) run on a daily basis for a period of 120 hours and obtain the differences between ON and X% runs with OFF mode to obtain the best performance emission reduction scenario for the next 48–72 hours. The X% emission reduction scenarios are simulated by applying this emission reduction over the last 48–72 hours. This operational architecture requires – as we said – cluster platforms. Our tests over a cluster with 20 nodes (2,4 GHz.) and one main PC (with 2,4 GHz) show a speed increase of about 10 times. This test was performed at a cluster in the University of Iowa (USA).

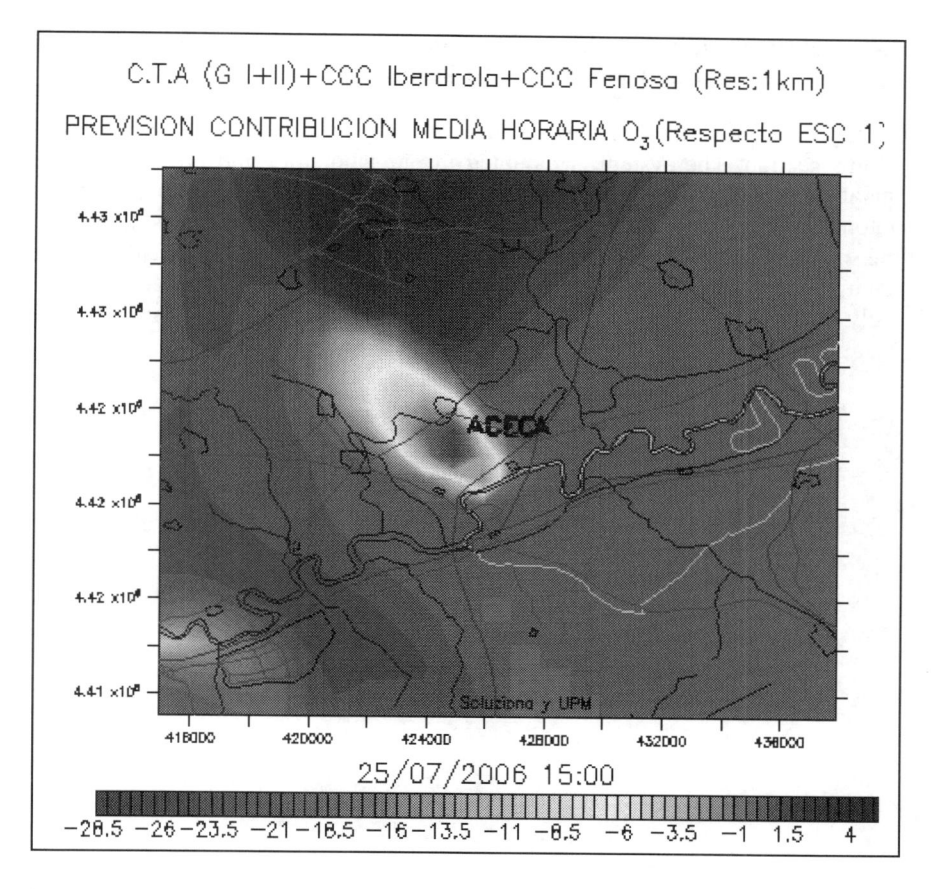

Figure 3: The map shows the impact of the emissions of the power plant by making the differences between ON and OFF scenarios for ozone on July 25, 2006 at 15:00.

2.2 Combined cycle power plant in Andalusia (South of Spain)

Similar cases have been performed for air quality impact assessment studies in Andalucia (Huelva area)

2.3 Incinerator in Basque Country (North of Spain)

In this particular air quality impact study we had to modify the CMAQ model to implement a dioxin/furna models together with the implementation of metals and B[a]P. Figure 4 shows an example of the calibration process in Basque Country (Spain). The comparison shows a good pattern between both data sets.

2.4 Real-time and forecasting application: urban application in Las Palmas de Gran Canaria (Canary Islands, Spain)

Finally, a real-time and forecasting application by using MM5-CMAQ is shown in this section. The system is mounted in our laboratory and provides the air quality forecasts through the Internet under daily basis by using a specific script automatic programme. In Figure 5 we observe the internal web presentation for the city of Las Palmas de Gran Canaria, which is accessed internally by the environmental experts in the Municipality of Las Palmas de Gran Canaria under daily basis. The model and the web interface are located in our laboratory in Madrid (Spain).

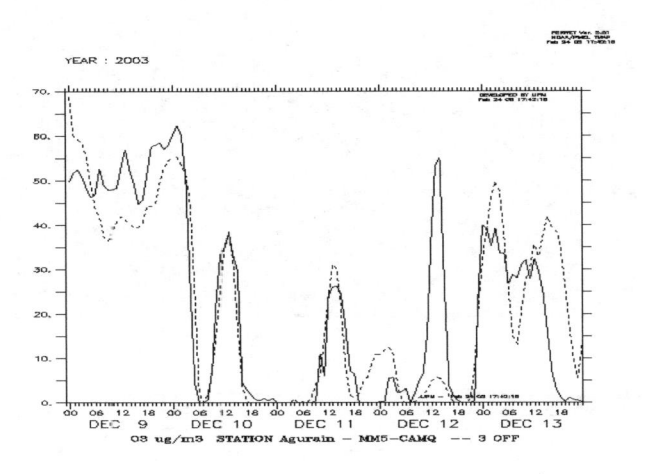

Figure 4: Comparison between O_3 observed and modeled data for Agurain monitoring station (Basque Country, Spain).

3 Conclusions

In this contribution we have shown several applications and studies by using the sophisticated MM5-CMAQ modeling system. The system has been proved to be very robust and reliable. The results assure us that it is possible to have in real-time and forecasting mode tools over the Internet that provide air quality impact forecasts for different industrial plants and urban areas and take emission reduction actions on time. Further work is currently under development to determine the best strategy to identify the best emission reduction strategies based on air quality forecasts.

In the case of industrial plants the complete switch off of the emissions for a period of 24–48 hours is the best possible solution assuming that the impact of the emissions of the industrial plant is the main cause of exceedance of the EU legislative concentration limits (or any other world legislation). In the case of urban areas, the situation is much more complex since different emission sources and spatial locations should be studied and identify to take the optimal emission reduction strategy decision. This can only be accomplished by increasing the number of model runs by using massively the cluster approach.

The system has been proved to be reliable and suitable to identify the impact in space and time of different air pollutants in real-time and forecasting mode. Further work should be done to improve the quality of the emission inventory to optimize the agreement between observations and simulations.

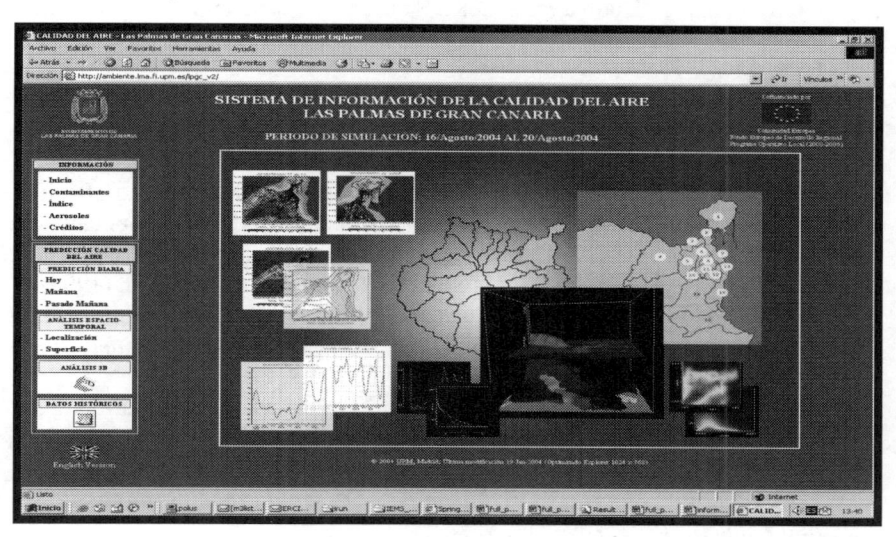

Figure 5: Home portal of the air quality forecasting system for the city of Las Palmas (Canary Islands, Spain).

Acknowledgements

We would like to thank Professor Dr. Daewon Byun formerly at Atmospheric Modeling Division, National Exposure Research Laboratory, U.S. E.P.A., Research Triangle Park, NC 27711 and currently Professor at the University of Houston, Geoscience Department for providing full documentation of CMAQ and help. We would also like to thank to U.S.E.P.A. for the CMAQ code and PSU/NCAR for MM5 V3.0 code. Also, we would like to thank Soluziona S.A. for funding this project.

References

[1] Collins W.J., D.S. Stevenson, C.E. Johnson and R.G. Derwent, Tropospheric ozone in a global scale 3D Lagrangian model and its response to NOx emission controls, *J. Atmos. Chem.* **86** (1997), 223–274.
[2] Derwent R., and M. Jenkin, Hydrocarbons and the long-range transport of ozone and PAN across Europe, *Atmospheric Environment* 8 (1991), 1661–1678.

[3] Gardner R.K., K. Adams, T. Cook, F. Deidewig, S. Ernedal, R. Falk, E. Fleuti, E. Herms, C. Johnson, M. Lecht, D. Lee, M. Leech, D. Lister, B. Masse, M. Metcalfe, P. Newton, A. Schmidt, C Vandenberg and R. van Drimmelen, The ANCAT/EC global inventory of NOx emissions from aircraft, *Atmospheric Environment* **31** (1997), 1751–1766.

[4] Gery M.W., G.Z. Whitten, J.P. Killus and M.C. Dodge, A photochemical kinetics mechanism for urban and regional scale computer modelling, *Journal of Geophysical Research* **94** (1989), D10, 12925–12956.

[5] Grell, G.A., J. Dudhia and D.R. Stauffer, A description of the Fifth-Generation Penn State/NCAR Mesoscale Model (MM5), NCAR/TN- 398+ STR. *NCAR Technical Note*, 1994.

[6] Jacobson M.Z. and R.P. Turco, SMVGEAR: A sparse-matrix, vectorized GEAR code for atmospheric models, *Atmospheric Environment* **28**(1994), 2, 273–284.

[7] Langner J., R. Bergstrom and K. Pleijel, European scale modeling of sulfur, oxidized nitrogen and photochemical oxidants. Model development and evaluation for the 1994 growing season, *SMHI report RMK No. 82*, Swedish Met. And Hydrol. Inst., SE-601 76 Norrkoping, Sweden, (1998).

[8] Roemer M., G. Boersen, P. Builtjes and P. Esser, *The Budget of Ozone and Precursors over Europe Calculated with the LOTOS Model*. TNO publication P96/004, Apeldoorn, The Netherlands, 1996.

[9] San José R., L. Rodriguez, J. Moreno, M. Palacios, M.A. Sanz and M. Delgado, Eulerian and photochemical modelling over Madrid area in a mesoscale context, *Air Pollution II, Vol I., Computer Simulation, Computational Mechanics Publications, Ed. Baldasano, Brebbia, Power and Zannetti*, 1994, 209–217.

[10] San José R., J. Cortés, J. Moreno, J.F. Prieto and R.M. González, Ozone modelling over a large city by using a mesoscale Eulerian model: Madrid case study, *Development and Application of Computer Techniques to Environmental Studies, Computational Mechanics Publications, Ed. Zannetti and Brebbia*, 1996, 309–319.

[11] San José, R., J.F. Prieto, N. Castellanos and J.M. Arranz, Sensitivity study of dry deposition fluxes in ANA air quality model over Madrid mesoscale area, *Measurements and Modelling in Environmental Pollution*, Ed. San José and Brebbia, 1997, 119–130.

[12] Schmidt H., C. Derognat, R. Vautard and M. Beekmann, A comparison of simulated and observed ozone mixing ratios for the summer 1998 in Western Europe, *Atmospheric Environment* **35** (2001), 6277–6297.

[13] Stockwell W., F. Kirchner, M. Kuhn and S. Seefeld, A new mechanism for regional atmospheric chemistry modeling, *J. Geophys. Res.* **102** (1977), 25847–25879.

[14] Walcek C., Minor flux adjustment near mixing ration extremes for simplified yet highly accurate monotonic calculation of tracer advection, *J. Geophys. Res.* **105** (2000), 9335–9348.

Section 3
Emission studies

Air quality in the vicinity of a governmental school in Kuwait

E. Al-Bassam[1], V. Popov[2] & A. Khan[1]
[1]*Environment and Urban Development Division,*
Kuwait Institute for Scientific Research, Kuwait
[2]*Wessex Institute of Technology, Southampton, UK*

Abstract

There is a growing concern in Kuwait for the air quality in the vicinity of schools. The problem exacerbates due to the peak time congestions, which adversely affect the traffic flow, and air quality. Several exceedances of certain primary pollutants have been observed during the peak periods in the country.

Air quality in the vicinity of a governmental school was assessed in March 2006 for a period of two weeks using air pollution monitoring station which recorded continuously various pollutants' concentrations and meteorological variables in five minute intervals. The results show that during the weekdays, the measured pollutants emitted from the road traffic next to the selected school, such as carbon monoxide (CO) and nitrogen dioxide (NO_2), were always under the allowable limits for Kuwaiti air quality standards except for a single exceedance of NO_2 concentration at morning hours. On the other hand, the values of non-methane hydrocarbon pollutants were found to be several times above the Kuwaiti air quality standards throughout the investigated period. The suspended particulates (PM_{10}) concentrations have exceeded twice the limits of Kuwaiti air quality standards. A traffic counter was used to record the number of cars in the main road next to the school in fifteen minute intervals for ten days during the monitoring period for air quality. Statistical analysis was used in order to test whether there is any correlation between variations in the CO concentrations and the traffic frequency during working days' morning and afternoon periods. A relation was developed for predicting the necessary reduction in traffic based on the necessary reduction in CO concentrations.
Keywords: air pollution, Kuwait, schools, statistical analysis.

1 Introduction

In Kuwait, the urban population is growing at about 3.4% per year (Institute of Banking Studies, 2004). This increase in population in addition to the development of urban areas has in turn resulted in massive increase in the demand for transport. Motor vehicles and buses are the only means of road transportation in Kuwait. Road vehicles have increased as shown in Figure 1 with an average growth of 3.0% annually. The number of buses has not increased from year 1993 till year 2002 (Ministry of Planning, 2003), and its annual growth rate is negligible. Presently, there are 377.2 vehicles for every 1000 persons, which indicates that there are 2.65 persons per car (Institute of Banking Studies. 2004). Motor vehicles and buses cause environmental pollution due to exhaust emissions and tyres abrasion which depend on driving cycles, engine design and condition, fuel composition and air to fuel ratio. The vehicular emissions constitute harmful pollutants that affect the health adversely such as carbon monoxide, particulate matter, nitrogen oxides, and lead. A large proportion of urban pollution is mainly due to road traffic.

According to the Ministry of Education (MOE) in Kuwait statistical data and Ministry of Planning statistics, the school buses are serving approximately 17 to 18% of students in the governmental schools. Based on 2003/2004 statistics, there are 23,302 students using buses out of 131,597 total students. The rest of students mostly depend on private transportation.

According to various reports (The Ashdon Trust, 1994), it was proved medically that the vehicle air pollutants such as nitrogen dioxide (NO2), carbon monoxide (CO) and particulates (PM10) have pronounced effect on human health as shown in Table1.

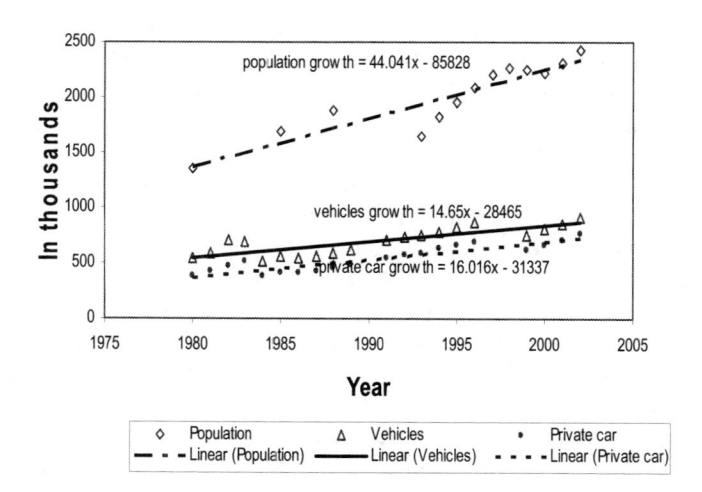

Figure 1: Vehicles in use and the growth of population (after Institute of Banking Studies, 2004).

Table 1: Health effects of vehicle air pollution.

Pollutant	Source	Health Effects
Nitrogen dioxide (NO_2)	One of the nitrogen oxides emitted in vehicle exhaust	May exacerbate asthma and possibly increase susceptibility to infections
Particulates PM10, Total Suspended Particulates, Black smoke	Includes a wide range of solid and liquid particles in air. Those less than 10μm in diameter (PM10) penetrate the lung fairly efficiently and are most hazardous to health. Diesel vehicles produce proportionally more particulates than petrol vehicles	Associated with a wide range of respiratory symptoms. Long-term exposure is associated with an increased risk of death from heart and lung disease. Particulates can carry carcinogenic materials into the lungs
Carbon monoxide (CO)	It is mainly produced from petrol car exhausts	Lethal at high doses. At low doses can impair concentration and neuro-behavioral function. Increases the likelihood of exercise related heart pain in people with coronary heart disease. May present a risk to the fetus.
Ozone (O_3)	Secondary pollutant produced from nitrogen dioxides and volatile organic compounds in the air	Irritates the eyes and air passages. Increases the sensitivity of the airways
Volatile organic compounds (VOCs)	A group of chemicals emitted from the evaporation of solvents and distribution of petrol fuel. Also present in vehicle exhaust	Benzene has given most cause for concern in this group of chemicals. It is a cancer causing agent which can cause leukemia at higher doses than are present in the normal environment

[Reproduced from "How Vehicle Pollution Affects Our Health" © The Ashden Trust 1994, p2.]

1.1 Study area

The governmental school which was selected for the study is located at the Mishref area in flat and homogeneous terrain region without any major local air pollution sources. This school is surrounded by Road 57 from north as shown in Figure 2 (www.municipality) which is considered a main street and from the east there is another school under construction. From west and south there are minor streets. The school is surrounded by residential houses and other governmental schools. The schools are adjacent to each other in one lane and there are no school buses at morning or afternoon for students.

The school area is 18,000 m² and has a parking in the front of the school entrance gate. About 985 students attended this school for year 2005/2006 during the time of monitoring the air quality.

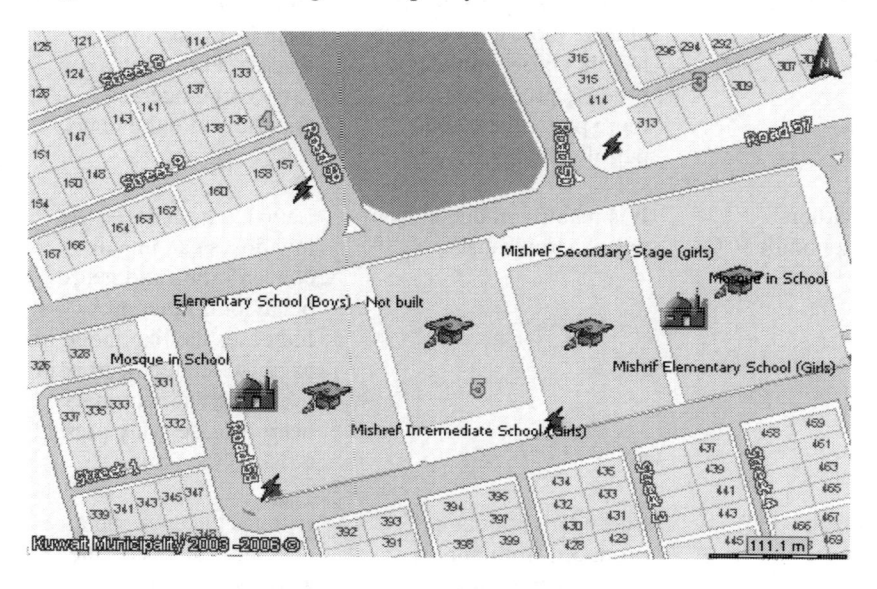

Figure 2: Location of the governmental school at the Mishref area.

2 Methodology

Air quality and weather data were recorded at sampling intervals of 5 minutes by Kuwait Institute for Scientific Research (KISR) air monitoring station as shown in Figure 3 for two weeks. The measured data included the concentration of different pollutants such as carbon monoxide (CO), carbon dioxide (CO_2), methane (CH_4) and non-methane hydrocarbons, nitrogen oxides (NO_x), nitrogen dioxide NO_2, and suspended particulates (PM_{10}). In addition, the measured data included wind speed, wind direction, solar radiation and ambient temperature. The monitoring station was parked in Mishref area next to the governmental school entrance. A traffic counter was used to record the number of cars in the roads as shown in Figure 4 next to the selected school for every 15 minutes throughout the study period. The measurements were taken in March 2006 including weekdays and weekend holidays. Statistical analysis of the recorded data was performed to establish whether there is any correlation between working days' variations in the levels of CO and the traffic frequency in the vicinity of the governmental school.

3 Discussion and results

3.1 Traffic

The hourly average weekday and weekend traffic flow profile is shown in Figure 5. The profile of the traffic indicates two peaks during the working days which are related to the opening time of the school and start of working hours in the morning and closing time in the afternoon and end of working hours. At weekends there are no sharp peaks and traffic flow gradually increases followed by slight decrease at afternoon time then minor increase in the evening.

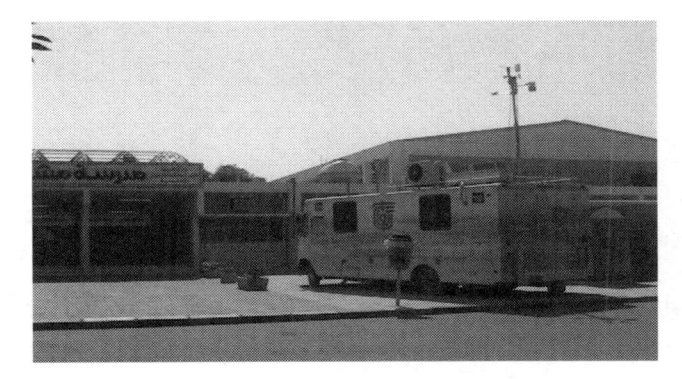

Figure 3:   Air monitoring station next to school.

Figure 4:   Cars counters on the selected road.

3.2 Air quality

All the measured pollutants' concentrations in the vicinity of the selected school for a period of two weeks were compared with the allowable levels according to Kuwait's air quality standards. The Air Quality Standards (A.A.Q) in the residential areas for Kuwait, Federal US and California states are presented in Table 2. The mean concentration and the maximum and minimum level of CO, NO_2, and PM_{10} pollutants are shown in Figures 6 to 8. The CO concentrations are always under the allowable limits. The average non-CH_4 concentrations are always above the specified limits as shown in Figure 9. NO_2 concentration had exceeded the allowable limits 15 times (do you mean on fifteen occassions?) during the study period. The NO_2 exceedances are mainly due to road traffic since these values were associated with the increase of CO levels. Regarding PM_{10} levels it has exceeded the limits of A.A.Q (on two occassions) during the time of recording.

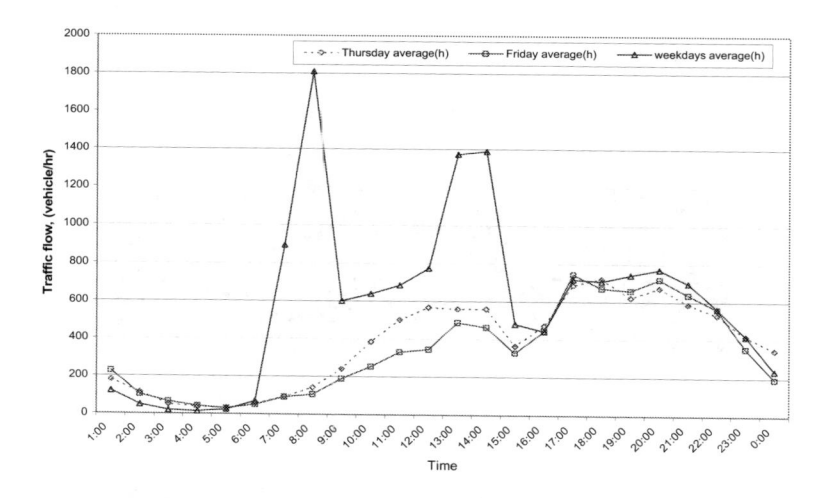

Figure 5: Hourly traffic flow for weekdays and weekend days at the site.

Table 2: The hourly air quality standards for Kuwait, Federal US and California State.

Pollutant	Kuwaiti Standard*	Federal Standard	California State standard
Ozone $\mu g/m^3$	157	235	180
CO (ppm)	30	35	20
NO2 (ppm)	0.1	-	0.25
PM10 $\mu g/m^3$	350 (24 hours)	150 (24 hours)	50 (24 hours)
Non methane HC	0.24 ppm for a period of 3 hours (6-9 AM)	-	-

*Al-Kuwait Al-Youm. 2001

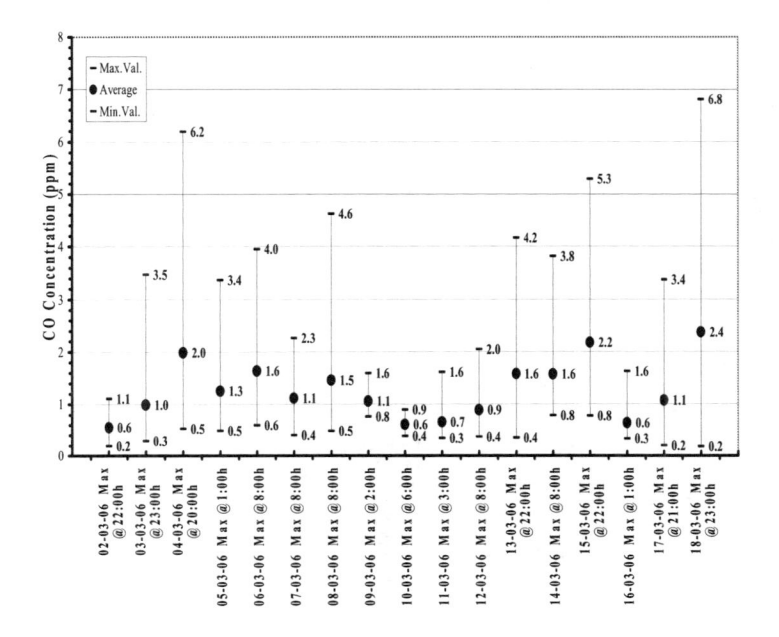

Figure 6:   Mean, maximum, and minimum level of CO concentrations.

3.3 Statistical analysis

The recorded data for CO concentration and cars counts on every 15 minutes during the study period were analyzed taking into consideration the morning hours from 5:00hr -10:00hr and afternoon hours from 11:00hr -16:00hr. This strategy was adopted in order to decrease the influence of traffic in the surrounding area and to focus mainly on the traffic in the vicinity of the school, which is the main objective of this research. For the selected time periods, the measured CO concentrations are plotted against car counts for the 15 minutes intervals. A strong correlation is found in mornings showing 4.4 ppb car while in the afternoon the CO emissions were 1.3 ppb car depending on the traffic flow as shown in Figure 10. The dispersion of pollutants is slower in morning times than afternoon due to prevailing meteorological conditions, temperature, wind, and inversion layer. The equation, which was obtained from the morning trend of cars versus CO concentration, was used to predict the effect of reducing the number of cars according to the desired level of CO concentrations. Figure 11 presents the dependency of CO concentration reduction as a function decrease in number of cars on the road in the vicinity of the school. This correlation is very important and can be used to regulate the traffic according to desired reduction in air pollution. In this case 40% reduction in traffic leads to 32% reduction in CO concentration.

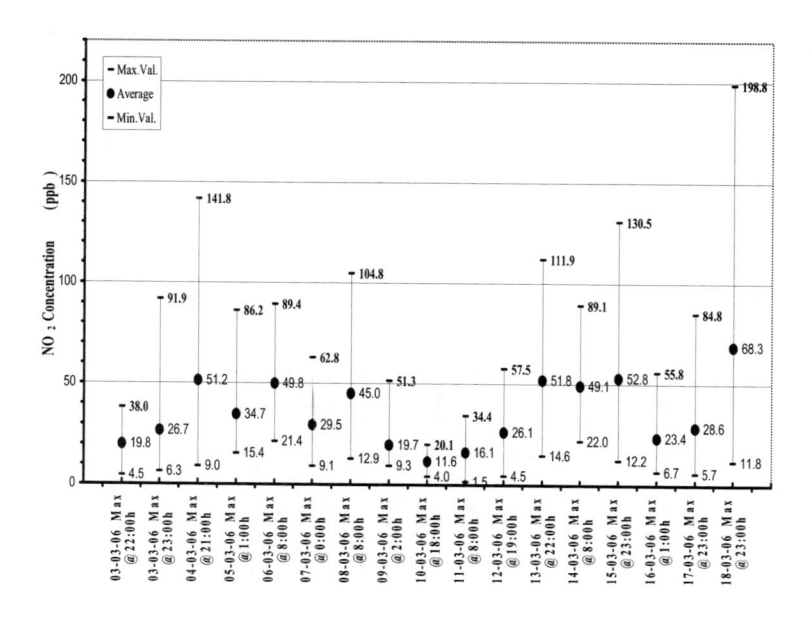

Figure 7: Mean, maximum, and minimum level of NO_2 concentrations.

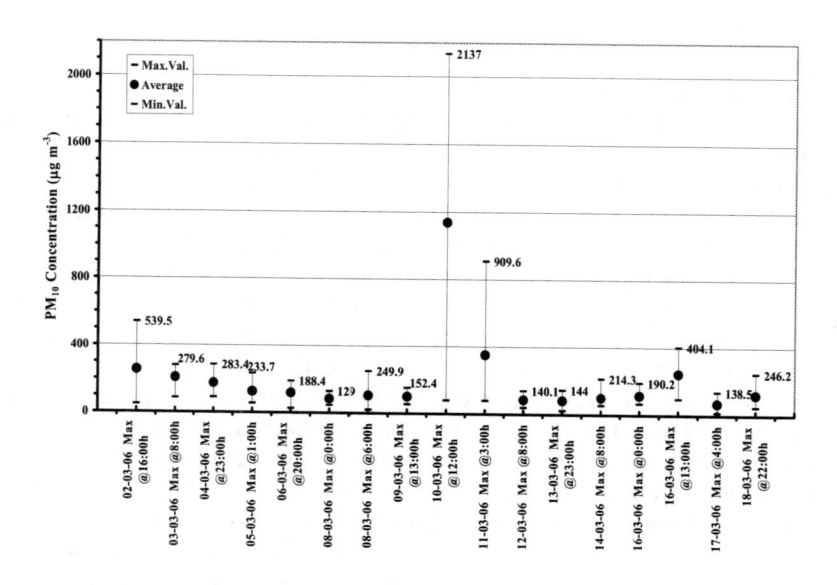

Figure 8: Mean, maximum, and minimum level of PM_{10} concentrations.

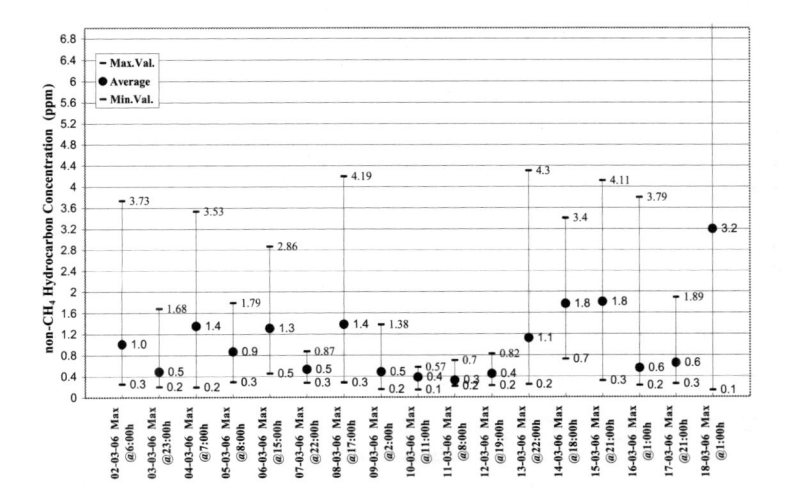

Figure 9: Mean, maximum, and minimum level of non-CH4 Hydrocarbon concentrations.

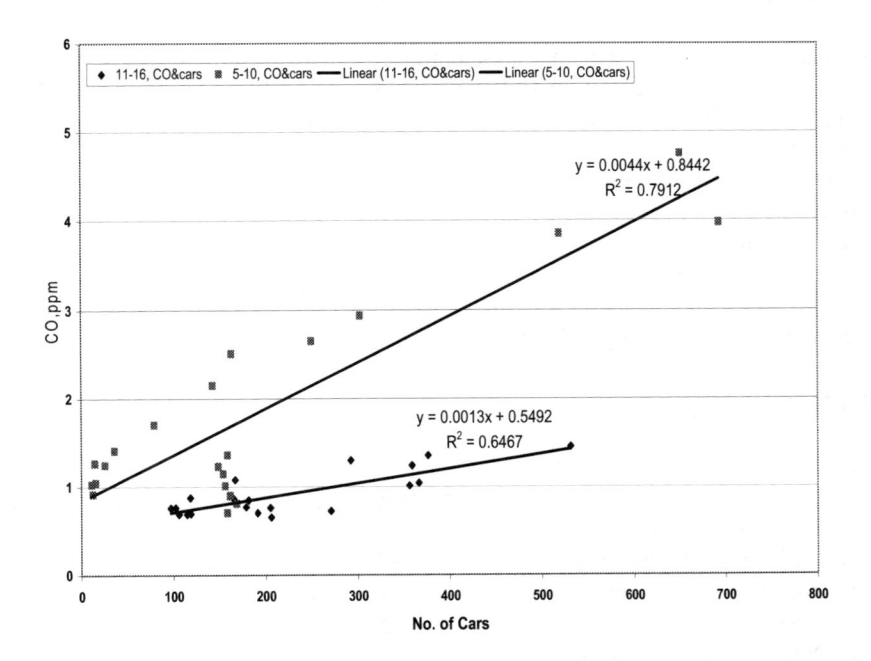

Figure 10: The correlation between CO pollutant and the number of cars (5 – 10 = morning hours; 11 – 16 = afternoon hours).

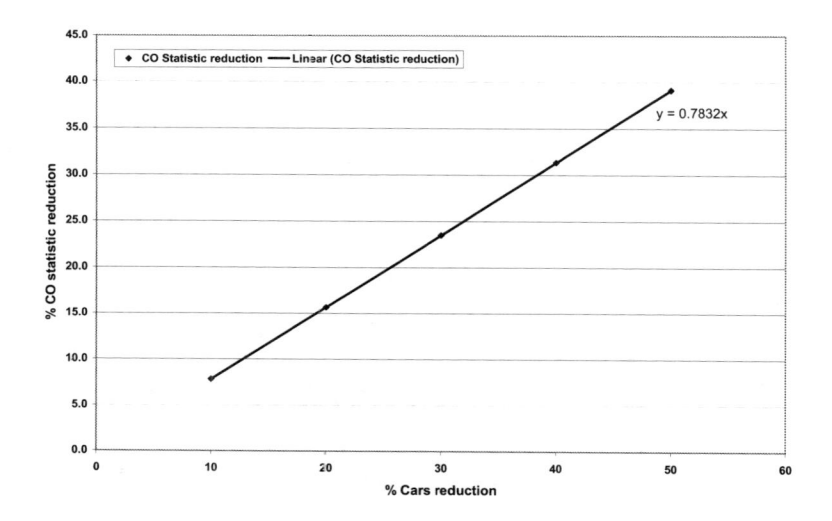

Figure 11: The predicted effect of decreasing the number of cars on CO concentration.

4 Conclusion and recommendation

It is important to maintain high standards of air quality around the schools in order to reduce the effect of traffic pollutants on health of children and their performance. High levels of pollution and traffic conjunctions are recognized as health risk.

Kuwait government should consider public transportation for the governmental schools students to abate traffic conjunction and associated air pollution problems in the country.

Protective measures such as introduction of school buses using superior quality fuel to combat high pollutants emissions are required to achieve good ambient air quality in the country.

References

[1] Ministry of Planning, 2003
[2] Institute of Banking Studies. 2004. Economic and Financial Data base for Bankers, Research Unit, Kuwait.
[3] "How Vehicle Pollution Affects Our Health" © The Ashden Trust 1994, p2
[4] http://www.municipality.gov.kw/gis/ebase_map.htm.
[5] Al-Kuwait Al-Youm. 2001. Annexure No. 533, 2 October 2001, year 47. Ministry of Information (Arabic).

Emissions of nitrogen dioxide from modern diesel vehicles

G. A. Bishop & D. H. Stedman
*Department of Chemistry and Biochemistry,
University of Denver, Colorado, USA*

Abstract

The traditional on-road emissions remote sensing technique has been enhanced with the ability to monitor not only CO, CO_2, HC and NO, but also SO_2, NH_3 and NO_2. Modern diesel powered vehicles, particularly in Japan and Europe are equipped with diesel particle filters (DPFs). These traps are frequently designed to be self regenerating by means of a prior catalytic oxidation of the exhaust NO to NO_2. Because of this intentional conversion to NO_2, the ratio of NO_2 to NO in modern diesel vehicles is very different from older vehicles. The detailed results of measurements from several thousand vehicles measured in Japan, Austria and Sweden in 2006 and 2007 will be compared to results from the USA where there are very few vehicles equipped with DPFs. The overall picture is vehicles without DPFs in the USA emit between 5 and 10% NO_2 while vehicles with DPFs are often observed at 50% NO_2 by moles. This, together with the overall NO_x reductions has strong negative implications for local photochemical ozone production.

Keywords: nitrogen dioxide, automobile emissions, on-road remote sensing.

1 Introduction

In his review of diesel exhaust emissions control strategies, Lemaire [1] suggests that nitrogen dioxide (NO_2) was forgotten as a separate component of the NO_x emissions from diesel vehicles. He further suggests that this omission has caused the observed increases in NO_2 and ozone in a number of cities in Europe. For instance, he shows a very strong correlation between the imposition of diesel particle filters (DPFs) on London buses with the increase in the NO_2/NO_x ratio in London. Carslaw et al. [2] draw the same conclusion.

The DPF is an extremely effective means to achieve particle reduction goals. However, these traps are bound to become clogged with carbon particles in normal use and must be regenerated in order to maintain a useful in-service lifetime. The most economical method of regeneration makes use of the fact that NO_2 can oxidize carbon (soot) to gaseous products (CO and CO_2) at typical diesel exhaust system temperatures. This regeneration is however not adequate with the normal NO_2/NO_x ratio in diesel exhaust (between 6% ad 9% [3]). As a result, an intentional catalytic conversion of NO to NO_2, using the excess oxygen from the engine, is carried out in the exhaust system of these DPF equipped vehicles in order to enhance the NO_2/NO_x ratio. Some diesel vehicles not equipped with DPF, nevertheless are equipped with an oxidation catalyst which also has the effect of increasing the NO_2/NOx ratio.

The reasons for concern about nitrogen dioxide emissions arise in part from observations of ambient air concentrations in urban core areas [2]. These observations generally show decreasing concentrations of total NO_x (NO + NO_2) but increasing concentrations of NO_2. Both of these observations, but particularly the NO_2 increase, are of concern to air pollution photochemistry because NO_2 leads directly, through solar photolysis, to ozone formation while NO initially removes ozone in the process of becoming NO_2. Lowering NO_x concentrations enhances urban ozone [4] while directly emitting NO_x as NO_2 further enhances ozone and many other manifestations of photochemical smog. An attempt to simplify the complexity of these photochemical processes was published by Stedman [5] and in a poster [6]. The two preceding references are available from the web site www.feat.biochem.du.edu and click on "Publications List". Our work and others suggests that modern gasoline powered vehicles emit less than 1% of their NO_x as NO_2. For instance, Heeb *et al.* measured emissions of twenty Euro III and Euro IV gasoline powered vehicles in a dynamometer based study [7]. Using high-speed sampling and a mass spectrometric analytical method, they arrived at the same conclusion.

1.1 On road remote sensing

On road remote sensing enables fuel-based emission measurements to be made from a very large number of vehicles driving in a real on-road situation. Typically the systems can monitor the emissions of between 3000 and 10000 vehicles per day. The system measures emissions in ratio to fuel burned. Results are frequently reported in gm of pollutant per kg of fuel. The recent development of on-road emission remote sensing for SO_2, NH_3 and NO_2 is described by Burgard *et al.* [8]. The first two gases are measured in the deep ultraviolet (200-220 nm) using the same diode array spectrometer used for NO at 230 nm. Nitrogen dioxide is monitored by means of its visible absorption at 430 nm using an identical spectrometer but set to monitor at these longer wavelengths. The spectroscopic matrix deconvolution process by means of which these analyses are carried out has been described by Bishop and Stedman [9]. In order to monitor heavy duty diesel vehicles in the USA, the same system which is normally used to monitor passing light duty vehicles in under one second each is placed upon scaffolding with the optical beam approximately three meters above the ground. A lower level optical trigger is added to initiate plume measurement.

The instrumentation with this new capability has been used on-road in the U.S.A., Japan, Austria and Sweden. In the USA very few diesel vehicles are equipped with either DPF or oxidation catalysts. The comparison study, also reported herein, [10] was carried out on a relatively small fleet of diesel school buses some of which had recently been retrofitted with either DPF or diesel oxygenation catalysts (DOC). Japan, Austria and Sweden, where some of these reported remote sensing observations have been made, have required new model year diesel vehicles to significantly lower their particle emissions using DPF technology.

2 Experimental methods

More than 40,000 measurements were made in Japan in May and June of 2006. We did not obtain vehicle model year (MY), but did obtain a vehicle classification number from the license plate. Vehicles classified in the 100 series are mostly more or less heavy duty diesel vehicles. Vehicles in the 500 series are mostly gasoline fuelled automobiles.

More than 15,000 Measurements were made in Sweden in June 2007 [11] while a similar number of measurements were made in Austria shortly afterwards.

Figure 1: Photograph from Colorado showing the necessary scaffolding and the three calibration cylinders.

Figure 1 shows the instrumentation as set up to monitor truck tractor emissions in Colorado. The three gas cylinders contain the necessary calibration gas mixtures. The main calibration gas cylinder contains CO, CO_2, NO, Propane and SO_2 with the balance nitrogen. The two secondary cylinders contain in one, a known mixture of NO_2, and CO_2 balance air (stored NO_2 is incompatible with both propane and NO) and, in the second, a known mixture of NH_3 and propane in nitrogen because NH_3 and CO_2 are incompatible. All calibration gases are supplied by Scott Specialty Gases of Longmont CO.

3 Results and discussion

Figure 2 shows, in histogram form, the data from the 2006 study in Japan. It is quite noticeable that the gasoline powered vehicles (class 5, mean NO_2 0.1 and median 0.05 g/kg) have lower NO_2 as well as lower total NO_x than the diesel (class 1, mean NO_2 1.9 and median 1.4 g/kg) and a very much lower NO_2/NO_x ratio. Several of the diesel trucks emit more than 50% by mass of their NO_x as NO_2. For these comparisons NO_x is given the mass of NO_2 although the majority gas emitted is most often NO.

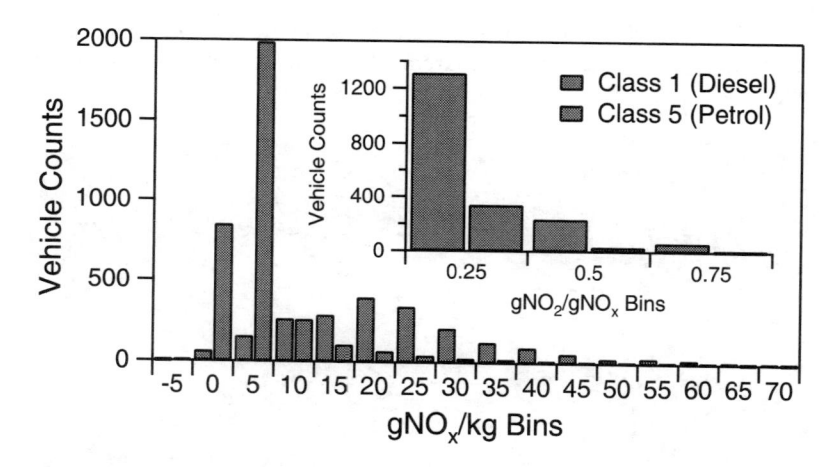

Figure 2: Histogram plots from Japan showing that the class 5 vehicles (cars) have very significantly less NO_2 and NO_x emissions than the class 1 (trucks).

Figure 3 shows a comparison between NO_x emissions from light duty gasoline powered vehicles measured in Sweden [11] and NO_x emissions measured in Denver in the same year [12] and using approximately the same model year groupings. The measured NO_2 emissions (not shown) from both fleets are not statistically distinguishable from zero. A conservative error analysis suggests less than 2%. The other phenomenon, very noticeable from this Figure, is the amazingly successful reduction of NO_x emissions over the about 20 year period shown. This observation is cause for congratulations all round, because the results not only point to the success of the European and U.S. regulations in reducing automobile emissions, but also to the success of on-road remote sensing in quantitatively demonstrating these phenomena. If the on-road data were significantly contaminated with noise, then these very significant differences would not be observed. The reductions in the U.S., although also dramatic, seem to be somewhat less successful than in Europe over the same time period.

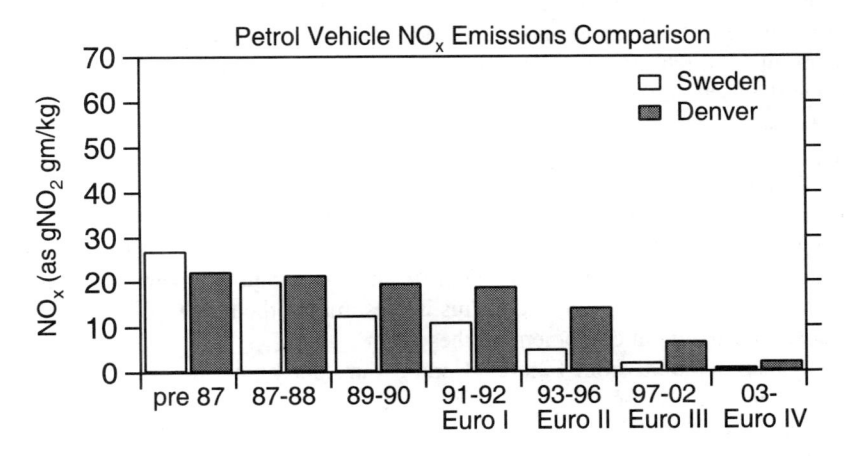

Figure 3: NO_x emissions from petrol (gasoline) light duty vehicles in 2007.

Figure 4 shows NO_x and % of NO_x as NO_2 from medium duty diesel bus vehicles in Sweden compared to a recently reported study from Washington State on retrofitted School Buses [10]. It is very noticeable that the highest NO_x

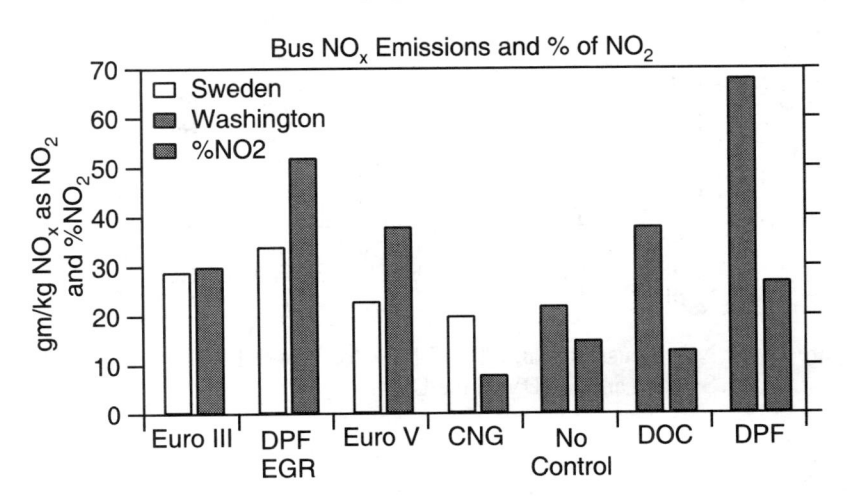

Figure 4: The 2007 NO_x emissions and NO_2 % for buses in the USA and Sweden.

emissions in this entire manuscript (by almost a factor of two) is the DPF equipped school buses in Washington State. It was reported that while the retrofits on these school buses were underway, there were several cases of filter clogging and this caused many of the DPFs to be removed (as worthless) and replaced with DOC on several fleets. This problem may have arisen because the duty cycle on school buses is relatively light load and frequent stops. The system may rarely become hot enough even for NO_2 to regenerate. We speculate that the retrofit contractor, becoming aware of this problem chose to solve it by increasing the NO_x emissions (and possibly the vehicle power) and thus generating enough excess NO_2 that adequate regeneration could take place despite the light duty cycle.

The last Figure 5 shows heavy duty diesel truck NO_x emissions and $\%NO_2$ compared to a U.S. study [3] and shows NO_x and % of NO_x as NO_2 from light duty diesel vehicles in Sweden. Comparing Figures 3 and the last three bars in Figure 5. it is apparent that the modern fleet of light duty diesel vehicles in Europe emit very significantly more NO_x than their gasoline (petrol) fuelled counterparts. Furthermore, a very significant percentage of their emissions is observed to be in the form of NO_2. This is presumably the result of DPF or DOC exhaust aftertreatment equipment on these cars.

It is also apparent from Figures 4 and 5 that, per kg of fuel, the trucks and buses in Sweden do not have very different NO_x emissions although the % of NO_2 is quite different. The Euro 2,3 and 4 standards for the trucks do not seem to have resulted in significant on-road NO_x reductions.

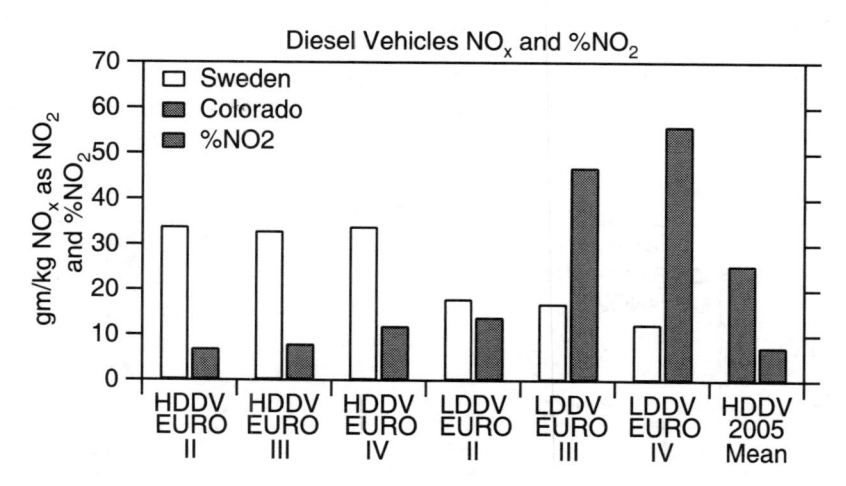

Figure 5: NO_x emissions and % of NO_2 for diesel HDDV and LDDV in Sweden and HDDV in the USA.

The fact that DPF equipped diesel vehicles show evidence of very large NO_2/NO ratios compared to pre-control on-road vehicles is apparent from these data. Our expectation was that the U.S.A. retrofit systems would also have shown evidence of reduced total NO_x emissions. At least in the case of the

Washington State school buses, it seems that the DPF fleet have higher overall NO_x emissions as well as a much larger NO_2/NO ratio.

NO and NO_2 emissions were monitored at three sites in Austria in the summer of 2007. The truck fleet showed relatively low NO_2/NO_x ratios, for instance 50% of the fleet showed a ratio lower than 0.1. Only at the highest load site was the highest 5% of the vehicles measured with $NO_2/NO_x > 0.5$. By contrast, and not unlike the situation in Sweden, the 85% diesel car fleet showed very high NO_2/NO_x ratios. At all three sites the top ten percent of the vehicles were all measured with a NO_2/NO_x ratio of above 0.7 and at the highest load site 50% of the whole fleet (gasoline included) were measured above a ratio of 0.5.

References

[1] Lemaire, J., How to select efficient diesel exhaust emissions control strategies for meeting air quality targets in 2010. *Österreichische Ingenieur und Architekten-Zeitschrift* **152** . pp1–12 Heft 1-3, 2007.

[2] Carslaw, D.C., Beevers, S.D. & Bell, M.C, Risks of exceeding the hourly EU limit value for nitrogen dioxide resulting from increased road transport emissions of primary nitrogen dioxide. *Atmospheric Environment*, **41** pp. 2073–2082, 2007.

[3] Burgard, D.A., Bishop, G.A., Stedman, D.H., Gessner, V.H., & Daeschlein C., Remote Sensing of In-Use Heavy-duty Diesel Trucks, D.A. Burgard, , *Environ. Sci. Technol.*, **40,** pp. 6938–6942, 2006.

[4] Chow, J.C. Introduction to special topic: weekend and weekday differences in ozone levels *EM* **July** pp. 16–25 2003.

[5] Stedman, D.H. Photochemical ozone formation, simplified,, *Environ. Chem.*, **1**, pp. 65–66, 2004.

[6] Stedman, D.H., The Weekend Effect: the science suggests that we are embarking on an expensive policy which will harm the environment, Poster Presented at the 2006 Diesel Engine-Efficiency and Emissions Research (DEER) Conference, Detroit, MI, August 20 - 24, 2006 available from www.feat.biochem.du.edu/reports.

[7] Heeb, N.V., Saxer, C.J., Forss, A-M., & Brühlmann, Trends of NO-, NO_2-, and NH_3-emissions from gasoline-fueled Euro-3- to Euro-4-passenger cars *Atmospheric Environment*, **42,(10)**, pp. 2543–2554, 2008.

[8] Burgard, D.A., Dalton, T.A., Bishop, G.A., Starkey, J.R., & Stedman, D.H. Nitrogen dioxide, sulfur dioxide, and ammonia detector for remote sensing of vehicle emissions, *Rev. Sci. Instrum.*, **77**, pp. 014101.1–014101.5, 2006.

[9] Bishop, G.A. & Stedman, D.H. Signal to Noise Improvements in On-road NO, SO_2, and NH_3 Measurements using a Multi-component Classical Least Squares Fitting Approach, Poster presented at the 18th CRC On-road Vehicle Emissions Workshop, San Diego, CA, March 31- April 2, 2008, available from www.feat.biochem.du.edu/reports.

[10] Provinsal, M.N., & Burgard D.A., Examining gaseous emissions from diesel school buses in Washington State. *Proceedings of the 18th CRC On-Road Vehicle Emissions Workshop*, March 31-April 2, 2008.

[11] Jerksjö, M., Sjödin, Å., Bishop, G.A., Stedman, D.H., On-road emission performance of a European vehicle fleet over the period 1991-2007 as measured by remote sensing. *Proceedings of the 18th CRC On-Road Vehicle Emissions Workshop*, March 31-April 2, 2008.
[12] Bishop, G.A. & Stedman, D.H., On-Road Remote Sensing of Automobile Emissions in the Denver Area: Year 6, January 2007, www.feat.biochem.du.edu/publications.

Errors in model predictions of NO_x traffic emissions at road level – impacts of input data quality

R. Smit
Pacific Air & Environment, Australia

Abstract

This study investigates the effects of three important input variables on the prediction accuracy of average speed emission models. These variables are average speed, basic traffic composition (proportion of heavy-duty vehicles) and model choice (COPERT, QGEPA). Sensitivity analysis (conditional NRSA) is used to determine to what extent the possible range in these input variables influences model outcomes (i.e. NO_x emissions for road links), and hence accuracy. It is shown that maximum errors can be large (up to a factor of about 3.5). Moreover, they are a function of the level of congestion with errors generally increasing with the level of congestion. Traffic composition is shown to most strongly affect NO_x emissions (29-241%), followed by average speed (2-168%) and model choice (0-177%). The results were similar for arterial roads and freeways. These results can be used to provide direction to the collection of model input data, further model development and model application. The external errors found in this study appear to be of the same order of magnitude as internal errors that have been reported from (partial) road validation studies. This implies that in terms of further improvements of traffic emission modeling, focus should be on both the quality of input data (application) and the quality of the actual emission models (model development). Given the relevance of these results, it would be worthwhile to extend and refine this work by including other air pollutant and greenhouse gas emissions, and to use more complex traffic and emission models.

Keywords: accuracy, error, road traffic emission, modeling, sensitivity analysis, NO_x, congestion.

1 Introduction

Road traffic is an important global source of air pollution and greenhouse gas emissions and its significance is increasing. As emissions are a complex function of many variables, impacts and solutions are commonly evaluated using multidisciplinary combinations of transport, emissions and dispersion models at different scales, ranging from local road projects to entire urban or regional transport networks and even national or global emission inventories. There is an increasing need for valid and accurate modeling results as many national and local authorities are faced with difficulties in meeting air quality standards and other environmental policy targets (e.g. National Emission Ceilings). There is however limited knowledge about the reliability of calculated emissions from road traffic. Testing the overall accuracy of road traffic emission models (model validation) is difficult as "true" emission values are unknown and cannot practically be determined by measurement.

A review of current literature showed that available validation studies are restricted to specific models (or model versions) and specific situations. Some studies report on modeling results that are close to observed values, but most studies indicate that errors in emission predictions can be quite substantial. Two types of validation studies can be distinguished, namely area and road level studies:

- Validation of traffic emission models at area level is possible by using ambient air pollutant concentration data collected downwind of these areas. In the US and Europe, a number of studies have compared ambient air sampling data or emission fluxes to the results from combined emission and dispersion modeling [1–4]. For NO_x, differences varying between a factor (predicted/observed) of 1.0 [1] to 2.2 [4] have been found.
- Validation at road level is possible for specific traffic situations during relatively short time periods and they include tunnel studies (e.g. [5–9]), near-road air quality monitoring (e.g. [10–14]) and remote-sensing studies (e.g. [15–18]). For NO_x, differences varying between a factor (predicted/observed) of 0.4 [19] to 4.2 [20] have been reported.

The accuracy of emission model predictions is affected by both internal and external errors. Internal errors are associated with the emission model itself (e.g. emission factors). External errors are associated with the errors in model input variables. Road level validation studies commonly use measured input data for key variables such as vehicle kilometers travelled (VKT, i.e. traffic volume multiplied with road length), travel speeds and traffic composition (e.g. [21–23]). As a result, these studies tend to quantify internal errors. In contrast, area level validation studies tend to quantify both internal and external emission modeling errors, since the model predictions are often based on the combination of traffic, emission and dispersion models. Similarly, the use of ambient concentration data in road level validation studies often requires the use of dispersion models (e.g. [24, 25]). As a consequence, these studies validate the model chain and do not directly assess the accuracy of emission modeling. This complicates explanation of the discrepancies between ambient and modeled data. For instance, errors due to dispersion modeling may have offset or amplified emission modeling errors, but the magnitude and direction of these errors in the validation studies are unknown.

Natural variation in traffic emissions may also complicate emission model validation. For instance, emissions from a traffic stream may vary substantially due to random fluctuations in the number of high-emitters, the number of cold-start vehicles, etc. [26]. These factors cannot be controlled and their proportion in the traffic stream is often unknown. Moreover, fleet characteristics continuously change in time and this significantly affects emissions observations [27]. Thus, a model may have performed well a number of years ago, but this may no longer be the case for the current situation. Finally, model validation is not possible for situations for which there is a lack of empirical data (e.g. future years).

In conclusion, validation of road traffic emission models is difficult and only limited information is available. Further work to increase our understanding in model accuracy is thus required. The quality of emission model input data is obviously an important factor for the accuracy of emission predictions. In particular the impact of input data accuracy on emission predictions seems to be an area where further work would be valuable. Identification of the most important input data can be used to provide guidance and direction to data acquisition (e.g. new emission testing focused on critical aspects), further emission model development (e.g. focus efforts on critical aspects) and model application (e.g. focus efforts on collecting input data that are most relevant).

This study seeks to quantify maximum errors due to changes in selected important input variables on prediction accuracy of two selected models, and to assess its relevance. Although more complex emission models and input data can be used, as will be discussed later, this study presents a first-order assessment of possible emission prediction errors to assess the relevance of input data accuracy in relation to reported results from validation studies.

2 Methodology

In addition to model verification and model validation, a model itself can be used to examine uncertainty in the predictions of traffic emissions. Sensitivity analysis (SA) can be used for this purpose, as it is able to apportion prediction variability to specific inputs [28]. There are various SA methods [29], but mathematical SA is well-suited to quantitatively assess the sensitivity of a model output to the (possible) range of variation of an input. An important limitation of traditional SA is that only a small portion of the possible space of (combined) input values is investigated. To address this limitation, conditional one-at-a-time (OAT) nominal range sensitivity analysis (NRSA) will be used. In addition, graphical methods will be used to clarify the results where useful.

NRSA is applicable to deterministic models and evaluates the effect of model outputs exerted by individually varying only one of the model inputs (OAT), while holding all other inputs at constant values. Conditional NRSA conditions the sensitivity on specific sets of input values (scenarios). These inputs are varied across their entire range of plausible values (two extreme values), which are derived from either test data, expert judgment or literature review. For each scenario the impact on the model output is then evaluated. The sensitivity of the model to a scenario is represented as a positive or negative percentage change compared to the nominal situation:

$$S_{i,alt} = \frac{E_{alt,i} - E_{nom,i}}{E_{nom,i}} \times 100\%. \qquad (1)$$

$E_{nom,i}$ and $E_{alt,i}$ are the predicted total traffic emissions (kg/h) for the nominal and alternative scenarios (minimum, maximum) for traffic situation i, respectively. Here traffic situation is defined in terms of road type (arterial, freeway) and level of congestion (V/C ratio). As will be discussed below, $E_{nom,i}$ and $E_{alt,i}$ are determined from computation of 21 different speed-congestion relationships and associated emission factors for two models (COPERT, QGEPA) and three basic traffic compositions (defined as proportion heavy duty vehicles). The maximum absolute error for traffic situation i (e_i) is then computed as:

$$e_i = MAX\left(\left|S_{i,\min}\right|, \left|S_{i,\max}\right|\right). \qquad (2)$$

$S_{i,\min}$ and $S_{i,\max}$ represent the sensitivity for the predicted minimum and maximum scenarios.

2.1 Emission model selection

Many road traffic emission models exist around the world. Of these models, so-called average speed models are most commonly applied in practice [35]. Although these models are complex with respect to the number of model categories (e.g. vehicle classes, number of pollutants, emission types), the overall computation process is straight forward. Road link emissions are computed by multiplying a composite emission factor for a pollutant (g/km) with total vehicle kilometers of travel (VKT). The composite emission factor for a link presents the "mean traffic stream emission factor" and it is equal to the sum of the emission factors for all vehicle classes and the VKT-weighted proportion of these classes in the traffic stream. These emission factors are computed as a function of average link speed, but can also be corrected for other factors such as road gradient, air conditioning and ambient temperature. As a consequence, total link (and thus network) emissions, are determined by three basic variables, namely VKT, traffic composition and traffic conditions (congestion, road gradient, etc.).

Given their common use and their relative simplicity of application, two average speed emission models were selected for this study, namely COPERT III [30] and QGEPA 2002 [31]. COPERT is (and has been) extensively used for emission modeling in Europe and other parts of the world. QGEPA is an Australian model that has been developed using Australian test data in combination with information from other models such as MOBILE6 [32]. Using these two models, speed-dependent composite NO_x emission factors were computed for three basic traffic composition scenarios (0, 5 and 20% heavy duty vehicles). The results are shown in Figure 1.

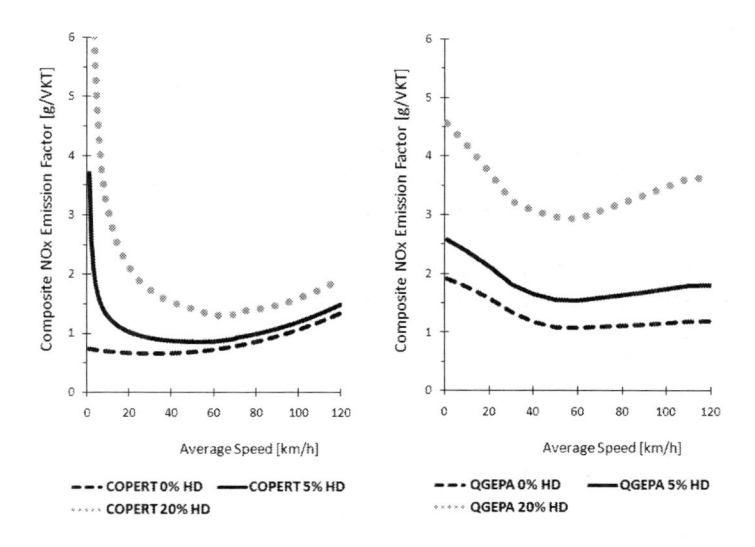

Figure 1: Composite emission factor curves for two emission models.

The composite emission factors were developed in different steps. Firstly, speed-dependent emission factors were computed for 32 vehicle classes, which are defined in terms of vehicle type (car, articulated truck, bus, etc.), fuel type (petrol, diesel, LPG) and technology type (legislative emission standards, type of catalyst, etc.). Secondly, these detailed emission factors were weighted using 2003 Brisbane fleet composition data that was taken from Smit [33]. Thirdly, the three basic traffic composition scenarios were developed by weighting the proportion of light-duty and heavy-duty vehicles accordingly.

2.2 Variable selection for simulation

This study will focus on three basic emission model variables, namely VKT, basic traffic composition (proportion of light-duty and heavy duty vehicles) and traffic conditions (level of congestion, expressed as volume-to-capacity ratio). It will also include, to some extent, internal errors by using two different emission models. The two models will reflect differences in emissions test data (e.g. due to country-specific differences in emission control technology and calibration of the engine and emission control systems), modeling approach and development (e.g. choice of driving cycles, statistical modeling), presence (or absence) of national inspection and maintenance programs and possibly systematic differences in measurement results between laboratories.

The use of traffic field data as input has a clear advantage in terms of accuracy when compared to modeled data. Moreover, data from traffic models may be the only source that can be (feasibly) used. In the simulations, the focus will be on variables for which field data are relatively scarce, i.e. traffic composition and average speed, but not VKT as will be discussed below.

2.2.1 Vehicle kilometers travelled

Efforts to improve the quality of emission model input data should focus on variables that have been shown to have a large effect on emission predictions. In this respect, the amount of travel (VKT) is a particularly important input variable as errors in VKT are proportionally propagated into emission predictions [33]. Therefore, particular attention should be directed at obtaining accurate information on traffic volumes. Compared to other input variables such as average speed and traffic composition, accurate VKT estimates for roads are relatively easy to obtain as traffic count data are commonly measured at various points (e.g. automatic detection, manual counting surveys) in road networks. Errors in VKT input, and thus emission prediction, are relatively small compared to traffic composition and average speed (as will be shown later), and this variable is therefore not included in the simulations.

2.2.2 Basic traffic composition

Basic vehicle classification data (e.g. light vehicle, heavy vehicle, perhaps a few heavy vehicle sub classes) is usually available for major roads. However, more comprehensive classification data, needed for emission estimation, is more difficult to obtain since they are usually collected by less common manual classified counting surveys or video image surveys [34]. For a detailed breakdown of (mean) traffic composition, additional data is commonly derived from other sources such as the National Bureau of Statistics and fleet turnover modeling. Following analysis of the Brisbane road network [32], a minimum, nominal and maximum value for the proportion of heavy-duty vehicles in the traffic stream of 0%, 5% and 20%, respectively, was determined for use in the sensitivity analysis.

2.2.3 Level of congestion

Average speed is needed as input to the selected emission models. Average speed, however, is not an adequate congestion indicator in certain speed intervals (between about 15 and 60 km/h) as the relationship between average speed and level of congestion is road-type specific [35]. As average speed models are based on emissions tests using driving cycles that typically run for about 10 minutes, the definition of average speed needs to be carefully considered when input speed data are collected. For instance, speed data measured at certain points in the network (e.g. by dual-loop detectors) can only be used when they represent average speeds for traffic conditions that are relatively homogeneous and stable over some distance of road (e.g. free-flow freeway driving away from on- and off ramps). On the other hand, average speeds measured on specific segments of road or entire routes using travel time studies [36] would align with the spatial resolution of driving cycles, but are only available to a limited extent.

There are several indicators for congestion but volume-to-capacity ratio (V/C) is a good one, since it combines the two principal causes of congestion (traffic demand and capacity) into one variable. Because of the availability of volume and capacity figures for network links and widespread acceptance by most transport agencies, V/C has been widely used as a fundamental congestion indicator [33]. To assess the relationship between prediction errors and congestion, a mathematical relationship between average speed and traffic conditions is needed and congestion functions can be used for this purpose. Congestion functions are used extensively in (macroscopic) traffic modeling and they are often calibrated using experimental data. They have evolved from relatively simple functions to more complex (sets of) equations by incorporating, for instance, traffic flow theory (e.g. queuing theory). The variables of traffic volume, road capacity and (mean) free-flow speed are fundamental to all congestion functions. The Akçelik function [37] is given as an example:

$$\overline{T}^* = \overline{T}_{ff}^* + 0.25\tau \left\{ (V/C - 1) + \sqrt{(V/C - 1)^2 + \frac{8 J_A V/C}{\tau C}} \right\}$$
(3)

where $\overline{T}^*$ represents mean unit travel time (min/km), which is (approximately) the reciprocal of average speed, $\overline{T}_{ff}^*$ represents free-flow unit travel time (min/km), V/C represents volume-to-capacity ratio, τ represents the time period over which traffic flow exceeds capacity (min), C is the road capacity (veh/h) and J_A is a delay parameter which is a function of road characteristics (e.g. signal density, signal coordination).

The posted speed limit on a road is often assumed to approximate the free-flow speed. Drivers tend to comply, on average, within certain margins above and below the speed limit in free-flowing traffic conditions. Therefore, the maximum mean speed in free-flow conditions is set to the speed limit plus 15 km/h for arterial roads and to the speed limit plus 25 km/h for freeways [38]. The minimum mean speed is set to 5 km/h below the speed limit for both arterials and freeways, and this can occur in specific conditions such as roads with strict radar control [35]. Using seven different congestion functions from a literature review [33] in combination with the three possible free-flow speeds (speed limit, minimum, maximum), an envelope of plausible mean speeds by level of congestion, including nominal speeds, has been computed for two basic road types (arterial, freeway).

The results are presented in figure 2 (next page). It shows that congestion has a large effect on average speed, and that congestion functions exhibit an inverted S-shape relationship between volume-to-capacity ratio and mean speed. The largest difference between congestion functions (50 km/h) occurs when traffic demand is near road capacity (V/C about unity).

For the sensitivity analysis is has been assumed that the range of average speed predictions by the different congestion functions represent the range of plausible values. The Akçelik function (eqn. 3) is the most complex function and its parameters have been calibrated using the aaSIDRA model, which is commonly used by traffic engineers around the world. Therefore this congestion function was taken to present the nominal situation as it provides probably the most accurate prediction of mean speed when the various functions are compared.

In the sensitivity analysis the three speed-congestion curves (minimum, nominal, maximum) for each road type, as depicted in figure 2, are used.

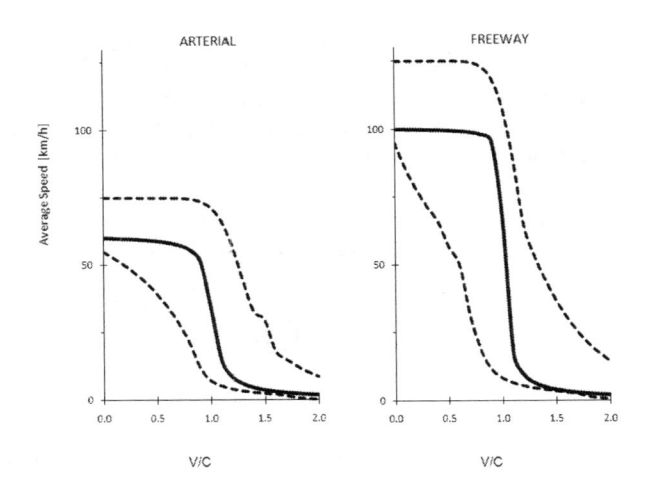

Figure 2: Plausible mean speed range and nominal speed by road type.

3 Results

Figure 3 (next page) presents the envelope of computed maximum absolute errors for average speed as a function of level of congestion and road type. It can be seen that errors can be quite substantial with a factor of up to 2.7 higher NO_x emissions compared to the nominal emissions value in freeway conditions, which was computed using the Akçelik function (eqn. 3). Errors are also dependent on congestion level, where relatively low errors (< about 30%) are computed for uncongested traffic conditions (V/C < 0.7). Errors peak when traffic flow approaches road capacity (V/C about 1), after which they are reduced but can still be substantial (< 60%). Although there are some differences (e.g. in maximum error value), road type does not seem to be an important factor in the relationship between prediction error and congestion level. The maximum values are consistently computed for one scenario, i.e. the COPERT model with 20% HDVs (denoted as COPERT/20%). However, the minimum values are computed for various scenarios depending on road type and congestion level, but include COPERT/0%, QGEPA/0% and QGEPA/20%. Interactions with traffic composition and model choice were observed in the simulations (not shown). For COPERT, errors generally increased with proportion HDVs; whereas, for QGEPA, errors generally decreased with proportion HDVs.

The extent of prediction error is dependent on two factors, i.e. the shape of the composite emissions curve and the difference in predicted average speeds. The location of the minimum value and the degree of non-linearity of both legs (left and right of the minimum value) of the parabolic curve are most relevant in this respect. For instance, figure 2 showed that for congested freeway conditions (V/C = 1) predicted average speeds can vary between 8 and 109 km/h with a nominal value of 68 km/h. Table 1 presents computed NO_x emission factors, sensitivities and maximum absolute errors for all scenarios (model, basic traffic composition). The strongly increasing and non-linear shape of the COPERT/20% composite emission factor curve at lower speeds, as was shown in Figure 1, results in a large increase in the emission factor for the nominal situation, and subsequently in a large sensitivity and error. In contrast, QGEPA/20% has substantially less non-linearity and also relatively large emission factors, which results in lower sensitivity and the lowest (maximum) error for this traffic situation.

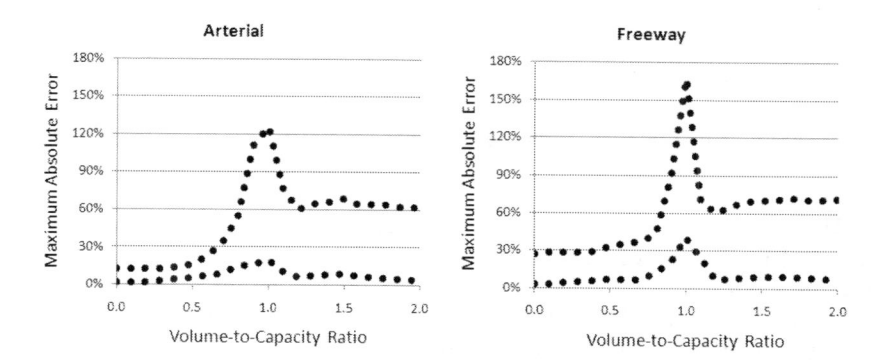

Figure 3: Envelope of maximum absolute errors for mean speed.

Table 1: Composite emissions factors, sensitivities and errors for traffic situation "Freeway, V/C = 1.0" for all six scenarios.

Average Speed	Composite NO$_x$ Emission Factors QGEPA			Composite NO$_x$ Emission Factors COPERT		
	0% HDV	5% HDV	20% HDV	0% HDV	5% HDV	20% HDV
(km/h)	(g/km)	(g/km)	(g/km)	(g/km)	(g/km)	(g/km)
8	1.80	2.42	4.26	0.70	1.40	3.49
68	1.09	1.58	3.04	0.76	0.90	1.31
109	1.19	1.79	3.61	1.18	1.32	1.72
$S_{i,min}$	65%	53%	40%	-9%	55%	167%
$S_{i,max}$	8%	13%	19%	55%	46%	32%
e_i	65%	53%	40%	55%	55%	167%

Figure 4 presents the envelope of computed maximum absolute errors for basic traffic composition as a function of level of congestion and road type. It can be seen that errors can be large with a factor of up to 3.4 higher NO$_x$ emissions compared to the nominal emissions value (5% HDV) in both arterial freeway conditions. Errors are again dependent on congestion level, where smaller errors (50% to 100%) are computed for relatively uncongested traffic conditions (V/C < 0.7 for arterial, V/C < 1.0 for freeways). For more congested conditions, errors increase with congestion level. Road type does not seem to be an important factor in the relationship between prediction error and congestion level, although errors can be slightly higher in arterial driving conditions.

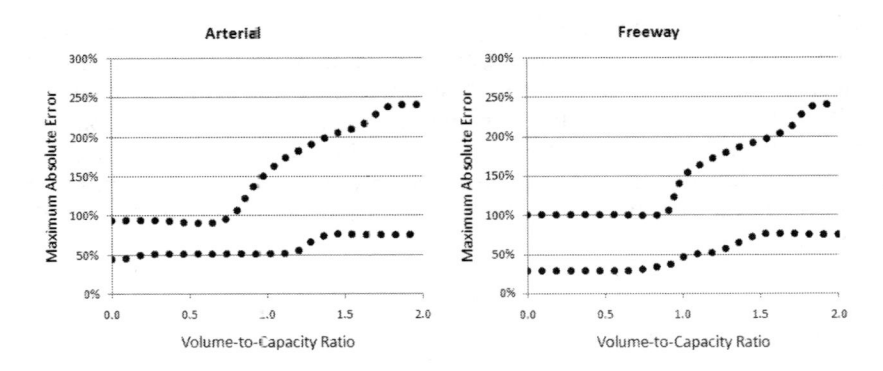

Figure 4: Envelope of maximum absolute errors for basic traffic composition.

Maximum error values are computed for QGEPA/max (maximum speed scenario) for V/C ratios less or equal to 0.8 (mean speeds higher than 55 km/h and 100 km/h for arterials and freeway, respectively) and for COPERT/max for V/C ratios larger than 0.8. In fact, error values computed for the QGEPA model are relatively stable and vary between 70% and 100%, where errors are reduced when congestion level increases. COPERT, on the other hand, is much more sensitive to congestion with errors starting from about 30-50% at free-flow conditions and then consistently increasing when V/C ratios exceed 0.5 (arterial) or 0.8 (freeway) up to errors between 120-240%.

As for average speed, the extent of prediction error is dependent on two factors, i.e. the shape of the composite emissions curve and the difference in predicted average speeds. In addition to the degree of non-linearity of the parabolic curve, the relative difference between composite emission factors for the nominal traffic composition and the minimum and maximum traffic compositions is relevant in this respect. Figure 1 shows that the relative difference between QGEPA/5% and QGEPA/20% is large and only varies slightly (75-100%) with congestion. In contrast, the relative difference between COPERT/5% and COPERT/20% is small at high speeds (30%) but consistently increases after that (up to 240%), having a larger difference than QGEPA at mean speeds below 35 km/h, which explains the increase in maximum error.

Figure 5 presents the envelope of computed maximum absolute errors for model choice as a function of level of congestion and road type. Although not an error in a strict scientific sense (we do not know which model is more accurate, so no baseline values were computed), this term is applied to be consistent with previous discussions. Nevertheless, model comparison provides a sense of possible internal errors that may arise from the arbitrary choices that were made and test data that were used in the development phase of the models.

It can be seen that errors can be large in highly congested conditions with a factor of up to 2.8 higher NO_x emissions in both arterial freeway conditions. Errors are to some extent dependent on congestion level, where relatively stable errors for arterials (approximately 30-60%) and freeways (approximately 10-

60%) occur until traffic conditions reach capacity. After this point, errors remain stable except for arterials where errors can be close to zero. Only at very congested conditions (V/C > 1.7, mean speeds < 25 km/h), maximum errors increase further to a maximum value of about 180%.

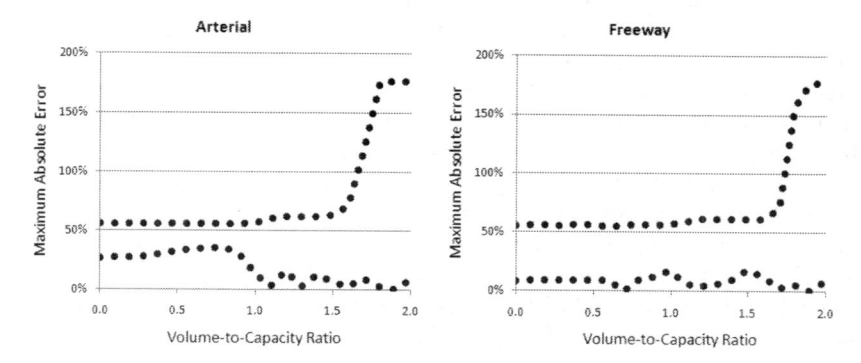

Figure 5:   Envelope of maximum absolute errors for model choice.

The extent of prediction error is dependent on two factors, i.e. the shape of the two composite emissions curves for a particular basic traffic composition and the difference in predicted average speeds. The relative difference between composite emission factors for both models is clearly relevant in this respect. Comparison of COPERT and QGEPA composite emission factor curves (refer to figure 1) reveals that QGEPA generally predicts higher emission factors than COPERT and that for mean speeds smaller or larger than 36 km/h relative differences are largest for a basic traffic composition of 0% HDV or 20% HDV, respectively. For low speeds (< 25 km/h), where largest errors occur, the relative differences between QGEPA and COPERT composite emission factors vary between factors of 2.3 to 2.6, which explain the large error for these traffic situations.

4 Discussion and conclusions

This study has shown that emission predictions at road level are sensitive to possible errors in key input data consisting of traffic activity (VKT), mean speed and basic traffic composition, and model choice. The magnitude of possible errors for mean speed, traffic composition, and model choice were found to be dependent on level of congestion. It was also shown that interaction effects exist. The magnitude of these external errors can be substantial (up to a factor of 3.4). Importantly, they appear to be of the same order of magnitude as internal errors that have been reported from partial road validation studies. This implies that in terms of further improvements of traffic emission modeling, focus should be on both the quality of input data (application) and the quality of the actual emission models (model development).

One limitation of this study is its focus on NO_x. Given the results of this work, it seems valuable to examine the relationships between prediction errors and level of congestion for other air pollutants (e.g. PM_{10}, speciated hydrocarbons and greenhouse gas emissions).

There are some accuracy issues that have not been addressed in this work. A primary issue is that average speed models do not explicitly take driving dynamics into account [38]. This may introduce substantial errors in the emission predictions. For instance, NO_x emissions from an average Euro 3 petrol car could vary between about –80% to +200% around the COPERT estimate for an average speed of 60 km/h [39]. However, Smit et al [35] showed that driving dynamics are implicitly included as lower mean speeds in the real-world are naturally the result of, for example, more speed fluctuation and idle time.

Another issue is the use of a single (mean) speed for all vehicles on a section of road. In reality a distribution of average speeds would apply to a traffic stream. Smit et al [38] showed that this can potentially lead to substantial errors (up to 75%) in road link emissions. In order to address these issues, the work presented in this paper could be extended and refined by using more complex emissions models like VERSIT+ [39], PHEM [40] and DIVEM [41] and by using vehicle-specific driving behavior data in the simulation process, which could be sourced from microscopic simulation models [42].

References

[1] Fujita, E.M., Croes, B.E., Bennett, C.L. & Lawson, D.R., Comparison of emission inventory and ambient concentration ratios of CO, NMOG and NOx in California's South Coast Air Basin. *J. Air Waste Manage. Assoc.*, **42**, pp. 264–276, 1992.

[2] Namdeo, A., Mitchell, G. & Dixon, R., TEMMS: an integrated package for modelling and mapping urban traffic emissions and air quality, *Environmental Modelling & Software*, **17(2)**, pp. 179–190, 2002.

[3] Panitz, H.-J., Nester, K. & Fiedler, F., Mass budget simulation of NOx and CO for the evaluation of calculated emissions for the city of Augsburg (Germany), *Atmospheric Environment*, **36**, pp. S33–S51, 2002.

[4] Mensink, C., Validation of urban emission inventories, *Environmental Monitoring and Assessment*, **65(1-2)**, pp. 31–39, 2000.

[5] Hwa, M., Hsieh, C., Wu, T. & Chang, L.W., Real-world vehicle emissions and VOC profile in the Taipei tunnel located at Taiwan Taipai area, *Atmospheric Environment*, **36**, pp. 1993–2002, 2002.

[6] Hausberger, S., Rodler, J., Sturm, P. & Rexeis, M., Emission factors for heavy-duty vehicles and validation by tunnel measurements, *Atmospheric Environment*, **37**, pp. 5237–5245, 2003.

[7] John, C., Friedrich, R., Staehelin, J., Schläpfer, K. & Stahel, W.A., Comparison of emission factors for road traffic from a tunnel study (Gubrist tunnel, Switzerland) and from emission modeling, *Atmospheric Environment*, **33**, pp. 3367–3376, 1999.

[8] Colberg, C.A., Tona, B., Catone, G., Sangiorgio, C., Stahel, W.A., Sturm, P., Staehelin, J., Statistical analysis of the vehicle pollutant emissions

derived from several European road tunnel studies, *Atmospheric Environment*, **39**, pp. 2499–2511, 2005.
[9] Rodler, J. et al., *ARTEMIS Validation Report WP 1200*, DGTREN Contract 1999-RD.10429, 2005.
[10] Vogel, B., Corsmeier U., Vogel, H., Fiedler, F., Kühlwein, J., Friedrich, R., Obermeier, A., Weppner, J., Kalthoff, N., Bäumer, D., Bitzer, A. & Jay, K., Comparison of measured and calculated motorway emission data, *Atmospheric Environment*, **34**, pp. 2437–2450, 2000.
[11] Reynolds, A.W. & Broderick, B.M., Development of an emissions inventory model for mobile sources, *Transpn. Res.-D*, **5(2)**, pp. 77–101, 2000.
[12] Mensink, C., Bomans, B. & Janssen, L., An assessment of urban VOC emissions and concentrations by comparing model results and measurements, *Int. J. Environment and Pollution*, **16(1-6)**, pp. 345–356, 2001.
[13] Parish, D.D., Critical evaluation of US on-road vehicle emission inventories, *Atmospheric Environment*, **40**, pp. 2288–2300, 2006.
[14] Mellios, G., Van Aalst, R. & Samaras, Z., Validation of road traffic urban emission inventories by means of concentration data measured at air quality monitoring stations in Europe, *Atmospheric Environment*, **40**, pp. 7362–7377, 2006.
[15] Ekstrom, M., Sjodin, A. & Andreasson, K., Evaluation of the COPERT III emission model with on-road optical remote sensing measurements, *Atmospheric Environment*, **38**, pp. 6631–6641, 2004.
[16] Kuhns, H.D., Mazzoleni, C., Moosmüller, H., Nikolic, D., Keislar, R.E., Barber, P.W., Li, Z., Etyemezian, V. & Watson, J.G., Remote sensing of PM, NO, CO and HC emission factors for on-road gasoline and diesel engine vehicles in Las Vegas, NV, *The Science of the Total Environment*, **322**, pp. 123–137, 2004.
[17] Chan, T.L. & Ning, Z., On-road remote sensing of diesel vehicle emissions measurement and emission factors estimation in Hong Kong, *Atmospheric Environment*, **39**, pp. 6843–6856, 2005.
[18] Guo, H., Zhang, Q., Shi, Y. & Wang, D., On-road remote sensing measurements and fuel based motor vehicle emission inventory in Hangzhou, China, *Atmospheric Environment*, **41**, pp. 3095–3107, 2007.
[19] Peace, H., Owen, B. & Raper, D.W., Comparison of road traffic emission factors and testing by comparison of modelled and measured ambient air quality data, *Science of the Total Environment*, **334-335**, pp. 385–395, 2004.
[20] Matzoros, A., Results from a model of road traffic air pollution, featuring junction effects and vehicle operating modes, *Traffic Engineering + Control*, **31(1)**, pp. 24–35, 1990.
[21] Pierson W.R., Gertler A.W., Robinson N.F., Sagebiel J.C., Zielinska B., Bishop G.A., Stedman D.H., Zweidinger R.B. & Ray W.D., Real-world automotive emissions - Summary of studies in the Fort McHenry and

Tuscarora mountain tunnels, *Atmospheric Environment*, **30**, pp. 2233–2256, 1996.
[22] Rogak, N., Pott, U., Dann, T. & Wang, D., Gaseous emissions from vehicles in a traffic tunnel in Vancouver, British Columbia, *J. Air & Waste Manage. Assoc.*, **48**, pp. 604–615, 1998.
[23] Mukherjee, P. & Viswanathan, S., Carbon monoxide modeling from transportation sources, *Chemosphere*, **45**, pp. 1071–1083, 2001.
[24] Park, J.Y., Noland, R.B. & Polak, J.W., Microscopic model of air pollutant concentrations – Comparison of simulated results with measured and macroscopic estimates, *Transportation Research Record*, **1773**, pp. 32–38, 2001.
[25] Marmur, A. & Mamane, Y., Comparison and evaluation of several mobile-source and line-source models in Israel, *Transpn. Res.-D*, **8**, pp. 249–265, 2003.
[26] Marsden, G., Bell, M. & Reynolds, S., Towards a real-time microscopic emissions model, *Transpn. Res.-D*, **6**, pp. 37–60, 2001.
[27] Pokharel, S.S., Bishop, G.A. & Stedman, D.H., An on-road motor vehicle emissions inventory for Denver: an efficient alternative to modelling, *Atmospheric Environment*, 36, pp. 5177–5184, 2002.
[28] Saltelli, A., Chan, K. & Scott, E.M. (Eds.), *Sensitivity Analysis*, John Wiley & Sons Ltd., Chichester, 2000.
[29] Frey, H.C. & Patil, S.R., *Identification and Review of Sensitivity Analysis Methods*, NCSU/USDA Workshop on Sensitivity Analysis Methods, 2001.
[30] European Environment Agency, *COPERT III Computer Programme to Calculate Emissions From Road Transport - Methodology and Emissions Factors (Version 2.1)*, by Ntziachristos, L., Samaras, Z., Eggleston, S., Gorissen, N., Hassel, D., Hickman, A.-J., Joumard, R., Rijkeboer, R., White, L. & Zierock, K.-H., Technical Report No. 49, 2000.
[31] Queensland Government Environmental Protection Agency (QGEPA), *South-East Queensland Motor Vehicle Emissions Inventory – Final Report*, Prepared by Adam Pekol Consulting, Parsons Australia, Pacific Air & Environment, Computing in Transportation and Resource Coordination Partnership, 2002.
[32] Pekol, A., Anyon, P., Hulbert, M., Smit, R., Ormerod, R. & Ischtwan, J., South East Queensland motor vehicle fleet air emissions inventory 2000, 2005, 2011, *Proc. of National Clean Air Conference*, Newcastle, Australia, 2003.
[33] Smit, R., *An Examination of Congestion in Road Traffic Emission Models and Their Application to Urban Road Networks*, Ph.D. Dissertation Griffith University, Brisbane, Australia, 2006.
[34] Taylor, M.A.P., Bonsall, P.W. & Young, W., *Understanding Traffic Systems: Data, Analysis and Presentation*, 2nd Edition, Ashgate Publishing Ltd., England, ISBN 0 7546 1248 1, 2000.
[35] Smit, R., Brown, A.L. & Chan, Y.C., Do air pollution emissions and fuel consumption models for roadways include the effects of congestion in the

roadway traffic flow?, *Environmental Modelling & Software*, accepted for publication, 2008.
[36] Bullock, P., Stopher, P., Pointer, G. & Jiang, Q., *GPS Measurements of Travel Times, Driving Cycles, and Congestion*, Institute of Transport Studies (ITS), University of Sydney, Australia, Working Paper ITS-WP-03-07, ISSN 1440 3501, 2003.
[37] Akçelik, R., Travel time functions for transport planning purposes: Davidson's function, its time-dependent form and an alternative travel time function, *Australian Road Research*, 21(3), pp. 49–59, 1991.
[38] Smit, R., Poelman, M. & Schrijver, J., Improved road traffic emission inventories by adding mean speed distributions, *Atmospheric Environment*, **42**, pp. 916–926, 2008.
[39] Smit, R., Smokers, R. & Rabé, E., A new modelling approach for road traffic emissions: VERSIT+, *Transp. Res.-D*, **12**, pp. 414–422, 2007.
[40] Zalinger, M., Ahn, T.L., Hausberger, S., Improving an instantaneous emission model for passenger cars, *Proc. of the 14th Symposium Transport and Air Pollution*, Graz, Austria, pp. 167–176, 2005.
[41] Atjay, D.E., Weilenmann, M., Soltic, P., Towards accurate instantaneous emission models, *Atmospheric Environment*, **39**, 2443–2449, 2005.
[42] European Commission, *SMARTEST*, Transport RTD Programme of the 4th Framework Programme, Contract No. RO-97-SC.1059, 2000.

Air pollution from traffic, ships and industry in an Italian port

G. Fava & M. Letizia Ruello
Department Fisica & Ingegneria dei Materiali e del Territorio, Università Politecnica delle Marche, 60100 Ancona, Italy

Abstract

In the present study, the levels of NO_x, primary and secondary formed particulate matter (PM10, PM2.5) originating from the intensive commercial and industrial activities within the port zone of Ancona (Italy) were analysed. The data have been evaluated, calculating by a Weibull distribution approach, the hourly, or using a recoded time, statistics of the pollutants concentrations and meteorological parameters measured in an area situated in a zone of influence of the harbour area. This gave the opportunity to study the dependence of the processes on the local scale as well as the impact of the harbour area on the urban scale. The highest concentrations of PM generally occurred in the harbour area whereas the NO_2 was found to be higher in the urban area. The pollutant transport from the harbour and related to the local dispersive ventilation capacity can become important in episodic situations even in the facing urbanised area.
Keywords: PM10, PM2.5, NO_2, harbour area, urban air.

1 Introduction

Ancona is an Italian town surrounded by a busy sea and hilly streets, with a port for agricultural and industrial goods, traffic and passengers, and for fishing boats. Mineral and vegetable dust coming from the intensive commercial and industrial activities on vegetable seeds such as soybean, sunflower and castorbean add to the gaseous Volatile Organic Compounds (VOC), SO_2 and NO_x from heating and ships sailing around the harbour. The urban area is characterized by high population density and economic development. The resulting pollutant emissions place an increasing pressure on the air quality of this area. Whereas in the past the major reasons for poor air quality were industrial activity and domestic heating, nowadays as a result of the rapid increase in mobility an important amount of air pollutants comes from road traffic. This is particularly significant for the harbour area, where rapid rates of growth in heavy vehicle numbers is expected simply because transportation is an important infrastructural facility and plays a very important role in the overall development of the community. As the number of vehicles is increasing rapidly, it has started hindering the atmospheric purity correlated with the emission loads and meteorological conditions. The lognormal distribution has been widely used to represent the type of air pollutant concentration distribution and the parent frequency distribution gives good result for evaluating the mean concentration of pollutants (Kao and Friedlander [4]; Taylor et al. [6]; Horowitz and Barakat [3]; Yu and Chang [7]). In the present study, we relate the dispersal of PM10, Pm2.5, NO_x and SO_2 originating from industry, ships and traffic in the harbours of Ancona (Italy), relying on local data of meteorological conditions with the aim of assessing the dispersive capacity for this place which is environmentally fragile and health sensitive.

2 Characteristics of the study area, emission data and assumptions

The city of Ancona with about one hundred thousand inhabitants is located at the Adriatic side of the Marche Region in Italy, with longitude between 43°37'00'' and 43°36'21'', and latitude between 13°30'2'' and 13°27'21''; its urban extension is approximately 2x2 square km. The city is surrounded by hills of approximately 300–500m height with only the North and East ends open to the Adriatic Sea. This natural topography has influence on the meteorological conditions determining the air pollution condition. The meteorological station furnishing the data used is located in the centre of the city. For the years 2005–2006 the annual rainfall reached 835 mm while the highest monthly rainfall was 112 mm during September 2006. January was the coolest month with a daily average temperature of 3°C, while in July the daily average reached 28°C in the year 2006. The annual mean wind speed was 3.5m/s with predominant wind directions of WNW, W, and NE.

Manufacturing of vegetable oil seed is the most air-polluting sector together with small and medium–sized manufacturing firms of towboats and yachts present in the harbour and its surroundings. In Fig. 1 a pictorial map of the area is shown, including the main industrial zones and monitoring stations.

Aerial emissions in the harbour originate from manoeuvring ships, activity at the dock, traffic and industrial manufacturing. Dock activity requires all the machinery used for loading and unloading of goods (mainly vegetable seeds, minerals such as kaolin and coal) and road traffic of light and heavy trucks. Annual statistics of embarkment and disembarkment at the Ancona Port are shown in Table 1. The strengths and weakness related to the harbour activities together with the economical opportunities characterise the threats and must be evaluated with a view to the sustainability of the entire port system.

Table 1: Annual statistics of embarkment and disembarkment for the years 2005–06 (tons x 10^3).

Goods	2005		2006	
	Embark	Disembark	Embark	Disembark
Petrol	1177	3688	907	3844
Coal	-	394	5	480
Cereals	10	210	23	265
Kaolin	4	287	3	264
Iron&Steel	36	33	45	25
Total	2692	6517	2542	6689
TEU	42	44	48	49
TIR+TRAILER	104	93	102	94
Passengers	500	1000	600	1000

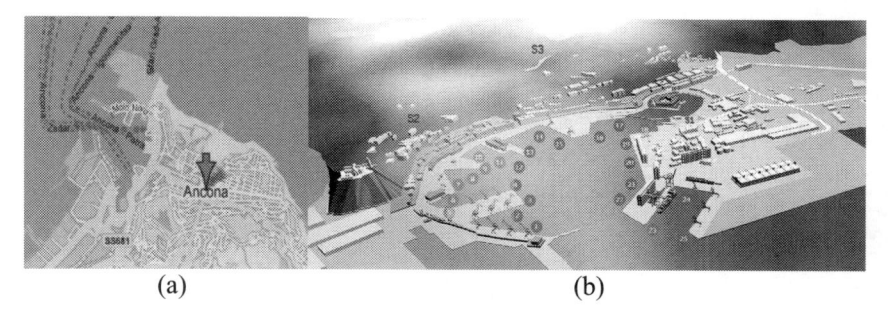

Figure 1: (a) Map of Ancona town (Italy) and (b) layout of its harbour with the sampling locations S1 and S2.

3 Meteorological data

In the present study, hourly and 4-hourly (00:00–06:00, 06:00, 0600–12:00, 12:00–18:00, 18:00–00:00) meteorological data of wind speed, wind direction, for the year 2005–2006, has been used. Table 2 gives the meteorological data for January 2005 through to December 2006. Fig. 2 reflects the statistics of hourly wind speed within the 16 sectors of provenience and for the selected time interval. It is possible to observe low winds especially in the summer season and at night time.

4 Monitoring locations

In the urban area, the municipality continuously monitors NO_x and particulate matter (PM10, PM2.5) levels. The monitoring network contains four permanent stations. Fig. 1 shows two locations of these stations in the harbour and in the city (S1, S2). Simultaneous weather and gaseous pollutants data (including SO_2, CO, NO and NO_2) were also available for data analyses. A monitoring campaign was performed between January 2005 and December 2006 in these two sites characterised by a different exposure to urban (S2) and harbour (S1) sources. A third monitoring station (S3) was located at a rural site far from the industrial and traffic emissions.

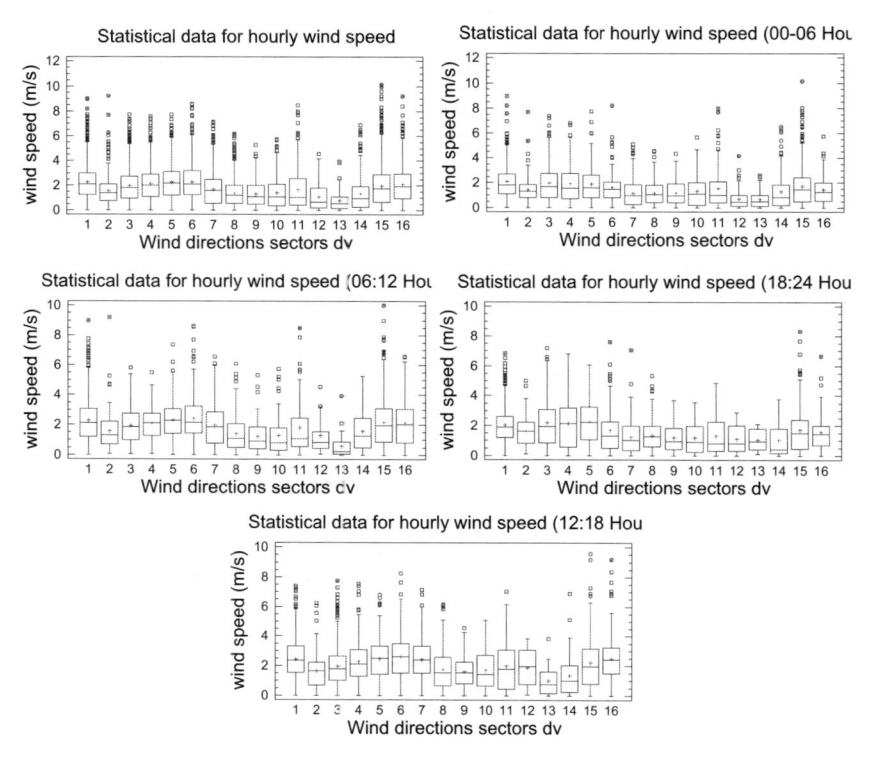

Figure 2: Statistical data for wind speed within the provenience sector and hourly segment (sector equals 22.5°).

A short description of the S1 site and of its exposure to the traffic emissions, dock activity and the duration of the period when the campaigns took place are reported in Table 1.

5 Results and discussion

The concentrations of NO_x, PM10 and PM2.5 caused by the ships, road traffic ad the local industrial activity at the S1 site were analysed and are reported in Fig. 3. At the S1 location PM10 has shown a downward trend since 1998, with a 40% decrease in the year 2006 compared to the year 1998. Within the two years of interest the data shown in Fig. 3 proves that the PM2.5 and to a minor degree also NO_2, decreased significantly in summer compared to winter, whereas the same trend is not observed looking at the coarser PM10 particulates. The distributions of the hourly time series of NO_2, PM10 and Pm2.5 at the stations S1 and S2 also deserve some attention.

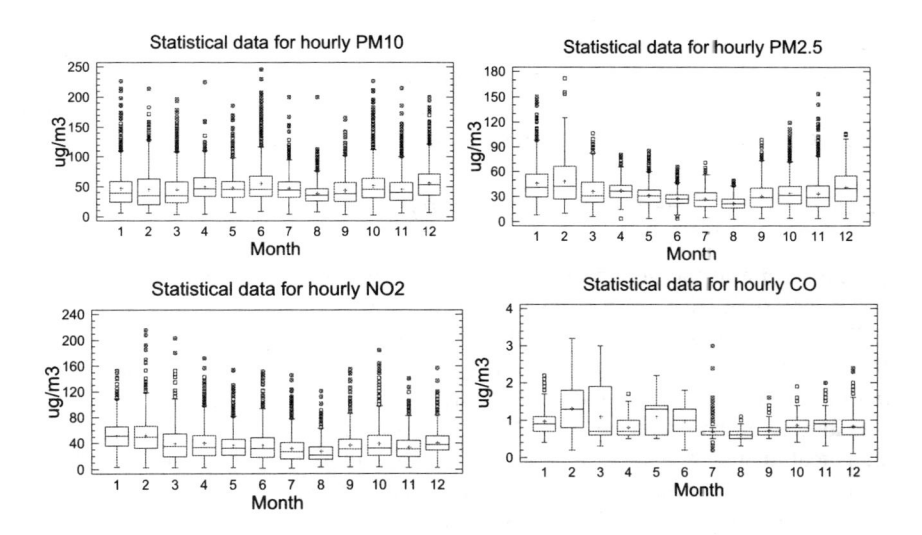

Figure 3: Statistics for PM10, PM2.5, NO_2.

Two theoretical distributions, namely lognormal and Weibull, were used to fit the measured PM10. PM2.5 and NO_2 data. The equations of these distributions were found and the distributional parameters estimated by the method of maximum-likelihood for the theoretical distributions (Georgopoulos and Seinfeld [2]; Lu [5]). Overall the Weibull distribution was found to be the most appropriate to represent all the parent distributions and used to fit the entire measured data. Fig. 4 shows the shape and scale factors for PM10. PM2.5 and NO_2 for the locations S1, S2, S3. The standard deviations of the annually averaged parameters can be considered moderate but require evaluation. We have used the atmospheric Turbulence and Diffusion Laboratory (ATDL) model already adopted to predict long term average pollution concentrations for a variety of urban situations (Gifford and Hanna [8]).

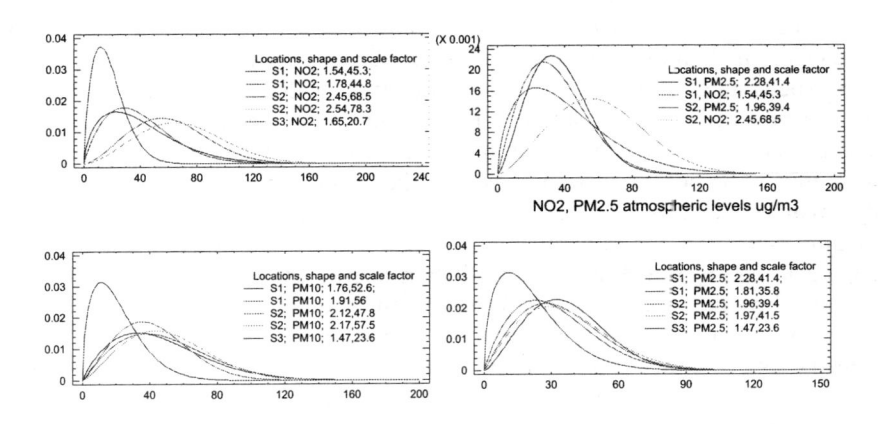

Figure 4: Weibull shape and scale factor for PM10, PM2.5 and NO_2.

In its simplest form the ATDL model assumes the following simple relationship between air pollution concentration χ and wind speed u

$$\chi = C*Q/u \quad (1)$$

where Q is the average source strength and C is an atmospheric stability factor given by $C = \left(\sqrt{\frac{2}{\pi}} * x^{1-b}\right) * (a*(1-b))^{-1}$ and x is the distance from a receptor point to the upwind edge of an urban pollution source. The constants a and b are defined by the vertical diffusion length, σz=axb. Daly and Steele (1975) used a modified version with χ=K/u where K=C*Q. Bencala and Seinfeld also made the important point that if χ and u are related as shown than the air pollution data are lognormally distributed if wind speed data are lognormally distributed. Simpson et al. (1983) have also shown that irrespective of the statistical distributions of air pollution data and wind speed the cumulative frequency distributions F(χ) and G(u) of air pollution and wind speed are related, then it would appear that if χ=K/u is applicable to all the data, the K value can be calculated from F(χ) =1- G(u) where F(χ)=prob(X≥ χ) anf G(u)=prob(U≥u). This in essence implies that the χ value corresponding to the p-percentile χp in F(χ) and the u value corresponding to the (100-p) percentile u100-p are related by K= χp* u100-p (equation (1)).

Looking to the relationship between the wind speed and air pollution levels we considered the data as hourly average, six hours average and data collected at the same hour of each day, the hours chosen being 06:00, 13:00 and 20.00 which should roughly correspond to stable, instable and neutral conditions respectively, even if the sea breeze effects make the general characterisation of atmospheric stability really difficult. The pollutant and wind speed cumulative distributions using the p-percentiles 30, 35, 40, 45, 50, 55, 60, 65, 70 have been used to see how well equation (1) applies by fitting a regression model of the form y=b*x where y = χp and x=1/(u100-p)n. The relationships come out to be best represented by an exponential relation of the type C(ug/m^3)=K*un. The K and n values are reported in table 2.

It would appear from Table 2 that the relationship suggested by equation (1) as a representation of the relationship between non extreme percentile values of wind speed and pollutant concentrations may be widely applicable but with a wind exponent varying from 0.5 to 1 depending upon the locations S1, S2, S3. The percentiles of the wind speed in the cumulative frequency distribution refer to a mix of atmospheric stabilities and the wind exponent n could refer to distinguishing and specific local dispersion conditions.

For estimating the assimilative capacity of the atmosphere, a different approach based on a dispersive ventilation coefficient (VC) was tried. VC is computed by the product of mixing height and the average wind speed for the two seasons, winter and summer. The Atmospheric Environment Services, Canada, has classified that high pollution potential occurs when the afternoon ventilation coefficient is less than 6000 m2s-1 and mean wind speed does not exceed 4ms-1, and during morning hours, the mixing height is less than 500m and mean wind speed does not exceed 4ms-1 (Stack Pole [12]; Gross [13]).

Table 2: K and n values for S1, S2 and S3 site locations.

Site Location S1							Site Location S2							Site Location S3					
PM_{10}		$PM_{2.5}$		NO_2			PM_{10}		$PM_{2.5}$		NO_2			PM_{10}		$PM_{2.5}$		NO_2	
K	n	K	n	K	n		K	n	K	n	K	n		K	n	K	n	K	n
60	0.70	41	0.63	50	0.74		55	0.5	44	0.63	80	0.51		35	0.84	26	0.83	-	-

Site Location S1 — Hour 06.00

PM_{10}		$PM_{2.5}$		NO_2	
K	n	K	n	K	n
37	0.6	30	0.57	31	0.67

Hour 13.00

PM_{10}		$PM_{2.5}$		NO_2	
K	n	K	n	K	n
111	0.86	67	0.87	85	0.93

Hour 20.00

PM_{10}		$PM_{2.5}$		NO_2	
K	n	K	n	K	n
49	0.65	34	0.56	43	0.80

Thus, the ventilation coefficient, boundary layer height for winter (December) and summer (July) at Ancona are shown in Fig. 5, which reveal that morning and evening mixing/inversion height values in both months are low compared to day time values.

The diurnal variation of mixing height follows a similar trend in both these months, since the wind speed is higher in the daytime during the summer season compared to morning and night time. Due to low wind speed and mixing height, the ventilation coefficient is lower in winter compared to the summer season. The maximum value of the ventilation coefficient is recorded as about 2300 m^2s^{-1} in the month of July and 1200 m^2s^{-1} in the month of December, which indicates that no season satisfies the Atmospheric Environment Services criteria. On this basis of this criteria it can be observed that no safe emission hours can be advised throughout the year, even if this approach is not sufficient to draw a conclusion because it does not include sources of pollutants.

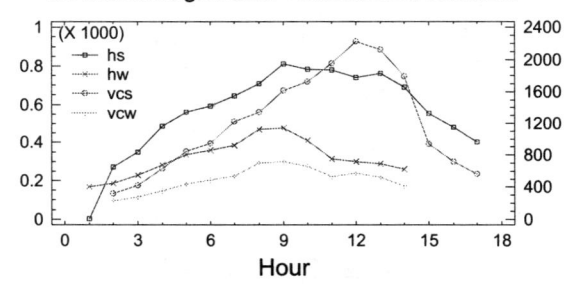

Figure 5: Boundary layer height and ventilation coefficient in winter and summer.

6 Conclusions

The model $\chi=C*Q/u$ as a representation of the relationship between the opposing percentile-values in the statistical distribution for the 30 to 70 percentile requires an exponent to wind speed ranging from 0.3 to 1. The agreement was good for PM10, PM2.5 or NO_2 on an hourly basis or daily or for any specific hour of the day. The empirical derived C-factor ($K(=C*Q)$) in the ATDL model is due to the detailed temporal relationship between χ and u and therefore not to the atmospheric stability, which seems more related to the wind speed exponent n.

Assimilative capacity within the harbour area, determined based on a ventilation coefficient, reveals that the VC is always less than 2400 $m^{-2} s^{-1}$, which indicates that the studied area has high pollution potential implying that assimilative capacity is low. An assessment of this approach would enable one to draw some plausible operational schedule for the sources. The results also reveal that the assimilative capacity is better in the summer season than in winter. In the present study the influence of topographical features and other complex terrain related forcings have not been considered.

References

[1] Sturm, P., Almbauer, R., Sudy, C., Pucher, K., 1997. Application of computational methods for the determination of traffic emissions. Air and Waste Management Association 47, 1204–1210.
[2] Georgopoulos, P.G., Seinfeld, J.H., 1982. Statistical distribution of air pollutant concentration. Environmental Science and Technology 16, 401A–416A.
[3] Horowitz, J., Barakat, S., 1979. Statistical analysis of the maximum concentration of an air pollutant: effects of autocorrelation and non-stationarity. Atmospheric Environment 13, 811–818.
[4] Kao, A.S., Friedlander, S.K., 1995. Frequency distributions of PM10 chemical components and their sources. Environmental Science and Technology 29, 19–28.
[5] Lu, H.C., 2002. The statistical characters of PM10 concentration in Taiwan area. Atmospheric Environment 36, 491–502.
[6] Taylor, J.A., Jakeman, A.J., Simpson, R.W., 1986. Modeling distributions of air pollutant concentrations-I. Identification of statistical models. Atmospheric Environment 20, 1781–1789.
[7] Yu, T.Y., Chang, L.W., 2001. Delineation of air-quality basins utilizing multivariate statistical methods in Taiwan. Atmospheric Environment 35, 3155–3166.
[8] Gifford F.A and Hanna S.R (1973) Modeling urban Air Pollution. Atmospheric Environment 9. 267–271.
[9] Daly N.J and Steele L.P (1975) Air Quality in Canberra Report to Dept. Capital Territory Camberra Australia.
[10] Simpson R.W, Daly N.J, and Jakeman AJ. (1983) The Prediction of Maximum air Pollution Concentration for TSP and CO using Larsen's Model and the ATDL model. Atmospheric Environment 17, 2497–2503
[11] Calabrese M, 2005 Doctoral Thesis.Università Politecnica delle Marche. Ancona Italy
[12] StackPole, J.D., 1967. The Air Pollution Potential Forecast Programme, ESSA Tech, Memo.,WBTMNMC 43.
[13] Gross, E., 1970. The National Air Pollution Forecast Programme, ESSA Technical Memo., WBTM-NMC 47, US Department of Commerce.

Fugitive dust from agricultural land affecting air quality within the Columbia Plateau, USA

B. S. Sharratt
USDA Agricultural Research Service, Pullman, Washington, USA

Abstract

Windblown dust originating from agricultural land has contributed to poor air quality within the Columbia Plateau region of the Pacific Northwest United States. In fact, the United States Environmental Protection Agency (US EPA) national ambient air quality standard for PM10 (particulates ≤10 µm in diameter) is exceeded each year in the Columbia Plateau due to fugitive dust emitted from agricultural land. Winter wheat - summer fallow is the conventional crop rotation employed on >1.5 million ha within the region. During the 13-month summer fallow period, multiple tillage operations are performed to conserve soil water and control weeds; these tillage operations also create erosive soil conditions due to burial of crop residue and degradation of soil aggregates. Instrumentation was installed to measure sediment flux and PM10 concentration at the windward and leeward positions in fields maintained in summer fallow. Soil loss resulting from singular high wind events ranged from 0 to 2317 kg/ha over a four-year observation period. The corresponding loss of PM10 during these high wind events ranged from 0 to 212 kg/ha. For those events with measurable soil loss, PM10 comprised 5 to 12% of the total soil loss. Although loss of PM10 during high wind events is relatively small compared to total soil loss, such quantities are sufficient to degrade air quality. In fact, under atmospheric conditions, which accompany high wind events within the Columbia Plateau, a loss of only 5 kg/ha of PM10 from all agricultural land in summer fallow is sufficient to raise ambient PM10 concentration above the US EPA standard. Therefore, alternative tillage or cropping practices are sought for reducing the loss of topsoil and PM10 from fields managed in summer fallow during high wind events.

Keywords: PM10, wind erosion, fugitive dust, agriculture, particulate matter, high winds, dust storms, windblown dust.

1 Introduction

Fine soil particles emitted into the atmosphere during high winds have contributed to the exceedance of the United States Environmental Protection Agency (US EPA) national ambient air quality standard for PM10 (particulate matter ≤10μm in diameter) in the western US. In fact, PM10 emitted from agricultural land has been a major contributor to poor air quality within the Columbia Plateau region of the Pacific Northwest (Sharratt and Lauer [1]) where the 24-h PM10 standard (150μg/m3) is exceeded on average twice annually (fig. 1). The semiarid climate, occasional high winds, fragile and fine-textured soils, and large extent of land managed in a conventional winter wheat – summer fallow rotation (>1.5 million ha) promotes wind erosion that contributes toward poor air quality within the region.

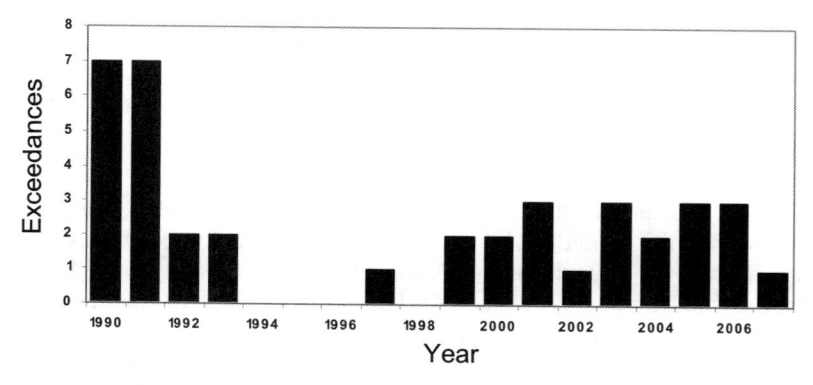

Figure 1: Number of days the US EPA 24-hour PM10 air quality standard was exceeded from 1990-2007 at Kennewick, Washington, USA.

Wind erosion and air quality studies conducted in the Columbia Plateau region over the past decade have focused on ascertaining the magnitude of soil and PM10 transport across agricultural landscapes (Kjelgaard et al [2]) and developing a model capable of simulating horizontal sediment flux and emission of PM10 from soils (Saxton et al [3]). The wind erosion and PM10 emission model developed by Saxton et al [3] attempts to predict vertical PM10 flux as a function of the estimated horizontal sediment transport. Although the aim of these previous studies was to ascertain horizontal sediment transport and vertical PM10 flux from large fields managed primarily in a winter wheat – summer fallow rotation, the experimental design lacked the rigor to assess the loss of soil and PM10 from agricultural fields.

Farm conservation programs such as the United States Department of Agriculture (USDA) Conservation Reserve Program are designed to conserve soil resources and to protect air and water resources. In an effort to conserve soil resources, quantitative data are needed regarding soil loss from farm land because biomass production cannot be sustained where the rate of soil erosion exceeds the rate of soil regeneration. To protect air resources, quantitative data are needed regarding PM10 loss from farm land because loss of fine particulates to the atmosphere must not jeopardize attainment of PM air quality standards. Future farm conservation programs will likely regulate the loss of PM10 from agricultural fields. In fact, the USDA is currently seeking information about tolerable limits of PM10 loss from agricultural fields to protect air quality during high winds within the Columbia Plateau. This study attempts to determine the loss of PM10 from agricultural fields during high winds and implications of PM10 loss on regional air quality.

2 Field campaigns

Soil erosion and loss of PM10 were assessed from agricultural fields located within the Columbia Plateau of the Pacific Northwest United States (fig. 2). Soils within the Columbia Plateau were formed from loessial deposits. Wind erosion is particularly acute in areas with <300 mm of annual precipitation and where winter wheat-summer fallow is the predominate dryland cropping system. This cropping system is typified by a 13-month fallow period that begins after harvest of wheat in July. Soils are typically cultivated on numerous occasions during the fallow period to control weeds and conserve soil water.

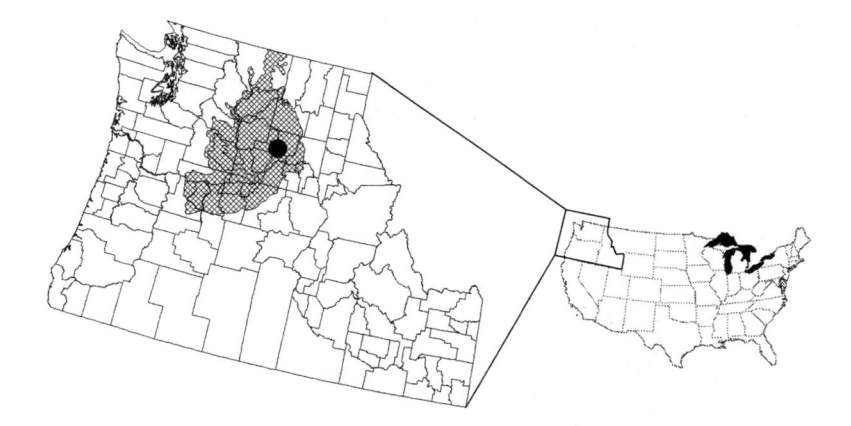

Figure 2: Location of field sites (dot) within Columbia Plateau (shaded area) of the Pacific Northwest United States.

Loss of soil and PM10 was assessed from 2003 to 2006 at field sites (2 to 9 ha) maintained in summer fallow. Soils at the field sites were classified as silt loam with a geometric mean particle diameter of 25 to 35 μm. A nonerodible surface was maintained upwind (south and west) of the sites. Each field site was managed by the cooperator using conventional tillage practices which included disking in the spring after wheat harvest and then rod-weeding at monthly intervals prior to sowing winter wheat in late August. PM10 samplers and creep and saltating/suspended sediment collectors were deployed at the windward and leeward positions in the field to respectively assess the influx and efflux of PM10 and soil. PM10 samplers were mounted at three heights above the soil surface and activated when the wind speed exceeded 6.4 m/s at a height of 3 m for 10 consecutive minutes. This threshold, designated as a high wind event, is that required to initiate movement of soils across the Columbia Plateau [3]. Filters for the PM10 samplers were equilibrated to standard laboratory conditions before and after deployment. Six sets of saltating/suspended sediment collectors (Big Spring Number Eight or BSNE collectors) were deployed in the field with each set consisting of 5 BSNE collectors mounted on a pole at heights from 0.1 to 1.5 m. Collection of field data was periodic due to the remoteness of the field sites and generally occurred immediately after a high wind event or series of high wind events with persistent SW winds. An instrumented field site is portrayed in fig. 3.

Total horizontal soil flux at the windward and leeward positions in the field was equivalent to the sum of creep and BSNE catch. The vertical distribution of saltating/suspended sediment captured by the BSNE collectors was described using the equation:

$$q = az^{-b} \qquad (1)$$

where q is sediment catch (kg/m^2), z is height (m) of the opening of the BSNE collector above the soil surface, and a and b are fitted parameters (Zobeck and Fryrear [4]). Saltating and suspended sediment flux was then determined by integrating eqn. (1) from 0.025 to 5 m (approximate height where integrated flux for all high wind events approached maximum value). Net soil loss from the field for each high wind event was calculated as the difference between total horizontal soil flux at the leeward and windward positions in the field.

Figure 3:   Instrumentation to measure soil and PM10 loss from a field during summer fallow within the Columbia Plateau.

Loss of PM10 from the field site was assessed for each high wind event by subtracting horizontal PM10 flux at the windward position from that at the leeward position in the field. Horizontal PM10 flux (PM10$_{hf}$) at the windward and leeward field positions was determined by:

$$PM10_{hf} = \int C_z u_z t \, dz \qquad (2)$$

where PM10$_{hf}$ is in μg/m, C_z is PM10 concentration at height z, u_z is wind velocity at height z, and t is the duration of an event. The integral was evaluated from 0.025 m to the height of the PM10 plume. The PM10 concentration profile was obtained from BSNE catch and PM10 samplers. PM10 concentration based upon BSNE catch was derived from PM10 mass in the BSNE collector and volume of air moving through the BSNE collector during a high wind event. Plume height was obtained by extrapolating the PM10 concentration profiles at the windward and leeward positions in the field.

3 Observed soil and PM10 loss

Over the four years of this study, soil and PM10 loss from field sites in summer fallow was assessed for 12 high wind events. These events were characterized by persistent SW winds whereby a nonerodible boundary was maintained upwind of the field site for the duration of the event. No erosion or PM10 loss occurred during two high wind events (23 Jun – 8 Jul 2005 and 19-25 Jul 2006) because of the presence of a crust on the soil surface during these events. Characteristics of high wind events during which soil and PM10 loss was observed are reported in Table 1. PM10 loss comprised from 5 to 12% of total soil loss.

Table 1: Characteristics of high wind events observed between sample dates from 2003 to 2006 within the Columbia Plateau, USA.

Year	Sample period	Time (h)	Maximum 3-m wind speed (m/s)	Soil loss (kg/ha)	PM10 loss (kg/ha)
2003	12-22 Sep	26	12.5	43	5
	3-15 Oct	47	14.4	118	10
	15-27 Oct	41	11.9	44	5
	27-29 Oct	14	17.6	2317	212
2004	6-23 Aug	20	10.3	138	16
	23 Aug – 9 Sep	43	10.7	1604	163
2005	31 Aug – 14 Sep	24	13.0	120	7
2006	14-19 Jul	11	9.3	29	2
	29-30 Aug	17	12.0	401	50
	30 Aug – 6 Sep	11	11.6	258	17

The maximum PM10 concentration (based upon 10 minute averages) observed at a height of 5 m equalled 8535 μg/m^3 and occurred during the 27-29

October 2003 high wind event. The mean 5-m PM10 concentration during this 14-hour event was 790 µg/m^3.

4 Implication of PM10 loss for air quality

Loss of PM10 from agricultural land influences air quality within the Columbia Plateau. To better understand the impact of fugitive dust emissions from agricultural fields on regional air quality, the Columbia Plateau Wind Erosion / Air Quality Project (http://pnw-winderosion.wsu.edu/) has supported the development of a regional air quality model. Although a regional air quality model has been developed for the Project and tested with varying degrees of success (Sundram et al [5]), the model is based upon a highly empirical wind erosion and PM10 emission algorithm. Current efforts are now underway to incorporate the Wind Erosion Prediction System or WEPS (WEPS is a processed based wind erosion model as described by Hagen [6]) into the Air Information Report for Public Access and Community Tracking or AIRPACT (AIRPACT can be accessed online at http://lar.wsu.edu/airpact-3/introduction.html) system to better understand PM10 emissions and atmospheric transport processes within the Columbia Plateau.

For the purpose of this paper, we present an analytical approach to determining the impact of fugitive dust emissions on regional air quality. The importance of PM10 loss observed over the four years of this study on air quality can be illustrated by considering a field with dimensions of 100 x 100 m that is maintained in summer fallow. Also assume that PM10 emissions occur from the field on a day when winds of 8 m/s are sustained for 10 hours (Stetler and Saxton [7]), PM10 is emitted into an affected area (cross-sectional length of 100 m) immediately downwind of the field and uniformly mixes to a height of 400 m within the atmospheric boundary layer (Claiborn et al [8]), and winds are calm prior to and following the high wind event. Based upon these field and atmospheric characteristics, emission of 5 kg PM10 during the high wind event will effectively increase atmospheric PM10 concentration by 0.43 µg/m^3 within the affected area. Thus, the atmospheric PM10 concentration would exceed the US EPA national ambient air quality standard of 150 µg/m^3 when >348 fields (>348 ha) immediately upwind of the affected area are each emitting 5 kg PM10 during the event. Similarly, the national ambient air quality standard for PM10 would be exceeded when >9 fields (>9 ha) immediately upwind of the affected area are each emitting 200 kg PM10 during the event. Furthermore, assume that regional storms with predominately SW winds affect PM10 concentrations across a 200 km swath of the Columbia Plateau (Claiborn et al [8]). Based upon the lateral dimension of regional storms, PM10 concentrations within the affected area would exceed the national ambient air quality standard when >0.7 million ha immediately upwind of the affected area are emitting 5 kg PM10/ha or when >17300 ha upwind of the affected area are emitting 200 kg PM10/ha. Although nearly 0.7 million ha are maintained in summer fallow in any given year, the latter case would represent about 2% of all fields in summer fallow within the Columbia Plateau. Therefore, regional air quality can be compromised even during seemingly small events when all land in summer fallow is emitting PM10.

5 Controlling fugitive dust emissions to improve air quality

A close linkage exists between wind erosion and fugitive dust emissions from agricultural lands. Thus, fugitive dust emissions can be suppressed by implementing practices that control wind erosion. A volume of literature exists that describes methods for controlling wind erosion (Nordstrom and Hotta [9])). These methods attempt to enhance cover of nonerodible material (e.g. straw) on the soil surface, roughen the soil surface using tillage implements, and decrease the fetch by using barriers such as tall wheatgrass. Nonerodible materials, however, have little application in the low precipitation zone of the Columbia Plateau where the economic plight of farming in this region precludes additional input costs associated with applying nonerodible material to the soil surface. In addition, the extent of land managed by a single farmer and the non-uniformity in size of farm machinery poses challenges in using wind barriers.

The low precipitation zone of the Columbia Plateau has an annual precipitation of <0.3 m; about 70% of this is received during winter (October – March). These precipitation characteristics necessitate maintaining the land in fallow every other year. For the past 120 years, farmers have used "dust mulch" tillage to conserve soil water during summer fallow. This type of tillage results in a layer of finely dispersed soil particles that is about 0.1 m deep. The "dust mulch" layer is effective in creating an enriched zone of moisture at a depth of 0.1 m (seed depth) when winter wheat is sown in late August. This enriched zone of moisture allows wheat to germinate and emerge through otherwise parched top soil; any delay in sowing past August will reduce yield.

Recent efforts to control wind erosion within the Columbia Plateau have focused on annual cropping systems. Annual crops protect the soil surface during the growing season, but these cropping systems lack the economic and productive stability of the conventional wheat-fallow crop system (Schillinger et al [10]). Reduced tillage systems have attained some success (Schillinger [11]); these systems protect the soil surface by retaining crop residue or nonerodible aggregates on the soil surface. Recently, the USDA Natural Resource Conservation Service has promoted the use of an undercutter tillage implement to reduce wind erosion and fugitive dust emissions (Burnham [12]). While conventional tillage implements (e,g, plow, disk) partially or completely invert the soil during primary tillage, the undercutter implement creates little soil disturbance due to the low pitch of the wide V-blades on the tool bar (fig. 4). Currently, the undercutter tillage implement is being evaluated as a wind erosion control method in the field. Measurements of soil and PM10 loss to date indicate that fugitive dust emissions during high wind events can be reduced by 30 to 70% using the undercutter versus the conventional disk implement during primary tillage.

Figure 4: The top image is an undercutter tillage implement with 0.8-m wide V-blades and the bottom image is a conventional disk implement.

References

[1] Sharratt, B.S. & Lauer, D., Particulate matter concentration and air quality affected by windblown dust in the Columbia Plateau, *Journal of Environmental Quality*, **35**, pp. 2011–2016, 2006.
[2] Kjelgaard, J., Sharratt, B., Sundram, I., Lamb, B., Claiborn, C. & Saxton, K., PM10 emission from agricultural soils on the Columbia Plateau:

comparison of dynamic and time-integrated field-scale measurements and entrainment mechanisms, *Agricultural and Forest Meteorology*, **125**, pp. 259–277, 2004.

[3] Saxton, K., Chandler, D., Stetler, L., Lamb, B., Claiborn, C. & Lee, B., Wind erosion and fugitive dust fluxes on agricultural lands in the Pacific Northwest, *Transactions of the American Society of Agricultural Engineers*, **43**, pp. 623–630, 2000.

[4] Zobeck, T.M. & Fryrear, D.W., Chemical and physical characteristics of windblown sediment I. Quantities and physical characteristics, *Transactions of the American Society of Agricultural Engineers*, **29**, pp. 1032–1036, 1986.

[5] Sundram, I., Claiborn, C., Strand, T., Lamb, B., Chandler, D. & Saxton, K., Numerical modeling of windblown dust in the Pacific Northwest with improved meteorology and dust emission models, *Journal of Geophysical Research*, **109**, D24208, doi:10.1029/2004JD004794, 2004.

[6] Hagen, L.J., A wind erosion prediction system to meet user needs, *Journal of Soil and Water Conservation*, **46**, pp. 106–111, 1991.

[7] Stetler, L.D. & Saxton, K.E., Wind erosion and PM10 emissions from agricultural fields on the Columbia Plateau, *Earth Surface Processes and Landforms*, **21**, pp. 673–685, 1996.

[8] Claiborn, C., Lamb, B., Miller, A., Beseda, J., Clode, B., Vaughan, J., Kang, L. & Newvine, C., Regional measurements and modeling of windblown agricultural dust: the Columbia Plateau PM10 program. *Journal of Geophysical Research*, **103**, pp. 19753–19767, 1998.

[9] Nordstrom, K.F. & Hotta, S., Wind erosion from cropland in the USA: a review of problems, solutions and prospects, *Geoderma*, **121**, pp. 157–167, 2004.

[10] Schillinger, W.F., Kennedy, A.C. & Young, D.L., Eight years of annual no-till cropping in Washington's winter wheat – summer fallow region, *Agriculture, Ecosystems and Environments*, **120**, pp. 345–358, 2007.

[11] Schillinger, W.F., Minimum and delayed conservation tillage for wheat-fallow farming, *Soil Science Society of America Journal*, **65**, pp. 1203–1209, 2001.

[12] Burnham, T.J., Study finds savings in using undercutter, *Western Farmer – Stockman*, May issue, p. 6, 2007.

Emission inventory for urban transport in the rush hour: application to Seville

J. Racero, M. Cristina Martín, I. Eguía & F. Guerrero
School of Engineering, University of Seville, Spain

Abstract

Energy and transport are indispensable ingredients for economic and social development. At the same time conventional forms of energy production, distribution and consumption as well as sustainable transport and mobility patterns are linked to environmental degradation.

The goal of the work is to develop a decision support system able to help local administrators in reducing the impact of air pollution due to urban traffic. The method designed is a framework, which included a transportation planning tool and a comprehensive model to estimate pollutants emissions. The hourly traffic flow data is obtained from transportation planning, and the emission model integrated is based on COPERT methodology. Detailed traffic data have been collected and analyzed from the city of Seville to test the methodology.

Keywords: CORINAIR, emission inventory, traffic assignment model, transportation planning.

1 Introduction

Among the sources of air pollution, road traffic is widely recognised to be the most important and increasing source. Carbon dioxide (CO_2) and water vapour (H_2O) are the most significant transport emissions to the atmosphere. When combustion is incomplete, fuel is oxidised to carbon monoxide (CO) with some volatile hydrocarbons. Impurities such as sulphur are oxidised mostly to sulphur dioxide (SO_2) or to sulphate during the combustion process. And finally, at high combustion temperatures, nitrogen is oxidised to nitric oxide (NO) and small quantities of nitrogen dioxide (NO_2) or nitrous oxide (N_2O). UNECE/EMEP [7] estimated road traffic emissions to account for about 54% of CO emissions, 47%

of NO_x emissions and 27% of hydrocarbon emissions within the European Union.

One of the main applications of road traffic emission estimation is to build an atmospheric emission inventory, a collection of data related to the emissions of pollutants into the air, often with spatial resolution [3]. As governments have the ability to influence the reduction of emissions by the application of different strategies, such as improved fuels and vehicle technology [4], then the emission inventories are important for helping in the identification of zones with similar pollution problems and for the analysis and evaluation of different strategies, which is necessary prior to any implementation. They are also important as input data for atmospheric dispersion models. Finally, these inventories are essential to quantify the population exposed to the pollution levels measured by an air quality monitoring station.

In the last decade, the European Environment Agency [7] has developed some air pollution projects. The aim of those projects was to collect, maintain, manage and publish information on emissions into the air from all relevant sources of environmental problems. All these projects are compiled using the CORINAIR (CORe Inventory AIR) methodology [7]. The pollutant emissions from road transport depend on driving speed and vehicles types, the estimation of those emissions need traffic related data, including the transport activity and driving conditions per category of vehicle.

In the context of an integrated modelling approach, emission models can be considered as tools allowing the calculation of air pollution emissions from a traffic network. Two approaches can be adopted to quantify the atmospheric emissions: the bottom-up and the top-down methodologies [19].

The bottom-up approach is the usual and detailed way to obtain a traffic emission inventory. The emission inventories are obtained directly by measurement of each individual source [13,17]. Total emissions for a geographical area can be obtained by the aggregation of the individual ones. To estimate the urban traffic inventory, the bottom-up approach starts with the calculation of the traffic flow for each road segment using a traffic model based on origin-destination surveys or traffic survey/counters [1]. Then, the results are used as input data of an emission model to finally obtain the estimates emissions for every road of the urban area studied.

The top-down approach starts with aggregated data of total pollution activities throughout the whole geographical area. Spatial desegregation is then performed by assuming that local emissions are proportional to same variable. This approach is usually used to build national emissions inventories, for example, using total petrol national sales as input data, average emission of pollutant per litre of petrol consumed and per category of vehicles, and finally population density.

COPERT-Computer Programme to Estimate Emissions from Road Traffic-[14] is the mathematical model recommended by the European Environment Agency for the compilation of CORINAIR emission inventories. COPERT is based on a database with information on the national automotive fleet, speed-dependent emission functions or average speed for each type of vehicle.

In this paper a bottom-up method to estimate the air pollutant and greenhouse gas emissions from road transport in urban areas is proposed, integrating the COPERT methodology into an assignment traffic model. An application to the city of Seville is also showed.

2 Description of the methodological approach

The emission inventory developed is based on a bottom-up methodology where the pollutants emissions are calculated by a peak day's hour. In this work, air pollutant emission for urban transport has been calculated using a transportation planning software, integrated with EMEP/CORINAIR guide [8]. Using this method, the traffic control operators and authorities can evaluate strategic rules to improve the pollutants emission previous to their implementation.

In this study, an assignment model is applied to forecast the traffic flow and speed in the street. The assignment model is part of classic transport model, which does not solve transport problems, but they can be used within a decision process adapted to the chosen decision-making [15].

The methodology proposed is divided into two phases. The first phase is oriented to forecast the traffic flow and the average speed on the streets or links (Fig. 1). The second phase consists of estimating the pollutants emission classified by vehicles categories (Fig. 2). The traffic flow forecasting is based on solving the equilibrium traffic assignment problem. The assignment model input data are:

- A detail description of road traffic network
- The traffic demand characteristics, expressed in term of origin-destination (O-D) matrix of road vehicles. This matrix shows the hourly mobility between zones. The matrix is updated by a calibration method and traffic surveys/counters located in the network.

One of mainly different between Top-Down and Bottom-Up methods is the speed calculation. The emission equations use the speed as input data, in the most cases the speed is obtained by surveys in specific location and the model applied an average speed for all network. The assignment model resolution provides, by the cost function application, the average speed on each link, that it is more detailed than surveys estimation.

2.1 Traffic assignment model

The assignment model used to forecast the traffic flow is an extension of Beckman's model [2]. This is a non-linear model, based on the Wardrop [18] equilibrium conditions and can be solve by the Frank-Wolfe algorithm [10]. This algorithm allows the estimation of traffic flows within each link of the road network from the knowledge of the network features and traffic demand.

The Frank-Wolfe algorithm computes, by iterative form, the minimum path between all pair Origin-Destination zones and then it assigns trips (Origin-Destination matrix) to the links. In each step the algorithm obtains an optimal factor, which is used to distribute the trips on all paths. The trip time on the links is calculated by applying parametric cost/time functions. All links have associated a cost/time function that depends on links characteristics and traffic flow. The functions have a saturation level that it limits the traffic flow. Finally, with the traffic flow on links, the delay functions are applied in each link to obtain the average speed.

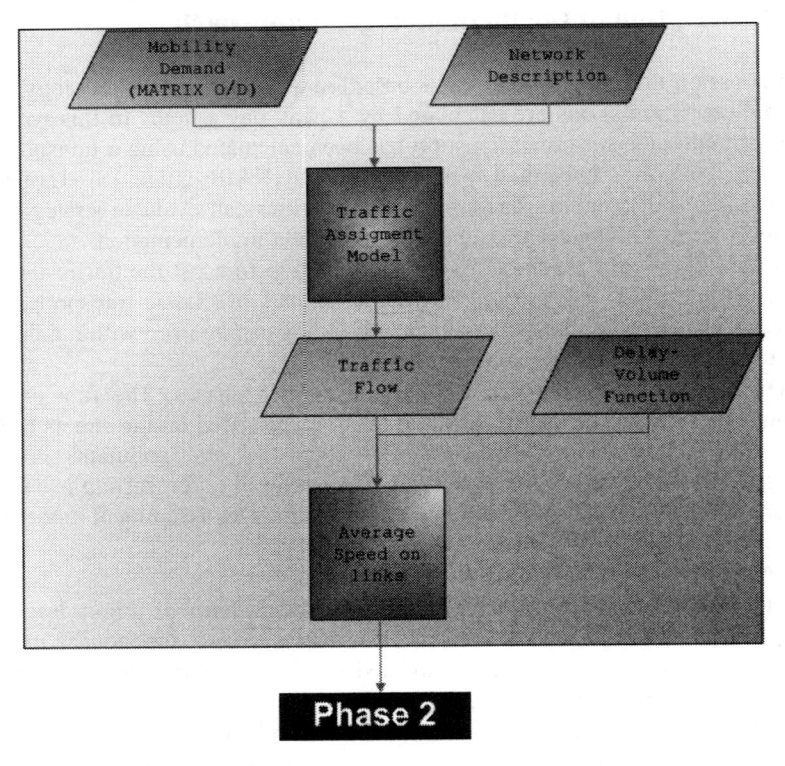

Figure 1: Structure of methodology, phase 1.

2.2 Pollutant emission model

The second phase is oriented to forecast the pollutants emission on all links, based on the average speed calculated in previous phase and using the mathematical equations proposed by COPERT which depends only on the vehicle speeds. The expressions for emissions factors provided in COPERT are based on analysis of a very large amount of data from several European vehicles. EMEP/CORINAIR guide classifies the vehicles in categories according to the type of vehicle, class and legislation. Different pollutants equations are defined for each category [6]. Then, the work focuses on statistical analysis of vehicle fleet. The goal is to compose a percentage or ratio tree that shows the composition of each category.

The final forecast is calculated for each link. It depends on the traffic flow, transport mode allowed and percent of vehicles categories.

3 Application

The methodology has been applied from 8pm to 9pm in the city and metropolitan area of Seville.

The first phase, applied in Seville, can be divided in three tasks: updating the traffic network, obtaining the traffic flow and calculating the average speed. The first one is to update the traffic network using transportation software. TRAMOS is a software developed for this study (an integrated Geographical Information System and Transportation planning, [16]) and used to specify the data associate to each link: transport mode allowed, length, number of lanes and the delay-volume function. In this study about 30 types of functions are used, which depend on typology of the street or avenue, pavement, saturation level, width and speed limit. A volume-delay function has been assigned to each link according to its own features.

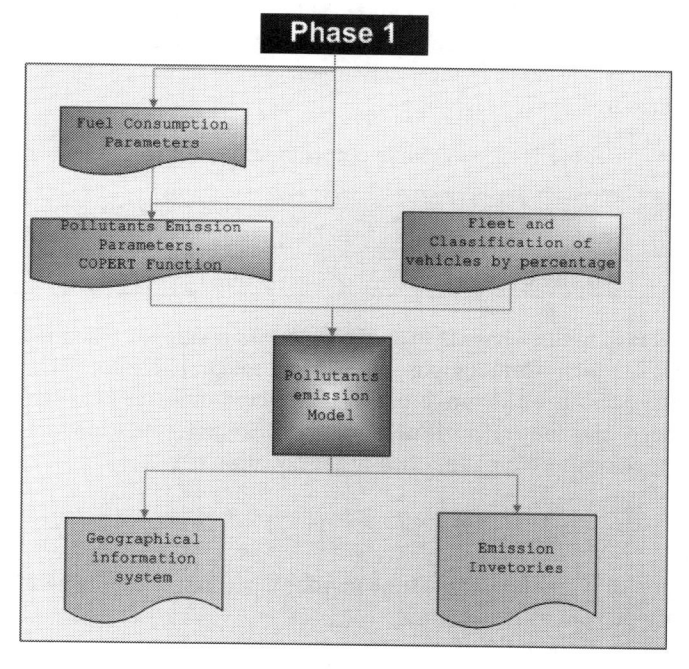

Figure 2: Structure of methodology, phase 2.

The hourly mobility demand between each zone in the study area is grouped in the Origin-Destination Matrix. The Matrix data is obtained from the Transport Planning Studies and it is updated by sensor counters located in some links. In resume, the city model presents about 5000 links, 1200 nodes and 220 zones and the mobility are from 9:00h to 10:00h. Once traffic demand data are provided and the network is updated, it is possible to simulate the traffic flow behaviour within each link of the network. TRAMOS includes a traffic assignment algorithm, which implements a Frank-Wolfe Algorithm. The input data necessary to run the algorithm are the Origin-Destination matrix (Updated) and the network characteristics. Once traffic flows over each links that have been assigned by the traffic model, it is possible to evaluate their relative mean speed by delay-volume function. The average speed on links is calculated by applying the expression (1):

$$Speed_l = Length_l / F_l(I) \qquad (1)$$

where $Speed_l$ is the average speed estimate in the link l, $Length_l$ is the length of the link and $F_l(I)$ is the volume-delay function assign to link l. I parameters are the link attributes as lanes, length and traffic flow.

The next phase is centred on classifying the vehicles into categories. The vehicle fleet information is colleted from vehicle registration in the general traffic administration [5].

The statistical analysis studies vehicles based on engine size, fuel type, emission control technology (registration year) and vehicle class. In each case, this is achieved by firstly determining the percentage according to the total vehicles, from statistical source data (Table 1). Then the percentage tree is determined, and the last level shows the category percentage in relation to total, and the intermediate node is the relative percentage in the category. In Seville, the percentage tree has 79 categories.

Table 1: Vehicle percentage regarding total.

Motorcycle	Car	Van Truck	Bus	Industrial Vehicles	Others Vehicles
0,0605	0,7473	0,1632	0,0023	0,0067	0,0199

The next task in this phase is to apply the emission model. The model is based on COPERT which formulations have been integrated into TRAMOS. The estimation process is an iterative procedure, where on each link the length, the average speed and the traffic flow is used to estimate the pollutants emission according to the expression described in the equation (2):

$$E_{lp} = \sum_{i=1}^{N} P_i \times \delta_{il} \times T_l \times F_{pi}(v_l) \times Length_l \quad \forall l, p \qquad (2)$$

E_{lp}: Emission of pth pollutant in a time period (measured in grams/hour).
P_i: Category percent of ith category
$Length_l$: Length of the link l (kilometre)
N: Number of categories
T_l: Traffic flow on the link (vehicles/hour)
δ_{il}: Link Parameter. 1 if the ith category is allowed on the lth link and 0 other case.
v_l: Average speed in the lth link (kilometre/hour).
F_{pi}: Emission equation of pth pollutant in the ith category (gram/kilometre).

The emission equation is described in CORINAIR guide. The guide presents a set of equations according to the pollutant and speed to estimate road traffic emission. The inventory emission is obtained in disaggregated form (by links and categories, Table 2), the results can be shown in aggregated form, and for example inventories are obtained according to the categories (Table 3). Then, the

Table 2: A category emission in a link (g/h).

Category	Class	Legisl.	NOx	CO	CO_2	N_2O
Private car	Gas<1,6l	Bef. 1985	3,5	130,4	1084,7	15,3
		...	...	...	...	...
		1998-2002	0,6	18,20	1068,8	229,09
	Gas 1,6l-2,0	Bef. 1985	5,0	99,54	1024,8	11,71
		...	...	...	...	...
		1998-2002	0,5	14,48	1105,8	174,61
	Gas>2,0l	Bef. 1985	0,9	21,97	280,14	2,58
		...	...	...	...	...
		1998-2002	0,1	4,43	319,11	38,57

Table 3: Total emissions by category.

Categories	NOx [Kg NO_2/h]	CO [Kg/h]	CO_2 [Kg/h]	N_2O [Kg/h]
Car gasoline < 1.6 l before 1985	65,657	1364,829	12971,420	291,911
...	...	...	...	...
Car gasoline < 1.6 l 1998-2002	12,140	204,310	18355,730	3 547,516
Car gasoline 1.6 l -2.0 l before 1985	125,882	1041,356	11722,904	222,726
...	...	...	...	...
Car gasoline > 2.0 l before 1985	19,361	229,836	3 112,539	49,158
...	...	...	...	...
Car gasoline > 2.0 l 1998-2002	2,617	54,419	5 079,652	597,367
Car diesel < 2.0 l	152,844	202,209	54 176,564	7 380,179
...	...	...	...	...
Motorcycles > 74 c.c.	3,912	194,500	18 643,038	125,161
Total (T/h)	4,103	10,325	416,367	24,367

Table 4: Total emissions in Seville between 8h–9h.

NOx [t $NO2$/h]	CO [t/h]	COV [t Ch1.85/h]	SO2 [t/h]	CO_2 [t/h]	N_2O [t/h]
4,103	10,325	2,479	0,068	416,367	24,367

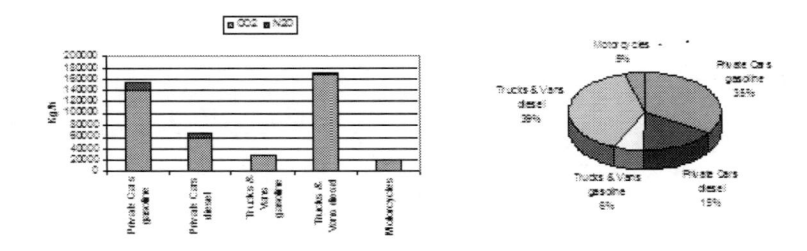

Figure 3: Greenhouse gas emission.

emissions could be aggregated to obtain global results (Table 4). Also, analysis of each category and pollutants group could be calculated (Fig. 3) and total emission can be obtained. Finally, the link data emissions (aggregated or disaggregated) could be used by the traffic control operators to analyze and compare different strategies.

An aggregate analysis (Fig. 3) shows that the private car is the main greenhouse gases contributor in the metropolitan area (50%) at 8–9 a.m. A detail link analysis can be used to design strategies to improve the environmental impacts. Although, the number of cargo vehicle is low their emission are very high (about 40% of total of greenhouse gases).

4 Validation

The Atmospheric Emission Inventory Guidebook [11] provides a general overview of verification procedure showing the tasks that can be applied to demonstrate the applicability and reliability of emission inventory data. The validation method used is based on the precision of the emission factors derived from COPERT [12]. The methodology has been compared with software last version of COPERT (COPERT IV) to probe that the new methodology is equivalent to disaggregate the EMEP/CORINAIR guide. The relation among methods is compared to probe that changes are minimal.

Table 5: Comparison with COPERT.

	NOx (t NO2·h-1)	CO (t·h-1)	CO_2 (t·h-1)
(A) COPERT	3,95	15,025	503 775
(B) New Approx.	4,103	10,325	416,367
Relation (A/B)	0,96	1,45	1,20

The new methodology proposed uses the same vehicles categories as COPERT IV. The average speed used as input data in COPERT IV is the diary mean. This speed is different according to the zone. The comparison shows the differences between the COPERT IV software and the estimation using the new local methodology (Table 5). The ratios between the bottom-up method and top-down method (COPERT) vary in the range 0.8–1.45 depending on the pollutants considered. The approximation indicates that both methods are similar, only to emphasize that the bottom-up method is a bit more accurate than Top-Down method.

5 Conclusion

A new methodology for estimating, predicting and evaluating air pollution due to road traffic in urban areas has been proposed.

The main contribution of the methodology is that it makes an integrated system for use by the local administration in order to forecast alert pollutants levels in urban atmosphere and to analyse and evaluate local and global strategies before applying them in local areas.

This method is a useful evaluator tool in the context of transport planning and local air quality management. The policy advisory tool will aid Local Authorities in trying to asses the implications of their traffic and pollution management policies and will assist in the integration of these activities within air quality management strategies.

Acknowledgements

This work is has been developed into the project "Estimation and improvement methods for power efficiency and environmental air quality in ATMS-ATIS systems" subsidized by the Public Works Ministry with number T30/2006.

References

[1] Baldasano, J. Guidelines and Formulation of an Upgrade Source Emission Model for Atmospheric Pollutants. 1998. Air Pollution Emissions Inventory.
[2] Beckman, M. J., C. B. McGuire, and C. B. Wisten. 1956. Studies in the Economics of Transportation. Ed: New Haven: Yale University Press.
[3] Bellasio, R., Bianconi, R., Corda, G., Cucca, P., 2007. Emission inventory for the road transport sector in Sardinia (Italy). Atmospheric Environment 41 (4), 677–691.
[4] Covile R.N., Hutchinson E.J. and Warren R.F. 2001. The transport sector as a source of air pollution. Atmospheric Environment, 35,1537–1565.
[5] DGT.2003. Dirección general de tráfico. Statistic vehicle registration.
[6] Eggleston H.S., Gaudioso D., Gorissen N., Jourmard R. Rijkeboer R.C., Samaras Z. and Zierock K.H. 1992. CORINAIR Working Group on Emission Factors for Calculating Emissions from Road Traffic. Volume 1: Methodology and Emission Factors.
[7] European Environment Agency (EEA) 2004. Environmental signals 2004. European Environment Agency. Copenhagen. ISBN: 92-9167-669-1

[8] European Environment Agency (EEA) EMEP/CORINAIR, 1999. Atmospheric Emission Inventory GuideBook.
[9] EPA. 2002. *User's Guide to MOBILE6.0. Mobile Source Emission Factor Model*. EPA420-R-02-001. United States Environmental Protection Agency.
[10] Frank, M. and P. Wolfe. 1956. An algorithm for quadratic programming. Naval research logistics quarterly 3 (1-2), 95-110.
[11] Mc Innes G. 1996, Atmospheric emission inventory guide book, A joint EMEP/CORINAIR Production, EEA, B710/9-11, Copenhagen.
[12] Mensink C. 2000 Validation of urban emission inventories. Environmental Monitoring and Assessment 65: 31–39.
[13] Negrenti E., 1995, Bottom-up traffic Emission models, COST 319. Estimation of pollutant emissions from transport. Brussels.
[14] Ntziachristos L. and Samaras Z. 2000. COPERT III Version 2.1: Methodology and emission factors, European Environmental Agency.
[15] Ortuzar J, Willumsen L., 2004. Modelling Transport. Ed: John Wiley & Sons. ISBN 0-471-86110-3. pp. 23-31
[16] Racero J. 2003. Transportation planning and simulation techniques to analysis traffic problems in urban context, thesis. University of Seville.
[17] Teng H., Yu L., and Qi Y., 2002. Statistical microscale emission models incorporating acceleration and deceleration, in Proceedings of the 81st Annual Meeting of the Transportation Research Board, Washington D.C.
[18] Wardrop, J. G. 1952. Some theoretical aspects of road traffic research. Proceedings of the Institution of Civil Engineers, Part II, Vol. 1, 325–378.
[19] Zachariadis, T. and Z. Samaras, 1995, Comparison of microscale and macroscale traffic emission estimation tools: DGV, COPERT and KEMIS, COST 319 – Estimation of pollutant emissions from transport, Pre-proceedings of the workshop on 27-28 November 1995 at ULB - Brussels.

Modeling carbon emissions from urban land conversion: gamma distribution model

A. Svirejeva-Hopkins[1] & H.-J. Schellnhuber[2]
[1]SIM, Faculty of Science, University of Lisbon, Portugal
[2]PIK Potsdam, Germany

Abstract

In this paper we examine the method used for calculating regional urban area dynamics and the resulting carbon emissions (from the land-conversion only) for the period 1980–2050 for the eight world regions. This approach is based on the fact that the spatial distribution of population density is close to the two-parametric gamma-distribution. The developed model provides us with the scenario of urbanization, based on which the regional and world dynamics of carbon emissions and export from cities, and the annual total urban carbon balance is estimated. According to our estimations, world annual emissions of carbon as a result of urbanisation increased to 1.25 GtC in 2005 and began to decrease afterwards. If we compare the emission maximum with the annual emission caused by deforestation, 1.36GtC per year, then we can say that the role of urbanised territories (UT) in the Global Carbon Balance is of a comparable magnitude. Regarding the world annual export of carbon from UT, we observe its monotonous growth by three times, reaching 505 MtC. The latter is comparable to the amount of carbon transported by rivers into the ocean (196-537 MtC). The current model shows that urbanization is inhibited in the interval 2020–2030, and by 2050 the growth of urbanized areas would almost stop. Hence, the total balance, being almost constant until 2000, then starts to decrease at an almost constant rate. By the end of the XXI century, the total carbon balance will be equal to zero, with the exchange flows fully balanced, and may even be negative, with the system beginning to take up carbon from the atmosphere, i.e., becomes a "sink".
Keywords: urban area, carbon emissions, cities' growth, urbanisation, population density, distribution, carbon source and sink, vegetation, carbon cycle.

1 Introduction

In this paper we continue to consider how the urbanisation process influences the Global Carbon Cycle (GCC), using the model of urban area growth based on the two-parametric gamma-distribution of population density (Kendall and Stuart [1], Vaughn [9]). Note that we shall not take into account the clear role of urbanisation in all anthropogenic emissions of CO_2. We are calculating the emissions from the conversion of natural ecosystems and landscapes unto urban lands that take place when cities are sprawling, with additional "natural" lands becoming "urbanised". Some part of the growing city remains "green" and continues to function as an "urbanised" ecosystem. Its characteristics and functioning though become very different, i.e. it is now an "urbanised" ecosystem, where not only the quantitative fluxes of carbon change, but most importantly the qualities of carbon fluxes change significantly (Svirejeva-Hopkins and Schellnhuber [7]). The estimation of the "green" area depends on the type of urbanisation, for example, the presence or lack of planning for city growth, municipal regulations and laws, the attractiveness of a city for a rural population and "*favelas*" phenomenon, i.e. the growth of informal settlements in the Third World.

At the present time, the total area of urbanised territories is relatively small, compared to the total territory participating in the GCC, and is estimated to be a little more than 1% of the total land area (Miller [2]). However, if we take into account the paradoxes of exponential growth so that a factor being negligibly small these days could rapidly become significantly important in the near future, urban territories are good candidates for becoming a leading actor in land use. As compared to our earlier approach, Svirejeva-Hopkins *et al* [5], that was based on the linear regression of urban area to urban population and statistically predicted an area's dynamics until the year 2020, here we have enriched our database by up to 1248 cities and added the generalised statistics for a dynamic case by implementing the multi-regional demographic model (Svirezhev *et al* [8]) that allowed us to predict the dynamics of urban areas further into the future, until 2050, for the eight world regions: Economies in Transition (ET); Highly industrialised/Europe (HI); USA, Canada, Australia and New Zealand (UCA); China (Cn); Africa (Afr); Latin America and the Caribbean (LAC); Arab States (Ar); Asia and the Pacific (AsP). Comparing the new estimations of the regional urban area growth to the previous statistical predictor [5], one could note that the estimations of the regional urbanised areas, although qualitatively coinciding in dynamics, differ in their values, i.e. gamma-model predicts higher numbers for UT.

1.1 Population density

One of the most important variables when determining the total contribution of urbanised territories to the GCC – at global, regional and local levels – is their area. However, although there are many prognoses in the literature, they are mostly for the population values. Nevertheless, there is one standard demographic variable that contains both spatial and population information: population density. It is obvious that if we know the density threshold that determines when a territory becomes urbanised then we may estimate the percent of urbanised area. There is some difficulty, however, in how to define the urban.

After many attempts have been made to find universal definitions, based on density, and none having been totally successful, we have decided to use the "directive" definition of the threshold density when this value is given *ad hoc*. The point is that to set the exact boundary between urban and rural territories with respect to population density is a difficult task, while the absence of a strong, formal definition generates fuzziness in the boundary. As a result, territories with very low densities are often accounted in national statistics as urban ones. It is important to take into consideration that when we estimate urban areas, we operate within the tails of the sampling distributions, which are as a rule determined with low accuracy; while the situation is not significantly improved if we change from the normal to log-normal distribution of densities.

1.2 Population density distribution

Generally speaking, for our purposes, it is more convenient to use another distribution, $p(D)$, where the value $p(D)\Delta D$ is the relative area (percent with respect to the total area, S_t) of such a domain, the population density of which lies between D and $D + \Delta D$. It is obvious that the normalizing condition below must hold:

$$\int_0^\infty p(D)dD = 1. \tag{1}$$

An urbanised area is then defined as:

$$S_u = S_t \cdot \int_{D^*}^\infty p(D)dD. \tag{2}$$

where D^* is the threshold density of population for urbanised territories. The mean population density is equal to:

$$\hat{D} = N_t / S_t = \int_0^\infty Dp(D)dD. \tag{3}$$

If the proportion of a population inhabiting urbanised territories is known (k_u) then:

$$k_u \hat{D} = \int_{D^*}^\infty Dp(D)dD. \tag{4}$$

So, if the distribution $p(D)$ belongs to the class of two-parametric distribution, and its type is known, then there is no problem to estimate the urbanised area if we know the total population, N_t, the percent of urban population, k_u, and the density threshold, D^*, for a given region.

In their classic monograph Kendall & Stuart [1] have indicated that the spatial distribution of population density for different species (including *Homo sapiens*) is close to the two-parametric gamma-distribution:

$$p(D) = \frac{1}{\Gamma(\alpha)\beta}\left(\frac{D}{\beta}\right)^{\alpha-1} \exp(-D/\beta). \qquad (5)$$

where $\Gamma(\alpha)$ is Euler's gamma-function, $\alpha > 0$ determines the form of distribution and β is a scale factor, see fig 1. Note that Vaughn [9] has successfully applied this distribution for the description of population densities in a city centre.

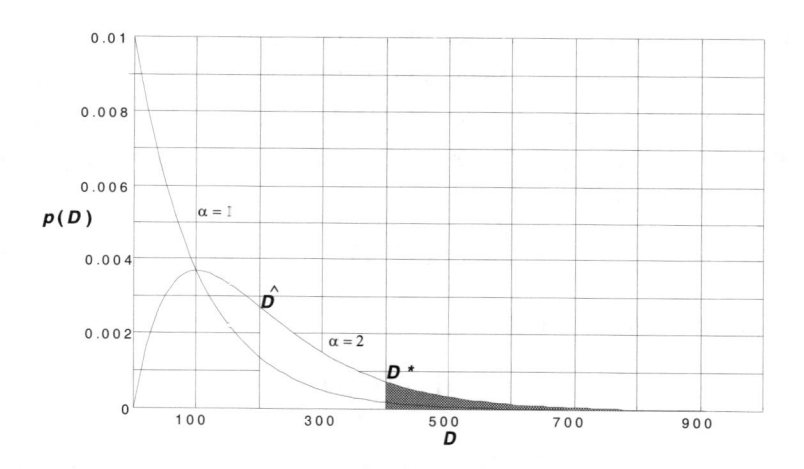

Figure 1: Gamma-distribution as the distribution with density $p(D)$.

Equalities (4) and (5) can be used to estimate these parameters. After such calculations, we obtain:

$$\hat{D} = \alpha\beta \text{ and } k_u = \frac{\Gamma(\alpha+1;\alpha\lambda)}{\Gamma(\alpha+1)}, \quad \lambda = \frac{D^*}{\hat{D}}. \qquad (6)$$

where $\Gamma(\alpha+1;\alpha\lambda)$ is the incomplete Euler function. Finally, knowing the parameters α and β for each region, the percentage of urbanised territory in this region is calculated as:

$$s_u = S_u / S_t = \frac{\Gamma(\alpha; \alpha\lambda)}{\Gamma(\alpha)}. \qquad (7)$$

For instance, the area of the grey domain in fig. 1 is equal to s_u.

Hence, as shown above, urbanised areas may be described through the quantiles of this distribution. Since its parameters depend upon such demographic characteristics as the mean population densities and the density threshold corresponding to an urban territory, by the same token we will obtain the dependence of urban area on urban population.

1.3 Data and demographic prognoses: scenario

As a scenario we use the UN regional prognoses for the total and urban population with some correction, which was necessary to introduce more accuracy to prognoses of the total regional populations. Therefore we use the statistical data and prognoses for the total country population from UN [10] and [8]. In order to estimate the percentages of urbanised populations for the world's regions, we use data from [10] with some corrections and extrapolations.

2 Method: Construction of a model for urban areas' growth based on the gamma-distribution

2.1 Parametric distribution

As it was shown in a previous section, In order to construct a functional relationship connecting the area of urbanised territory and its population, we can use some parametric distribution, the parameters of which are some function of urban population, and, possibly, other demographic characteristics.

We assume that the distribution of population density is described by the formula (5). This determines a fraction of the total area populated with a density range between D and $D + \Delta D$. This share is equal to $p(D)\Delta D$. As already shown, the parameters α and β are determined by using demographic characteristics.

We describe this further using:

$$k_u = N_u / N_t = \frac{\Gamma(\alpha+1; \alpha\lambda)}{\Gamma(\alpha+1)} \quad \text{and} \quad \hat{D} = N_t / S_t = \alpha\beta. \qquad (8)$$

where $\lambda = D^* / \hat{D}$ and $\Gamma(\alpha+1; \alpha\lambda)$ is the incomplete Euler function (Ryzhik and Gradstein [3]). Finally, knowing the parameters α and β for each region, the share of urbanised territory (in relation to the total area) in this region is calculated as:

$$s_u = S_u / S_t = \frac{\Gamma(\alpha, \alpha\lambda)}{\Gamma(\alpha)}. \tag{9}$$

or, using the various relations for gamma-functions from [3] we obtain:

$$s_u = k_u - \frac{(\alpha\lambda)^\alpha e^{-\alpha\lambda}}{\Gamma(\alpha+1)}. \tag{10}$$

If the values of λ and k_u are known, then the values of α are found as solutions of the functional eqn. (6, left). We presume that the values of the density threshold do not change with time, since within each region urban lifestyle, traditions, structure etc. do not change rapidly over the course of considered time. Substituting the inferred values of α into eqn. (9) or eqn. (10), we obtain the relative areas of urbanised territory (% of the total regional area) for each region, fig. 2. Considering the dynamics of the relative regional urban areas we can say that regions such as Afr, Ar, Cn and AsP manifest fast increasing areas of urbanised territory. For instance, urban area in the Africa region will increase 3–4 times during the 80 years (between 1980 and 2050), while, during the same time, the HI, ET and UCA regions will experience insignificant increases in urban area. Fig.3 shows the world dynamics of relative urbanised area. We see that the percentage of the world's urban area in 1985 is about 3%, which is very close to the estimation by Small and Cohen [4].

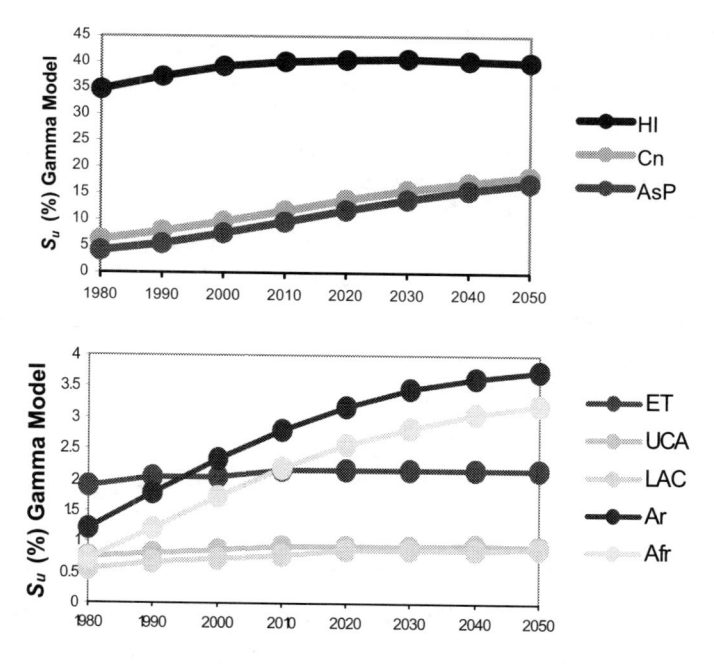

Figure 2: Dynamics of the regional relative urban areas (% of the total regional area).

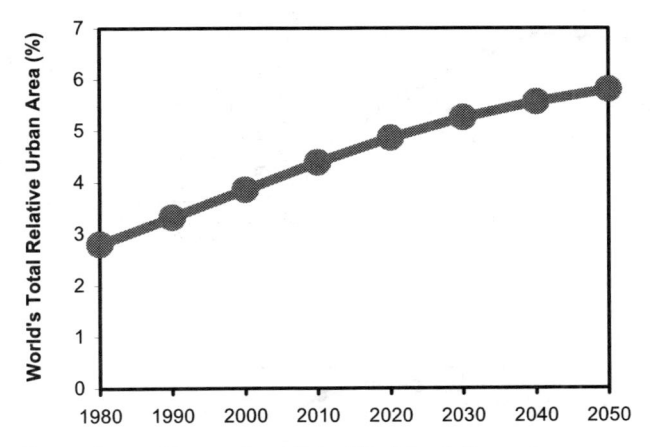

Figure 3: Dynamics of the World relative urban area.

2.2 Emissions, export and the total carbon balance in urbanised areas

As the next step we calculate the dynamics of all three carbon flows [7], i.e. emission, export and the total balance, by using data describing the dynamics of an urbanized area obtained by means of the gamma-model and applying the model of non-random cities distribution, i.e. taking into account that cities are not distributed randomly within the region, but are located within the certain biomes. These calculations were performed for the eight World regions (Svirejeva-Hopkins and Schellnhuber [6]), but we will show here the results for the World in fig. 4; where a result of interaction between the annual carbon losses due to land conversion accompanying the process of urbanisation, $dC_l(t)$ and the annual export of carbon, $dC_e(t)$, is the total carbon balance of the exchange fluxes between the atmosphere and urbanised territory, $dC_{tot}(t)$.

3 Discussion

The World as a whole is an integrator of all-regional tendencies, so let us see what kind of tendencies are observed at the global level.

For the annual emission flow out of the world urbanised territories, fig. 4, we can say that it will slowly increase from 1.12GtC per year in 1980 up to 1.25GtC per year in 2005, after which an increasingly accelerated decrease starts, such that by the year 2050 emissions will have fallen to 623MtC. If we compare the emission maximum, 1.25GtC per year, with the annual emission caused by the process of deforestation, 1.36GtC per year in 1980, then we can say that the role of urbanised territories in the GCC is of a comparable magnitude as the role of deforestation.

World dynamics of the annual "horizontal" export of carbon out of urbanised territories is the following: there is an almost linear growth from 249MtC in

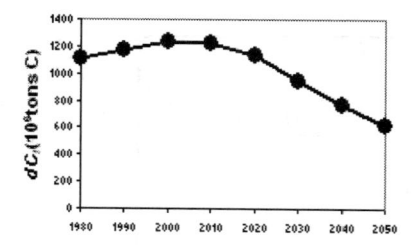

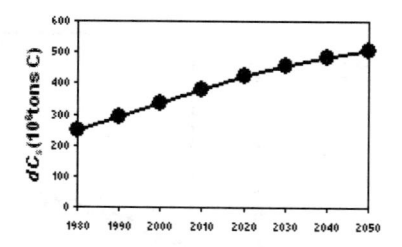

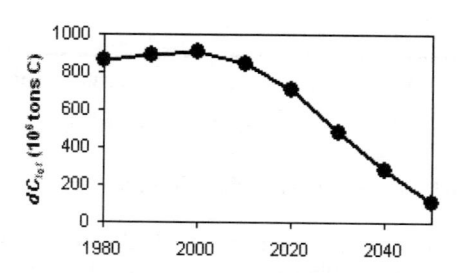

Figure 4: World annual emission of carbon as a result of land conversion (upper left); annual export of organic carbon out of "urbanised" ecosystems into neighbouring territories (upper right) and dynamics of the total carbon (bottom centre).

1980 up to 505MtC in 2050, i.e., during those seventy years export increases two fold. Basically, the transport power of urbanised territories is comparable to the amount of carbon transported by rivers into the ocean (196–537MtC per year, [8]).

Finally, the total carbon balance being almost constant until 2000, then starts decreasing at an almost constant rate. If its maximal value in 2000 is 905MtC, then by 2050 this value has fallen to 118MtC, i.e., by almost eight times. By extrapolating this graph into the future, we can say that by the end of the XXI century, the total carbon balance will be equal to zero, and may even be negative (note that we are talking only about land conversion here). The picture is quite different for the regional dynamics, for example China and ASP regions are continuing to be major sources, while HI region has switched from the source state to the sink one around 2006. However, it is necessary to note that the formation of "sinks" in urbanised territories is accompanied by the appearance of "sources" in other locations. In general, this situation is functionally similar to the case when rivers transport organic carbon to other locations, where it is then decomposed and emitted into the atmosphere.

4 Conclusions

Gamma-model infers dynamics of urbanisation that are very similar to realistic one. The reason for this is that this dynamics is, on the one hand, a result of the generalization of concrete statistical data, while on the other, is a consequence of statistical extrapolation to a near future. It is natural that the use of a model based on the two-parametric gamma-distribution is more attractive then using statistical model, since these parameters are integral indicators that combine demographic characteristics, such as the threshold density, which in turn is connected with life quality and also with a city structure. Therefore, as a future development, by changing the threshold density in accordance with some scenarios, we can study the dynamics of urbanised areas as a function of such scenarios. To estimate the role of UT in the global balance of carbon, we have calculated the total balance of carbon between the atmosphere and urbanised territories, when UT act either as sink or source of carbon. When the growth of UT in a given region happens faster, then the shift towards a "source" state is greater. The "through-pumping" of atmospheric carbon through an "urbanised" ecosystem shifts its state to a "sink". This depends on the green area: the larger it is, the greater is the shift towards a "sink" state. The model forecasts that the peak of urbanisation will happen in a decade. In this light it becomes especially important to consider dynamics of UT and to include urbanisation as a component in the Global Carbon Models.

References

[1] Kendall, M. G. & Stuart, A., *The Advanced Theory of Statistics*, v. 1.2. Academic Press, New York. 1958.
[2] Miller, G.T., *Living in the Environment*, 6th ed. Woodswort, Belmont, CA., 1988.
[3] Ryzhik, I. S. &. Gradstein, I. M., *Tablicy intiegralov, sum, riadow i proizvedenij*, Nauka, Moskva, (in Russian), 1971.
[4] Small, C. & Cohen, J.E., Continental physiography, climate and the global distribution of human population. *Proc. of the International Symposium on Digital Earth*. Chinese Academy of Science, Beijing: 965–971, 1999.
[5] A. Svirejeva-Hopkins, H.-J.Schellnhuber, V.L. Pomaz, Urbanised territories as a specific component of the Global Carbon Cycle. *Ecological Modelling*, 173, pp. 295–312, 2004.
[6] A. Svirejeva-Hopkins & H.-J.Schellnhuber, Urbanised territories as a specific component of the Global Carbon Cycle. *PIK Report No. 94*, ISSN 1436.0179, 2005.
[7] Svirejeva-Hopkins, A. & Schellnhuber H.-J., Modelling carbon dynamics from urban land conversion: fundamental model of city in relation to a local carbon cycle. *Carbon Balance and Management*, 1:8, 15 August. http://www.cbmjournal.com 2006.

[8] Svirezhev, Yu. M. et al. New version of the Moscow Global Biosphere Model. PIK Core Project BBM/Gaia. *Final scientific report.* Potsdam-Institut für Klimafolgenforschung, Potsdam – Moscow: 834pp., 1997.
[9] Vaughn, R., *Urban Spatial Traffic Patterns*, Pion, London, 1987.
[10] United Nations (UN), *The state of the World Cities 2001*. Centre for Human Settlements, UNCHS: 121pp., 2001.

Reduction of CO_2 emissions by carbonation of alkaline wastewater

M. Uibu, O. Velts, A. Trikkel & R. Kuusik
Laboratory of Inorganic Materials,
Tallinn University of Technology, Estonia

Abstract

In energy production based on combustion of fossil fuels, particularly the low-grade ones (lignite, sulphur-rich coal, oil shale etc.), huge amounts of alkaline wastewater are formed. This situation is characteristic for the Republic of Estonia where 59% of the need in primary energy was covered by local oil shale in 2006. Due to the calcareous origin of the mineral part of oil shale, its combustion yields lime-containing ashes (5–6 million tons annually, containing 10–25% of lime as free CaO) as well as enhanced amounts of CO_2 (up to 29.1 t C per TJ energy produced) as compared to coal combustion. To transport the ash into wet open-air deposits, a hydraulic system is used. As a result tens of millions of cubic meters of alkaline water (pH level 12-13) saturated by Ca^{2+}-ions is recycling between the plant and sedimentation ponds. Before directing excessive water to nature it should be neutralized to a pH level accepted by environmental regulations (< 9). Related to this, the current work was targeted to develop possibilities for neutralizing wastewater with CO_2-containing flue gases, leading to reduction of CO_2 emissions. Intensification of respective heterogeneous gas–liquid reactions and characterization of calcium carbonate (PCC) forming were other objectives of the study.

A cycle of laboratory-scale neutralization experiments using industrial wastewater samples and a model gas of flue gas composition has been carried out. Two different reactors – barboter-type columns and a dispergator-type phase mixer – have been comparatively examined. Sedimentation of PCC particles of *rhombohedral* crystallic structure has been shown and their main characteristics have been determined. A neutralization method having up to fifty times higher specific intensity has been proposed. According to calculations 970–1010 tons of CO_2 can be trapped and 2200–2300 tons of PCC per million cubic meters of water can be obtained.

Keywords: carbon dioxide, alkaline wastewater, Ca-compounds, carbonation, precipitated calcium carbonate.

1 Introduction

In several process industries, for example, in energy processing based on combustion of fossil fuels (coal, oil shale, brown coal etc) huge amounts of highly alkaline (pH 10–13) technological and wastewaters are formed. Such technological water is the circulating water used to carry ash formed during combustion to ash-fields. If limestone is used for SO_2 capture or when a limestone-containing fuel (such as oil shale) is burnt, the ash formed contains free lime and its transport water has a high concentration of Ca^{2+}-ions. As a rule, before directing these wastewaters to natural water basins they have to be neutralized, decreasing their pH to the level fixed in environmental regulations (usually, below 9). Neutralization with strong mineral acids such as hydrochloric or nitric acid is well-known. The disadvantage of this method is usage of expensive reagents, release of notable quantities of acid anions to natural water and corrosion of equipment. Besides, due to the character of the neutralization curve (sharp decrease of pH around the equivalent point), a risk of reagent overdose exists. These disadvantages can be avoided by carbonation i.e. by using carbon dioxide for wastewater neutralization, which forms a weak carbonic acid at dissolution. In addition, calcium carbonate is formed in the process, which can be used as a separate product.

Carbonation of alkaline water is a heterogeneous process taking place via gas – liquid reactions. Carbonation methods can be divided into two groups. In the first case gaseous CO_2 is directed into water that is to be carbonized. As the gas-bubbles move through the water-layer, CO_2 is dissolved. This process is slow and to achieve appropriate rates, long residence time is needed, which requires big dimensioned barboter-columns. The rate can be increased by enlarging the area of phase contact, using different methods for dispersing gas bubbles. For example, specific rotating dispergator elements are used inside columns [1]. Carrying out the carbonation process under pressure is also well-known and used widely in food-processing industries [2].

The other group of methods uses injection of liquid phase into a chamber filled with CO_2 or into an adsorption column with rising gas flow. In the last case atomizers are used to intensify the process increasing the contact surface by reducing the liquid particle size. Industrial absorption columns are big dimensioned apparatus having a diameter of 5–6 m and height of 30–40 m. Neutralizing processes for alkaline wastewaters using liquid carbon dioxide as the neutralizing agent are also known, such as the SOLVOCARB®-process and respective apparatus of the company "Linde Gas" [3]. However, using commercial carbon dioxide at power plants where the flue gas contains 12–15% of CO_2 is not reasonable.

The method described in [4] is based on a gas–liquid reaction taking place in a column filled with coal ash (contains free lime), water slurry or eluate into which CO_2-containing flue gas is directed through a distributor. One product of the process is calcium carbonate. Carbonation takes place at bubbling flue gas through the ash–water slurry requiring prolonged duration of phase contact (20 min) and, hereby, a big dimensioned reactor.

Formation of highly alkaline wastewater is also characteristic for the industry of the Republic of Estonia, where more than 90% of electric energy is produced by combusting local carbonate-rich fuel–oil shale [5]. This brings along the formation of CO_2 emissions and waste ash, which contains 10–30% of free Ca and Mg oxides [6], in the annual amounts of 11–13 and 5–6 million tons, respectively. Sedimentation ponds of the hydrosystem for water separation and transportation have a volume 15–20 million m^3 covering an area of about 20 km^2 [7]. The water that reaches the ash fields is oversaturated with Ca^{2+}-ions (1000–1400 mg/L), after sedimentation the Ca^{2+}-ion content in circulation water has dropped to 400–800 mg/L [8]. Taking this into account, the aim of the current research was to reduce CO_2 emissions, to clarify intensification possibilities of carbonation of alkaline wastewater with CO_2-containing flue gases and to give the first characterization of the formed PCC.

2 Materials and methods

2.1 Theoretical considerations and calculations

Neutralization of alkaline wastewater, which contains Ca2+-ions with CO2-containing flue gases, results in the following heterogeneous reactions:

$$Ca^{2+}(aq) + 2OH^-(aq) + CO_2(g) \rightarrow CaCO_3(s) + H_2O \quad (1)$$

$$CO_2(g) + OH^-(aq) \rightarrow HCO_3^-(aq) \quad (2)$$

Dissolution of CO_2 in water depends on the pH, because forming H_2CO_3 (CO_2*H_2O) dissociates to CO_3^{2-} and HCO_3^- and, herewith, the overall concentration of dissolved carbon C_{CO2} increases together with the amount of HCO_3^- and CO_3^{2-}-ions formed. Fractional amounts of carbonate species in solution depending on its pH are presented in Fig 1. This distribution diagram shows that at higher pH levels more CO_2 can be dissolved due to the formation of hydrogen carbonate or carbonate ions. Around the targeted pH value of wastewater (below 9), CO_2 is found in the solution mainly as HCO_3^-. Therefore, the amount of CO_2 necessary to neutralize the alkaline wastewater can be calculated according to Eq (2).

As the wastewater also contains Ca^{2+}-ions (300-1200 mg/L), determining the actual demand of CO_2 is more complex. Contact between Ca^{2+}-ions and CO_2 leads to precipitation of $CaCO_3$, which is practically insoluble at pH $\geq$ 9. Along with the formation of calcium carbonate, alkalinity of the solution also diminishes according to Eq. (1). In the case of high Ca^{2+} concentrations (600-1200 mg/L) and lower pH values (pH<12.4), the stoichiometric amount of CO_2 needed for $CaCO_3$ precipitation is also sufficient for wastewater neutralization. Higher pH values of the initial solution require more CO_2 to neutralize extra OH^--ions according to Eq. (2). So, the overall CO_2 demand can be calculated as the combination of CO_2 amounts needed for precipitation of Ca^{2+}-ions and neutralization of OH^--ions (Fig. 2).

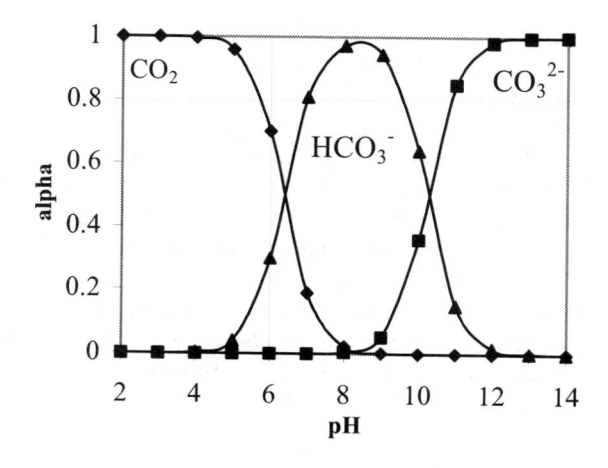

Figure 1: Dependence of the fractional amounts of all carbonate species on the pH of the solution.

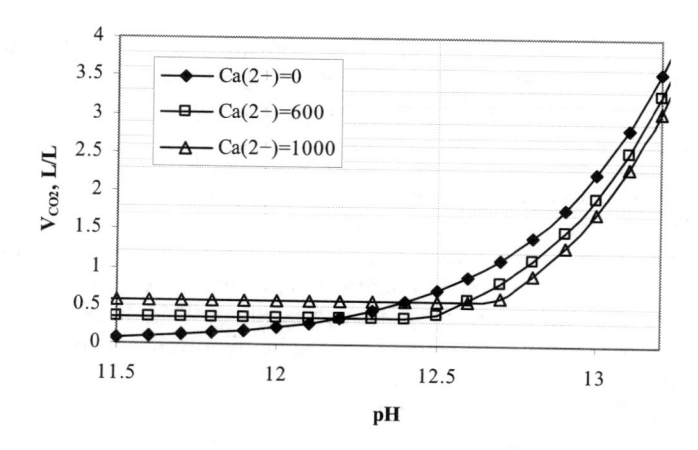

Figure 2: The amount of CO_2 needed for precipitation of Ca^{2+}-ions and neutralization of OH^--ions.

2.2 Dispergator-type phase mixer

Recirculation water from the ash transportation system of an Estonian Power Plant was used for laboratory experiments. The wastewater contained 800–1020 mg/L Ca^{2+}-ions; and its pH was 13.0–13.1. Neutralization of alkaline ash transportation water was carried out in a laboratory dispergator-type batch device (Figure 3a), which provides intensive stirring and good mechanical mixing of gas and liquid phases to increase interfacial contact surface [9]. The initial ash transportation water (flow-rate 82 L/h) was treated with model gases (flow-rate 1497.4 L/h) having flue gas CO_2 (SO_2) content: FG 1 (15% CO_2 in air) or FG 2

(15% CO_2 and 0.07% SO_2 in air). It was determined that the pH of the treated wastewater decreased to 10 after one carbonation cycle at CO_2 excess (amount of CO_2 passed per stoichiometric amount according to Eq (1) and (2)) N_{CO2}=1.5 and to 8 after three cycles at N_{CO2}=4.5. The experiments were carried out at an ambient temperature under atmospheric pressure and the contact time of phases was 1.25-3.75 sec. After separation of the liquid and gas phase the gas was directed into the atmosphere and the liquid phase was either led back into the reactor or to filtration, where the solid reaction product was separated and dried at 105°C.

2.3 Barboter-type column

In order to compare different technological approaches, a conventional barboter-type column (Figure 3b) was also used. Carbonation of alkaline wastewater with model gases FG 1 and FG 2 (flow-rate 100 L/h) was carried out in the laboratory absorber (water column height 500 mm, volume 0.6 L) equipped with a magnetic stirrer (for achieving better interfacial contact) and a sintered gas distributor (pore diameter 100 μm). The experiments lasted until the pH of the treated wastewater reached 10 or 8. The gas phase was analyzed by a gas analyzer (TESTO 350) to determine the residual concentration of SO_2. After treatment the suspension was filtered and solid residue dehumidified at 105°C.

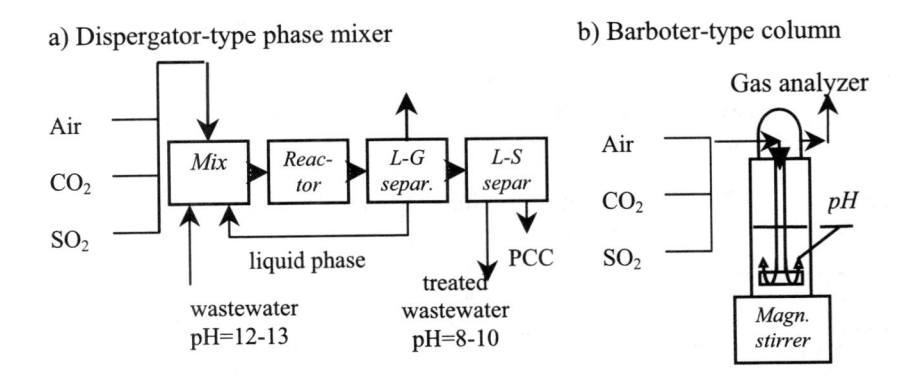

Figure 3: Experiment setup for alkaline wastewater carbonation.

In the liquid phase, pH, the Ca^{2+} content [10], SO_4^{2-} content (spectrophotometer Spectrodirect Lovibond), TDS (microprocessor conductivity meter HI9032) and alkalinity [11] were determined. Solid samples were analyzed for chemical composition as well as for surface properties and particle size distribution. Surface observation of particles was carried out using the scanning electron microscope Jeol JSM-8404A. The laser diffraction analyzer Beckman Coulter LS 13320 was used for particle size distribution analysis.

3 Results and discussion

3.1 Carbonation of wastewater

Comparative experiments with dispergator- and barboter-type reactors indicated that alkaline ash transportation waters (pH=13.0-13.1, Ca^{2+} content 800-1020 mg/L) react readily with CO_2-containing model gases FG1 and FG2, decreasing the content of Ca^{2+}-ions and pH value (Fig. 4 and Table 1). In the case of

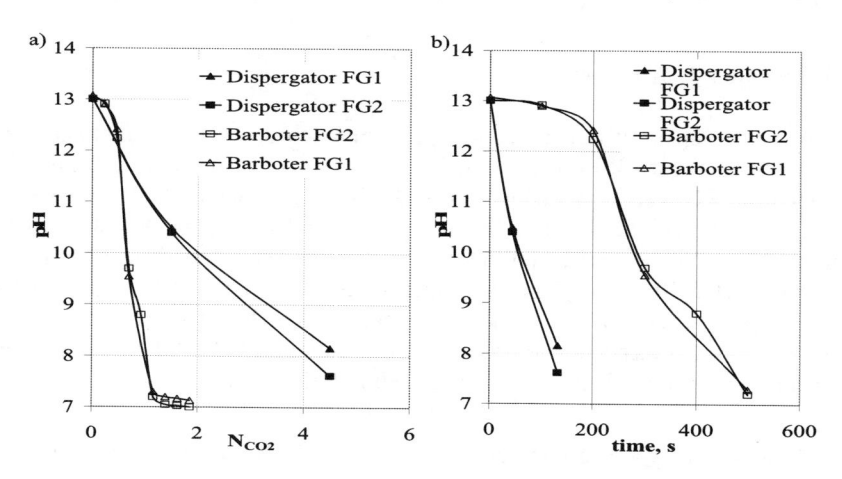

Figure 4: Changes in wastewater pH during carbonation as a function of a) CO_2 excess N_{CO2} and b) time (calculated on 1 L of alkaline wastewater).

Table 1: Wastewater treatment with CO_2 and SO_2-containing model gases: composition of solid and liquid phase.

N_{CO2}	pH	TDS	Ca^{2+}	SO_4^{2-}	OH^-	CO_3^{2-}	HCO_3^-	Solid phase CaCO$_3$	PCC sample
		g/L	mg/L			mmol/L		%	
1. Wastewater treatment in dispergator type phase mixer with FG 1 (CO_2+air)									
1.5	~10	5.61	10	1980	2.0	24.0	0.0	86.01	PCC1
4.5	~8	6.33	30	2210	0.0	6.0	25.0	90.40	PCC2
2. Wastewater treatment in dispergator type phase mixer with FG 2 (CO_2 + SO_2+air)									
1.5	~10	6.43	10	2180	0.0	27.0	1.5	94.16	PCC3
4.5	~8	6.49	100	2340	0.0	7.0	24.5	93.22	PCC4
3. Wastewater treatment in barboter type column with FG 1 (CO_2+air)									
0.7	~10	6.30	5	960	0.0	22.5	6.5	87.95	PCC5
1.04	~8	6.54	135	1200	0.0	0.0	32.5	84.77	PCC6
4. Wastewater treatment in barboter type column with FG 2 (CO_2 + SO_2+air)									
0.7	~10	6.39	10	2540	0.0	14.0	14.75	91.20	PCC7
1.04	~8	6.48	160	1922	0.0	0.0	42.5	88.25	PCC8

the dispergator-type phase mixer the content of Ca^{2+}-ions dropped notably – to 10 mg/L after 1 carbonation cycle (N_{CO2}=1.5) and the pH value decreased to the acceptable level (about 8) after 3 carbonation cycles (N_{CO2}=4.5). The dispergator-type reactor promotes dissolution of CO_2 and the following dissociation of H_2CO_3, which raises the rate of neutralization (Fig. 4b), but due to the short contact time of phases (1.25 s per cycle) dissolution of CO_2 is not completed and higher CO_2 excess N_{CO2} is needed (Fig. 4a) as compared to the conventional barboter-type column. Nevertheless, by using the dispergator-type reactor an increase in specific intensity (q, $m^3/m^3 \cdot h$) of up to 50 times, calculated as the volume of processed wastewater per unit of volume of the reactor, was achieved. It was found that the filtered solid samples (PCC1–PCC8, see Table 1) contained predominantly $CaCO_3$ (85–94%). The distribution of sulphate ion between the liquid and solid phases needs further analysis.

3.2 Characterization of PCC

The PCC samples obtained were analyzed for particle size distribution and surface properties. Results showed that the end-point pH value of carbonation does not influence the size of PCC particles formed in dispergator type phase mixer (PCC1 and PCC2, Table 2). The samples can be characterized by homogeneous particle size distribution: median size ~5 µm and mean/median ratio ~1. The presence of SO_2 in the gas mixture considerably influenced the size of particles (PCC3 and PCC4, Table 2): the median size of particles increased noticeably due to agglomeration of particles reaching 8–9 µm. Additionally, the agglomeration effect escalated with decreasing pH. The sample of PCC4 was not homogeneous, containing particles in the size range of 3–310 µm. The PCC samples from the barboter-type reactor (PCC5–PCC8, Table 2) had somewhat bigger particle sizes, which was also influenced by the extent of carbonation: median size was ~6 µm and 8–10 µm at pH~10 and pH~8, respectively. The presence of SO_2 had a minor influence on the size distribution characteristics of PCC.

Table 2: Particle size distribution of PCC samples.

	Dispergator-type phase mixer				Barboter-type column			
	PCC1	PCC2	PCC3	PCC4	PCC5	PCC6	PCC7	PCC8
Mean, µm	5.263	5.225	9.163	27.38	6.074	9.524	5.734	7.715
Median, µm	5.184	5.103	5.205	8.381	6.169	9.632	5.773	7.883
Mean/Median	1.1015	1.024	0.995	3.267	0.985	0.989	0.993	0.979
	Average size, m^{-6}							
<10%	0.632	0.567	0.880	1.196	0.374	0.789	0.384	0.623
<25%	2.315	2.111	4.895	4.001	3.273	5.728	2.937	4.454
<50%	5.184	5.103	9.205	8.381	6.169	9.632	5.773	7.883
<75%	7.797	7.885	13.32	15.01	8.923	13.54	8.457	11.16
<90%	9.972	10.1	16.71	96.55	11.23	16.77	10.69	13.78

Shape and surface observations confirmed the results of the particle size distribution analysis. The shape of PCC particles from the dispergator-type reactor was influenced noticeably by the extent of carbonation as well as the presence of SO_2. Regularly structured particles of PCC were formed at the high pH region in a SO_2-free environment (Fig. 5a). Extending the carbonation process led to cleaving of particles (Fig. 5b). In the presence of SO_2 most of the individual particles were merged forming agglomerates (Fig. 5c).

The PCC formed in the barboter-type column can be characterized by more homogeneous particles with a distinctive regular rhombohedral crystal structure as compared to the particles formed in the dispergator-type reactor. Shape and surface characteristics were not noticeably influenced by the extent of carbonation or presence of SO_2 in the model gas (Fig. 5d–f).

Figure 5: SEM images of PCC samples formed at different conditions in the dispergator-type phase mixer: a - PCC1, b - PCC2, c - PCC4 and the barboter-type column: d - PCC5, e - PCC6, f - PCC8.

According to thermodynamic calculations [12] precipitation of $CaCO_3$ in the system wastewater–flue gas can be expected at the starting stage of interfacial contact at high pH values. At pH 7.5–9 the binding capacity of the liquid phase is almost utilized and the increase in CO_2 binding is mainly due to the increase in the amount of dissolved CO_2 and formed HCO_3^-. Also, at deeper carbonation to pH<9, $CaCO_3$ becomes soluble in water and the concentration of Ca^{2+}-ions in the solution starts to rise again as is shown in Table 1. Therefore, to avoid agglomeration of the particles and redissolution of $CaCO_3$, the alkaline wastewater neutralization process with production of PCC should be divided into two stages: first, $CaCO_3$ precipitation and separation at higher pH values and second, decreasing the residual alkalinity of wastewater to an acceptable level (pH~8-8.5). Producing PCC of specific properties is a complicated process, which needs further research to obtain better understanding of the influence of different conditions on the crystallization processes.

4 Conclusions

According to calculations 970–1010 tons of CO_2 can be trapped and 2200–2300 tons of PCC per million cubic meters of wastewater can be obtained.

Comparative experiments with dispergator- and barboter-type reactors indicated that alkaline ash transportation waters with pH=13.0–13.1 and Ca^{2+} content 800-1020 mg/L) react readily with the CO_2-containing model gas of flue gas composition. Rapid neutralization accompanied by forming of useful by-product $CaCO_3$ with the following decrease in pH value to an acceptable level (pH<9) was shown. In the case of the dispergator-type phase mixer, which provides intensive stirring and good mechanical mixing of gas and liquid phases increasing interfacial contact surface, an increase in specific intensity of up to 50 times was achieved as compared to the barboter.

In the case of the dispergator-type reactor the shape and structure of PCC particles were influenced by the end-point pH value of carbonation – particles formed at a higher pH value were characterized by a more regular crystal structure and homogeneous size distribution (median size ~5 μm). The presence of SO_2 in the gas mixture influenced the size and shape of particles: due to extensive agglomeration, which escalated with decreasing pH, the median size of the particles increased noticeably. The PCC from the barboter-type reactor was characterized by more homogeneous particles of distinctive regular rhombohedral crystal structure and somewhat larger particle size as compared to samples from the dispergator-type reactor. The size and shape of particles were not influenced by the presence of SO_2 in the gas mixture.

According to the preliminary results, Ca-rich wastewater from the oil shale ash transport system can be used as a sorbent for reduction of CO_2 emissions as well as a calcium resource for PCC production.

Acknowledgements

The authors express their gratitude to Dr Valdek Mikli for performing SEM measurements and to Prof. Juha Kallas from Lappeenranta University of Technology for organizing particle size measurements. The financial support of the ESF (grant G7379) and SC Narva Elektrijaamad is also highly appreciated.

References

[1] Takeuchi, T., Yoshida, S. and Kawai, K. Gas-liquid contacting apparatus, Patent, US-4519959, 28.05.1985.
[2] Alistar, S., Fluid treatment, Patent, EP-0301169 A1, 01.02.1989.
[3] Neutralization of alkaline waste water: A simple and environmentally friendly process using carbon dioxide, http://www.linde-gas.com/International/Web/LG/COM/likelgcom30.nsf/repositorybyalias/pdf_neutralisation/$file/NeutralisationWasteWater_e.pdf
[4] Teruo, N., Takashi, K., Yoshihiro, K., Kouji, A. Carbon dioxide absorption and fixation method for flue gas, Patent, US 2004/0228788 A1, 18.11.2004.

[5] Veiderma, M. Estonian oil shale – resources and usage. *Oil Shale*, **20(3S)**, pp. 295–303, 2003.
[6] Kuusik, R. Uibu, M. and Kirsimäe, K. Characterization of oil shale ashes formed at industrial-scale CFBC boilers. *Oil Shale*, **22(4S)**, pp. 407–420, 2005.
[7] Prikk, A. Oil shale ash - formation, environmental problems and utilization: research report. Tallinn University of Technology: Tallinn, 2004.
[8] Arro, H., Prikk, A. and Pihu T. Reducing the environmental impact of Baltic Power Plant ash fields. *Oil Shale*, **20(3S)**, pp. 375–382, 2003.
[9] Kuusik, R. et al. Method for neutralization of alkaline waste water with carbon dioxide consisting in flue gas, Patent, EE200600041, 22.12.2006.
[10] Vilbok, H., Ott, R. *Volumetric analysis. Instruction for Practical Works*, Tallinn Polytechnical Institute: Tallinn, 1997.
[11] Water Quality. Determination of Alkalinity. International Standard ISO 9963-1:1994(E).
[12] Kuusik R., Türn L., Trikkel A., Uibu, M. Carbon dioxide binding in the heterogeneous systems formed at combustion of oil shale. 2. Integrations of system components – thermodynamic analysis. *Oil Shale*, **19(2)**, pp. 143–160, 2002.

Section 4
Monitoring and measuring

Characterisation of inhalable atmospheric aerosols

N. A. Kgabi[1], J. J. Pienaar[2] & M. Kulmala[3]
[1]Department of Physics, North-West University, Mmabatho, South Africa
[2]School of Chemistry, North-West University, Potchefstroom, South Africa
[3]Department of Physical Sciences, University of Helsinki, P. O. BOX 64, FIN-00014, Helsinki, Finland

Abstract

A better understanding of the chemical constituents of ambient particles is fundamental in bridging the knowledge gap between the air quality and its health effects. In this study, Inductively Coupled Plasma Mass Spectroscopy (ICP-MS), Atomic Absorption Spectrometry (AAS) and Scanning Electron Microscopy coupled with Energy Dispersive Spectrometry (SEM/EDS) were used to characterise the inhalable atmospheric particulate matter (PM10). About 30 elements were identified using ICP-MS and the mean concentrations of Cr, Ni, V and Pb for the sites RMINE and RCBD were found to be 2.55(±1.42) and 0.18(±0.08), 1.41(±0.73) and 0.13(±0.07), 0.28(±0.12) and 0.03(±0.01), and 0.35(±0.15) and 0.48(±0.28) µg/m^3 respectively. SEM/EDS yielded information on most crustal elements and their total oxides but could not detect Ni, V and Pb. The Cr concentrations at the two sites in the Rustenburg municipality were measured as 2.28(±0.64) and 0.14(±0.04) µg/m^3. AAS yielded the toxic metal concentrations of 1.03(±0.54) and 1.52(±0.56) for Cr, 0.29(±0.09) and 0.58(±0.23) for Ni, 1.37(±1.02) and 1.38(±0.59) for V, and 1.32(±0.27) and 1.31(±0.57) µg/m^3 for Pb. All the methods used could not give unambiguous information on specific oxides of metals.

Keywords: particulate matter, ICP-MS, SEM/EDS, AAS.

1 Introduction

Characterisation of inhalable airborne particulates is becoming increasingly important to governments, regulators and researchers due to their potential impacts on human health [1], transnational migration and influence on climate forcing and global warming [2] and on the ecosystems [3]. Chemical speciation is essential for establishing more specific relationships between particle concentrations and measures of public health. Monitoring of both PM mass and chemical composition is also important for identification of emission sources, determination of compliance, and development of effective control programs.

Major components of atmospheric aerosols include sulphate, nitrate, ammonium, and hydrogen ions; trace elements (including toxic and transition metals); organic material; elemental carbon (or soot); and crustal components [4]. Each aerosol species is formed in the atmosphere separately by different processes and are mixed together to form particles of mixed composition.

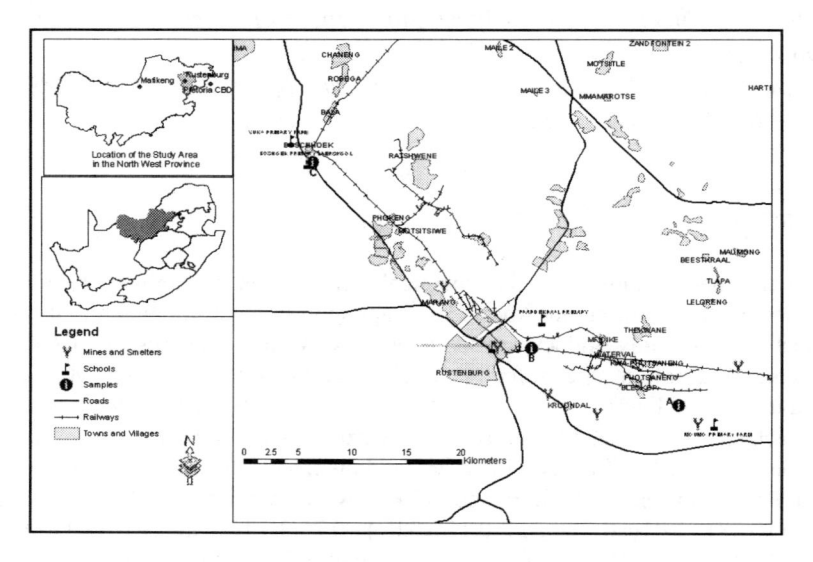

Figure 1: A map showing the location of the study sites RMINE and RCBD given as A and B respectively, in the Rustenburg municipality.

Insoluble aerosols act as foreign bodies stimulating defence reactions in living organisms, e.g. crystalline silica (quartz) relative toxic to cells. Soluble aerosols are easily transferred to blood and other parts of the body, e.g. lead, cadmium. The molecular form (e.g. oxidation state) is very important in metal-containing aerosols [5].

The sampling of the atmospheric particulate aerosols was performed using the TEOM series 1400a incorporated with the PM_{10} inlet at two sites A, hereafter referred to as RMINE with latitude $25^0 43' 03,0'' $ E and longitude

$27°23'57,8''$ S, and B, which is referred to as RCBD with latitude $25°40'01,3''$ E and longitude $27°16'38,5''$ S. These sites are representative of well-defined environments, exposure situations or source activities like remote areas, urban background, traffic, and industry. The map in Figure 1 shows the location of RMINE and RCBD.

The objective of this study was to determine the composition of the inhalable particulate aerosols and hence evaluate the efficiency of the Inductively Coupled Plasma Mass Spectroscopy (ICP-MS), Atomic Absorption Spectrometry (AAS) and Scanning Electron Microscopy coupled with Energy Dispersive Spectrometry (SEM/EDS).

2 Experimental

The ambient air samples used for ICP-MS, SEM/EDS and AAS analysis were collected on 15mm Teflon Coated Borosilicate glass fibre filters at the flow rate of 3.0 L/min with the total flow of 16.7 L/min where the 13.7 L/min was the bypass flow. The filters were changed monthly, or when they are approximately 80% full, using the forceps and placed in the plastic containers with caps. The filters were scanned using SEM/EDS, followed by extraction for ICP-MS. A portion of the extracted solution from ICP-MS was stored for further analysis using AAS. Thus each sample was analysed using the three methods.

The ESEM FEI QUANTA 200 coupled with the OXFID ENCA 200 EDS was used for elemental analysis of samples. The requirements for sample analysis by SEM are that the sample has to be stable under the vacuum conditions, it has to be conductive to allow electron beam irradiation, and lastly the substrate of sample should fit well into the sample stage. The samples were analysed at high vacuum, with a voltage of 15 kV and the working distance of 10 mm. A dead time of forty percent, which corresponds to the live time of 100 seconds, was used during the analysis of samples. The filters were fixed onto sample studs to ensure good electrical connection between the specimen and the microscope stage. The samples were not coated. The filters were scanned several (10) times (1 scan per 10 second) to ensure that the representative portion of the sample is covered. No extraction was performed.

The samples collected on filters were also analysed using the Agilent ICP-MS 7500c with the operating conditions and measurement parameters given as 3 replicates, Rf power of 1.6 kW, Rf Matching of 1.64 V, carrier gas flow rate of 1.09 L/min, wash time and rinse time of 5 s and 60 s respectively, sample uptake rate and stabilization of 55 s and 45 s respectively. Air particulates collected on filters were extracted into a dilute nitric acid solution by sonication with heating and the extracted solution was stored at room temperature until analysis by ICP-MS.

AAS is the quantitative technique in which the optical absorption of atoms in the ground state is measured when the sample is irradiated with the appropriate source that is; the absorption of optical radiation by atoms in the gaseous state is measured. The elemental analysis depends on the measurements made on the analyte (solution) that is transformed into free atoms.

A limitation associated with the ICP-MS and AAS analysis is the difficulty in assessing the efficiency of extracting the PM from different filters. This has been reported before by researchers who estimated an efficiency of 20–50% of PM removed from filters [6] and those who estimated an efficiency of 10–30% [7]. Attempts were made to assess the extraction efficiency by measuring the extracted mass. The other limiting factor in these procedures may be the possibility of contamination as the extraction solution for ICP-MS was further treated and used for AAS analysis.

3 Discussion of results

The chemical composition of samples that were compared was collected at the sites described in section 2. Sample 1, 2 and 3 were obtained at RMINE during the sampling periods 20 February – 12 March 2004, 22 April – 15 May 2004 and 15 May – 8 June 2004 respectively, and Sample 4, 5, 6, 7, 8, 9, 10 and 11 were collected at RCBD during the periods 8 – 29 June 2004, 29 June – 23 July 2004, 23 July – 19 August 2004, 19 August – 10 September 2004, 10 – 28 September 2004, 28 September – 26 October 2004, 26 October – 12 November 2004 and 12 – 23 November 2004 respectively. The composition of these samples was determined using SEM/EDS, ICP-MS and AAS.

3.1 SEM/EDS

Figure 2 gives the morphology of Sample 1. The dark area indicates the carbon content, the greyish area is mainly the silica and the white areas are the Cr content [8]. Greyish particles are predominantly geological elements that include the iron, silica and manganese. This can be attributed to re-suspension of soil particles during wind episodes and emissions from the industry during mineral processing. Angular fractured grains are produced as a result of mining and milling processes. The irregularly shaped particles may be linked to the results of anthropogenic abrading processes. The white particles that are predominantly chromium and its oxides can be associated with the industrial

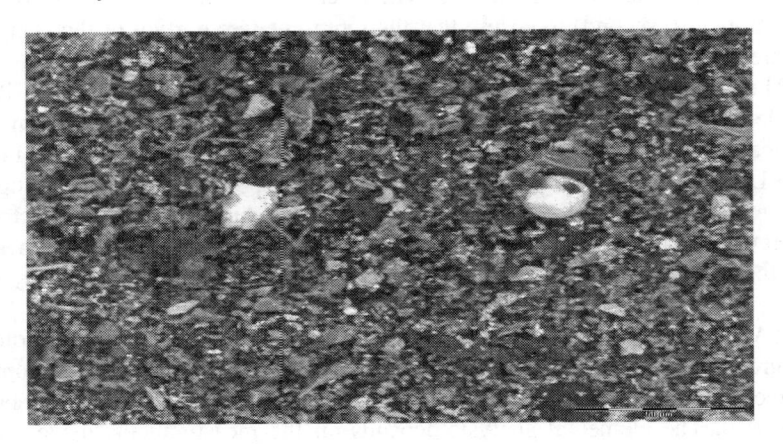

Figure 2:   Micrograph of Sample 1 obtained at RMINE.

emissions particularly the ferrochrome smelter located about 15 km from the study area.

Spherical particles shown in Figure 2 are typically associated with particles that have been formed in a high-temperature furnace, such as a coal-fired boiler. These are typically alumino-silicates, often with significant concentrations of iron that come from pyrites and other iron-containing minerals in coal [9]. Smelting operations may also generate spherical particles, including elements specific to the ores. This simply means that the ferrochrome smelter in the study area is expected to emit Cr and Fe that are predominantly spherical in shape. Elemental analysis indicates that a large portion of the spherical particles contain elevated concentrations of carbon, oxygen, and silicon and lesser quantities of magnesium, sodium, aluminium, sulphur, calcium, and silicon.

Table 1 gives the concentrations of the oxides of the elements identified at RMINE. For all the samples described above, the data was obtained from analysis of Teflon-coated borosilicate fibreglass filters.

Table 1: Concentrations (in μgm^{-3}) of oxides of the elements identified by using SEM/EDS during autumn and winter at RMINE.

Sample	PM10	Si	Fe	Al	Ca	Mg	K	Na	Ti	Cr	C	Cl	S	F	O
1	131.10	7.08	3.28	4.06	1.44	1.70	0.52	1.05	0.40	3.28	18.22	-	6.42	8.39	74.73
2	99.74	14.16	3.49	5.69	3.59	3.29	0.60	0.70	-	1.10	8.08	0.30	4.29	-	54.66
3	94.36	14.44	5.57	5.76	3.96	3.02	0.57	0.57	0.28	2.45	6.89	0.28	1.60	-	45.39
MEAN	108.4	11.89	4.11	5.17	3.00	2.67	0.56	0.77	0.23	2.28	11.06	0.19	4.10	2.80	58.26
SD	11.46	2.41	0.73	0.56	0.79	0.49	0.02	0.14	0.12	0.64	3.59	0.10	1.39	2.80	8.66

The oxides on the main elements identified at this site are Si, Fe, Al, Ca, Mg, K, Na, Ti, Cr, C, Cl, S, F and O. There is an increase in concentrations of crustal elements Si, Fe, Al, Ca, Na and Mg from the autumn to the winter season. The S and C concentrations also show a decrease from autumn to winter. South Africa has dry winter seasons and the concentration of S shown in Table 1, at the beginning of winter (15 May to 8 June) is 1.60 μgm^{-3}, 4.29 μgm^{-3} (22 April to 15 May) and 6.40 μgm^{-3} in autumn (20 February to 12 March). The rainfall reported by the South African Weather Services as given in Table 2 are 175.6, 11, 27.4, 0.2, and 5.2 mm for the months of February, March, April, May and June respectively. The levels of the oxides of sulphur are very important in a study that deals with toxic trace metals, because most of the metals occur in ambient air as sulphates.

The potentially toxic metal of interest identified at RMINE is Cr, with concentrations of 3.28, 1.10, and 2.45 $\mu g.m^{-3}$ for the three samples analysed at this site. This has serious implications since for Sample 1 and 3; the limit of 1.5 $\mu g.m^{-3}$ for Cr, set by the Asia Pacific Centre for Environmental Law (APCEL) is exceeded [10]. The limit of 1000 ng/m^3 (1 $\mu g.m^{-3}$) set by the National Institute for Occupational Safety and Health (NIOSH) was exceeded for the whole sampling period at this site.

The elements identified at RCBD, as shown in Table 2 include Si, Fe, Al, Ca, Mg, K, Na, Ti, Cr, C, Cl, S, F, V, Pb, and N. For most samples, the oxides of Pb, V, N, P and F could not be detected.

Table 2: Concentrations (in μgm^{-3}) of oxides of the elements identified by using SEM/EDS during winter, spring and summer at RCBD.

Sample	PM10	Si	Fe	Al	Ca	Mg	K	Na	Ti	Cr	C	Cl	S	F	P	V	Pb	O
4	48.73	1.41	-	0.68	0.15	0.39	0.19	0.24	-	-	11.26	0.10	0.68	-	-	-	-	33.67
5	49.28	2.12	0.30	0.99	0.25	0.39	0.35	0.15	-	0.10	10.50	0.10	0.84	-	0.05	-	-	33.17
6	48.39	2.41	0.19	0.77	0.24	0.34	0.44	0.34	0.05	-	9.19	0.10	0.44	4.65	-	-	-	29.28
7	47.10	3.96	1.27	-	0.61	0.57	0.47	0.33	0.09	0.28	7.58	0.19	0.99	-	0.09	-	-	29.11
8	45.51	2.78	0.68	1.18	0.41	0.41	0.46	0.36	0.05	0.14	8.33	0.09	0.68	1.55	-	-	-	28.35
9	44.22	2.39	0.62	0.97	0.31	0.4	0 35	0.27	-	0.13	7.03	-	1.81	-	-	-	-	28.88
10	43.06	2.50	0.60	0.99	0.34	0.56	0.26	0.34	-	0.17	8.01	0.13	1.25	-	-	-	-	27.90
11	46.46	3.58	1.02	1.49	0.37	0.51	0.23	0.23	0.09	0.28	7.43	-	1.81	-	-	0.02	0.24	29.08
MEAN	46.52	2.64	0.59	0.88	0.34	0.45	0.34	0.28	0.04	0.14	8.67	0.09	1.06	0.78	0.02	0.00	0.03	29.93
SD	0.90	0.29	0.15	0.15	0.05	0.03	0.04	0.03	0.01	0.04	0.54	0.02	0.18	0.59	0.01	0.00	0.03	0.78

The oxides linked with Cl were identified in almost all samples and P was identified in the winter (June – July) and Spring (August – September) samples. The occurence of oxides of P, Cl at RCBD may suggest the presence of a Pb phosphate mineral (Pb-P-Ca and/or Cl) in the samples [5].

The concentrations of S identified in this study ranged from 0.44 to 1.81 $\mu g.m^{-3}$. Sulphur-bearing particles are very common in ambient PM samples. They are found in virtually all airborne PM samples [11–13].

The potentially toxic trace metals identified in the samples include Cr, V and Pb, though V and Pb were only be identified the samples collected in summer. This could not be explained since it can be expected that the levels of Cr, V, and Pb are related mainly to mining and/or industrial activities and not meteorology. There is no clear trend in the levels of Cr within the study site. The levels ranged from 0.10 to 0.28 $\mu g.m^{-3}$ for Cr, and the concentrations of Pb and V were measured as 0.24 and 0.02 $\mu g.m^{-3}$ respectively.

3.2 ICP-MS

Table 3 shows the metals identified using ICP-MS and their concentration levels. About 15 main metals were identified. The crustal elements identified included Si, Fe, Al, Ca, Mg, K, Na, Ti, and toxic trace elements identified are Cr, Ni, V, Pb, Cu, Zn, and Mn.

There are no clear relations between the toxic trace metal concentrations (Cr, Ni, V, Pb, Cu, Zn, and Mn) and the seasons or months of the year. K increases from autumn to winter and crustal elements Ti, Fe, Ca decrease from autumn to winter. This may be because of the increase in biomass burning during winter since K is a tracer for biomass burning.

Table 3: Concentrations (in µg.m^{-3}) of the elements identified using ICP-MS during autumn and winter at RMINE.

Sample	PM10	Si	Fe	Al	Ca	Mg	K	Na	Ti	Cr	Ni	V	Pb	Cu	Zn	Mn
1	131.10	2.12	25.00	4.43	7.52	3.08	0.46	1.44	0.32	2.10	1.10	0.40	0.50	0.19	0.38	0.21
2	99.74	2.50	8.48	5.20	5.30	3.56	0.54	2.02	0.12	0.36	0.34	0.04	0.06	0.26	0.38	25.00
3	94.36	1.54	5.02	1.92	1.92	1.54	0.57	1.40	0.08	5.20	2.80	0.40	0.50	0.21	0.26	0.21
Mean	108.40	2.05	12.83	3.85	4.91	2.73	0.52	1.62	0.17	2.55	1.41	0.28	0.35	0.22	0.34	8.47
SD	11.46	0.28	6.16	0.99	1.63	0.61	0.03	0.20	0.07	1.42	0.73	0.12	0.15	0.02	0.04	8.26

It is worth noting that the different techniques used in this study show different sensitivities to different metals thus, the elements that could not be identified using SEM/EDS are Ti, Cu, Zn, and Mn. The levels of Mn are very high for Sample 2 and this could not be explained. It can however be an analysis error associated with the technique because the value is of three orders of magnitude higher than Sample 1 and 3 and thus clearly out of acceptable range. The Zn and Cu identified have concentration range of 0.26 to 0.36 µg.m^{-3} and 0.19 to 0.26 respectively. Table 4 shows the average monthly concentrations determined at RCBD during the study.

Table 4: Concentrations (in µg.m^{-3}) of elements identified by ICP-MS during winter, spring and summer at RCBD.

Sample	PM10	Si	Fe	Al	Ca	Mg	K	Na	Ti	Cr	Ni	V	Pb	Cu	Zn	Mn
4	48.73	10.20	0.20	3.10	2.30	3.20	2.30	3.10	0.01	0.02	0.00	0.05	0.00	0.08	0.10	0.40
5	49.28	1.60	5.20	2.00	2.00	1.60	0.60	1.50	0.80	0.50	0.30	0.04	0.05	0.20	0.30	0.20
6	48.39	9.10	10.70	1.90	3.20	1.30	0.20	0.60	0.10	0.09	0.05	0.02	0.02	0.08	0.17	0.09
8	45.51	3.30	11.30	6.90	7.10	4.8	0.70	2.70	0.17	0.50	0.46	0.05	0.08	0.35	0.50	0.33
9	44.22	0.20	8.50	3.10	3.70	2.70	2.20	2.90	0.07	0.05	0.03	0.02	1.40	0.02	0.35	0.51
10	43.06	0.93	10.00	4.20	3.70	3.60	1.80	4.40	0.07	0.06	0.05	0.04	1.70	0.03	0.49	0.62
11	46.46	-	0.93	3.00	1.70	1.40	0.41	0.82	0.09	0.03	0.03	0.01	0.10	0.61	0.40	0.04
Mean	46.52	4.22	6.69	3.46	3.39	2.66	1.17	2.29	0.19	0.18	0.13	0.03	0.48	0.20	0.33	0.31
SD	0.90	1.77	1.75	0.64	0.69	0.50	0.34	0.52	0.10	0.08	0.07	0.01	0.28	0.08	0.06	0.08

There seem to be no clear trends for most of the elements identified except for a trend observed during spring for Na, Pb, Mn and K in Sample 6, 8 and 9. These show an increase in concentration as temperature increases as shown in Table 5, where the temperatures 23.8, 25.7 and 28.8 ^{0}C were measured for the months of August, September and October respectively.

Table 5: Monthly meteorological data for the Rustenburg area supplied by the South African Weather Services (SAWS) [14].

Season	Month	Temperature (°C)		Rainfall (ml)	Wind Speed (m/s)	Wind Direction (Degrees from true North)
		Min	Max			
Autumn	Mar -04	14.7	25.9	11.0	15.5	347 (NNW)
	Apr -04	12.4	25.4	27.4	16.3	337 (NNW)
Winter	May –04	7.0	24.3	0.2	8.7	337 (NNW)
	Jun –04	2.7	19.4	5.2	12.2	354 (NNW)
	Jul –04	1.0	19.6	0.0	13.3	349 (N)
Spring	Aug –04	6.5	23.8	1.6	15.7	349 (N)
	Sep –04	7.5	25.7	0.2	16.7	340 (NNW)
	Oct –04	12.5	28.8	39.4	9.4	176 (S)
Summer	Nov –04	15.3	31.5	38.2	10.2	165 (SSE)
	Dec –04	16.0	29.2	83	10.4	163 (SSE)
	Jan – 05	17.4	29.4	160	9.0	204 (SSW)
Autumn	Feb - 05	16.5	29.7	78.6	8.5	161 (SSE)
	Mar - 05	14.1	27.3	42.4	8.8	190 (S)
	Apr – 05	11.5	23.2	91	7.3	196 (SSW)
Winter	May - 05	6.4	23.5	0.8	6.7	188 (S)
	Jun – 05	4.4	22.6	0.0	6.8	205 (SSW)
	Jul - 05	2.8	22.0	0.0	6.0	164 (SSE)

The peak in the K concentrations for Sample 4 and 9 may indicate activities related to biomass burning. This can be expected during winter (Sample 4) since there are more forest fires during this season. The peak in spring (Sample 9) may also indicate the effect of forest fires but from regional pollution, since the spring season is characterised by high wind speeds of 15.7 $m.s^{-1}$ in August, 16.7 $m.s^{-1}$ in September and 9.4 $m.s^{-1}$ in October 2004 (shown in Table 5) blowing from regions known for biomass burning.

The Cr and Ni concentrations are lowest for Sample 4 and 11. This could not be explained since the wind directions corresponding to the samples are oppositely directed, that is, NNW in June and SSE in November 2004, and the temperatures are also varied i.e. 19.4 and 31.5°C for June and November respectively. The wind speeds are however, roughly the same (12.2 and 10.2 $m.s^{-1}$) for June and November. The levels of Ca, Mn and V are also lowest for Sample 11.

The Pb concentrations show a maximum for Sample 9 (1.4 $\mu g.m^{-3}$) and 10 (and 1.7 $\mu g.m^{-3}$). The wind speeds for Sample 9 and 10 are lower (9.4 and 10.2 $m.s^{-1}$ for October and November) than for the samples obtained between June and September, which rules out the possibility of contribution from soil dust. The wind direction (S and SSE) corresponding to these samples can be responsible for the Pb levels, since they imply that the wind was blowing from the central business district, where there are more activities related to traffic.

3.3 AAS

The AAS analysis was performed to determine only Cr, Ni, V and Pb since these were the only elements for which the lamps were available. The results discussed in this section therefore only covers the four trace metals. Figure 3 gives the concentrations of these elements for Sample 1 to 7.

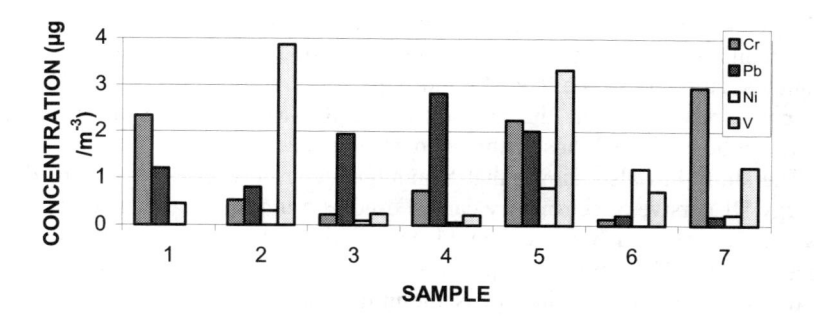

Figure 3: Concentrations of trace metals at RMINE (1, 2 and 3) and RCBD (4, 5, 6 and 7).

Chromium (Cr) and Nickel (Ni) concentrations decrease as the season changes from autumn to winter. This is probably because of the decrease in temperature, since it is warmer in autumn than in winter. It is documented that the winter nocturnal surface inversion has a depth of 400 – 600 m and strength of 5 – 7 ^{0}C over southern Africa as a whole [15]. This implies that the pollution released into a stable inversion layer is seldom able to rise through it and disperses slowly in clearly defined plumes.

The metal concentrations of Cr and Ni decrease from Sample 1 to 3, this is surprising since it was expected that concentrations should be higher in Sample 3 collected during winter than Sample 1 during autumn, because during winter the community burn coal and woods as a source of fuel for heating and cooking. This however, may suggest the source of high metal concentrations to be the industrial activities and not domestic and natural activities. The large variations in concentrations and differences in the ratio between the metals are indicative of the diversity of emission sources at this site.

The average concentration of Pb obtained from data over a 3 month period at RMINE is 1,32 µg.m^{-3} and this value is above the 3 months average value of 0.20 µg.m^{-3} set by Ministry for the Environment and the Ministry of Health (ME and MH) [16].

The Pb concentration decreases from winter to spring at RCBD. This may be due to the fact that, Pb may occur in the atmosphere in the form of Lead sulphate, Lead carbonate and Lead-phosphate; and AAS cannot successfully measure the total carbon in particulate matter.

The levels of Cr, Ni, V, and Pb were successfully determined and the concentration levels (in general) were in the order of decreasing abundance V, Pb, Cr, Ni for RMINE, and Cr, V, Pb, Ni for RCBD.

4 Conclusions

The concentrations of the metals identified at RCBD are generally lower than at RMINE. The average concentrations of Cr obtained at RMINE and RCBD were identified as 2.55(±1.42) and 0.18(±0.08) µgm^{-3} for ICP-MS and 2.28(±0.64) and

$0.14(\pm0.04)$ µgm^{-3} for SEM/EDS. The relatively good agreement between the two techniques shows that the SEM/EDS technique is a suitable non-destructive characterisation tool. The ICP-MS and AAS concentrations for Cr, Ni, V and Pb are not comparable. The discrepancy in the ICP-MS and AAS determinations could not be explained thus further studies are needed.

The study has also shown that Scanning Electron Microscopy coupled with Energy Dispersive Spectrometry can be used as a tool to characterise particulate matter. The method however, could not yield all the required information on trace metals. The ability to perform non-destructive analysis of individual particles by SEM/EDS is indispensable in the characterisation of PM10 samples because it allows the same sample to be chemically speciated by other spectroscopic methods.

ICP-MS was shown to be a relevant characterization tool for particulate aerosols. The technique provides valuable information, which when properly analysed, can contribute significantly to studies on seasonal variations and source apportionment, and ultimately to exposure assessment studies. A large number of metals, both crustal and trace metals were identified using ICP-MS. It is suggested however, that determination of elemental composition of PM using ICP-MS for the purpose of identifying and quantifying major source contributions, be coupled with chemical speciation of ammonium, sulphate, nitrate, organic carbon, and elemental carbon. This is necessary because atmospheric particulate aerosols, in general, consist of sulphates, nitrates, sea-salt, mineral dust, organics, carbonaceous components.

Acknowledgements

The authors acknowledge Dr L Tiedt and Mr P. Janse van Rensburg for the role they played in the analysis of filters. The financial support from the Finnish Environment Institute (SYKE) and the National Research Foundation (NRF) is greatly acknowledged.

References

[1] Dockery, D.W., Pope, C.A., Xu, X., Spengler, J.D., Ware, J.H., Fay, M.E., Ferris, B.G. & Speizer, F.E. *New England Journal of Medicine.* 329, p. 1753, 1993.

[2] IPCC. The Third Assessment Report of Working Group I of the Intergovernmental Panel on Climate Change: Technical Summary, Lead Authors, Albritton DL (USA), Meira Filho LG (Brazil), Shanghai, pp. 17–20, January 2001.

[3] McLaughlin, S.B. Effects of air pollution on forests: a critical review. *Journal of the Air Pollution Control Association,* 35 (5), pp. 512–534, 1985.

[4] Khlystov, A. Quality Assurance Project Plan for Pittsburgh Air Quality Study (PAQS), Department of Chemical Engineering, Carnegie Mellon University, 2001.

[5] Goldstein, J., Newbury, D., Joy, D., Lyman, C., Echlin, P., Lifshin, E., Sawyer, L. & Michael, J. *Scanning Electron Microscopy and X-Ray Microanalysis,* Third Edition, Kluwer Academic Publishers, New York, 2003.
[6] Li, X.Y., Gilmour, P.S., Donaldson, K. & MacNee, W. Free radical activity and pro-inflammatory effects of particulate air pollution (PM10) in-vivo and in-vitro, *Thorax,* 51, pp. 1216–1222, 1996.
[7] Gilmour, P.S., Brown, D.M., Lindsay, T.G., Beswick, P.H., MacNee, W. & Donaldson, K. Adverse health effects of PM10 particles: involvement of iron in generation of hydroxyl radical, *Occupational Environment and Medicine.* 53, pp. 817–822, 1996.
[8] Herbst, J.A. *Control 84:* Mineral/Metallurgical processing, American Institute of Mining, Metallurgical, and Petroleum Engineers, Inc. New York, 1984.
[9] Li, W-W., Bang, J.J., Chianelli, R.R., Yacaman, M.J., & Ortiz, R. Characterization of Airborne Particulate Matter in the Paso del Norte Air Quality Basin, *Morphology and Chemistry*, 2000.
[10] Asia Pacific Centre For Environmental Law (APCEL), National University of Singapore, http://www.nus.edu.sg
[11] Buseck, P.R. & Pósfai, M. Airborne minerals and related aerosol particles: effects on climate and the environment. *Proc. of the National Academy of Sciences of the United States of America*, 96, pp. 3372–3379, 1999.
[12] Zhang, D., Shi, G-Y., Iwasaka, Y. & Hu, M. Mixture of sulfate and nitrate in coastal atmospheric aerosols: individual particle studies in Qingdao, *Atmospheric Environment,* 34, pp. 2669–2679, 2000.
[13] Pósfai, M., Simonics, R., Li, J., Hobbs, P.V. & Buseck, P.R. Individual aerosol particles from biomass burning in southern Africa: Compositions and size distributions of carbonaceous particles. *Journal of Geophysical Research-Atmospheres*, 108, SAF 19/1–SAF 19/13, 2003.
[14] South African Weather Services (SAWS), http://www.saws.co.za
[15] Preston-Whyte, R.A. & Tyson, P.D. *The atmosphere and weather of Southern Africa,* New York: Oxford University Press. Pp. 366, 1988.
[16] Ministry for the Environment and the Ministry of Health, Ambient Air Quality Guidelines, *Air Quality Report No.32*, New Zealand, ME number: 438, pp 9, May 2002.

Data handling of complex GC-MS signals to characterize homologous series as organic source tracers in atmospheric aerosols

M. C. Pietrogrande, M. Mercuriali & D. Bacco
Department of Chemistry, Ferrara University, Italy

Abstract

A description is given of a chemometric approach used to extract information on the characteristics of n-alkane and n-alkanoic acid homologous series as useful markers for PM source identification and differentiation. The key parameters of the homologous series – number of terms and Carbon Preference Index – are directly estimated by the Autocovariance Function (*EACVF*) computed on the acquired chromatogram. The homologous series properties – relevant as the chemical signature of specific input sources – can be efficiently extracted from the complex GC-MS signal thus reducing the labour, time consumption and the subjectivity introduced by human intervention.

Keywords: aerosol chemical composition/homologous series/GC-MS analysis/ signal processing/ multicomponent mixtures.

1 Introduction

Atmospheric aerosols consist of a complex mixture of hundreds of compounds belonging to many different compound classes: despite this complexity, in environmental monitoring and assessment studies, the sample chemical analysis is usually limited to selected compounds to adequately represent a chemical signature of the possible input sources [1–3]. Homologous series of n-alkanes and n-alcanoic acids are especially suited for use as molecular tracers: they are common to multiple sources and they give information relevant to differentiating aerosols of anthropogenic origin (i.e. associated with industrial and urban activities) from those of natural, biogenic origin [4–6]. The key parameters to characterize specific sources are the number of terms and the carbon preference index (*CPI*, i.e., the sum of the concentrations of the odd/even carbon number terms divided by the sum of the concentrations of the even/odd carbon number terms). This parameter makes it possible to identify the biogenic contribution (that exhibits a strong odd/even carbon number predominance and thus a high *CPI* value) versus petroleum-derived fuels (displaying *CPI* values close to 1).

Gas chromatography-mass spectrometry (GC-MS), the best analytical technique for these organics, generates extensive amounts of data when applied to such complex mixtures as polluted environmental samples, which are complicated by a vast amount of noise, artefacts, and data redundancy. This motivates the need for computer-assisted signal processing procedures to transform GC data into usable information by extracting all the analytical results hidden in the complex chromatogram [7].

In the present paper, a signal processing procedure based on the AutoCovariance Function (*ACVF*) is applied to GC-MS signals of atmosferic aerosols. The case of n-alkanes and n-alkanoic acids is discussed as useful markers for PM source identification and differentiation. As molecular marker, the key parameters of the homologous series – number of terms and the *CPI* value – are directly estimated from the *ACVF* computed on the acquired chromatogram, thus reducing the labour, time requirements and the subjectivity introduced by human intervention.

1.1 Theory

The chemometric approach studies the Autocovariance Function, *ACVF*, that can be directly computed from the whole experimental chromatogram acquired in digitized form, Experimental $ACVF_{tot}$, $EACVF_{tot}$ [7]. The $EACVF_{tot}$ is plotted vs. the interdistance between subsequent points in the chromatogram, Δt, to obtain the $EACVF_{tot}$ plot (inset in Figure 1 shows the $EACVF_{tot}$ plot computed on the chromatogram of Figure 1). Theoretical models have been developed to extract information on sample complexity and chromatographic separation from the $EACVF_{tot}$. The mathematical description is reported elsewhere [7–9]: here the main parameters relevant for environmental analysis are discussed:

1. Information on sample complexity and separation performance is contained in the first part of the $EACVF_{tot}$ plot: the number of compounds present in the mixture is estimated from the $EACVF_{tot}(0)$ value, and the mean separation performance, σ, from the $EACVF_{tot}$ peak width at half height [7].

2. Information on the separation pattern is contained in the second part of the $EACVF_{tot}$ plot. In particular, the $EACVF_{tot}$ plot is specifically useful to single out the presence of ordered sequences of peaks appearing in the chromatogram [7]. This is the case of homologous series: if *n* compounds belonging to a homologous series are present in the sample, they will appear in the chromatogram as an ordered sequence of *n* peaks located at a constant interdistance value between subsequent terms in the series, e.g., $\Delta t=b$ where *b* is the CH_2 retention time increment (signed by arrows in the chromatogram of Figure 1) in GC analysis under linearized temperature programming conditions [7]. In this case, the $EACVF_{tot}$ computed on the acquired signal displays well defined deterministic peaks located at the interdistances $\Delta t=bk$, where $k=1,2,....n-1$ (arrows in the inset of Figure 1): their appearance identifies the presence of the series, even if the corresponding chromatographic peaks are hidden within the complex signal [7].

3. Number of terms of the homologous series. The height of the $EACVF_{tot}(bk)$ peaks ($EACVF_{tot}$ values at $\Delta t=bk$) can be quantitatively related to the abundance of the terms of the homologous series, i.e., the combination of the number of terms in the series, n, and their concentration in the sample, according to the following equation:

$$EACVF_{tot}(bk) = \frac{\sqrt{\pi}\sigma a_h^2(n-k)}{X}\left[\frac{\sigma_h^2}{a_h^2}+1\right] k = 0,1,2,.....n-1 \quad (1)$$

where all the reported parameters can be directly estimated from the chromatographic signal: X is the total chromatogram time span, σ_h^2/a_h^2 is the peak height dispersion ratio describing the relative abundance distribution of the n terms of the series: it derives from the mean, a_h^2, and the variance, σ_h^2, of peak height computed from the observed peak maxima in the chromatogram [7].

4. Abundance distribution of the homologous series terms. A random distribution of the series terms (no odd/even prevalence) yields a monomodal distribution of the subsequent $EACVF_{tot}(bk)$ peaks. If the terms of the series display an odd/even prevalence, the obtained $EACVF_{tot}(bk)$ peaks show a bimodal height distribution with lower values at $\Delta t=bk$ for odd k values and higher values at even k values. This pattern is the basis for extracting quantitative information on the odd/even prevalence of the terms by computing the preference index CPI [9]. Such a parameter can be related to the $EACVF_{tot}(bk)$ values at $\Delta t=b$ and at $\Delta t=2b$ according to the equation:

$$\frac{EACVF_{tot}(b)}{EACVF_{tot}(2b)} = \frac{\frac{2}{CPI}(n-1)}{\left(1+\frac{1}{CPI^2}\right)(n-2)} \quad (2)$$

This is a quadratic equation, and can be solved to estimate CPI. The CPI_{tot} value is obtained from $EACVF_{tot}$ by evaluating all the series components, i.e., the C_{12}-C_{35} n-alkane range. Otherwise, the CPI index can be calculated on selected terms in order to describe specific source contribution to the sample, i.e., the CPI_{plant} parameter is computed on the heavier C_{25}-C_{35} n-alkanes to describe the contribution of n-alkane plant waxes. CPI_{plant} is directly estimated from the $EACVF_{plant}$ computed on the partial region of the chromatogram containing the selected terms [9].

All these key parameters, used to characterize the homologous series as source chemical signature, can be directly obtained from the $EACVF_{tot}$ computed on the acquired chromatogram, thus reducing the labour, data handling time and removing the subjective step of peak integration. The big advantages of the present procedure becomes obvious when compared with the traditional procedure which requires identification of the homologous series terms by comparison with retention times and MS spectra of the reference standards, integration of the identified peaks, and computation of CPI from the concentrations of the odd and even carbon numbered terms. It must be underlined that labour and time saving in GC-MS signal processing is especially relevant for environmental analysis requiring high-throughput chemical monitoring.

2 Experimental

The aerosol samples ($PM_{2.5}$ and PM_{10}) were collected daily on quartz-fibre filters in urban (city centre of Bologna, Italy) and rural sites (San Pietro Capofiume, located on a flat, homogeneous terrain of harvested fields, about 40km north east of Bologna) during Spring 2008.

The PM filters were submitted to the traditional approach of solvent extraction and GC-MS analysis for n-alkane determination (procedure reported in [8]). Then the solution was submitted to the derivatization procedure for n-alkanoic acid analysis: 30 µL of bis(trimethylsilyl) trifluoroacetamide (BSTFA) plus 1% trimethylchlorosilane (TMCS) were added to form trimethylsilyl (TMS) derivatives (reaction at 70 °C for 2h) [7]. The GC-MS system was a Scientific Focus-GC (Thermo-Fisher Scientific Milan, Italy) coupled with PolarisQ Ion Trap Mass Spectrometer (Thermo-Fisher, Scientific, Milan, Italy). The column used was a DB-5 column (L=30m, I.D.=0.25mm, d_f=0.25µm) (J&W Scientific, Rancho Cordova, CA, USA). Proper temperature program conditions were selected for n-alkanes and n-alkanoic acids to obtain linearized temperature programming conditions, i.e., constant CH_2 retention time increment. The mass spectrometer operated in EI mode (positive ion, 70eV). Three different samples were analyzed for each PM type: the obtained mean values are reported (Table 1) and discussed below.

3 Results and discussion

3.1 n-alkane series

The aliphatic hydrocarbons present in the PM samples were identified from the SIM (Selected Ion Monitoring) signal using the typical fragments of these compounds at m/z=57+71+85 (Figures 1 and 2 for urban and rural samples, respectively). The investigated n-alkanes showed a distribution profile resulting from the contribution of vehicular exhaust and lubricant residues (C_{24} or C_{25} n-alkanes) and inputs of biological sources (C_{27}, C_{29}, and C_{31} terms displaying odd carbon number preference).

To extract information on the PM chemical composition, the $EACVF_{tot}$ was computed on the whole chromatographic signal ($EACVF_{tot}$ plots reported in insets of Figures 1 and 2: solid lines). The $EACVF_{tot}$ plots show well-defined deterministic peaks at Δt=1.9min and multiple values that are diagnostic for the presence of the n-alkane homologous series (b=1.9min in these experimental conditions). The number of n-alkanes present in the mixture, n, can be estimated from the $EACVF_{tot}$(1.9min) values (eqn (1)): the same value n=16 is obtained from both the chromatograms (Table 1, $EACVF$ estimation).

The $EACVF_{tot}$ values of subsequent peaks give quantitative information on the distribution of the odd/even terms: both the plots show a monomodal distribution of the $EACVF_{tot}$ peak heights suggesting a homogeneous distribution of the odd/even terms. Such a pattern can be quantively described by computing CPI_{tot} (eqn (2)): CPI_{tot}=1.1 and CPI_{tot}=1.6 were estimated for urban and rural samples,

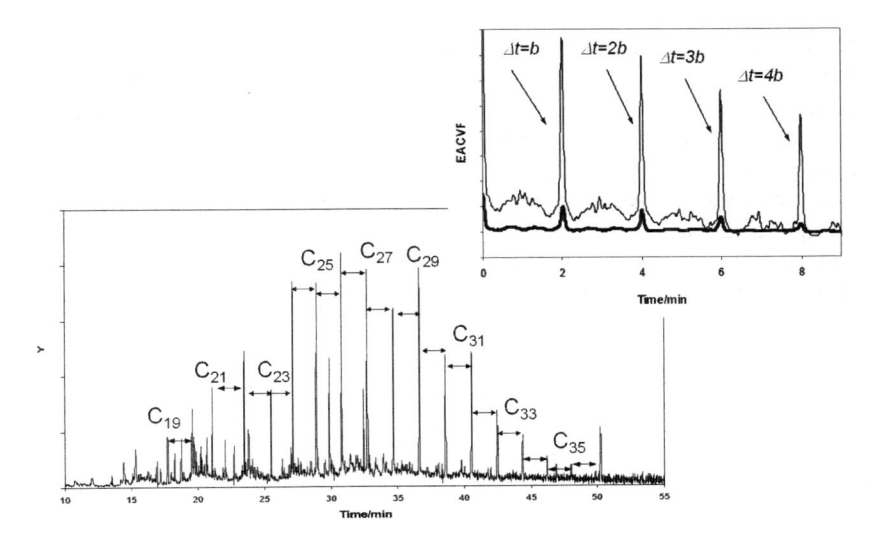

Figure 1: n-alkanes in urban $PM_{2.5}$: GC-MS chromatogram (SIM at $m/z=57+71+85$); inset: $EACVF_{tot}$ plot (solid line) and $EACVF_{plant}$ plot (bold line).

respectively. These values close to 1 suggest, for both the samples, a major contribution from petroleum residues derived from vehicular emissions as compared to biological inputs.

For all the studied chromatograms, the $EACVF_{tot}$ plots clearly show diagnostic peaks: this behaviour highlights the power of the $EACVF$ procedure in extracting information on homologous series, singling them out from the involved signal of the complex chromatograms. In fact, the $EACVF_{tot}$ pattern is independent of the concentration level of n-alkanes, i.e., total concentrations of n-alkanes in the urban $PM_{2.5}$ are nearly four times higher than those in the rural $PM_{2.5}$ sample, and nearly three times lower than those in the urban PM_{10} [4]. Moreover, the chromatographic signal of urban PM samples is further affected by a cluster of unresolved peaks (*UCM* band) (Figure 1): the $EACVF_{tot}$ of the urban sample (inset in Figure 1, solid line) retains the shape of the *UCM* band, but clearly displays the $EACVF_{tot}$ peaks characteristic of the homologous series.

To distinguish the role played by the biogenic vs. anthropogenic sources on the atmospheric n-alkanes, the $EACVF_{plant}$ was separately computed on the chromatographic region where the biogenic C_{27}-C_{35} n-alkanes are eluted (t=32-55min). For both samples, the number of terms n_{plant}=9 is estimated from the $EACVF_{plant}$ values at Δt=1.9min ($EACVF_{plant}$ plots in insets of Figures 1 and 2, bold lines). The differences in plant contribution to the two samples can be simply identified by visual inspection of the $EACVF_{plant}$ plots obtained. For the rural sample, the $EACVF_{plant}$ (inset of Figure 2, bold line) shows a bimodal distribution of subsequent peak heights that is diagnostic for the presence of odd/even prevalence, as revealed by the high estimated value of CPI_{plant}=2.4 that characterizes the contribution of biogenic sources (i.e., higher plant waxes). Otherwise, a lower CPI_{plant}=1.3 value is estimated for the urban sample, as typical for urban environments. The $EACVF$ value at $\Delta t=2b=3.8$min is related to the total amount of the terms of homologous series (eqn. (1)): therefore, for each sample, the ratio between $EACVF_{tot}$(3.8min) and $EACVF_{plant}$(3.8min) can be used to estimate the relative contribution of plant waxes ($EACVF_{plant}$) to the overall n-alkane components ($EACVF_{tot}$). Such a contribution was quantified as percentages of plant wax fraction in the total n-alkanes: 23% and 10% for the rural and urban samples, respectively.

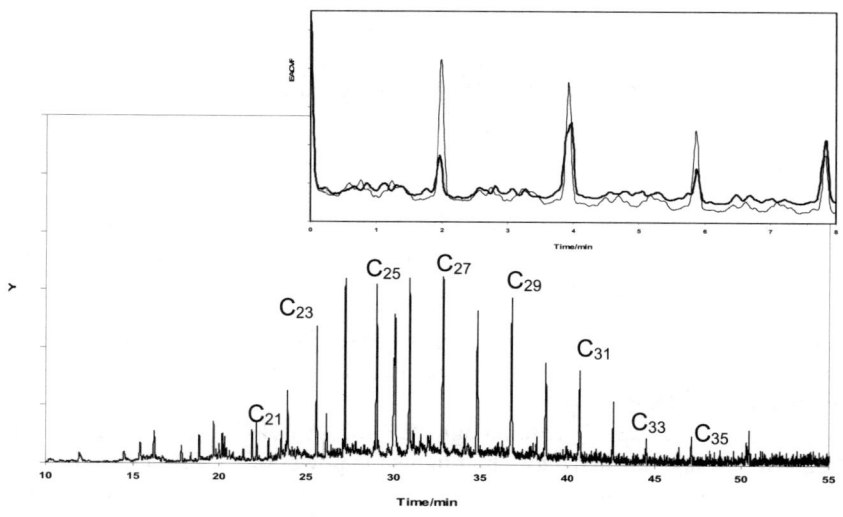

Figure 2: n-alkanes in rural $PM_{2.5}$: GC-MS chromatogram (SIM at m/z= 57+71+85); inset: $EACVF_{tot}$ plot (solid line) and $EACVF_{plant}$ plot (bold line).

To check the accuracy of the results obtained (Table 1, 1st-4th columns, $EACVF$ estimation), the traditional procedure, based on computation on the integrated peaks, was applied to the PM chromatograms (Table 1, 5th-8th columns, traditional calculations). A comparison between the independently computed values show a close agreement, validating the reliability of the information obtained by the $EACVF$ procedure. This result confirms the usefulness of the $EACVF$ mehod as a simple, time saving approach to characterize the n-alkane series as molecular marker in complex environmental samples.

3.2 n-alkanoic acid series

The $EACVF_{tot}$ method was also applied to characterize n-alkanoic acids, as another homologous series of organics useful in discriminating the relative

Table 1: *CPI* and *n* parameters estimated by using the *EACVF* method (1st-4th columns, *EACVF* estimation) and traditional calculations (5th-8th columns: traditional method).

Sample	EACVF Estimation				Traditional method			
	n	CPI_{tot}	n_{plant}	CPI_{plant}	n	CPI_{tot}	n_{plant}	CPI_{plant}
n-alkanes $CPI_{tot}=\Sigma(C_{13}-C_{35})/\Sigma(C_{12}-C_{34})$; $CPI_{plant}=\Sigma(C_{25}-C_{35})/\Sigma(C_{24}-C_{34})$								
PM 2.5 urban	16.2	1.1	8.8	1.3	8.8	1.1	9	2
PM 2.5 rural	15.6	1.6	9.2	2.4	9.2	1.8	9	2.5
PM 10 urban	16.4	0.9	8.6	1.2	8.6	0.9	9	1.8
n-alkanoic acids $CPI_{tot}=\Sigma(C_{14}-C_{30})/\Sigma(C_{13}-C_{29})$; $CPI_{plant}=\Sigma(C_{20}-C_{30})/\Sigma(C_{19}-C_{29})$								
PM 2.5 urban	13.1	6.9	5.6	8.1	14	6.2	6	7.8
PM 2.5 rural	13.6	9.8	7.5	18.7	14	9.4	8	17.2
PM 10 urban	14.2	6.1	5.4	7.9	14	5.4	6	7.5

extent to which various sources contribute to the aerosol burden of organics: the lower molecular weight n-alkanoic acids (< C_{20}) are mainly emitted by petroleum based sources, while the heavier C_{20}-C_{30} terms, which display a strong even-to-odd carbon number preference, are mostly derived from plant waxes [6].

After derivatization, the urban and rural PM samples were submitted to GC-MS analysis: the n-alkanoic acids present in the samples were identified in the SIM signal by monitoring the typical fragments of the TMS derivatives at $m/z=75+147$ (Figure 3a: rural sample). Under the experimental conditions used, the retention increment for subsequent n-alkanoic acids is b=2.5min. The $EACVF_{tot}$ was computed on the whole signal (Figure 3b: solid line): deterministic peaks at Δt=2.5min and multiple values are diagnostic for the presence of this homologous series. All the data set to characterize the series are estimated (Table 1, 1st-4th columns, *EACVF* estimation) and compared to results obtained with the traditional procedure (Table 1, 5th-8th columns, traditional calculations).

The $EACVF_{tot}$ plot shows a marked bimodal distribution with a predominant peak at $\Delta t=2b$=5min: this is consistent with predominant contribution of hexadecanoic (C_{16}) and octadecanoic (C_{18}) acids that are known to be the most abundant species in most of the PM samples [3,6]. The even/odd prevalence of acid isomers was confirmed by high CPI_{tot}=9.8 and CPItot=6.9 values found for rural and urban samples, respectively (Table 1).

To extract information on the biological sources of n-alkanoic acids, the selected chromatographic region containing the C_{20}-C_{26} terms (35-60min) was separately investigated by computing $EACVF_{plant}$. The obtained $EACVF_{plant}$ plot

(Figure 3b, bold line) clearly identifies the contribution of biogenic sources, since it displays the strong bimodal distribution ($EACVF_{plant}(bk)$ peaks are low for $k=1, 3, 5$ and high for $k=2, 4$) characteristic of a strong odd/even prevalence. This is confirmed by the high CPI value ($CPI_{plant}=18.7$) computed from subsequent $EACVF_{plant}$ peaks, reflecting the stronger vascular plant wax signatures. Otherwise, a lower $CPI_{plant}=8.1$ value was obtained for the urban PM, indicating that plant waxes make a weaker contribution (Table 1).

The contribution of biogenic n-alkanoic acids in PM samples can also be directly estimated by the ratio between $EACVF_{tot}(5min)$ and $EACVF_{plant}(5min)$ computed on each chromatogram: the plant fraction ($\geq C_{20}$ congeners) accounted for about 25% and 8% of the total measured n-alkanoic acids levels in rural and urban samples, respectively.

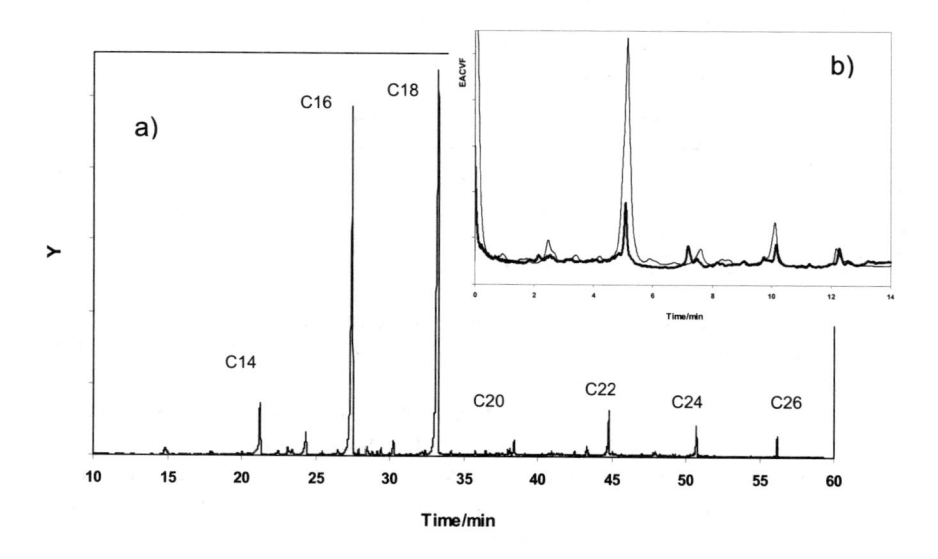

Figure 3: n-alkanoic acids in rural $PM_{2.5}$: a) GC-MS chromatogram (SIM at $m/z= 75+147$); b) $EACVF_{tot}$ plot (solid line) and $EACVF_{plant}$ plot (bold line).

4 Conclusions

The described results reveal the effectiveness of the $EACVF_{tot}$ procedure for handling complex GC-MS data of PM samples in order to characterize the homologous series as molecular marker to trace the origin and fate of atmospheric aerosols. The key parameters – number of terms and the odd/even prevalence – are efficiently extracted from the $EACVF$ computed on the acquired chromatogram, with low labour and time consumption. This seems a promising method for high-throughput analysis of the large data sets generated by chemical monitoring in environmental analysis: the obtained chemical information can serve as useful tracers for source apportionment and processes involving organic carbonaceous aerosols when coupled with receptor models.

Acknowledgement

The Emilia Romagna Regional Environmental Agency (ARPAER) is acknowledged for providing the aerosol samples.

References

[1] Simoneit, B.R.T., Characterization of organic constituents in aerosols in relation to their origin and transport: a review. *International Journal of Environmental Analytical Chemistry,* **23**, pp. 207–237, 1986.

[2] Schauer J.J., Rogge W.F., Hildemann L.M., Mazurek M.A., Cass G.R., Simoneit B.R.T., Source apportionment of airborne particulate matter using organic compounds as tracers. *Atmospheric Environment,* **30**, pp. 3837–3855, 1996.

[3] Park S.S., Bae M., Schauer J.J., Kim Y.J., Cho S.Y., Kim S.J., Molecular composition of $PM_{2.5}$ organic aerosol measured at an urban site of Korea during the ACE-Asia campaign. *Atmospheric Environment,* **40**, pp. 4182–4198, 2006.

[4] Wang G., Liming Huang L., Zhao X., Niu H., Dai Z., Aliphatic and polycyclic aromatic hydrocarbons of atmospheric aerosols in five locations of Nanjing urban area, China, *Atmospheric Research,* **81**, pp. 54–66, 2006.

[5] Cheng Y., Li S.-M., Leithead A., Brook J.R., Spatial and diurnal distributions of n-alkanes and n-alkan-2-ones on PM2.5 aerosols in the Lower Fraser Valley, Canada, *Atmospheric Environment,* **40**, pp. 2706–2720, 2006.

[6] Oliveira C., Pio C., Alves C., Evtyugina M., Santos P., Goncalves V., Nunes T., Silvestre J.D., Palmgren F., Wahlinc P., Harrad S., Seasonal distribution of polar organic compounds in the urban atmosphere of two large cities from the North and South of Europe, *Atmospheric Environment,* **41**, pp. 5555–5570, 2007.

[7] Pietrogrande M.C., Zampolli M.G., Dondi F., Identification and Quantification of Homologous Series of Compound in Complex Mixtures: Autocovariance Study of GC/MS Chromatograms, *Anal. Chem.* **78**, pp. 2579-2592, 2006.

[8] Pietrogrande M.C., Mercuriali M., Pasti L., Signal processing of GC–MS data of complex environmental samples: Characterization of homologous series, *Analytica Chimica Acta,* **594**, pp. 128–138, 2007.

[9] Pietrogrande M.C., Mercuriali M., Pasti L., Dondi F., Data handling of complex GC-MS chromatograms: characterization of n-alkane distribution as chemical marker in organic input source identification, submitted to publication.

Monitoring of trace organic air pollutants – a developing country perspective

P. B. C. Forbes[1] & E. R. Rohwer[2]
[1]Natural Resources and the Environment, CSIR, South Africa
[2]Department of Chemistry, University of Pretoria, South Africa

Abstract

Air pollutants arise both from natural sources and from various anthropogenic activities, and are of concern due to their environmental impacts, including human health effects. In developing countries, atmospheric monitoring has largely focused on inorganic pollutants, such as sulphur dioxide, nitrogen oxides and ozone. Organic air pollutants, however, are monitored infrequently, because of factors such as the cost of equipment required; necessary expertise of monitoring personnel; and the trace levels at which such pollutants are usually present in the atmosphere. This is of concern since organic air pollutants, such as polycyclic aromatic hydrocarbons (PAHs) and dioxins, are emitted from combustion processes, which are often employed for domestic heating and cooking purposes in developing countries.

This paper focuses on the current status of organic air pollutant monitoring in southern Africa, and discusses developments in this regard. Screening methods and monitoring of indicator compounds, which allow for more widespread sampling and analysis of samples for spatial and temporal trend determinations, are discussed. A laser induced fluorescence (LIF) technique, for example, has been developed by the authors, based on sampling onto a novel silicone rubber trap. This will allow for the rapid screening of air samples for the presence of PAHs, prior to comprehensive, quantitative analysis of samples of interest by gas chromatography-mass spectrometry (GC-MS) (particularly those with PAH levels above background concentrations). LIF provides sufficient selectivity for a screening procedure, without the need for sample clean-up and separation processes prior to analysis. The technique also lends itself to real-time monitoring and "chemical fingerprinting" via the fluorescence spectra obtained.

Keywords: trace organic air pollutants, developing countries, combustion emissions, polycyclic aromatic hydrocarbons, dioxins, laser induced fluorescence.

1 Introduction

1.1 Air pollution in developing countries

The air quality of developing countries is deteriorating due to industrial development, economic growth and large-scale migration of rural residents to urban areas. Consequences of this include a decline in food security, an increase in respiratory related illnesses, and a degradation of both the quality of life and the environment [1]. Comprehensive and reliable air monitoring data is key to the improvement of air quality, as without this information it is impossible to identify and apportion emission sources [2].

Whilst it is acknowledged that significant progress has been made in organic air pollutant monitoring in many developing countries in Asia and South America, this paper focuses on southern Africa. The geographic delineation for the region was used, i.e. the portion of Africa south of the Cunene and Zambezi rivers. This incorporates the countries of Botswana, Lesotho, Namibia, Mozambique, South Africa, Swaziland and Zimbabwe, as shown in Figure 1.

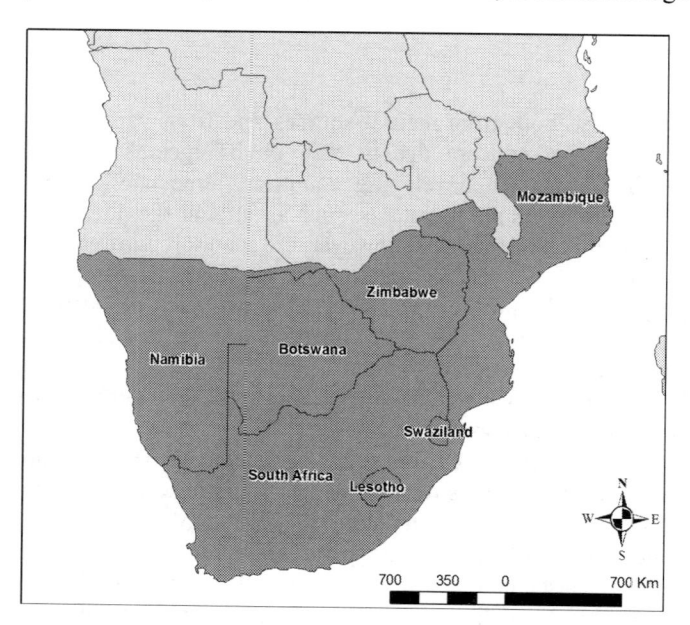

Figure 1:   Southern African countries.

In addition to industrial emissions, major sources of air pollutants in developing countries include dependence on fossil fuels, and increasing traffic densities in urban areas, mainly involving an aged vehicle fleet without catalytic converters. Besides the relatively commonly monitored inorganic pollutants which are released from these combustion processes, for example, nitrogen oxides, sulphur dioxide and carbon dioxide; organic pollutants may also be released in the gaseous form or associated with particles. Hydrocarbons and polycyclic aromatic hydrocarbons (PAHs) are examples of such pollutants, as are polychlorinated dibenzodioxins and furans which may be generated when chlorinated material is combusted. Although the organic pollutants would generally be present at lower concentrations than the inorganic species, they are nevertheless of environmental significance due to their potential impacts (including health effects) at low levels.

1.2 Hindrances to air pollutant monitoring in developing countries and means by which these are being addressed

Organic air pollutant monitoring has been minimal in many developing countries to date, particularly those on the African continent. This is due to a number of reasons, some of which are the result of socio-political priorities. In addition, a lack of resources is often the most prominent hindrance, where funding, skilled human capital, and suitable equipment may be unavailable.

Various initiatives have been established in order to address these issues. For example, the Air Pollution Information Network for Africa (APINA) is a sub-Saharan regional network of scientists, policy-makers and non-governmental organizations, and its member countries include Botswana, Mozambique, South Africa and Zimbabwe. APINA aims to address air pollution problems, as part of the RAPIDC (Regional Air Pollution in Developing Countries) programme. RAPIDC is funded by the Department of Infrastructure and Economic Cooperation (INEC) of SIDA, the Swedish International Development Cooperation Agency and includes initiatives in both Asia and Africa [3].

The Southern and Eastern Africa Network of Analytical Chemists (SEANAC) was established to assist with capacity building and collaboration in the region, specifically in the areas of health, food security and environmental monitoring. One of the problems cited as being faced by African analytical chemists is that many of them received training abroad in techniques which are much more appropriate to the developed world [4].

The Organisation for Economic Co-operation and Development (OECD) has investigated the impact of monitoring equipment on air quality management in developing countries [2]. Capacity building is achieved through case studies, where participant countries are assisted with monitoring equipment and training. The case studies conducted to date do not include southern African countries, and have focused on inorganic pollutants, with the exception of one study involving fuel testing in India, where PAHs were monitored.

2 Overview of organic air pollutant monitoring in southern Africa

It is relevant to consider southern Africa as a region in terms of air quality, as the climate and air circulation of the region south of northern Angola is dominated by a gyre centred on Botswana. The air rotates anti-clockwise, completing one revolution approximately every week and the gyre thus formed remains in place for several weeks, particularly in winter, which leads to a build up of various trapped atmospheric pollutants, resulting in deterioration in visibility and air quality over parts of the subcontinent [5].

Monitoring activities are largely driven by legislation, thus an overview of the existing air quality legislation pertaining to the region is included in this section. It is generally accepted that environmental legislation in many African countries is outdated, or is poorly enforced due to a lack of capacity or poor institutional organisation. In some cases, more modern legislation still allows industries which were in existence before promulgation to continue operating outdated equipment, even if this leads to excessive pollution.

Some monitoring is undertaken by local government municipalities, industries, and research institutions in the different countries of interest, but in most cases such monitoring is currently restricted to inorganic air pollutants. Publicly accessible published organic air pollutant monitoring activities, conducted prior to 2007, are summarised in the following sections. It is acknowledged that additional monitoring may have taken place for the purpose of Environmental Impact Assessments (EIAs), for example.

2.1 Botswana

Routine air quality monitoring began in Botswana in the mid 1970s, upon the promulgation of the Atmospheric Pollution Prevention Act in 1971. Continuous monitoring of hydrocarbons has been conducted in Botswana at one site since 1999 [6], but most of the monitoring sites focus on inorganics (SO_2, NO_x, O_3, CO and particulate matter) and associated meteorological data. Ambient air quality objectives for Botswana do not include organic air pollutants. A National Environmental Laboratory was established in 2002, however, which has the capabilities to analyse organic pollutants [1].

2.2 Lesotho

The Environment Act 2001 provides for the management of the environment and all natural resources of Lesotho. The Act makes provision for the establishment of environmental quality standards and environmental monitoring, but it is unclear as to whether such provisions have been enacted [7].

2.3 Mozambique

Mozambique has an Environmental Law of 1997, but few air pollutant monitoring studies have been conducted in this country [1].

2.4 Namibia

The Namibian Directorate of Environmental Affairs has noted that air pollution is a less serious problem than in many other countries, although the need to update the environmental legislation to enable effective law enforcement is acknowledged [8]. No systematic air quality monitoring had occurred in Namibia, as of 2001 [6].

2.5 South Africa

A new era in air quality management in South Africa began with the promulgation of the National Environmental Management: Air Quality Act in 2004 (Act No. 39 of 2004) (AQA), which replaced the Air Pollution Prevention Act of 1965 (Act No. 45 of 1965). This has resulted in a shift from an emission control focus to an airshed approach, which culminated in the establishment of national standards for permissible ambient concentrations of air pollutants [9].

This has implications in terms of air quality monitoring, as methods are required which are suitable for monitoring pollutants at ambient concentrations, which are generally lower than emission levels. Alternative methods, which allow for pre-concentration of analytes of interest and which have lower detection limits, are therefore of importance. Sampling methods employed in ambient and emission air monitoring are also different. Development of the capacity and capabilities to perform ambient air monitoring (as well as in air quality management) is therefore receiving attention in South Africa, particularly at local government level.

At present, benzene is the only organic air pollutant for which ambient standards have been set (an annual average of 5 $\mu g.m^{-3}$), although the AQA makes provision for the Minister of Environmental Affairs and Tourism to make notice of additional air pollutants, as necessary.

In terms of monitoring, local municipalities in South Africa currently monitor Volatile Organic Compounds (VOCs) (specifically benzene, toluene, ethylbenzene and xylenes) at nine sites across three provinces. Methane is also monitored at four stations in two provinces, one of which is the WMO Global Atmospheric Watch (GAW) site at Cape Point [10]. Non-methane hydrocarbons are also monitored monthly at the GAW site, by means of grab sampling into canisters which are analysed in Europe. These samples have yielded interesting results in the context of biomass burning episodes, for example [11].

VOCs have been studied in various contexts, including that of emissions from spontaneous combustion of coal [12] and the Cape Town brown haze, which forms in winter months under inversion conditions [13].

In addition to monitoring campaigns, fundamental research has been conducted in South Africa in terms of novel means of sampling and analysing organic air pollutants. Ortner and Rohwer [14] developed thick film multi-channel silicone rubber traps, which have been applied in semi-volatile organic air pollutant sampling. These traps serve as pre-concentrators and can be thermally desorbed for GC analysis, thereby negating the need for solvent extraction and concentration. Burger *et al* [15] has also developed a high capacity, polydimethylsilicone rubber sample enrichment probe (SEP), which can be thermally desorbed for GC analysis. A novel thermal modular array for comprehensive two-dimensional gas chromatography has also been produced by these researchers [16].

2.6 Swaziland

Air Pollution Control Regulations have been drafted for Swaziland, which include air quality objectives for a number of inorganic air pollutants only [17].

2.7 Zimbabwe

Zimbabwe has an Atmospheric Pollution Control Act (No. 33 of 1971 as amended by Act No. 22 of 2002) and Atmospheric Pollution Prevention Regulations of 1975. The Environmental Management Act was enacted in December 2002, which will repeal the existing air legislation in due course, and will provide for the establishment of Air Quality Standards [1].

Some random air monitoring studies have been conducted in Zimbabwe, and the Air Pollution Control Unit of the City of Harare Health Department has carried out routine air pollution monitoring of SO_2, NO_2 and soot at eight sites over the past 20-30 years. Other parameters which have been monitored include methane and VOCs, in addition to inorganic analytes (Pb, NH_3, HCl and particulate matter) [1].

2.8 Other African studies

Few published organic air pollutant monitoring campaigns have been conducted in other African countries, such as the monitoring of PAHs from charcoal burning in Kenya (using liquid chromatography with fluorescence detection) [18]. The majority of these studies were carried out in collaboration with developed countries, which assisted primarily with analyses. Studies of organic pollutants, particularly pesticides, in other environmental media [19], also indicate that there is some existing capacity that could provide a basis for air monitoring. An example is the determination of PAHs in surface runoff and sediments in Nigeria [20].

2.9 Involvement of southern Africa in international conventions

In terms of organic pollutant monitoring, the Stockholm Convention on Persistent Organic Pollutants (POPs) is of relevance. The majority of southern African countries have become signatories to this convention, and as such have a responsibility to manage and monitor POPs. Existing monitoring progammes, activities and datasets have recently been compiled [21], and are summarised in Table 1. Although concentrations of POPs were determined in ambient air in Durban, South Africa, all analyses were performed in the USA [22]. It is evident that capabilities in the region require further development.

2.10 Southern African air monitoring campaigns

The Southern Africa Fire-Atmosphere Research Initiative of 1992 (SAFARI -92) and the Southern African Regional Science Initiative of 2000 (SAFARI 2000) provided scientists in the region with the opportunity to participate in large research projects with international experts. SAFARI-92 investigated the role of savanna fires in atmospheric chemistry, climate and ecology, where organic monitoring included CH_4 and non-methane hydrocarbons [23]. More than 150 scientists from 14 countries were involved, including the southern African countries of Botswana, Namibia, South Africa, Swaziland and Zimbabwe.

Table 1: Existing national monitoring programmes, activities and datasets for southern African countries [21].

Country	Monitoring activities
Botswana	No data
Lesotho	No data
Mozambique	No data
Namibia	Limited data
South Africa	No existing national POPs information gathering activities. Academic research studies have been conducted and published to assess POPs in various media (Note: air was not included).
Swaziland	No data
Zimbabwe	Limited data

SAFARI 2000 addressed a broad range of phenomena related to land-atmosphere interactions and biogeochemical functioning of the southern African system. Here biogenic volatile organic compounds (BVOCs), VOCs, oxygenated VOCs, and semivolatile organic compounds (SVOCs) were also monitored [24].

3 Use of alternative monitoring methods

In light of the hindrances experienced in developing countries, as discussed under section 1.2, alternative monitoring methods to the standard comprehensive methods may be more appropriate. Such methods should be cost effective and simple, in order to allow for more widespread monitoring in the region. We now present a few options in this regard, with respect to organic air pollutants.

3.1 Passive sampling

Passive air samplers provide a useful means of monitoring in developing countries due to their simplicity, cost effectiveness and non-reliance on provision of a power source. Passive sampling may provide pre-concentration of analytes, thereby increasing analytical sensitivity, and may reduce or eliminate solvent consumption for sample preparation purposes. Sampling is based on diffusion or permeation of the analyte into the sampler, as reviewed by Seethapathy [25]. Passive samplers have found widespread application in airborne POPs monitoring, for example [26, 27].

Besides the use of evacuated canisters for passive organic air pollutant sampling, passive samplers of interest include solid phase microextraction (SPME), polyurethane foam based samplers, and samplers containing adsorbents such as graphitised charcoal or Tenax.

Studies have been conducted in South Africa to assess the chronic health impacts of exposure to VOCs (as well as naphthalene), by means of passive sampling, using 3M badges [28] and IVL passive samplers [29]. There is significant scope to implement passive sampling more widely in the field of organic air pollutant monitoring in southern Africa, and to utilise this technique for a wider range of organic analytes. It should be noted, however, that passive samplers may not provide the time resolution required in some monitoring campaigns, due to the relatively long sampling intervals usually employed (one to several weeks).

3.2 Indicator compounds

The monitoring of by-products (indicator compounds) of reaction pathways which generate the analyte of interest has been successfully utilised for various classes of chemicals. A relevant example is the monitoring of chlorobenzenes and chlorophenols as indicators for dioxins and furans, where correlation coefficients are used to relate the indicator compound concentrations to that of the target analyte [30]. Precursor compounds may be similarly utilised where intermediates in the synthesis of the target analyte are monitored.

The choice of indicator compound is based on the existence of a correlation with the target compound(s), ease of sampling, as well as other characteristics which simplify analysis, such as a higher concentration and existence of fewer congeners compared to the target analyte(s). The monitoring of indicator or precursor compounds for organic air pollutant monitoring therefore has great potential for application in developing countries, where resources and capacity are limited.

3.3 Screening methods

Screening methods find application in large sampling campaigns, in that numerous samples can be taken and analysed in order to determine whether more comprehensive analysis (by standard accepted methods) is required, such as samples which screen positive. A screening method should ideally be fast, simple and of low cost, yet meet the sensitivity and selectivity requirements of the application.

The authors have developed a laser induced fluorescence method as a screening tool for PAHs sampled onto multi-channel silicone rubber traps [14] by means of a portable GilAir sampling pump. An excimer laser (Lambda Physik EMG201) was used to optically pump a dye laser equipped with Rhodamine 6G dye. An intra-cavity grating allowed for wavelength tuneability over the gain bandwidth of the dye, and ensured a narrow bandwidth output (0.1 cm^{-1} at 495 nm). In the case of naphthalene, for example, the 584 nm output beam was frequency doubled through second harmonic generation in a non-linear crystal, resulting in a 292 nm excitation laser wavelength (~500 µJ pulse energy in a 30 ns pulse, with a 5 mm spot size), which was directed onto the sample trap.

The fluorescence of the sample was optically collected, and resolved with a scanning double monochromator (Kratos, Schoeffel Instruments), equipped with a photomultiplier tube. Results were recorded on a PC linked to an oscilloscope (Tektronix TDS 360).

Different silicones were tested in order to minimize the background signal of the substrate and thereby enhance the detection limit of the method. The best results were obtained for the silicone which had the least aromatic moieties in the final product, as would be expected due to their chromophoric properties. Losses of naphthalene as a result of volatilization and photodegradation during laser irradiation were found to be less than 10 % over a typical analysis period of 5 minutes. Initial experiments using standards yielded promising results, and the proof of concept of the method was demonstrated by the monitoring of PAHs arising from a sugar cane burn in KwaZulu Natal, South Africa.

The method is rapid, and has acceptably low limits of detection. LIF also provides sufficient selectivity for a screening procedure, without the need for sample clean-up and separation processes prior to analysis. Interface with GC-MS for further quantitative analysis post LIF analysis is also possible, due to the non-destructive nature of the fluorescence measurement. The method is being further optimized, and may find application in a centralized environmental laboratory for the southern African region.

4 Conclusion

Currently most analytical technology used in the developing world for environmental purposes, is imported from the developed nations [2]. Significant capacity building is needed, but should be undertaken in the context of the differing requirements of developing countries. Alternative organic air pollutant monitoring strategies, such as the use of passive samplers and screening methods, or the monitoring of indicator compounds, have the potential to provide viable, cost effective alternatives to established methods in southern Africa.

References

[1] Mmolawa, M.D., Scoping report on existing monitoring activities in the 7 APINA member countries, Issued as part of Phase III Activity 2.3 of the RAPIDC Programme 2005-07, Botswana, 2006.
[2] Hight, J. & Ferrier G., *The impact of monitoring equipment on air quality management capacity in developing countries*, Organisation for Economic Co-operation and Development, Joint Working Party on Trade and Environment, Report COM/ENV/TD(2006)7/FINAL, 2006.
[3] Air Pollution Information Network for Africa (APINA), www.sei.se/rapidc/apina.htm
[4] Southern and Eastern Africa Network of Analytical Chemists (SEANAC), www.seanac.org
[5] Scholes, R.J. & Biggs, R. (eds), *Ecosystem services in Southern Africa: A regional assessment*, CSIR, Pretoria, South Africa, ISBN 0-7988-5527-4, pp. 55, 2004.
[6] Scholes, R., *Regional Implementation Plan for Southern Africa, Global Terrestrial Observing System GTOS-21*, CSIR internal report ENV-P-R 2001-002, South Africa, February 2001.
[7] Department of Tourism, Environment and Culture, Republic of Lesotho, Environment Act, 2001 Online. www.faolex.fao.org/faolex

[8] Namibia Directorate of Environmental Affairs, www.met.gov.na/dea
[9] Department of Environmental Affairs and Tourism, South Africa, National Environmental Management: Air Quality Act, 2004, Government Notice No. 528, Government Gazette No. 28899, 9 June 2006.
[10] Department of Environmental Affairs and Tourism, South Africa, *Technical Compilation to inform the Initial State of Air Report*, National Air Quality Management Programme, May 2007.
[11] Brunke, E.-G., Labuschagne, C. & Scheel, H.E., Trace gas variations at Cape Point, South Africa, during May 1997 following a regional biomass burning episode. *Atmospheric Environment*, **35**, pp. 777–786, 2001.
[12] Pone, J.D.N., Hein, K.A.A., Stracher, G.B., Annegarn, H.J., Finkleman, R.B., Blake, D.R., McCormack, J.K. & Schroeder, P., The spontaneous combustion of coal and its by-products in the Witbank and Sasolburg coalfields of South Africa. *International Journal of Coal Geology*, **72**, pp. 124–140, 2007.
[13] Burger, J.W., Pienaar, J.J., Fourie, L. & Jordaan, J.H.L., Identification of volatile organic compounds in Cape Town brown haze. *Air Pollution 2004*, WIT Press, UK, pp. 631–640, 2004.
[14] Ortner, E.K. & Rohwer, E.R., Trace analysis of semi-volatile organic air pollutants using thick film silicone rubber traps with capillary gas chromatography. *Journal of High Resolution Chromatography*, **19**, pp. 339–344, 1996.
[15] Burger, B.V., Marx, B., le Roux, M. & Burger, W.J.G., Simplified analysis of organic compounds in headspace and aqueous samples by high-capacity sample enrichment probe. *Journal of Chromatography A*, **1121**, pp. 259–267, 2006.
[16] Burger, B.V., Snyman, T., Burger, W.J.G. & van Rooyen, W.F., Thermal modulator array for analyte modulation and comprehensive two-dimensional gas chromatography. *Journal of Separation Science*, **26(1-2)**, pp. 123–128, 2003.
[17] Department of Tourism, Environment and Communications, Swaziland, Air Pollution Control Regulations, 1999 Online. www.ecs.co.sz
[18] Gachanja, A.N. & Worsfold, P.J., Monitoring of polycyclic aromatic hydrocarbon emissions from biomass combustion in Kenya using liquid chromatography with fluorescence detection. *Science of the Total Environment*, **138**, pp. 77–89, 1993.
[19] Torto, N., Mmualefe, L.C., Mwatseteza, J.F., Nkoane, B., Chimuka, L., Nindi, M.M. & Ogunfowokan, A.O., Sample preparation for chromatography: An African perspective. *Journal of Chromatography A*, **1153**, pp. 1–13, 2007.
[20] Ogunfowokan, A.O., Asubiojo, O.I. & Fatoki, O.S., Isolation and determination of polycyclic aromatic hydrocarbons in surface runoff and sediments. *Water, Air, and Soil Pollution*, **147**, pp. 245–261, 2003.
[21] United Nations Environmental Programme, Stockholm Convention on Persistent Organic Pollutants (POPs), *Compilation of existing national*

monitoring programmes, activities and datasets, Document UNEP/POPS/GMP/TWG-2/6, Geneva, 2007.
[22] Batterman, S., Chernyak, S., Gounden, Y. & Matooane, M., Concentrations of persistent organic pollutants in ambient air in Durban, South Africa. *Organohalogen Compounds*, **68**, pp.1111–1114, 2006.
[23] Lindesay, J.A., Andreae M.O., Goldammer, J.G., Harris, G., Annegarn, H.J., Garstang, M., Scholes, R.J. & van Wilgen, B.W., International Geosphere-Biosphere Programme/International Global Atmospheric Chemistry SAFARI-92 field experiment: Background and overview. *Journal of Geophysical Research*, **101(D19)**, pp. 23521–23530, 1996.
[24] Swap, R.J., Annegarn, H.J., Suttles, T., King, M.D., Platnick, S., Privette, J.L. & Scholes, R.J., Africa burning: A thematic analysis of the Southern African Regional Science Initiative (SAFARI 2000). *Journal of Geophysical Research*, **108(D13)**, pp. SAF 1-1 – 1-15, 2003.
[25] Seethapathy, S., Górecki, T. & Li, X., Review: Passive sampling in environmental analysis. *Journal of Chromatography A*, **1184**, pp. 234–253, 2008.
[26] Shoeib, M. & Harner, T., Characterization and comparison of three passive air samplers for persistent organic pollutants. *Environmental Science and Technology*, **36**, pp. 4142–4151, 2002.
[27] Jaward, F.M., Farrar, N.J., Harner, T., Sweetman, A.J. & Jones, K.C., Passive air sampling of PCBs, PBDEs, and organochlorine pesticides across Europe. *Environmental Science and Technology*, **38**, pp. 34–41, 2004.
[28] John, J., *Research to quantify atmospheric volatile organic compounds in the major metropolitan areas of South Africa*, CSIR internal report ENV-P-C-98016, South Africa, January 1998.
[29] Wichmann, J., *Human health risk assessment case study – value addition of passive sampler generated data*, CSIR internal report ENV-P-I-99008, South Africa, October 1999.
[30] Kaune, A., Lenoir, D., Schramm, K-W., Zimmermann, R., Kettrup, A., Jaeger, K., Rückel, H.G. & Frank, F., Chlorobenzenes and chlorophenols as indicator parameters for chlorinated dibenzodioxins and dibenzofurans in incineration processes: influences of various facilities and sampling points. *Environmental Engineering Science*, **15(1)**, pp. 85–95, 1998.

NIST gas standards containing volatile organic compounds in support of ambient air pollution measurements

G. C. Rhoderick
Analytical Chemistry Division, Chemical Science and Technology Laboratory, National Institute of Standards and Technology (NIST), Gaithersburg, Maryland, USA

Abstract

Since the late 1970s the National Institute of Standards and Technology (NIST), Gaithersburg, Maryland, USA, has been developing and supplying reference materials (RMs) in support of gas measurements for measuring ambient air pollution. These RMs have been developed for federal and state governments in the United States (US), such as the US Environmental Protection Agency (EPA) and the State of California Air Resources Board (CARB), as well as academia, research laboratories such as the National Center for Atmospheric Research (CARB), and other laboratories around the world. Many of these volatile organic compounds (VOCs) pose health concerns and are monitored at ground level to assist U.S. government agencies in assessing the need for emission studies and controls. Ozone is also a ground level pollutant of great concern. Hydrocarbons are ozone precursors and are measured to track, in particular, automobile emissions. RMs are split into two groups containing either multicomponent mixtures of VOCs or non-methane hydrocarbons (NMHCs). Typically these RMs are produced at the 0.02–10 nmol/mol (ppb) level. The research into the preparation and stability of such chemical compounds contained in gas mixtures has led to the development of two Standard Reference Materials (SRMs); a 30-component VOC (SRM 1804) and 18-component NMHC (SRM 1800) at 5 nmol/mol per component. NIST has participated in international comparisons with other National Metrology Institutes (NMIs) through bilaterals and CCQM key and pilot comparisons for many of these VOCs and NMHCs. Results of these comparisons have shown equivalency among these NMIs, helping to underpin their measurement and standards claims. Development of these SRMs and RMs, as well as the international comparisons will be discussed.

Keywords: standard reference material (SRM), reference material (RM), volatile organic compounds (VOCs), non-methane hydrocarbons (NMHC), primary standard mixtures (PSMs), air pollution measurements.

1 Introduction

Over the past 30 years, the need and demand for accurate standards containing trace levels [<10 nmol/mol (ppb)] of toxic organic vapor in a matrix gas has increased. Federal and state environmental programs have been implemented requiring the monitoring of compounds from automobile exhaust, industrial emissions, hazardous waste sites and incineration to name a few. In the United States (US) the US Environmental Protection Agency (US-EPA) has identified many volatile organic compounds (VOCs) that represent possible health risks [1]. Several of these compounds are documented as potential carcinogens [2,3]. Included in the classification of VOCs are non-halogenated, non-methane hydrocarbons (NMHC) primarily from automobile exhaust. These hydrocarbons (NMHCs) are precursors to the formation of ozone (O_3) and are major contributors to photochemical smog [4]. This has increased the interest in measuring the levels of VOCs and NMHCs at ground level and the upper atmosphere. The measurement data aids regulators in determining the level of reduction in ambient concentrations required to achieve national ambient air quality standards for ozone [5,6]. Measurement data aids in developing control measures and as input for urban atmosphere modelling [7,8].

The ability to intercompare adequately the host of laboratories that analyze VOCs and NMHCs depends on the provision of standards with certified concentrations with low levels of uncertainty. Primary and secondary standards are necessary to correlate data accurately, a prerequisite for adequate regulation of toxic organic pollution. In addition, specialty gas companies preparing VOC/NMHC gas mixtures for customers prefer to have a traceability link to a nationally recognized standard. The amount-of-substance fraction (concentration) of a chemical species in a gas mixture can be accurately determined by comparing the unknown mixture with a primary standard mixture (PSM) prepared by an absolute method such as gravimetry. Comparative techniques referenced to PSMs must be used to certify other mixtures as Standard Reference Materials (SRMs).

There have been several techniques developed and used to prepare gas standards, or calibration standards, for VOCs. A procedure for the gravimetric preparation of volatile organic compounds in a gas matrix was previously developed at the National Institute of Standards and Technology (NIST) [9,10]. Up to 30 - 40 liquid analytes (if one is extremely careful) can be accurately weighed into a 30 L aluminium gas cylinder. More than 70 carbon based compounds have been studied for feasibility and stability in gas mixtures. Standards have been developed containing varying numbers of compounds (1–30) and concentrations 10 pmol/mol to 10 μmol/mol. NIST has used this procedure to develop more than 200 in-house primary standard mixtures (PSMs). These PSMs have then been used to certify over 200 Reference Materials (RM) for the US EPA, the state of California Air Resources Board (CARB), and many other laboratories and university research groups in the USA and internationally. A total of 475 Standard Reference Materials (SRMs) have been developed and certified. This paper will discuss two of the most complex VOC/NMHC SRMs developed and a complex RM. Results from international comparisons with other National Metrology Institutes (NMIs) for some of the VOC/NMHCs will also be discussed.

2 SRM development

Twelve different NIST SRMs for VOCs and NMHCs have been developed over the past 20 years of which there are currently 2 available. They include SRM 1804, a 30 component VOC and SRM 1800, a 19 component NMHC, both at 5 nmol/mol per component. The most often requested VOC gas standard is for EPA Method TO-14 (organics, semi-volatile and volatile). The TO-14 list varies from 39-45 organic analytes. NIST settled on a subset of 30 analytes so as to create a simple chromatogram using one analytical column which would result in no co-elution of compounds.

2.1 SRM 1804 – VOCs

The 30 component VOC SRM 1804 was developed and certified in 2003. A detailed description of the development of this SRM has been published [11]. Using the previously mentioned techniques and procedures for preparing gravimetric standards, 4 PSMs were developed and used to certify SRM 1804. Table 1 gives the concentration and expanded uncertainty (95% confidence interval) in nmol/mol, for each of the 30 VOCs in a representative mixture. The majority of the components have expanded uncertainties of $\leq \pm 5.0\%$. The analytical data have larger uncertainties for the higher molecular mass components resulting in larger certified uncertainties. Each of the samples in this SRM 1804 lot was individually certified.

A key component of the quality assurance program at NIST includes stability checks of SRMs at appropriate time intervals. This SRM was recertified in 2007 after stability analysis were preformed using the original and newly prepared PSMs. Figure 1 shows the 4 year stability data for the SRM lot standard (LS) 1804-C-01. The LS is a cylinder prepared at the same time as the SRM lot and is retained at NIST. All of the cylinders comprising the lot are compared to the LS. The LS is assumed to be representative of the SRM lot.

2.1.1 SRM 1800 – NMHCs

The original 15 component NMHC SRM 1800 was developed in 1993, the current 18 component SRM was certified in 2003. The same techniques and procedures for PSM development and SRM preparation and certification were used for both SRM 1800 lots and a detailed description has been published [12]. Five PSMs were developed and used to certify the current SRM 1800. Table 2 lists the compounds with their certified concentrations and expanded uncertainties at the 95% confidence interval. All but one of the components has an uncertainty of $\leq \pm 2.9\%$ with decane having an uncertainty of $\pm 5.6\%$.

Table 1: Certified concentrations and uncertainties for a representative SRM 1804 sample.

Analyte	Concentration and Uncertainty in nmol/mol (95%)
Dichlorodifluoromethane	4.43 ± 0.09 (2.0%)
Chloromethane	4.83 ± 0.10 (2.1%)
1,2-Dichlorotetrafluoroethane	4.25 ± 0.09 (2.1%)
Vinyl chloride	3.42 ± 0.27 (7.9%)
Chloroethane	5.14 ± 0.10 (1.9%)
Trichlorofluoromethane	5.26 ± 0.11 (2.1%)
1,1-Dichloroethene	5.27 ± 0.11 (2.1%)
Dichloromethane	5.41 ± 0.11 (2.0%)
1,1-Dichloroethane	5.42 ± 0.11 (2.0%)
cis-1,2-Dichloroethene	5.30 ± 0.11 (2.1%)
Chloroform	5.39 ± 0.11 (2.1%)
1,1,1-Trichloroethane	5.34 ± 0.11 (2.1%)
Carbon tetrachloride	5.34 ± 0.11 (2.1%)
Benzene	5.33 ± 0.11 (2.1%)
Trichloroethylene	5.42 ± 0.11 (2.0%)
1,2-Dichloropropane	5.35 ± 0.11 (2.1%)
Toluene	5.38 ± 0.11 (2.0%)
1,1,2-Ttrichloroethane	5.30 ± 0.21 (2.1%)
Tetrachloroethylene	5.37 ± 0.11 (2.1%)
Chlorobenzene	5.33 ± 0.11 (2.1%)
Ethylbenzene	5.32 ± 0.11 (2.1%)
para-Xylene	5.42 ± 0.11 (2.0%)
ortho-Xylene	5.03 ± 0.20 (4.0%)
1,1,2,2-Tetrachloroethane	5.51 ± 0.28 (5.1%)
1,3,5-Trimethylbenzene	5.34 ± 0.22 (4.1%)
1,2,4-Trimethylbenzene	5.52 ± 0.22 (4.0%)
1,3-Dichlorobenzene	5.52 ± 0.28 (5.1%)
1,2-Dichlorobenzene	5.48 ± 0.28 (5.1%)
1,2,4-Ttrichlorobenzene	4.88 ± 0.49 (10.0%)
Hexachloro-1,3-butadiene	4.97 ± 0.50 (10.1%)

Three major stability analyses have been performed on the original SRM 1800 certified in 1993. The data has shown 11 years of continued stability for this SRM. Figure 2 illustrates the stability of LS 1800-01-A (original SRM 1800). The spread in the data points for any given hydrocarbon is about 2%, which is within the uncertainty. An interesting observation is that the last stability analysis in 2004 results in concentrations that are closer to the original value than at any other time.

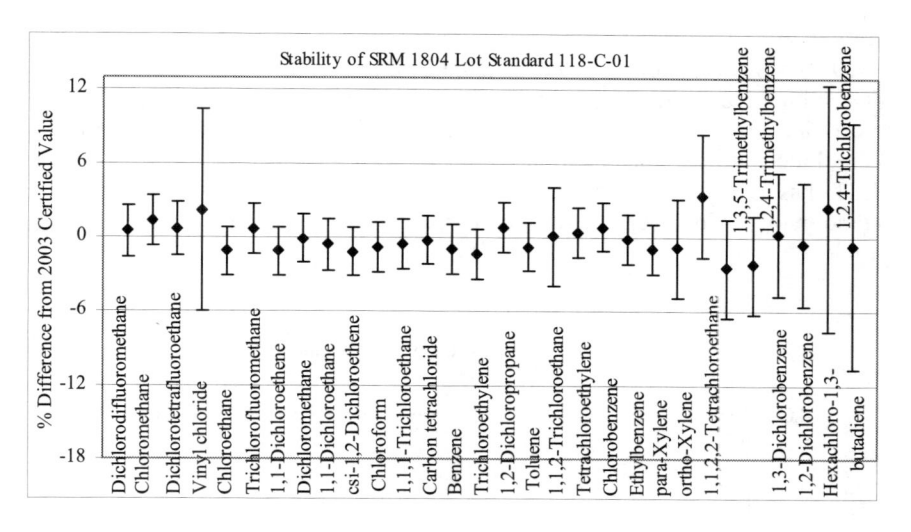

Figure 1: Stability Data for SRM 1804 Lot Standard 1804-C-01 from 2007 analysis.

Table 2: Certified concentrations and uncertainties for a representative SRM 1800 sample.

Analyte	Concentration and Uncertainty in nmol/mol (95%)
Ethane	5.35 ± 0.11 (2.0%)
Propane	5.64 ± 0.11 (2.0%)
Propene	5.40 ± 0.11 (2.0%)
iso-Butane	5.68 ± 0.11 (1.9%)
n-Butane	5.54 ± 0.11 (2.0%)
iso-Butene	5.68 ± 0.12 (2.1%)
iso-Pentane	5.24 ± 0.11 (2.1%)
n-Pentane	5.31 ± 0.11 (2.0%)
1-Pentene	5.13 ± 0.15 (2.9%)
n-Hexane	5.30 ± 0.11 (2.1%)
n-Heptane	5.39 ± 0.11 (2.1%)
Benzene	5.50 ± 0.11 (2.0%)
iso-Octane	5.32 ± 0.11 (2.1%)
n-Octane	5.14 ± 0.11 (2.1%)
Toluene	5.23 ± 0.15 (2.9%)
Nonane	4.98 ± 0.14 (2.8%)
ortho-Xylene	4.96 ± 0.12 (2.4%)
Decane	4.82 ± 0.27 (5.6%)

2.1.1.1 Multi-component RM Many of the gas mixtures for ambient air pollution monitoring are prepared as RMs. They can be very complex with very different concentrations. One such VOC gas mixture used for ambient toxic calibration by CARB contains 27 components with concentrations varying from 0.020 to 5.0 nmol/mol as listed in table 3, along with stability data. Several VOCs, highlighted in bold, have shown decreases outside or just at the original uncertainty limits. Although carbon tetrachloride has demonstrated stability at 5 nmol/mol, at this concentration there has been a major decrease of > 51%. NIST has observed good stability for the major percentage of VOCs at 5 nmol/mol. However, there are a few that consistently exhibit stability problems.

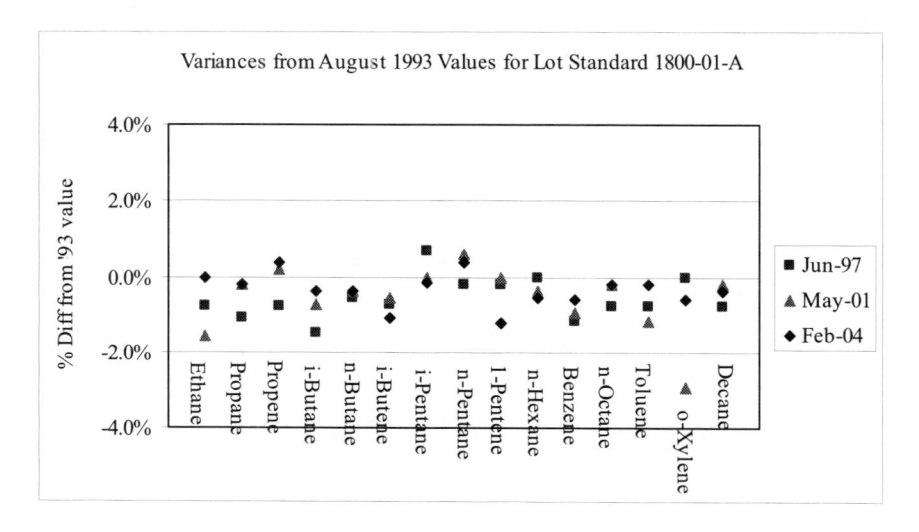

Figure 2: Stability Data for SRM 1800 Lot Standard 1800-01-A.

2.1.1.1.1 Comparison of standards between National Metrology Institutes (NMIs)
Only a few NMIs have or are currently developing standards and RMs of VOC/NMHCs. The National Physical Laboratory (NPL) in the United Kingdom and the National Metrology Institute (NMi) in the Netherlands analyzed a representative NIST PSM for the original SRM 1800. Figure 3 illustrates the results of that comparison. The agreement was within the uncertainty for almost all of the analytes and demonstrates equivalency among these NMIs. Results of other comparisons for VOC gas mixtures between NMIs have been published [13–15].

3 Conclusions

These RM and SRM mixtures containing nmol/mol multi-component VOC/NMHC have allowed laboratories to use traceable gas mixtures to determine the levels of these components in ground and atmospheric environments. In general, most of these compounds have exhibited good stability in gas mixtures contained in treated aluminium gas cylinders, and those compounds that have shown decreases over time are limited. Many combinations of compounds and concentrations are possible, and concentration demands for gas mixtures will continue to trend lower.

Table 3: Certified concentrations and uncertainties for a complex RM VOC gas mixture.

Analyte[a]	Concentration and Uncertainty in nmol/mol (95%)	
	Original Value February 2004	Recertified Value February 2007
Dichlorodifluoromethane	1.01 ± 0.03	1.01 ± 0.02
Vinyl chloride	0.60 ± 0.01	0.60 ± 0.02
1,3-Butadiene	**1.49 ± 0.12**	**1.36 ± 0.14**
Bromomethane	**2.03 ± 0.17**	**1.89 ± 0.20**
Trichlorofluoromethane	1.77 ± 0.06	1.77 ± 0.07
Dichloromethane	1.86 ± 0.11	1.86 ± 0.11
Chloroform	0.25 ± 0.01	0.25 ± 0.02
1,1,2-Trichlorotrifluoroethane	0.53 ± 0.02	0.53 ± 0.02
1,1,1-Trichloroethane	1.33 ± 0.04	1.33 ± 0.04
Carbon tetrachloride	**0.17 ± 0.01**	**0.072 ± 0.04**
1,2-Dichloroethane	0.44 ± 0.02	0.44 ± 0.02
Benzene	2.59 ± 0.07	2.59 ± 0.05
Trichloroethylene	0.44 ± 0.02	0.44 ± 0.02
cis-1,3-Dichloropropene	**2.68 ± 0.03**	**2.37 ± 0.45**
trans-1,3-Dichloropropene	**2.68 ± 0.09**	**1.64 ± 0.67**
Toluene	4.38 ± 0.08	4.38 ± 0.15
Tetrachoroethylene	0.52 ± 0.03	0.52 ± 0.02
1,2-Dibromoethane	**0.95 ± 0.03**	**0.93 ± 0.11**
Chlorobenzene	1.81 ± 0.04	1.81 ± 0.05
Ethylbenzene	2.76 ± 0.07	2.76 ± 0.06
para-Xylene	2.62 ± 0.06	2.62 ± 0.06
meta-Xylene	2.65 ± 0.12	2.65 ± 0.07
ortho-Xylene	2.70 ± 0.09	2.70 ± 0.06
Styrene	**3.7 ± 0.2**	**2.3 ± 0.6**
1,3-Dichlorobenzene	**2.59 ± 0.09**	**2.50 ± 0.21**
1,4-Dichlorobenzene	**2.75 ± 0.07**	**2.63 ± 0.19**
1,2-Dichlorobenzene	**2.69 ± 0.07**	**2.54 ± 0.28**

[a]Analytes and concentrations in bold represent those compounds that are exhibiting decreases (either marginally or significantly) in concentration over time.

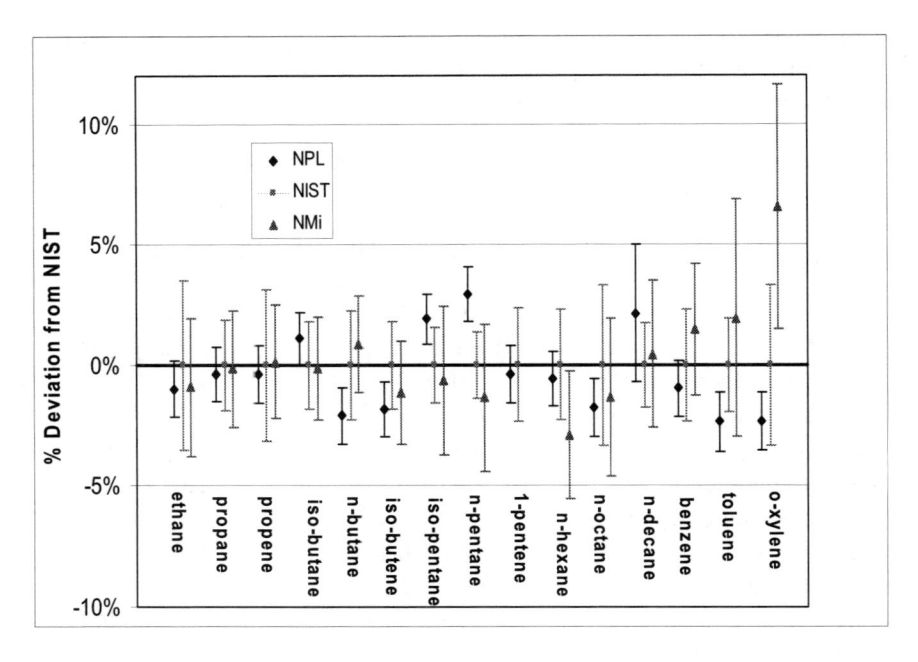

Figure 3:   Analytical results for comparison of SRM 1800 representative standard with NIST, NPL and NMi.

References

[1] IRAC Monographs on the Evaluation of Carcinogenic Risk of Chemicals to Man: International Agency for Research on Cancer. Health Organization: Geneva, Switzerland, **1**, pp. 53-65, **1972** and **7**, pp. 203–216, 291-305, **1974**.
[2] von Lehmden, D.J., Conference on Recent Developments in Methods for Toxics in the Atmosphere, Boulder, CO, **1987**.
[3] New health data may affect clean air rules, *C&EN*, **68(4)**, pp. 5, **1990**.
[4] Haagerr-Smit, A.J., Chemistry and Physiology of Los Angeles Smog, *Industrial and Engineering Chemistry*, **44**, pp.1342–1346, **1952**.
[5] Guidance for Collection of Ambient Non-methane Organic Compounds (NMOC) Data for Use in 1982 Ozone SIP Development and Network Design and Siting Criteria for the NMOC and NO_x Monitors, *EPA-450/4-80-011*, **1980**.
[6] Singh, H.B., Guidance for the Collection and Use of Ambient Hydrocarbon Species Data in Development of Ozone Control Strategies, *EPA-450/4-80-008*, **1980**.
[7] Derwent, R.G., Hov, O., Computer Modeling Studies of the Impact of Vehicle Exhaust Emission Controls on Photochemical Air- Pollution Formation in the United Kingdom, Environmental Science and Technology, **14**, pp. 1360–1366, **1980**.

[8] Whitten, G.Z., Hugo, H., Killus, J.P., The Carbon-Bond Mechanism-A Condensed Mechanism for Photochemical Smog, *Environmental Science & Technology*, **14**, pp. 690-700, **1980**.

[9] Schmidt, W.P., Rook, H.L., Preparation of Gas Cylinder Standards For Measurement of Trace Levels of Benzene and Tetrachloroethylene, *Analytical Chemistry*, 55, pp.290–294, **1983**.

[10] Rhoderick, G.C., Zielinski, W.L., Jr., Preparation of Accurate Multicomponent Gas Standards of VOCs in the Low-Parts-per-Billion Range, *Analytical Chemistry*, **60**, pp. 2454–2466, **1988**.

[11] Rhoderick, G.C., Development of a NIST Standard Reference Material Containing Thirty Volatile Organic Compounds at 5 nmol/mol in Nitrogen, *Analytical Chemistry*, **78(9)**, pp. 3125-3132, **2006**.

[12] Rhoderick, G.C., Development of a 15 Component Hydrocarbon gas Standard Reference Material at 5 nmol/mol in Nitrogen, *Fresenius Journal of Analytical Chemistry*, **359**, pp. 477–483, **1997**.

[13] Guenther, F.R., Rhoderick, G.C., et al.; Key Comparison CCQM-K7: Benzene, Toluene, Ethylbenzene, m-Xylene and o-Xylene in Nitrogen, *Metrologia*, **39, 2002**.

[14] Guenther, F.R., Rhoderick, G.C., et al, Key Comparison CCQM-K10: Benzene, Toluene, o-Xylene in Nitrogen, *Metrologia*, **39, 2002**.

[15] Kato, K., Maruyama, M., Kim, J.S., Heo, G.S., Kim, Y-D., Guenther, F.R., Rhoderick, G.C., van der Veen, A.M.H., Baldan, A., Milton, M.J.T., Vargha, G., Brookes, C., Konopelko, L., Kustikov, Y., Vishnyakov, I., Final Report on key comparison CCQM-K22: Benzene,chloroform,dichloromethane,trichloroethylene, tetrachloroethylene,1,2-dichloroethane,1,3-butadiene and vinyl chloride in nitrogen, *Metrologia* **44**, 08006, **2007**.

A comparison of EPA and EN requirements for nitrogen oxide chemiluminescence analyzers

J. Barberá[1], M. Doval[1], E. González[1], A. Miñana[1] & F. J. Marzal[2]
[1]*Departamento de Ingeniería Química, Universidad de Murcia, Spain*
[2]*Departamento de Ingeniería Térmica y de Fluidos,
Universidad de Cartagena, Spain*

Abstract

Nitrogen oxides analyzers currently used for measuring air quality in European countries meet the standard ISO 7996:1985, which does not establish any special feature for such equipment, except the analytical technique implemented. Nevertheless, most of them are designated as "Reference methods" by the Environmental Protection Agency of the United States. Future European legislation in this matter establishes as reference method for measuring these pollutants that described in EN 14211:2005, where a number of tests for chemiluminescence analyzers are described. In this paper, we compare the requirements of both documents, evaluating the suitability of each and their approach.

Keywords: chemiluminescence, nitrogen oxides, EPA, type approval tests, EN 14211:2005.

1 Introduction

Following the "Better Regulation" initiative for updating and simplying community legislation, the *Proposal for a Directive on ambient air and cleaner air for Europe* [3] aims at integrating in a single document the Directives 96/62/EC [4], 99/30/EC [5], 2000/69/EC [6], 2002/3/EC [7] and the Decision 97/101/EC [8]. Moreover, the Proposal establishes the complete control of particulate matter ($PM_{2.5}$) and also stipulates new reference documents for the measurement of regulated pollutants in air, as shown in table 1.

The standards of the Proposal of Directive, except for those related to particulate matter and lead, include a new and wide-ranging section about

Table 1: Reference documents currently used and those set by the Proposal of Directive for measuring pollutants.

Pollutant	Analytical Method	Normas In force	Proposal of Directive
Sulfur dioxide	Ultraviolet fluorescence	ISO 10498	EN 14212 : 2005
Particulate lead	Atomic absorption spectrometry or Inductively coupled plasma mass spectroscopy	EN 12341 : 1999	EN 14902 : 2005
Benzene	Gas cromatography	-------------	EN 14662 : 2005
Carbon monoxide	Non-dispersive infrared spectroscopy	-------------	EN 14626 : 2005
Particulated matter PM_{10}	Gravimetric determination	EN 12341 : 1999	EN 12341 : 1999
Particulate matter $PM_{2.5}$	Gravimetric determination	-------------	EN 14907 : 2005
Ozone	Ultraviolet photometry	ISO 13964:1998	EN 14625 : 2005
Nitrogen oxides	Chemiluminescence	ISO 7996:1985	EN 14211 : 2005

requirements to be met by the analyzers, known as "Type-Approval Test", which aims at assuring data quality from the networks measuring ambient air pollution. "Type- Approval Tests" include a number of tests in the laboratory and field and the calculation of the expanded uncertainty, which must be lower than that specified in the corresponding legislation.

Independently, the Environmental Protection Agency from the US certifies the analyzers for measuring air pollutants as Reference or Equivalent Methods [9]. In the next section and regarding nitrogen oxides analyzers, the main differences, advantages and drawbacks of the proposed tests in both documents are discussed.

2 EPA and EN requirements for chemiluminescence nitrogen oxides analyzers

2.1 Performance parameters and criteria required by EPA

The performance parameters and their criteria described in the Code of Federal Regulations by EPA are compiled in table 2.

2.2 Performance parameters and criteria required by EN 14211:2005

Regarding the performance characteristics required to obtain the "Type-Approval Test" certificate, characteristics evaluated in laboratory (table 3) and those tested in field (table 4) must be distiguished.

Table 2: Performance parameters and criteria for nitrogen oxides analyzers established by EPA. All the tests are carried out with NO_2.

Performance parameter	Performance criteria
Range	0-500 ppb
Noise	
At zero	≤ 5 ppb
At 80% URL	≤ 5 ppb
Lower detectable limit	≤ 10 ppb
Interference equivalent	
Each interferant	≤ 20 ppb
Water vapour	
Sulfur dioxide	
Nitric oxide	
Ammonia	
Total interferant	≤ 40 ppb
Zero drift, 12 and 24 hours	≤ 20 ppb
Span drift, 24 hours	
20% of URL	$\leq 20\%$
80% of URL	$\leq 5\%$
Lag time	≤ 20 min
Rise time	≤ 15 min
Fall time	≤ 15 min
Precision	
20% of URL	$\leq 20\%$
80% of URL	$\leq 30\%$
NO_2 efficiency converter[1]	$\geq 96\%$

[1]From EPA [10].

2.3 Comparison between EPA and EN requirements in laboratory tests

2.3.1 General considerations about performance parameters, performance criteria and experimental procedures

As can be deduced from tables 2 to 4, tests targets are, in some cases, similar but neither performance criteria nor experimental procedures are.

Firstly, it is noticeable that EPA does not carry out any field tests, unlike some of the tests proposed in EN 14211:2005.

On the other hand, tests established by EPA are focused on NO_2 whereas EN tests are designed for NO, except in the case of converter efficiency, which is carried out for NO_2 and response time and averaging effort, for both NO and NO_2. This is explained by the fact that NO_2 is measured by reduction to NO so, if conversion efficiency is appropriate, the measurement procedure is common to both species. Evaluating the performance characteristics with NO_2, instead of NO, avoids introducing the standard uncertainty of the reduction process but could lead to mistakes in the NO_2 channel when evaluating certain performance characteristics.

As regards performance criteria, the European standard is much more demanding than EPA. In most cases, these criteria are defined by absolute figures so it is necessary to express them as relative values, since the concentrations of the pollutants used for tests are not the same, in order to establish a correct comparison. The performance parameters of both documents that can be compared are displayed in table 5 with their respective performance criteria in relative value, where it can be easily seen that EPA is much more

Table 3: Laboratory performance parameters and their criteria established in EN 14211:2005 for chemiluminescence nitrogen dioxide analyzers.

Performance parameter	Performance criteria	
	NO	NO$_2$
Range	0-962 ppb	0-261 ppb
Response time for NO channel:		
Rise response time	≤ 180 s	
Fall response time	≤ 180 s	
Difference between rise and fall response times	≤ 10 s or ≤ 10 % relative difference	
Response time for NO$_2$ channel:		
Rise response time		≤ 180 s
Fall response time		≤ 180 s
Difference between rise and fall response times		≤ 10 s or ≤ 10 % relative difference
Repeatability		
at zero level	≤ 1 ppb	
at the hourly limit value	≤ 3 ppb	
Lack of fit to linear		
Largest residual different from zero	≤ 4 %	
Zero residual	≤ 5 ppb	
Sensitivity to sample gas pressure	≤ 8 ppb/kPa	
Sensitivity to sample gas temperature	≤ 3 ppb/K	
Sensitivity to surrounding temperature	≤ 3 ppb/K	
Sensitivity to electrical line voltage	≤ 0.3 ppb/V	
Cross-interferences		
Water vapour	≤ 5 ppb	
Carbon dioxide	≤ 5 ppb	
Ammonia	≤ 5 ppb	
Ozone	≤ 2 ppb	
Averaging effort	≤ 7 %	≤ 7 %
Short term drift, 12 hours		
At zero	≤ 2 ppb	
At the span level	≤ 6 ppb	
Differences in response between sample and calibration port, if applicable	≤ 1 %	
NO$_2$ efficiency converter		≥ 98 %

Table 4: Field performance parameters and their criteria established in EN 14211:2005 for chemiluminescence nitrogen dioxide analyzers.

Performance parameter	Performance criteria	
	NO	NO$_2$
Long term drift, 3 months		
At zero	≤ 5 ppb	
At the span level	≤ 5 %	
Reproducibility		≤ 5 %
Period of unattended operation	3 months or less if manufacturer indicates a shorter period	> 90 %
Period of availability	> 90%	

permissive than EN. The response time and interferences criteria are especially striking.

Differences in test procedures are also found. Whereas EN establishes the number of individual or independent measurements to be taken in each test without specifying any special sequence for their execution, EPA requires at

Table 5: Performance criteria comparison for analogous EPA and EN tests.

Performance parameter	Performance criteria	
	EPA	EN
Response time for NO_2		
Rise response time	≤ 35 min (20 min lag time + 15 min rise time)	≤ 180 s
Fall response time	≤ 35 min (20 min lag time + 15 min fall time)	≤ 180 s
Repeateability (EN) or noise (EPA)		
At zero level	≤ 5 ppb	≤ 1 ppb
At span	≤ 1.25%	≤ 0.6 %
Interferences		
Water vapour[1]	≤ 20 %	≤ 1 %
Ammonia[2]	≤ 20 %	≤ 1 %
Total interferences	≤ 40 %	≤ 3.4 %
Short term drift		
Zero, 12 hours	≤ 20 ppb	≤ 2 ppb
Span		
12 hours	-	≤ 0.9 %
24 hours		
20% URL	≤ 20 %	-
80% URL	≤ 5 %	-

(1) Concentration used by EPA, 20000 ppm; concentration used by EN 19000 ppm.
(2) Concentration used by EPA, 100 ppb; concentration used by EN, 200 ppb.

least seven days for carrying them out, and establishes which tests should be repeated following a strict sequence.

In the next section, most significant differences between both documents are described.

2.3.2 Certification range

Regarding the certification range of both documents, that established in EN 14211:2005 for NO_2 (0 to 261 ppb) is the most coherent since it is 1.25 times the highest limit for this gas in the legislation, that is, the alert threshold (208 ppb). Nevertheless, as was mentioned before, NO_2 is hardly ever used in EN tests as most of them are described for NO, with the certification range in this case of 0 to 962 ppb. This range, as well as that proposed by EPA, is excessively high since in real environments, concentrations are not likely to exceed 200 to 300 ppb, being typically less than 100 ppb.

2.3.3 Response time

This test shows important differences from one document to the other. EN establishes 180 seconds as the maximum time to pass from zero concentration to 90% of the introduced concentration (770 ppb) and the same time to pass from 770 ppb to 10% of this value when changing to sample zero air.

As far as EPA is concerned, it defines the response time as the necessary to obtain a reading of 95% of the introduced concentration (400 ppb) from zero air and must be, at most, 35 minutes which is not admissible taking into account that there are commercially available nitrogen oxides analyzers with response times as low as 30 seconds.

2.3.4 Short term drift

This test also shows how EPA is less restrictive than EN. When sampling zero air, the short term drift performance criteria established by EPA is 20 ppb, whereas in EN it is 2 ppb, which is ten times below the permitted value of the American agency.

Something similar happens with the span concentration. Whereas EPA allows a 20% variation when using a concentration equal to 20% of the upper range limit (URL) (100 ppb) and 5% when using a concentration of 80 % of the URL (400 ppb), EN allows a maximum drift of 0.9% for a test concentration of 670 ppb. Leaving apart the discussion about whether EN drift criteria are adequate or not, a short term drift of 20% in 100 ppb (hourly limit value for the protection of human health) is not admissible since the uncertainty of the readings would be higher than that permitted in the legislation (±15%).

2.3.5 Interferences

Each document establishes four interferent gases, water vapour and ammonia being the two common interferences in both. Water vapour is a specie that is always present in the environment but ammonia is not likely to be found except in industrial fields.

It has been tested experimentally that certain NO converters are able to oxidize ammonia to nitric oxide giving, as a result, a positive signal in the NO_2 channel without interfering with NO. This interference is not taken into account in the EN standard since it evaluates only NO readings, which remain the same in the presence of ammonia, for which this part of the standard should be reviewed. On the other hand, EPA studies the interference of ammonia on NO_2 but only at zero concentration as it assumes that NO_2 and ammonia can react to produce NO.

As has been previously stated [11], water vapour is able to absorb part of the released energy in the chemiluminescence reaction because of its absorption strip in the infrared region. The greater or lesser degree of interference depends on the type of photomultiplier tube used but, can only be totally eliminated with great difficulty since the excited NO_2 emission strip and the water vapour absorption strip practically overlap.

Carbon dioxide and ozone are the two gases that complete the interferences established in the EN standard. Carbon dioxide absorbs infrared radiation but at higher wavelengths, which makes it possible to totally eliminate the interference using a suitable photomultiplier tube.

As regards ozone, it is well-known that this gas reacts with NO to produce NO_2 so if the NO reading diminishes when mixing with ozone, it is not because of interference but due to a reaction between them. This reaction is commonly used to produce NO_2 standard concentrations by means of gas phase titration. The EN standard needs to be reviewed in this respect.

Nitric oxide and sulfur dioxide are the other two gases that EPA specifies as interferences in the NO_2 measurement process, but this *a priori* should not be especially problematic.

2.3.6 Lower detectable limit

This performance parameter is only taken into account by EPA and gives relevant information about the analytical method. Since the annual limit value for the protection of human health is 21 ppb it is essential that analyzers give reliable readings at these low concentration levels.

2.3.7 Lack of fit

This test checks the capacity of the analyzer to give proportional readings to different concentrations, this being one of the basic features of any analytical method.

Although this test is not included among the tests set by EPA, the American agency establishes as a previous step before carrying out all the tests the calibration of the analyzer with, at least, seven points of concentration, correcting the readings by means of the calculated straight line.

2.3.8 Sensitivity coefficients to device characteristics and ambient parameters

One of the most clearest difference between both protocols is that EN 14211:2005 incorporates the evaluation of the analyzer response for a number of performance parameters, such as sensitivity to variations in the sample temperature and pressure, the surrounding air temperature and the line voltage. Even though the two last variables can be controlled in the networks by means of thermal conditioning and voltage stabilizators, respectively, the first can only be controlled with difficulty and, thus, it is necessary to know the influence they have on the measurement process to evaluate their contribution to the expanded uncertainity and decide whether corrections should be made.

3 Conclusions

We have compared the requirements of EPA and EN 14211:2005 for nitrogen oxides analyzers. The latter document will be the first one to impose a number of requirements to be met by this type of analyzers when the Proposal for a Directive on ambient air and cleaner air for Europe comes into force. The performance characteristics are more numerous than those set by EPA to certified the equipments as reference methods for measuring nitrogen dioxide. In cases where the tests are comparable, the performance criteria are much more demanding in EN than in the EPA document. All of this will involve important changes for most of the equipment used at present in European countries to meet the aforementioned requirements. Nevertheless, some of the performance parameters or their criteria must be reviewed and, when necessary, modified.

Acknowledgements

This work has been possible thanks to the financial support of the Fundación Séneca of the Autonomic Community of Murcia by means of the predoctoral grant 01210/fpi/03.

References

[1] ISO 7996:1985. Determination of the mass concentration of nitrogen oxides. Chemiluminescence method.
[2] EN 14211:2005. Ambient air quality. Standard method for the measurement of the concentration of nitrogen dioxide and nitrogen monoxide by chemiluminescence. ISBN 0580457206.
[3] European Union. Proposal for a Directive of the European Parliament and of the Council on ambient air quality and cleaner air for Europe, 2005. http://eur-lex.europa.eu/LexUriServ/LexUriServ.do?uri=COM:2005:0447:FIN:EN:PDF
[4] European Union. Council Directive 96/62/EC on ambient air quality assessment and management. Official Journal of European Communities, OJ L 296, pp 0055–0063, 21.11.1996.
[5] European Union. Council Directive 99/30/EC relating to limit values for sulphur dioxide, nitrogen dioxide and oxides of nitrogen, particulate matter and lead in ambient air. Official Journal of European Communities, OJ L 163, pp 0041–0060, 29.06.1999.
[6] European Union. Council Directive 2000/69/EC relating to limit values for benzene and carbon monoxide in ambient air. Official Journal of European Communities, OJ L 313, pp 0012–0021, 13.12.2000.
[7] European Union. Council Directive 2002/3/EC relating to ozone in ambient air. Official Journal of European Communities, OJ L 67, pp 0014–0030, 09.03.2002.
[8] European Union. Council Decision 97/101/EC establishing a reciprocal exchange of information and data from networks and individual stations measuring ambient air pollution within the Member States. Official Journal of European Comunities, OJ L 035, pp 0014 - 0022, 05.02.1997.
[9] United States Environmental Protection Agency. Ambient Air Monitoring Reference and Equivalent Methods. Code of Federal Regulations, Title 40, Part 53, pp 21–39, 2007. http://frwebgate3.access.gpo.gov/cgi-bin/PDFgate.cgi?WAISdocID=57073018495+56+1+0&WAISaction=retrieve
[10] United States Environmental Protection Agency. Technical Assistance Document for the chemiluminescence measurement of nitrogen dioxide. EPA-600/4-75-003, 1975. http://www.epa.gov/ttnamti1/files/ambient/criteria/reldocs/4-75-003.pdf
[11] Gerboles, M., Lagler, F., Rembges, D. & Brun, C. Assessment of uncertainty of NO_2 measurements by the chemiluminescence method and discussion of the quality objective of the NO_2 European Directive. *Journal of Environmental Monitoring*, 5, pp 529–540, 2003.

A procedure for correcting readings in chemiluminescence nitrogen oxide analyzers due to the effect of sample pressure

M. Doval[1], J. Barberá[1], E. González[1] & F. J. Marzal[2]
[1]*Department of Chemical Engineering, University of Murcia, Spain*
[2]*Department of Thermal Engineering and Fluids, University of Cartagena, Spain*

Abstract

The influence of sample pressure on chemiluminescence nitrogen oxides analyzers is studied in this paper. Although both of the employed analyzers comply with the requirements established in the standard EN 14211:2005 for this performance feature, it is advisable to make corrections to ensure a better quality of the data. The proposed procedure reduced to ± 1.5% the mistakes caused by the effect of sample pressure.
Keywords: EN 14211:2005, nitrogen oxides, chemiluminiscence, pressure, correction procedure.

1 Introduction

Networks measuring ambient air pollution within the Member States use, in general, analyzers certified by the Environmental Protection Agency of the United States as reference methods [2], although no European legislation in this matter obliges the analyzers to meet any special requirements, except as regards the analytical method employed.

In order to improve emission quality data, the European Commission has elaborated the Proposal for a Directive on ambient air and cleaner air for Europe [3], which will replace the Framework Directive 96/62/EC [4] currently in force. The Proposal establishes new reference documents for the measurement of the gaseous pollutants regulated in this field.

Nitrogen oxides (NO and NO_2) receive special attention due to their effects on humans and the environment and for being tropospheric ozone precursors. The reference method for the measurement of these pollutants established in the

Proposal of Directive is that described in EN 14211:2005 [1], which will replace the current document in force (ISO 7796:1985 [5]). The analytical method is the same in both documents, that is, the chemiluminescence reaction between NO and O_3 which has been widely studied in the literature [6-9], but the new standard sets a number of requirements to be met by the nitrogen oxides analyzers before they can obtain the so-called "Type-Approval" certificate. The aforementioned requirements consist of several tests in the laboratory and field and calculation of the expanded uncertainty, which has to be lower than that specified in the legislation (±15%).

The performance characteristics in the laboratory and field are multiple and, among them, the response time, short and long term drift or the influence of pressure and temperature samples are evaluated. In this paper, results concerning the influence of sample pressure are discussed. A correction procedure for the readings is proposed in order to improve data quality.

2 Experimental section

An in-house designed test chamber was used to reproduce the tests described in EN 14211:2005 regarding sample pressure, and those carried out to obtain a procedure for correcting the readings.

NO was used for the tests as indicated in EN 14211:2005 in order to eliminate the possible effect of converter efficiency on the results. Gaseous standards were generated by mixing known flow rates of zero air (produced in the laboratory by compression and purification of ambient air) with known flow rates of a mixture of 50.3 ppm of NO in N_2 matrix from a cylinder. Measurement and control of the flows were made possible by using mass flow controllers. The system is placed inside a thermally isolated chamber operating in the range of 0 to 30° C. Different pressures in sample line were achieved by means of regulating valves, fig. 1. The analyzers were calibrated at 20° C and ambient pressure (for both surrounding air and sample flow). Higher pressures than the ambient are obtained by partially closing the valve located in the main line (V1), whereas lower pressures are achieved by closing V2.

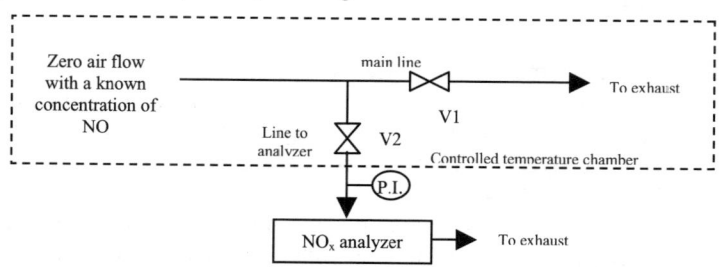

Figure 1: Basic diagram of the experimental device used for the tests.

2.1 Pressure influence test described in EN 14211:2005

First, tests regarding the calculation of the sensitivity coefficient to sample pressure, b_{gp}, were carried out. Two different chemiluminescence nitrogen dioxides analyzers were used. The test is carried out with 770 ppb NO in zero air (which corresponds to 80% of the upper range of NO certification set by the Standard, 961 ppb). The abovementioned mixture is introduced in the analyzer at an absolute pressure of 80 kPa and, afterwards, at 110 kPa, the readings being recorded in both situations. Sample and ambient temperatures are kept at 20° C ± 0.5° C and 20° C ± 2° C, respectively, during the tests.

From eqn (1), b_{gp} (ppb/kPa) can be calculated, where C_{P1} and C_{P2} are the average reading concentrations at sample pressure P_1 (80 kPa) and P_2 (11 kPa), respectively.

$$b_{gp} = \left| \frac{(C_{P1} - C_{P2})}{P_2 - P_1} \right| \qquad (1)$$

The performance criterion for this parameter is $b_{gp} \leq 8$ ppb/kPa.

2.2 Pressure influence at the hourly limit value

The concentration established in EN 14211:2005 (770 ppb) is much higher than that found in ambient air. This is why the pressure influence was evaluated at a concentration of NO and NO_2 close to the hourly limit value for NO_2 (105 ppb). The rest of the conditions of the tests remained the same.

2.3 Tests for obtaining a correction procedure of the readings

Correlations between NO readings and NO standard generated concentrations at different pressures were made. These tests were carried out at 0, 50, 100, 150, 200 and 250 ppb as concentrations higher than the last value are not expected to be found in ambient air. Relative sample pressure was set at 19.6, 14.7, 9.8, 4.9, 0, -4.9, -9.8, -14.7 and -16.6 kPa. For this set of experiments only Analyzer I was employed.

3 Results and discussion

3.1 Pressure influence test described in EN 14211:2005

Figure 2 shows the results obtained when reproducing the sample pressure test described in 2.1. In both cases, pressure was seen to influence the readings. The sensitivity coefficient, b_{gp}, for Analyzer I is 4.8 ppb/kPa, whereas that calculated b_{gp} for Analyzer II is 7.9 ppb/kPa, so both of them meet the performance criterion for this test (8 ppb/kPa). A summary of the calculated coefficients is given in table 1 together with the value of b_{gp} expressed in percentage of deviation/kPa.

3.2 Pressure influence at the hourly limit value

Table 2 shows the results when repeating the test for NO (108 ppb) and for NO_2 (108 ppb) with Analyzer I. The deviation in both channels is the same. The percentual variation of the readings with pressure does not depend on concentration as can be seen by comparing tables 1 and 2.

Table 1: b_{gp} coefficients calculated according to EN 14211:2005 for two different nitrogen dioxide analyzers.

ANALYZER I		ANALYZER II	
b_{gp} (ppb/kPa)	b_{gp} (% of change/kPa)	b_{gp} (ppb/kPa)	b_{gp} (% of change/kPa)
4.8	0.62	7.9	1.02

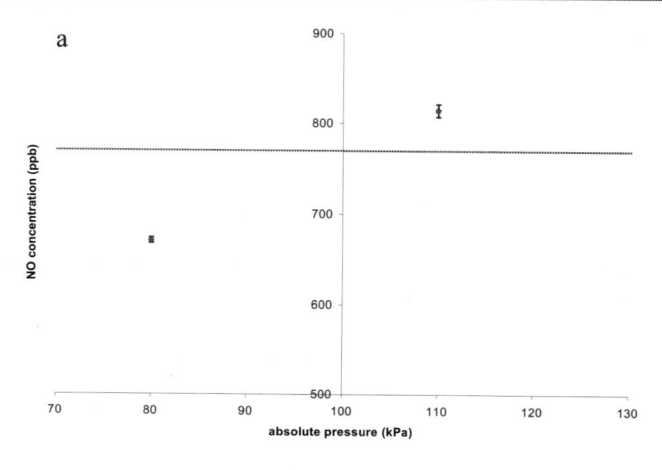

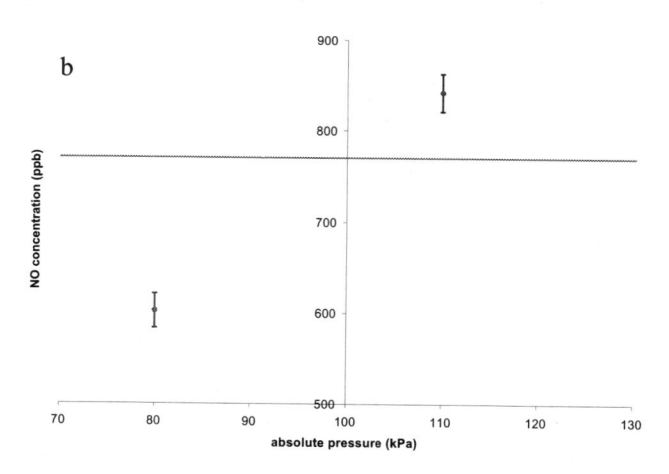

Figure 2: Deviation in NO readings through effect of pressure sample. The continuous line represents the standard generated concentration in the test chamber (770 ppb). (a) Analyzer I. (b) Analyzer II.

3.3 Correction procedure of the readings by effect of pressure

In practice, analyzers are calibrated *in situ* in the measuring ambient air pollution networks by means of pressure cylinders using an exhaust outlet to prevent the analyzers from overpressuring, although the sample will always be at a higher pressure than the ambient one. On the other hand, while analyzers are measuring the ambient air, a drop in pressure is created in the sample line. When line maintenance is not appropriate, it can become blocked for different reasons, and the depression will increase. All of this can lead to sample pressure differences from calibration to measurement ranging from a few tenths of kPa to more than the exit pressure of the pressure reducer in the worst situation.

Table 2: b_{gp} coefficients calculated using 108 ppb of NO and 105 ppb of NO_2.

ANALYZER I			
108 ppb de NO		102 ppb de NO_2	
b_{gp} (ppb/kPa)	b_{gp} (% of change/kPa)	b_{gp} (ppb/kPa)	b_{gp} (% of change/kPa)
0.63	0.58	0.59	0.58

As checked in previous tests, changes in sample pressure lead to concentration reading variations. For this reason, many analyzers have an option called "pressure compensation" in order to avoid these mistakes but, although data quality is improved, it is not enough to avoid differences of up to ±12%, in some cases.

A number of experiments were designed consisting of generating different NO concentrations ranging from zero to 250 ppb at different sample pressures and measuring them with the Analyzer I. Figure 3 shows the readings obtained when generating different NO concentrations at each pressure. The analyzer was calibrated at 20 °C and ambient pressure (for both surrounding air and sample flow).

When working at a different sample pressure from that of the calibration conditions and no equation for its respective line is available (eqn (2)) it can be estimated from the slopes and y-intercepts of the straight lines in figure 3. As the y-intercept does not follow any mathematical law with pressure, an averaged y-intercept, A, was calculated. Slopes at different pressures, B, are calculated from figure 4.

$$[NO]_{reading} = A + B \cdot [NO]_{standard} \quad (2)$$

It is convenient to correlate the slope *versus* an internal parameter of the analyzer, such as the chamber pressure or the sample line flow, which changes proportionally to sample line pressure in order to avoid incorporating a pressure meter in the line. Figure 5 shows the relation between the calculated slope and the sample line flow.

Once the slope is known, it is possible to correct the analyzer readings by eqn (3).

$$[NO]_{corrected} = \frac{[NO]_{reading} - A}{B} \quad (3)$$

Before using eqn (3), it is advisable to first correct the NO readings with the calibration curve of the analyzer calculated from at least 6 points in the calibration conditions (in this example, at 20 °C and ambient pressure), giving the [NO]' reading.

Table 3 shows the mistakes of the readings when no correction for the pressure effect is made, just the correction mentioned in the above paragraph, and those after correction has been made.

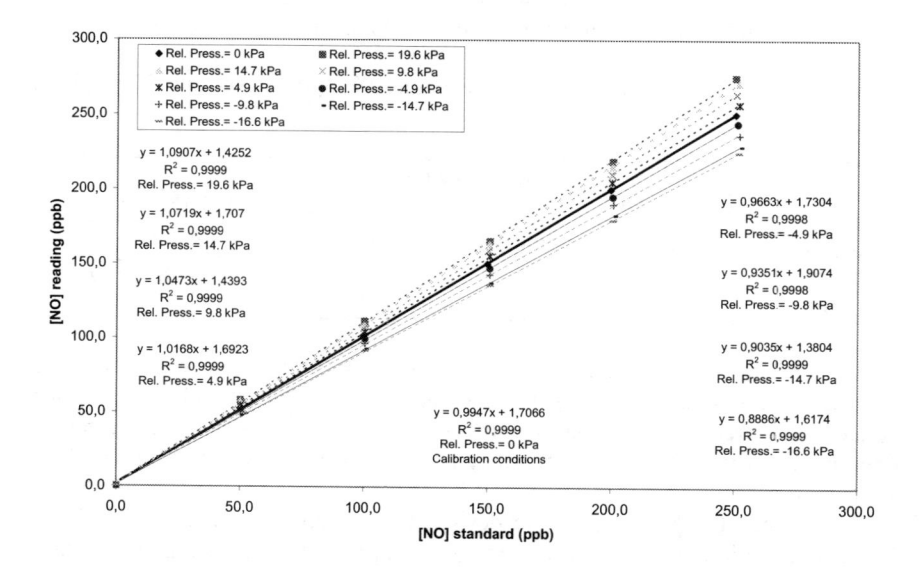

Figure 3: Reading changes as a function of sample pressure.

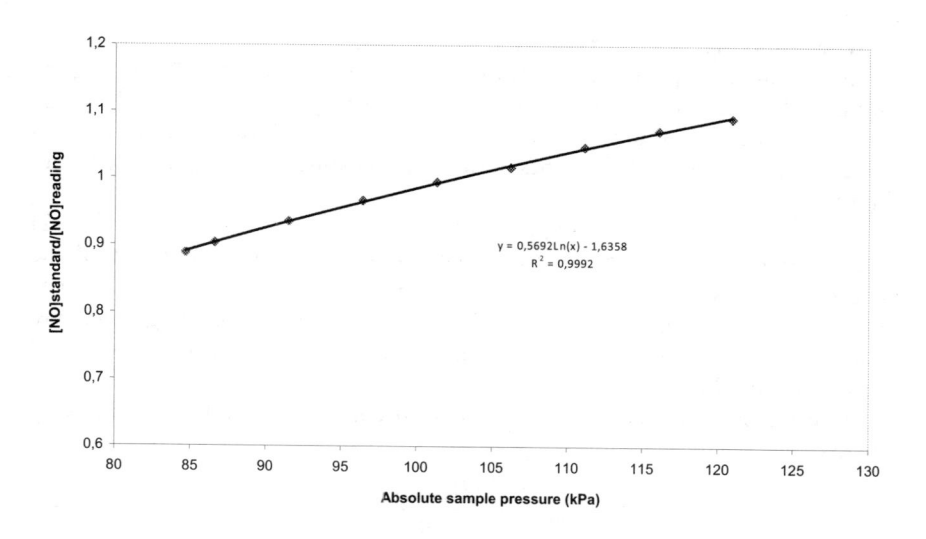

Figure 4: Estimation of the slope of eqn (2) from sample line pressure.

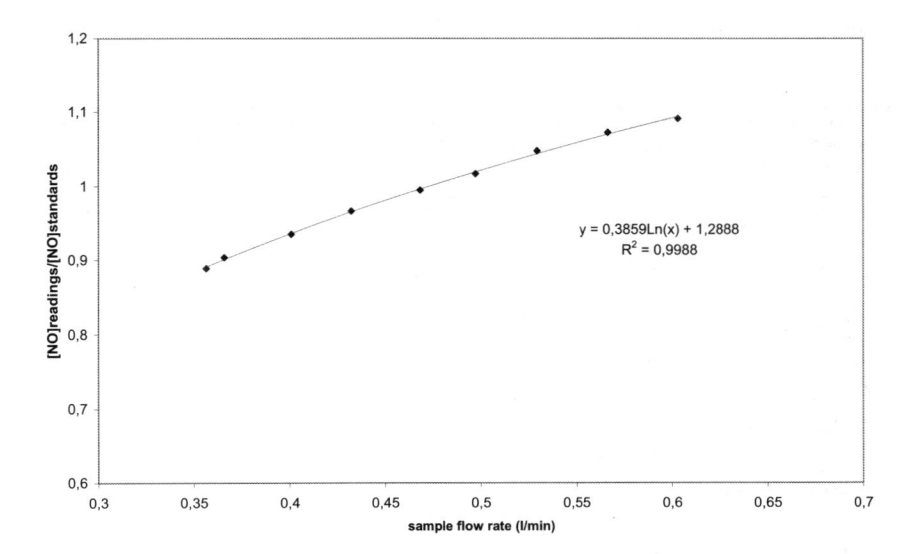

Figure 5: Estimation of the slope of eqn (2) from sample line flow.

Plot shows $y = 0{,}3859\mathrm{Ln}(x) + 1{,}2888$, $R^2 = 0{,}9988$.

The proposed correction improves the data in all cases except four (table 3, in bold) and maximum deviation is below ±1.5%. Figure 6 shows a chart where the mistakes due to the effect of sample pressure are represented depending on the concentration level. The dashed lines delimit the region where correction is not necessary. Outside this region, correction is highly recommended.

Deviations from the standard generated concentration are constant at high concentrations for a fixed sample pressure but at lower concentrations the deviation can be higher or lower. This is due to the intrinsic mistakes of the calibration curve in the calibration conditions, which are higher and positive in the example at low concentrations. Higher sample pressures than the calibration pressure give higher responses than the "real" ones, increasing the mistake at low concentrations. When sample line pressure is lower than that used for calibration, the readings are lower than "real" ones, partially compensating the calibration mistake in these concentration levels.

4 Conclusions

In this paper the influence of sample line pressure on chemiluminescence nitrogen oxides analyzers has been studied. When the pressure difference between calibration and operating conditions is about 15 kPa, mistakes of ±10% are found. These differences of pressure occur as a result of the calibration procedure (overpressure) and the operation conditions (depression).

The proposed correction for the data regarding pressure consists on the following steps:
 1. Calibrate the analyzer in known conditions (e.g. 20° C and atmospheric pressure). Introduce at least 6 values of concentration (zero included).

Table 3: Comparison of mistakes with and without correction due to sample pressure.

Sample pressure (kPa)	Sample line flow (l/min)	[NO]$_{standard}$ (ppb)	[NO]$_{reading}$ (ppb)	[NO]'$_{reading}$ (ppb)	Deviation of [NO]'$_{reading}$ (%)	[NO]$_{corrected}$ (ppb)	Deviation of [NO]$_{corrected}$ (%)
101.325	0.468	250.4	250.5	250.1	-0.10	249.5	**-0.35**
101.325	0.468	199.7	200.3	199.6	-0.06	198.8	**-0.47**
101.325	0.468	149.8	150.3	149.4	-0.30	148.3	**-0.98**
101.325	0.468	99.9	101.8	100.6	0.71	99.4	-0.52
101.325	0.468	50.0	52.9	51.5	3.05	50.1	0.20
101.325	0.468	0.0	0.3	-1.4	-	-3.1	-
120.925	0.603	250.3	274.9	274.7	9.71	249.6	-0.28
120.925	0.603	200.6	219.3	218.7	9.02	198.5	-1.06
120.925	0.603	150.8	165.5	164.7	9.21	149.1	-1.13
120.925	0.603	100.5	111.5	110.4	9.88	99.5	-1.01
120.925	0.603	50.3	58.0	56.6	12.59	50.3	-0.01
120.925	0.603	0.0	0.1	-1.7	-	-3.0	-
116.065	0.567	251.5	271.2	270.9	7.72	251.8	0.11
116.065	0.567	200.5	216.0	215.4	7.45	199.9	-0.29
116.065	0.567	150.6	163.5	162.6	8.00	150.6	-0.03
116.065	0.567	100.4	109.4	108.3	7.89	99.7	-0.64
116.065	0.567	50.2	57.2	55.8	11.32	50.7	1.06
116.065	0.567	0.0	0.2	-1.5	-	-2.9	-
111.135	0.529	250.7	264.0	263.7	5.18	251.2	0.18
111.135	0.529	200.3	210.4	209.8	4.71	199.5	-0.42
111.135	0.529	150.3	159.3	158.4	5.40	150.3	-0.02
111.135	0.529	100.3	106.9	105.7	5.37	99.8	-0.57
111.135	0.529	50.1	55.6	54.1	7.98	50.3	0.39
111.135	0.529	0.0	-0.1	-1.8	-	-3.2	-
106.225	0.497	251.8	257.1	256.7	1.96	250.3	-0.60
106.225	0.497	200.3	205.2	204.6	2.12	199.1	-0.61
106.225	0.497	150.7	155.5	154.6	2.57	150.1	-0.42
106.225	0.497	100.6	104.3	103.2	2.57	99.6	-0.95
106.225	0.497	50.3	54.5	53.0	5.48	50.4	0.32
106.225	0.497	0.0	0.0	-1.7	-	-3.3	-
96.425	0.432	251.2	244.2	243.8	-2.95	250.9	-0.11
96.425	0.432	200.6	195.3	194.6	-2.98	200.0	-0.31
96.425	0.432	150.8	147.2	146.3	-3.00	149.9	-0.62
96.425	0.432	100.6	99.6	98.4	-2.17	100.3	-0.31
96.425	0.432	50.3	52.2	50.8	0.92	50.9	1.22
96.425	0.432	0	0	-1.7	-	-3.5	-
91.515	0.401	251.9	236.4	235.9	-6.33	250.3	-0.62
91.515	0.401	200.9	190.3	189.6	-5.63	200.8	-0.04
91.515	0.401	150.7	142.9	141.9	-5.81	149.9	-0.52
91.515	0.401	100.5	96.5	95.3	-5.18	100.1	-0.42
91.515	0.401	50.2	50.6	49.2	-2.08	50.8	1.16
91.515	0.401	0	0	-1.7	-	-3.6	-
86.585	0.366	252.2	229.0	228.5	-9.40	251.9	-0.13
86.585	0.366	201.1	182.9	182.2	-9.42	200.4	-0.33
86.585	0.366	150.8	137.5	136.5	-9.47	149.8	-0.69
86.585	0.366	100.6	92.9	91.7	-8.87	100.0	-0.62
86.585	0.366	50.3	48.0	46.5	-7.48	49.9	-0.86
86.585	0.366	0	0.16	-1.6	-	-3.5	-
84.655	0.357	251.7	225.0	224.5	-10.81	250.2	-0.60
84.655	0.357	200.8	179.5	178.7	-10.99	198.8	-0.98
84.655	0.357	151.0	136.0	135.0	-10.59	149.7	-0.83
84.655	0.357	100.6	91.9	90.7	-9.87	100.0	-0.63
84.655	0.357	52.3	49.3	47.8	-8.51	51.9	-0.78
84.655	0.357	0	0.16	-1.6	-	-3.6	-

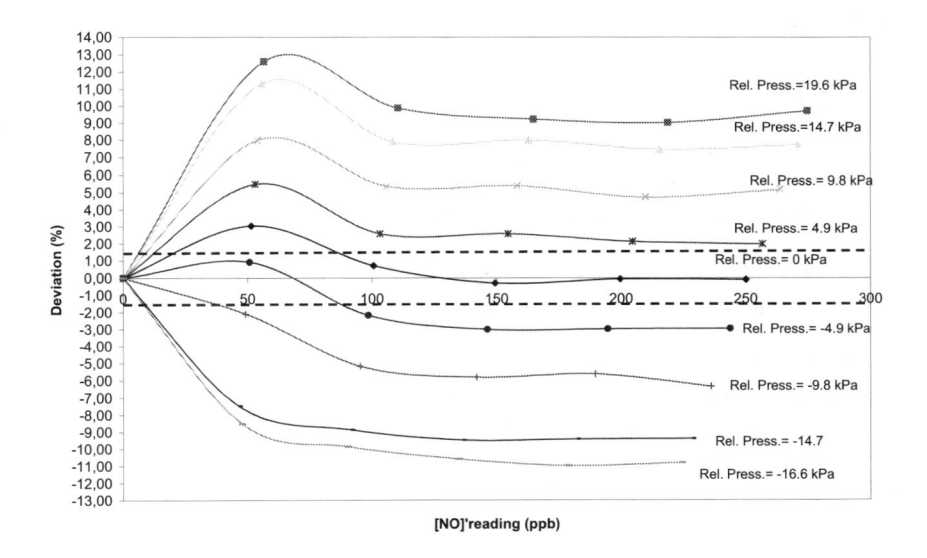

Figure 6: Chart for determining the convenience of the reading correction.

2. Generate different known concentrations of NO (advisable from 0 to 250 ppb) at different known sample relative pressures (from −17 to +20 kPa) and record the responses. Represent the analyzer responses vs. generated concentration. Calculate an average y-intercept from those lines.
3. Represent the slopes of the above lines (NO concentration measured by the analyzer/NO generated concentration) vs. line pressure or sample flow rate.
4. Knowing the line pressure (or sample flow) it is possible to obtain a straight line, using the above graph to calculate the slope, and the average y-intercept.
5. Data obtained with the analyzer should be corrected by means of the calculated line in 4. To improve quality of the data it is advisable to correct the analyzer readings with the calibration carried out in 1 before using the correction in 4.

The results have shown that the mistakes introduced through the effect of pressure can be reduced to ±1.5%. A chart is included to decide whether the correction is worth carrying out.

Acknowledgements

This work was carried out thanks to the financial support of the Consejería de Desarrollo Sostenible y Ordenación del Territorio and the Fundación Séneca of the Autonomic Community of Murcia.

References

[1] EN 14211:2005. Ambient air quality. Standard method for the measurement of the concentration of nitrogen dioxide and nitrogen monoxide by chemiluminescence. ISBN 0580457206.
[2] United States Environmental Protection Agency. Ambient Air Monitoring Reference and Equivalent Methods. Code of Federal Regulations, Title 40, Part 53, pp 21-39, 2007.
[3] European Union. Proposal for a Directive of the European Parliament and of the Council on ambient air quality and cleaner air for Europe, 2005. http://eur-lex.europa.eu/LexUriServ/LexUriServ.do?uri=COM:2005:0447:FIN:EN:PDF
[4] European Union. Council Directive 96/62/EC on ambient air quality assessment and management. Official Journal of European Communities, OJ L 296, pp 0055-0063, 21.11.1996.
[5] ISO 7996:1985. Determination of the mass concentration of nitrogen oxides. Chemiluminescence method.
[6] Clough P.N., Thrus B.A. Mechanism of chemiluminescence reaction between nitric oxide and ozone. Trans. Faraday Soc. (1967), 63, 915–925.
[7] Mathews R.D., Sawyer R.F., Schefer R.W. Interferences in chemiluminescent measurement of NO and NO_2 emissions from combustion systems. Environ. Sci. Technol. (1977), 11, 1092–1096.
[8] Mehrabzadeh A.A., O'Brian R.J., Hard T.M. Optimization of response of chemiluminescence analyzers. Anal. Chem. (1983), 55, 1660-1665.
[9] Winer A.M., Peters J.W., Smith J.P., Pitts J.N. Response of commercial chemiluminescent $NO-NO_2$ analyzers to other nitrogen-containing compounds. Environ. Sci. Technol. (1974), 8, 1118–1121.

New measures of wind angular dispersion in three dimensions

P. S. Farrugia & A. Micallef
Department of Physics, Faculty of Science, University of Malta, Malta

Abstract

Wind is a 3D vector quantity. However, it is frequently found that the vertical component is much smaller in magnitude than the horizontal ones and acts over shorter distances so that its effect is considered negligible. As a result of this, wind is often treated as a two dimensional circular variable. In fact, most of the statistical treatment used to derive the wind angular dispersion, a parameter that is very important in determining the rate of diffusion of pollutants, is based on this assumption. Nevertheless, the reduction of a dimension might not always be justified. A case in point occurs when considering air flow in a street canyon whereby all three directions of air flow can be significant. In such a situation, a 3D measure of angular dispersion would be much more appropriate. For this reason, this work concentrates on establishing consistent 3D measures of angular dispersion. Thus a detailed geometric analysis of an existing and widely used measure of 3D angular dispersion was carried out. This showed that this measure accounts for dispersion mainly along one direction. In order to improve on this, a new set of measures for angular dispersion that account for all three directions is derived using some recently proposed measures of angular dispersion. It is also shown that these new measures can be further extended to any dimension.

Keywords: spherical variables, three dimensional measure of angular dispersion, spherical variance, projection variance, geometric variance, diffusion of pollutants in three dimensions, angular dispersion in any dimension.

1 Introduction

Air motion, can occur in any of the three spatial dimensions. However, it is common practice to ignore the vertical component since this is frequently much smaller in magnitude than the other two and acts over shorter distances. This procedure has the effect of reducing wind from a spherical to a circular variable.

One of the consequences of this reduction in the number of dimensions is that the angular standard deviation is calculated using equations from circular statistics. These can be either the simple ones proposed by Mardia [1] and Batschelet [2] that were obtained by linearization of the angles using some type of mapping, or else, more often, by employing complex, multi-step algorithms like those of Essenwanger [3], Irwin [4], Nelson [5], Skibin [6] and Yamartino [7]. Once the result is obtained, it is then employed to determine the rate of diffusion of pollutants. The important thing to note is that the resultant pollutant diffusion rate applies only to the horizontal plane in which the x- and y-axis lie and gives no information about the vertical diffusion.

A more complete analysis would show that even though the procedure of limiting the number of dimensions that are considered is often justified, it cannot be applied indiscriminately in the general case. A very important example where this approximation is not appropriate is provided by air motion in a street canyon whereby all three components can be significant due to the vortices that can be induced by the vertical walls. Another case arises in connection with the emission of a hot plume, such as the exhaust from motor vehicles, where the buoyancy effect can make the vertical component of air motion sizeable. For such situations, a three dimensional measure of angular dispersion would seem a more appropriate parameter to use in order to calculate the diffusion of air pollutants.

Such considerations lead to the need to investigate 3D measures of dispersion that can be used in 3D pollutant dispersion models. To the knowledge of the authors, the only 3D measure of dispersion available so far in the literature is the spherical variance given by Mardia [1],

$$s^2_{\text{Mardia 3}} = 1 - R_3 \qquad (1)$$

where R_3 is the magnitude of the vector to the centre of gravity of the 3D system $\mathbf{R}_3$ (see eqn (15)). This equation is a generalisation of the circular variance that Mardia [1] proposed in the case of circular variables. However, this extension to a higher number of dimensions is not straight forward, and no indication was foreseen in the work of how the derivation used for the circular variables can be extended directly to the 3D case. Justification for doing so comes from the fact that the equation has the same form as that of circular variance and that R_3 is a measure of the concentration of the angles. In fact for a unimodal distribution the values of R_3 range from zero (when the angles are highly dispersed) to one (when all the angles in the sample are along one single direction).

The problem encountered by Mardia [1] to extend his measure of angular dispersion to higher dimensions is also encountered in the case of other equations that have been proposed for the angular variance. In fact, looking at the other algorithms and the way in which they were derived, there does not seem to be an intuitive way of including an additional angle and proceed to derive a consistent measure of angular dispersion. This means that the transition from circular to spherical variables for these equations is not straight forward even though circular variables should, in theory, be a subset of spherical variables.

An alternative pathway is now possible through the geometric approach taken by Farrugia and Micallef [8]. In this work, using the fact that in general the most important thing is the distance of the angle from the average angle of the sample θ_a (see eqn (6)), a suitable line on which to map the angles or their difference from θ_a was chosen. This led to the development of two new measures of 2D angular dispersion, termed the projection variance,

$$s_{P2}^2 = 1 - R_2^2 - s_{S'}^2 \tag{2}$$

and the geometric variance,

$$s_{G2}^2 = 2(1 - R_2) - s_{S'}^2 \tag{3}$$

where R_2 is the magnitude of the vector to the centre of gravity of the system in the 2D case $\mathbf{R}_2$ (see eqn (8)), while $s_{S'}^2$ gives the variance of the sines of the angles in a direction rotated such that $\mathbf{R}_2$ is aligned with the x'-axis (see eqn (11)).

The real important difference that stems out between these and the linear equations proposed by Mardia [1] (eqn (9)) and Batschelet [2] (eqn (10)) is the term $s_{S'}^2$. In the work by Farrugia and Micallef [8] it was shown that, from a geometric perspective, while R_2 gives a measure of angular concentration along the direction of $\mathbf{R}_2$, $s_{S'}^2$ gives a measure of angular concentration along a line perpendicular to $\mathbf{R}_2$. Thus R_2 and $s_{S'}^2$ together give a 2D picture of the concentration of the angles along two orthogonal directions. In this way, their negative will give a measure of dispersion in two 2D. Farrugia and Micallef [8] also showed that $s_{S'}^2$ emerges naturally if the Taylor or Fourier expansions are used to calculate the variance of the sample angles.

Further analyses on circular distributions carried out in the course of the work presented here reveals that $s_{S'}^2$ is also one of the terms that can emerge when calculating the angular variance of standard distributions. For example, the angular variance of the von Mises distribution with mean zero and concentration parameter κ obtained by Fisher [9], can be written in the form,

$$\mathrm{var}[\theta] = \left(\frac{\pi^2}{3} + 1\right) - 4\rho - 2\sigma_{S'}^2 + 4\sum_{n=3}^{\infty} \frac{(-1)^n I_n(\kappa)}{I_0(\kappa)} \tag{4}$$

where $I_n(\kappa)$ is a modified Bessel function of the first kind and order n, while ρ and $\sigma_{S'}^2$ are the population parameters corresponding to R_2 and $s_{S'}^2$. In a similar fashion, using the Fourier expansion of the wrapped normal distribution [1] and that for the minimum angular distance squared [8, 9] it is a matter of going through the algebra to show that for this distribution the variance is given by,

$$\mathrm{var}[\theta] = \left(\frac{\pi^2}{3} + 1\right) - 4\rho - 2\sigma_{S'}^2 + 4\sum_{i=3}^{\infty} \frac{(-1)^i \rho^{(i^2)}}{i^2} \tag{5}$$

Additionally, in these particular cases, the first few terms of eqns (4) and (5) are actually the same, even though the distributions are different. This gives further indication of the need for using both R_2 and $s_{S'}^2$ in order to obtain a consistent two dimensional measure of angular dispersion.

Another important feature of the projection and geometric variance is that the 2D derivation can be extended to three and higher order dimensions. This is the main scope of this work. In the process it will also be shown that the expression in k-dimensions contains k terms that can be identified as measure of dispersion in k orthogonal directions. However, before proceeding any further, in the next section, some important two dimensional parameters are defined.

2 Some important two dimensional parameters

In this section some important 2D parameters are defined. The first one is the mean angle of the sample θ_a, which for a sample of angles θ_i of size n is defined as,

$$\theta_a = \tan^{-1}\left(\overline{S}/\overline{C}\right) \tag{6}$$

where $\overline{S}$ and $\overline{C}$ are the averages of the sines and the cosines of the angles respectively i.e.,

$$\overline{S} = \frac{1}{n}\sum_{i=1}^{n}\sin(\theta_i) \text{ and } \overline{C} = \frac{1}{n}\sum_{i=1}^{n}\cos(\theta_i). \tag{7}$$

Note that when evaluating θ_a, care needs to the taken of the quadrant in which $\overline{S}$ and $\overline{C}$ will locate the angle.

This definition for the mean angle of the sample was obtained by mapping each angle θ_i to the unit vector $_2\hat{\mathbf{r}}_i = \left[\sin(\theta_i), \cos(\theta_i)\right]$. Subsequently, θ_a was derived as the angle that the mean of the vectors $\mathbf{R}_2$, termed the vector towards the centre of mass of the system, makes with the positive x-axis.

The magnitude or length of $\mathbf{R}_2$, given by,

$$R_2 = \sqrt{\overline{S}^2 + \overline{C}^2} \tag{8}$$

is another important parameter. Its significance stems from the fact that for a unimodal distribution, its value gives an indication of the dispersion of the angles along a single direction. In fact its value ranges from zero (when the angles exhibit extensive dispersion) to one (when all the angles are pointing in the same direction). For this reason, $-R$ is often taken as a measure of dispersion, as can be seen by looking at the equations used as measures of angular dispersion such as the circular variance given by Mardia [1],

$$s_{\text{Mardia 2}}^2 = 1 - R_2 \tag{9}$$

and angular variance proposed by Batschelet [2],

$$s_{\text{Bat}}^2 = 2(1 - R_2) \tag{10}$$

Consider a rotation of the coordinate system by an angle θ_a such that in the new $x'y'$ coordinate system the new mean angle of the sample $\theta'_a = 0$. In such a system $\mathbf{R}_2$ is aligned along the x' direction, so that R_2 becomes a measure of dispersion along this direction.

However, provided that not all the angles are aligned in the same direction, then there should also be some "concentration" about the y' direction and as so it should be possible to find some measure of concentration along this direction. This can be provided by the variance of the sines along the y'-axis [8],

$$s_{S'}^2 = \frac{1}{n}\sum_{i=1}^{n}\sin^2\left(\theta_i - \theta_a\right) \tag{11}$$

(Note that the biased form of the variance is being adopted in order to simplify the equations derived.) The reason for this is that $s_{S'}^2$ takes values ranging from zero, when all the angles are aligned along the x'-axis, to one, when all the angles are aligned along the y'-axis. Naturally, the upper limit cannot be attained because of the alignment of $\mathbf{R}_2$ with the x'-axis. Thus, in a similar way as was done for R_2, it is possible to consider $s_{S'}^2$ as a measure of dispersion along the y' direction.

It can be seen that in 2D it is possible to find two parameters that represent measures of concentration along two orthogonal directions so that their negatives can be taken as measures of dispersion along these same directions. In the next section it will be shown how it is possible to obtain three parameters of concentration/dispersion for spherical variables.

3 Derivation of measures of dispersion in three dimensions

In this section the derivations of the projection and geometric variance as given in [8] will be generalised to 3D. As in the 2D case, the first thing that needs to be done is to find a suitable geometric measure, by which it is meant that a line will be found such that the projections on this line can be representative of the 3D system.

Thus, given a sample (ϕ_i, θ_i) of 3D wind directions (see fig. 1), these can be mapped on the unit vector,

$$_3\hat{\mathbf{r}}_i = \begin{pmatrix} X_i \\ Y_i \\ Z_i \end{pmatrix} = \begin{bmatrix} \sin(\theta_i)\cos(\phi_i) \\ \sin(\theta_i)\sin(\phi_i) \\ \cos(\theta_i) \end{bmatrix} \tag{12}$$

At this point, since the general form of the components will always be the same in any coordinate system, without loss of generality, the projections on the x, y and z-axis will be analysed to assess the potential use of each coordinate axis as the line that represents the angles. Once an adequate axis is chosen, then it is possible to rotate the coordinate system so as to optimise the orientation of the axis.

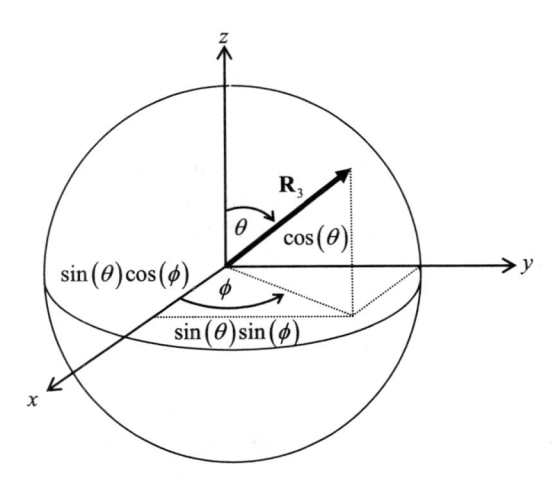

Figure 1: The spherical coordinate system representation in 3D.

For an axis to be able to represent adequately all three special directions, it is necessary that the component along it contains terms with both ϕ_i and θ_i. This means that the z-component cannot be taken. As for the other two axes, in order to discriminate between them, use can be made of the same criteria that has been adopted in the two-dimensional case i.e. that the cosine term increases monotonically when the angle is in the range $[0,\pi]$, while the sine term increases monotonically only when the angle is in the range $[0, 0.5\pi]$. Thus, mapping on the x-axis will retain much more information than mapping on the y-axis. Hence, it can be concluded that the x-axis is the most suitable candidate for the geometric measure.

Even though ideally all terms should involve cosines so that they retain maximum information, given the way in which ϕ_i and θ_i are defined this is not possible since $\cos(\theta_i)\cos(\phi_i)$ is not a component along any fixed line.

At this stage the question arises weather it makes sense to take as the geometric measure $1 - X_i$ ($= 1 - \sin(\theta_i)\cos(\phi_i)$) as was done in the 2D case, or whether X_i should be used. In response to this question it is worth noting that to develop the geometric measure in 3D, before mapping the angles on a line, they are being projected on a circle (through the term $\sin(\theta)$). Once the sphere has been projected on the circle, the argument used in the 2D case [8] can be adopted. One may recall that the argument used was that for the unit circle, the arc length is equal to the subtended angle measured in radiance and hence given that $1 - \cos(\theta)$ is a good measure of the arch length then it is also a good measure of the angle. It follows that $1 - X_i$ should be used.

Once the appropriate axis and measure have been established, the coordinate system can be rotated so as to optimise its orientation. The adopted criterion is once again similar to the one used in the 2D case, i.e. an orientation will be chosen such that it minimises the expected mean direction. Thus we need to minimise,

$$\frac{1}{n}\sum_{i=1}^{n}\left[1-\sin(\theta_i-\beta)\cos(\phi_i-\alpha)\right], \tag{13}$$

with respect to both α and β. This can be done by differentiating with respect to both α and β, setting the results equal to zero,

$$\frac{1}{n}\sum_{i=1}^{n}\cos(\theta_i-\beta)\cos(\phi_i-\alpha)=0 \text{ and } \frac{1}{n}\sum_{i=1}^{n}\sin(\theta_i-\beta)\sin(\phi_i-\alpha)=0 \tag{14}$$

and solving the two simultaneous equations. To solve this system of equations, consider the vector to the centre of gravity in three dimensions $\mathbf{R}_3$. If the coordinate system is rotated such that $\mathbf{R}_3$ is aligned with the new x'-axis, $\mathbf{R}_3$ can be written simply as,

$$R_3 = \frac{1}{n}\sum_{i=1}^{n}\sin(\theta_i-\theta_a)\cos(\phi_i-\phi_a), \tag{15}$$

where θ_a and ϕ_a give the mean direction [1, 10]. If both sides of eqn (15) are differentiated with respect to θ_a and ϕ_a, remembering that R_3 is a constant, then eqns (14) would be recovered if α and β are equated to ϕ_a and θ_a. This means that the geometric measure is optimised if the x-axis is aligned with $\mathbf{R}_3$.

Note that aligning the x-axis with $\mathbf{R}_3$ would not minimise all the possible moments of any distribution. In order to obtain this, the distribution from which the sample has been taken needs to be highly symmetric and n has to be large.

Once the geometric measure has been established, the projection variance can be obtained by equating the system of angles (ϕ_i, θ_i) to the corresponding $1 - X'_i$ ($= 1 - \sin(\theta_i-\theta_a)\cos(\phi_i-\phi_a)$) and then calculating the variance,

$$s_{P3}^2 = \frac{1}{n}\sum_{i=1}^{n}X'_i - \left\{\frac{1}{n}\sum_{i=1}^{n}X'_i\right\}^2, \tag{16}$$

where in deriving this expression use has been made of the identity $\text{var}[ax+b] = a^2\text{var}[x]$, with a and b being constants. On the other hand the 3D geometric variance can be obtained by equating the system of angles ($\phi_i-\phi_a, \theta_i-\theta_a$) with $1 - X'_i$. In this case the resulting equation would be,

$$s_{G3}^2 = \frac{1}{n}\sum_{i=1}^{n}\left[1-\sin(\theta_i-\theta_a)\cos(\phi_i-\phi_a)\right]^2. \tag{17}$$

After some mathematical manipulation using trigonometric relations, it is possible to rewrite eqns (16) and (17) as,

$$s_{P3}^2 = 1 - R_3^2 - s_{y'}^2 - s_{z'}^2 \text{ and } s_{G3}^2 = 2(1-R_3) - s_{y'}^2 - s_{z'}^2. \tag{18}$$

where $s_{y'}^2$ and $s_{z'}^2$ are the variances of the components of the system of angles in the $x'y'z'$ coordinate system:

$$s_{y'}^2 = \frac{1}{n}\sum_{i=1}^{n}\sin^2(\theta-\theta_a)\sin^2(\phi-\phi_a) \text{ and } s_{z'}^2 = \frac{1}{n}\sum_{i=1}^{n}\cos^2(\theta_i-\theta_a) \tag{19}$$

In analogy with the 2D case the terms R_3 and $s_{y'}^2$ can be easily identified with the measures of concentration along the x' and y'-axis that have been identified for the circular variables, namely R_2 and $s_{S'}^2$ respectively. Hence their negative constitutes a measure of angular dispersion in the $x'y'$-plane. The really new term is $s_{z'}^2$, which is again a measure of concentration, this time along the z'-axis. Thus it needs to be subtracted from the angular variance. In this way the two measures of dispersion given by eqn (18) constitute an improvement over the measure of dispersion given by Mardia [1] for 3D spherical variables (eqn (1)). It can also be noted that the maximum value for the projection and geometric variance in the three-dimensional case is the same as for the circular case (one and two respectively for the biased form).

A further point that can be noted is that like in the 2D case [8] the mean bias (MB) of the geometric measure is equal to the mean absolute bias (MAB), given that $1 - X'_i \geq 0$ for any angle, and is equal to the spherical variance (eqn (1)):

$$\text{MB} = \text{MAB} = \frac{1}{n}\sum_{i=1}^{n}[1 - X'_i] = 1 - R_3. \qquad (20)$$

This provides a direct derivation for the measure of dispersion proposed by Mardia [1]. However, in linear statistics, the mean bias and mean absolute bias are usually discarded in favour of the variance. Such a consideration confirms the superiority of the geometric or projection variance as compared to Mardia's spherical variance as a measure of 3D angular dispersion.

4 Results for the *k*-dimensional case

One of the most important aspects, at least from the academic point of view, of the derivation of the projection and geometric variance is that it can be extended to any dimension k. The steps to follow in the most general case parallel exactly those of the 3D case. Thus, taking a sample of angles in k-dimensions given by $(\theta_{1,i}, \theta_{2,i}, \theta_{3,i}, ..., \theta_{k-1,i})$ this can be mapped to the unit vector,

$$_k\hat{\mathbf{r}}_i = \begin{pmatrix} X_{1,i} \\ X_{2,i} \\ X_{3,i} \\ \vdots \\ X_{k,i} \end{pmatrix} = \begin{pmatrix} \sin(\theta_{k-1,i})\sin(\theta_{k-2,i})...\sin(\theta_{3,i})\sin(\theta_{2,i})\cos(\theta_{1,i}) \\ \sin(\theta_{k-1,i})\sin(\theta_{k-2,i})...\sin(\theta_{3,i})\sin(\theta_{2,i})\sin(\theta_{1,i}) \\ \sin(\theta_{k-1,i})\sin(\theta_{k-2,i})...\sin(\theta_{3,i})\cos(\theta_{2,i}) \\ \vdots \\ \cos(\theta_{k-2,i}) \end{pmatrix}. \qquad (21)$$

Since it is desired that the derived geometric measure represents adequately all angles, the x_1-axis needs to be chosen. The geometric measure will then take the form $1 - X_{1,i}$ since as in the 3D case, the k-spherical variable is being mapped successively onto a lower dimensional systems and when it is mapped on a circle the same criteria used for the 2D case will apply.

The next thing to do is to orient the coordinate system so as to optimise the orientation. Using as a condition the need to minimise the moments, and focusing on the minimisation of the mean given by,

$$\frac{1}{n}\sum_{i=1}^{n}\left[1-\sin\left(\theta_{k-1,i}-\alpha_{k-1}\right)\ldots\sin\left(\theta_{2,i}-\alpha_{2}\right)\cos\left(\theta_{1,i}-\alpha_{1}\right)\right], \qquad (22)$$

in a similar fashion to the 3D case, it can be determined that this occurs when α_i's are equal to the $\theta_{a,i}$'s that give the direction of the vector towards the centre of gravity $\mathbf{R}_k$, in the k dimensional system. Thus the resulting geometric measure would be given by, $1 - X'_{1,i} (= 1 - \sin(\theta_{k-1,i} - \theta_{a,k-1})\ldots \sin(\theta_{2,i} - \theta_{a,2})\cos(\theta_{1,i} - \theta_{a,1}))$

The projection variance can now be found by equating the set of angles $(\theta_{1,i}, \theta_{2,i}, \theta_{3,i}, \ldots, \theta_{k-1,i})$ to the geometric measure and then determining the variance. The geometric variance would be given by equating $(\theta'_{1,i}, \theta'_{2,i}, \theta'_{3,i}, \ldots, \theta'_{k-1,i})$ where, $\theta'_{j,i} = \theta_{j,i} - \theta_{a,i}$ to the geometric measure and then determining the variance. The final equations obtained using some trigonometric identities would be given respectively by,

$$s_{Pk}^2 = 1 - R_k^2 - \sum_{j=2}^{k-1} s_{x'_j}^2 \text{ and } s_{Gk}^2 = 2(1 - R_k) - \sum_{j=2}^{k-1} s_{x'_j}^2, \qquad (23)$$

where $R_k = |\mathbf{R}_k|$, given by,

$$R_k = \frac{1}{n}\sum_{i=1}^{n}\sin\left(\theta'_{k-1,i}\right)\ldots\sin\left(\theta'_{2i}\right)\cos\left(\theta'_{1i}\right) = \frac{1}{n}\sum_{i=1}^{n} X'_{1,i}, \qquad (24)$$

and the $s_{x'_j}^2$'s are the variances of the components along the x'_j-axis,

$$s_{x'_j}^2 = \frac{1}{n}\sum_{i=1}^{n} X'^2_{j,i}. \qquad (25)$$

Considering now the mean bias (MB) and the absolute bias (MAB) of this k-dimensional measure of dispersion, it can be easily shown that,

$$\text{MB} = \text{MAB} = \frac{1}{n}\sum_{i=1}^{n}\left[1 - X'_{1,i}\right] = 1 - R_k, \qquad (26)$$

which is Mardia's spherical variance for the k-dimensional system.

5 Conclusion

This work considers the need for consistent 3D measures of dispersion. The problematic issue considered here is that it is not straight forward to generalise the existing 2D measures of angular dispersion to include a higher dimension. An analysis of a known 3D measure of dispersion, that given by Mardia [1], reveals that while this measure has been obtained through a generalisation of Mardia's circular variance, it does not seem to have been justified through a direct generalisation of the derivation.

In order to address the situation, two newly proposed measures of dispersion for circular variables, the projection and the geometric variance, have been generalised to three and higher dimensions. This has been accomplished by finding a suitable line on which to map the angles such that as much information as possible about the directional data is retained.

The generalised projection and geometric variance obtained were shown to have terms that represent dispersion along each orthogonal axis that is needed to represent the directional data. This is a positive feature, which is retained from the 2D case. In the 3D case, this feature makes the variances superior to Mardia's spherical variance, which according to the present analysis, can only represent dispersion along one direction. Furthermore, from the point of view of the geometric measure proposed, it was shown that Mardia's spherical variance is equivalent to the mean bias or absolute mean bias, these being measures of dispersion that are usually set aside in favour of the variance. In all cases, it is another positive feature of the geometric measure that it allows a direct derivation for Mardia's spherical variance, which in turn justifies its use as a measure of dispersion.

A look at the generalised projection and geometric variance would reveal that they have the same basic structure. The only real difference between them is that s_{Gk}^2 emphasises the measure of dispersion along the x_1-axis more than any other axis by not squaring it and multiplying it by two. Thus, it is not an easy task to discriminate between the two in order to decide which one should be adopted. However, the use of these two measures of dispersion is likely to give researchers a superior tool for modelling angular dispersion in three (and higher) dimensions and hence to obtain a better insight on the way in which pollutants will diffuse.

References

[1] Mardia, K.V., *Statistics of Directional Data*, Academic Press: New York, 1972.
[2] Batschelet E., *Circular Statistics in Biology*, Academic Press: New York, 1981.
[3] Essenwanger, O.M., *World Survey of Climatology, Vol. 1, B: General Climatology 1 B: Elements of Statistical Analysis*, Elsevier: Amsterdam, 1985.
[4] Irwin, J. S., Dispersion Estimate Suggestion 9: Processing of Wind Data. (document for internal distribution) Environmental Application Branch, Meteorology and Assessment Division, ESRL, U.S. Environmental Protection Agency, Research Triangle Park, N. C., 1980.
[5] Nelson, E. W., A simple and accurate method for calculation of the standard deviation of horizontal wind direction. *Journal of the Air Pollution Control Association*, 34, pp. 1139–1140, 1984.
[6] Skibin, D., Simple Method for Determining the Standard Deviation of Wind Direction. *Journal of Atmospheric and Oceanic Technology*, 1, pp. 101–102, 1983.

[7] Yamartino R.J., A comparison of several "single pass" estimators of the standard deviation of wind direction. *Journal of Climate and Applied Meteorology*, 23, pp. 1362–1366, 1984.
[8] Farrugia, P. S. & Micallef, A., Derivation of a new measure of angular dispersion for circular variables using geometrical description. *Meteorological Journal*, 7(3), pp. 111–117, 2004.
[9] Fisher, N. I., Problems with the Current Definitions of the Standard Deviation of Wind Direction. *Journal of Climate and Applied Meteorology*, 26, pp. 1522–1529, 1987.
[10] Fisher, N. I., Lewis T. & Emblet B.J.J., *Statistical analysis of spherical data*, Cambridge University Press: Cambridge, 1987.

Variation of air pollution with related meteorological factors in Tripoli (case study)

T. A. Sharif[1,2], A. K. El-Henshir[2] & M. M. Treban[2]
[1]Department of Atmospheric Science, Al-Fateh University, Tripoli, Libya
[2]Libyan Petroleum Institute, Tripoli, Libya

Abstract

Ground level concentrations of nitrogen oxides NO_x, ozone O_3 and some meteorological variables in downtown Tripoli were measured and analyzed as a case study. It is indicated that there are strong relationships between the air pollution gases such as nitrogen oxides NO_x, nitrogen monoxide NO, nitrogen dioxide NO_2, in addition to ozone O_3 and the meteorological factors such as wind speed and direction, solar radiation and temperature. These relationships are positive for some gases and negative for others.

Ozone and nitrogen dioxide are formed during photochemical processes by conversion of nitrogen monoxide by chain reaction, with participation of hydrocarbons and organic compound products. Low concentrations of nitrogen oxides and nitrogen monoxide were observed during the daytime while high concentrations of ozone and nitrogen dioxide were observed during the same period. In other words, these gases showed marked diurnal variation characteristics.

Pollutant emissions follow the cycles of human activity and traffic load with rapid rises near the start of each day's activities, various degrees of decrease or increase around midday, increases in early evening, and declines late at night. Such traffic load activity obviously depends on various characteristics of meteorological factors, and the meteorological factors have a significant impact on air quality.

Keywords: pollutant concentrations, meteorological factors, air pollution gases, traffic load, photochemical processes, diurnal variation.

1 Introduction

Air Pollution is a problem that can only be solved by a collaboration of scientists, the public, industry and politicians. For example, educating the public in order that they might let elected officials know they are willing to pay the price for cleaning the environment is, at least partially, a role of public-interest groups; while passing laws requiring reducing emissions is a political problem. In the scientific area, the actual reduction of emissions is an engineering problem, the transport and diffusion of pollutants in the atmosphere is a meteorological problem, while transformation of atmospheric pollutants is a chemical problem.

A general characteristic of the air quality records for any one of the common pollutants in urban areas is the variability of the concentrations. Concentrations are especially variable in time: from hour to hour, day to day, season to season, and year to year. In some cases the concentrations are highly variable over rather short distances. For those pollutants that react relatively slowly while they are airborne, their temporal and spatial variations in concentration are largely due to an interaction between similar variations in atmospheric transport/diffusion and pollutant emissions. For pollutants that more readily undergo reactions in the atmosphere (e.g., photochemical oxidants O_x formation or physical removal by precipitation) their concentrations also depend on the intensities of the phenomena involved (Liu *et al* [1]).

Temporal variations in transport/diffusion are tied to natural cycles of the weather and large spatial differences, to climatic effects (Georgii [2]). Variations in pollutant emissions are dictated by social/economic practices, and in some cases these are also influenced by the weather (Wang *et al* [3]; Niccum *et al* [4]). Although emissions, weather and atmospheric reactions may follow diurnal and seasonal cycles, their phases, amplitudes and periods are usually different (Holzworth [5]). This tends to confound explanations of observed pollutant concentrations.

2 Site description

The Green Square site at which the measurements were taken is located in the central area of Tripoli adjacent to the Mediterranean Sea at an elevation of 10m above mean sea level. This square extends about 400m in a north-south direction and 300m in an east-west direction and the surface is coated with green paint. Five main streets and a subway are branched from the square, two of the streets have two-way traffic and the other four have one-way traffic, as shown in fig. 1.

3 Data collection

The data utilized in this case study is part of the data that was acquired with the instrumented mobile laboratory of the Technical Center for Environmental Protection positioned at The Green Square in Tripoli for a few days during May 1997. The data measured includes emission and meteorological variables. The primary meteorological variables measured were: wind speed, wind direction, air temperature, global solar radiation, relative humidity, atmospheric pressure and rainfall. The emission variables measured include: NO, NO_2, NO_x, O_3, SO_2, TS, H_2S, THC, Non-CH_4, CH_4, CO, and dust. Only the data on 21st May 1997 will be used for illustrations and discussion.

In addition to the collection of meteorological and emission data, traffic load count was also done at the same position where the mobile laboratory was sitting. The count was taken for all streets that lead to the Green Square area for a weekday (Monday) and a weekend (Friday and Saturday).

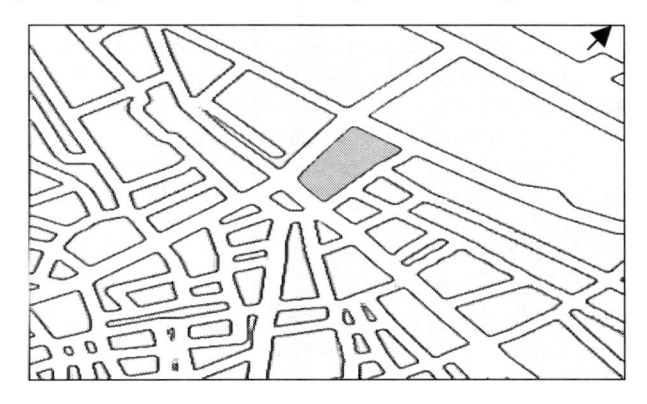

Figure 1:   Downtown Tripoli.

4 Methodology

One of the most important objectives of the study of air pollution is the development of a methodology capable of predicting the variation of pollutant concentrations under varying meteorological and emission conditions. Such a methodology will not only enhance the present understanding of various physical and chemical processes that take place in the atmosphere, but will also provide tools for national and local air pollution agencies to use in regional and urban planning.

Many studies have attempted to develop such a methodology. They can be classified according to the basic modeling approach used: fluid, numerical or statistical.

In the fluid modeling approach, a laboratory device is employed to simulate the behavior of the atmosphere. For instance, laboratory simulation of photochemically reacting atmospheric boundary layers (Hoffert et al [6]).

In the numerical modeling approach, a set of equations is employed as a mathematical representation of the atmospheric processes. Numerical solution of this set of model equations provides the desired concentration distributions. This approach takes into account temporal as well as spatial variations in wind speed, diffusivity, mixing depth, radiation intensity and emission rate. Liu et al [1] used numerical photochemical air pollution modeling to asses the effect of atmospheric parameters on concentration levels of both primary and secondary air pollution in an urban area.

In the statistical modeling approach, air quality data is first collected in conjunction with meteorological and emission data. This data is subsequently analyzed, and correlation relationships between air quality or emission and meteorological parameters are derived.

Statistical techniques are divided into the following three types: graphical, tabular, and summary statistics. Bornstien and Anderson [7] surveyed the statistical techniques used in validation studies of air pollution prediction models.

In this study observational data was used to graphically determine the relationship between the concentrations of gaseous pollutants such as NO_x, NO, NO_2 and O_3 and meteorological parameters such as wind speed, wind direction, temperature and solar radiation. Unfortunately the use of this approach requires a large data base which is not available. Furthermore, the relationships established cannot be extracted to other locales or to future times.

Graphical presentations of the measured pollutant concentrations and meteorological variables can be carried out using values at a single site over extended time period, as in our case, or values at various sites at a single time. Variable time series plots allow us to qualitatively evaluate how they will vary with time. We can also estimate:

1) the magnitude of extreme (maximum and minimum) values,
2) the time of occurrence of extreme values,
3) stable and unstable behaviors,
4) weekdays (high source strength) and weekends (low source strength) behaviors.

5 Results

Graphical comparisons between emission variables and meteorological factors were carried out using values at a single site over a time period of 10 hours. The diurnal variation of each meteorological variable (wind speed, wind direction, temperature and solar radiation) was superimposed on the time series of emission variables (NO, NO_x, NO_2, O_3).

Figure 2 shows the average hourly traffic load count for all streets that branch from the Green Square figure 1. The diurnal variations shown here, especially for weekends, are generally representative. The weekday curve shows traffic peaks at noon with a lull during the morning and evening hours. Weekend curves (Friday and Saturday) differ considerably from the weekday (Monday). A significant change is also feasible between Friday and Saturday. In addition to that, curves from holiday periods may have considerable changes from the weekdays and weekends curves. It is expected that there are some relationships between the traffic load and the concentrations of the different pollutants gases, these relationships will be discussed in the subsequent paragraphs.

The main cause of the variations in gaseous pollutant concentrations is the variation in the patterns of emissions, which at least in part is related to the weather, and variations in atmospheric transport/ diffusion. However, the dearth of available emission and meteorological information precludes the determination of quantitative relationships. Meteorology in detailed information is required in the immediate vicinity and upwind of the air quality station and downwind of the emissions come from large specific sources. In this study, relationships of the wind speed and direction, the air temperature and solar radiation with gaseous pollutants are discussed.

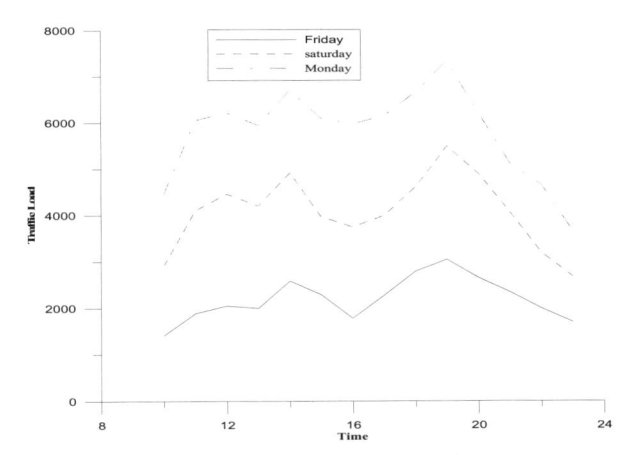

Figure 2: Hourly traffic count Green Square, Tripoli for Friday, Saturday and Monday.

Figure 3 shows the diurnal variations in concentrations of four major gaseous pollutants and wind speed on a weekday in Green Square. The variation of the wind speed with time is conversely proportional to the variation of NO and NO_2. The wind speed is lowest during the evening while the concentrations of NO and NO_2 have there lowest values during the afternoon and about midnight. The NO_x is flocculating more rapidly with its higher values after the sunset and lower values at about midnight. The variation of O_3 is very stable compared to NO_x with larger values during afternoon.

Figure 4 shows the diurnal variations in concentrations of nitrogen monoxide, nitrogen oxides, nitrogen dioxide, ozone, and wind direction on a weekday in Green Square. The wind direction starts northeasterly in the morning and then becomes gradually southwesterly; the peak concentrations of nitrogen monoxide, nitrogen oxides and ozone are when the wind direction becomes southeasterly. As expected the nitrogen monoxide and nitrogen oxides have about a similar diurnal variation. There are no pronounced relationships between the wind direction and ozone concentration.

Figure 5 shows the diurnal variations in concentrations of nitrogen monoxide, nitrogen oxides, nitrogen dioxide, ozone and temperature on a weekday in Green square. As expected the values of temperature is high during the afternoon except when there are some clouds. As usual it states a gradual decrease after sunset to reach low values near midnight. The maximum value of temperature is in agreement with the high values of the concentrations of

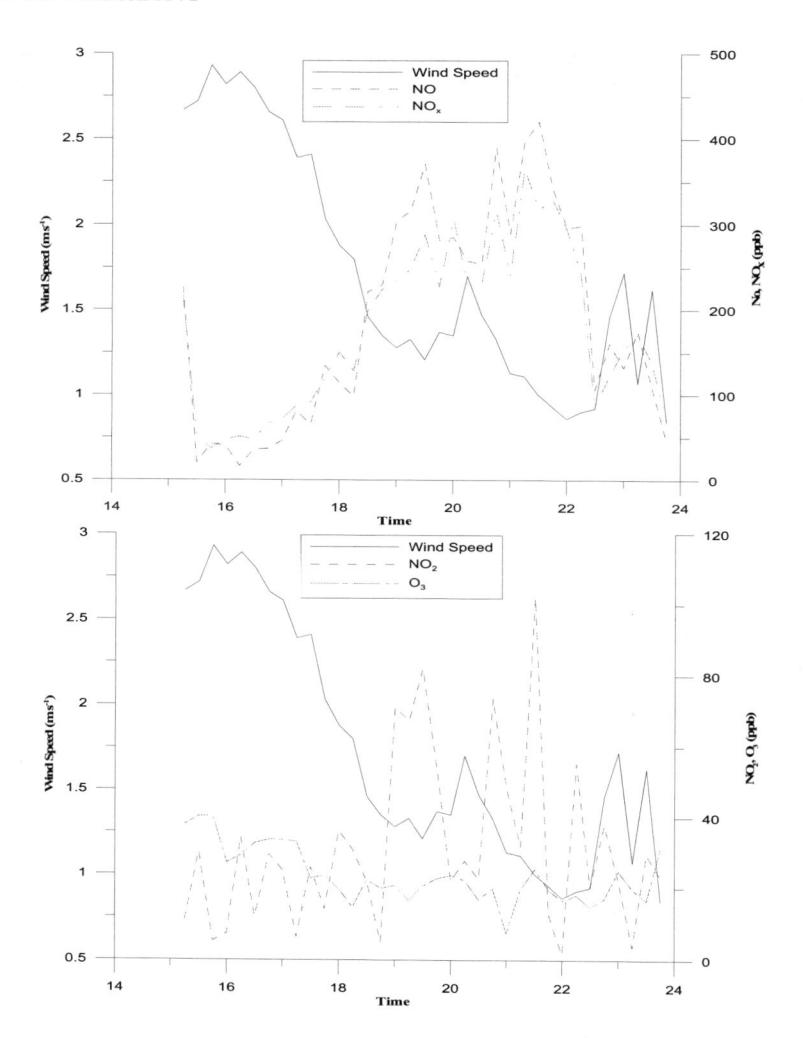

Figure 3: Diurnal variation of NO, NO_2, NO_x and wind speed.

nitrogen monoxide, nitrogen oxides, and nitrogen dioxide during the afternoon period. The opposite is true for the ozone where its values are low during the same period.

Figure 6 shows the diurnal variations in concentrations of nitrogen monoxide, nitrogen oxides, nitrogen dioxide, ozone and solar radiation on a weekday in Green Square. The solar radiation is a maximum at about 3:30pm then decreases gradually to become a minimum at sunset. The ozone values are high during the period where solar radiation values are high. These high values of ozone concentration are due to photochemical processes.

As seen from figures 3–6 each meteorological factor has a different characteristic of diurnal variation. The behaviors of the concentration curves have a meaningful variation in relation to the specific meteorological factor.

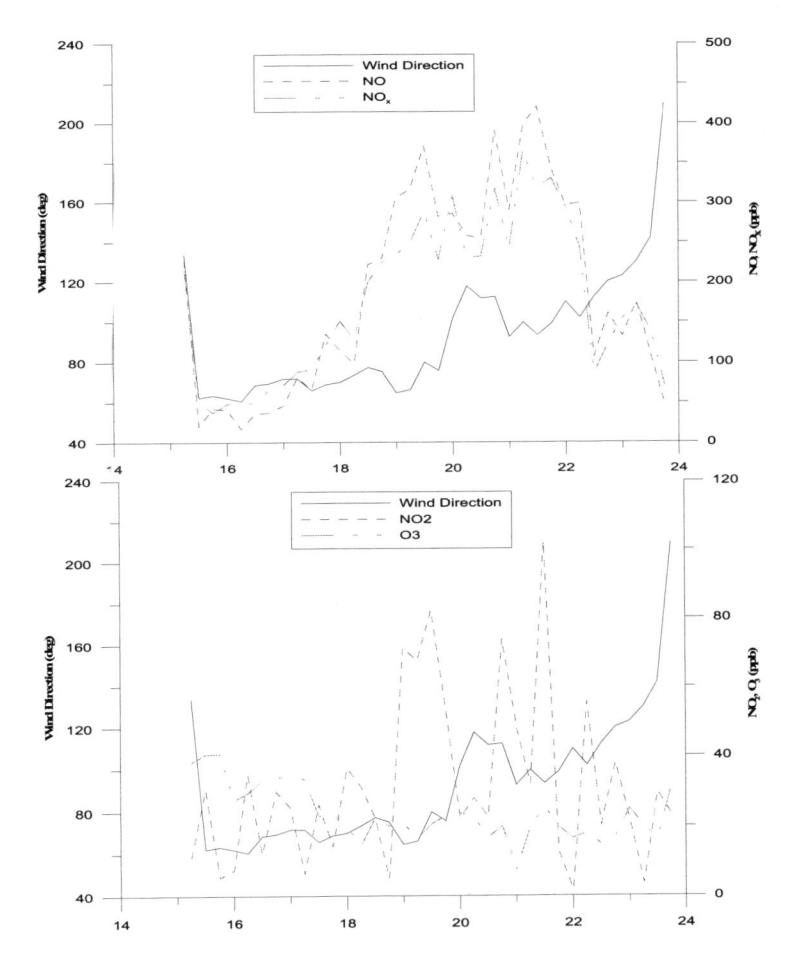

Figure 4: Diurnal ariation of NO, NO$_2$, NO$_x$ and wind direction.

6 Conclusion

In this paper we have used the technique that has been used by many meteorologists to find the relationship between the meteorological factors' variations with pollutant concentrations. The case study was for a week day over a period of 9 hours. A similar case study can extend to cover the weekend for a single station and/or area sources.

Graphical summaries were used to demonstrate qualitatively how the air pollution concentrations and meteorological variables vary over time for a single station. The relationships of the measured concentrations of the pollutant gases are realistic; the wind speed is a good example. Whenever the wind speed values are high, the concentration values are low.

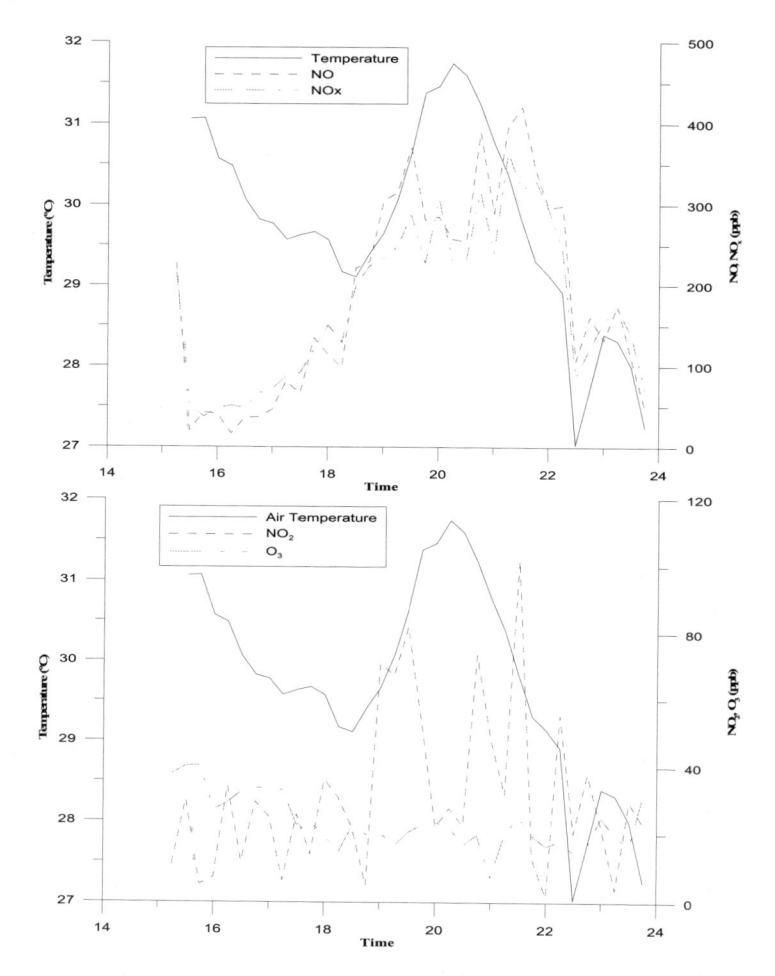

Figure 5: Diurnal variation of NO, NO_2, NO_x and temperature.

Very generally, pollutant emissions follow the cycles of human activity and load traffic with rapid rises near the start of each day's activities, various degrees of decrease or increase around mid-day, increases in early evening, and decline late at night. Such traffic load activity obviously depends on various characteristics of meteorological factors. And the meteorological factors have a significant impact on air quality.

The overall effect on air quality of the usual diurnal patterns of emission and dilution is the high concentrations at the beginning of a day's activities are mainly due to a morning surge in emissions; lower concentrations in the forenoon and afternoon are due to increased dilution and sometimes partly to decreased emissions; by both decreasing dilution and increasing emissions; and low concentrations at night are mainly due to less emissions. Ozone which depends on solar radiation for its formation in the atmosphere, is of course an exception.

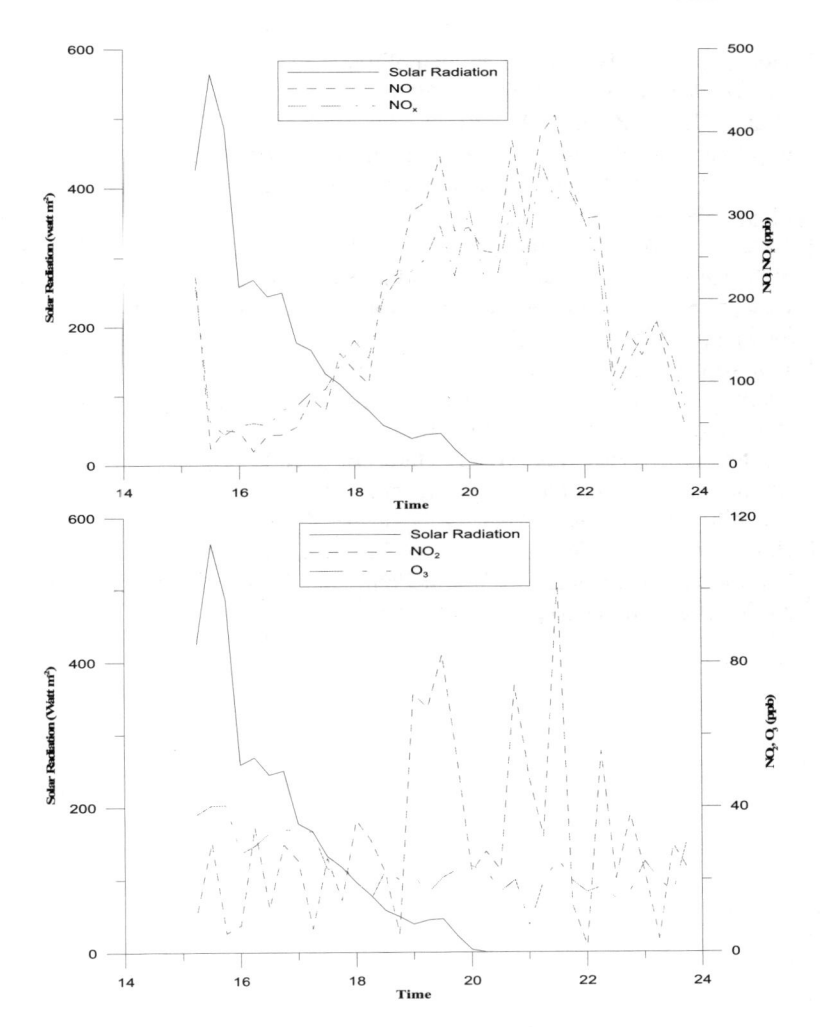

Figure 6: Diurnal variation of NO, NO$_2$, NO$_x$ and solar radiation.

Acknowledgements

We would like to express our deep thanks to the Technical center for Environmental Protection for supplying the emission data and the Libyan Petroleum Institute for supporting part of this work.

References

[1] Liu, M.K., Whitney, C. & Roth, P.M., Effects of atmospheric parameters on the concentration of photochemical air pollution. *J. Appl. Meteo.*, 15(8), pp. 829–835, 1976.

[2] Georgii, H.W., The effects of air pollution on urbane climates. *Bull. World Health Org.*, 40, pp. 624–635, 1969.
[3] Wang, X..J., Zhang, W., Li Y., Yang K.Z. & Bai, M., Air quality improvement estimation and assessment using contingent valuation method, a case study in Beijing. *Environmental Monitoring and Assessment*, 111(1-3), pp. 89–112, 2006.
[4] Niccum, E.M., Lehrman, D.E. & Knuth, W.R., The influence of meteorology on the air quality in the San Luis Obispo Country-southern western San Joaquin Valley region for 3–6 August 1992. *J. Appl. Meteo.*, 34(8), pp. 1834–1847, 1995.
[5] Holzworth, G.C., Variations of meteorology, pollutant emissions and air quality. *Second Joint Conference on Sensing of Environment Pollutants*: Washington D.C., December 10–12, pp. 247–255, 1973.
[6] Hoffert, M. I., Hoydysk, W. G., Hameed, S. & Lebedeff, S. A., Laboratory simulation of photochemically reacting atmospheric boundary layers: a feasibility study. *Atmos. Environ.*, 9, pp. 33–48, 1975.
[7] Bornstein, R.D. & Anderson, S.F., *A Survey of statistical Techniques used in Validation Studies of Air Pollution Prediction Models*, Technical Report No. 23, Stanford University, pp. 1–47, 1979.

Section 5
Urban air management

Prediction of air pollution levels using neural networks: influence of spatial variability

G. Ibarra-Berastegi[1], A. Elias[2], A. Barona[2], J. Sáenz[3], A. Ezcurra[3] & J. Diaz de Argandona[4]
[1]Department of Fluid Mechanics & N.I.,
University of the Basque Country, Spain
[2]Department of Chemical and Environmental Engineering,
University of the Basque Country, Spain
[3]Department of Applied Physics II,
University of the Basque Country, Spain
[4]Department of Applied Physics I,
University of the Basque Country, Spain

Abstract

This work focuses on the prediction of hourly levels up to 8 hours ahead for five pollutants (SO_2, CO, NO_2, NO and O_3) and six locations in the area of Bilbao (Spain). To that end, 216 models based on neural networks (NN) have been built. Spatial variability for the five pollutants has been assessed using Principal Components Analysis and different behaviour has been detected for the nonreactive pollutant (SO_2) and the rest (CO, NO_2, NO and O_3). This can be explained by the very local effects involved in the photochemical reactions. The inputs used to feed the NN models intended to predict forthcoming levels of these five pollutants, include a baseline based on autocorrelation plus a linear or nonlinear combination of different meteorological and traffic variables. The nature of these combinations is different depending on the sensor thus showing the importance of the spatial variability to build the models. The number of hourly cases, due to gaps in data predictions, can have a possible range from 11% to 38% depending on the sensor. Depending on the pollutant, location and number of hours ahead the prediction is made, different types of models have been selected. The use of these models based on NNs can provide Bilbao's air pollution network originally designed for diagnosis purposes, with short-term, real time forecasting capabilities. The performance of these models at the different sensors in the area range from a maximum value of $R^2=0.88$ for the prediction of NO_2 1 hour ahead, to a minimum value of $R^2=0.15$ for the prediction of ozone 8 hours ahead. These boundaries and the limitation in which the number of cases that predictions are possible represent the maximum forecasting capability that Bilbao's network can provide in real-life operating conditions.
Keywords: PCA, neural networks, fluid mechanics, air pollution forecasting, air quality network, traffic network, Bilbao, photochemistry, chemical engineering, applied physics.

1 Introduction

This work describes the results of a study carried out in the Bilbao area corresponding to years 2000 and 2001, in which data from the three existing networks in this city (air quality, meteorological and traffic) have been analyzed jointly to see if short-term, real time hourly forecasts can be obtained for SO_2, CO, NO_2, NO and O_3.

For the period analyzed in this study (2000–2001) the main sources of SO_2 are small domestic heating systems and to a much lower extent, traffic. CO emissions are mainly due to traffic and, in winter, also domestic heating. As far as NO_2 and NO emissions are concerned, a report to the Basque Government suggests that NOx emissions in the area are mainly due to traffic. All these emissions are scattered throughout the whole area and do not show an important spatial variability.

The underlying assumption for this work has been that if the system formed by the three existing networks can properly describe the joint evolution of air pollution, meteorology and traffic, a thorough analysis of their historical records can detect and recognize patterns and relationships among them and as a result, lead to the prediction of forthcoming air pollution levels. These relationships correspond to well known complicated fluid mechanics and photochemistry mechanisms, which are not always easy to model. However, the links between inputs (current and past values of air pollutants, meteorology and traffic) and outputs (future values of air pollution) can be modelled using statistical techniques. Since the mechanisms involved are known to be highly nonlinear, neural networks have been widely used [1–3]. In this work the effect that spatial variability has on this type of model is analyzed.

2 Database

Data to build and test the models were obtained from the historical records of the three networks existing in the area of Bilbao (Spain) corresponding to years 2000 and 2001. In Fig. 1, the six air pollution sensors are labelled from #1 to #6 and the three meteorological sensors as A, B and C. Hourly values of SO_2, CO, NO_2 and NO are measured at the six sensors in the area while O_3 is only measured at locations #1, #2 and #3.

Temperature and relative humidity are measured at the three meteorological sensors, wind speed only at locations B and C, while atmospheric pressure and radiation only at location A. Since sensor B is nearly at sea level and C is 200 m.a.s.l., the difference of temperature between them can be considered as a descriptive estimator of the true vertical thermal gradient.

Traffic is monitored at 181 locations throughout the central area of Bilbao (black zone in Fig. 1) with sensors located under the streets. At each of them, a variable, which represents the number of vehicles (NV) passing above each sensor every ten minutes, is measured. Hourly averages of NV were calculated and mean hourly values for the whole area were computed using measurements from the 181 sensors.

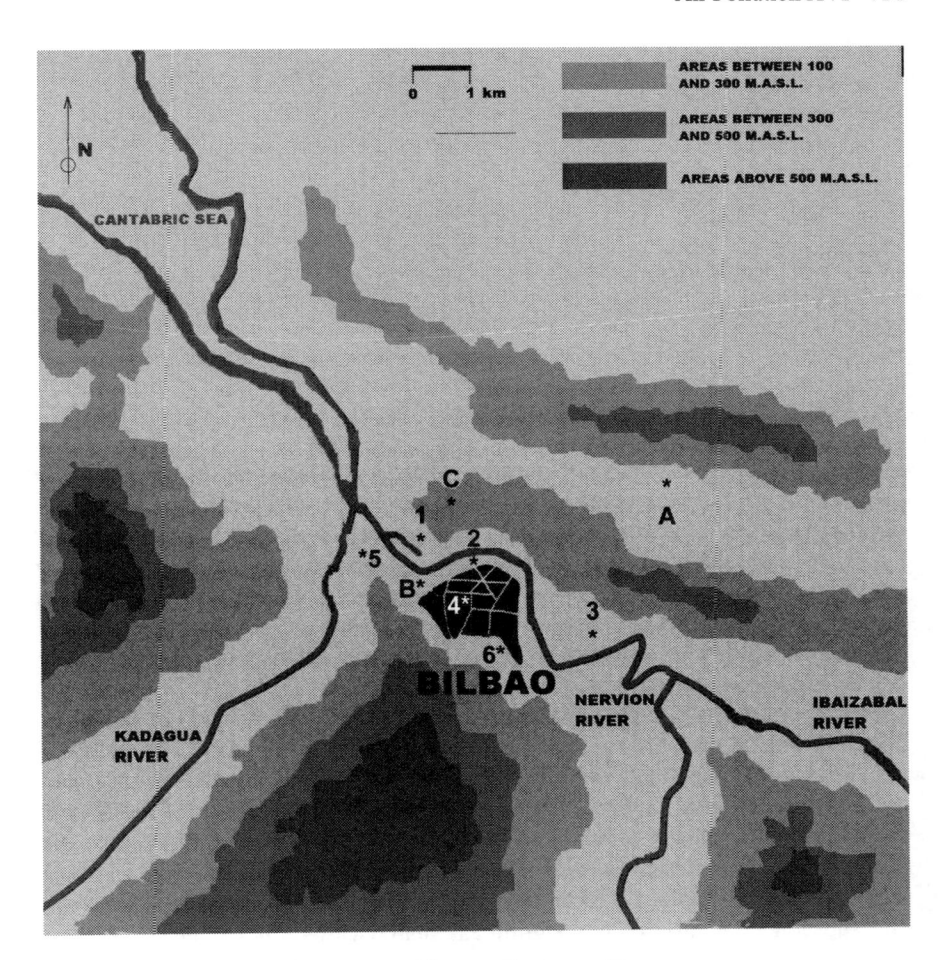

Figure 1: Bilbao and surrounding area.

Finally, all these variables exhibit daily, weekly and yearly cycles, so the sine and cosine functions of these periodicities were calculated and considered as additional candidate inputs. Apart from traffic, additional unknown emission sources were expected to show similar periodicities, so the sine and cosine functions could also be understood as surrogate input variables associated to these unknown pollutants' emissions.

For this study, historical hourly records of the above mentioned variables corresponding to years 2000 and 2001 were available [4–7]. Data of year 2000 were used to build the different groups of candidate prognostic models while data belonging to year 2001 were reserved to test and select the best model. Each case consisted of a complete set of values corresponding to the output and candidate input variables. In principle, for year 2000, 8760 hourly cases with data from the three networks were available. However, due to the gaps and missing data existing in the historical records and also those produced after data pre-processing, the total number of cases available of year 2000 ranged from

1496 in sensor n#2 to 4372 in sensor n#6 and for year 2001, (test set) ranged from 971 in sensor n#2 to 3337 cases in sensor n#5. The goal of this work was to build a group of statistical prognostic models to forecast SO2, CO, NO2, NO and O_3 hourly levels at six locations (Fig. 1) in the city of Bilbao (Spain). The statistical tools employed have been several types of neural networks (NNs). These NNs obtained can be easily incorporated into the air pollution network management activities to obtain hourly forecasts [8,9,10]. The NNs are the core of the BISTAPOF (BIlbao Short-Term Air POllution Forecast) model. A demo of BISTAPOF is available at no cost from http://www.ehu.es/eolo/software/bistapof_demo/index.html

3 Results

For the five pollutants analyzed, predictions are made using all the types of NNs [4–10]. However, it was necessary to analyze the effect that the spatial variability of pollutants' emissions have on the different models used at each location.

To that end, the only pollutant that in the space and time frame of this study could be considered inert -SO_2.-was analyzed separately. The rest of pollutants are involved in complex photochemical reactions that are highly site-dependant like the availability of VOCs or the NOx/VOC ratio and as a consequence, they can be expected to have a higher spatial variability

The tool used to detect spatial variability in the emission fields of SO_2, CO, NO_2, NO and O_3 measured at each of the six sensors during years 2000 and 2001 was Principal Component Analysis (PCA). For the five pollutants, the results of the PCA show that 2 factors are enough to account for fractions of the overall variability ranging form 79% to 92%. For each pollutant, if the measurements in each sensor are represented on the factor plane corresponding to the two main factors, it can be graphically detected clusters of sensors with similar factor loads and subsequently, little or non-significant spatial variability. Inversely, sensors with different factor loads will appear in the graph far from each other, thus suggesting relevant spatial variability (Fig. 2a-2e). In the case of SO_2, the representation of the six sensors on the two-factor plane (Fig. 2a; ~80% of the overall variability) shows similar factor loads, and the sensors tend to cluster close to each other. Therefore, it can be concluded that the six sensors are capable of capturing the same main SO_2 regimes in the area.

If the emissions are scattered throughout the whole area and the SO_2 sensors "see" nearly the same, it can be concluded that the dispersion in the area of Bilbao covered by the six sensors of a non-reactive pollutant like SO_2, can be described following the single box model where the SO_2 is almost perfectly mixed in the boundary layer throughout the whole area. The single box model is based on the mass conservation of pollutants inside a Eulerian box. In the area above the 6 sensors the two mountain ranges (Fig. 1) form a box in which measured SO_2 levels are due to emissions inside the box plus the transportation from or to nearby areas (advection). These SO_2 apportions are associated to the main circulations in the area which take place forth and back along the river axis (SE-NW).

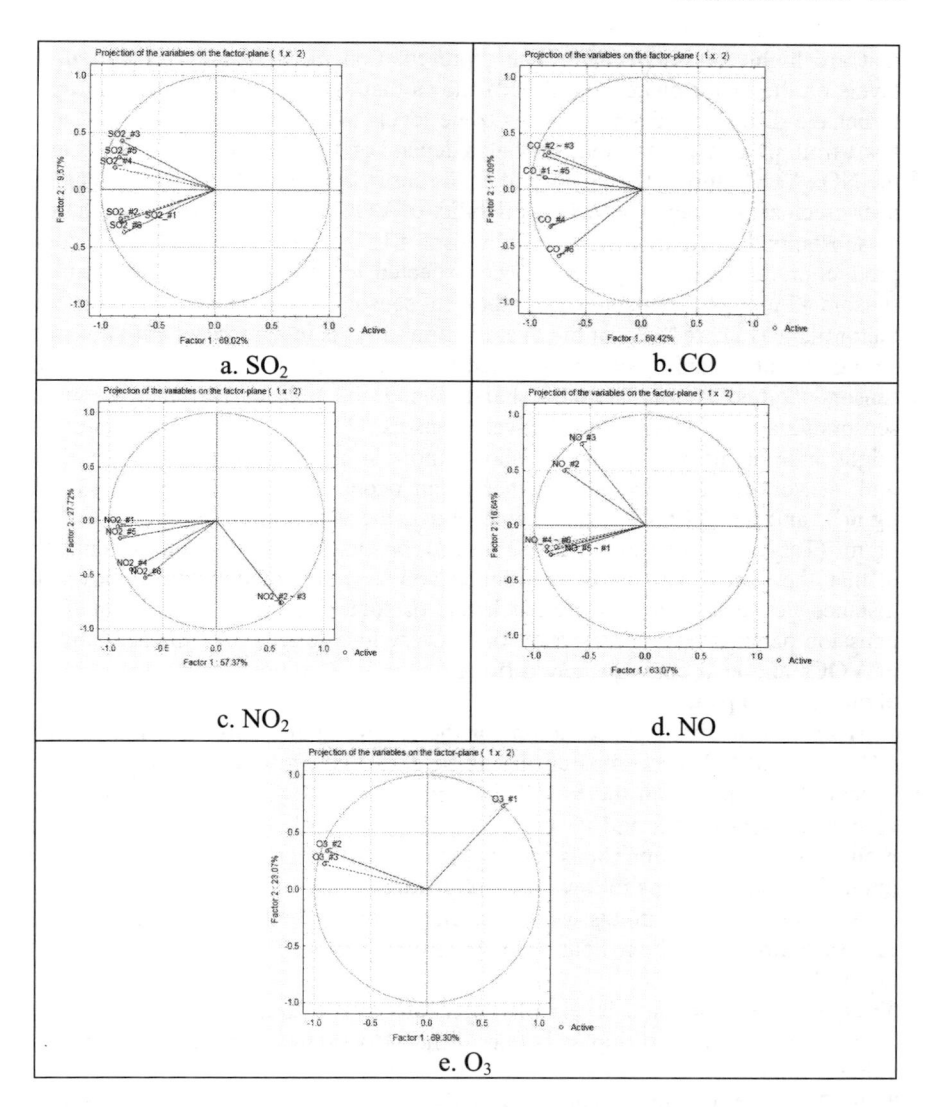

Figure 2: a–e. Factor plane for the five pollutants and six sensors.

In the case of CO, emissions are primarily originated by traffic throughout the whole area and, although involved in the photochemical reactions, CO emission levels are mainly associated to traffic cycles. The traffic flows in the whole area are similar, with low spatial variability, so emissions can also be expected to be similar. The results of the PCA applied to the six CO sensors show quite a similar behaviour (Fig. 2b), which is in agreement with the emission pattern originated by traffic. However, the representation in the factor plane shows that sensors #2 and #3 are particularly near to each other. NO_2 and NO emissions are also mainly due to traffic [4,8,10] and these emissions do not show strong spatial variability. However, both pollutants are involved in photochemical reactions that are highly site-dependant. The representation of NO_2 and NO measured levels on the factor plane (Fig. 2c-2d) shows that again, sensors #2 and #3 tend to behave differently from the rest. Ozone levels are highly site-dependant, and apart from precursor emissions they also depend on the availability of VOCs and the NOx/VOC ratio in the vicinity of each sensor. Sensors #2 and #3 are located near green areas where the local availability of VOCs produced by the vegetation is substantially higher than that in sensor #1. This explains that the ozone production/destruction regimes can be expected to be significantly different in sensors #2 and #3 if compared with those of sensor #1. The representation in the factor plane (Fig. 2e~92% of the overall variability) is in agreement with this and strong spatial variability can be detected between sensor #1 on the one hand, and sensors #2 and #3 on the other hand. Being the physical distance between sensors #2 and #1 smaller than between sensors #2 and #3 (Fig. 1), the question might arise about why sensor #2 shows more similarities with sensor #3 than with sensor #1. This includes higher mean ozone levels in sensor #2 (37.33 $\mu g/m^3$), and #3 (34.65 $\mu g/m^3$) – both above the area's average value of 33.4 $\mu g/m^3$ (Table 1) – than in #1 (28.32 $\mu g/m^3$), below the average The explanation is that the spatial variability mentioned above is not related to the physical distance between sensors, different levels of solar radiation, or different NOx emission patterns near each sensor, but to very local effects like the availability of VOCs and as a consequence, different NOx/VOC ratios leading to different photochemical patterns.

If the two-factors representation for the reactive pollutants (CO and mainly NO_2, NO and O_3) are compared, it can be seen that sensors #2 and #3 tend to behave differently from the rest. This is in agreement with the fact that in the vicinity of sensors #2, and #3 the availability of VOCs can be expected to be higher due to the emissions from green areas nearby. As a result, the photochemical reactions follow different patterns.

Six major types of inputs were identified in the 216 NNs finally selected for the prediction at a given hour H of a certain pollutant's levels H+K=(1,...8) hours ahead:

Type 1: The pollutant's levels measured at current hour H. The information contained in this type of inputs is that of the autocorrelation function at a lag of value K hours.

Type 2: The pollutant's levels measured at hour H-Z, being Z the number of hours before H that the variable has been measured.

Type 3. In the case of pollutants involved in photochemical reactions, the rest of reactive pollutants – measured at hour H or H-Z- usually also appear as inputs.

Type 4: Traffic measured at time H and H-Z.

Type 5. Meteorological variables, mainly radiation and wind speed at time H and H-Z.

Type 6. Sine and cosine functions corresponding to the daily, weekly and yearly cycles.

Very often the typical value of Z =24-K, thus indicating the strength that the 24 h cycle has on all the forecasts. A sensitivity analysis applied to the 216 NNs showed that in all cases, the inputs belonging to type 1 were the most relevant to explain changes in the outputs. This indicates that predictions are made using a baseline, which is the autocorrelation function, that is, current values of the pollutants measured in the network (input type 1). In some cases (31.9%) this is the best option (persistence) while in others, the information corresponding to additional input variables (types 2-3-4-5-6) is also incorporated and combined in the frame of linear (51.4%) or nonlinear (16.7%) models. The rather high amount of models in which persistence is not outperformed indicates that although in some of these cases, linear or MLP models perform equally, there is no need to select a complicated model if a simple one (like in this case, persistence) is enough [10].

In the case of SO_2 (the only non-reactive gas in the frame of this study's time and space scale), persistence is enough in as many as 68.8% of cases. It can be seen that for this pollutant, up to 4h ahead, in most sensors persistence is the best option. From 5h ahead onwards, at least in half the sensors more elaborated models (linear or not), which incorporate the rest of input types, are used. This suggests that being SO_2 a pollutant whose emission levels are closely linked to emissions, four hours can be understood as the average period of time needed to detect changes in emissions and/or transportation from/to nearby areas. During nightly hours, perhaps persistence could be the best option for more than 4 hours ahead. However, in other periods of the day changes in emissions take place more rapidly and the persistence model might not work so well. However, the same model is used for daily and nightly hours so 4 hours can be considered as an average period for the validity of the persistence models in the case of SO_2.

For the CO predictions up to 2 hours ahead, persistence is the best option in four sensors. This gas is reactive but its emission levels are mainly guided by the evolution of traffic. The mean autocorrelation function for traffic up to 2-4 hours lag, shows quite high values which explains why persistence (based exclusively on type 1 inputs), is the best option. Performance of persistence models was aided by the prevalence of moderate CO levels in the area. After 4 hours ahead more elaborated models (linear or not) incorporating additional information corresponding to the rest of input types are needed to predict CO levels.

For the rest of reactive pollutants like NO_2, NO and ozone, autocorrelation (persistence) cannot be used beyond 1h-2h ahead and in some cases, like ozone, not even that. The reason is that concentrations are continuously varying, not only due to changes in the emissions but also owing to their participation in the photochemical processes. Therefore, predictions need to be calculated incorporating (in a linear or nonlinear way) additional information corresponding to the rest of inputs.

The spatial variability described above for reactive pollutants is not reflected in the selected model's architecture but mainly in the relative relevance that the six types of inputs have on the final prediction. A sensitivity analysis shows that, in sensors #2 and #3, the most relevant inputs belong to type 1, followed by type 3 and type 5. In the rest of sensors, after type 1 inputs, it is type 2 and type 4 inputs that contribute most to build the predictions. As said before, sensors #2, and #3 are affected by local emissions of VOCs and the information that needs to be added to the prediction baseline (input type 1) must also include the past behaviour of the rest of reactive pollutants (input type 3) and meteorology (input type 5). In the rest of sensors, there is a major single source of precursors (traffic) and therefore, 24h cycle corresponding to the different pollutants, constitutes the most powerful signal after autocorrelation.

4 Conclusions

Air quality networks are usually designed for diagnosis purposes, being a key feature of a good network that it has enough time and space resolution to follow the evolution of the most important pollutants.

The air pollution network of Bilbao was originally designed as a diagnosis tool to describe in real time the evolution in the area of several pollutants and meteorological variables. The traffic network was also intended to follow the evolution of the traffic flow in the area of Bilbao. Bringing together the information from these networks, statistical models based on NNs to obtain short-term forecasts of air pollution levels can be built. The use of these models can provide the air pollution network with new forecasting capabilities. Very local effects have a great influence in the mechanisms of production for reactive pollutants. This results in an important spatial variability though the NNs can capture this variability and performance is not affected by this fact.

Acknowledgements

This work has been financially supported by the University of the Basque Country UPV-EHU (Spain) (contract #1/UPV 00149.345-E-15398/2003 and research group GIU05/12), the Spanish Ministry of Science and Education (contract CGL2005-06966-C07-05/CLI-INVENTO National R+D+I Project) and project EKLIMA21 ETORTEK (Basque Autonomous Government's meteorological agency, EUSKALMET).

The authors also wish to thank the Basque Government and Bilbao's Municipality for providing data for this study.

References

[1] Gardner, M.W., Dorling, S.R. Artificial Neural Networks (The Multilayer Perceptron). A review of applications if the atmospheric sciences. *Atmospheric Environment,* **32**, pp. 2627–2636, 1998.
[2] Gardner, M.W., Dorling, S.R. Neural network modelling and Prediction of hourly NO_x and NO_2 concentrations in urban air in London. *Atmospheric Environment,* **33**, pp. 171–176, 1999.
[3] Gardner, M.W., Dorling, S.R. Statistical surface ozone models: an improved methodology to account for nonlinear behaviour. *Atmospheric Environment,* **34**, pp. 21–34, 2000.
[4] Ibarra-Berastegi G., Elias, A., Agirre, E., Uria, J. Long-term changes in ozone and traffic in Bilbao. *Atmospheric Environment,* **35**, pp. 5581–5592, 2001.

[5] Ibarra-Berastegi, G., Madariaga, I., Agirre, E., Uria, J. Short-term real time forecasting of hourly ozone, NO_2 and NO levels by means of multiple linear regression modelling. *Environmental Science and Pollution Research,* **8**, pp. 250, 2001

[6] Ibarra-Berastegi G., Elias, A., Agirre, E., Uria, J. Traffic congestion and ozone precursor emissions in Bilbao (Spain). *Environmental Science and Pollution Research,* **10**, pp. 361–367, 2003.

[7] Elías, A., Ibarra-Berastegi. G. Arias, R., Barona, A. Neural networks as a tool for control and management of a biological reactor for treating of hydrogen sulphide. *Bioprocess & Biosystems Engineering*, **29**, pp. 129–136. 2006.

[8] Ibarra-Berastegi, G. Short-term prediction of air pollution levels using neural networks. Air Pollution XIV. pp. 23–32. ISBN 1-84564-165-5. WIT Press. Southampton. UK. 2006.

[9] R. Arias, A. Barona, G. Ibarra-Berastegi, I. Aranguiz and A. Elías. Assessment of metal contamination in dredged sediments using fractionation and Self-Organizing Maps. *Journal of Hazardous Materials* **151**, 78–85. 2008

[10] G Ibarra-Berastegi, Ana Elias, Astrid Barona, Jon Saenz, Agustin Ezcurra, Javier Diaz de Argandoña. From diagnosis to prognosis for forecasting air pollution using neural networks: air pollution monitoring in Bilbao (Spain). *Environmental Modelling & Software.* **23,** pp. 622–637, 2008

[11] A. Ezcurra, J. Sáenz, G. Ibarra-Berastegi, J. Areitio. Rainfall yield characteristics of electrical storm observed in the Spanish Basque country area during the period 1992-1996. *Atmospheric Research. Accepted,* 2008.

Environmental planning and management of air quality: the case of Mexicali, Baja California, Mexico

E. Corona-Zambrano & R. Rojas-Caldelas
Faculty of Architecture, Autonomous University of Baja California, Mexico

Abstract

There is a worldwide increase in the number of people living in cities. As a consequence, this brings a number of social, economic and environmental problems relating to a decline of the environment and the quality of life. Air quality stands out among the problems of medium and large cities, mainly due to the increase of industrial activity, urban mobility, mobility from one city to another and of urbanization levels. That is why human settlements face important challenges in order to lower the levels of emission of pollutants into the atmosphere that are considered aggressive and hazardous for the built up environment, ecosystem and human health. In this context, the city of Mexicali stands out in the country for the problems relating to air quality it faces, mainly pollutants such as CO, PM10 and O_3. This report presents the results of research which has the purpose of designing a system of indicators for the study of air quality in the city of Mexicali, based on the pressure-state-response (PSR) model from OECD. This system is complemented with information of other urban variables, as well as population, traffic infrastructure, transportation, and land uses among others, trying to identify which strategies need to be applied in the city. Lastly, an urban assessment about pollution and strategies that can be applied in different fields involved in air quality is presented.

Keywords: urban air quality, environmental management of air quality, atmospheric pollution, sustainable urban planning.

1 Introduction

The management and strategies applied in Mexico to protect health and decrease the pollution levels in human settlements, has taken place through establishing quality regulations for each air pollutant. More recently, monitoring has been established in the main cities and the programs to improve air quality known as PROAIRES, which include concrete measures to decrease and control pollutant emissions. A characteristic of these programs is that government authorities participate (federal, state and local).

However, industrial regulations are inefficient (SEMARNAP [11]), through the urban scope, planning is still focused on land use, reducing the environmental dimension, along with the scarce information available for citizens about the industrialization and urbanization phenomenon and their impact on health, specifically on respiratory diseases.

The previous gets complicated with the actual trend, like the worldwide increase of the urbanization process in developing countries. According to United Nations (UNEP [12]), currently almost half of the world population lives in urban areas, expecting it to be 90% by the year 2030. In Mexico, 75% of its total population live in urban areas, where cities like Chihuahua, Aguascalientes, Mérida, Hermosillo and Mexicali among others which have around 500 000 to 1'000 000 million inhabitants, are part of the main urban system, and considered also strategic cities of economic and demographic corridors important in the region.

Based on that, this report has a double purpose; on the one side it shows the problem of air quality in Mexicali through state indicators and their relationship with land uses in the city, and on the other side to show the strategies used to decrease pollution from the point of view of different agencies. Lastly, to present some strategies dealing with urban planning and management topics to tackle air quality problems.

In the case of Mexicali, the economy of the city-region is supported mainly by the services sector, which employs 54.48% of the working population, while the industrial sector employs 35.46% and the primary sector a 10.06% (XVIII City Council of Mexicali-IMIP [2]), where agriculture and geothermal electric production as well as the sands and gravel extraction for construction work. These activities show certain patterns of territorial distribution and at the same time show a concentration of pollutants which affect the health of the inhabitants of the city. Mexicali is located north in the Mexican Republic in the state of Baja California (32°40'N, 115°27'W), adjoins north with the state of California, USA (Figure1), and it is known for its extreme dry weather and extreme temperatures. During summer temperature rises up to 54°C and in winter to 7°C. The average precipitation is 54.2 mm; the city is located in an arid region named Lower Colorado Delta belonging to the physiographic region of the Sonora desert. Also, the Colorado Delta has supported the agricultural areas from both sides of the border, along with its problems in air pollution for agricultural fires and spraying crops. On the other hand, car ownership is high, with one vehicle for every three inhabitants in the city of Mexicali. In the same way, the manufacturing industry has always been one of the motors of the local economy, activities that have worsen air quality, an issue that due to its local impact has become of binational concern.

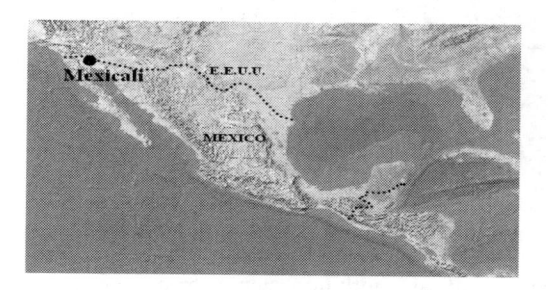

Figure 1: Location of Mexicali.

2 Methods

The assessment of air quality was made through the interpretation of the state indicators of environmental concentration of criteria pollutants recorded by the local monitoring system, composed by 6 stations located in the city, from which 4 monitor Ozone (O_3), Carbon Monoxide (CO), Sulfur Dioxide (SO_2), Nitrogen Dioxide (NO_2), and Particulate Material (PM10). In the case of Progreso and CONALEP stations, they only measure PM10, CBTIS station stopped working to be relocated south of the city in 2005, under the name Campestre. The monitoring project derived from an agreement between United States and Mexico governments through the Border XXI Program, which later became Border 2012 Program. Access to such information was through the Air Quality System database (AQS) 1997-2005 (AQS-UABC [1]) from the environmental protection agency (EPA) from the United States. For this case and in order to simplify the management of information, measuring of air quality was made by the number of days in which the respective standard was violated. To determine the characteristics of the immediate environment around the stations, an analysis of the land uses was made within a range of 1,500 meters, using the geographic information system Mapinfo vr. 7.8. For the proposed strategies, the base documents were checked and authorities in charge of the programs were interviewed.

3 Results

3.1 Air quality in Mexicali

Monitoring stations that integrate the local monitoring system are generally located inside the facilities of a school or in a public sector's building like the public health center of Progreso residential development. The urban features in land uses vary according to location (Figure 2).

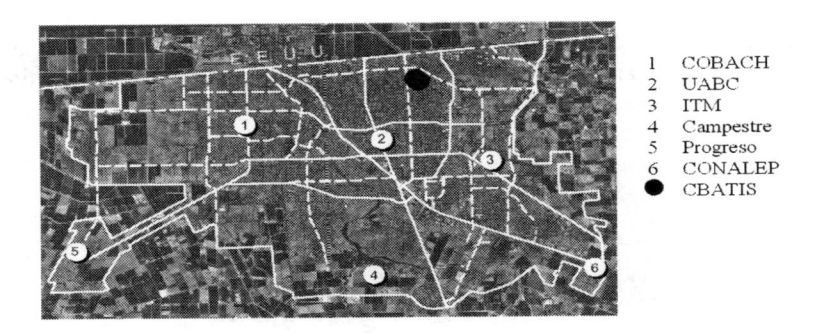

Figure 2: Monitoring stations.

Legend:
1 COBACH
2 UABC
3 ITM
4 Campestre
5 Progreso
6 CONALEP
● CBATIS

The main causes of air pollution come from energy production, motor vehicles, industry, open dumps (sanitary landfills now) dust storms which are very characteristic in the region, car traffic on unpaved roads, as well as crops burning on both sides of the border (Mendoza, et al., [6]; SEMARNAP [10]).

Results obtained from monitoring between 1997 and 2005, in the city of Mexicali show the dynamics of the city regarding the economic activities that support it and the environmental conditions related to its geographic location. Current data generated by the monitoring system match with previous reports made in 1997 and 1998 by SEMARNAP [10], where the main problem seemed to be PM10 particles, CO and O_3 pollutants which mainly concentrated in COBACH, Progreso, CBTIS and UABC stations. (SEMARNAP [10]). However, data from 2005 reports a significant decrease even though currently levels of concentration of these exceed permissible levels of PM10 and O_3.

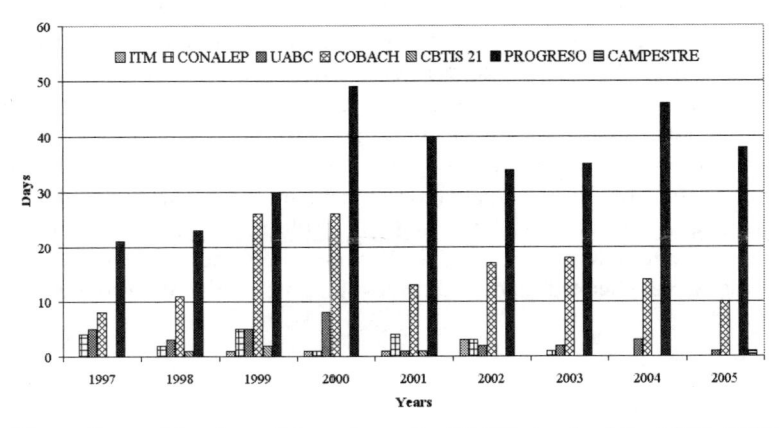

Figure 3: Number of days above the PM10 standard for 1997–2005.

In the case of PM10, a rising tendency from 1997 to 2000 for all the stations can be observed, reverting from 2000 to 2005 except for Progreso and COBACH stations, which currently exceed the corresponding quality standard. This is because these stations are located in the periphery of the city, where due to the low urbanization levels, traffic dust of unpaved roads, smoke from burning crops or derelict lands eroded by wind affect productive agricultural areas. The case of UABC station presents an exception, for being in an area of good urbanization, and a lower percentage of unpaved roads, a case that would be necessary to study in greater detail (figure 3).

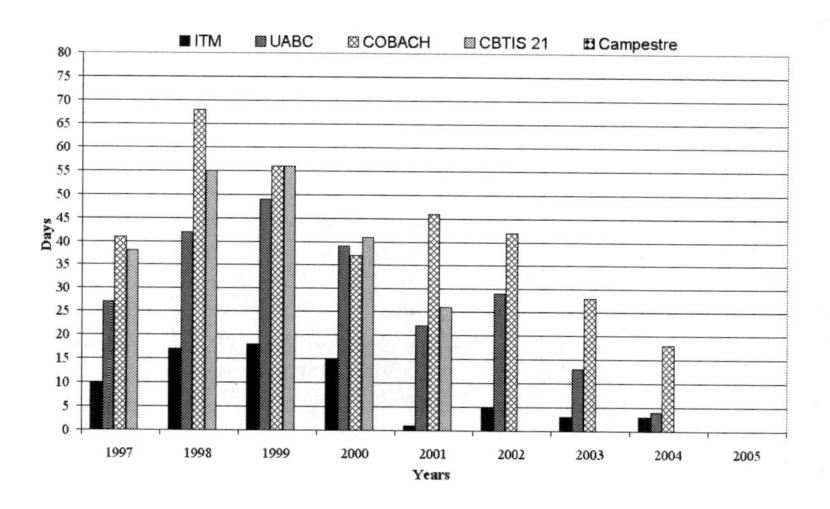

Figure 4:   Number of days above the CO standard for 1997–2005.

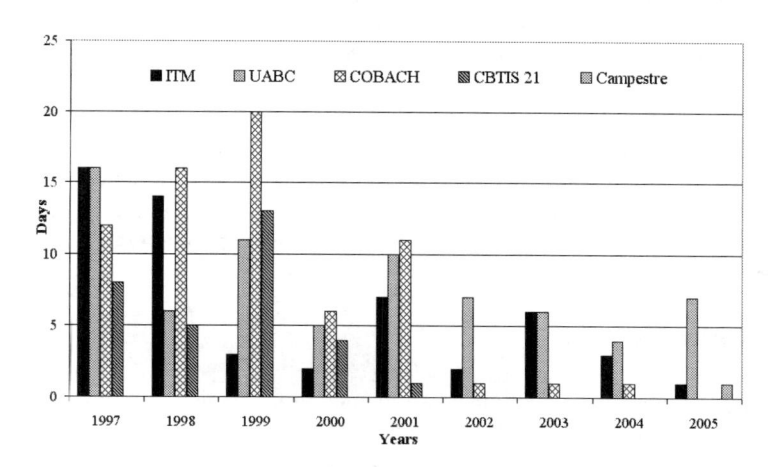

Figure 5:   Number of days above the O_3 standard for 1997–2005.

The CO for the analysis period shows a rising increase in the first three years, to decrease considerably and disappear in 2005. Although during this period stations like COBACH, UABC and Campestre reported violations to the standard also but with a smaller number of days that year. During 2005 no more violations to the standard were registered (figure 4).

O_3 pollutant shows a general trend to decrease on most of the stations, despite its recovery in 1999 and 2001, by 2005 the number of violations to the standard decreased considerably. The presence of O_3 continues on UABC, COBACH and ITM stations, the last two continued exceeding the standard during 2005, though in a lower proportion. These stations represent mostly urban areas, with moderated to high population densities with concentration groups of urban public services of urban and regional importance, or industrial activities that involve great volume of traffic caused by the transportation of people and goods in the city (Figure 5).

3.2 Implemented actions

As an answer to air pollution problems in Mexicali, during 2000 and 2005 programs and actions were put into practice by different authorities and levels of the government, mostly the Ecology and Environment sector, like the ones made by the federal government for the improvement of air quality in Mexicali 2000-2005 (SEMARNAP [10]), trying to reduce the emissions of different pollutants that exceeded the standard. Strategies were aimed toward: vehicles, urban management and transportation, ecology recovery research and national agreements, from which 27 actions resulted.

According to the 2006 results only 5 specific actions were put into practice (Gordillo [3]; IIS-JUEBC [4]; Paez [8]), among these were: the transference of the monitoring net by the United States, who financed and operated it at first; the Integrated Pavement and Air Quality Program (PIPCA), that worked during 2003-2004, improving the levels of urbanization with the pavement work that benefited about 20 residential areas from the city, with a total surface of 214 hectares. Integrated study on public transportation and its improvement, suggests replacing a part of the fleet with new and modern buses; and the creation of a system of automated red-light on the most important roads in the city; environmental culture and city forestry, this program is operated in elementary school and recreational areas; vehicles, on this topic, the implementation of the program of vehicle checking, oriented to emission control that public and private transportation cause, a project which by 2006 was being elaborated and subject to approval by the city council.

4 Discussion

A series of strategies and actions instrumented to tackle the air pollution problem from the point of view of the Environment and natural resources department have been commented. However management of air quality has not been integrated to the urban local planning, neither an integral planning that joins other sectors which would need to be planned together to help solve the problems of pollution (Leitmann [5]; Mitchell [7]). It is necessary to remember that air quality is also competence of regional urban planning, because several situations may be prevented, avoided or minimized using urban and environmental planning instruments that offer a national, state and municipal legal framework.

It is also clear that Mexicali presents problems of scarce planning or no planning at all, related to topics such as the growing number of vehicles, most of them imported from United States whose characteristics do not comply with the environmental quality standards; the public transportation that covers one third of the population; pavement work necessary for almost 60% of the streets in the city; industrial development that consumes 50% of the electric energy that is produced and that in terms of users it represents 1.14% (Rojas [9]). Added to all of this, there are no local codes and institutions are not capable to control the banks of material around the city. Incompatible mixed land uses of residential and industrial developments, places for disposal of solid wastes and hazardous wastes. Agricultural land that have become derelict and plots of land that lack vegetation and get eroded, and with a deficit of 49.9% of green areas in the city (XVIII City Council of Mexicali-IMIP [2]).

Previous information shows that even though it is true that implementation of programs partially successful like pavement of roads, transportation and intelligent red-lights have contributed to reduce pollutants significantly such as CO and in less degree PM10 and O_3, it has not been enough to keep under the standard these air pollutants with the urban and environmental characteristics that prevail in the area of influence of the monitoring stations.

It is important to note that in the case of industry, even though the levels of NO_2 and SO_2 have not presented violation to the standard since 2003. From the point of view of preventing environmental pollution also, it is necessary to consider the importance of this sector in the economic development, because if the trend to increase the number of industries planned for the industrial project of binational impact named Silicon Border continues, pollution tendencies for NO_2 and SO_2 will change for sure. Most of all because among the currently installed are the chemical, metallurgical, paper and food industries, considered as the ones that affects the environment the must.

4.1 Environmental planning and air quality management: strategies for the improvement of air quality

Urban planning needs to consider aspects like the following in order to face important challenges of the cities that have to do with air quality and the health of its population.

- **Industry**: joint land use planning with strategic environmental assessment; the implementation of norms in authorizing and monitoring of industry operation; publishing and providing communication to the general public about the characteristics of industrial emissions; promoting industries to join environmental certification programs; and discourage the setting of industries in the fields of electricity, metallurgic, chemical, paper and food, to reduce the participation of industries identified as those who generate considerable amounts of emissions and that are considered highly polluting.
- **Agriculture**: controlling and monitoring agricultural field burning; the application of agrochemical substances and pesticides in productive areas and the setting up of green buffer zones.

- **Material extraction sites**: regulating and monitoring the emissions during their operation.
- **Sanitary land fills**: defining an adequate and timely site for disposing solid waste, hazardous waste, as well as their emissions.
- **Transportation and traffic**: promote through the regulation of land uses, a compact pattern of the city with higher population densities that tends to reduce distances between home and work. There should also be an increase in the primary traffic infrastructure and better pavement coverage. Provide incentives for an increase public transportation, with newer and more adequate units appropriate to local weather. Also, it is of outmost importance to regulate the vehicles in use, and the ones that are 10 years old or more; verify emissions and promote economic or fiscal incentives to substitute old vehicles.
- **Public green areas**: reduce the deficit of public parks and promote its contribution to improve air quality, through adequate species for the region.
- **Energy**: promote the quality of fuels used for transportation, but also the generation and use of alternative sources of energy for industrial, domestic and commercial use in order to reduce the burning of fossil fuels.
- **Health and quality of life**: systematic follow up of health problems related to respiratory diseases due to high pollution levels, along with a preventive program that provides information and guides the population to preserve good health.
- **Policy and urban environmental standards**: defining policy, as well as standards, that respond to the local situation are required in order to define and instrument strategies and programs oriented towards improving the present urban environmental problems.
- **Production of environmental data**: developing pressure and response indicators that can be complemented by the state indicators (present air quality conditions), providing information based on the monitoring systems and the emission inventories that influence the design of public policies to measure the performance of programs provided by other sectors.
- **Information and communication:** it is recommended that this is implemented so that the policies, programs, and actions are not only known by the government authorities who work in the design and implementation of strategies and successful programs for air quality improvement, but that the general public has the adequate and timely information about air quality and the atmosphere conditions when precautions must be taken. These aspects will contribute considerably in the promotion of environmental and cultural education.
- **Environmental education**: this is an important issue, as it can promote interest and awareness of the population about environmental problems. In a particular manner about the state and conditions that local air quality has. Due to the importance and magnitude of the issue, it is convenient not only to depend on the strategies implemented by the government. The educational sector needs to incorporate in its curriculum, subjects related with the environment and the perspective it presents for modern times.

5 Conclusion

City planning must count with tools that help it make an environmental assessment not only of air quality, but also other resources to have influence over them. It is also necessary to stop thinking that air quality in the city, is only a responsibility of the ecological and energetic sectors, it also needs the participation of other sectors, like urban development, in which it would be necessary to develop suitable methodologies for the treatment of environmental aspects that improve the air quality of the cities.

References

[1] AQS-UABC, Database: pollutants PM10, CO, NO_2 SO_2, and O_3, Technical research report: UABC, Mexicali, B.C., Mexico, 2006.
[2] XVIII Mexicali City Council and IMIP, Urban development programme for Mexicali, B.C. 2025.: Council of Mexicali, Mexico, 2006.
[3] Gordillo, C. Personal communication 19 October 2006, Department of ecology of the city council of Mexicali, B.C.
[4] IIS-UABC and JUBC Assessment of integrated programme for the improvement of quality of life in the state of B.C. Technical research report, UABC: Mexicali, B.C., Mexico, 2006.
[5] Leitmann, J. *Rapid urban environmental assessment. Lessons from cities in the developing world. Methodology and preliminary findings. Urban management programme.* World bank: Washington, D.C. 1994.
[6] Mendoza, A., García, M. and Pardo, E., Air quality information catalogue for the Mexicali-Imperial Valley border region: Mexicali-Imperial Valley air quality modeling and monitoring programme in support of LASPAU's border ozone reduction and air quality improvement program. *Draft final report* LASPAU-*ITM:* Mexico, 2004
[7] Mitchell, G. Forecasting urban futures: a systems analytical perspective on the development of sustainable urban regions (Chapter 5). *Exploring sustainable development: Geographical perspectives, ed.* M. Purvis and A. Grainger, Earthscan: UK-USA, 2004.
[8] Paez, E. Personal communication 20 January 2007, City council institute of research and urban planning (IMIP).
[9] Rojas, R., Urban environment assessment: methodological proposal and application to a study case. Mexicali, Baja California. Doctoral thesis in urbanism, faculty of architecture, UNAM, Mexico D.F. 2000.
[10] SEMARNAP, Government of Baja California, Health department, city council of Mexicali Programme *for air quality improvement in Mexicali 2000–2005.* SEMARNAP, Mexico, 1999.
[11] SEMARNAP, *Integrated system of regulation and environmental management on industries in Mexico.* SEMARNAP,-INE: Mexico, 1997
[12] United Nations Environment Programme (UNEP) Urban outdoor pollution, in global environmental outlook year book 2006. Online.
[13] http://www.unep.org/geo/yearbook/yb2006/057.asp

High-resolution air quality modelling and time scale analysis of ozone and NOx in Osaka, Japan

K. L. Shrestha, A. Kondo, A. Kaga & Y. Inoue
Graduate School of Engineering, Osaka University, Japan

Abstract

The ozone pollution in the Osaka region is mainly influenced by the long-range transport of tropospheric ozone or its precursors from outside in winter but nearly 50% of the total ozone is chemically produced by regional emissions in summertime. The effects of the regional transport and the chemical production of ozone by regional emissions in summertime can be analysed by spectral decomposition of the time series and studying the inherent intra-day, diurnal and longer-term time scales in the time series of ozone and NOx. The relative contributions of these time scale components associated with different physical processes are also used to evaluate the performance of air quality models in simulating ozone and NOx. We used the Japanese national emission data with resolution of 10-km and 1-km resolution in the MM5-CMAQ modelling system for the nested simulation in Osaka with the finest domain having 1-km grid size. Among the different time scales, diurnal time scale has the highest correlation for ozone, and longer time scale has the highest correlation for NO and NO2. Spatial variation in the simulation of NOx is evident from the study of correlation of diurnal and longer-term time scales, with higher correlations in the longer-term time scale. Overall, the time scale analysis of ozone and NOx time series has proved useful for detailed quantitative analysis of the air quality modelling in Osaka.
Keywords: time scale analysis, air quality evaluation, urban air quality, MM5, CMAQ.

1 Introduction

Since the ozone concentration is affected by cyclic patterns of meteorological forcings as well as the anthropogenic influences such as traffic emission of NO_x

and VOCs [1], the study of time scale components can be used to analyze such influences. There are different components of time scales used in the analysis of meteorological variables, ozone and its precursors. The highest and lowest resolvable frequencies are determined by the sampling interval and the length of the data record [2]. Since our simulation time period is one-month, we can only resolve time scales of about 10 days. Considering this limitation, only the intra-day scale, diurnal scale and longer-term scale have been used in our study. The intra-day scale has frequencies less than 12 hours, diurnal scale has frequencies around 24 hours and longer-term scale contains all the frequencies above the diurnal scale. The frequency bands were determined [2, 3], and the clean separation of the intra-day, diurnal and longer-term scales was confirmed from the periodograms using the Fast Fourier Transform (FFT) technique (not shown here). The Kolmogorov-Zurbenko (KZ) filter method is a widely used technique for the separation of different time scales of meteorological variables [4]. Besides having a powerful and efficient separation characteristics, KZ filter method can be applied to datasets having missing observation data as well. This method has also been successfully used for the analysis of different scales of ozone time series data so that the temporal information hidden in the time series data can be extracted cleanly by separating the time series into different scales of motion [1,3,5]. In order to determine the time scale frequencies, logarithmic scale was used for ozone and NOx while the meteorological data were analyzed without using the logarithmic scale [2, 3]. This work presents the results of time scale analysis of ozone, and Nox in Osaka region for July, 2002.

1.1 Domain and grid structures

The Osaka prefecture region was selected as the target area for evaluating the MM5-CMAQ modelling system (Fig. 1). The area around Osaka was used as the finest domain area with a grid size of 1 km (Domain 3). This domain was nested inside a coarser domain with a grid size of 3 km (Domain 2). These domains were further nested in the coarsest domain of 9 km grid size covering nearly all of Japan (Domain 1).

1.2 Meteorological modelling

The Pennsylvania State University/National Center for Atmospheric Research Mesoscale Modelling System (MM5) is a limited-area, non-hydrostatic, terrain-following sigma coordinate model designed to simulate or predict mesoscale and regional-scale atmospheric circulation [6, 7]. In this study, the national mesoscale grid point value data (GPV-MSM) provided by Japan Meteorological Agency was used to initialize the MM5 model. The grid size of each pressure level of this data is $0.2° \times 0.25°$. The simple ice microphysics, Dudhia's longwave and shortwave radiation scheme, Grell cumulus parameterization scheme and Medium-Range Forecast (MRF) PBL scheme [8] with multi-layer soil model [9] were used in MM5 simulation.

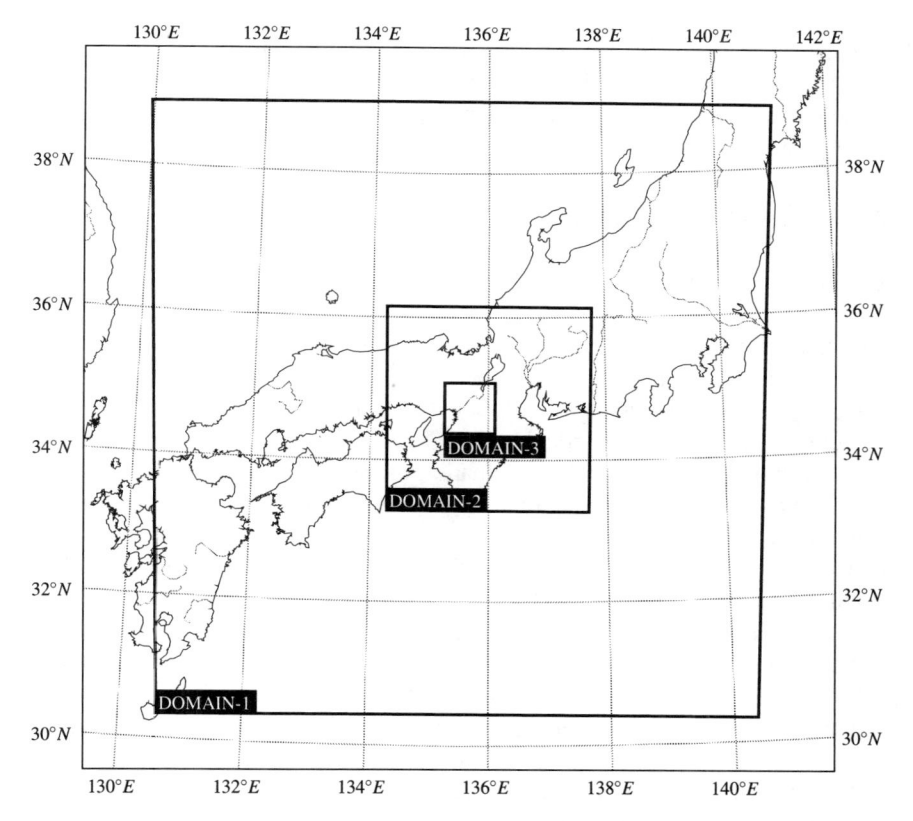

Figure 1: Nesting of model domains for CMAQ.

1.3 Emission modelling

The emission source data was obtained from Japan Clean Air Program (JCAP). For the 9-km domain (Domain 1) and 3-km domain (Domain 2), the national emission data with the spatial resolution of 0.125° by 0.0833° was used. For 1-km domain (Domain 3), the data covering the Kansai region with 0.0125° by 0.00833° resolution was used. This emission data is available as the average data for the month of July. Another feature of this data is the availability of weekday and weekend mobile emission data. Similarly, the biogenic emission has also been incorporated as the average of July.

1.4 CMAQ modelling

Community Multi-scale Air Quality (CMAQ) model is a Eulerian-type air quality model that simulates concurrently the atmospheric and land processes affecting the transport, transformation, and deposition of air pollutants and their precursors, on both regional and urban scales [10]. CMAQ is configured with the chemical mechanism of Statewide Air Pollution Research Center (SAPRC-99) mechanism, which is a detailed mechanism for the gas-phase atmospheric reactions of volatile organic compounds (VOCs) and oxides of nitrogen (NOx) in urban and regional atmospheres.

2 Results and discussion

2.1 Time scale analysis

For ozone as well as meteorological variables, the relative contribution of the different temporal components to the total time series can be examined from the variances of the components [11]. It can assist in examining the accuracy of the model in predicting the individual time scale components that are associated with different physical processes. For evaluation of the simulated meteorological and air quality variables, the observation data were obtained from the "Air Pollution Continuous Monitoring Network Data Files" provided by Environmental Pollution Control Center, Osaka Prefecture. The following results were obtained from simulation results of Domain 3 (1-km grid) of MM5-CMAQ system.

Figure 2 shows the contribution of different time scale components of time series of near-surface air temperature and wind speed using the percentage contribution of variance of each component to the total variance. The contributions of intra-day, diurnal and longer-term components are differing from the observed values by 0.95, 1.01 and 0.99 times respectively for the temperature time series and by 0.88, 1.07 and 0.97 times respectively for the wind speed time series. It indicates a strong diurnal influence on air temperature with the intra-day component differing the most from the observed values. Though one would expect the intra-day component to have larger underpredictions than this result [11], the use of very high resolution (1-km) grid and high resolution meteorological analysis data could have been one of the factors behind this improvement. The wind speed is also being influenced by diurnal forcings but the contribution of intra-day component is nearly 6 times than that of air temperature. Ohsawa et al. [12] reported that the maximum wind speed due to southerly sea breeze are underpredicted by MM5 even in fine domains of nearly 1 km grid size and that a better representation of diurnal variation is necessary in the summertime in regions with complex terrain. This situation is improved in our results because four dimensional data assimilation (FDDA) is not used and so a significant loss in the contribution of diurnal variations is avoided. In addition, the contribution of intra-day component has increased due to the absence of smoothing of intra-day variations by FDDA, and the local wind circulations being captured at very fine resolution of 1 km.

The contribution of intra-day, diurnal and longer-term components are differing from the observed values by 2.95, 0.84 and 1.05 times respectively for ozone, 1.69, 1.80 and 0.66 times respectively for NO_2, and 0.99, 1.91 and 0.38 times respectively for NO time series (Fig. 3). The intra-day component of ozone has the highest difference in contribution (nearly 3 times) when compared with the observed data, and this might have contributed to the slightly less contribution in diurnal

(a) Temperature

Simulated
- diurnal: 61.37
- intra: 3.13
- longer: 35.50

Observed
- diurnal: 60.68
- intra: 3.29
- longer: 36.04

(b) Wind speed

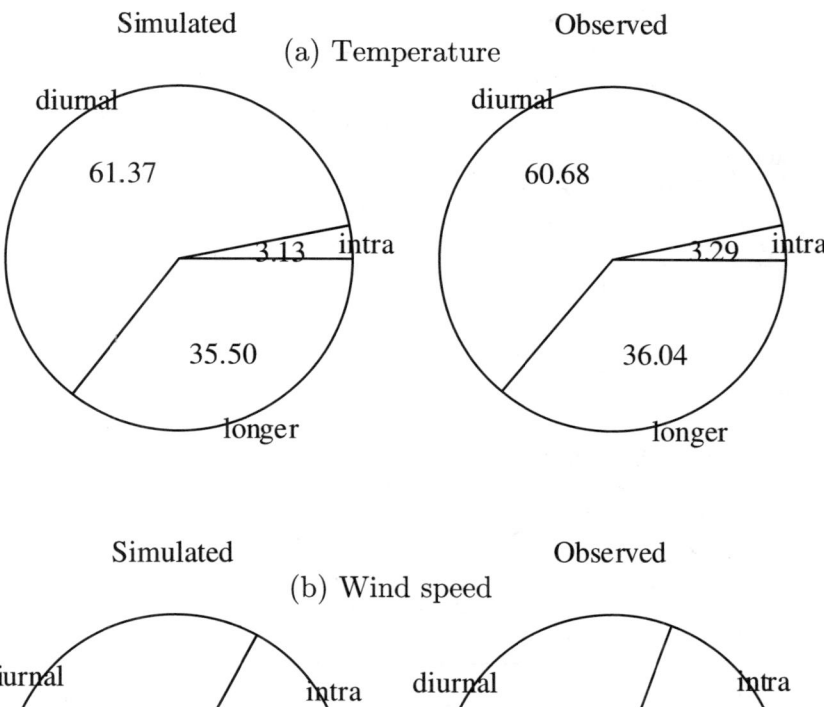

Simulated
- diurnal: 45.62
- intra: 17.05
- longer: 37.33

Observed
- diurnal: 42.38
- intra: 19.30
- longer: 38.32

Figure 2: Contribution of different time scales to the total variance of (a) near-surface air temperature and (b) wind speed. The values are shown as percentage of contribution of time scale component to the total variance.

component. The overestimation of intra-day component of ozone may be partially attributed to the overestimation of the intra-day component of its precursors, and also to the fact that the intra-day fluctuations of ozone are significantly affected by the intra-day wind speed fluctuations. Moreover, the uncertainties in the local emissions of ozone precursors also affect the intra-day fluctuations of ozone. The intra-day component is well-captured by MM5-CMAQ system for NO than its longer-term and diurnal components. For NO_2, the results show overpredictions in the contribution of variability from intra-day and diurnal components and under-

(a) Ozone

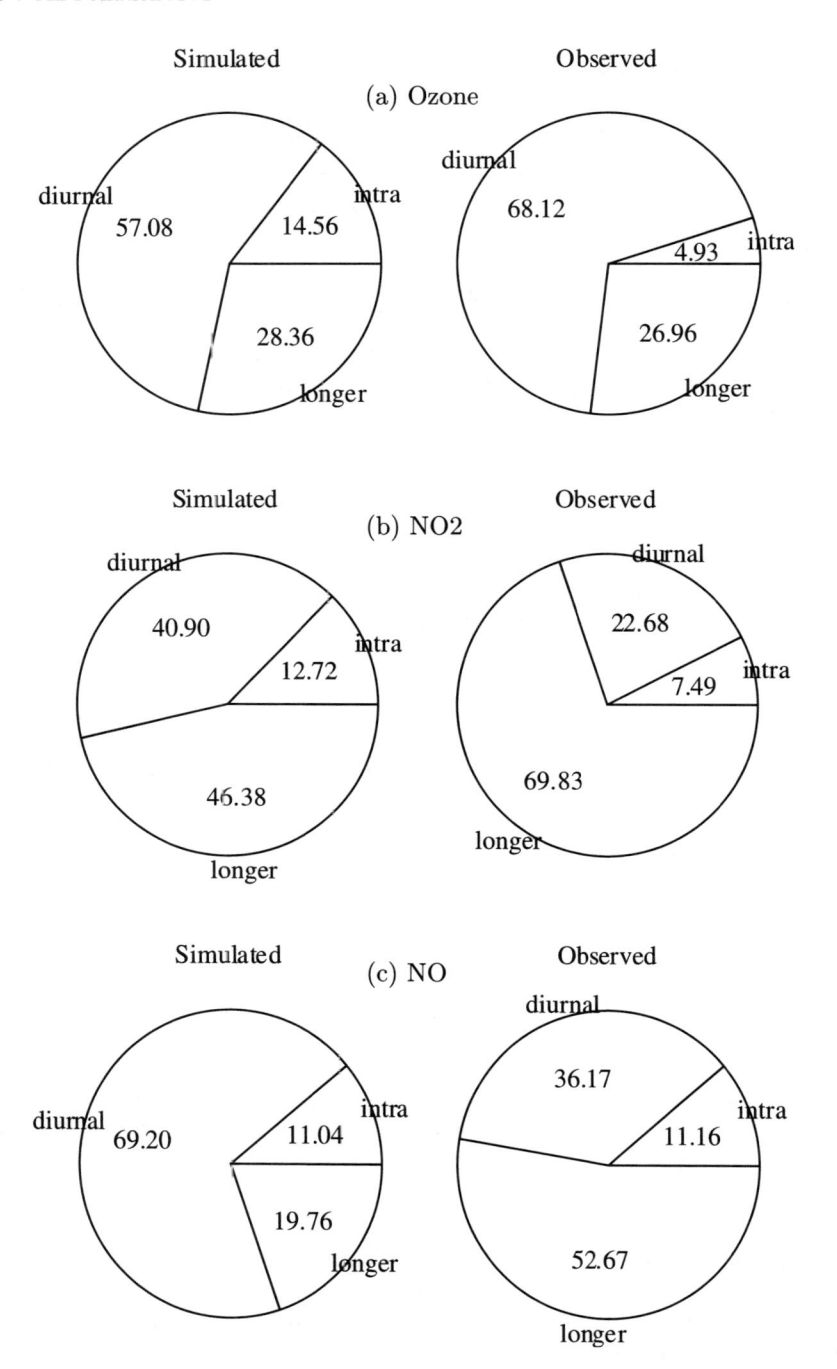

Simulated: diurnal 57.08, intra 14.56, longer 28.36

Observed: diurnal 68.12, intra 4.93, longer 26.96

(b) NO2

Simulated: diurnal 40.90, intra 12.72, longer 46.38

Observed: diurnal 22.68, intra 7.49, longer 69.83

(c) NO

Simulated: diurnal 69.20, intra 11.04, longer 19.76

Observed: diurnal 36.17, intra 11.16, longer 52.67

Figure 3: Contribution of different time scales to the total variance of (a) ozone, (b) NO_2 and (c) NO. The values are shown as percentage of contribution of time scale component to the total variance.

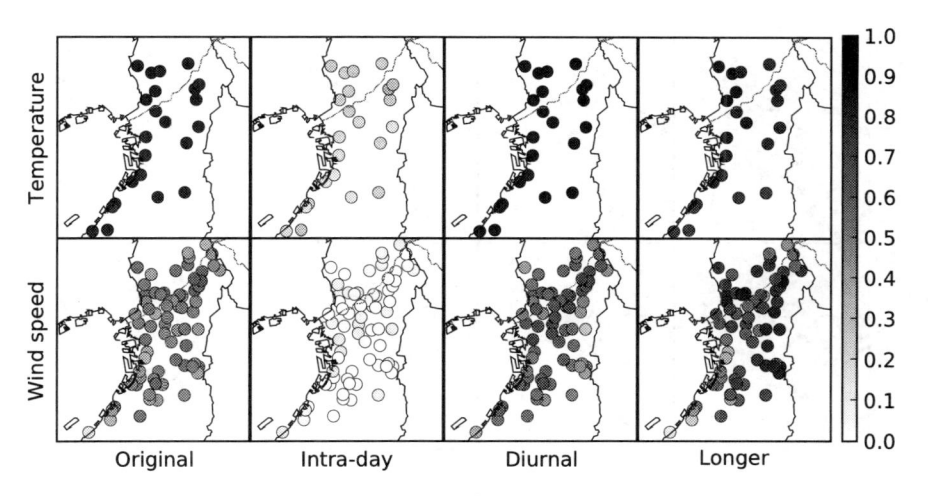

Figure 4: Correlation between simulated and observed temperature and wind speed for different time scales. The light gray line from the top right corner is Yodo River.

prediction in the longer-term component. Though the diurnal processes seem to be the major contributors in determining the time series of NO_2 and NO, the longer-term component is apparently contributing the most (69.83% and 52.67% respectively for observed NO_2 and NO). This apparently suggests the prevalence of synoptic pattern in NO_x that may have been caused by the unique and complex terrain and geography of the region as well as the lack of sufficient sensitivity to diurnal fluctuations in the observed data.

The correlation coefficient between the simulated and the observed time scale components is helpful in identifying the spatial information of the model performance. The correlation coefficients for near-surface temperature (Fig. 4) are consistently high over 0.8 over all the observation stations of Osaka. Diurnal component is the best simulated time series component with correlation coefficients higher than 0.9. The longer-term component has also high values (> 0.7) but the intra-day component has small correlation coefficients less than 0.3, especially in the stations near the west coast. The correlation coefficients for wind speed (Fig. 4) are very small in all the stations (< 0.2) and the longer-term component has better coefficient values among the time series components. This result is typical of summertime wind, which is affected by model representation of the boundary layer mixing processes and PBL growth and transition periods [13]. The diurnal component of wind speed shows some relatively higher correlation coefficients (0.6 to 0.7) in the stations around the Yodo River, which, among other reasons, may be due to complex land use pattern in the region.

Similarly, the correlation coefficients for ozone (Fig. 5) for the diurnal component have higher values (0.6–0.8) in the northern and southern parts of Osaka. The central city and the eastern regions have relatively smaller correlation coefficients (0.5–0.6) for the diurnal component. On the other hand, the longer-term compo-

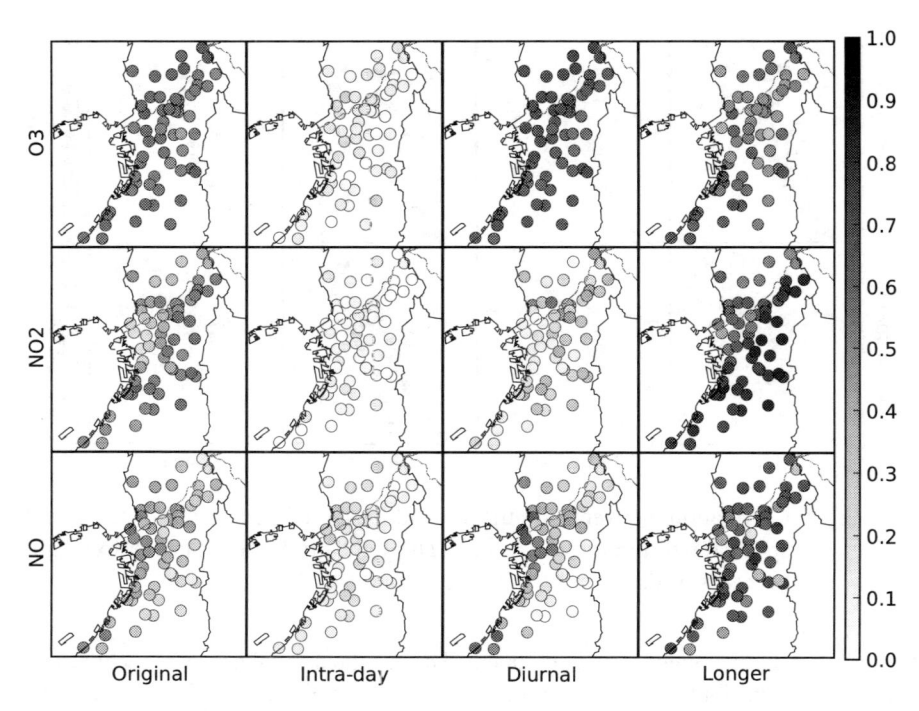

Figure 5: Correlation between simulated and observed ozone, NO_2 and NO for different time scales. The light gray line from the top right corner is Yodo River.

nent has coefficients less than the diurnal component but contains some higher values (around 0.6) along southern Osaka. From these results the higher correlation coefficients for peak ozone concentrations (not shown here) can be explained by the higher correlations in the diurnal and longer-term components. Similarly, the poor correlation in the intra-day component as well as the larger ratio of variability in intra-day time scale (Fig. 3) contributes to the overall gross error.

The correlation coefficients for NO_2 and NO (Fig. 5) have also very small values similar to that of ozone but the overall effect of the components of time scales is spatially different for both of them. The correlation coefficients for NO_2 are very low (< 0.2) in the downstream region of Yodo River consisting of highly urbanized area, which may have been caused by relatively lower values for the diurnal and longer-term components of NO_2. The longer-term component of NO_2 has very high correlation coefficients (0.7–0.9) in almost all stations.

In the case of NO, the downstream region of Yodo River has higher overall correlation coefficients (0.4–0.5) as compared to other regions, which may be attributed to relatively higher values for the diurnal component of NO. In both the cases of NO_2 and NO, longer-term components show higher correlation coefficients than the other components though the overall correlation coefficients for NO are comparatively lower than NO_2.

3 Summary

High resolution GPV-MSM meteorological data was used for driving the nested domains having 9-km, 3-km and 1-km grid sizes in MM5, and then CMAQ model was run using the meteorological input from MM5 and emission data based on Japanese emission database. Then the time scale components of the meteorological fields (temperature and wind), ozone and NOx were evaluated against the observation data in the Osaka prefecture region using the MM5-CMAQ system.

The time scale analysis reveals a strong influence of spatial distribution on both the meteorological and air quality model outputs. Moreover, the intra-day scales for wind, ozone and NO_x are relatively less well-represented than the other time scales by the MM5-CMAQ system, and thus, a more accurate represented land use and vegetation can be recommended for the terrains of Osaka.

The contribution of intra-day component to the overall variability of air temperature was found to be well simulated though correlation coefficient was low (< 0.4) compared to the best simulated diurnal component. Overall, the ozone predictions show the largest contributions from the diurnal component and the longer-term component was the best-simulated time scale for NO_x. Similarly, the meteorological fields were also simulated with an adequate level of accuracy.

References

[1] Sebald, L., Treffeisen, R., Reimer, E. & Hies, T., Spectral analysis of air pollutants. Part 2: ozone time series. *Atmospheric Environment*, **34(21)**, pp. 3503–3509, 2000.

[2] Hogrefe, C., Rao, S., Kasibhatla, P., Hao, W., Sistla, G., Mathur, R. & McHenry, J., Evaluating the performance of regional-scale photochemical modeling systems: Part II - Ozone predictions. *Atmospheric Environment*, **35(24)**, pp. 4175–4188, 2001.

[3] Rao, S.T., Zurbenko, I.G., Neagu, R., Porter, P.S., Ku, J.Y. & Henry, R.F., Space and time scales in ambient ozone data. *Bulletin of the American Meteorological Society*, **78(10)**, pp. 2153–2166, 1997.

[4] Eskridge, R.E., Ku, J.Y., Rao, S.T., Porter, P.S. & Zurbenko, I.G., Separating different scales of motion in time series of meteorological variables. *Bulletin of the American Meteorological Society*, **78(7)**, pp. 1473–1483, 1997.

[5] Rao, S., Ku, J., Berman, S., Zhang, K. & Mao, H., Summertime characteristics of the atmospheric boundary layer and relationships to ozone levels over the eastern united states. *Pure and Applied Geophysics*, **160(1)**, pp. 21–55, 2003.

[6] Dudhia, J., Gill, D., Manning, K., Wang, W. & Bruyere, C., *PSU/NCAR Mesoscale Modeling System Tutorial class notes and users' guide: MM5 Modeling System Version 3*, 2005.

[7] Grell, G.A., Dudhia, J. & Stauffer, D.R., A description of the fifth-generation Penn State/NCAR mesoscale model (MM5). NCAR Technical Note NCAR/TN-398+STR, 1994.

[8] Hong, S.Y. & Pan, H.L., Nonlocal boundary layer vertical diffusion in a medium-range forecast model. *Monthly Weather Review*, **124(10)**, pp. 2322–2339, 1996.
[9] Dudhia, J., A multi-layer soil temperature model for MM5. Preprints, The Sixth PSU/NCAR Mesoscale Model Users' Workshop, 1996.
[10] Byun, D.W. & Ching, J.K.S., (eds.) *Science Algorithms of the EPA Models-3 Community Multiscale Air Quality Model (CMAQ) modeling system*. US Environmental Protection Agency, Office of Research and Development, Washington, DC, 1999.
[11] Hogrefe, C., Rao, S., Kasibhatla, P., Kallos, G., Tremback, C., Hao, W., Olerud, D., Xiu, A., McHenry, J. & Alapaty, K., Evaluating the performance of regional-scale photochemical modeling systems: Part I - Meteorological predictions. *Atmospheric Environment*, **35(24)**, pp. 4159–4174, 2001.
[12] Ohsawa, T., Hashimoto, A., Shimada, S., Yoshino, J., de Paus, T., Heinemann, D. & Lange, B., Evaluation of offshore wind simulations with MM5 in the Japanese and Danish coastal waters. Proceedings of European Wind Energy Conference EWEC: Milan, 2007.
[13] Gilliam, R.C., Hogrefe, C. & Rao, S., New methods for evaluating meteorological models used in air quality applications. *Atmospheric Environment*, **40(26)**, pp. 5073–5086, 2006.

Practical problems associated with assessing the impact of outdoor smoking on outdoor air quality: an Edinburgh study

D. G. Snelson[1], A. J. Geens[1], H. Al-Madfai[1] & D. Hillier[2]
[1]Faculty of Advanced Technology, University of Glamorgan, UK
[2]Faculty of Health, Sports & Science, University of Glamorgan, UK

Abstract

Smoking in the United Kingdom is now banned in public places, subject to certain exempt situations. The last country to introduce the ban being England on the 1st July 2007. People who smoke now congregate outside premises either sitting at tables or standing while smoking their cigarettes. Previous studies have concluded that measured outdoor pollutants are proportional to and correlated with smoker count and not with motor vehicle traffic. This paper reports on a case study conducted in Edinburgh and identifies lessons learnt from this and earlier studies. The authors having conducted this field work suggest that this is not necessarily the case and they have concluded that further work is required to establish a more robust methodology.

Keywords: smoking in public places, outdoor air quality, particulates, carbon monoxide.

1 Introduction

An experimental study was completed by Klepeis et al. [1] where outdoor communal areas were monitored, in the presence of smokers, to ascertain the levels of environmental tobacco smoke (ETS) in close proximity to smokers with burning cigarettes. For this study a number of similar venues in Edinburgh were used and further location types were added. In this study, levels of particulates (PM 2.5) and carbon monoxide (CO) have been recorded. Carbon dioxide (CO_2) Temperature (°C), relative humidity (%), rain (mm), sun (hrs) and wind speed (knots) were also recorded as they may have an impact on the particulate concentration levels in the outdoor environment where ETS is being measured.

The background pollutant level concentrations are taken from the data supplied by the St. Leonards Air Quality monitoring station in the South of Edinburgh, part of the UK wide monitoring network managed by the Department for Environment, Food & Rural Affairs (DEFRA). These background levels will be used to normalise locally monitored data in proximity to smokers. The particulates and gaseous pollutant threshold / limit values for air quality are set out in European Directives. In the United Kingdom the National Air Quality Standards [2] define levels which are believed to avoid significant risks to health. This paper compares the results from three Edinburgh venues, used to ascertain the levels of pollutant in close proximity to cigarette smokers, to those of the Klepeis et al. [1] study.

2 Methodology

The authors measured particulates (PM 2.5) and gaseous emission (CO) in the outdoor environment at three outdoor seating areas.

2.1 On-site monitoring visits

Three outdoor seating areas were measured (Last Drop Tavern, Beehive and The White Hart Inn) to investigate the levels of CO and PM 2.5 where smoking occurs in the outdoor environment. The sampling devices used in this study were

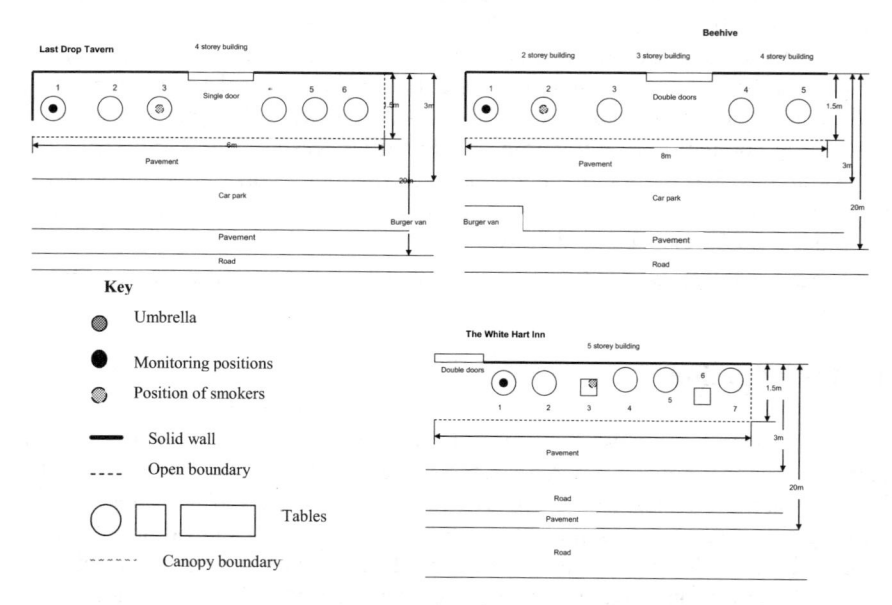

Figure 1: Schematic diagram of measurements venues and sites.

the Dustrak Aerosol Monitor Model 8520 by TSI Inc, using the 2.5 μm inlet conditioner and a flow rate of 1.7 l/min, and the Q-Trak Plus IAQ Monitor Model 8554 by TSI Inc. A venue plan (Fig. 1), the characteristics (Table 1), the position of smokers and the duration of cigarette being smoked were recorded (Table 2). The sampling devices were located in the outdoor seating areas at a height approximating to the breathing zone, on a suitable table marked on the venue plan. Monitoring periods ranged from 60 minutes to 76 minutes.

Table 1: Characteristics of outdoor monitoring locations.

Venues	Monitoring date	Length of monitoring (mins)	Monitoring location	Width (m)	Depth (m)	Building height (m)	Distance to building (m)	Distance to road (m)	Number of Tables	Seating capacity	comments
Last Drop Tavern	Thursday 20th September 2007	76	Table 1	6	1.5	4 storeys	0.5	20	6	22	Situated in the Grassmarket Car park opposite seating area French Connection food van 8 metres from outdoor seating area
Beehive	Thursday 3rd August 2007	60	Table 1	8	1.5	3 storeys	0.5	20	5	21	Situated in the Grassmarket Car park opposite seating area French Connection food van 6 metres from outdoor seating area
The White Hart Inn	Thursday 20th September 2007	63	Table 1	8	1.5	5 storeys	0.5	20	7	18	Situated in the Grassmarket Car park opposite seating area French Connection food van 22 metres from outdoor seating area

3 Background measurements

Information on measurements recorded at the Edinburgh St. Leonards air quality monitoring station (Table 2) is outlined below.

3.1 Carbon Dioxide

Carbon Dioxide is a product of respiration and occurs naturally in the atmosphere. It is also worth noting that Carbon Dioxide is present in the Earth's atmosphere at approximately 375 ppm [3] by volume and so, unlike other indicators, Carbon Dioxide will not tend towards zero when no pollution sources are active.

3.2 Temperature

The daily and seasonal solar cycles will essentially control the temperature profile of the lower atmosphere [4]. During daylight hours, the temperature in the lower atmosphere typically decreases with height. As evening approaches, the suns heating effect decreases and the earth's surface loses heat to the atmosphere. As the earth surface and the lower levels of the atmosphere cool the vertical temperature profile of the atmosphere is now in reverse of the daytime situation [4]. The increasing temperature with height is called a temperature inversion. The region of air that extends from the earth's surface to the base of the temperature inversion is referred to as the mixing layer [4]. The lower the depth of the mixing layer the less volume that is available to mix the pollutant with fresh air. The mixing layer is important in cities where high volumes of pollutants are released near ground level [4].

3.3 Wind speed

Changes in pressure and temperature in the atmosphere cause the movement of air. The speed of the wind will determine how long it takes for air pollutants to travel from their source to the measuring device. Also at higher wind speeds the pollutants released at or near ground level will disperse more rapidly into the surrounding atmosphere. Pollutants released when the wind speeds are low at or near ground level will disperse at a lower rate [5] for example vehicle exhaust. Wind speed data was supplied by defra from Edinburgh St. Leonards air quality monitoring site (see Table 2). Table 2 refers to the time period when the outdoor smoking measurements were recorded by the authors.

Table 2: Edinburgh air quality data for the 20th September 2007.

Time	20/09/2007							
	PM_{10} (mg/m^3)	CO (mg/m^3)	Temp (°C)	Rel Humidity (%)	Rain (mm)	Sun (hrs)	Wind - Mean Speed (knots)	Wind - Mean Dir
11:00	7	0.2	13.7	72.1	0.0	0.0	13	240
12:00	10	0.2	13.8	72.1	0.0	0.0	13	240
13:00	12	0.2	13.8	65.8	0.0	0.3	15	240
14:00	12	-	15.6	61.9	0.0	0.9	17	240
15:00	14	0.2	14.9	62.9	0.0	0.8	19	250

4 Smokers

The data outlined below relates to outdoor smoking activities at venues (Table 1 and Fig. 1) where CO and PM 2.5 levels were measured.

4.1 Data on smokers

The approximate times when a cigarette was lit and extinguished, the position of the smoker and the distance to the monitoring equipment were recorded (Table 3). This will give an indication of the distance at which the cigarette smoke can be detected by the monitoring equipment.

Table 3: Details of smoking activities.

Venues	Monitoring date	Length of monitoring (mins)	Monitoring location	Smokers table	Number of smokers	Number of cigarettes smoked	Time and duration of smoking activity	Distance from monitoring location
Last Drop Tavern	Thursday 20th September 2007	76	Table 1	Table 3	One	One	13.05-13.12	2.5m
Beehive	Thursday 3rd August 2007	60	Table 1	Table 2	Three	Three	16.41-16.47	1m
The White Hart Inn	Thursday 20th September 2007	63	Table 1	Table 3	Two	Two	13.55-14.02	1.5m

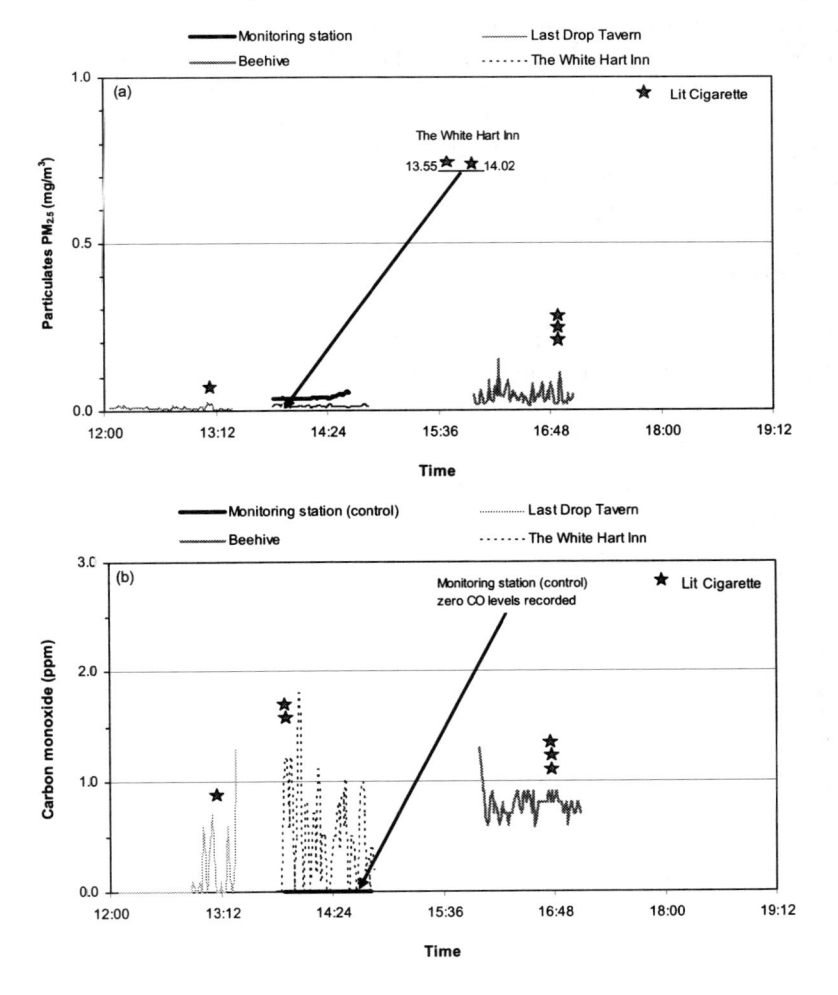

Figure 2: Graphs of CO and PM 2.5 levels measured at all sites.

4.2 Concentration levels of CO and PM 2.5

Figure 2 shows the levels of CO and PM 2.5 recorded at all sites. The maximum concentration of CO recorded at the individual venues is shown in Figure 3. The monitoring station (St. Leonards) gives the background pollutant level concentrations. Higher levels of CO were recorded at a bus stop from traffic exhaust fumes than from the cigarette smoke at the three venues. This was probably due to the maintainace work being carried out on the road causing the stagnation of the traffic increasing the concentration of the CO; this is further discussed in section 5.

4.3 Distance to the smoker

As the cigarette smoke plume is released into the atmosphere the concentration will be diluted by the surrounding air. The distance at which the CO and PM 2.5 can be detected is difficult to establish. The authors have conducted two tests; a walk by test one metre from a lit cigarette and another sitting opposite a smoker at a table, but these results are not presented in this paper as they are inconclusive. The weather conditions at the time of monitoring will have a bearing on the distance from the source that the CO and PM 2.5 can be detected by the monitoring equipment.

5 Discussion

The investigating team has found that whilst the background levels of a wide range of airborne contaminants can be successfully monitored, as demonstrated

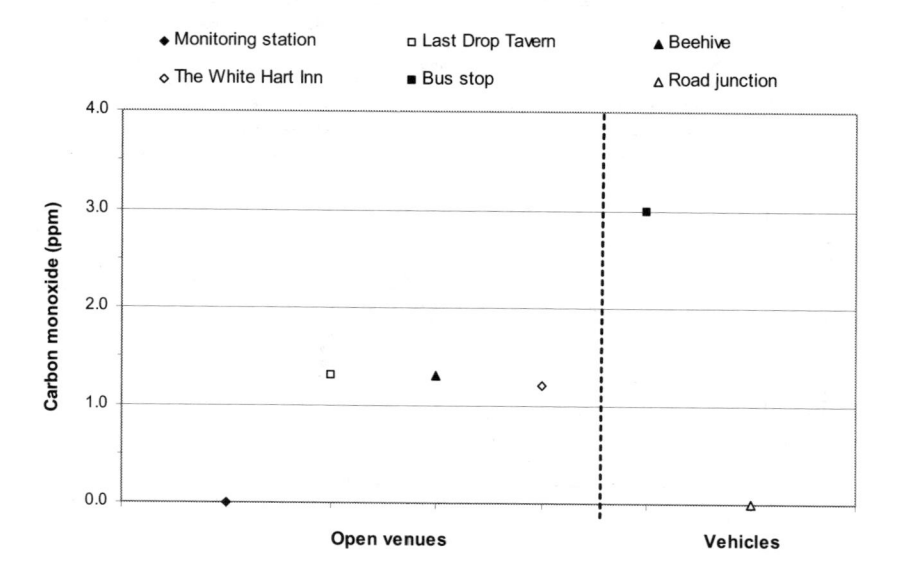

Figure 3: Graphs of maximum CO levels measured at all sites.

by the defra monitoring stations, it is very difficult to reliably monitor what is happening at a local level (Fig. 3). For example, the emissions from a bus exhaust can be observed to decay rapidly with distance and yet carbon monoxide can be detected as a smoker walks by. Although sources are very variable, (size/speed of vehicle, brand of cigarette), the rate of carbon monoxide production (l/s) from a vehicle is of the order of ten times the rate from a cigarette. The levels that are monitored are very dependent on distance. Furthermore, in assessing health risk, duration is as significant as level. Weather factors as well as degree of enclosure by adjacent structures are further variable factors. Some theoretical modelling (Gaussian plume) of the effect of the wind speed and direction would be useful in determining some of the critical factors, before moving to further field tests.

6 Conclusions

With the number of variables present and the subjectivity of the monitoring methods, studies such as the one reported by Klepeis et al. [1] and repeated in Edinburgh by Snelson et al, can only provide broad observations and perhaps identify areas of concern that may justify a more considered and lengthy treatment. Much further work is required to establish a robust methodology for this type of investigation as well as to establish the important or significant criteria before any sound conclusions can be drawn.

Acknowledgement

This study was commissioned by the Scottish Licensed Trade Association with funding support from the UK Tobacco Manufacturers' Association.

Reference

[1] Klepeis, N., Ott, W. & Switzer P., Real-Time Measurement of Outdoor Tobacco Smoke Particles, *Journal of Air & Waste Management Association*, **57, (5)**, pp. 522–534, 2007.
[2] The Air Quality Strategy for England, Scotland, Wales and Northern Ireland. Department for Environment Food and Rural Affairs in Partnership with the Scottish Executive, Welsh Assembly Government and Department of the Environment Northern Ireland, Volume 2, The Stationary Office 2007, http://www.defra.gov.uk/environment/airquality/strategy/pdf/air-qualitystrategy-vol2.pdf
[3] Whorf, T.P. & Keeling, C.D., Atmospheric CO_2 records from sites in the SIO air sampling network, 2005.
[4] Temperature, 2006, http://casadata.org/whatis/temperature.asp
[5] Wind, 2006, http://casadata.org/whatis/wind.asp

Role of leaf- and rhizosphere-associated bacteria in reducing air pollution of industrial cities in Saudi Arabia

M. A. Khiyami
General Directorate of Research Grants Programs,
Directorate of Research Follow-up,
King Abdulaziz City for Science and Technology, Riyadh, Saudi Arabia

Abstract

With the establishment of industrial cities such as Jubail and Yanbu, Saudi Arabia has been increasing its efforts to protect the country from various environmental hazards, several air quality monitors and meteorology network stations are operated throughout the kingdom to monitor parameters such as sulfur dioxide, ozone, nitrogen oxide, and hydrogen sulfide. As a result, air pollution in the Saudi cities is the lowest in the Middle East.

Microorganisms play a major role in removing pollutants from water, soil and air. Leaf- and rhizosphere-associated bacteria have an impact in reducing air pollution. Plant species growing in the industrial cites were studied by scanning electron microscopy (SEM) to observe the bacteria. Seventeen bacterial colonizations of leaves and rhizosphere were identified by API, biochemical tests, and PCR. The ability of the isolates to biodegrade different air and soil pollutants such as phenol, sulfate and nitrogen compounds was examined. Furthermore, the ability of leaf isolates to produce Extracellular polysaccharide (EPS) was examined.
Keywords: industrial cities, emission, pollutants, bacteria, rhizosphere, biodegradation, extracellular polysaccharide.

1 Introduction

Saudi Arabia is strategically located in the southwest corner of Asia, at the crossroads of Europe, Asia and Africa. Jubail and Yanbu are two major industrial cities in Saudi Arabia on the eastern and western coasts. The Royal

Commission (RC), the administrator of the two cities promotes and encourages the investors in these areas to invest in various oil and petroleum-related activities and businesses including the manufacture of organic chemicals, fertilizers, plastics, and synthetic materials. The RC made a conscious decision to balance industrial productivity with environmental quality, capitalizing on the lessons learned from other countries.

Several early measures were implemented to protect the environment, including setting up standards, regulations, and design review criteria aimed at restricting industrial air emissions and effluent discharges to levels consistent with the best available control technologies. In the beginning, the RC established comprehensive programs in Jubail and Yanbu to monitor ambient and source factors influencing the environment. Also, the RC maintains environmental control units at Jubail and Yanbu. Staffed with environmental scientists, engineers and other specialists, these groups monitor and enforce the standards and regulations in the industrial cities. On the other hand, the RC operates a number of sophisticated air monitoring stations in Jubail and Yanbu to ensure and maintain that facilities meet national air quality standards. These stations record parameters such as sulfur dioxide, inhalable particulates, ozone, nitrogen oxides, carbon monoxide, and hydrogen sulfide.

Beside the sophisticated air monitoring stations operated in Jubail and Yanbu, the RC built a network of greenways surrounding the cities. These greenways become a great source for recreation, being used for activities such as fitness, walking, running, skating, and cycling. Furthermore, greenways serve as natural conservation from the evolving gases that might escape from the air quality filters in the factories.

Plant leaves are commonly colonized by large bacterial populations, as well as by other microorganisms including yeasts, mycelia fungi, and algae [4]. Each leaf has its own ecosystem - an ecosystem in which microorganisms: a) battle among themselves for limited resources; b) forge strategies for surviving making rapid changes in environmental conditions, and; c) actively change their own microenvironment by instigating changes in their host plant. Among the bacterial species found on leaves, some can induce disease, other can fix atmospheric nitrogen, and some can degrade airborne pollutants that have collected on the leaf surface. In most cases, the potential for bacteria to perform these functions depends on the size of bacteria population that survives the environmental conditions and its ability to colonization [5]. In this study we have isolated bacterial populations (Epiphytic and Endophytic bacteria) that colonize the leaves and rhizosphers of trees to define their role in reducing air and soil pollutants in industrial city of Yanbu in Saudi Arabia.

2 Material and methods

2.1 Study location

Yanbu is located on the red sea, about 350 km from Jeddah. Yanbu city occupies about 185 km^2. Jubail and Yanbu industrial cities have established a range of 228 support industries and the infrastructure has provided encouragement to establish

41 primary and 50 secondary industries. Most of these industrial units are oil refineries and petrochemicals. Jubail and Yanbu Industrial cities are surrounded by a network of greenways to protect the environment in the residential areas. Trees in the greenway were the target of this study to collect samples and isolates in an attempt to identify the bacteria that can degrade the evolving gases.

2.2 Gases and particles in city sky

Usually most of studies on air population in all cities report gases that include SO2, NOx, CO, O3 and carbon hydrogen compounds except CH_4 [3]. However, international studies reported the emission of numerous hydrocarbon gases from oil refinery and petrochemical complexes [7, 9, 15, 17]. A study funded by King Abdulaziz City for Science and Technology [4], conducted on air pollutants in Yanbu industrial city has reported the emission of gases from three different stations [2, 10], as shown in table 1.

Table 1: Monthly mean pollution concentration for station 3 at Madinat Yanbu Al-Sinaiyah (units are microgram per cubic meter, except CO).

Month	NO_x	NO	NO_2	SO_2	O_3	CO^*	THC	TSP
May 1981	40.9	14.8	32.3	40.1	62.6	1.1	DI	***
June	34.8	9.0	26.1	25.6	59.3	1.5	1198	***
July	16.4	3.1	13.4	4.2	62.3	0.9	1398	***
August	17.0	4.3	12.8	12.1	59.9	2.2	1048	153***
September	51.5	11.2	40.3	27.3	39.2	1.3	1028	228
October	35.8	14.4	22.1	34.3	29.9	1.0	DI	151
November	26.6	11.8	28.8	27.4	37.8	1.2	DI	190
December	29.0	DI^{**}	DI	21.4	32.3	1.3	DI	142
January	15.9	DI	DI	12.7	44.2	1.0	1298	189

* Milligrams per cubic meter
** DI- data insufficient
*** Particulate sampling began on August 26

2.3 Samples collation

Leaves and soil samples were collected during the summer of 2006. The leaf samples were collected from different plants. The samples were kept in clean sacks. The soil samples were collected from the rhizosphere region at a depth of 5cm below the soil surface. The samples were kept in clean sacks. All samples were stored at 4°C for chemical and microbiological analysis [3].

2.4 Scanning electron microscopy (SEM) study

Different parts of leaf trichomes at stomata and at the epidermal cell wall and the depressions in the cuticle, were scanned using SEM to isolate bacteria

2.5 Bacterial isolation

The leaf samples were cleaned from dust and individual leaves were placed in large tubes containing 20 ml of 10 mM, pH 7.0 potassium phosphate buffer, sonicated for 7 min, and vortexed vigorously [6]. Appropriate suspensions were prepared and plated on different media to estimate bacterial cell numbers. Each leaf was weighed to allow the bacterial population size to be normalized for the amount of leaf tissue.

For rhizosphers soil sample the dilution plate method, originally described by Johnson et al [13] was employed with some pH modification. Appropriate soil suspensions was prepared and plated on different media to estimate bacterial cell numbers.

2.6 Bacterial identification

The colonies and isolates were identified by morphological, Gram Stain and biochemical methods using the API 20 system. Finally the isolates were subjected to Rapid PCR and 16 sRNA, the work is still in progress.

2.7 Degradation of air pollution

For preliminary screening experiments aimed at testing the ability of isolates to degrade air pollution, the minimal medium contained 8.0 g of $NH_4H_2PO_4$, 0.2 g of yeast extract, 2.0 g K_2HPO_4, 0.5 g of $MgSO_4 \cdot 7H_2O$, 0.5 g of Na_2SO_4, 0.5 g of NaCl, 10 mg of $ZnCl_2 \cdot 2H_2O$, 8.0 mg of $MnSO_4 \cdot 7H_2O$, 10 mg of $FeSO_4 \cdot 7H_2O$, 50 mg of $CaCl_2$ and 1.5 g% agar in 1 liter of distilled water. The media was autoclaved at 121°C for 15 min and after cooling different concentration of phenol (10.0, 20.0, 30.0, 40.0, 50.0, 70.0 100.0 mM) were added as filter sterilization. All isolates were screened to phenol degradation by inculcation on the plates with different phenol concentration and incubated at 30°C for 7 days. The ability of isolates to degrade phenolic compounds was also examined in a minimal medium and with high phenol concentrations including 100.0, 150.0, 200.0, 250.0, and 300.0 mM. The phenol degradation measured by phenolic assay is based on the oxidation of phenolate ion where ferric ions are reduced to the ferrous state which was detected by the formation of the Prussian blue complex ($Fe_4[Fe(CN)_6]_3$) with a potassium ferricyanide-containing reagent [16].

The assay mixture contained 25 ml of deionized water, 250 µl of sample, 3 ml of ferric chloride reagent and 3 ml of potassium ferricyanide reagent. The sample absorbance was measured at 720 nm. The phenolic contents in the samples were expressed as syringic acid equivalents (10 µg ml^{-1} gives A_{760} of 0.377/ml) [1].

2.8 Degradation of soil pollutant

Pollutants may accumulate in soil, and these pollutants stimulate microbial proliferate. Heavy metals, nitrogen and sulfate accumulate and contaminate soil environments and encourage special microbial groups to proliferate. To observe the role of pollutants on soil and their chemical activity effect on the microbial community, an enrichment technique for the growth of sulfate and nitrogen fixation bacteria was employed [8]. The isolates from study locations were compared to the isolates from the control location, which was out of the study zone.

2.9 Extracellular polysaccharide (EPS) production

Bacteria may modify their environment by producing a layer of EPS on leaf surfaces. The ability of isolates to produce the EPS was examined by using Plastic Compost Support (PCS) to develop biofilm. The PCS was produced as described by Ho *et al* [12]. The PCS tubes were cut into small pieces. 10.0 g of PCS was autoclaved at 121°C for 15 min in a 250 ml flask with distilled water. After PCS sterilized the distilled water it was removed aseptically and 100 ml of the minimal medium with 10.0% glucose was added. The flasks were inculcated with 0.5 ml of 620Abs 18 hr isolates bacteria and incubated at 30°C for 7 days. The suspended cell biomass was measured indirectly by absorbance at 620 nm. Gram staining was performed on suspended cell pellets obtained from centrifugation at 8816 x g for 15 min at 4.0°C of spent culture medium and of the sand of stripped cells from PCS biofilm [14]. Also, the flow cytometer measurement was employed to determine the bacterial cells.

3 Results and discussion

Seventeen bacteria leaves and rhizosphere were isolated and identified from leaves and soil samples, however, conformation of bacteria identification is still in progress. Table 2 showed the pre-identification of the bacteria.

The phenol degradation experiment was performed to represent the degradation of volatile compounds, VOCs, fig. 1. Previous study showed that evolving VOCs were resulted from refineries and petrochemical complexes [9, 11]. The VOCs were measured by using Multivariate receptor modeling which was applied to hourly observations of total nonmethane organic carbon (TNMOC) and 54 hydrocarbon compounds from C-2 to C-9. It should be mentioned that "in spite of the best efforts of government and industry, the emissions from refineries and chemical plants are notoriously hard to determine. Most of the emissions are so-called fugitive emissions from leaking valves, pipes, or connectors, of which there are tens of thousands in a large facility" [11].

Table 2: Bacterial strain isolated and pre-identified from leave and soil.

No	Bacteria	Source
1	*Pseudomonas* sp_1	Leaf
2	*Pseudomonas* sp_2	Leaf
3	*Pseudomonas* sp_3	Leaf
4	*Pseudomonas* sp_4	Leaf
5	*Erwinia* sp	Leaf
6	*Thiobacillus* sp_1	rhizosphere
7	*Thiobacillus* sp_2	rhizosphere
8	*Thiobacillus* sp_3	rhizosphere
9	*Thiobacillus* sp_4	rhizosphere
10	*Thiobacillus* sp_5	rhizosphere
11	*Thiobacillus* sp_6	rhizosphere
12	*Nitrobacter* sp_1	rhizosphere
13	*Nitrobacter* sp_2	rhizosphere
14	*Azatobacter* sp	rhizosphere
15	*Rhizobium* sp_1	rhizosphere
16	*Rhizobium* sp_2	rhizosphere
17	*Rhizobium* sp_3	rhizosphere

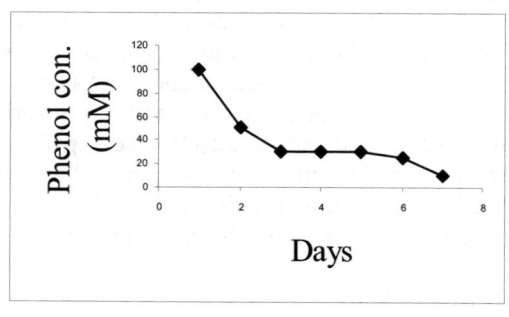

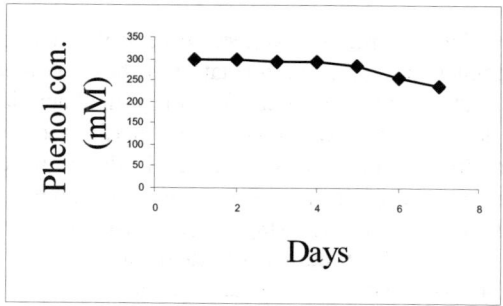

Figure 1: Degradation of different phenol concentration, 100 and 300 mM by *Pseudomonas* sp_2.

The ability of bacterial rhizosphere to degrade the air pollution that accumulates in soil in the form of nitrogen and sulfate compounds was determined by isolating the sulfate reducing-oxidizing bacteria and nitrogen fixation bacteria. The formation of elemental sulfur and various other higher oxidation states of sulfur by microbial oxidation of more reduced forms of sulfur is a common activity in nature. There are a variety of types of microorganisms capable of such activities including chemoautotrophs, chemoheterotrophs, and photoautotrophs. Nitrification and symbiotic nitrogen fixation are especially sensitive to disruption by pollutants, probably in part due to the small numbers of species involved in these processes. In this study, the increment of bacteria that metabolized sulfur and nitrogen compounds was used as an indicator for contamination of soil by the sulfur or nitrogen compounds evolving from oil refinery and other petrochemical units. The results showed that total numbers of sulfate and nitrogen bacteria were high compared to the total bacterial count in the control location, see fig. 2.

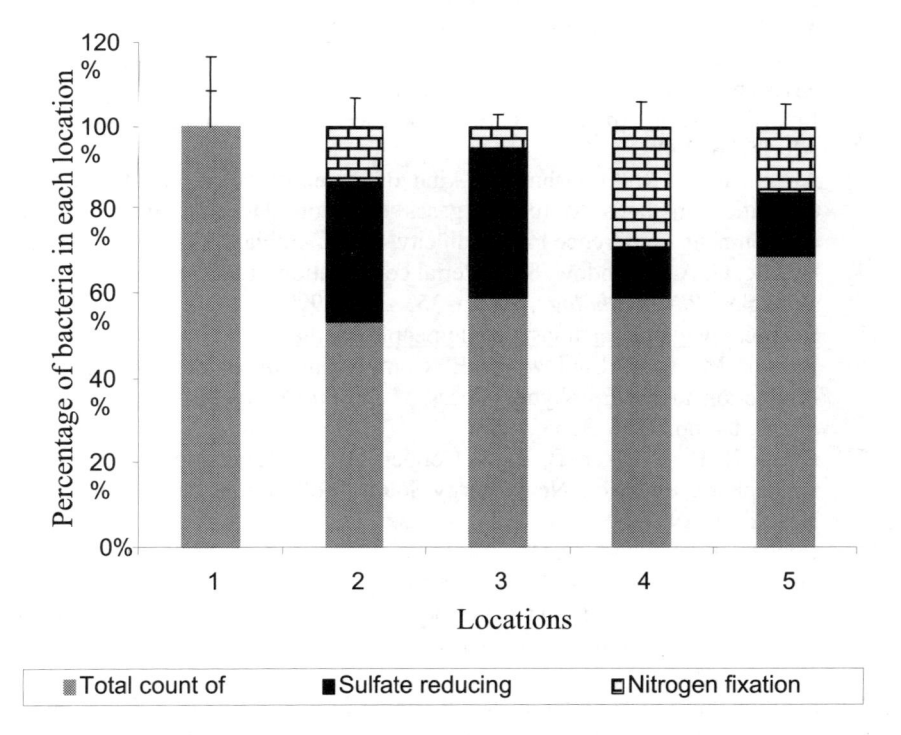

Figure 2: The percentage of sulfate reducing and nitrogen fixation bacteria to the total count bacteria in five different locations.

The ability of leaf surface isolates to produce the EPS was examined. Two out of five bacteria showed the ability to produce EPS. *Pseudomonas* sp_3 and *Pseudomonas* sp_4 could grow on PCS within 7 days and CFU were 2.5×10^{-5} and 2.1×10^{-6} respectively.

4 Conclusions

In spite of the efforts of the Royal Commission the emissions from refineries and chemical factories are notoriously hard to determine. Greenways around the industrial cities seem to serves as natural conservation from the gases that might escape from the air filters in the factories. Leaf and rhizosphere associated bacteria play a significant role in removing the pollutants. However, not all kinds of trees can develop leaf and rhizosphere associated bacteria, and it is important to select appropriate plants for industrial cities.

References

[1] Aggelis, G., Ehaliotis, C., Nerud, F., Stoychev, I., Lyberatos, G., & Zervakis, G. Evaluation of white-rot fungi for detoxification and decolorization of effluents from the green olive debittering process. *Appl. Microbial. Biotechnol*, **59**, pp. 353–360, 2002.

[2] Al-Radady, A. Study of air pollutants and their effects on the environmental and public health in the Yanbu Industrial City, Project funded by King Abdulaziz City for Science and Technology, Final report, AR-15-26, 2006.

[3] Al-senosy, I., & Al-nashmi M. kind of air emission in Riyadh city and evaluation methods to reduce gases emission. Develop and effect on environment, conference in Riyadh city, Saudi Arabia, 1998.

[4] Beattie, G. A. & Lindow, S. Bacterial colonization of leaves: A spectrum of strategies. *Phytopathology*, **89**, pp. 353–359, 1999.

[5] Beattie, www.plantpath.iastate.edu/people/beattie

[6] Brandl, M. T. & Lindow, S. E. Contribution of Indole-3-Acetic Acid Production to the Epiphytic Fitness of *Erwinia herbicola*. *Appl. Environ. Micro*. **64**, pp. 3256–3263, 1998.

[7] Byers, R. L., Crocker, B. B., & Cooper, D. W. Dispersion and control of atmospheric emission New-Energy-Source pollution potential Engineers, New York, 1997.

[8] Cooper, R. Jenkins, D. & Young, L. *Aquatic microbiology laboratory manual*. University of California, Berkeley. 1976.

[9] DeLuchi, M. A., Emission from the production, storage, and transport of crude oil and gasoline. *J. Air and West Mange. Assoc,* **43**, pp. 1486–14995, 1993.

[10] Ecology and Environment of Saudi Arabia Co. Ltd. (EESAL). Final report. Environmental monitoring program at Madinat Ynabu Al-Sinaiyah, prepared for the Royal Commission for Jubail and Ynbau under contract GST E-4021, 1984.

[11] Henry, R. Spiegelman, C.Collins, J. & Park, E. Reported emissions of organic gases are not consistent with observations. *Proc.Natl.Acad.Sci.*, **94**, pp. 6596–6599, 1997.

[12] Ho, K., Pometto III, A., Hinz, P., Dickson, N., J. & Demirci, A. Ingredient selection for plastic composite supports for $_L$-(+)-lactic acid biofilm

fermentation by *Lactobacillus casei* subsp. *rhamnosus*. *Appl. Environ. Microbiol,* **63**, pp. 2516–2523, 1997.
[13] Johanson, L. E., Curl, E. A., Bond, J. H., & Fribourg, H. A. Methods for studying soil microflora plant disease relationship. Burgess Publ. Co. Minn. USA. 1959.
[14] Ping-Shing, M. Isolation and characterization of thermophilic and hyperthermophilic microorganisms from food processing facilities. Thesis at Iowa State University, Ames, 2003.
[15] Sax, N. I., Industrial pollution, Van Nostrand Reinhold, New York, 1974.
[16] Waterman, P., & Mole, S. *Analysis of phenolic plant metabolites.* (ed.), Methods in ecology: J. H. Lawton & G. E. Likens Blackwell Scientific, Oxford. 1994.
[17] WHO, Rapid assessment of source of air, water and land pollution (WHO), Publisher, WHO, offset publication, **62**, Geneva, 1982.

Section 6
Indoor air pollution

Indoor concentrations of VOCs and ozone in two cities of Northern Europe during the summer period

J. G. Bartzis[1], S. Michaelidou[2], D. Missia[1], E. Tolis[1], D. Saraga[4], E. Demetriou-Georgiou[2], D. Kotzias[3] & J. M. Barero-Moreno[3]
[1]*University of West Macedonia, Greece*
[2]*State General Laboratory, Cyprus*
[3]*Institute for Health and Consumer Protection, JRC, Italy*
[4]*Environment Research Laboratory, NCSR "DEMOKRITOS", Greece*

Abstract

Major sources of indoor organic compounds are, commonly, building materials including vinyl tiles and coverings, carpets, wood based panels, paints etc. in new or recently renovated buildings as well as human activities indoors such as cleaning or cooking. Ozone, which has both indoor (photocopiers and other) and outdoor (due to ventilation and infiltration systems) sources, is a highly reactive oxidizing agent. This study was conducted in the frame of the BUMA project (Prioritization of Building Materials Emissions). Herein is presented one week's indoor and outdoor VOCs and ozone concentration measurements from field campaigns at two urban cities in Northern Europe, Dublin and Copenhagen, during a cold period. Sampling was conducted inside and outside four buildings. The concentrations of hazardous compounds (formaldehyde, benzene, acetaldehyde, toluene and xylenes) ranged from 5.9–42.7, 0.6–3.4, 2.3–41.6, 2.2–15 and 0.4–6 $\mu g/m^3$, respectively. Ozone levels were significantly higher outdoors that indoors.

Keywords: indoor air quality, VOCs, formaldehyde, ozone, passive sampling.

1 Introduction

Human activity studies have shown that people spend on average more than 85% of their time inside buildings. This proportion can be analysed more specifically as 66% in residential buildings and another 5% inside vehicles [1]. In recent years, extensive research effort has been invested in examining the relationship between indoor air quality and the application of building materials in new or recently renovated buildings. A variety of studies have demonstrated that building materials with large surfaces are meaningful emission sources of organic compounds and influence the concentration levels in indoor environments. Temporary pollution events such as painting, cleaning, cooking or smoking can contribute to indoor air quality even after the the application has stopped. Sakr et al [2] pointed out that even building materials and the installation of new furnishings that are designed to have low emissions can play a significant role in polluting the indoor environment through sorption and subsequent desorption of pollutants.

Building materials and human activities indoors are major sources of Volatile Organic Compound (VOC) emissions. VOCs have both indoor and outdoor sources and they are of particular interest due to their potential impact on human health [3]. Formaldehyde and benzene, for example, are considered the most studied pollutants since they are classified in Group 1 of human carcinogens by the International Agency for Research on Cancer because of their carcinogenicity [4]. On the other hand, ozone also has both indoor (photocopiers and other office equipment) and outdoor (due to ventilation processes) sources in indoor environments. Ozone can easily react with terpenes and other organic compounds forming ultrafine particles and irritating gaseous organic compounds [5].

For many of these chemicals, the risk on human health and comfort is almost unknown and difficult to predict because of the lack of toxicological data. In the framework of the INDEX project the existing knowledge worldwide has been assessed on type and levels of chemicals in indoor air as well as the available toxicological information. Thus, the INDEX project concluded in a priority ranking of 14 chemicals assigned to three groups [6].

The present work was conducted in the framework of the BUMA project (Prioritization of Building Materials Emissions) and aims to thoroughly assess the human exposure to air hazards emitted from building materials. This study focuses on compounds belonging to the first two priority groups of the INDEX project such as benzene, toluene, xylenes, formaldehyde and acetaldehyde.

2 Materials and methods

2.1 Campaign organization

This study was carried out in 2007 from 28 of May to 3 of June and 24 of June to 1 of July for Dublin and Copenhagen, respectively. Measurements were conducted in four buildings in order to evaluate the indoor air VOCs and ozone. The study design included the selection of the buildings in which passive samplers were installed. The buildings employed in the present study were selected according to the following criteria: (1) the age (less than two years), (2) the last reconstruction or renovation and (3) the position of the building (urban sites were preferred). There are four buildings in every case, one public building, one school and two private houses. In addition, temperature and relative humidity (RH) data loggers were used inside the tested rooms. Finally, the tracer gas technique was used for ventilation measurements, where it was possible.

Moreover, questionnaires were filled in, giving valuable information regarding sampling sites and activities taken place during sampling. At indoor locations, the passive sampling equipment was placed on sites on the wall approximately 1.5 m above the ground, or on tables or other furniture, wherever possible. Outdoor sampling locations were chosen to avoid significant point sources of pollution, such as building exhaust vents.

2.2 Sampling and analysis

Indoor and outdoor measurements of BTEX, carbonyls and ozone were conducted using passive samplers named Radiello in each tested room and outside for one week. The samplers used for BTEX were Activated Charcoal Cartridges (CS_2-desorption) for GC-analysis (code 130), for Aldehydes DNPH-covered cartridges (acetonitrile desorption) for HPLC-VIS (code 165) and for ozone 1, 2- di- (4-pyridyl) ethylene covered cartridges (MBTH (3-Methyl – 2-Benzothiazolinone Hydrazone) solution desorption), for UV-VIS (code 172).

The analysis of BTEX was carried out by GC/FID after desorption of the analytes with CS_2 and included determination of benzene, toluene, xylenes, ethylbenzene, 1,2,4 trimethylbenzene, d-limonene and a-pinene. The determination of all analytes was confirmed by GC-MS. The analysis of carbonyls and ketones was carried out using HPLC-VIS after desorption of the analytes with acetonitrile and included determination of formaldehyde, acetaldehyde, acetone, propanal and hexanal. Finally, the analysis of ozone was conducted using spectrophotometer-VIS after desorption with (MBTH) solution. The analysis of the samples was conducted in the State General Laboratory (SGL) of Cyprus.

With tracer gas technique, air exchange rates were estimated, only for Dublin's office building, using NORDTEST METHOD NT VVS 118. This method can be used in types of buildings, dwellings, offices, schools etc. The testing of ventilation is performed by using homogeneous emission of tracer gas at a constant rate in the ventilated system and subsequent analysis of the steady state concentration of that tracer gas in different parts of the system [7].

3 Results and discussion

3.1 Volatile Organic Compounds (VOCs)

Summary statistics for the concentrations of all measured compounds in indoor air are given in table 1. The most prevalent VOCs in buildings were formaldehyde, acetaldehyde, acetone, hexanaldehyde and a-pinene. Indoor concentrations usually exceeded outdoor levels. High priority compounds constituted large proportion of sum of VOCs in both cities' schools, fig. 1, lower than those reported in Michigan classrooms by Godwin et al [8]. Formaldehyde in Copenhagen school exhibited the same levels observed in schools at Porto [9] and in Shangai, China [10].

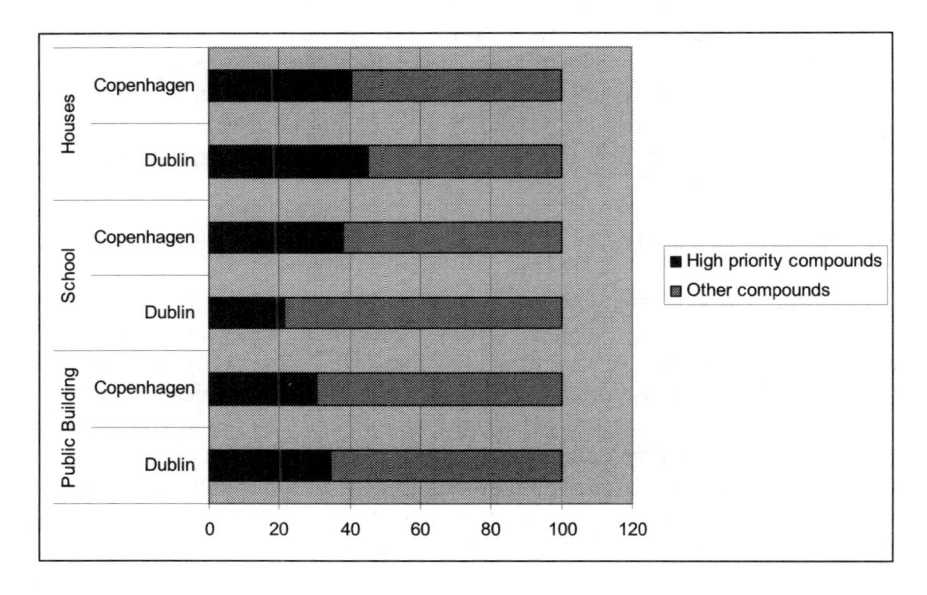

Figure 1: Average percentage of sums of high priority compounds in selected buildings.

Reported public buildings' indoor concentrations of individual VOCs are generally below 50 µg/m^3. Levels of aldehydes in Dublin's office building are lower than those observed in Copenhagen except for formaldehyde. Mean concentrations of the majority of VOCs in both cities are below 10 µg/ m^3, a common trend for both European and American countries [11]. Low concentrations of VOCs in Dublin's office building can be associated with the large air flow rate, which was estimated to be 3.66 h^{-1}.

Acetaldehyde and hexanaldehyde levels are similar to those found for private houses and dwellings in Paris [12, 13]. Such observations suggest probably the absence of indoor sources for acetaldehyde since acetaldehyde is mainly emitted from combustion processes. Propionaldehyde concentrations in houses were measured lower than those found in dwellings in Paris [13]. The levels of formaldehyde did not exceed the WHO guideline value of 100 µg/m^3, which may cause nose and throat irritation in humans after short-term exposure [14].

It is worth to notice that concerning high priority compounds, except for benzene and toluene in some cases, indoor to outdoor ratios (I/O) are substantially greater than one (>1) suggesting important indoor sources for these VOCs. More specifically, for hexanaldehyde I/O ratio is up to 37.8, for acetaldehyde 26.1, for acetone 14.6, and for formaldehyde 15.3.

Table 1: Mean concentrations of VOCs and ozone in µg/m³.

	Public Buildings		Schools		Houses	
	Dublin	Copenhagen	Dublin	Copenhagen	Dublin	Copenhagen
Benzene	1.4	0.7	1.1	0.9	1.7	0.9
Toluene	4.2	2.3	4.1	3.5	9.5	6.6
Ethylbenzene	0.8	1.5	1.5	0.6	0.8	0.6
m,p- xylene	2.5	5.3	4.4	1.6	2.3	1.6
o-xylene	1	2.1	2	0.7	0.8	0.7
a-pinene	0.5	0.5	0.6	7.6	2.9	14
1,2,4- TMB	1.4	0.7	3.1	1.1	1.7	1.2
d-limonene	2.1	0.2	1.1	2.2	10.4	22.6
Formaldehyde	21.2	15.8	7.6	19.9	18	24.4
Acetaldehyde	4.1	4.5	2.5	11.7	10.1	23.5
Acetone	4.5	5.5	2.7	11.5	8.7	9.5
Proprionaldehyde	1.7	1.9	1.5	8	2.6	3.2
Hexanaldehyde	9.6	11.9	7.6	37.8	25.5	34
Ozone	1	11.7	6.2	6.8	1.8	3.2

Table 2: Indoor to outdoor ratios.

	Public Buildings		Schools		Houses	
	Dublin	Copenhagen	Dublin	Copenhagen	Dublin	Copenhagen
Benzene	1	1.2	1.8	1.1	2.6	1.4
Toluene	1	1.3	2.4	1.5	5.9	3.2
Ethylbenzene	0.9	3.8	3.8	1.5	2	1.5
m,p- xylene	1	5.3	3.4	1.2	1.8	1.7
o-xylene	1	5.3	4	1.8	2	2
a-pinene	1.3	-	-	15.2	-	-
1,2,4- TMB	1.1	0.7	6.2	1.6	2.8	2.7
d-limonene	0.9	-	5.5	-	52	-
Formaldehyde	2.4	6.9	5.1	10.5	11.6	15.3
Acetaldehyde	1.3	3.5	3.1	9	11.9	26.1
Acetone	1.4	6.9	3.9	10.5	13.4	14.6
Proprionaldehyde	1.3	6.3	7.5	40	10.4	16
Hexanaldehyde	1.6	9.2	12.6	27	39.2	29.6
Ozone	0.03	0.2	0.2	0.1	0.03	0.02

3.2 Ozone (O_3)

The importance of measuring ozone in indoor environments comes from its ability to react with high molecular organic compounds and specifically with terpenes forming ultrafine particles and free radicals. As it is observed, outdoor concentrations are significantly higher in contrast with indoor levels. The indoor to outdoor ozone concentration ratio generally ranged between 0.03 and 0.2 indicating ozone – indoor chemistry relationship as mentioned by Nicolas et al [15].

4 Conclusions

The concentration data show a considerable diversity due to the different indoor emission sources, ventilation rates and outdoor environments concentrations. Aromatic compounds' levels in all buildings are lower than those expected in indoor environments. The sum of high priority compounds consist a large proportion of the TVOC in all selected buildings. The relatively high I/O ratios for carbonyls and ketones indicate strong indoor emissions sources. Ozone outdoor concentrations seem to be reduced substantially inside; indicating relatively strong indoor ozone sinks.

Acknowledgement

The BUMA project is funded by the European Community, Public Health Executive Agency (PHEA).

References

[1] Ogulei D., Hopke P.K. & Wallace L.A., Analysis of indoor particle size distributions in an occupied townhouse using positive matrix factorization. *Indoor Air,* **16(3)**, pp. 204–215, 2006.
[2] Sakr W., Weschler C.J. & Fanger P.O., The impact of sorption on perceived indoor air quality. *Indoor Air* **16(2)**, pp. 98–110, 2006.
[3] Marchland C., Bulliot B., Le Calve S. & Mirable Ph., Aldehyde measurements in indoor environments in Strasbourg (France). *Atmospheric Environment* **40**, pp. 1336–1345, 2006.
[4] IARC, Overall evaluation of Carcinogenicity to Humans, Formaldehyde [50-00-0], *Monographs Series*, **88**, International Agency for Research on Cancer, Lyon, France.
[5] Molhave L., Kjaergaad S.K., Sigsgaard T., Lebowitz M. Interaction between ozone and airborne particulate matter in office air. *Indoor Air*, **15(6)**, p.p. 383–392, 2005.
[6] Kotzias D., Koistinen K., Kephalopoulos S., Schlitt C., Carrer C., Maroni M., Jantunen M., Cochet C., Kirchner S., Lindvall T., McLaughlin J., Mølhave L., De Oliveira Fernandes E. & Seifert B., ''The INDEX project''

Critical appraisal of the setting and implementation of indoor exposure limits in the EU. Final Report. (2005)
[7] Ventilation: Flow rate, total effective – by single zone approximation. NORDTEST METHOD NT VVS 105, 1994.
[8] Godwin C., Batterman S., Indoor air quality in Michigan schools. *Indoor Air*, **17(2)**, pp. 109 – 121, 2007.
[9] Samudio M.J, Ventura Silva G., Oliveira Fernandes E., Guedes J., and Vasconcelos M.T.S.D., A Detailed Indoor Air Study in a School of Porto. Proc. of Healthy Buildings Conf. eds. E. de Oliveira Fernandes, M Gameiro da Silva, J. Rosado Pinto, pp. 345–349, 2006.
[10] Mi Y.H., Norback D., Tao J., Mi Y.L., Ferm M., Current asthma and respiratory symptoms among pupils in Shangai, China: influence of building ventilation, nitrogen dioxide, ozone and formaldehyde in classrooms. *Indoor Air*, **16(6)**, pp. 454–464, 2006.
[11] Wolkoff P., Wilkins C.K., Clausen P.A., Nielsen G.D., Organic Compounds in office environments – sensory irritation, odor, measurements and the role of reactive chemistry. *Indoor Air*, **17(1)**, pp. 7–19, 2006.
[12] Marchand C., Bulliot B., Le Calve S., Mirable Ph., Aldehyde measurements in indoor environments in Strasbourg (France). *Atmospheric Environment*, **40**, pp. 1336–1345, 2006.
[13] Clarisse B., Laurent A.M., Seta N., Le Moullec Y., El Hasnaoui A., and Momas I., Indoor aldehydes: measurement of contamination levels and identification of their determinants in Paris dwellings. *Environmental Research*, **92**, pp. 245–253, 2003
[14] WHO (2001). World Health Organization, Air quality guidelines for Europe 2000. WHO Regional Office for Europe.
[15] Nicolas Melanie, Ramalho Olivier, Maupetit Francois, Reactions between ozone and building products: Impact on primary and secondary emissions. *Atmospheric Environment*, **41**, pp. 3129–3138, 2007.

Comparative study of indoor-outdoor exposure against volatile organic compounds in South and Middle America

O. Herbarth[1], A. Müller[2], L. Massolo[3] & H. Tovalin[4]
[1]*Department Environmental Medicine and Hygiene, Faculty of Medicine, University of Leipzig, D-04103 Leipzig, Germany*
[2]*UFZ-Centre for Environmental Research, Department of Human Exposure Research and Epidemiology, D-04318 Leipzig, Germany*
[3]*Centro de Investigaciones del Medio Ambiente, Facultad de Ciencias Exactas, Universidad Nacional de La Plata, Argentina*
[4]*División de Estudios de Posgrado e Investigación, FES-Zaragoza, Universidad Nacional Autónoma de México (UNAM), México D.F., Mexico*

Abstract

Volatile organic compounds (VOCs) play an important role in indoor and outdoor air pollutants. In the present study, samples were analyzed from indoor (schools and houses) and outdoor air in urban, industrial, semi-rural and residential areas from Argentina (La Plata region) and Mexico (Mexico City region) to consider VOC exposure in different types of environments. VOCs were sampled using a passive sampling method with passive 3M monitors. Samples were extracted with CS_2 and analyzed by GC/MS detectors.

The results show significant differences in concentration and distribution between indoor and outdoor samples, depending on the study area. Most VOCs predominantly originated indoors influenced by local outdoor emissions (traffic and industry).

Keywords: VOCs, indoor-outdoor ratio, urban and industrial burdens, indoor sources.

1 Introduction

Volatile organic compounds (VOCs), arising from natural and anthropogenic processes represent an important group of indoor and outdoor air pollutants, because they are ubiquitous and associated with increased long-term health risks [1, 2], for the general population and those people that stay outdoors for a long time [3]. Vehicular and industrial emissions are the major sources of outdoor VOCs. It has been estimated that 35% of total VOC emissions are due to vehicle exhaust and evaporative losses [4–6].

Indoor VOC exposure is thought to be of greater concern in the community because indoor concentrations of many pollutants are often higher than those typically encountered outdoors. At the same time, indoor VOC concentration may be the dominant contributor to personal exposure because most people spend over 80% of their time indoors, either in the home or in the work place [7, 8]. Measurement of VOCs in the indoor environment has received substantial research attention for several years because indoor VOC levels may pose potential health effects to occupants of dwellings. While some VOCs may be present at concentrations that are not considered acutely harmful to human health with short-term exposure, long-term exposure may result in mutagenic and carcinogenic effects.

The occurrence and concentrations of VOCs in residences can be affected by outdoor atmospheric conditions, indoor sources, indoor volume, human activities, chemical reactions, ventilation rates, and seasonal factors [9, 10].

Indoor sources are quite numerous including combustion products, cooking, construction materials, furnishings, paints, varnishes and solvents, adhesives and caulks, office equipment and consumer products [11–13].

To consider the VOC exposure in the different types of environments, measurements of VOC concentrations were carried out indoors and outdoors in areas influenced by industrial emissions, in an urban area influenced by traffic, as well as in semi-rural and residential areas as control and as an area influenced by emissions coming from the (mega)-city. The indoor-outdoor ratios were calculated to get an impression regarding the sources of VOC exposure and the impact of outdoor-related air pollution on the indoor environment.

2 Methods

2.1 Measurement sites

The study region, La Plata and neighboring areas, located northeast of Buenos Aires province (Argentina), has a population around 700,000 inhabitants. Four sampling zones were considered: industrial (I), urban (U), semi-rural (SR) and residential (R) areas (Figure 1).

The industrial area holds the country's main oil refinery (total crude oil distillation capacity: 38,000 m^3/day^{-1}) located next to six petrochemical plants producing diverse compounds such as aromatics (benzene, toluene, xylenes), aliphatic solvents (n-pentane, n-hexane, n-heptane), polypropylene, polybutene, maleic anhydride, cyclohexane, methanol, methyl tertiary butyl ether, and petroleum coke. This industrial complex is approximately 10 km north-northeast of the main urban sector of the city.

The urban area is characterized by heavy traffic; the number of vehicles registered is approximately 200,000. The semi-rural and residential sites, places with low traffic, were considered control areas.

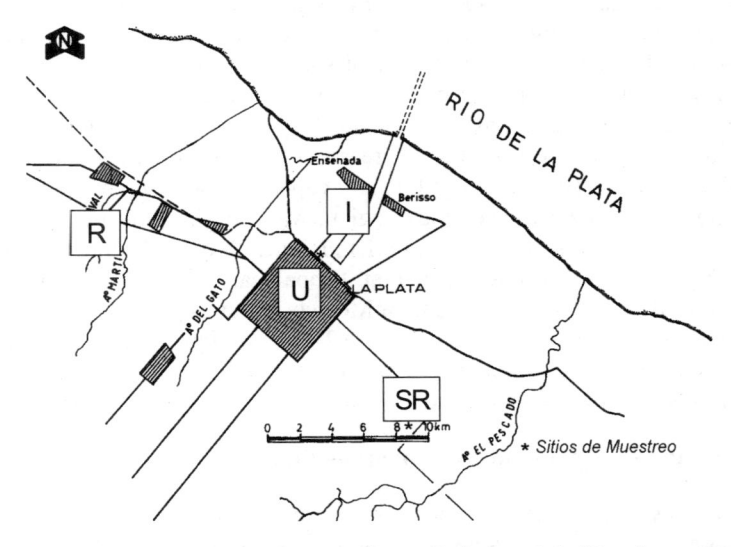

Figure 1: Sampling points in La Plata (I, industrial; U, urban; SR, semi-rural; R, residential).

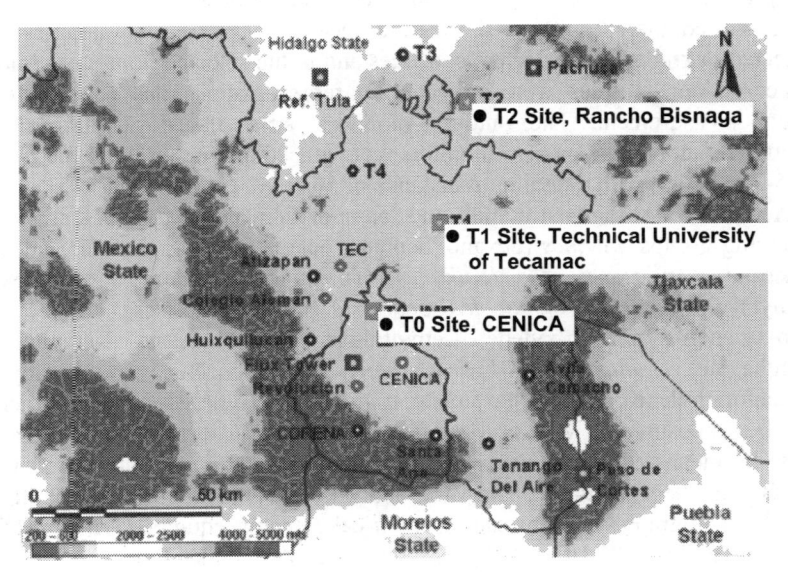

Figure 2: Sampling points in Mexico (T0 urban, T1 suburban, T2 rural).

The Mexico study is part of the MILAGRO project (Megacity Initiative: Local and Global Research Observations) [16]. The air pollution problem in megacities is influenced by several factors that include the topography, meteorology, demographic growth, industrial growth and urban expansion.

The Mexico City Metropolitan Area (MCMA) – the second largest megacity in the world – has an estimated population of 18 million people and covers an urbanized area that totals 1,500 km2, encompassing the 16 delegations of the Federal District, 37 municipalities of the State of Mexico and 1 municipality of the State of Hidalgo. In the MCMA, emissions of pollutants reach millions of tons per year, and atmospheric concentrations of ozone and particulate matter routinely exceed the standards recommended by the World Health Organization [6]. As a result, there has been an increase in diagnosed incidences of chronic bronchitis, asthma, reduction of pulmonary capacity, and in premature mortality rates among the citizens. The study region, Mexico City and the neighboring areas in main wind direction (SW to NE) includes three different sites (T0, T1, T2): T0 – within MCMA located at Iztapalapa, T1 – at the Universidad Tecnológica de Tecámac in the State of Mexico, T2 in Rancho La Bisnaga, north of Tizayuca in the State of Hidalgo (Figure 2). The designations refer to transport of the urban plume to different points in space and time. The selected VOC represent alkanes, cycloalkanes and aromatics, chloroaromatics, terpenes.

2.2 Sampling design and measurement method

Sampling was carried out in indoor and outdoor air using passive sampler (3M OVM 3500, 1986) [14]. A sampling period of four weeks was selected, since an integrative sampling described a mean load relative to personal exposure. Using this procedure, it can be assumed that the human exposure is more selectively described compared with a short time measurement. Indoor samples were collected in kindergartens, schools, offices and in the selected homes of children, while outdoor samples were collected in representative sites parallel to the indoor measurement. The outdoor monitors were placed at rain-protected positions. Indoors, the passive monitors were placed in the middle of the room at a 1.5-2 m height with a minimum distance of 50 cm to the ceiling [7, 15].

According to their structure and composition the VOC belong to the following groups: *alkanes* (hexane, heptane, octane, nonane, decane, undecane, dodecane, tridecane); *cycloalkanes* (methylcyclopentane, cyclohexane, methylcyclohexane); *aromatic hydrocarbons* (benzene, toluene, ethylbenzene, m+p-xylene, styrene, o-xylene, 4-ethyltoluene, 3-ethyltoluene, 2-ethyltoluene, naphthalene); *chlorinated hydrocarbons* (chlorobenzene, trichlorethylene, tetrachlorethylene); *terpenes* (a-pinene, b-pinene, 2-carene, 3-carene, limonene).

After exposure, the VOCs were desorbed from the adsorption layers (charcoal pads) by means of 1.5 ml of carbon disulfide (with low benzene, from Merck) containing 1% of methanol. The VOC analysis was performed on a PerkinElmer gas chromatograph with a masspectrometrical detector, equipped with an RTX-1 column (60 m 0.32 mm I.D., 1.0 µm film thickness; Restek). The oven temperature was held at 43 °C for 5 min, and then programmed to 200 °C at a rate of 2.5 °C min^{-1}. The injector temperature was held at 250 °C as well as the transfer line temperature. An electron energy of 70 eV was used for ionization. The source temperature was held at 200 °C. A sample volume of 1 μl was splitless injected. Integrated areas of selected fragment ions from each of the 29 VOCs were obtained with the software Turbomass, Version 4.4 (PerkinElmer) [7, 10].

2.3 Quantification and statistical analysis

Recovery coefficients were determined by direct injection of a known amount of the standard into a 3M sampler and subsequent extraction with carbon disulfide containing 1% methanol. The recovery was between 98% and 102%. For the studied VOCs, the background amounts contained in the monitors and in the carbon disulfide were determined for each charge of the samplers and chemicals and were subtracted from the sample results. The detection limits for the components were estimated as the three-fold standard deviation of five replicate measurements of monitor blanks. For components with blank values too low to be registered, it is usual practice to use the three-fold standard deviation of replicate measurements of a low-level standard solution [15]. The average concentration of each component over the sampling interval (in $\mu g\ m^{-3}$) was calculated according to the equation adopted from 3M [14], using the adsorption coefficients in charcoal. The detection limits were found to be between 0.01 and 0.05 mg m^{-3} regarding a sampling interval of 4 weeks.

A non-parametric method (Mann–Whitney U-Test) was used for statistical analysis of results, according to the non-symmetrical distribution of the data. Statistical analysis and plots of median values were prepared using Statistica 7.1 [17]. A p-value below 0.05 was regarded to be statistically significant.

3 Results and discussion

3.1 Total VOCs load

The concentration levels of the total indoor and outdoor VOC content per study area is shown in Figure 3a and 3b.

Significantly higher concentrations of indoor VOCs were observed independent from the study and local area.

3.2 Indoor-outdoor ratio

Since most people spend over 80 percent of their time indoors [10], indoor VOCs are thought to be of greater concern considering potential health effects because the indoor concentrations of these compounds are often higher than those typically encountered outdoors. On the one hand the indoor-outdoor ratio (I/O) is generally used to infer penetration to indoor environments and indoor sources [18]. On the other hand this ratio will be determined also by the use of the buildings. It makes a difference whether the building is mainly used for apartments or for public demands. An I/O ratio lower and close to one indicates more outdoor sources. Significant differences between indoor and outdoor concentrations indicate mainly indoor sources for the VOC compounds.

The I/O ratio of the apartments in the urban and suburban areas is similar (between 2.5 and 2.7).

The ratio is lower for the schools/offices than for the apartments. This reflects the lower indoor exposure in the public building coming from a reduced use of indoor exposure causing materials and maybe from a reduced redecoration activity but a better air exchange rate.

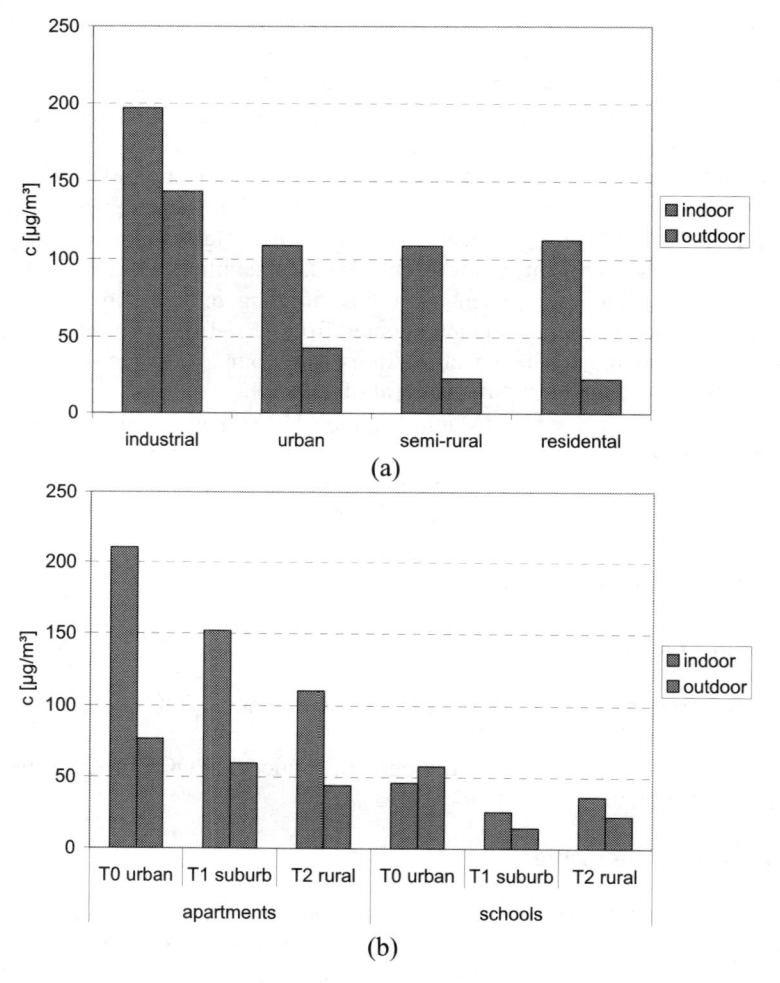

Figure 3: (a) Indoor and outdoor concentration of VOC (sum of measured VOC) in La Plata and surroundings (mean values). (b) Indoor and outdoor concentration of VOC (sum of measured VOC) in different building types in Mexico City and surroundings (mean values).

The main sources are the industrial complex and the traffic. The highest outdoor exposure was measured in the industrial area of La Plata following by the traffic dominated exposure in the urban area of Mexico City. The lowest load was recorded in the rural respectively semi-rural sites of both study areas.

4 Conclusions

Ambient air in the vicinity of the industrial and urban site is significantly different than the air quality found in the semi-rural and suburban residential areas. Whereas traffic was identified as the main source of outdoor VOC exposure in the urban and residential areas, in the industrial area, outdoor as well indoor air is strongly affected by emissions of the local industry.

Total VOCs in the urban, residential and semi-rural areas are equivalent to those reported for other major cities worldwide, indicating comparable burdens, although non-equivalent distribution patterns could be a result of different lifestyles and habits.

Indoor/outdoor ratios suggest higher indoor burdens in all places and demonstrate an exceptional importance of indoor exposure.

The high outdoor VOC content in the industrial and "high" traffic loaded area, together with the lifestyle quality, very strongly affected the indoor exposure for these inhabitants.

Further research will be necessary to evaluate the environmental impact of VOCs and the assessment of effects of the different volatile compounds on human health.

Acknowledgements

Parts of this study were supported by the International Bureau of the German Federal Ministry of Education and Research, the Secretary of Science and Technology and Province Research Council of the Province of Buenos Aires (CICPBA), the National Research Council of Argentina (CONICET), the Metropolitan Environmental Commission of México City and the Molina Center for Energy and the Environment. The authors would also like to thank Brigitte Winkler for her excellent assistance in the chemical analyses, and Jephte Cruz for her assistance during the monitoring.

References

[1] WHO (World Health Organization). Protection of the Human Environment. The health effects of indoor air pollution exposure in developing countries, Geneva, 2004.
[2] Brown, S.K. Volatile organic compounds in new and established buildings in Melbourne, Australia. Indoor Air. 12:55–63, 2002.
[3] Tovalin, H., Whitehead, L. Personal exposure to volatile organic compounds among outdoor and indoor workers in two Mexican cities. Science of the Total Environment. 376, 60–71, 2007.

[4] Herbarth, O., Rehwagen, M., Ronco, A. The Influence of Localized Emittants on the Concentration of Volatile Organic Compounds in the Ambient Air Measured Close to Ground Level. Environ Toxicol Water Qual. 12:31–37, 1997.

[5] Chan, C., Chan, L., Wang, X., Liu, Y., Lee, Y., Zou, S., Sheng, G., Fu, J. Volatile organic compounds in roadside microenvironments of metropolitan Hong Kong. Atmos Environ. 36 (12):2039–2047, 2002.

[6] SMA - Secretaria del Medio Ambiente. Inventario de contaminantes tóxicos del aire en la ZMVM (MCMA Air toxics inventory). Gobierno del Distrito Federal. México. 2006.

[7] Rehwagen, M., Schlink, U., Herbarth, O. Seasonal cycle of VOCs in apartments. Indoor Air 13:1–9, 2003.

[8] Guo, H., Lee, S., Chan, L., Li, W. Risk assessment of exposure to volatile organic compounds in different indoor environments. Environ Res. 94:57–66, 2004.

[9] Son, B., Breysse, P., Yang, W. Volatile organic compounds concentration in residential indoor and outdoor and its personal exposure in Korea. Environ Int. 29:79–85, 2003.

[10] Schlink, U., Rehwagen, M., Damm, M., Richter, M., Borte, M., Herbarth, O. Seasonal cycle of indoor-VOCs: comparison of apartments and cities. Atmos Environ. 38: 1181–1190, 2004.

[11] Jones, A. Indoor air quality and health. Atmos Environ. 33:4535–4564, 1999.

[12] Watson, J., Chow, J., Fujita, E. Review of volatile organic compound source apportionment by chemical mass balance. Atmos Environ. 35:1567–1584, 2001.

[13] Guo, H., Lee, S., Li, W., Cao, J. Source characterization of BTEX in indoor microenvironments in Hong Kong. Atmos Environ. 37:73–82, 2003.

[14] 3M Deutschland GmbH Technische Informationen 3M Monitore, 1986

[15] Begerow J, Jermann E, Kelles T, Ranft U, Dunemann L. Passive sampling for Volatile Organic Compounds (VOCs) in air at environmentally relevant concentration levels. Fresenius Journal of Analytical Chemistry 351: 549–554, 1995.

[16] MILAGRO-Factsheet 2006: http://www.eol.ucar.edu/projects/milagro/media/MILAGRO-Factsheet-Final.pdf

[17] StatSoft, Inc. (2005). STATISTICA für Windows [Software-System für Datenanalyse] Version 7.1. www.statsoft.com

[18] Tang, J., Chan, C.Y., Wang, X., Chan, L.Y., Sheng, G., Fu, J. Volatile organic compounds in a multi-storey shopping mall in Guangzhou, South China. Atmos Environ. 39:7374–7383, 2005.

Sampling of respirable particle PM_{10} in the library at the Metropolitana University, Campus Azcapotzalco, Mexico City

Y. I. Falcon, E. Martinez, A. Cuenca, C. Herrera & E. A. Zavala
Department of Energy, Metropolitana University, Mexico City, Mexico

Abstract

This study presents the results of a sampling campaign of PM_{10} respirable particles, carried out at the library located in the Metropolitana University, Campus Azcapotzalco, through seven months: March, April, May, June, July, September and October, 2005. We collected a total of 84 samples, considering only the three main levels of the Library and selecting a total of nine sampling points. The other levels are located at the mezzanine and the main entrance, where there is no rug and the floor is cleaned up daily with a wet cloth. Samplings were carried out for nine weeks, three weeks for each level, so finally, we ended up with 18 samples for level 1, 27 for level 2, and 36 for level 3; plus three reference samples. Samples were collected using a set of sampling pumps SKC for indoor sampling, from 10:00 a.m. to 6:00 p.m. on Monday, Wednesday, and Friday. Three reference samples were collected on Thursday, when there is less activity in the Library. In level 1, where there is a study area, the minimum mass concentration of R.P. was 15.27 $\mu g/m^3$. This area has no rug and it is near the main entrance. The maximum mass concentration of R.P. was 131.94 $\mu g/m^3$ for the same level, in a sampling point located in a corner far from the entrance. For level 2, the minimum concentration of R.P. was 8.33 $\mu g/m^3$, on the shelves T-J, and the maximum concentration of R.P. was 157.77 $\mu g/m^3$, on the shelf Q. This difference in concentration between the two sampling points could have been originated due to the type of books on the shelves, which are demanded more by the students. It is important to point out that this level has an old rug. In level 3, the minimum concentration of R.P. detected was 6.94 $\mu g/m^3$, in the sampling point of the F shelf, and the maximum concentration, on the shelves H-T, was 197.22 $\mu g/m^3$. The difference in concentrations may be due to the subjects of the books on each shelf and the interest of students for those books - so there are more people going in and out - also, in this level there is an old rug that is usually cleaned using a vacuum cleaner, but not very often. This type of cleaning is the same for level 2. It is important to point out that the library building does not have any air conditioning or air movers.

Keywords: indoor air pollution, respirable particles, library indoor air pollution.

1 Introduction

Describing indoor air quality is not a simple thing to do. It is a constantly changing interaction of complex factors that affect the types, levels, and importance of pollutants in indoor environments. These factors include: sources of pollutants or odors; design, maintenance, and operation of building ventilation systems; moisture and humidity; and occupant perceptions and susceptibilities. In addition, there are many other factors that affect comfort or perception of indoor air quality [7].

The rate at which outdoor air replaces indoor air is described as the air exchange rate. When there is little infiltration, natural ventilation, or mechanical ventilation, the air exchange rate is low and pollutant levels can increase [8].

Inadequate ventilation can increase indoor pollutant levels by not bringing in enough outdoor air to dilute emissions from indoor sources and by not carrying indoor air pollutants out of the building. High temperature and humidity levels can also increase concentrations of some pollutants [8].

Immediate effects may show up after a single exposure or repeated exposures. These include irritation of the eyes, nose, and throat, headaches, dizziness, and fatigue. Such immediate effects are usually short-term and treatable. Sometimes the treatment is simply eliminating the person's exposure to the source of the pollution, if it can be identified. Symptoms of some diseases, including asthma, hypersensitivity pneumonitis, and humidifier fever, may also show up soon after exposure to some indoor air pollutants. [8].

Typically, indoor $PM_{2.5}$ consists of ambient (outdoor) particles that have infiltrated indoors, particles emitted indoors (primary), and particles formed indoors (secondary) from precursors emitted both indoors and outdoors [1].

Fine particles smaller than 2.5 μm in aerodynamic diameter ($PM_{2.5}$) are of the greatest concert owing to their size and transportability in the human body [2].

In the present project, samplings of respirable particles were made at the inside of the library of the Metropolitana University - Campus Azcapotzalco, on all the three levels, during a school trimester. The results concluded by this research will be used to suggest changes at the interior of the library, such as: getting rid of the rugs, open new spaces for ventilation, and installing or replacing air extraction equipments -among other proposals- in order to improve the inside ambiance of the building and to prevent health problems of users and library staff.

2 Methodology

2.1 Selection of sampling points

The sampling points were chosen after a careful inspection of the facilities of the library in the Metropolitana University, aimed at making a qualitative evaluation of the place, identifying the spots with greater activity, and to observe their physical characteristics (like student inflow intensity, the kind of furniture, if there is a rug, ventilation, or whether there are curtains, air currents, etc.).

Keeping all these in mind, it was possible to select the right spots where to place the sampling equipment, preferably on those shelves located at more than 1.5 meters high from the floor.

With that information, it was decided to divide the interior of the Library in three levels: LEVEL 1: with two sample points, LEVEL 2: with three sample points, LEVEL 3, with three sample points, and one sample point at the thesis and internet section; in such a way, as to include the biggest possible area of the Library.

2.2 Library characteristics

The study area has an air conditioning system, which is out of order. In spite of the fact that there are many large windows, they cannot accomplish their function of ventilating, because they cannot be opened; they only work as a source of light. The only ventilation is the one provided by the main entrance.

2.3 Sampling period

The sampling period was established considering certain factors, like: the trimester school term at the Metropolitana University (eleven weeks), the service schedule offered by the Library, and the rush hours, among others. This way, it was decided that the sampling period would be of three months in a row, starting on April the 25th, 2005, and finishing on July the 15th, in the same year. There was a three week sampling period made for each level.

During the nine-week sampling period, samples were taken every three days, on Mondays, Wednesdays, and Fridays, and each sample-taking lasted eight hours - at rush hours - from 10:00 AM to 18:00 PM.

3 Analysis

A gravimetric analysis was made to the filters. They were conditioned at constant temperature and humidity, before and after the sampling. The weighing of the filters was made with a Cahn Electronic Scale at the Atmospheric Monitoring Direction Lab of Mexico City's Hall.

4 Results and discussion

The selected sampling points for each level were classified and distributed as follows: two in Level 1: PM1 at the section for documentation and PM2 in the study area; three in Level 2: PM3 at the area for rare books, PM4 at the shelf area for books classified as Q, and PM5 at the shelf area for books classified as JT; four in Level 3: PM6 at the shelf area for books classified as HD, PM7 at the shelf area for books classified as HT, PM8 at the shelf area for books classified as F, and PM9 at the shelf area for books classified as PQ. It is important to point out that Levels 2 and 3 have a rug and they do not get an adequate cleaning. This nomenclature for the sampling points will be used from now on at the forthcoming discussion of results.

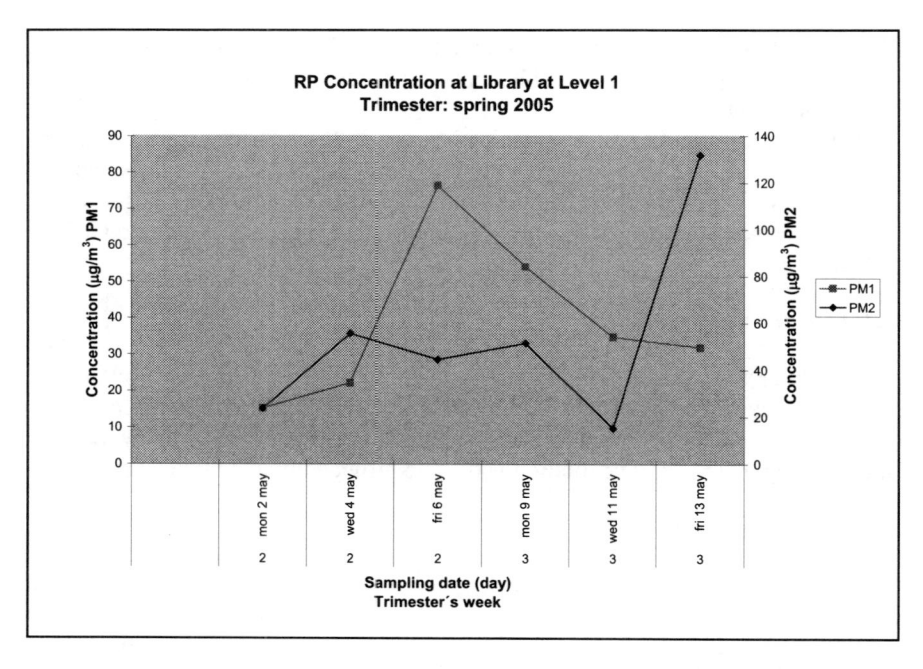

Figure 1: Concentration of respirable particles (RP) at library level I.

The maximum concentration of RP at Level 1 in the Library was measured at both sampling points on a Friday. Point PM1, located in the section for documentation, got a value of 76.9 µg/m^3, in the second week of the trimester. This RP concentration is low, what reflects the activity at this time of the school term, since students are only starting to attend the library. For point PM2, located at the study area, the maximum concentration value of RP was 131.94 µg/m^3, measured at the third week of the trimester. The RP concentration value is bigger than the previous week because of the increase of students visiting the library, since they are getting ready for their first partial exam then. The minimum value for both points matches with those registered on Wednesday, during the second week. As it can be seen in Figure 1, they got a value of 15.28 µg/m^3.

The maximum RP concentrations for sampling points PM3 and PM5 were measured on a Friday. At PM3, located in the shelf area for rare books, the value obtained was 125 µg/m^3, during the fifth week of the school term. The value registered for PM5, located in the JT bookshelf area, was 116.67 µg/m^3, in the sixth week of the trimester. The maximum concentration at PM4, located in the Q bookshelf area, was 152.78 µg/m^3, in the seventh week of the trimester, on Monday. These results match the activities of the school term, since the second partial exam takes place at this time, and lots of students come to the library to get ready for their test in these weeks. The minimum concentrations for the Q (PM4) and JT (PM5) bookshelf areas were measured on a Wednesday, registering 69.44 µg/m^3 at PM4 and 12.50 µg/m^3 at PM5. For the rare books

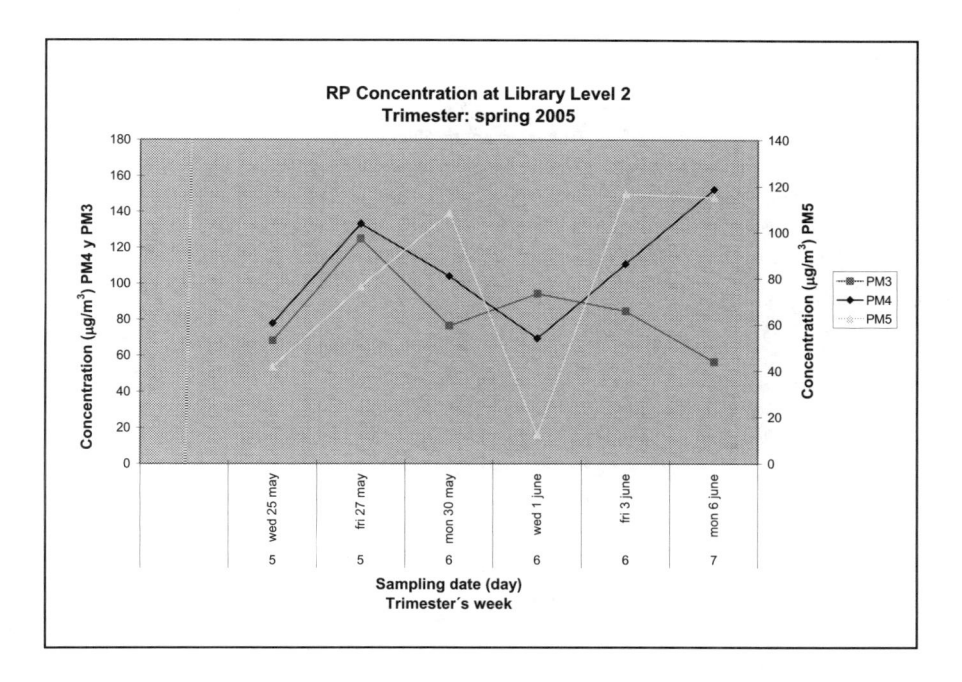

Figure 2: Concentration of respirable particles (RP) at library level 2.

shelf area (PM3), the minimum concentration was 56.50 µg/m³ and it was registered on a Monday (see Figure 2). It is important to point out that the concentration values at the different sampling points also show the shifts in the demand for the different kinds of books located on different shelves. In general, the books located in Level 1 are theses; in Level 2, related to Basic Science and Engineering; and in Level 3, Social Science and Arts.

The maximum RP concentration at sampling points PM6, PM7, PM8, and PM9, was registered during the tenth week of the school term, on Wednesday. The concentration obtained for PM6, located in the HD bookshelf area, was 156.94 µg/m³. For PM7, located at the HT bookshelf area, the value registered was 197.22 µg/m³. For PM8, placed at the F book shelf area, was 184.72 µg/m³, and for PM9, located in the PQ book shelf area, was 200 µg/m³, being the latter the highest registered during the whole sampling period.

The concentrations obtained at this stage of the sampling period were the highest registered by the present study, and match the school activities, since the three-month academic calendar schedules the third partial exam at the eleventh week. Thus, the Library is visited by a large amount of students from the Campus at this time, and most of them ask to borrow the books to take home, anticipating the final exams (global evaluations) scheduled for the twelfth week of the school term.

The minimum concentration for all the sampling points at this Level took place on a Wednesday, with the difference that for points PM6, PM7, and PM8, it occurred at the eighth week; while for PM9, at the seventh. The minimum RP

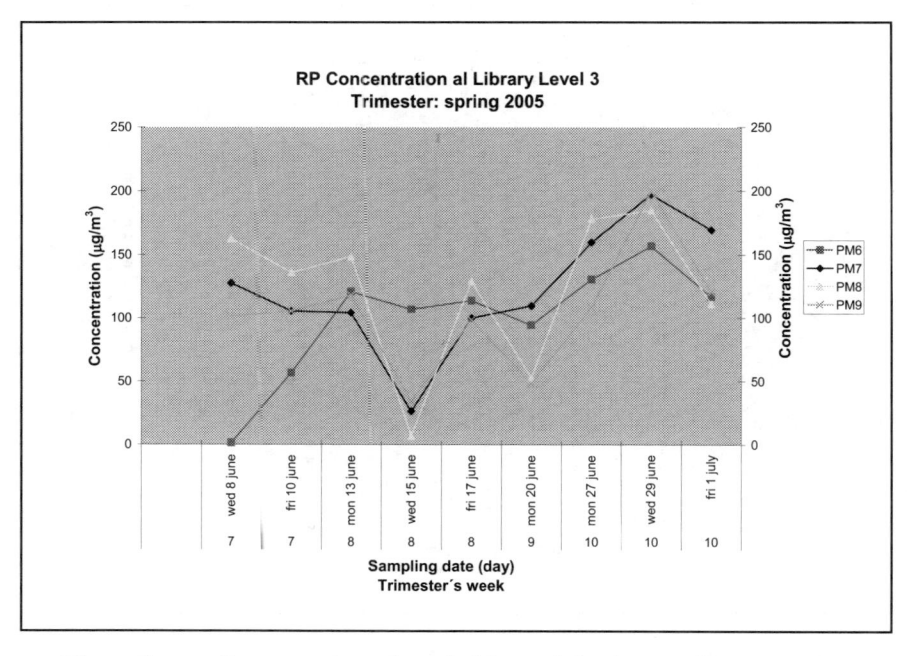

Figure 3: Concentration of respirable particles (RP) at library level 3.

concentration value measured at PM6 was 1.38 µg/m³, and we have to draw attention to the fact that this was the lowest registered in all the sampling period. This particular point was located at an area that had no air currents. The values for the minimum concentrations at all the other sampling points were: 26.39 µg/m³, at PM7, 6.94 µg/m³, at PM8, and 34.72 µg/m³, at point PM9 (see Figure 3).

The minimum concentration at all the sampling points in this Level took place on a Wednesday. The difference is that for points PM6, PM7, and PM8, it occurred at the eighth week; while for PM9, at the seventh. The minimum RP concentration registered for PM6 was 1.38 µg/m³. It is important to point out that this is the lowest concentration registered throughout all the sampling period, and that this particular point was located in a place where there are no air currents. The minimum concentration values for all the other sampling points were: 26.39 µg/m³ at PM7, 6.94 µg/m³ at PM8, and 34.72 µg/m³ at point PM9 (see Figure 3).

5 Conclusions

As it can be seen in the Results, the maximum RP concentration in Level 1 was 131.94, µg/m³, measured in the study area on Friday, May 13th, 2005; while the minimum RP concentration was 15.28 µg/m³, registered at the area for documentation, on Monday, May the 2nd, 2005.

Based on the results obtained throughout all the sampling period, we can assume that, since there is no carpet in this Level, the RP have no place where to deposit and accumulate, adding up the fact that the main entrance provides an air current that makes it possible to force the RP out of the library.

In Level 2, the maximum concentration was 152.78 µg/m^3, reported at the Q bookshelf area, on Monday, June the 6th, 2005. The minimum concentration was 12.50 µg/m^3, registered at the JT bookshelf area, on Wednesday, June 1st, 2005. The maximum concentration value in this Level is higher than in Level 1. This may be because it has a carpet, what gives the RP a place where to deposit and accumulate. Also, it lacks adequate ventilation that could force them out or disperse their concentration in the air inside. It is also necessary to consider that there is a high attendance of students to this Level, what makes the RP deposited on the rug be re-suspended in the air.

In Level 3, the maximum RP concentration was 200 µg/m^3, registered at PQ bookshelf area on Wednesday, June 29, 2005. The minimum concentration was 1.38 µg/m^3, reported at the HD bookshelf area, on Wednesday, June the 8th, 2005. It is important to consider that, besides having a rug, there are no air currents in this Level, and that the activity in the library is increasing at this time, because students are preparing their final exams; what makes the end of the school term.

The maximum RP concentration registered in the Library during the whole sampling period was reported in Level 3 (200 µg/m^3). We assume that this high RP concentration is produced by an accumulation of dust as a result of the bad cleaning of rugs, and because there are no air currents in this level that can provide adequate ventilation.

6 Recommendations

- To include a ventilation system capable of providing enough oxygen and to disperse the pollutants at the inside of the occupied spaces (dispersing of polluting particles emissions at the source).
- To do the cleaning of clothed furniture and rugs by means of a vacuum cleaner.
- To do the cleaning at a time when there is no personnel, neither users, present at the library.
- If possible, to get rid of the rugs and replace them with some other economical and easy to clean kind of floor.
- To replace the clothed furniture with a plastic one, to avoid dust accumulation.
- To avoid using ammonia-based cleaners and/or germicides, which contribute to Volatile Organic Compounds (VOCs).
- To use air deodorants, wax for furniture, or paint, the least as possible, since they are considered toxic aerosols.
- If the rug is not replaced, the cleaning should be done more frequently.

- In the case that a ventilation system cannot be acquired, we suggest to, at least, open a window at each Level, so that an air current can come through.

References

[1] A. Polidori, M. Arhami, C. Sioutas, R. J. Delfino and R. Allen. Indoor/Outdoor Relationships, Trends, and Carbonaceous Content of Fine Particulate Matter in Retirement Homes of the Los Angeles Basin. Journal of the Air & Waste Management Association, U.S. pp 366-379, March 2007.
[2] K. Zhu, J. Zhang and P.J. Lioy. Evaluation and Comparison of Continuous Fine Particulate Matter Monitors for Measurement of Ambient Aerosols. Journal of the Air & Waste Management Association, U.S. pp 1499-1506, December 2007.
[3] Wadden, A.R. "Contaminación del Aire en Interiores" LIMUSA 1987.
[4] SKC Inc. "Universal Flow Sample Pump Model 224-FCXR3. Operating Instructions". USA.
[5] Wark & Warner, "Contaminación del Aire, Origen y Control" LIMUSA, 2001.
[6] Heinsohn & Cimbala. "Indoor Air Quality Engineering", Marcel Dekker, Inc., New York, 2003.
[7] An Office Building Occupant's Guide to Indoor Air Quality; U.S. Environmental Protection Agency www.epa.gov/iaq/pubs/occupgd.html
[8] An Introduction to Indoor Air Quality; U.S. Environmental Protection Agency www.epa.gov/iaq/ia-intro.html
[9] The Inside Story: A Guide to Indoor Air Quality; U.S. Environmental Protection Agency www.epa.gov/iaq/pubs/insidest.html
[10] Indoor Air Quality, Basic Information; U.S. Environmental protection Agency www.epa.gov/iaq/is-build2.html

Impacts of ventilation: studies on "environmental tobacco smoke"

A. J. Geens, H. Al-Madfai & D. G. Snelson
Faculty of Advanced Technology, University of Glamorgan, Wales, UK

Abstract

A number of legislative bodies in Europe have already made or are currently considering making policy decisions on the issue of smoking in public places. Policy alternatives have been discussed in Town & Country Planning (2004 and 2008). Scientific evidence relating to this debate has been reported in a diverse range of publications such as the British Medical Journal, Indoor Air and the Chartered Institution of Building Services Engineers Journal. On inspection much of this reporting concludes negatively on the performance of ventilation systems. In this paper a critical review is undertaken of three "Environmental tobacco smoke" study papers, to supplement the overview provided by the authors in their paper in the International Journal of Innovative Computing, Information and Control (IJICIC) in 2007.
Keywords: ventilation, environmental tobacco smoke, environmental chamber.

1 Introduction

A number of legislative bodies in Europe have already made or are currently considering making policy decisions on the issue of smoking in public places. Policy alternatives have been discussed in Town & Country Planning [1]. Scientific evidence relating to this debate has been reported in a diverse range of publications such as the BMJ, Indoor Air and the CIBSE Journal. On inspection much of this reporting concludes negatively on the performance of ventilation systems [2–7].

In the UK the smoking ban has allowed a number of exemptions, and it is important that these spaces are ventilated using the best techniques available in order to protect both user groups and staff employed in these buildings for example residential care homes, hospices and mental health units where patients are held in secure conditions for more than six months [8]. The most immediate health and safety concern from smoking in this type of building is probably that of fire with the risk of smokers falling asleep in their rooms whilst smoking. This risk is reduced by providing a smoking room which is more easily monitored than individual rooms. The same strategy facilitates easier management of longer term health and safety concerns about the exposure of staff to environmental tobacco smoke (ETS). The use of ventilation to prevent migration of ETS through the building and to dilute ETS in the smoking room is more easily and economically managed if smoking is limited to one room. Ironically, many in the medical profession have dismissed the role of ventilation in limiting exposure to ETS in their campaign for the introduction of smoking bans, although this debate has highlighted the case that many hospitality venues do not use ventilation systems effectively, and that not all ventilation systems are equally effective. Ventilation systems are now being installed in hospitality venues to reduce smells that were originally masked by the tobacco smoke after the smoking ban came into force for example stale beer and food odours.

As a result of the negative reporting on ventilation in the debate leading up to the introduction of the ban, there is a possibility that the potential contribution from ventilation systems in managing such risks may be ignored. It would appear that the UK government unquestioningly accepted the argument that adequately ventilated rooms were not an alternative to a complete ban. Consequently it is now difficult for the government to offer advice to exempt building operators on how to ventilate their buildings to comply with Health and Safety requirements. Many of these buildings are government controlled and regulated.

2 Environmental tobacco smoke studies

To illustrate the dismissive behaviour towards the use of ventilation, three studies into environmental tobacco smoke are reviewed.

2.1 Impact of various air exchange rates on the levels of (ETS) components [9]

This is a report on experiments carried out in an environmental chamber. The chamber has a volume of 30 m^3. Measurements were taken at a number of air change rates. For this chamber these air change rates can be analysed as shown in Table 1. In other words the ventilation rate of the chamber at 2 air changes per hour was 16.67 l/s, adequate for 2 non smoking occupants. An experiment was conducted for air change rates of 0.2, 0.5 and 1, with 5 cigarettes being burnt in the chamber. A further experiment was conducted with 2 air changes per hour and 10 cigarettes being burnt. With 5 cigarettes being burnt in the hour long experiment, and allowing for 2 cigarettes per hour per smoker, this equates to 10 people in a room with 25% smoking room which according to CIBSE Guide B, Table 2.11, [10] requires 16 l/s/p or 160 l/s, and 1 air change per hour is 8.3 l/s.

The report states in its opening summary that changes in ventilation rates simulating conditions expected in residential and commercial buildings during

Table 1:   Analysis of chamber air change rates.

Air change rate (air change/h)	0.2	0.5	1	2
Supply rate (l/s)	1.67	4.17	8.3	16.67
Number of people (no smoking 8 l/s/p)	0.21	0.52	1.04	2.08
Number of people (some smoking 16 l/s/p)	0.1	0.26	0.52	1.04

smoking do not have a significant influence on ETS levels. The air flow rates in the chamber underestimate likely rates in a mechanically ventilated building by a factor of approximately 20. The only place that these ventilation rates and ETS levels are likely to occur is in a domestic dwelling (with only infiltration to dilute ETS) with 5 or ten cigarettes being smoked per hour in one room. One useful outcome of this study, although not commented on by the author is that the results show that all contaminants measured behaved in the same way, demonstrating Dalton's Law of Partial Pressures [11], negating the need to measure large numbers of different ETS markers. The key points from the paper are summarised in Table 2 below.

2.2 Environmental tobacco smoke exposure in public places of European cities [7]

This paper reports that nicotine levels are lower in no smoking areas than where smoking is permitted (see Table 3). In the abstract the authors argue that policies should be implemented that would effectively reduce levels of tobacco smoke in public places. The authors do not make any policy suggestions, but improved ventilation would substantially meet many of their demands/suggestions. Despite the scale of the study the authors make no strong conclusions and refer to the work as a pilot study pointing the way for further investigation. The key points from the paper are summarised in Table 3 below.

2.3 An international study of indoor air quality, ventilation and smoking activity in restaurants: a pilot study [12]

This paper offers an attempt at estimating ventilation rates and delivering a consistent methodology across a large number of studies, however there are a great many assumptions, and unnecessary variations in the methodology to be overly confident in the analysis and the findings. For example different cigarette counting methods were used in different locations. The key points from the paper are summarised in Table 4 below.

Table 2: Summary of Nebot et al [9] 2005 study.

Reference	Setting	Air tightness of building measured	Type of ventilation	Description of venues	Length of Measurements	Measured pollutants	Measured outdoor air quality	Weather conditions recorded	Number of active smokers recorded	Measured area/volume of venue	Findings	Remarks
Nebot et al. (2005)	Europe against cancer initiative - seven European cities: Vienna (Austria), Paris (France), Athens (Greece), Florence (Italy), Oporto (Portugal), Barcelona (Spain), and Orebro (Sweden). Public places that were sampled are an airport, train station, hospital, restaurant, university and disco. The study was carried out from October 2001 to October 2002.	Not specified	Yes: The data was recorded but not included in this paper	Sampling location and smoking policy	4 hours, 2 days, 7 days and 14 days	Nicotine vapour phase - ETS passive samplers (diameter 37 mm) comprising of a plastic cassette (with a windscreen on one side) containing a filter treated with sodium bisulfate. Samplers placed in both non-smoking and smoking areas. The samplers had to hang freely in the air, not be placed within 1 metre of an area where smokers regularly smoke, where air does not circulate, or under a shelf, or buried in curtains. The sampler used for personal samples had to be clipped to a shirt collar or lapel, with the windscreen facing away from the clothes. The study has a limitation regarding the placement of the samplers, which may result in differences between countries unrelated to actual exposure. The filters were analysed at the laboratory of the Public Health Agency of Barcelona, by gas chromatography / mass spectrometry (GM/MS) method.	No	No	No	Yes: The data was recorded but not included in this paper	In areas where smoking is prohibited, concentrations of nicotine are lower than in areas where smoking is allowed but they are not zero. The study showed that 22% of the samples had nicotine concentrations greater than 6.8 µg/m3; concentrations associated with lung cancer risk of one in 1000 assuming 45 years of working life, this is equivalent to the "significant harm" action level defined by the US Occupational Safety and Health administration. The results indicate that well implemented smoke-free polices are necessary to eliminate exposure to tobacco smoke in public areas.	In the abstract the authors argue that policies should be implemented that would effectively reduce levels of tobacco smoke in public places. The authors do not make any policy suggestions, but improved ventilation would substantially meet many of their demands/suggestions.

Table 3: Summary of Kotzias et al [7] 2004 study.

Reference	Setting	Air tightness of building measured	Type of ventilation	Description of venues	Length of Measurements	Measured pollutants	Measured outdoor air quality	Weather conditions recorded	Number of active smokers recorded	Measured area/volume of venue	Findings	Remarks
Kotzias (2004)	INDOORTRON walk-in type environmental chamber, temperature control (15-40 °C), relative humidity (20-90%) and air exchange rates 0.1-2 ach ("climate model"). Under non-controlled climatic conditions ("rising mode") air exchange rates can be increased up to 5 ach (air exchange rates per hour). Commercial smoking machine was used in these experiments. Cigarettes had nicotine content of 0.6 mg and tar of 7.0 mg.	Air exchange rates were set as the experiment was carried out in a walk-in environmental chamber	Dilution ventilation. Air exchange rates were determined by using a tracer gas. Stagnant air conditions for Experiment one 3 different ventilation rates 0.2, 0.5 & 1 and experiment two five different ventilation rates 0.5, 1, 2, 3.5 & 5. Homogeneity is 100% in climate mode and 75% in the rising mode.	Yes: walk-in environment chamber	100 minutes	Volatile Organic Compounds (VOC): benzene, toluene, pyridine, m+p-xylene, limonene and nicotine (1st and 2nd series of experiments at stagnant air conditions, 0.5, 1 and 2 ach). Carbonyl compounds: formaldehyde and acetaldehyde 2nd series of experiments at 0.5, 1 &5 ach). Inorganic gases: NOx (NO + NO2) and carbon monoxide (CO) all experiments. Air samples were taken at distinct time intervals to follow changes occurring in concentrations of the compounds formed during the burning of the tobacco. VOC concentrations were measured using TENAX TA tubes, analysis was done by thermal desorption and gas chromatography with Mass Selective Detector. Carbonyl compounds concentrations were measured using Sep-Pak DNPH-Silica cartridges.	No	No	Yes: Experiment 1 - five cigarettes (RH at 50%, temperature 23 °C) and experiment 2 - five cigarettes smoked simultaneously five times (RH at 50% (at 5 ach dropped down to 23%) and temperature 23 °C)	Yes (volume 30-m³)	The chemicals (volatile hydrocarbons, carbonyls, polycyclic aromatic hydrocarbons, inorganic gases and particles) emitted by smoking cigarettes are not rapidly and substantially eliminated from the indoor atmosphere, even when higher air exchange rates are applied. These results show that "wind tunnel"- like rates or other higher rates of dilution ventilation would be expected to be required to achieve pollutant levels close to ambient air limit values.	The report states in its opening summary that changes in ventilation rates in residential and commercial buildings during smoking do not have a significant influence on ETS levels. The air flow rates in the chamber underestimate likely rates in a mechanically ventilated building by a factor of approximately 20. The only place that these ventilation rates and ETS levels are likely to occur is in a domestic dwelling (with only infiltration to dilute ETS) with 5 or ten cigarettes being smoked per hour in one room. One useful outcome of this study, although not commented on by the author is that the results show that all contaminants measured behaved in the same way, demonstrating Dalton's Law of Partial Pressures, negating the need to measure large numbers of different ETS markers.
	Modelling of NOx and CO to show build-up and decay up to 120 minute experiment at different air exchange rates. An attempt was made to calculate at which air exchange rates CO and NOx concentrations reach levels comparable to those in ambient air [NO2: 200 ug/m³ (one hour), CO: 10 mg/m³ (8-hour average)].		Yes: modelling of conditions inside a walk-in environment chamber		Up to 120 minutes	Carbon monoxide (CO) and (NOx) A linear ODE (ordinary differential equation) was used to simulate mathematically the experimental setup. The concentration change of NOx and CO was attributed to emissions from smoking device, removal due to exchange and introduction of outdoor polluted air into the chamber (for the experiments in "rising mode"). No other sink source or sink terms for the two pollutants was considered. The assumption was that the chamber gases were well mixed.			Yes: The same emission rate was used to simulate both the first and second series of experiments, multiplied by 4 in the latter case.		Model and experimental data agree fairly well. The correlation coefficient between measured and calculated time series stays above 99% in all cases while the normalized bias is below 5% in all but one dataset.	

Table 4: Summary of Bohanon et al [12] 2003 study.

Reference	Setting	Air tightness of building measured	Type of ventilation	Description of venues	Length of Measurements	Measured pollutants	Measured outdoor air quality	Weather conditions recorded	Number of active smokers recorded	Measured area/volume of venue	Findings	Remarks
Bohanon et al. (2003)	Uniformed protocol used in 34 medium-priced restaurants where smoking took place in six countries (France, Korea, Japan, Switzerland, United Kingdom, and United States). Air samples and questionnaires were obtained during the lunch and/or dinner period on high occupancy days.	Not specified	Both natural and mechanical ventilated systems are mentioned in the paper but no further details are given. Two methods were used to determine the rate of outside air supply to the test space. An engineering estimation of air exchange rates from the available mechanical ventilation parameters and in-duct air flow measurements combined with counts of people in the space were used to estimate air exchange rates. The use of CO_2 to estimate restaurant ventilation rates is based on CO_2 exhalation at a rate of 31l/min-person.	Sampling location and questionnaire that sort basic information from the occupants. The questionnaire assessed the perception of indoor environmental conditions: noise, temperature, draft, odours humidity, freshness, environmental tobacco smoke, and overall IAQ and/or indoor air quality (IEQ). The questionnaires were normally distributed by the waiting staff after they took the orders for the meal.	In duplicate over 3 to 4 hours and in most cases over 1 to 2 days	A protocol was devised to obtain quantitative information on the environmental conditions within the restaurants. Respirable suspended particulate matter, ultraviolet particulate matter, fluorescing particulate matter, solanesol particulate matter, nicotine, 3-ethenylpyridine, carbon dioxide, carbon monoxide, temperature and relative humidity were measured in this study. Two sampling locations were identified in each restaurant. The sampling equipment was at least 50 cm from any walls, and located approximately at head height of a seated person. The sampling would not be influenced by fans or ventilation system or direct exposure to sidestream or exhaled mainstream smoke plumes. Particulate matter samples were collected using 37-mm filters with opaque filter holders to ensure the stability of the solanesol collected on the filters. Gas-phase samples were collected with XAD-4 cartridges.	Yes: Outdoor concentration of CO_2 for the estimated ventilation rates but no details in the paper	No	Yes - Number of smokers were estimated from an average count every 30 minutes. Two methods were employed, the first method was collecting and counting the cigarette butts about every 30 minutes. The second method was frequent periodic visual observations and tabulation of occupant smoking. Regardless of the method used, the number of cigarettes smoked per hour was estimated.	Yes: Volume of room for the estimated ventilation rates but no detail in the paper	It is not necessary to measure a large number of constituents to gain insight into ETS levels in a restaurants facilities. It is necessary to measure at least one constituent in each of the vapour and particle phases. The most beneficial measurements are nicotine or 3-EP and solanesol or Sol-PM. Carbon dioxide generated by occupants was found to be a viable tool to determine ventilation rates	This paper offers an attempt at estimating ventilation rates and delivering a consistent methodology across a large number of studies, however there are a great many assumptions, and variations in the methodology to be overly confident in the analysis and the findings.
	Indoor air modelling a simple model was used derived from the basic model, (ventilation estimation model) to calculate the steady-state concentrations of an indoor air pollutant. For 28 sessions in five Swiss restaurants nicotine and 3-EP concentrations were calculated.	Not applicable	(Q²F) becomes an effective ventilation rate	Not applicable	Not applicable	ETS yields of nicotine and 3-EP from cigarettes were taken to be 1585 and 334 µg/cig.	Not applicable	Not applicable	Yes - The number of cigarettes smoked per hour (cig/h).	Not applicable	The results appear to predict the measured concentrations, especially at higher concentrations. The simplicity and the potential uncertainties associated with the input parameters, the model predicts nicotine and 3- EP concentrations surprising well. Experimental errors in counting the number of cigarettes smoked are a potential factor. The study does demonstrate that there is significant potential for the use of carbon dioxide measurements to be used to estimate ventilation rates.	

3 Conclusion

In introducing smoking bans it can be argued that insufficient consideration has been given to the use of ventilation systems to control levels of environmental tobacco smoke or to provide segregation by pressurization / de-pressurization of zones. Effective use of ventilation is not straightforward and the evidence from the scientific community has not been helpful, however well intentioned and executed.

The summary of the Bohanon paper confirms the complexity of the problem, which is likely to deter the policy makers from further investigation, whilst the Nebot paper recommends further work, a point apparently overlooked by policy makers. The Kotzias paper provides technically concise and accurate findings, and it is unsurprising therefore that this paper is widely quoted as evidence that ventilation is ineffective in controlling environmental tobacco smoke. This is unfortunate, as although the Kotzias work is accurate and reliable, it was mainly testing using air exchange rates expected in non-mechanically ventilated buildings as those were the rates specified in the project brief.

It is perhaps unreasonable to expect policy makers to have spotted this simple but fundamental weakness in the experimental methodology in the past, but future decisions should now be better informed.

Acknowledgement

This study was commissioned by the Scottish Licensed Trade Association with funding support from the UK Tobacco Manufacturers' Association.

References

[1] Jones, P., Geens, A.J., Hillier, D. & Comfort, D., Smoking in public places. *Town and Country Planning*, **73**, pp. 328–330, 2004.

[2] Mulcahy, M. & Repace, J.L., Passive smoking exposure and risk for Irish bar staff. Proceeding of the 9th International Conference on Indoor Air Quality and Climate, Monterey California, 30 June -5 July 2002. *Indoor Air*, **2**, pp. 144–149, 2002.

[3] Carrington, J., Watson, A.F.R. & Gee, I.L., The effect of smoking status and ventilation on environmental tobacco smoke concentrations in public areas of UK pubs and bars. Proceeding of the 9th International Conference on Indoor Air Quality and Climate, Monterey, California, 30 June-5 July 2002. *Indoor Air*, pp. 495–499, 2002

[4] Carrington, J., Watson, A.F.R. & Gee, I.L., The effect of smoking status and ventilation on environmental tobacco smoke concentrations in public areas of UK pubs and bars. *Atmospheric Environmental*, **37**, pp. 3255–3266, 2003.

[5] Gee, I.L., Watson, A.F.R. & Carrington, J., The contribution of environmental tobacco smoke to indoor pollution in pubs and bars. *Indoor and Built Environment*, **14**, pp. 301–306, 2005.
[6] Gee, I.L., Watson, A.F.R. & Carrington, J., Edwards, P.R., van Tongeren, M., McElduff, P. & Edwards, R.E., Second-hand smoke levels in UK pubs and bars. Do the English Public Health White Paper proposals go far enough? *Journal of Public Health Medicine*, **28**, pp. 17–23, 2006.
[7] Kotzias, D., Geiss, O., Leva, P., Bellintani, A. & Arvanitis, A., Impact of various air exchange rates on the levels of Environmental Tobacco Smoke (ETS) Components. *Fresenius Environmental Bulletin*, **13**, pp. 1536–1549, 2004.
[8] Geens, A. J., Breath of fresh air. *Nursing Standard*, **22**, pp. 22–23, 2008.
[9] Nebot, M., Lopez, M. J., Gorini, G., Neuberger, M., Axelsson, S., Pilali, M., Fronseca, C., Abdennbi, K., Hackshaw, A., Moshammer, H., Laurent, A. M., Salles, J., Georgouli, M., Fondelli, M. C., Serrahima, E., Centrich, F. & Hammond, S. K., Environmental tobacco smoke exposure in public places of European cities. *Tobacco Control*, **14**, pp.60–63, 2005.
[10] The Chartered Institution of Building Services Engineers (CIBSE), Ventilation and air conditioning. CIBSE Guide B, The Chartered Institution of Building Services Engineers, London, 2001.
[11] The Columbia Encyclopedia, *Dalton's Law*. New York, Columbia University Press, 2001.
[12] Bohanon, H. R., Piade, J-J., Schorp, M. K. & Saint-Jalm, Y., An international survey of indoor air quality, ventilation, and smoking activity in restaurants: a pilot study. *Journal of Exposure Analysis and Environmental Epidemiology*, **13**, pp.378–392, 2003.

PCB contamination in indoor buildings

S. J. Hellman[1], O. Lindroos[1], T. Palukka[1], E. Priha[2], T. Rantio[2] & T. Tuhkanen[1]

[1]*Tampere University of Technology, Institute of Environmental Engineering and Biotechnology, Finland*
[2]*Finnish Institute of Occupational Health, Finland*

Abstract

PCBs can still be found in open applications such as additives in paints or elastic sealants used in buildings built in the 1960s and 1970s. The objectives of this study were to analyze the occurrence of PCBs in paints used in buildings. PCBs were applied especially in chlorinated rubber paints, cyclorubber paints and in vinyl paints, which have been used widely, especially in industrial buildings. The results from the research show that 60% of buildings studied contain PCB over 20 mg/kg in indoor paints. From all the samples collected almost every fourth contained PCBs over the limit value of 50 mg/kg given for hazardous waste. The highest concentration (102 900 mg/kg) was found in the concrete floor of a school building's basement. Wipe samples taken from paint surfaces also contained high amounts of PCBs. The highest amount (83 000 µg PCB/m^2) was found in the surface of a building, which had not yet been renovated. The most common method for removing existing paint is sandblasting. After sandblasting the sand contains high concentrations of PCBs and has a large surface area thus PCBs leaching capacity may be significant. This may cause occupational and also residential exposure to PCBs. In buildings that had been renovated by using sandblasting for paint removal, PCBs were found in the surface, which had not originally contained PCBs. The secondary contamination was often over the Finnish limit value of 100 µg PCB/m^2 as the highest concentration detected was 1100 µg PCB/m^2. The congener profiles from the samples reminded the profile of Aroclor 1260 or Clophen 60 profiles in almost all samples. The profiles showed that the PCBs most probably originate from paints and not from other PCB source such as capacitors.
Keywords: polychlorinated biphenyls (PCB), paint, indoor dust.

1 Introduction

Polychlorinated biphenyls (PCBs) are a class of Persistent Organic Pollutants (POPs) that are highly toxic, resistant to degradation and have a high bioaccumulation potential. PCBs are transported through air, water and migratory species across international boundaries and deposited far from their place of release, where they accumulate in terrestrial and aquatic ecosystems (Gjessing et al [1]). The commercial production of technical mixtures of PCBs with different degrees of chlorination started in 1929. PCBs were soon used for various purposes; in closed systems in industrial applications such as heat exchange fluids in electrical transformers and capacitors (liquid PCBs) but also in various open applications such as softeners or additives in paints, sealants and varnishes (non-liquid PCBs). NLPCBs in buildings pose a possible risk from a residential, occupational and also an environmental point of view. The amount of PCBs produced worldwide from 1929–1977 is estimated to be about 1.5 million metric tons (Breivik et al [2]). The serious risks associated with PCBs were noticed more than 30 years ago when Sören Jensen detected PCBs in pike from Sweden (Jensen [3]). Since then their use began to be restricted but PCBs are still found in use in previous applications.

In Finland PCBs can be found in buildings constructed from 1956–1975. PCBs are also found in soil in the vicinity of buildings (Hellman [4]). The use of PCBs in new building materials was banned in 1989 (VNp, 1071/1989 [5]). Even though the production of PCBs was stopped 30 years ago, PCBs continue to be detected in environmental samples and in the built environment around the world. Estimates of the volume of PCBs in sealants in Finland vary from 130 t to 270 t (Pentti and Haukijärvi [6]). The amount of PCBs used in paints and varnishes is approximately six times the amount of PCBs used in sealants (PCB-committee report [7]).

In this study PCB containing materials investigated were paints in various applications. Also paint dust caused by aging of paints or sandblasting was examined. The objective was to find out to what extent PCBs exist in buildings built before the year 1975. The painted surfaces such as floors, walls and staircases were studied to find out where PCB containing paints were applied. The indoor air of industrial buildings as well as school buildings was sampled in order to find out about the presence of PCBs in the indoor air in different types of buildings.

The study adds up to the studies conducted earlier in Finland about PCBs in buildings [4, 8–10].

2 Materials and methods

2.1 Sampling targets

Samples were collected in different types of industrial and official buildings primarily according to their year of construction. Buildings were mostly located in the Tampere region and were built before the year of 1970. Some targets were undergoing renovations as they were constructed for residential use, for day-care centers or for offices. Firstly the industrial buildings and warehouses were studied. After that the sampling was extended as schools and military buildings were studied. Most of the paint and surface samples were taken from concrete floors but some were also taken from doors, doorsteps and walls.

All together 114 material samples were collected from 23 different buildings (8 schools, 7 old industrial buildings, 8 buildings used for military purposes). 21 surface wipe samples were taken from four different industrial buildings. Indoor air was sampled in ten targets, where 18 volatile samples and 14 particle-bound samples were taken. Also the penetration of PCBs from paint into the concrete underneath was studied by three drilling samples.

2.2 Sampling methods

The material samples of paints were taken by using metallic knife, scraper of chisel. The tools were cleaned after each sample with ethanol. The samples were packed in capped brown glass bottles and taken to the laboratory at the Tampere University of Technology (TTY). For wipe samples 1 m x 1 m areas were marked as sampling locations. Two 20 cm x 20 cm areas were measured inside the marked sampling area. Clean cotton tissue was moisturized with 10 ml of ethanol and each area was wiped twice with the same cotton. Cotton was then sealed in a glass tube and taken to the Institute of Occupational Health in Tampere (FIOH) for analysis. The air samples were collected with OVS tubes containing XAD-2 sorbent and a glass fiber filter. Suction velocity in pumps was about 21 l min^{-1}. Sampling time was about 2 hours. For drilling samples three holes were drilled to the surface, which was known to have PCB concentration of 23 000 mg/kg. Before drilling the paint was removed carefully. Upper depths were from surface to 2.5 mm, 3.5 mm and 5 mm. The deeper wholes were drilled further up to 4,5 mm, 6 mm and 7.5 mm respectively.

2.3 PCB-analyses

Paint samples were extracted with hexane in an ultrasonic bath using 2,4,6-trichlorobiphenyl (PCB 30) as an internal standard. For wipe samples 2,2′,3,3′,4,4′,5,5′,6,6′-decachlorobiphenyl (PCB209) was also added as an internal standard. The extracts from paint and wipe samples were cleaned up with concentrated sulphuric acid as needed. PCBs were determined by using Agilent Technologies gas chromatograph (6890) with an electron capture detector or mass selective detector (5971A or 5973). HP-5MS capillary column (30m long, 250µm internal diameter and 0.25µm of 1µm film thickness) was used with helium as a carrier gas (1ml/min for MSD, 2 ml/min for ECD). Oven temperature at TTY started at the temperature 80°C, was held for 1 minute and then increased to 150°C at 30 °C/min, to 250° at 5°C/min and to 300°C at 30°C/min. In IHO starting temperature was 70°C, held for 1 minute and then increased to 250° at 15°C/min where hold for 20 minutes. The total PCB concentration in paint and wipe samples was calculated according to the PCB profile found in a sample. Sample profiles were compared to the profiles of technical mixtures Aroclor 1254 or Aroclor 1260. The individual PCB congeners used for quantification were 28, 31, 52, 77, 101, 105, 118, 126, 128, 138, 153, 156, 169, 179 and 180 as their concentrations were determined. For air analysis both volatile and particulate PCBs were measured. Volatile gaseous phase PCBs were collected by using XAD-2 sorbent equipped with a Reciprotor-pump. The suction velocity was between 3-5 l/min and sampling time was 24 hours. Particulate phase PCBs were collected by using high-volume dust collector with a glass fiber filter (e.g. Staplex TFAGF41, diameter 10,16cm), suction velocity was about 0.7 m^3/min and sampling time about an hour. XAD-2 sorbent and glass fiber filters were analyzed together. The total PCB concentration in air samples was calculated by using the method presented by Benthe *et al* [11]. The method adds up the concentrations of PCB congeners 28, 52, 101 and 138 and multiplies the sum with six. To assure the quality during the analysis, internal standards in samples, blank samples and calibration mixtures together with blank samples were used.

3 Results

3.1 PCBs in paints

Target buildings were chosen according to the year of construction or known reconstruction time so that the buildings were painted 1940–1972. The paint samples were collected from 8 schools (45 samples), 7 buildings known for previous industrial use (49 samples) and two military areas (20 samples). From 18 locations sampled, 11 sampling sites were found to contain PCBs over 20 mg/kg. From 114 samples 33 had PCB-concentrations over 20 mg/kg. The percentages of paint samples containing PCBs more than 20 mg/kg are summarized in table 1. The highest concentration (102 900 mg/kg) was analyzed at a schools boiler room where the sample was taken from the painted concrete floor. Over 10% of the dry weight of the paint was shown to consist of PCBs. Although, this was not the case in most of the samples the concentrations were still close to the regulatory limit value of 50 mg/kg for hazardous waste. The limit value was exceeded in 23 samples. Most of the samples that had high PCB concentrations were painted during 1950s.

Table 1:   Percentage of paint samples containing more than 20 mg/kg of PCB in different building types.

Building type	%
School	31
Industrial	24
Military	35
Total	29

Most of the high concentrations were on concrete basement floors in elementary school buildings. The congener profiles resembled mostly of the profile of Aroclor 1260 or Clophen 60. Aroclor 1254 was found in few paint samples.

3.2 PCBs in surfaces

Wipe samples were taken from painted surfaces known to contain PCBs. Seven targets were sampled and five of them exceeded the guideline limit of 100 $\mu g/m^2$, which is given for the control of occupational exposure surfaces in Finland (PCB-committee report, 1983). The highest surface concentration analyzed with known paint concentration was 75 000 $\mu g/m^2$. It was the spot that had the highest paint concentration 102 900 mg/kg as well so it indicates that PCB concentrations in dust correlate with PCB concentrations in paint. Results are shown in table 2.

Table 2: PCB-concentrations in surfaces vs. PCBs in paints.

PCB-concentration in surface ($\mu g/m^2$)	PCB-concentration in paint (mg/kg)
120	70
68	70
44	530
92	530
2 600	48 000
2 100	23 600
25 000	29 000
660	11 000
< 8	< 20
230	280
59 000	69 300
10 000	69 300
75 000	**102 900**
5 900	69 300
380	400
2 000	1 990
4 100	1 990

Samples taken from the vicinity of PCB-containing paint showed secondary contamination, which is remarkable since the cleaning may not be done properly after renovations, if the presence of PCBs is unknown.

3.3 PCBs in air samples

Ten targets were sampled; 18 air samples for volatile PCBs and 14 particle phase air samples for particulate PCBs. From 18 samples in six volatile PCBs were found to exceed the detection limit. The concentrations of particulate PCBs were detected in three samples out of 14. Summary of the samples is shown in table 3.

Table 3: Volatile (left) and particulate (right) PCBs in the indoor air samples.

Air volume (m^3)	PCB (ng/m^3)	Air volume (m^3)	PCB (ng/m^3)
4,5852	< 50	58,50	< 4
4,6100	< 50	50,03	< 5
4,5319	< 50	45,68	< 5
6,1280	38	45,75	< 5
6,1820	33	46,40	< 5
0,6955	< 100	36,00	< 7
1,0993	< 60	48,68	< 5
4,7580	56	38,75	18
4,7920	39	34,80	5,5
5,6308	350	60,18	< 4
4,1122	330	41,85	< 6
6,2097	< 40	45,68	< 5
5,1665	< 50	45,00	5,3
2,8819	< 80	39,00	< 6
2,6197	< 90	41,85	< 6
0,5270	290		
5,8950	< 40		
0,7547	320		
0,5270	290		
5,8950	< 40		

3.4 PCBs in concrete

The penetration of PCBs from paint into the concrete underneath was studied by drilling samples from concrete surfaces. It was shown that PCBs penetrate into concrete in concentrations that are significant if thinking about the reuse of concrete.

4 Discussion

This study shows that PCBs have been added to various paints, which have been applied in various types of Finnish buildings, mostly indoors. The concentrations in paints exceed often the hazardous waste limit value of 50 mg/kg. In Finland there is no legislation that especially requires testing old paints before renovation or demolition of buildings. Though, it is said that the construction company need to be aware of the safety requirements during construction work. If paints containing PCBs are removed by using sandblasting, it is evident that without sufficient knowledge there may be health risks associated with e.g. poor safety gear and the technique used. Also the question about demolition waste becomes important. This study suggests that demolition waste from buildings built and painted especially in late 1950s may contain significant amounts of PCB. If present, PCB concentrations in paints most probably will exceed the given limit value for hazardous waste and should be treated as hazardous waste.

There was no clear trend between the PCB concentrations in paints and the wipe sample concentrations in surfaces. An interesting thing was shown in a school building, as the surface concentrations were order of magnitude lower compared to the concentrations of other surfaces versus paint. The samples were taken from the floor, which was painted in 1951. Also another target showed the same trend as the painting year was 1954. This suggests that PCBs had been eroded away from the floor paint. This may be due to the cracking and erosion of an old paint.

Another question is the concrete under the painted surface. The limited investigations in this study suggest that PCBs pass the paint surface and are found also in underlying concrete. The results suggest that this should be further examined when testing concrete from demolished buildings for reuse purposes.

As the results show more attention should be paid on paint removal techniques as well as the safety requirements. The attention should be paid also to secondary contamination. While sandblasting is the most commonly applied technique in paint removal, special attention should be paid in order to meet the safety requirements and also to prevent the exposures of workers (cf. exposure to lead or asbestos). Also the cleaning, especially after the paint removal, is important. Especially, if buildings are renovated from industrial to residential use, the clean-up is important in preventing the PCB exposure. Further the risk assessment is to be done by using three types of scenarios: residential, occupational and environmental.

References

[1] Gjessing, E.T., Steiro, C., Becher, G. and Christy, A. Reduced analytical availability of polychlorinated biphenyls (PCBs) in colored surface water. *Chemosphere*, 2006, doi:10.1016/j.chemosphere.2006.07.086.
[2] Breivik, K., Sweetman, A., Pacyna, J. M. and Jones, K. C., 2002. Towards a global historical emission inventory for selected PCB congeners – a mass balance approach 1. Global production and consumption. *The Science of the Total Environment.* Vol. 290, pp. 181-198, 2002.
[3] Jensen, S. Report of new chemical hazard. New Scientist. Vol. 32, p. 612, 1966.

[4] Hellman, S. Saumausmassojen PCB-yhdisteet elementtitalojen pihamaiden ongelmana. MSc Thesis. Tampereen teknillinen korkeakoulu, Tampere. 104 p, 2000.
[5] VNp, 1071/1989. Decision of the Council of State 1071/1989. Government decision on restricting the use of PCBs and PCTs. Ministry of Environment, Finland.
[6] Pentti, M. and Haukijärvi, M. Betonijulkisivujen saumausten suunnittelu ja laadunvarmistus. Tampereen teknillinen korkeakoulu, Talonrakennustekniikka. Publication no 1000, 2. edition. Tampere. 88 p. + app. 2000.
[7] PCB-committee report, 1983. *Committee report 1983:47*. Government printing center. Helsinki. Finland.
[8] Pyy, V, and Lyly, O. PCB elementtitalojen saumausmassoissa ja pihojen maaperässä [PCB in apartment building sealants and soil]. Summary in English. Helsinki City Environment Centre, Helsinki, Finland, 1998.
[9] Hellman, S. and Puhakka, J. Polychlorinated Biphenyl (PCB) contamination of apartment building and its surroundings by construction block sealants. *Geological Survey of Finland*, Special Paper Vol. 32, pp. 123–127, 2001.
[10] Priha, E., Hellman, S. and Sorvari, J. PCB contamination from polysulphide sealants in residential areas – exposure and risk assessment. *Chemosphere* Vol. 59, pp. 537–543, 2005.
[11] Benthe, C., Heinzow, B., Jessen, H., Mohr, S. and Rotard, W. Polychlorinated biphenyls in indoor air due to Thiokol-rubber sealant in office buildings. *Chemosphere* 25, pp. 1481–1486, 1992.

Evaluation of Indoor Air treatment by two pilot-scale biofilters packed with compost and compost-based material

M. Ondarts, C. Hort, V. Platel & S. Sochard
Université de Pau et des Pays de l'Adour, Laboratoire de Thermique, Energétique et Procédés, Département Génie des Procédés, Equipe "Traitement des effluents gazeux" Site de Tarbes, France

Abstract

The interest in Indoor Air Quality (IAQ) has increased this past decade. Indeed, many studies focused on Indoor Air analysis have pointed out many potential health risks and non-negligible associated costs. But only a few studies have dealt with the adaptation of industrial processes like sorption or photocatalysis and with development of new processes. Moreover, the use of these processes is still limited by the characteristics of this pollution: lots of components with different properties (Volatile Organic Compounds (VOC), aldehyde, inorganic compounds), competition phenomena between these pollutants, and low concentrations.

Biofiltration is currently used to treat high flow rate effluents with low concentrations and various pollutants, so this technology seems adapted for IA treatment. This study focuses on the evaluation of the performance of biofilters packed with compost and compost-based material (a mixture of compost/activated carbon). Indeed, compost is a natural material that possesses a lot of microorganisms, as well as good physical properties (water retention, pH) and nutriment content. The model effluent is constituted of ten compounds (aldehyde, aromatic, chlorinated, inorganic...) at low concentrations (sub-ppmv), which have been chosen for their ubiquity in indoor environments, their different physical and chemical properties (solubility, vapor pressure, biodegradability) and their potential health risks in chronic exposures.

The pilot scale is constituted by a gas generator (permeation module) feed with high quality zero air and two biofilters with compost or compost/activated carbon (AC) based-bed material. VOCs are analysed by Gas Chromatography/Mass Spectrometry coupled with a cryogenic preconcentrator.

The first part of this study has demonstrated the possibility of generating a continuous sub-ppmv pollutants mixture. A simple analysis method that demonstrates this adaptation to this range of concentration is presented.
Keywords: Indoor Air, VOCs, biofiltration, sub-ppmv level, model effluent.

1 Introduction

During the past decade, the interest in Indoor Air Quality (IAQ) has increased. Indeed, the progress in analytical chemistry methods and the multiplication of Indoor Air (IA) studies as RIOPA (Relationship between Indoor, Outdoor, Personal Exposure), NHEXAS (National Human Exposure Assessment Survey) in United States or Expolis in Europe have allowed an increase in knowledge of IA pollution. Indoor Air pollution presents two specificities. On one hand, IA pollution is constituted by a great number of various pollutants: few hundred compounds, including hydrocarbon compounds as alkanes, aromatic compounds, chlorinated compounds as trichloroethylene (TCE), pesticides, or inorganic compounds as ammonia, nitrogen dioxide, were detected in large studies as NHEXAS [1] or Expolis [1]. On the other hand, concentrations of these pollutants are generally very low and vary between 0.1 $\mu g\ m^{-3}$ to 100 $\mu g\ m^{-3}$ for the most concentrated [1–3]; these levels are higher than in outdoor: the ratio between indoor and outdoor pollutant concentrations is generally greater than 1 and can reach 20 or more [4]. There are lots of pollution sources: typically, building materials and use of cleaning products are the main sources [4]. Use of pesticide (acarina treatment, wood treatment), use of combustion apparatus and tobacco smoke are additional sources which can be specific from one country to another, and specific of practices of each country too [5].

Since the time spent in the indoors is close to 80%, the chronic exposure to IA pollutants is the major source of exposure to pollution for human being. Moreover, different studies have observed or proposed relations between some indoor pollutants and some diseases: lung cancer could be linked to radon exposure, others cancers to pesticides, disorders of the reproduction to glycol ether and, at last, Sick Building Syndrome could be linked to multiple Volatile Organic Compounds (VOCs) [6]. Thus IAQ becomes a public health matter priority for public health organization as World Health Organization (WHO), United States Environmental Protection Agency (US-EPA) or governmental action as National Health Environment Plan in France [7].

Three ways can be investigated to remove IA pollution:
- suppression of pollutants sources: for example bio-materials, like tannin resin, are studied to replace formaldehyde in particle boards but this strategy cannot be applied to all buildings;
- increase the ventilation rate to dilute the pollution: this solution is not compatible with energy policy;
- use processes to remove pollutants.

Processes studied for IA treatment are current industrial processes that have to be adapted to IA pollution specificities. Indeed, concentrations are 1000 times lower than classical concentrations found in industrial effluents. Moreover, industrial treatments are adapted to specific chemical and physical properties, and so are limited for the treatment of various pollutants. Despite the present poor number of studies about IA treatment, many of classical processes have been studied in this aim. On one hand, a part of processes are based on the transfer of pollutants to another phase, which include adsorption [8], absorption

[9], membrane process [10] and electrostatic separation [11]. Most of the studies and present commercial cleaners are based on adsorption process, most often on AC [8]. On the other hand, some processes are based on the degradation of pollutants, generally by oxidation, which include catalytic oxidation [12], UV photocatalysis [8], plasma and ionisation technologies [13], phytoremediation [14] and bioprocesses [15, 16].

All processes can present good efficiency, for example phytoremediation and plasma technology can respectively remove 75% and 50% of total VOCs (TVOC) [14, 13], UV photocatalysis can treat 83% of inlet nitrogen oxide [17].

However, many limitations were found during these processes studies. The first limitation, for treatments based on pollutant transfer is the necessity to regenerate the cleaning phase or to change it [8]. The fact that most of these processes are selective constitutes a second typical limitation. For example, adsorption on activated carbon is not adapted for compounds with low molecular weight as formaldehyde, which is a ubiquitous, and toxic compound in indoor. Membrane process presents by definition selectivity due to the choice of the membrane (hydrophobic or hydrophilic, pore size). The most observed others limitations are [8]:

- competition or inhibitory effect between different components;
- inhibition in presence of humidity;
- creation of by-products more dangerous or generating inhibition of the process.

The bioprocess has been studied for IA treatment since 1980 with Wolverton's work [16]. This work deals with potted plant biofilter for 3 compounds: formaldehyde, TCE, toluene. The efficiency of the process, which depends on the used plant, is closed to 80% for toluene. More recently, Darlington studied a trickling biofilter and observed great removal for aromatic compounds at low concentration [15].

In fact, biofilter process has properties that seem to be adapted to IA pollutants. First, lots of compounds present in indoor are successfully treated by biofiltration: efficiency closed to 100% can be obtained for biodegradable compounds as alcohol, acetate and efficiency from 40% to 100% can be obtained for more recalcitrant compounds like chlorinated compounds, aromatic or terpene [18]. Moreover, the biofiltration have been used for the treatment of waste air from publicly owned wastewater, which is effluent with low concentrations of pollutants [18]. This process is recommended for the treatment of large flow rate with low concentration [19].

Biofiltration is also a technology that is friendly for environment without chemical add products and which is relatively cheap. Thus biofiltration that seems to be an adapted technology has been chosen for IA pollutants removal. In order to evaluate this technology, the determination of the most representative experimental conditions is required: the composition of the model effluent, the concentrations of its pollutants and the flow rate have to be discussed. This paper describes the establishment of a pilot-scale, which allows one to obtain these conditions.

2 Protocol

With the expansion of the IA cleaners market, especially for particular pollution, some organizations have established normalisations in order to test and qualify these cleaners. Different protocols can be differentiated according to the nature of the target pollution: particular pollution, biological pollution or chemical pollution. Attention is largely paid to particular pollution. Only a Japan protocol, JEM 1767, has been established and adopted as far as we know for chemical pollution [20]. It deals with odorous compounds linked to tobacco smokes. Two other protocols has been proposed but not adopted yet by the ASHRAE (American Society of Heating, Refrigerating and Air-Conditioning Engineers) [8] and the association EDF/CETIAT [20]. Table 1 presents the principal characteristics of these 3 protocols.

Table 1: Normalizated and non-normalizated protocol for chemical IA cleaners.

Organisation	Date	Configuration test	Target Compounds	Efficient criterion
JEM 1467 Japan	March 1995	chamber	ammonia acethaldehyde acetique acid	Lifespan
ASHRAE USA	2005	chamber	16 VOCs	CADR[a]
EDF/CETIAT France	May 2006	open system	toluene	CADR

a :Clean Air Delivery Rate

Three main differences can be noted in these 3 methods: efficient criterion, configuration of the test chamber and model pollutants. Two different efficient capacity criteria are used. Japan's protocol uses a criteria based on the lifespan of the cleaner calculated with a series of tests made until cleaner saturation is reached. The lifespan is expressed as a function of the number of cigarettes whose smoke can be treated. The 2 others protocols use the "Clean Air Delivery Rate" (CADR), product of flow rate (Q) and efficiency (RE) of the cleaner. The last one seems to be more adapted to different compounds.

The choice of model pollutants is a critical point for a good qualification of the cleaner. Indeed, all processes can have a good efficiency with compounds that present a good affinity for the process. However, one of the difficulties encountered in IA treatment is the presence of lots of chemical families with various properties.

This point is clearly proposed by ASHRAE's protocol: 16 compounds with different properties (alkanes, TCE, aromatic) are taken as model components.

Usually, only few compounds are chosen in studies dealing with IA treatment; a low number of studies use more than three compounds as model pollutants.

We have chosen to take 10 compounds into account in our model effluent. The compounds usually chosen are formaldehyde and toluene as aromatic compound. Indeed, these compounds are ubiquitous in the indoors and their toxicity is clearly established. These two points constituted the first part of our choice of model compounds. Moreover, the various chemical and physical properties of the 10 pollutants have been taken into account to evaluate the capacity of biofiltration to treat a broad range of pollutants. This latest point is necessary to answer to the typical problem of IA pollution: multi-exposure.

Public health organizations have already established lists of priority compounds in indoor based on ubiquity and toxicity of pollutants. These lists are in agreement with criteria retained for the constitution of the model effluent we will generate. Indeed, three organizations lists were retained to establish the model effluent: WHO's [21] US-EPA's lists [22] (reference organizations for public health matter); and OIAQ list (Observatory of Indoor Air Quality) in order to take into account the national specificities of indoor air pollution [23].

The IA compounds properties can be very different from a component to another: the Henry coefficient of limonene is 10^6 times higher than this of nitrogen dioxide, the biodegradability of 1-butanol is very good whereas TCE is known to be a xenobiotic compound [19].

Moreover, compounds concentrations are very different. Final concentrations are not obviously defined. Actually, only a few normalizations define good IA quality characteristics; only radon, for chemical pollution, is concerned by normalization. Table 2 presents some selected compounds median indoor concentrations in a French study and some toxicological reference concentrations recommended by WHO, US-EPA, ATSDR (Agency for Toxic Substances and Disease Registry) and Canadian directive.

Table 2: Medium IA concentration in France [23] and toxicological concentrations.

Priority Compound in Indoor	Concentrations mg m^{-3}	Toxicological Ref.[a] mg m^{-3}
Formaldehyde	$2.40.10^{-2}$	$1.0.10^{-2}$ ATSDR $1.0.10^{-1}$ Canada
Toluene	$1.56.10^{-2}$	$4.0.10^{-1}$ EPA $3.0.10^{-1}$ ASTDR
TCE	$8.60.10^{-4}$	$2.0.10^{-2}$ ASTDR
NO$_2$	$4.30.10^{-2}$	$4.0.10^{-2}$ OMS $1.0.10-2$ Canada
Dichlorvos	$4.55.10^{-4}$	$5.0.10^{-4}$ ATSDR $5.0.10^{-4}$ EPA

a: Chronic exposition

All the concentrations of the different compounds do not behave in the same manner with regard to toxicological threshold: some medium concentrations are lower than toxicological concentrations and some medium concentrations are higher than recommendations. However, only the Canadian directive includes the multi-exposure indoors. In this directive, the recommended value for formaldehyde depends on the presence and on the concentration of acetaldehyde and acroleine. Any other toxicological concentration is defined for multi-exposure. Owing to this fact, and as it's recommended by NIOSH (National Institute for Occupational Safety and Health) for carcinogenic compounds, the treatment have to decrease the indoor pollutant concentrations as low as possible.

The last point discussed in indoor cleaner protocol is the configuration of the test unit. Chamber test is currently used and has been retained in JEM protocol and by the ASHRAE. However, the results can be strongly influenced by the sample point in the chamber. Moreover, for low pollutant concentrations, high sample volume is required and so can disturb the system. Open system with continuous feed seems to be more adapted and simplify the analytic part.

3 Installation of pilot bench scale

3.1 Design of the biofilter

Despite the recent progress in the comprehension of phenomena involved during biofiltration and so despite recent development of new biofilter models, the design of biofilter is still often empirical since these phenomena and then these models are very complex. "Empty Bed Time Retention" (EBRT) is usually used to design biofilter. Recommended EBRT varies between 20 s and 60 s respectively for low and high pollutant concentration [19]. Moreover, Darlington observed that elimination capacity of his bioprocess for indoor air treatment increases with the increase of the flow rate. The mass loading is then increased and tends towards classical mass loading in biofiltration. So, biofilter is designed for an EBRT of 20 s. The second parameter is the depth of the bed. The recommended depth varies between 0.5 and 2 m [24] to minimize the pressure drop and the compaction of the bed. For the pilot bench scale, depth of the bed is 0.5 m. The flow rate adapted to this depth, to the diameter of the column, and to the EBRT is 10 l mn^{-1}.

The pilot bench scale can be divided into 3 parts:
1. generation of model effluent;
2. biofilter;
3. analytical apparatus.

3.2 Effluent generation

The concentration of each component varies between 30 µg m^{-3} to 100 µg m^{-3}. This effluent is produced by a calibration gas generator (PUL200, Calibrage society, France) constituted with 2 permeation ovens with capacity of 5 components by oven. The range of concentration generated by the gas generator, with accuracy about 5% (according to the pollutant), is close to hundreds µg m^{-3}. The generator is fed by zero air with maximal concentration in carbonate compounds about 50 ppbv. In order to decrease the effluent concentration, a system enabling the dilution of the effluent (ratio 1:6) on one hand and the split of the gas generator outlet (ratio 1:10) on the other hand has been implemented. All regulations are performed with Brook Instrument flow meter with an accuracy of 1% of the flow range.

3.3 Biofilter design

Two biofilter supports are tested for this application: compost and mixture of compost and activated carbon (AC). Compost is a natural support known for its good degradation capacity: it presents natural microorganisms and nutriments content [19]. Moreover, compost is relatively cheap and allows waste valorisation. The mixture compost/AC has already been studied for its good removal capacity and its bulking agent property [19].

The two Biofilters are two columns of 100 mm inner diameter in borosilicate glass, which is a material with low adsorption capacity. Some distribution Teflon meshes are arranged at different levels of the biofilter in order to avoid gas preferential paths. A bubbler unit enables water saturation of the effluent. All pipes are in Teflon (pfa), which is inert.

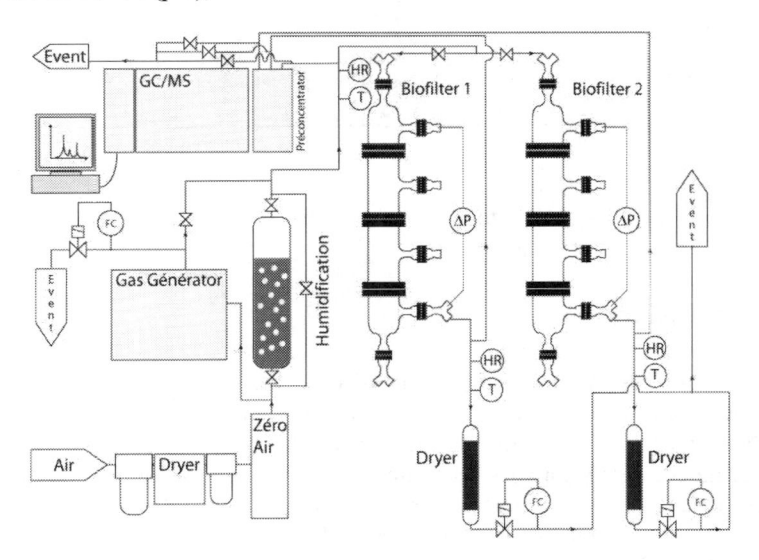

Figure 1: Schematic diagram of biofiltration unit.

3.4 Analytical apparatus

VOCs are analysed by Gas Chromatography/Mass Spectrometry (GC/MS, Trace MS Plus, Thermoelectron SA). First a cryogenic preconcentration removes water and carbon dioxide, which are responsible for interference in the analysis, and concentrates the effluent at the same time. Preconcentration is necessary because the MS sensibility is too low. The temperature program applied was 35°C during 5mn, 8°C until 100°C and 15°C until 200°C. The detection mode applies is Single Ion Monitoring (SIM).

Figure 2 shows analytical result for a mixture of toluene, acetate butyl, limonene, undecane, TCE.

The nitrogen dioxide is analysed by an apparatus based on chemiluminescence with a detection limit about the ppbv level.

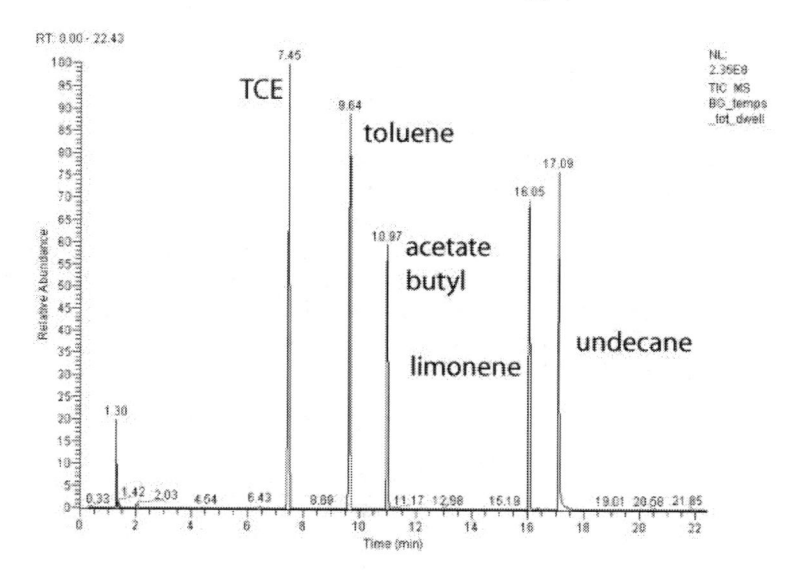

Figure 2: Typical chromatograph of effluent with concentration of 100 μg m^{-3} for each compound.

4 Conclusion

The focus on IAQ, which started in the last decade, has increased in recent years. On one hand, public health organizations are more and more concerned with IAQ. On the other hand, lots of institutions as governments or industries focus on IAQ. Indeed, the poor Indoor Air Quality has been pointed out in numerous European studies and the implementation of treatment solutions seems to be unavoidable. Note that space industry is also concerned since the IA of cabin has to be treated. Moreover, the same problem holds for car industry. At the present time, treatment processes are not enough efficient: processes are always limited by inhibition, selectivity or by-products problems. Moreover, the present knowledge is limited by tests with poor number of pollutants with high concentration with regard to real IA constitution.

This work has proposed a protocol based on the constitution of a representative effluent: the pollutants were selected for their occurrence in indoor, their toxicity and their different chemical and physical properties. The concentrations of pollutants are relatively close to IA concentrations. These different aspects have needed original technical solutions, which have been described. The generation of effluent with low concentration has been done. The analysis apparatus has been optimized and is efficient for sub-ppmv analysis.

The biofiltration technology seems to be adapted to IA treatment. Biofiltration is currently recommended for treatment of large effluent flow rate with low concentration of pollutants as shown in some past studies. The biofilter has been designed in agreement with literature and seems to be adapted.

Acknowledgements

This work was supported by ADEME (Environment and Energy Control Agency, France) and Hautes Pyrénées General Council, France.

References

[1] L. Mosqueron and V. Nedellec. Revues des enquêtes sur la qualité de l'air intérieur dans les logements en europe et aux etats-unis. Ethnical report, Observatoire de la Qualité de l'Air Intérieur, 2004.

[2] C.P. Weisel, J. Zhang, B.J. Turpin, M.T. Morandi, S. Colome, T.H. Stock, and D.M. Spektor. Relationships of indoor, outdoor and personal air (riopa), part i. collection methods and descriptive analyses. Technical report, Health Effects Institute, 2005.

[3] K. Saarela, T. Tirkkonen, J. Laine-Ylijoki, J.Jurvelin, M.J. Nieuwenhuijsen, and M. Jantunen. Exposure of population and microenvironmental distributions of volatile organic compound concentrations in the expolis study. Atmospheric Environment, 37: 5563–5575, 2003.

[4] R.D. Edwards, J. Jurvelin, K. Koistinen, K. Saarela, and M. Jantunen. Voc source identification from personal and residential indoor, outdoor and workplace microenvironment samples in expolis-Helsinki, Finland. Atmospheric Environment, 35: 4829–4841, 2001.

[5] M.S. Zuraimi, C.-A. Roulet, K.W. Tham, S.C. Sekhar, K.W.D. Cheong, N.H. Wong, and K.H. Lee. A comparative study of VOCs in Singapore and European office buildings. Building and Environment, 41: 316–329, 2006.

[6] A.P. Jones. Indoor air quality and health. Atmospheric Environment, 33: 4535–4564, 1999.

[7] AFSSE. Rapport de la commission d'orientation du plan national santé environnement. Technical report, Agence Française de Sécurité Sanitaire Environnementale, 2004.

[8] W. Chen, J.S. Zhang, and Z. Zhang. Performance of air cleaners for removing multiple volatile organic compounds in indoor air. ASHRAE Transactions, 111: 1101–1114, 2005.

[9] P.-C. Luo, Z.-B. Zhang, Z. Jiao, and Z.-X. Wang. Investigation in the design of a co2 cleaner system by using aqueous solutions of monoethanolamine and diethanolamine. Ind. Eng. Chem. Res., 42: 4861–4866, 2003.

[10] S. Aguado, A.C. Polo, M.P. Bernal, J. Coronas, and J. antamaria. Removal of pollutants from indoor air using zeolite membranes. Journal of Membrane Science, 240: 159–166, 2004.

[11] T. Ito, N. Namiki, M. Lee, H. Emi, and Y. Otani. Electrostatic separation of volatile organic compounds by ionization. Environmental Sciences and Technology, 36: 4170–4174, 2002.

[12] Y. Sekine. Oxidative decomposition of formaldehyde by metal oxides at room temperature. Atmospheric Environment, 36: 5543–5547, 2002.

[13] S.L. Daniels. "On the ionization of air for removal of noxious effluvia"(air ionization of indoor environments for the control of volatile and particle contaminants with nonthermal plasmas generated by dielectric-barrier discharge. IEEE Transaction on plasma science, 4: 1471–1481, 30.

[14] R. A. Wood, M. D. Burchett, R. Alquezar, R. L. Orwell, J. Tarran, and F. Torpy. The potted-plant microcosm substantially reduces indoor air voc pollution: I. office field-study. Water, Air, and Soil Pollution, 175: 163–180, 2006.

[15] A.B. Darlington, J.F. Dat, and M.A. Dixon. The biofiltration of indoor air: air flux and temperature influences the removal of toluene, ethylbenzene and xylene. Environmental Sciences and technology, 35: 240–246, 2001.

[16] B.C. Wolverton and A. Johnson. Interior landscape plants for indoor air pollution abatement. Technical report, National Aeronautics and Space Administration, 1989.

[17] C.H. Ao and S.C. Lee. Indoor air purification by photocatalyst tio2 immobilized on an activated carbon filter installed in an air cleaner. Chemical Engineering Science, 60: 103–109, 2005.

[18] R. Iranpour, H.H.J. Cox, M.A. Deshusses, and E.D. Schroeder. Literature review of air pollution control biofilters and biotrickling filters for odor and volatile organic compound removal. Environmental Progress, 24: 254–267, 2005.

[19] J.S. Devinny, M.A. Deshusses, and T.S. Webster. Biofiltration for air pollution control. Lewis Publishers, 1999.

[20] P. Blondeau, A. Ginestet, Dr F. Sqinazi, Pr F. de Blay, M. Ott, B. Ribot, and D. Frochot. Mise en place de protocoles de qualification des appareils d'épuration d'air, rapport final. Technical report, CETIAT, EDF RD, Mairie de Paris, Université de la Rochelle, Les Hôpitaux Universitaires de Strasbourg, 2006.

[21] WHO. Air Quality Guidelines - second Edition. WHO Regional Publications, 2000.

[22] P.K. Johnston, G. Hadwen, J. McCarthy, and J.R. Girman. A screening-level ranking of toxic chemicals at levels typically found in indoor air. In Indoor Air, 2002.

[23] L. Mosqueron and V. Nedellec. Hiérarchisation sanitaire des paramètres mesurés dans les bâtiments par l'observatoire de la qualité de l'air intérieur. Technical report, Observatoire de la Qualité de l'Air Intérieur, 2002.

[24] I. Datta and D.G. Allen. Biotechnology for odor and air pollution control. Springer, 2005.

Section 7
Aerosols and particles

The role of PM_{10} in air quality and exposure in urban areas

C. Borrego[1], M. Lopes[1], J. Valente[1], O. Tchepel[1], A. I. Miranda[1] & J. Ferreira[2]
[1]CESAM & Department of Environment and Planning, University of Aveiro, Aveiro, Portugal
[2]School of Environmental Sciences, University of East Anglia, Norwich, UK

Abstract

In recent years, there has been an increase of scientific studies confirming that long- and short-term exposure to particulate matter pollution leads to adverse health effects. The calculation of human exposure in urban areas is the main objective of the current work combining information on pollutant concentration in different microenvironments and personal time-activity patterns. Two examples of PM_{10} exposure quantification using population and individual approaches are presented. The results are showing important differences between outdoor and indoor concentrations and stressing the need to include indoor concentrations quantification in the exposure assessment.

Keywords: air pollution, respiratory diseases, exposure, particulate matter.

1 Introduction

Every day, a person breathing is exposed to different concentrations of atmospheric pollutants, as he moves from and to different outdoor and indoor places. Particulate matter, coarse and fine, is one of the pollutants of most concern in terms of adverse health effects [1]. Epidemiological studies point out tobacco smoke, indoor and outdoor pollutants as preventable risk factors of respiratory diseases such as cancer, allergic diseases, asthma and other chronic respiratory diseases [2]. Smoke from fuels, urban air pollution or occupational airborne particulates contribute to the increase of disability-adjusted life years (DALYs, one DALY represents the loss of one year of full health) either in developed and developing countries [3]. For both chronic and acute health effects, the elderly, children, and those suffering from respiratory or heart conditions seem to be most at risk.

Air pollution problems related to particulate matter are more frequent and severe at urban areas, with high population density, and where industrial and traffic particulate matter emissions are the major contributor to air pollution.

Health effects of air pollution are the result of a sequence of events, which include release of pollutants, their atmospheric transport, dispersion and transformation, and the contact and uptake of pollution before the health effects take place. The conditions for these events vary considerably and have to be accounted for, in order to ensure a proper assessment.

Studies of human exposure to air pollution have different applications, namely: i) impact assessment of population exposure, in connection with various types of management such as traffic and city planning; ii) comparison of the exposure of different specific population groups; iii) estimation of the average or peak exposures of the population in connection with, for example, health assessment; iv) identification of the most important sources of pollution exposure; v) identification of possible associations between exposure and health effects [4].

Exposure studies can be carried out to obtain estimates of the exposure of the individual (personal exposure) or for a larger population group (population exposure). The exposure can be obtained from direct measurements on individuals or can be determined from model calculations [4]. The former is used to access individual dose and the later for strategic studies, for example, to estimate the number of inhabitants exposed above limit value.

The general approach for exposure estimation can be expressed by:

$$Exp_i = \sum_{j=1}^{n} C_j t_{i,j} \qquad (1)$$

where Exp_i is the total exposure for person i over the specified period of time; C_j is the pollutant concentration in each microenvironment j and $t_{i,j}$ is the time spent by the person i in microenvironment j.

This paper explores two different approaches to analyse the impact of air pollution on human health, focused on the estimations of population exposure and individual exposure, applied in two different case studies.

2 Population exposure

The exposure of an entire population to PM_{10} air concentration is a useful parameter to evaluate air quality effects on human health. The assessment of measured ambient air particulate matter concentrations in Portugal has been carried out for the past few years to evaluate their effect on human exposure and health. Urban areas are particularly of concern due to usual high PM emissions and population density.

2.1 Study case

Porto is the second biggest city of Portugal, located on the coast. The municipalities in the region are characterized as urban (workplaces) and suburban (mainly industrial and residence areas) with different population density accordingly.

The analysis of the data from the air quality network of Porto, in the period 2001-2004, reveals that PM_{10} levels have exceeded the limit values (LV) established by legislation for the protection of human health for the daily average as well as for the annual mean.

Air quality modelling tools have been used to evaluate the air quality over the Metropolitan Area of Porto (MAP) showing also high levels of PM_{10} where air quality monitoring stations do not exist.

Therefore, the MAP was selected as a study case to estimate the population exposure to PM_{10} by the application of an exposure module linked to an air quality modelling system.

2.2 Methodology

To model human exposure, over a selected region, by a deterministic approach, three types of input data are needed: the population characterization (number of people and daily time-activity pattern), the spatial distribution of the microenvironments visited by the population and the temporal variation of PM_{10} concentrations in each microenvironment.

The Portuguese National Statistics Institute (INE), in 1999, 2000 and 2003, did some inquiries to the resident population (students over 15 and employees) of the MAP focusing on: mobility of the population, work/school-home displacements, displacements from and to the MAP presented as Origin-Destination matrixes (number of people and time spent in displacement by mean of transport). Those matrixes allowed calculating the number of people that enters and leaves each municipality of MAP and thus, the people presented during the day and during the night in each cell of the modelling domain. The time spent was also addressed in the INE national enquires which has permitted the definition of a daily time-activity pattern per microenvironment considered in the exposure module, on an hourly basis, for the students and employed population of Portugal. According to the detail of information gathered it was possible to consider four different microenvironments – outdoor, home, other indoors and in vehicle.

The exposure module calculates the indoor concentrations using the outdoor concentrations simulated by the application of MM5-CAMx modelling system for the year 2004 and for a domain of 3 km horizontal resolution covering the MAP [5] and indoor/outdoor relations obtained experimentally (Table 1) [6,7].

Figure 1 presents the input data prepared for the exposure modelling application based on all the compiled and treated information, namely the population presented in the study domain during the day and night, the time-activity pattern for outdoor, home and other indoor microenvironments, the daily

Table 1: Indoor-outdoor relations considered in the exposure module for PM_{10}.

Home	Other indoors	In vehicle
$C_{it}(day) = 48 + 0.51 C_{out}$	$C_{in}(day) = 48 \cdot (1 - 0.14) + 0.51 C_{out}$	$C_{vehicle} = 13.1 + 0.83 C_{out}$
$C_{in}(night) = 20 + 0.52 C_{out}$	$C_{int}(night) = 20 \cdot (1 - 0.14) + 0.52 C_{out}$	

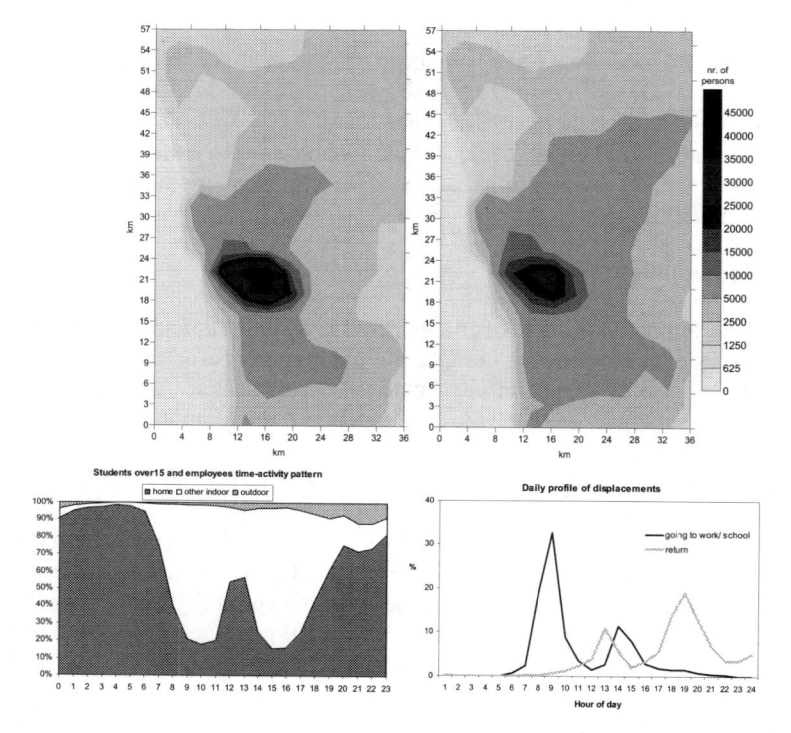

Figure 1: Input data for exposure module to estimate population exposure in the Metropolitan Area of Porto.

profiles of displacements to and from work/school and the PM_{10} annual average concentration field simulated by the MM5-CAMx modelling system [8,9].

2.3 Results

The application of the air quality-exposure modelling system has permitted the estimation of the spatial distribution of the exposure to PM_{10} for the population subgroup selected and for weekdays (no information for the weekend time-activity pattern was available). Figure 2 presents the annual averages of simulated PM_{10} concentration and individual exposure fields for the study domain.

The PM_{10} concentration field shows high levels of PM_{10} in Porto urban area, tending to decrease with the increase of radius distance to the city. The spatial distribution of exposure levels follows the concentration field, however with lower levels.

Considering the annual limit value for PM_{10} of 40 $\mu g.m^{-3}$, concentration results reveal a non-accomplishment of the legislation in terms of air quality possibly leading to significant impacts that are minimized regarding human health as exposure results show.

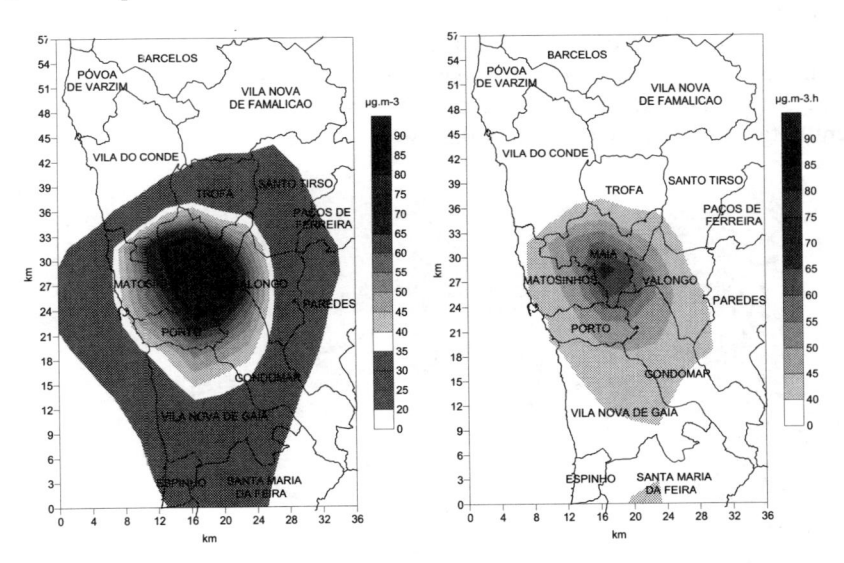

Figure 2: Annual average fields of simulated PM_{10} concentration and individual exposure for the Metropolitan Area of Porto study domain.

To evaluate the behaviour and applicability of the exposure module applied to MAP it would be important to validate the obtained results. However, previous similar studies have not been performed for Portugal and there are no exposure measurements to compare the modelling results with.

3 Individual exposure

Individual exposure is quantified for single individuals as they represent some population subgroups. Different methodologies could be applied for this purpose using: (i) direct measurements, or; (ii) estimations based on exposure concentration data and the time of contact.

3.1 Study case

One of the Portuguese middle-size towns currently not presenting significant air pollution problems was selected in this study. The town of Viseu is located in the central Portuguese mainland region, near important road transport networks and a high urban development is expected in a near future.

Asthmatic children were identified as the population subgroup more susceptible to air pollution and therefore they are the focus of the exposure estimation. The work was developed as a part of SaudAr (The Health and the Air we breath) project, which main objective is to contribute for the urban sustainable development by preventing air pollution problems and health related diseases in the future due to expected economical development.

In total, 4 schools were selected, including 2 located in town centre – *Massorim* and *Marzovelos* – and 2 in city suburbs – *Ranhados* and *Jugueiros* (Figure 3). The ISAAC (International Study of Asthma and Allergies in Childhood) questionnaire was applied to 805 children between ages 6 and 12 to identify those presenting respiratory disease.

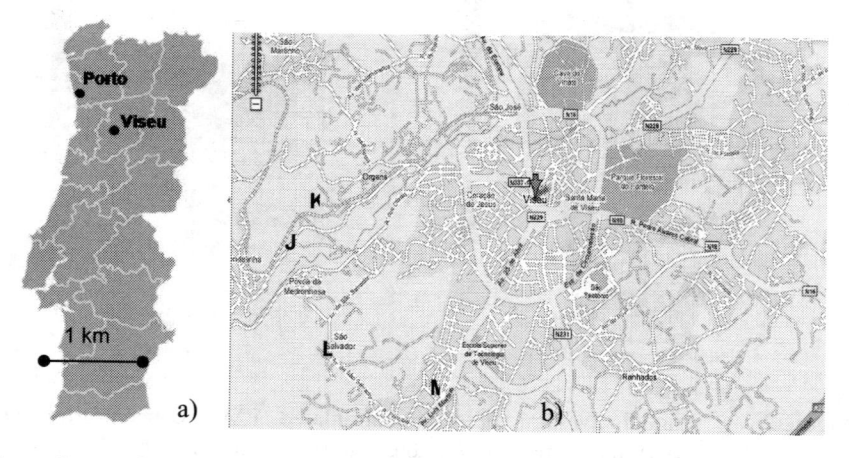

Figure 3: Geographic location of Viseu in Portugal (a) and location of the schools in Viseu (b).

3.2 Methodology

Similarly to the population exposure, the individual exposure was estimated using the microenvironment approach and calculated according to Equation 1. The input data required for the exposure quantification are determined separately for each individual under the study. For this purpose two main tasks were carried out: a) the definition of the daily activity profile of each child for a typical winter and summer school week, which allowed the identification of the microenvironments frequented by those children and the time spent in each one; and b) the air quality characterisation of those microenvironments. The daily activity profile was established through personal interviews of parents and child during the medical consulting hour. The air quality evaluation in the identified microenvironments, both outdoor and indoor, was performed using a multi-strategy approach: measurements during field campaigns and air quality modelling simulations.

Campaigns and model simulations were performed both in winter (14–28 January 2006) and summer time (19-26 June 2006) to take into account seasonal variability of pollutant levels. In microenvironments where outside measurements where not possible, PM_{10} concentrations where obtained through air quality modelling. These microenvironments were geo-referenced and outdoor concentrations where obtained directly from modelling while for indoor concentrations the relations presented in Table 1 were used. With these data daily personal exposure to PM_{10} was calculated for each child for a week in summer and winter time.

3.3 Results

The analysis of the time activity profiles obtained shows that the school children spend more than 95% of their time indoors. This percentage slightly decreases in summer and in the weekend. Also, the children living in a suburban location tend to spend more time outdoors than urban children. Figure 4 schematically represents the typical winter and summertime weekday and weekend of the children participating in the study.

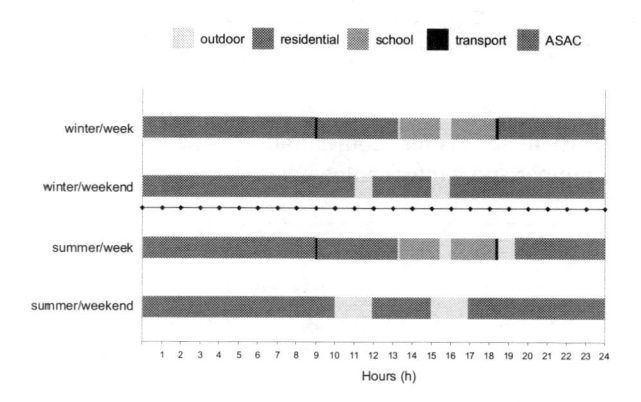

Figure 4: Typical winter and summertime weekday and weekend of the children that participated in the study (ASAC – after school activity centre).

The air quality measurements made at schools outdoor and in the classrooms are shown in Figure 5. During weekdays, the PM_{10} concentrations in the classrooms are significantly higher than in the outdoors for the same location. At urban school Marzovelos, the indoor PM_{10} levels measured during the winter campaign are higher than the double of the outdoor concentrations. At summer, the difference between indoor and outdoor pollution levels is lower. This fact can be explained by presence of indoor emission sources and low ventilation rates at wintertime, while open windows during summertime promote better conditions to indoor/outdoor air exchange. Along weekends, the indoor PM_{10} values are similar to the ones measured outdoor, confirming the existence of indoor PM sources on weekdays.

The model simulations performed are in agreement with the measurements made. They show that the legislated daily value of PM_{10} is often exceed, mainly in winter and in the most urbanised area of the town.

Exposure results are plotted in Figure 6. These results are divided in urban and suburban (according to the children address and school) and winter and summer.

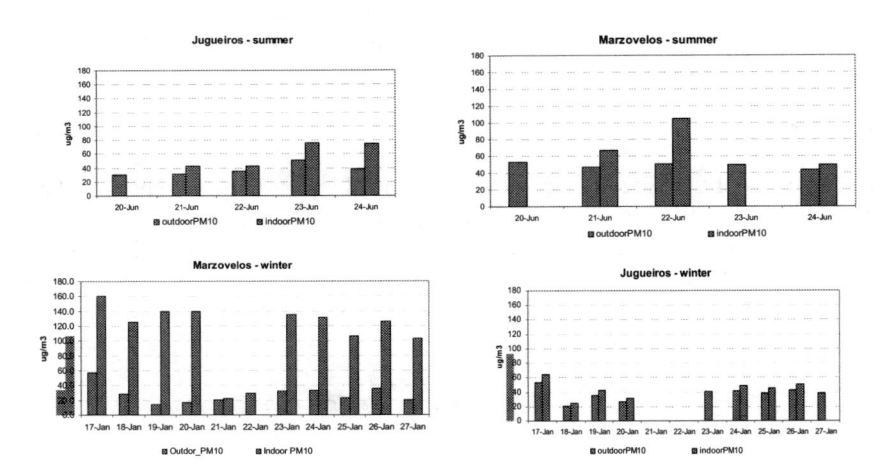

Figure 5: Particulate matter concentration values ($\mu g.m^{-3}$) measured in summer and winter campaigns.

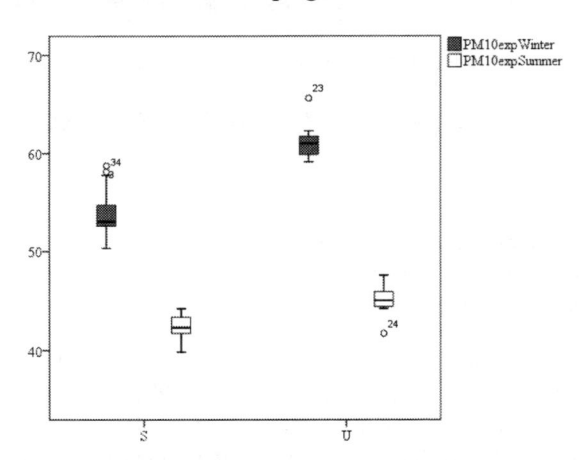

Figure 6: Weekly mean of hourly exposure to PM_{10} ($\mu g.m^{-3}.h$) for urban (U) and suburban (S) children both for winter and summer.

As it can be seen from Figure 6, exposure to PM_{10} is quite high attaining levels of concern, particularly in winter and for the urban location (61$\mu g.m^{-3}.h$). Although results are calculated for two weeks in a year, they clearly indicate that the annual mean value of 20 $\mu g.m^{-3}$, recommended by WHO [10] is easily exceeded by any of the studied children. The microenvironment that contributes the most to exposure is clearly the classroom.

4 Conclusions

This study intends to contribute for the general knowledge on PM_{10} exposure levels highlighting the importance of their estimation for the definition of air quality standards. Two approaches applied in this work to quantify population and individual exposure have demonstrated the significant contribution of indoor microenvironments to the human exposure by PM_{10} and, therefore, the importance to include indoor concentrations quantification in the exposure assessment. The results of the study show that indoor concentrations could be more than double in comparison with outdoor levels and the individuals are spending more than 90% of time indoors. Besides the contribution of indoor pollution sources, the ventilation of the buildings is also a relevant factor that has to be taken into account. The results indicate the characterisation of indoor pollution sources and indoor PM_{10} levels as a major source of the uncertainty of exposure quantification.

Acknowledgements

The authors are grateful to the Calouste Gulbenkian Foundation for the SaudAr project financial support. The financial support under the 3rd EU Framework Program and the Portuguese 'Ministério da Ciência, da Tecnologia e do Ensino Superior' for the Project PAREXPO (POCI/AMB/57393/2004) and the Ph.D. grant of J. Valente (SFRH/BD/22687/2005) and J. Ferreira (SFRH/BD/3347/2000) is also acknowledged.

References

[1] EEA – European Environmental Agency. Environment and Health, EEA report No. 10/2005. Office for Official Publications of the European Communities, 2005.
[2] Beaglehole R. et al. Preventing chronic diseases: a vital investment. Geneva, World Health Organization, 2005.
[3] Bousquet J and Khaltaev. Global surveillance, prevention and control of chronic respiratory diseases: a comprehensive approach. World Health Organization, 2007.
[4] Hertel, O.; De Leeuw, F.; Raaschou-Nielsen, O.; Jensen, S.; Gee, D.; Herbarth, O.; Pryor, S.; Palmgren, F.; Olsen E. (2001): Human Exposure to Outdoor Air Pollution. IUPAC Report. *Pure and Applied Chemistry* Vol. 73, No. 6, pp 933–958.
[5] Ferreira, J., 2007, Relation Air Quality and Human Exposure to Atmospheric Pollutants. PhD Thesis, University of Aveiro, Aveiro, Portugal.

[6] Gulliver, J., Briggs, D.J., 2004, Personal exposure to particulate air pollution in transport microenvironments, Atm. Env. 38, pp. 1–8.
[7] USEPA Air Quality Criteria for Particulate Matter, v.1, 1997.
[8] Dudhia, J. (1993). A nonhydrostatic version of the Penn State - NCAR Mesoscale Model: Validation tests and simulation of an Atlantic cyclone and cold front. Mon. Wea. Rev., 121, 1493–1513.
[9] ENVIRON (2004). Comprehensive Air Quality Model with Extensions – CAMx. Version 4.0, User's guide. ENVIRON International Corporation.
[10] WHO, Air Quality Guidelines – Global Update 2005. WHO Europe, June 2004.

Spatial distribution of ultrafine particles at urban scale: the road-to-ambient stage

F. Costabile[1], B. Zani[2] & I. Allegrini[1]
[1]*Institute for Atmospheric Pollution-National Research Council (CNR-IIA), Monterotondo-Rome, Italy*
[2]*ARPA Parma, Italy*

Abstract

The spatial variability of ultrafine particles (UFPs) is believed to be an important issue to assess urban air pollution fate and exposure in connection with traffic motorised emissions. In this work, the high-time resolution total number concentration of UFPs was measured at traffic-oriented and urban background locations in a middle-size city in Italy. The major objective was to study valuable connections with local traffic sources, as well as measurement sites' representativeness. On the one hand, it was found that the total concentration at the traffic site can be representative of vehicle exhaust sources in ambient air. On the other hand, it was possible to identify three prevailing contributions for the total UFPs number concentration at urban scale: a very low urban background concentration, a significant contribution due to local traffic sources, and a significant contribution due to secondary transformation processes closely linked to meteorology.

Keywords: urban air pollution, ultrafine particles, traffic emissions, background, representativeness, exposure.

1 Introduction

During the last decades a growing body of research has investigated worldwide the extremely vast subject of urban air quality [1]. Measuring any potential effect of any urban air pollutant requires the understanding of its variation and distribution in both space and time. Traffic-related pollution and its spatial variations are particular concerns; a comprehensive understanding is crucial [2–

8]. However, current actions are particularly hampered by a lack of knowledge when characterising the spatial variability within urban areas [9].

For particles, the spatial variability is particularly relevant, and depends on the size fraction. Particles smaller than 100 nm (the so-called ultrafine particles, UFPs) are more variable in space and time than fine particles as they have a higher dependence on particle sources, and a faster removal from the atmosphere [10–12]. Consequently, their spatial variability is believed to be an important issue to assess air pollution fate and exposure.

Internal combustion engines are known to be a major emission source of UFPs [13–17]. However, in spite of extensive laboratory studies on engine emissions, there are few investigations of how particle mobile emissions evolve and affect air quality establishing a link between sources and receptors [18–20]. It has been shown that the size distribution of emitted particles evolves substantially within a few hundred meters of emission [8, 21]. After release into the atmosphere, UFPs are subjected to complex dilution and transformation processes. Neither current models nor those that will be available in the near future are tough to be able to cover all the spatial and temporal scales that are involved from the emission (centimetres, milliseconds) to the urban/regional scale (kilometres, hours) [22].

In this work, high-time resolution UFP total number concentration was measured in a middle-size urban area of Italy (Parma). The major objective was to study valuable connections with local traffic sources, as well as the measurement sites' representativeness [23].

2 Experimental

A Water-Condensation Particle Counter (WCPC,TSI Mod. 3781, Figure 1 and Figure 2) [24–27] was used to measure the concentration of particles in air with a diameter larger than 6 nm (and smaller than 3 μm), N_6,

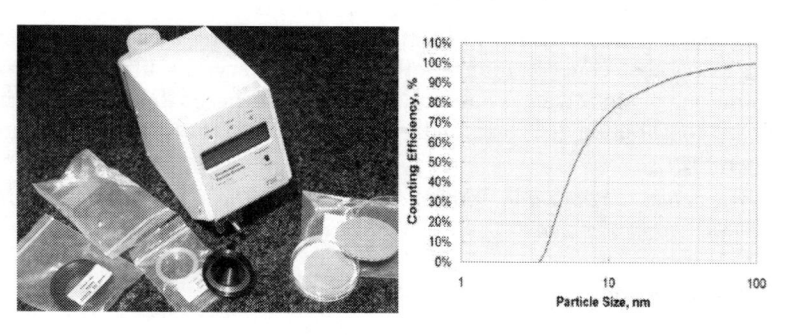

Figure 1: Picture and counting efficiency of the Water-Condensation Particle Counter (WCPC,TSI, Modello 3781) [28].

Data were collected every 2 seconds, and then averaged to generate 1-minute values. Measurements were taken during the winter (January, February and March) and summer (July, August, September) of 2007 (both weekdays and weekends). Instruments were located at two urban sites, 740 meters away from each other in Parma downtown: via Montebello, and Parco della Cittadella (Figure 3, Figure 4).

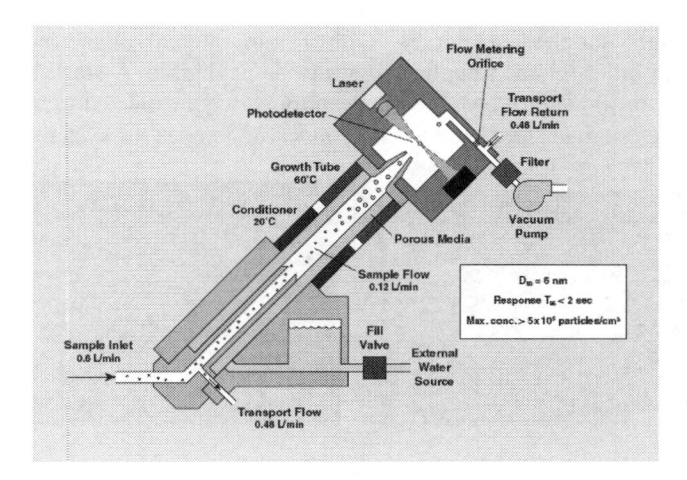

Figure 2: Working principle of the Water-Condensation Particle Counter (WCPC,TSI, Modello 3781) [28].

Figure 3: Picture by Google Earth of the two measurement sites in Parma: Parco della Cittadella (top) and via Montebello (bottom).

Representativeness criteria with respect to emission sources drove the selection of these two sites [29, 30]. Parco della Cittadella was selected as representative of urban background concentrations, the closest traffic emission source being around 100 meters away. Conversely, the traffic site in Via Montebello would reflect pollution concentrations strongly influenced by urban traffic emissions; measurements were taken less than 1 meter from the road (the road is south of the measurement point, Figure 4). The sampling probe was made by a straight stainless steel tube, with 7 mm internal diameter and length < 15 cm, in order for the sampling losses to be negligible. A small cyclone was used at the traffic site (where the concentrations were expected to be higher) to avoid particles larger than 1 µm to either block or damage the system.

Figure 4: Details by Google Earth of the two measurement sites in Parma: Parco della Cittadella (left) and via Montebello (right).

3 Results and discussion

3.1 From traffic site to urban background

The total number concentrations of UFPs measured (Figures 5–8) were found to be low, but comparable to similar works [31]: urban background locations have usually average concentrations of 10.000 #/cm^3, whereas traffic locations exhibit 30.000-50.000 #/cm^3. Since the total particle number is generally dominated by local emissions, it is particularly affected by primary particles generated by vehicle exhausts. Therefore, as found in previous studies [32], UFP total number concentrations measured at Via Montebello – close to traffic emissions – showed tendencies significantly different from the concentrations measured just 700 meters away in the park; representative of the urban background. Differences related mainly to two factors. The traffic site showed the UFP total number to have both much higher concentrations, and much more rapid fluctuations (Figures 5–8). Much faster and more intensive dilution processes could result in a more rapid reduction of UFP total number, as well as a trigger for other physical processes (such as nucleation) in connection to other parameters, such as solar radiation. Concentrations are comparable only at nighttime, from midnight to 4.00 a.m., when both the traffic flow and the concentration fluctuations at the Montebello station were very low.

Contrary to the urban background site, the total number concentrations at the traffic location showed a daytime trend shaped by two peaks (Figure 10). They can be likely related to traffic rush hours (7.00 a.m. and 18:00), and were also

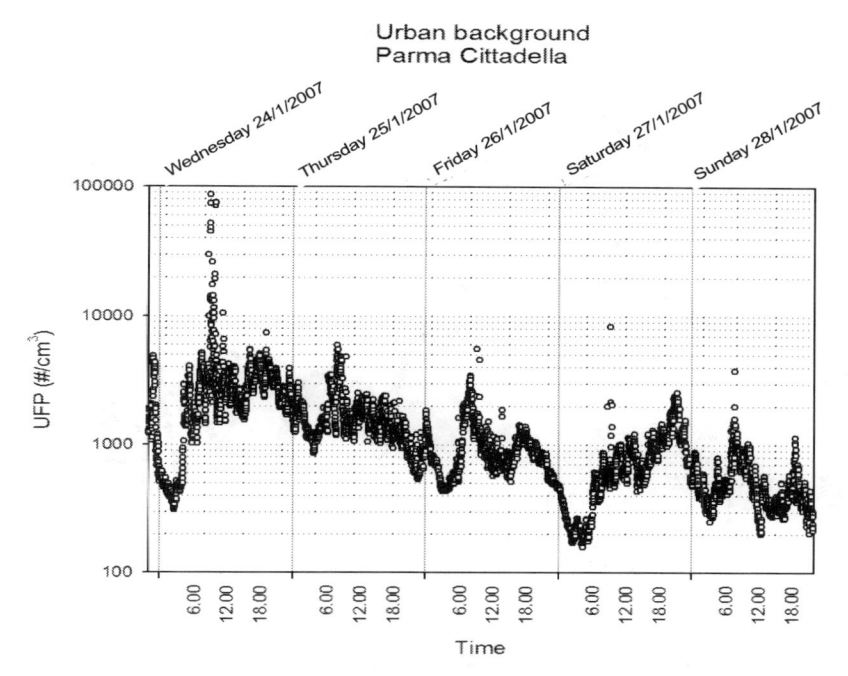

Figure 5: UFP number concentration during the winter campaign at an urban background station in Parma [23].

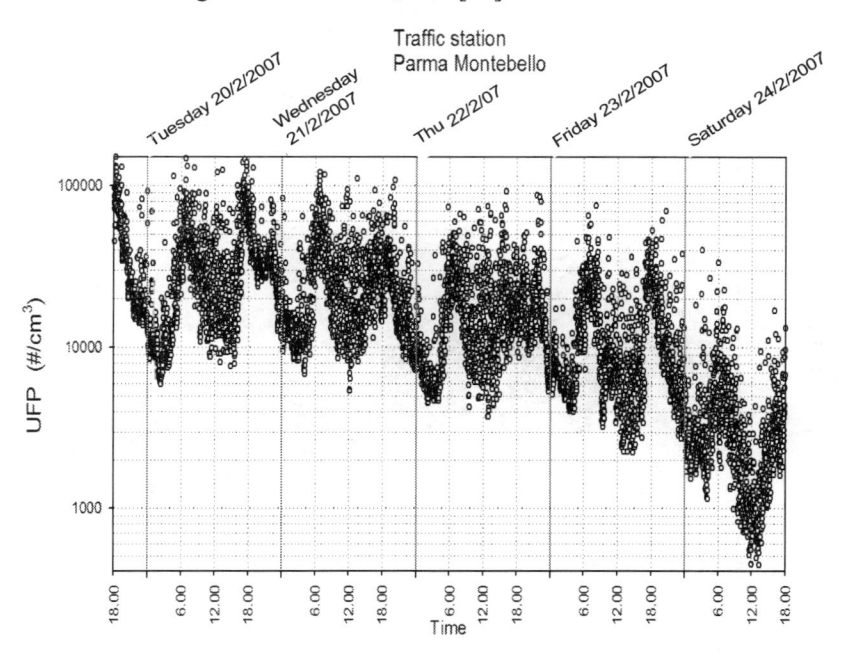

Figure 6: UFP number concentration during the winter campaign at a traffic station in Parma [23].

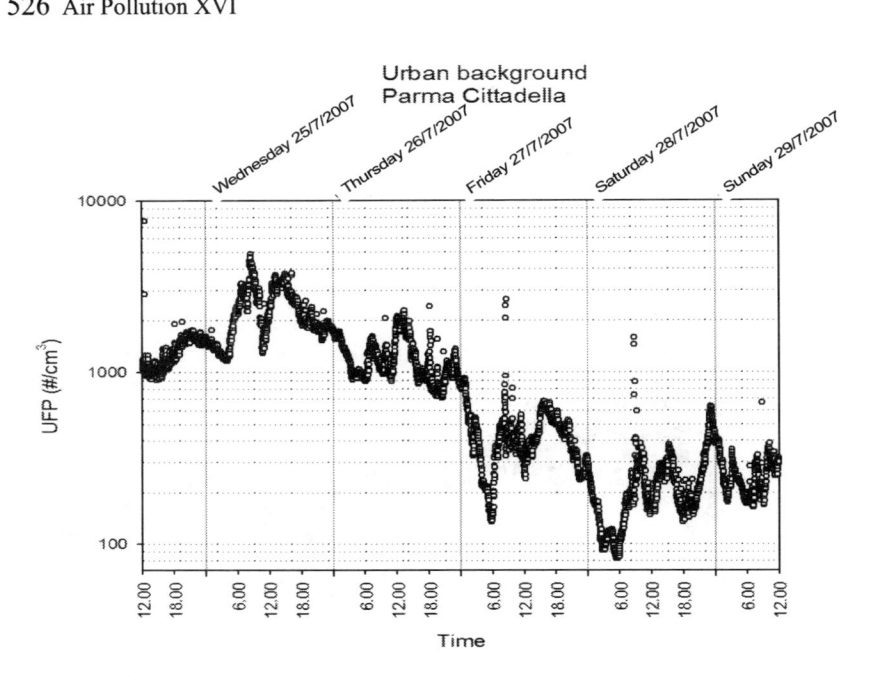

Figure 7: UFP number concentration during the summer campaign at the background station in Parma [23].

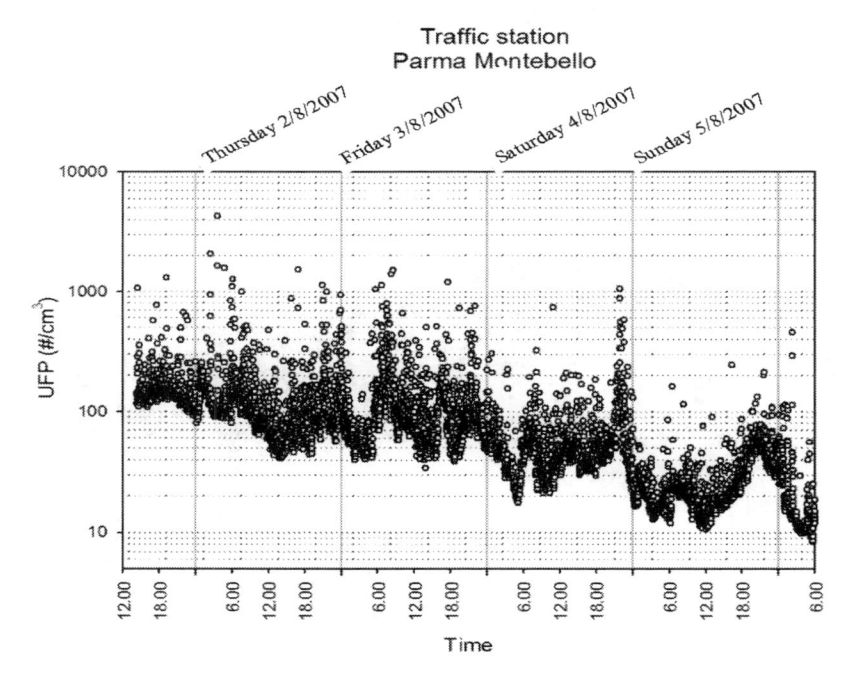

Figure 8: UFP number concentration during the summer campaign at a traffic station in Parma [23].

shown by other gaseous pollutants. This finding clearly indicates traffic emissions to be the prevailing source of the UFP total number [33].

3.2 Coupling with meteorology and gaseous pollutants

Variations due to meteorology can be more easily analysed at the urban background site, where traffic flows influence can, however, also be recognised. UFP total number concentration peaks (relative or absolute peaks) measured in the Cittadella often corresponded to (midday) solar radiation peaks (Figures 5 and 7). This suggests the secondary particle formation to be more significant at the background site [31]. The highest value of UFP total number concentration (not including spikes) at this site was measured on January 19 (Figure 9). This

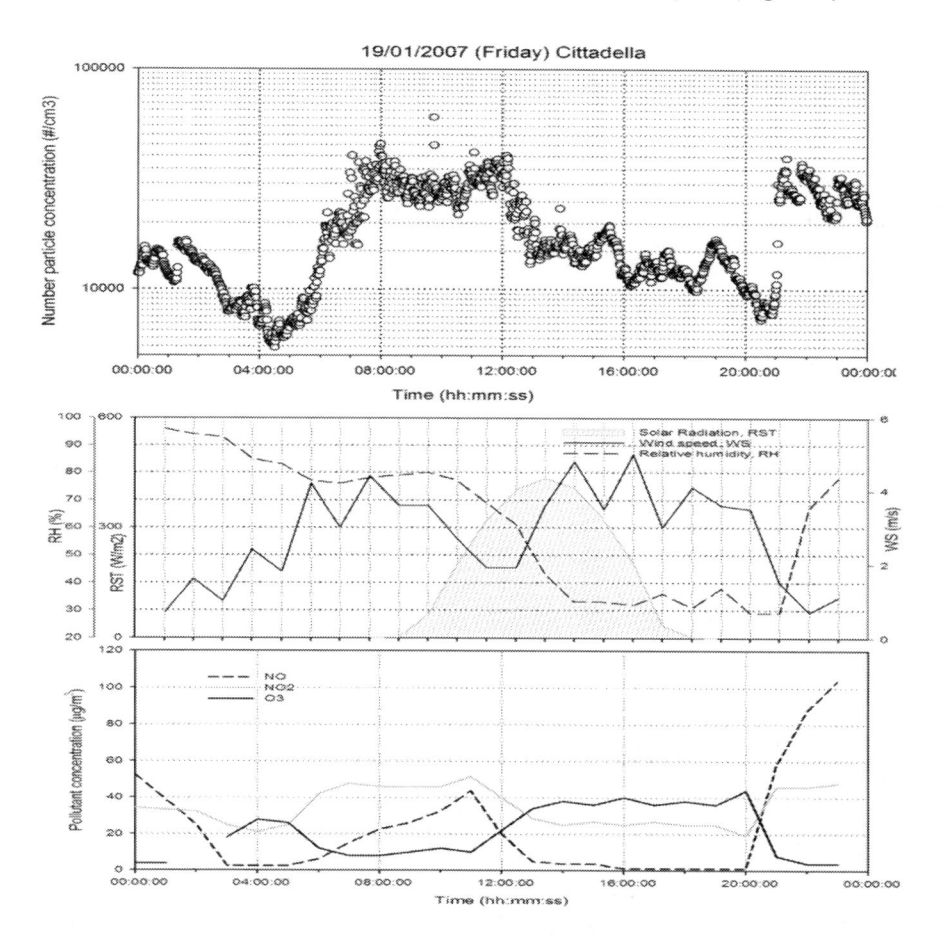

Figure 9: Total number concentration of UFPs measured at the background station (Parco della Cittadella) on Jan 19, 2007. Meteorological parameters and gaseous pollutant concentrations measured at the same site (ARPA Parma) are shown below.

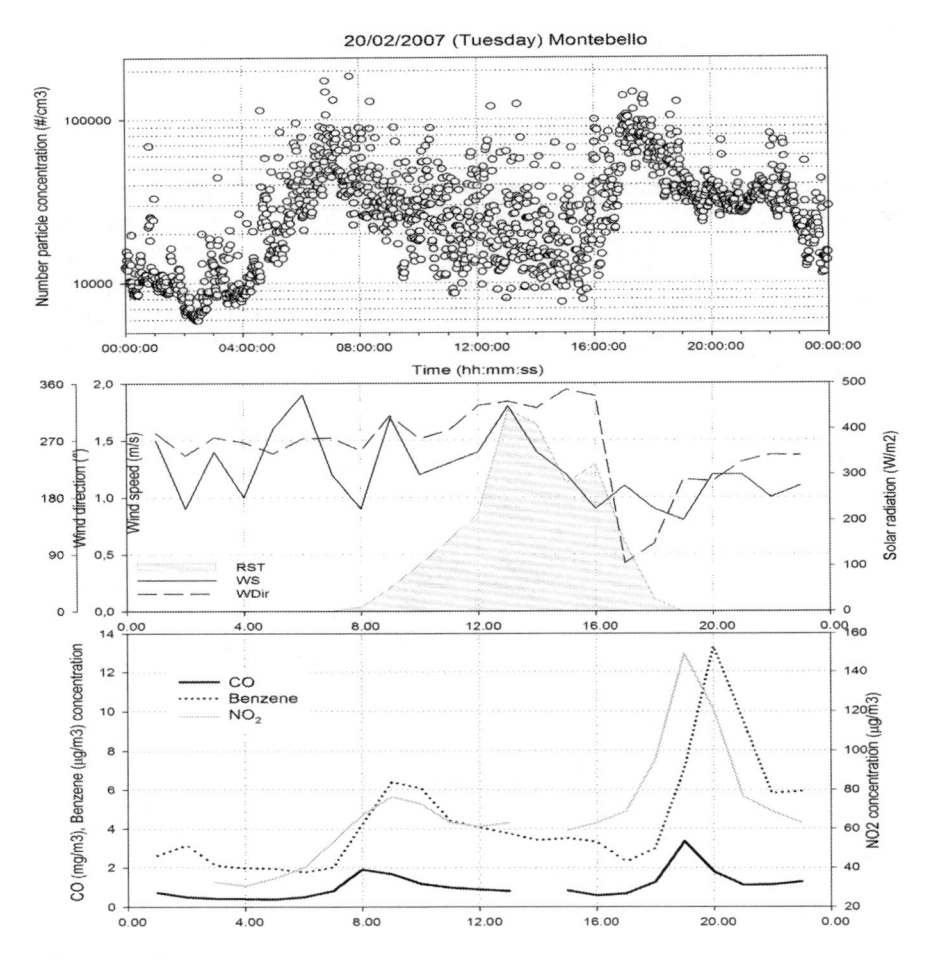

Figure 10: Total number concentration of UFPs measured at the traffic site (via Montebello) on Feb 20, 2007. Meteorological parameters and gaseous pollutant concentrations measured at the same site (ARPA Parma) are shown below.

was a weekday (Friday) with high solar radiation, temperature and wind speed, and low relative humidity.

Though less clear, these noontime peaks can also be seen at the traffic site (Figures 6, 8 and 10).

Beyond this effect, a correlation with wind speed was also found (Figure 9, Figure 10). It is not clear whether increased wind speeds can either induce significant transport phenomena of UFPs emitted/formed by nearby traffic sources [33], or enhance total number concentration by other phenomena (i.e., secondary new particle formation). This effect was less evident at the traffic site: local emitted particles effectively dominated measurements even when the site was upwind to the road nearby. It is not clear if the afternoon peak measured for both UFPs and gaseous pollutants on February, 20 in Montebello (Figure 10) should be related either to the wind direction change, or (more likely) to the lower wind speed (wind calm, <1m/s), the rush hours, and increased atmospheric stability. Interestingly enough, diurnal trends of both total UFPs number and primary gaseous pollutants (particularly NO and CO) were found to be quite similar (Figures 9 and 10) [34, 35]. Different time resolutions of data collected (1min versus 1h) are likely the major reason for the differences in the fluctuations, and the related time shifts. These findings indicate the total UFP number as a valuable marker of fresh emissions at traffic stations, significantly correlated with combustion-generated pollutants.

Finally, the comparison between the frequency distributions of UFPs measured at the two sites (Figure 11) shows a double FD mode only at the urban background location. This could likely suggest the existence of an (extremely low) urban background level (the lowest mode, 1000#/cm^3, which can be recognised at both the sites). The second mode in the background location could be related to UFPs concentrations (extremely diluted) transported from nearby sources and already aged because of meteorology-induced transformations. The FD at the traffic site looks like an Ln-FD, suggesting mixing and dilution of local traffic source emissions to be the prevailing process over time [29, 36]

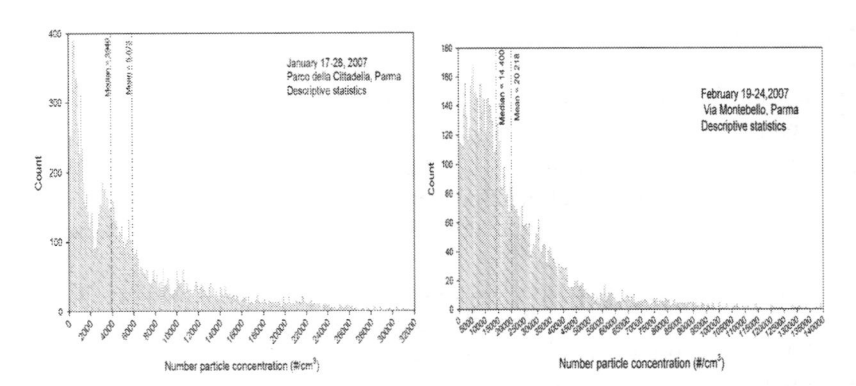

Figure 11: Frequency distribution of the total number concentration of UFPs measured at the background (Parco della Cittadella) and traffic (Via Montebello) stations.

4 Conclusions

The high-time resolution total number concentration of UFPs was analysed at traffic-oriented and urban background locations of a middle-size city in Italy. Total number concentration at the traffic site was found to be a clear indication of vehicle exhaust sources in ambient air. At an urban scale it was possible to identify three prevailing contributions for the total UFPs number concentration. Firstly, a very low urban background concentration (lower than 1000 #/cm^3) was particularly visible in summer. Secondly, a significant contribution due to local traffic sources, up to 100.000 #/cm^3 (traffic site in winter time). Finally, a significant contribution due to secondary transformation processes closely linked to meteorology, particularly solar radiation. Transport from sources nearby was also found to be significant, indicating significant UFP lifetime in the atmosphere. The conditions near mobile emission sources were found to differ from typical background conditions in that total particle number concentrations were much higher, and dilution processes were much faster and stronger. The stronger variability induced more rapid concentration fluctuations, probably due to faster and more intense dilution processes, which can reduce the number concentration more rapidly, as well as trigger physical processes other than dilution, such as nucleation, coagulation, and condensation.

Acknowledgements

Financial support mainly from the Italian Ministry of Environment, Land and Sea is gratefully acknowledged. As well, the authors would like to thank the ARPA-Parma (Regional Environment Protection Agency), and in particular Eriberto De Munari. Finally, our thanks are extended to our colleagues, F. De Santis, M. Montagnoli, and M. Giusto.

References

[1] Fenger, J., 1999. Urban air quality. Atmospheric Environment 33, 4877–4900

[2] Morawska, L. Vishvakarman, D. Mengersen, K. Thomas, S. 2002. Spatial variation of airborne pollutant concentrations in Brisbane, Australia and its potential impact on population exposure assessment. Atmospheric Environment 36, 3545–3555

[3] Briggs DJ. 2000. Exposure assessment. In: Elliott P, Wakefield JC, Best NG, Briggs DJ, editors. Spatial epidemiology: methods and applications. Oxford: Oxford University Press; p. 335–59

[4] Ito K, Xue N, Thurston G., 2004. Spatial variation of PM2.5 chemical species and source-apportioned mass concentrations in New York City. Atmos. Environ 38:5269–82.

[5] Kim E, Hopke PK, Pinto JP, Wilson WE, 2005. Spatial variability of fine particle mass, components, and source contributions during the Regional Air Pollution Study in St. Louis. Environ Sci Technol; 39:4172–9.

[6] Pinto JP, Lefohn AS, Shadwick DS., 2004. Spatial variability of PM2.5 in urban areas in the United States. J Air Waste Manage Assoc. 54, 440–9.

[7] Wilson JG, Kingham S, Pearce J, Sturman AP. A review of intraurban variations in particulate air pollution: implications for epidemiological research. Atmos Environ 2005;39:6444–62.

[8] Zhu, Y., Hinds, W.C., Kim, S., Shen, S., Sioutas, C., 2002. Study of ultrafine particles near a major highway with heavy-duty diesel traffic. Atmospheric Environment 36 (27), 4323–4335.

[9] Lebret, E., Briggs, D., van Reeuwijk, H., Fischer, P., Smallbone, K., Harssema, H., Kriz, B., Gorynski, P., Elliott, P., 2000. Small area variations in ambient NO2 concentrations in four European areas. Atmospheric Environment 34, 177–185.
[10] WHO, 2006. Air Quality Guidelines. Global update 2005. WHO Regional Office for Europe, Copenhagen, Denmark (http://www.euro.who.int/Document/E90038.pdf)
[11] Pekkanen, J. and Kulmala, M., 2004. Exposure assessment of ultrafine particles in epidemiologic time-series studies, Scandinavian Journal Work Environment and Health 30 (Suppl. 2), pp. 9–18
[12] Monn, C., 2001. Exposure assessment of air pollutants: a review on spatial heterogeneity and indoor/outdoor/personal exposure to suspended particulate matter, nitrogen dioxide and ozone. Atmospheric Environment 35, pp. 1–32.
[13] Weijers, E. P., Khlystov, A. Y., Kos, G. P. A., Erisman, J. W., 2004. Variability of particulate matter concentrations along roads and motorways determined by a moving measurement unit. Atmospheric Environment, 38 (19), 2993-3002.
[14] Charron, A., Harrison, R. M., 2003..Primary particle formation from vehicle emissions during exhaust dilution in the roadside atmosphere. Atmospheric Environment 37, 4109–4119
[15] Alam A., Shi Ping, J. and Harrison R. (2003), Observations of new particle formation in urban air. *J. Geophys. Res.* 108: D3, doi: 1029/2001/JD001417.
[16] Kittelson, D.B., Watts, W.F., Johnson, J.P., 2004. Nanoparticle emissions on Minnesota highways. Atmospheric Environment 38 (1), 9–19.
[17] Morawska, L., He, C., Hitchins, J., Mengersen, K., & Gilbert, D. , 2003. Characteristics of particle number and mass concentrations in residential houses in Brisbane, Australia. Atmospheric Environment, 37, 4195–4203.
[18] Abdul-Khalek, I. S.; Kittelson, D. B.; Brear, F. , 1999. Influence of dilution conditions on diesel exhaust particle size distribution measurements. *SAE Technol. Pap. Ser.*, no. 1999-01-1142
[19] Scheer, V., Kirchner, U., Casati , R., Vogt, R., Wehner, B., Philippin ,S., Wiedensohler, A., Hock, N., Schneider, J., Weimer , S., Borrmann, S., 2005. Composition of Semi-volatile Particles from Diesel Exhaust. SAE Paper 2005-01-0197
[20] Giechaskiel, B.; Ntziachristos, L.; Samaras, Z.; Scheer, V.; Casati, R.; Vogt, R.; Van Grieken, 2005. Formation potential of vehicle exhaust nucleation mode particles on-road and in the laboratory. Atmospheric Environment 39, 3191-3198
[21] Shi, J.P., Evans, D.E., Khan, A.A., Harrison, R.M., 2001. Sources and concentration of nanoparticles (10nm diameter) in the urban atmosphere. Atmospheric Environment 35, 1193-1202
[22] Ketzel, M, Berkowicz, R., 2004. Modelling the fate of ultrafine particles from exhaust pipe to rural background: an analysis of time scales for

dilution, coagulation and deposition. *Atmospheric Environment* 38, 2639–2652

[23] Costabile, F., Allegrini, I,., 2007b. Measurement and analysis in space and time of ultrafine particle number concentration in ambient air. The case of Parma. Proceeding of the International Conference UFIPOLNET, Dresden (Germany), 23-24 October 2007, pp.56-57

[24] Petäjä, T.; Mordas, G.; Manninen, H.; Aalto, P. P.; Hämeri, K.; Kulmala, M., 2006. Detection Efficiency of a Water-Based TSI Condensation Particle Counter 3785 Aerosol Science & Technology, 40(12):1090-1097

[25] Hering, S. V., Stolzenburg, M. R., Quant, F. R., Oberreit, D. R., and Keady, P. B., 2005. A Laminar-Flow, Water-Based Condensation Particle Counter (WCPC). *Aerosol Sci. Technol.* 39:659–672

[26] Hering, S. V., and Stolzenburg, M. R. (2005). A Method for Particle Size Amplification by Water Condensation in a Laminar, Thermally Diffusive Flow. *Aerosol Sci. Technol.* 39:428–436

[27] Biswas, S., Fine, P. M., Geller, M. D., Hering, S. V., and Sioutas, C., 2005. Performance Evaluation of a Recently Developed Water-Based Condensation Particle Counter. *Aerosol Sci. Technol.* 39, 419–427

[28] TSI, operating manual, 2006.

[29] Costabile, F., Bertoni, G., Desantis, F., Wang, F., Hong, W., Liu, F., Allegrini, I., 2006. A preliminary assessment of major air pollutants in the city of Suzhou, China. Atmospheric Environment 40 (33), 6380-6395.

[30] Costabile, F., Desantis, F., Wang, F., Hong, W., Liu, F. and Allegrini, I., 2006. Representativeness of Urban Highest Polluted Zones for Siting Traffic-Oriented Air Monitoring Stations in a Chinese City. JSME International journal-B 49 (1), 35-41

[31] Harrison, R.M., Jones, M., Collins, G., 1999. Measurements of the physical properties of particles in the urban atmosphere. Atmospheric Environment 33, 309- 321

[32] Zhang, K. M., Wexler, A.S., Zhu, Y.F., Hinds, W. C., Sioutas, C., 2004a. Evolution of particle number distribution near roadways. Part II: the 'Road-to-Ambient' process. Atmospheric Environment 38, 6655–6665

[33] Shi, J.P., Khan, A.A., Harrison, R.M., 1999. Measurements of ultrafine particle concentration and size distribution in the urban atmosphere. *The Science of Total Environment.* 235, 51-64..

[34] Kittelson, D.B., Watts, W.F., Johnson, J.P., Rowntree, C., Payne, M., Goodier, S., Warrens, C., Preston, H., Zink, U., Ortiz, M., Goersmann, C., Twigg, M.V., Walker, A.P., Caldow, R., 2006. On-road evaluation of two Diesel exhaust aftertreatment devices. Aerosol Science 37, 1140 – 1151

[35] Kittelson, D.B., Watts, W.F., Johnson, J.P., Schauer, J.J., Lawson, D.R., 2006b. On road and laboratory evaluation of combustion aerosol-Part 2: summary of spark-ignition engine results. Aerosol science 37, 931-949.

[36] Ott, W.R.A., 1990. Physical explanation of the lognormality of pollutant concentrations. Journal of Air and Waste Management Association 40, 1378–1383.

Electromagnetic and informational pollution as a co-challenge to air pollution

A. A. Berezin[1] & V. V. Gridin[2]
[1]*Department of Engineering Physics, McMaster University, Hamilton, Canada*
[2]*Department of Chemistry, Technion-Israel Institute of Technology, Haifa, Israel*

Abstract

The present focus on air pollution problems in a proper sense (emission of gaseous and particulate pollutants to the atmosphere) detracts, to a certain degree, attention from other growing problems of electromagnetic and informational pollution. At first glance, it appears that these problems are unrelated to the air pollution as such and may require completely different approaches for their solution (or at least, mitigation). We argue that the links between the above problems are more direct than is usually expected. Electromagnetic emissions in specific frequency bands, which are due to the exponentially increasing use of wireless communication technologies such as cell phones, satellite TV broadcasting, etc have selective effects at different elevation levels of the atmosphere and different geographical areas. Such known problems as holes in the ozone layer may be not be entirely due to chemical pollution, but may also be affected by high frequency electromagnetic effects. We discuss thermodynamical and quantum aspects of links between these three forms of pollution in the context of physics of chaotic and self-organizing systems.

Keywords: air pollution, electromagnetic pollution, informational pollution, mass communications, information density, Shannon entropy, resonance effects, quantum effects, order-disorder transitions, self-organized criticality, physics of chaos.

1 Introduction

With all the emphasis that is presently placed on air pollution (AP) in a proper sense, it remains one of several contributing factors to the general problem of pollution and the sustainability of the environment we are living in. The main target of this article is electromagnetic pollution (EMP) and informational pollution (IP). These, newly defined forms of pollution seem at first glance unrelated to AP. Yet, a generic analysis of various physical aspects of them undertaken in this (in essence interdisciplinary) article shows numerous possible connections between AP on one hand and EMP and IP on the other hand.

While traditional methods of AP control such as emission reduction and electrostatic precipitation are part of mainstream technology, new avenues were offered for conceptualization and possible experimental development. In [1] some links between AP and quantum physics were outlined. Here we propose some other links of that nature, placing special emphasis on the relationship between AP, EMP and IP. One of our central points is to draw attention to the fact that human-produced electromagnetic emissions have a structured nature and are informationally rich. This may result in pattern-organizing (catalytic) effects on various atmospheric processes directly related to AP and sustainability.

2 Conceptual framework of pollution discourse

Like other major social and economic issues, the pollution problem in general is multi-faced and prone to all sorts of controversies and clashes of opinions. At this point public environmental discourse is dominated by the Global Warming (GW) debate. The issue of human-generated GW is highly politicized and controversial with enthusiastic supporters at both ends of the debate [2,3]. This, to a certain degree, obfuscates and overshadows pollution problem(s) as such.

2.1 From atmosphere to infosphere

The fact that we live on an (approximately) spherical planet has historically created the notion of many spheres. The number of such conceptually defined "spheres" is now well over a dozen and keeps increasing. Without claiming completeness of such a list, we mention here *lithosphere*, *hydrosphere*, *atmosphere* (with a further subordination to *troposphere*, *stratosphere*, *ionosphere*), *biosphere*, and *magnetosphere*. These are, so to say, "naturally created" spheres. To that one can add an almost arbitrary number of "human created spheres" such as "*anthroposphere*", "*noosphere*" (sphere of intellectual life) and even such peculiar notions as "*pedosphere*" (sludge and pavement waste under our feet), "*garbagesphere*" or "*trashesphere*", etc.

The concept of infosphere is presently gaining momentum [4], especially with the advent of global communication networks and, particularly, the Internet. Its precursor can be seen in the notion of the "noosphere" introduced by Vladimir Vernadsky (1863-1945) and discussed by such visionaries as Pierre Teilhard de

Chardin (1881-1955) and Nikola Tesla (1856-1943). Vernadsky considered the dynamical sequence of geosphere (non-organic Earth) to biosphere and, finally, to the noosphere. While nominally the noosphere can be defined as a sphere of human thought, it presently to some degree is merging with the notion of *infosphere*. The "thickness" of noosphere is now greatly enhanced by human-generated electromagnetic (EM) emissions and satellites. This can produce some global effects in ecosphere, e.g., EM energy fluxes can affect migratory paths of birds, or lead to a variety of genetic and ecological modifications.

2.2 Major types of human-induced pollution

With an enormous volume of pollution-related literature and documentation, below is just a mere scope of major forms of human-induced pollution.

2.2.1 Air pollution
Without detailed discussion of AP, it is suffice to say that its relationship to human health is far from a simplistic formula that AP is "always bad". In reality its effects are highly non-linear and non-monotonic. Human immune system is known to exhibit hormesis effect, that is the adaptability of organisms to moderate pollution levels. In a sense, some forms of AP can "train" immune system to be more resistive. Medical inhalations and scent therapy is just one example. Absolutely sterile atmosphere can, in fact, be less health-friendly than moderately polluted. Non-organic (2.2.2.1) and organic (2.2.2.2) forms of pollution are key aspects of contamination from air, water and ground alike.

2.2.2 Water pollution
The problem of clean water availability becomes more and more acute with population growth and general contamination of the environment [3,5]. At the same time, various facets of water purification technology, such as seawater desalination plants, bottled water industries or transportation of polar icebergs to populated areas remain in the focus of social attention. We can point here to some novel quantum development related to informational content (memory effect) in water due to it quasi-crystalline structure [6]. The latter may potentially open a route of "self-purification" of water – the effect somewhat similar to natural fractionation of phases in heterogeneous mixtures (physical effect called *spinodal decomposition* which is a spontaneous separation of phases happening in thermodynamically non-equilibrium systems).

2.2.2.1 Non-organic pollution mostly refers to chemical spills containing such allegedly toxic elements as Cd, Hg, Co, Mn, etc. Curiously, some of them are part of typical poly-vitamin supplements and hence they are thought to have some positive health action. This again brings us to the concept of hormesis (see 2.2.2.3).

2.2.2.2 Organic compounds such as car exhausts, biological wastes, numerous food industry wastes, VOC (volatile organic compounds [7]), wastes from textile and other industries, etc all have numerous pollution control technologies associated to them. Quality, cost factors and efficiency vary in broad limits. Likewise, different geographical areas and countries have greatly different conditions and opportunities for applications of pollution control technologies.

2.2.2.3 Hormesis effect is a medically non-standard (and still controversial) effect of allegedly bringing benefits to the immune system which is "trained" by low toxicity levels. It may manifest itself in various kinds of pollution – air, water, low levels of radioactivity, etc.

2.2.3 Ground pollution
This type of pollution is often difficult and even impractical to separate from water pollution. Such specific forms as "brown fields" (former industrial grounds which can [sometime] be reclaimed for habitable use) have special means to deal with them.

2.2.4 Electromagnetic, noise and light pollution
These forms of "pollution" are specific in their "intangible" nature. Their interaction with other forms of pollution has many consequences. For example, patterns of birds migration are influenced by urban areas and, especially, by megalopolis sites (super-cities) which unstoppably grow across the world with exponential rate.

2.2.5 Informational pollution
Informational pollution constitutes an even more special category and at this point it is only gaining grounds in terms of its recognition as a form of pollution in some general sense. We mention it here for the purpose of consistency of the list of pollution items. More on it later in this article.

3   Electromagnetic emissions and electromagnetic pollution

Apart from some minor exceptions, almost all electricity produced by humans serves two purposes – to deliver power or to facilitate information and communication systems. The first group (electrical power engineering) includes power generation plants of various kinds (coal, hydro, nuclear, wind, solar), distribution systems (power lines, residential networks, etc) and user end (electrical appliances of all kinds). To that, to a certain degree, one can include autonomous small-scale electrical systems such as automobile or aircraft electrical equipment, etc. The second group (communication and computer engineering) includes computers, radio and TV broadcasting, phone lines, cells phones, satellite communications, etc. All this has numerous (and often non-trivial) links to a variety of physical phenomena and effects.

3.1 Physical aspects of electromagnetic pollution

While the abundance and ever-growing volume of human-generated electromagnetic waves (EMW) of various frequencies is obvious, a systematic study of electromagnetic pollution (EMP) is still relatively fragmentary. Because modern electrical, informational and communication technologies move towards atomic (and possibly soon to sub-atomic) scales [8], the quantum aspects of information [9] and its interaction with material environment are coming to the focus of environmental attention.

3.1.1 Power lines
Alleged health effects of power transmission lines is a hot topic of social discussions. While actual data on that are mixed, unreliable and controversial, there is an open (but largely unexplored) possibility that low frequencies (50 or 60 Hz) may have a complicated non-linear effect on human immune system. On a negative side there can be some increase of risk to trigger genetic-related processes (due to, e.g., polarization and/or magnetic effects on biological structures) and on a positive side it may lead to the above mentioned hormesis effect. Direct effects of power lines on AP can also take place in a variety of ways. One is the trapping effect of charge particulate due to eddy electric field as well as induced magnetism (Faraday induction) from changing alternating currents.

3.1.2 Communication satellites
Much of global communications is now based on satellites with equatorial geostationary orbits about 37, 800 km above the Earth's surface. They orbit Earth every 24 hours with orbital speed of 3.07 km/sec. Therefore, they "hang" over a fixed (equatorial) point on Earth's surface and appear immovable for an Earth's observer. Because of their distance from the Earth's surface the velocity of light limitation leads to a communication latency about 0.5 sec (both ways). One can ask what kind of a cumulative effect these satellites and their electromagnetic emissions can produce at a terrestrial scale and how these emissions can interact with AP and general environmental quality. Also, there are numerous satellites on lower orbits (a few hundred km over the ground) which rotate at about 90 min periods and may interact with ionosphere and Kennelly-Heaviside layer.

3.1.3 Magnetic effects
It is well known that the interaction of Earth's magnetic field with ion flux from the Sun produces noticeable effects such as *Aurora Borealis* (Northern Lights), disruption of radio communications, etc. One may question how the numerous satellites with their regular orbits can contribute to these effects. Because satellites have regular orbits and emit directed and informationally structured electromagnetic waves, they, in-spite of the weakness of their signals, may trigger various types of non-linear chaotic response in the ionospheric plasma as well as on dust particulates at the ground level.

3.1.3.1 Dust particles in magnetic fields are capable to exhibit a range of quantum effects. Special role of the magnetic component of the EMW may be caused by specific quantum effects characteristic for a dynamics of charged particles in a magnetic field, such as Larmour precession, Quantum Hall Effect and other macroscopic quantum phenomena. Most of a micron and sub-micron AP particulate usually has a net charge (negative or positive) of several elementary charges. This brings us to quantum regimes and the EMW-induced symmetry-breaking which can lead to a variety of chaos related effects, such as order-disorder structural phase transitions in suspended dust systems.

3.1.3.2 Accidental energy degeneration which was originally introduced in 1935 by one of the founders of quantum physics V.A.Fock (original German term: *Zufalligen Entartung*) refers to fact that in symmetrical systems, such as hydrogen atom (spherical symmetry), there are more energy levels having the same energy than can be expected from the geometry in a 3-dimensional space. Symmetry considerations involving higher (4-dimensional) space can explain that. Although at first glance accidental degeneration may be unrelated to AP problem, at a deeper level a possible link can be established through recognition that a very small perturbation in a symmetrical system can have a drastic effect (above mentioned "Butterfly effect"). The fact that Earth's gravitational field has (approximately) spherical symmetry can have a significant bearing on the enhancement of interaction between EMW and AP in a form pertinent to a new macroscopic quantum effect.

3.1.4 Order-disorder effects

Physically speaking, various forms of pollutions are non-linear systems. Response of such systems to external factors are far more rich in possible outcomes than the response of simplistic linear systems.

3.1.4.1 Ordered electromagnetic network can be specified as one producing EM noise which is different from random EM noise. Random EM noise in circuits (Johnson noise) is used in physics-based random number generators (RNG). The latter differ from common computer-based RNG which use some deterministic mathematical procedures to generate set of numbers which still (due to the way they are produced) contain some hidden correlations. For the topic of this paper (cross-effects between EMP and AP) the opposite claim may be the case, namely a possible detrimentality of *order* in human-generated IP. Random EMW (noise) produce effects which, so to say, average out and result in self-cancellation of the effects. Ordered EM signals emitted by satellites with fixed and periodic orbits may trigger pattern-formation phenomena in Earth's atmosphere. Non-linear "telegrapher's equations" introduced by Oliver Heaviside (1850-1925) for long electrical lines admit *soliton* (localized energy packs) solutions. Thus. low frequency (50 or 60 Hz) alternating currents in power lines and complicated connectivity of power grids can lead to formation of energy nodes (beats) or running localized solitons which, in turn, may result in large scale power disruptions propagating as domino effect. Likewise, charged particles from Solar Wind and/or galactic high-energy cosmic rays can acquire some focusing and directionality which can foster accumulation of dust particulate at some specific locations. This is a macroscopic analogue of the quantum effect of *Anderson localization*.

3.1.4.2 Anderson localization is a specific quantum effect of the formation of discrete (bound) quantum states in a periodical (regular) potential with some admixture of randomness (e.g., impurities in crystal lattice). In case of human-made EMW a similar situation happens. There is some regularity (described above) and there is also some disorder (randomness and noise). Because AP in general is a non-linear complex system, it can produce nodes of energy agglomeration which then can proliferate in a variety of ways.

3.1.5 Entropy argument

In its historical development biological life on Earth (biosphere) had adjusted itself to optimize its own survivability and proliferation One example is an apparent "self-healing" of the ozone layer holes. This effect not necessarily can be fully explained by the reduction of chlorofluorocarbons (CFC) emissions alone and most likely can point to some active participation of the biosphere itself.

From the systemic point of view self-regulation implies a negative feedback which is fed on ample availability of fluctuations. This exponentially enormous "library of fluctuations" [6] allows the system (biosphere in this case) to chose the optimal path of adjustability. In physical analogy this resembles the Least Action Principle or Entropy maximization principle introduced by Jaynes [10] in the context of Bayesian (conditional) probability. This is especially relevant for the case when different forms of pollution (here AP, EMP and IP) are in the state of dynamical interaction.

In the environmental/pollution context, the imposition of the ordered human-created EM network on the system can result in truncation (diminishing) of available choices. Therefore, EMP can limit degrees of freedom and impose an outside order on the biosphere. Instead of a free rein, the biosphere now faces additional constrains and may be derailed from its natural adaptational paths.

3.1.5.1 Ergodicity principle is an important aspect of thermodynamics which states that the phase trajectory of a system passes (eventually) through all isoenergetic (energetically available) microscopic states. In other words, all fluctuations are available and potentially can be tried by the system. The ordered (or partially ordered) EMP reduces space of available fluctuations. Ergodicity principle breaks down and the system no longer has an option of "visiting" all isoenergetic states. Some, otherwise available, channels for optimization now may be blocked with drastic consequences to the biosphere. The informational side of EMP can even further exacerbate this effect.

4 Informational pollution

At the present stage of civilization we are overwhelmed with all kind of information. It is generally recognized that many of us live in a state of informational and communicational overload. Thousands of TV channels and present-day public obsession with cell phones is a vivid illustration to that. Much of this information is in a digital form and is transmitted by EMW.

4.1 Entropy and information

Here we draw attention to the fact that EMP is not a random EM noise. The genuine randomness does not admit any hidden correlations or constrains. However, as we have seen, EMP is prone to some degree of order (though sporadic and usually local) and hence is conducive to produce a number of constraints which may deprive the eco-system from full-fledged adaptability.

4.1.1 Non-randomness versus non-equilibrium

The concepts of "randomness" and "non-randomness" are largely mathematical while the tandem "equilibrium" and "non-equilibrium" are defined for physical systems. And yet, there is some inner similarity between these two dichotomic pairs. For all practical purposes the physical (thermodynamic) equilibrium is equivalent to ergodicity principle, which implies an equal probability (and potential accessibility of *all* microscopic (statistical) states within a given macroscopic (thermodynamic) state of the entire system.

In non-equilibrium systems ergodicity principle breaks down. Different microstates laying on a fixed isoenergetic surface in the phase space are no longer having equal probability. In non-equilibrium systems certain sub-sets of states and dynamical regimes are becoming preferential, that in-essence means the departure from equiprobable randomness. This (non-equilibrium condition) makes system especially prone to the outside perturbations (like EMP in this case). They exhibit even higher sensitivity to initial conditions (greater "Butterfly effect") than systems in thermodynamic equilibrium. Instead of settling into their own quasi-stable states and dynamical regimes, they are now forced to adopt a behavior dictated by the external ordering forces. In other words, non-random EMP has a tendency to deprive the eco-system from its natural adaptability. While the latter may not be completely lost, its range (freedom of choices) gets truncated.

4.1.2 Boltzmann and Shannon entropy

Since the advent of concept of entropy by Rudolf Clausius (1822-1888) and Ludwig Boltzmann (1844-1906), it was felt that it has some connection to information. One popular metaphor for that is the idea of "Maxwell Demon" (MD) which has an extensive literature. Key quest here is how MD acquires the detailed (microscopic) information about the state of a large system and how this information is used, processed and discarded. Later Claude Shannon (1916-2001) introduced alternative definition of entropy within the framework of the information theory. These definitions put a bridge between thermodynamics and information. Information about the macroscopic system can be identified with its *negentropy*. In other words, when *our* information about a physical system increases, its entropy must decrease. A similar (but somewhat later) line of thinking has emerged on the basis of quantum theory of measurement and the idea of the *implicit order* introduced by David Bohm (1917-1992). Here, a fundamental premise is that the information is not just an abstract mathematical concept but has or may have a direct action on physical systems. In-spite that such ideas are facing opposition from some skeptics, we see them as central for our discourse on the interaction between EMP/IP and AP. One might expect the "Bohm effect" to be obscured if the electrons go around the region where a non-uniform (spatially and temporally) IP is superimposed onto the static magnetic flux.

4.1.3 Information bits as bosons

Quantum physics separates all elementary particles on two groups – fermions and bosons. From standpoint of quantum physics *bits* (quanta of information) are like bosons and hence, like physical bosons, they can exhibit the effect of the so called Bose condensation.

4.1.3.1 Bose condensation refers to the formation of the highly condensed agglomerate of bosons in macroscopic quantum systems such as superconductors, superfluids, and some biological systems. We suggest that in the modern global economy world, the said Bose condensation of high-density information bits in the informationally rich EM emissions (EMP) can amplify the pattern-forming effect on components of ecosystem and the environment (such as global AP) and lock them into some more detrimental modes. Another aspect is direct effects of EMP and IP on humans (e.g., cell phones). Collapse of communication could also be envisaged as the IP density keeps ever increasing in the modern global economy world.

4.1.3.2 Catalytic effect is another way to look at possible interaction between EMP/IP and AP. It is known that the catalyst is generally not spent in chemical reactions. Its action, therefore, is primarily informational. An example is formation of ozone holes by CFC pollutants. Here, our quest is how information carried by EMW (especially if it experiences Bose condensation of *info-bits*) can affect chemical processes in such pollutants as, say, VOC. It may appear in a form of beats (solitons), i.e., energy accumulation in localized packs when the EMW are correlated and/or coherent.

4.2 Chaos and self-organized criticality

With the advent of modern theory of chaos with all its ammunition of terms (bifurcations, Mandelbrot and Julia sets, strange attractors, sensitivity to initial conditions ["Butterfly effect"] and many others) we were exposed to the dichotomy of chaos and order [11]. There are windows of order within chaos. An example of it is bifurcation cascades revealed by iterations of the logistic equation (Mitchell Feigenbaum).

Another example of extreme sensitivity to initial conditions is the *self-organized criticality* (Per Bak) when a sudden breakdown in a large system (like a pile of sand) is triggered by a minute (but "critical") perturbation. This effect is somewhat akin to the sudden transitions in the Catastrophe Theory (Rene Thom). Such a triggering of sharp (phase) transitions by a very weak (but "threshold") perturbation is what one can expect in the interaction of IP with AP system. It should be noted that even a very small (but ordered) perturbation can often lead to drastic effects, like symmetry breaking in complex systems. One example is a weak interaction in beta-decay of some elementary particles. Likewise, a proper Fourier analysis can extract ordered information (message) buried under a much stronger noise. Thus, human-generated fluxes of EMW with its high informational content can lead to a variety of resonance-type effects in noosphere and ecosphere the consequences of which may be unexpected and unpredictable.

5 Conclusion

In this paper we have outlined some possible links regarding the interaction of "ordinary" pollution, such as AP with information-loaded electromagnetic emissions, which we tentatively label as EMP and IP. No form of pollution can be totally controlled and even less so eliminated. Calls for a "zero tolerance" to this-or-that form of pollution are naive and unworkable. Some reasonable level of accommodation (a generalized hormesis effect) is inevitable. We see a deepening need for a development of multi-dimensional, multi-sided interdisciplinary vision of the whole range of pollution problems in order to enhance our ability to understand all the important links in this global issue.

References

[1] Berezin, A.A. Quantum effects in electrostatic precipitation of aerosol and dust particles. *Air Pollution XIII*, ed. C.A. Brebbia, WIT Press: Southampton, pp. 509–518, 2005.
[2] Wilson, J.R. & Burgh, G. *Energizing our future: rational assessment of energy resources*, John Wiley & Sons, Inc.: Hoboken, New Jersey, 2008.
[3] Murray, I. *The really inconvenient truths*, Regency Publishing, Inc.: Washington, D.C., 2008.
[4] McDowell, S.D., Steinberg, P.E. & Tomasello, T.K. *Managing the infosphere: governance, technology, and cultural practice in motion*, Temple University Press: Philadelphia, 2008.
[5] Orrell, D. *Apollo's arrow; the science of prediction and the future of everything*, Harper Perennial (Harper Collins Publishers Ltd): Toronto, 2007.
[6] Berezin, A.A. Isotopic diversity in natural and engineering design. *Design and Nature II*, ed. M. Collins & C.A. Brebbia, WIT Press: Southampton, pp. 411–419, 2004.
[7] Vercammen, K.L.L., Berezin, A.A., Lox, F. & Chang, J.S. Non-Thermal Plasma Techniques for the Reduction of Volatile Organic Compounds in Air Streams: A Critical Review. *Journal of Advanced Oxidation Technologies*, 2(2), 312–329, 1997.
[8] Harms, A.A., Baetz, B.W. & Volti, R.R. *Engineering in time: the systematics of engineering history and its contemporary context*, Imperial College Press: London, 2004.
[9] Berezin, A.A. Quantum computing and security of information systems. *Safety and Security Engineering II*, eds. M. Guarascio, C.A. Brebbia & F. Garzia, WIT Press: Southampton, pp. 149–159, 2007.
[10] Jaynes, E. T. *Papers on probability, statistics and statistical Physics*, Reidel publishing Company: Dordrecht, Holland, 1983.
[11] Schroeder, M. *Fractals, Chaos, Power Laws*, W.H. Freeman and Company: New York, 1991.

Genotoxic and oxidative damage related to PM$_{2.5}$ chemical fraction

Sa. Bonetta[1], V. Gianotti[1], D. Scozia[2], Si. Bonetta[1], E. Carraro[1], F. Gosetti[1], M. Oddone[1] & M. C. Gennaro[1]
[1]*Dipartimento di Scienze dell'Ambiente e della Vita, University of Piemonte Orientale "A. Avogadro", Alessandria, Italy*
[2]*AMC, Azienda Multiservizi Casalese S.p.A., Italy*

Abstract

Many studies have pointed out a correlation between airborne PM quantitative exposure and health effects. The aim of this research is the investigation of the role of the PM$_{2.5}$ chemical fraction in the DNA damage induction in human cells (A549). Air samples (PM$_{2.5}$) were collected in different sites (urban, industrial and highway) using a high–volume sampler. Organic and water-soluble extracts of PM$_{2.5}$ were tested on A549 cells to evaluate genotoxic and oxidative damage using the Comet assay without and with formamido-pyrimidine-glycosylase (Fpg). Organic and water extracts were analysed for determination of PAHs by GC-MS methods and metals by the ICP-MS technique respectively. The PM$_{2.5}$ organic extract of all the samples caused a significant dose-dependent increase of the A549 DNA damage. The genotoxic effect was related to IPA PM$_{2.5}$ content and the highest effect was observed for the motorway site sample (65.03 CL/10m^3) while the oxidative damage was observed in PM$_{2.5}$ water extract of the industrial and motorway sites. The extent of the oxidative damage seems to be related to the type and concentration of metals present in these samples. The results of this study emphasize the importance of evaluating the PM chemical composition for the biological effect determination. This concern highlights the need for considering its qualitative composition in addition to its size and air concentration for PM health effect evaluation and exposure management.

Keywords: PM, genotoxicity, oxidative damage, Comet assay, PAH, metals.

1 Introduction

Particulate matter (PM) is an important environmental health risk factor for many diseases. Different studies show that long-term exposure to high concentrations of PM increases the risk of lung cancer, respiratory and cardiovascular diseases [1, 2]. Although epidemiological studies have consistently demonstrated adverse effects of PM exposure on human health, the physical and chemical properties of PM responsible for toxicity as well as the mechanisms underlying particle-induced carcinogenesis are still not fully known [3]. It is probable that different characteristics of PM are responsible for its adverse health effects. The particles size influences the capacity of PM fractions to reach the deepest sites of the respiratory system. Moreover, because of their large specific surface, PM can contain various organic substances that are known human mutagens and carcinogens, such as polycyclic aromatic hydrocarbons (PAHs) and nitroaromatic hydrocarbons (nitro-PAHs), as well as metals [4]. In particular, transition metal ions (Fe, Cu, Zn etc) are abundantly present in PM and have been shown to be potent inducers of oxidative DNA damage through the generation of reactive oxygen species (ROS) [5]. The health effects are mainly attributed to the small size particles with an aerodynamic diameter below 2.5 μm ($PM_{2.5}$) that penetrate deeply into the alveoli. Moreover the large and irregular surface areas favour the better adsorption of pollutants. The aim of this research is the investigation of the role of the $PM_{2.5}$ chemical fraction on the oxidative and genotoxic effects in human cells. With this scope the biological effects caused by $PM_{2.5}$ samples collected in sites with different emission sources were compared.

2 Materials and methods

2.1 Airborne particulate sampling

$PM_{2.5}$ samples were collected in different sites near the city of Alessandria (Piemonte): urban site (UR) (medium vehicular traffic density and home heating); highway site (HW) (high vehicular traffic density); industrial site (IN) (located near a foundry). The 24h sampling was performed in February 2007 using a high-volume air sampler (TISH Environmental, INC.) and glass fibre filters [6].

2.2 Extraction of PM components

The glass fibre filters were fractionated and one fraction (1/4) was added with 30mL of Milli-Q ultrapure water for 30min in an ultrasonic bath to extract water-soluble compounds and with 300 mL of dichloromethane to extract organic-extractable compounds. Soluble components were separated from the insoluble ones by centrifugation at 5000rpm for 10min. The supernatant was filtered with polypropylene membrane filters (0.45μm). PM extracts were separated into two aliquots (one for the chemical analysis and the other for Comet assay). Organic extracts were concentrated by rotary evaporation.

2.3 Chemical analysis

The PAHs level in the organic extracts was evaluated using a GC-MS Finnigan Trace GC Ultra-Trace DSQ (Thermo Scientific, San Jose, CA, USA) instrument with quadrupole mass analyzer. The column was a Thermo TR-5MS (60m x 0.25mm i.d.) coated with a 0.25µm film of 5% phenyl polysilphenylene-siloxane. The inlet temperature was 250°C and the splitless time was 1.00min. The column temperature was: 70°C (4min), then at 10°C min^{-1} from 70 to 120°C and then at 2°C min^{-1} from 120 to 300°C (21min). Helium was the carrier gas at a constant flow (1.0mL min^{-1}). Electron impact (EI+) mass spectra were acquired with an ionization energy of 70.0eV in full scan mode between 30-300amu (scan rate 500.0amu s^{-1}). The ion source and transfer line temperatures were at 250°C. The compounds identification was based on the comparison of their retention times and mass spectra with those of reference standards. The metal content in water extracts were measured by ICP-MS by ThermoFisher XSeries1 ICP-MS (Winsford UK), software PlasmaLab V2.5.4.289, equipped with an Apex-Q fully-integrated inlet system. The instrumental conditions were as follows: main run was: peak jumping, sweeps at 60ms, dwell time at 10000ms, channels per mass was 1, acquisition duration was 33750ms and channel spacing is 0.02amu. Resolution was standard.

2.4 Comet assay

The organic extracts were dissolved in DMSO for the biological test. A549 (human alveolar carcinoma) cells were maintained in nutrient mixture F-12 Ham supplemented with L-glutamine (200mM), 10% heat-inactivated newborn calf serum, 100U/mL penicillin, 100µg/mL streptomycin and 25mM Hepes. The cultures were incubated in a humidified incubator at 37°C with 5% CO_2 in air. The A549 cells were cultured for 24h before exposure to PM extracts. The proportion of living cells was determined by the trypan blue staining (overall > 90%). Cells were exposed (24h at 37°C) to serial dilutions of the PM extracts. After exposure, cell viability was checked again. The Comet assay was performed under alkaline conditions (pH>13) according to the method of Moretti *et al.* [7]. The slides were stained with ethidium bromide (20µl/mL) and examined with a fluorescent microscope (Axiovert 100M, Zeiss). One hundred cells per sample (two slides), randomly selected, were analysed using an image analysis system (CometScore). The Comet Length (CL) was selected as the parameter to estimate DNA damage. The data obtained were statistically evaluated by the Student's t-test (SYSTAT statistical package) to assess the significant differences (P≤ 0.05) between control and exposed cells.

2.5 Fpg-modified Comet assay

The Fpg-modified Comet assay was carried out as described above with the exception that, after lysis, the slides were washed 3 X 5 min with Fpg Buffer (40mM Hepes, 0.1M KCl, 0.5mM EDTA, 0.2mg/mL bovine serum albumin, pH 8). Then the slides were incubated with 1 unit of Fpg enzyme at 37°C for 1h.

Control slides were incubated with the buffer only. For each experimental point the mean comet length for enzyme untreated cells (CL) (direct DNA damage) and the mean comet length for Fpg-enzyme treated cells (CLenz) (direct and indirect DNA damage) were calculated.

3 Results and discussion

3.1 Gravimetric and chemical analysis

The obtained $PM_{2.5}$ air concentration and the abundance of different PAHs in PM extracts of the three sites investigated are summarized in table 1.

Table 1: $PM_{2.5}$ air concentration and $PM_{2.5}$ organic extract PAH concentration.

Parameter	UR	HW	IN
$PM_{2.5}$ (µg/m^3)	27.5	75.3	56.6
Naphthalene (ng/m^3)	n.d.	n.d.	0.1
Acenaphthylene (ng/m^3)	n.d.	n.d.	n.d.
Acenaphthene (ng/m^3)	n.d.	n.d.	n.d.
Fluorene (ng/m^3)	n.d.	n.d.	n.d.
Phenanthrene (ng/m^3)	0.5	1.5	0.8
Anthracene (ng/m^3)	n.d.	n.d.	n.d.
Fluoranthene (ng/m^3)	0.8	4.6	2.3
Pyrene (ng/m^3)	0.9	n.d.	2.5
Benzo(a)anthracene (ng/m^3)	1.4	n.d.	8.5
Benzo(b)anthracene (ng/m^3)	1.2	n.d.	2.7
Benzo(b)fluoranthene (ng/m^3)	1.1	11.0	8.2
Benzo(k)fluoranthene (ng/m^3)	1.1	5.0	n.d.
Benzo(a)pyrene (ng/m^3)	n.d.	10.4	4.3
Benzo(e)pyrene (ng/m^3)	2.2	5.8	2.5
Indeno(1.2.3cd)pyrene (ng/m^3)	n.d.	17.9	9.7
Dibenz(ah)anthracene (ng/m^3)	n.d.	n.d.	n.d.
Benzo(ghi)perylene (ng/m^3)	5.6	28.4	8.3
Benzo(ghi)fluoranthene (ng/m^3)	0.3	n.d.	0.5
Total mutagenic PAHs (ng/m^3)	10.9	77.3	43.8
Total PAHs (ng/m^3)	15.1	84.6	50.4
Total PAHs (ng/100µg $PM_{2.5}$)	54.4	112.5	89.0

The results of the gravimetric analysis showed that in all the sites the PM level exceeded the WHO guideline (10 µg/m^3), the future limit value proposed by the new European Directive (25 µg/m^3 mean annual level) and the standard value recently approved by the US EPA (15 µg/m^3 standard value for the annual level) [8]. The highest level of PM was observed in the HW site followed by the IN and UR sites. The great content of atmospheric $PM_{2.5}$ in the HW site is due to the high traffic density of the highway [9]. The chemical analysis of the PM organic extracts showed a variability of PAH composition in the three different sites. In the HW site the highest concentration of total and mutagenic PAHs was measured and the highest PAH concentration per µg of PM was found. The type and density of vehicular traffic likely affects PAH concentration. The chemical analysis (Table 2) of the PM water extracts points out the presence of 14 metals in all the PM samples investigated being the more abundant Fe, Cu, Zn, Sb and Ba.

Table 2: Metal concentrations in the $PM_{2.5}$ water extracts.

Parameter	UR	HW	IN
V (ng/m^3)	158	299	118
Cr (ng/m^3)	46	154	99
Mn (ng/m^3)	296	781	730
Fe (ng/m^3)	2127	7275	2747
Co (ng/m^3)	41	32	22
Cu (ng/m^3)	324	1236	3588
Zn (ng/m^3)	58906	72893	594011
As (ng/m^3)	21	137	74
Cd (ng/m^3)	21	64	644
Sn (ng/m^3)	n.d.	17,933	6321
Sb (ng/m^3)	92403	284229	233376
Ba (ng/m^3)	11160	10513	10805
Pt (ng/m^3)	25	69	21
Pb (ng/m^3)	213	460	839
Transition metals (ng/m^3)	61898	82671	601,314
Total metals (ng/m^3)	165741	396076	853394
Total metals (ng/µg $PM_{2.5}$)	6027	5260	15078

The highest concentration of total and transition metals was observed in the industrial site. Calculating the total load of soluble metals per unit mass of PM (µg) the IN site showed also the highest concentration of metals confirming the data expressed per m^3. The result is probably related to the presence in this area of a foundry that typically releases metals (Zn, Cu, Pb, Fe etc.). The presence of metals seems to be more affected by industrial emissions than by vehicular traffic.

3.2 Genotoxic damage and oxidative stress of PM extracts

The alkaline version of the Comet assay (sensitive to DNA strand breaks, direct oxidative DNA lesions and alkali-labile sites) was used to evaluate the genotoxic effect of organic extracts. The exposure of the A549 cells to $PM_{2.5}$ organic extracts showed a statistically significant (t-test) dose-dependent increase of the genotoxic effect with respect to the control cells in all the samples investigated (Figure 1).

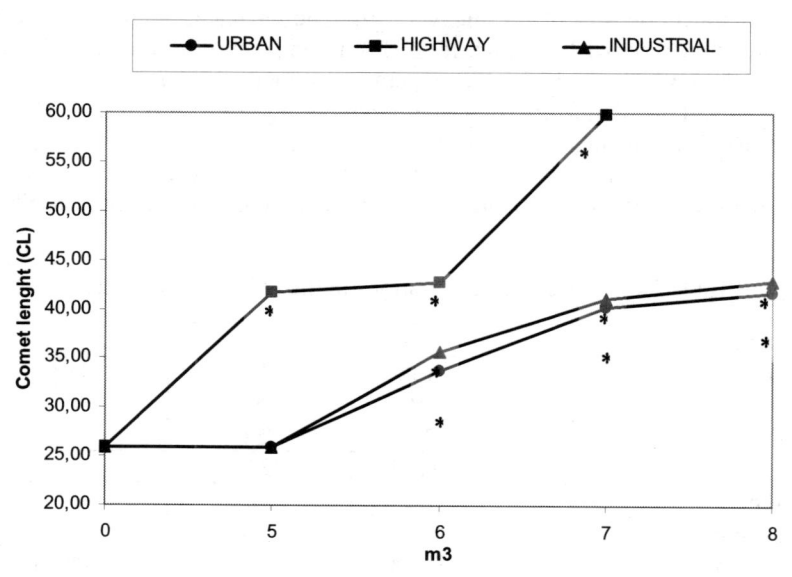

Figure 1: Effect of A549 cells exposure to organic extracts evaluated by alkaline version of the Comet assay (*: P<0.001).

In order to compare the results of the genotoxic effect the data obtained with $PM_{2.5}$ extracts sampled in the different sites were analysed as genotoxic parameter (CL) referred to 10 m^3 of air calculated from the dose-response curves by linear regression analysis (r^2> 0.7). The results obtained showed a variable degree of genotoxic damage in the monitored sites. The highest genotoxic activity was evidenced in the HW site (65.03 CL/10m^3) and the lower effect in the UR site (44.54 CL/10m^3). The biological effect seems to be related to PAH concentration observed in the sites investigated. In fact the HW site showed higher levels of total and mutagenic PAHs with respect to the IN and UR sites. Other studies showed the relationship between PM PAH concentration and the biological effects [10]. To evaluate the direct and oxidative DNA damage of water extracts the Fpg-modified Comet assay was used. The results obtained in the IN site showed the presence of a genotoxic effect both in enzyme untreated cells (CL) (direct DNA damage) and in enzyme treated cells (CLenz) (direct and indirect DNA damage) (Figure 2).

On the other hand, for the HW site a biological effect only using Fpg enzyme was observed. No genotoxic effect was showed in water extract of the UR site. For the IN site the DNA damage observed in enzyme untreated cells underlines the presence of pollutants with direct genotoxic effect. The subtraction of the mean CL from the relative CLenz value of the exposed cells (Clenz-CL) compared with unexposed cells at each experimental point provides the intensity of the oxidative damage. A significative oxidative damage was observed only in the IN site (39.40 CL/m^3) and a lower oxidative effect was revealed in the HW site (31.33 CL/m^3). The presence of the oxidative genotoxic damage could be related to the composition of metal in these sites. In fact in the UR site the lower

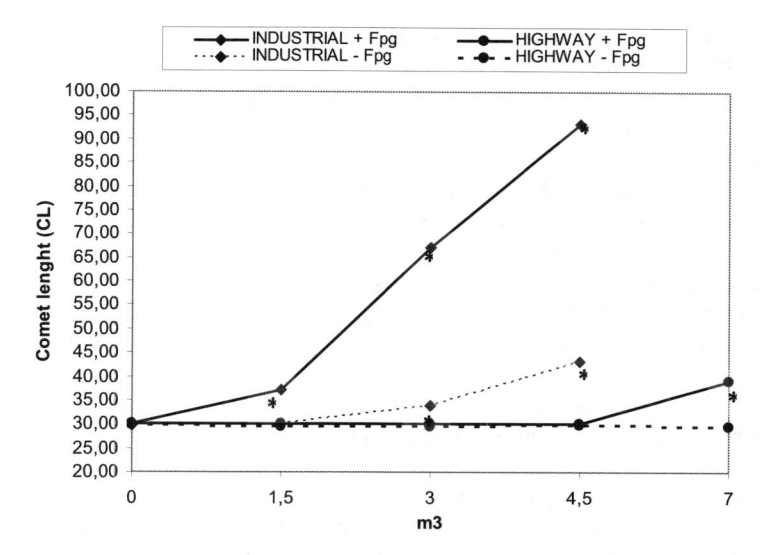

Figure 2: Effect of A549 cells exposure to water extracts evaluated by Comet assay and Fpg-modified Comet assay (*: P<0.001).

concentration of total and transition metals was observed while in the IN and HW sites a higher levels of these compounds was revealed. In particular the high level of oxidative damage in the IN site with respect to the HW site could be ascribed to the concentration of Cu, Zn e Cd that is 3-10 fold higher then in the PM sampled in the HW site. This finding supports the results obtained in other studies that showed the role of transition metals in reactive oxygen specie production and in oxidative stress induction [11, 12]. In different studies the genotoxic and oxidative activity of Cu and Zn are reported [13].

4 Conclusion

The results of this study highlight that emission sources characterized by different prevalent pollution (PAHs, metals) can significantly affect the intensity and type of the biological effect observed, emphasizing the importance of the PM chemical composition for the biological effect. This concern showed the need of considering the qualitative composition of the PM in addition to its size and air concentration for the PM health effect evaluation and the exposure management.

References

[1] Pope, C.A., Burnett, R.T., Thun, M.J., Calle, Krewski, D., Thurston, G.D., Lung cancer, cardiopulmonary mortality and long-term exposure to fine particle air pollution. *JAMA*, **287**, pp. 1132–1141, 2002.

[2] Jerrett, M., Burnett, R.T., Ma, R., Pope, C.A., Krewski, D., Newbold, K.B., Thurston, G., Shi, Y., Finkelstein, N., Calle, E.E., Thun, M.J., Spatial

analysis of air pollution and mortality in Los Angeles. *Epidemiology*, **16**, pp. 727–736, 2005.

[3] Knaapen, A.M., Shi, T., Borm P.J.A. & Shins, R.F.P., Inhaled particles and lung cancer. Part A. Mechanisms. *International Journal of Cancer*, **109**, 799–809, 2004.

[4] Claxton, L.D., Matthews, P.P. & Warren, S.H., The genotoxicity of ambient outdoor air, a review: *Salmonella* mutagenicity. *Mutation Research*, **567**, pp. 347–399, 2004.

[5] Knaapen, A.M., Shi, T., Borm, P.J.A. & Schins, R.F.P., Soluble metals as well as the insoluble particle fraction are involved in cellular DNA damage induced by particulate matter. *Molecular Cell Biochemistry*, **234/235**, pp. 317–326, 2002.

[6] D.M. Decreto Ministeriale 02/04/2002, n. 60. Recepimento della direttiva 1999/30/CE del Consiglio del 22 aprile 1999 concernente i valori limite di qualità dell'aria ambiente per il biossido di zolfo, il biossido di azoto, gli ossidi di azoto, le particelle e il piombo e della direttiva 2000/69/CE relativa ai valori limite di qualità dell'aria ambiente per il benzene ed il monossido di carbonio. Gazzetta Ufficiale n. 87, Suppl. n. 77, 13/04/2002.

[7] Moretti, M., Marcarelli, M., Fatigoni, C., Scassellati-Sforzolini, G. & Pasquini R., In vitro testing for genotoxicity of the herbicide terbutryn: cytogenetic and primary damage, *Toxicology In vitro*, **16**, pp. 81–88, 2002.

[8] Ballester, F., Medina, S., Boldo, E., Goodman, P., Neueberger, M., Iniguez, C., Künzli N., on behalf of the Apheis network, Reducing ambient levels of fine particulates could substantially improve health: a mortalità impact assessment for 26 European cities. *Journal of epidemiology and community health*, **62**, pp. 98–105.

[9] De Kok, T.M.C.M., Driece, A.L., Hogervorst, J.G.F., Briedè, J.J, Toxicological assessment of ambient and traffic-related particulate matter: A review of recent studies. *Mutation research*, **613**, pp. 103–122.

[10] Gilli, G., Pignata, C., Schilirò, T., Bono, R., La Rosa, A. & Traversi, D., The mutagenic hazards of environmental $PM_{2.5}$ in Turin. *Environmental Research*, **103**, pp. 168–175, 2007.

[11] Gábelová, A., Valovičová, Z., Lábaj, J., Bačová, G., Binková, B. & Farmer P.B., Assessment of oxidative DNA damage formation by organic complex mixtures from airborne particles PM_{10}. *Mutation research*, **620**, pp. 135–144.

[12] Gutiérrez-Castillo, M.E., Roubicek, D.A., Cebrián-García, M.E., De Vizcaya-Ruíz, A., Sordo-Cedeno, M. & Ostrosky-Wegman P., Effect of chemical composition on the induction of DNA damage by urban airborne particulate matter. *Environmental and Molecular Mutagenesis*, **47**, pp. 199–211, 2006.

[13] Shi, T., Knaapen A.M., Begerow, J., Birmili, W., Borm, P.J.A. & Schins, R.P.F., Temporal variation of hydroxyl radical generation and 8-hydroxy-2'-deoxyguanosine formation by coarse and fine particulate matter. *Occupational and Environmental Medicine*, **60**, pp. 315–321, 2003.

Preliminary results of aerosol optical thickness from MIVIS data

C. Bassani[1], R. M. Cavalli[1] & S. Pignatti[2]
[1]Institute for Atmospheric Pollution (IIA), National Research Council (CNR), Italy
[2]Institute of Methodologies for Environmental Analysis (IMAA), National Research Council (CNR), Italy

Abstract

Aerosols are the principal atmospheric constituents which the radiative field in the atmospheric window of solar spectral domain depends on. This dependence have to be considered when the radiative quantities of the atmosphere are simulated for solving the radiative transfer equation written for at sensor signal. The most important parameter for these simulations is the aerosol optical thickness at wavelength of $\lambda = 550$ nm, $\tau_a(\lambda = 550$ nm$)$, that is an input of the atmospheric radiative transfer codes used by the scientific community. The study carried out is based on the estimation of $\tau_a(\lambda = 550$ nm$)$ starting only from the MIVIS (Multispectral Infrared and Visible Imaging Spectrometer) remote sensing data. A minimization algorithm is applied to a quadratic cost function of the $\tau_a(\lambda = 550$ nm$)$ variable. In this paper are presented the preliminary results obtained from the MIVIS image processing in particulary pixels that satisfy both the conditions of lambertianity and spectral homogeneity of the surface.
Keywords: aerosol optical thickness, atmospheric radiative transfer, scattering, remote sensing.

1 Introduction

The aerosols are significant atmospheric constituents to estimate and predict direct component of climate forcing by scattering and absorption of solar and infrared radiation in the atmosphere [1]. Besides, Houghton *et al.* [1] showed that the aerosol mass and particle number concentrations are highly variable in space and time representing one of the largest uncertainties in climate change studies. In the last decades, the passive optical remote sensing data are observational support to estimate aerosol properties [2] since in atmospheric window the magnitude of at-sensor signal depends on the direct radiative forcing exerted by the aerosols on the solar radiation [3–5]. The most sensitive variable of the radiative field model in the atmosphere is the aerosol optical thickness, $\tau_a(\lambda)$. This is the fundamental basis of atmospheric study in optical remote sensing: the analysis of the aerosol contribution by simulation of the atmospheric state, in scattering and absorption terms, is achieved by atmospheric radiative transfer codes.

For istance, Kaufman et al. [6] developed a technique for the aerosol retrieval over land from MODIS data based on the aerosol radiative effects noticeable in the visible spectral domain and negligible in the near infrared.

In this work, the preliminary results of the aerosol optical thickness retrieved by using remote sensing data of AA5000 MIVIS (Multispectral Infrared and Visible Imaging Spectrometer) sensor are presented for the first time. In order to greatly reduce the computational time, the least complicated approach described by Guanter et al. [7], has been used for this study. Initially, a cost function is created that measures the deviation between the remote sensing data observed by the sensor and simulated by the computer code for atmospheric radiative transfer, namely 6S (Second Simulation of the Satellite Signal in the Solar Spectrum) [8]. The cost function is then minimized using the Powell algorithm described in [9]. The computed cost is dependent on the aerosol properties, mainly on optical thickness. Thus the minimum cost function provides the parameter set which fits the model most closely with the observed data.

2 MIVIS data

The Daedalus AA5000 MIVIS (Multispectral Infrared and Visible Imaging Spectrometer) is an airborne hyperspectral scanner [10] belonging to Italian National Research Council (CNR).

The MIVIS system is a whisk-broom scanner in which the light is collimated by the scan head and distributed by means of dichroic filters into four simultaneous recorder spectrometers operating in the visible, short-wave infrared and thermal spectral domain.

The sensor detects the downwelling radiation providing information on the earth-atmosphere system according to the characteristics shown in table 1. MIVIS sensor has proven to be highly suitable to perform surface detection into natural [11] and urban [12] environments by using a carefully selected spectral bands of the instrument.

3 Method

In the atmospheric window of visible spectral domain, the attenuation of electromagnetic wave through the atmosphere is caused by several factors, primarily the scattering of aerosol.

Table 1: MIVIS sensor technical details and spectrometer charasteristics.

	Channels Number	102
	Total Spectral Coverage	0.43 – 12.7 μm
	Istantaneous Field Of View	2.0 mrad
	Sample rate (angular step)	1.64 mrad
	Total Scan Angle (FOV)	71.059°
	Pixel per Scan-line	755

Spectrometer	Range	Channels	Bandwidth
I - VIS	0.43 – 0.83 μm	20	20 nm
II - NIR	1.15 – 1.55 μm	8	50 nm
III - SWIR	2.0 – 2.5 μm	64	8 nm
IV - TIR	8.2 – 12.7 μm	10	450 nm

Thus, the radiative contribution of scattering process is not negligible on the at-sensor radiance in the optical spectral range, this means that the sensor bands in the atmospheric window depends on the optical properties of the principal scatterers, the aerosols. The aerosol optical properties can be investigated from the at-sensor radiance by the inversion of the following radiative transfer equation written for the optical remote sensing data, [8]:

$$\rho_{sim}(\theta_s, \theta_v, \phi_s - \phi_v, z) = T_g(\theta_s, \theta_v, z)\left[\rho_{R+A}(z) + T^\uparrow T^\downarrow \frac{\rho_s}{1 - S\rho_s}\right] \quad (1)$$

In the equation 1, ρ_{sim} is the reflectance of the pixel observed by the sensor and obtained from simulation of radiative quantities; ρ_s is the pixel reflectance measured at ground. The other terms are spectral radiative quantities, functions of solar zenith angle, θ_s, view zenith angle, θ_v, relative azimuth, $\phi_{rel} = \phi_s - \phi_v$, and airborne altitude, z. $T^\downarrow = e^{-\tau/\mu_s} + t_d(\theta_s)$ and $T^\uparrow = e^{-\tau/\mu_v} + t_d(\theta_v)$ are the total transmission, direct and diffuse component sum, of the atmosphere along downwelling, $\mu_s = \cos(\theta_s)$, and upwelling, $\mu_v = \cos(\theta_v)$, directions respectively, and where $\tau(z)$ is the aerosol optical thickness in the layer between the ground and the z altitude. T_g is the gaseous transmission; ρ_{R+A} is the intrinsic reflectance of the molecular and aerosol layer; and S is the spherical albedo of the atmosphere.

There are several codes to simulate the spectral radiative quantities presented in the equation 1, providing specific atmospheric parameters. In particular, the aerosol optical thickness at 550 nm wavelength, $\tau_a(\lambda = 550$ nm$)$, and the four pure aerosol types (dust-like, water soluble, oceanic and soot) are the underlying variable parameters to control the aerosol radiative contribution on the radiative quantities of the equation 1 by means of the radiative transfer code.

In this study, the $\tau_a(\lambda = 550$ nm$)$ and the aerosol types have been retrieved by an inversion method of the at-sensor radiance from the MIVIS bands, L_{meas}^{MIVIS}, that are inside the atmospheric window of visible spectral domain assuming that the reflectance of the MIVIS pixel used, ρ_s, is known. The algorithm is based on the existing approach developed by Guanter et al. [7] specifically for aerosol retrieval from remote sensing data of CHRIS (Compact High Resolution Imaging Spectrometer) sensor on-board of the PROBA (Project for On-Board Autonomy) satellite.

The inversion algorithm is performed by means of the cost function minimization designed on the at-sensor signal of a specific pixel measured by the MIVIS sensor and simulated by the 6S (Second Simulation of the Satellite Signal in the Solar Spectrum) atmospheric radiative transfer code described in [8]. The 6S satisfies the requirement of lower computational cost preserving the computational accuracy for aerosol scattering effects, [8].

In order to match measured (L_{meas}^{MIVIS}) and simulated (ρ_{sim}^{MIVIS}) MIVIS data, the at-sensor radiance has been expressed in terms of equivalent reflectance by the following equation:

$$\rho_{meas}^{MIVIS} = \frac{\pi L_{meas}^{MIVIS}}{\mu_s E_s} \quad (2)$$

where E_s is the solar flux at the top of the atmosphere.

All the above-mentioned spectral quantities presented in the equation 1, namely Q_λ, and simulated by 6S code (spectral sampling is 2.5 nm), have been convolved in ith MIVIS band response filter, $f_i(\lambda)$, by:

$$Q_i^j(x, y, z) = \frac{\int_{\lambda_{min}}^{\lambda_{max}} Q_\lambda^j(x, y, z) f^j(\lambda) d\lambda}{\int_{\lambda_{min}}^{\lambda_{max}} f(\lambda) d(\lambda)} \quad (3)$$

where (x, y, z) are pixel coordinates on which the inversion is applied. The Q_i^j is the effective spectral value of jth simulated quantity referred to the ith MIVIS band. $[\lambda_{min}; \lambda_{max}]$ is the spectral range of filter function of ith's MIVIS band.

Once the pixel reflectance is retrieved from MIVIS acquisition, ρ_{meas}^{MIVIS}, and 6S simulation, ρ_{sim}^{MIVIS}, the aerosol properties are retrieved by the inversion of the equation 1 performing the minimization of the cost function described in [7]:

$$\delta^2 = \sum_{\lambda_i=1}^{20} \frac{1}{\lambda_i^2} [\rho_{sim}^{MIVIS} - \rho_{meas}^{MIVIS}]^2 \quad (4)$$

where λ_i are the central wavelengths of the selected twenty MIVIS bands inside the atmospheric window of the visible spectral domain; the λ_i^{-2} are the coefficients in order to weigh much more the MIVIS bands at the beginning of visible range where the aerosol scattering is higher and consequently the radiative contribution of aerosol on the at-sensor signal is considerable, after that the scattering decreases rapidly.

The inversion algorithm is performed by the Powell method [9] based on the minimization of cost function along each parameter space separately. Typically, design parameters that drive the inversion algorithm in the visible spectral domain and that are implemented in the 6S database, are the aerosol optical thickness and the four aerosol types (dust-like, water soluble, oceanic and soot) presented in [13].

4 Results

The inversion algorithm, that is the minimization of the cost function expressed in the equation 4, has been applied to MIVIS data acquired in Venice lagoon (45.4° N, 12.5° E, Italy) the August 22th, 2006 at 11:41 UTC. The MIVIS scene, shown in the figure 1, is a flat coastal region close to urban sites.

The selected study area, shown in the left of figure 2, includes pixels made up of concrete material only. The pixel reflectance, shown in the right of figure 2, satisfies the conditions of spectral homogeneity and reflectance isotropy. Both the conditions assure that the reference pixel reflectance, the ρ_s of the equation 1, is not composed by mixed materials and it is completely independent from the view and illumination angles (i.e., the bi-directional reflection function of the pixel is negligible). Besides, the clear sky at the MIVIS acquisition time was a particularly favorable radiative condition. The cloudless atmosphere implies that the radiative

Figure 1: MIVIS image (oriented to the North) acquired in Venice lagoon (45.4° N, 12.5° E, Italy) the August 22th, 2006 at 11:41 UTC. The study area is the bright coastal region at the northwestern part of the image.

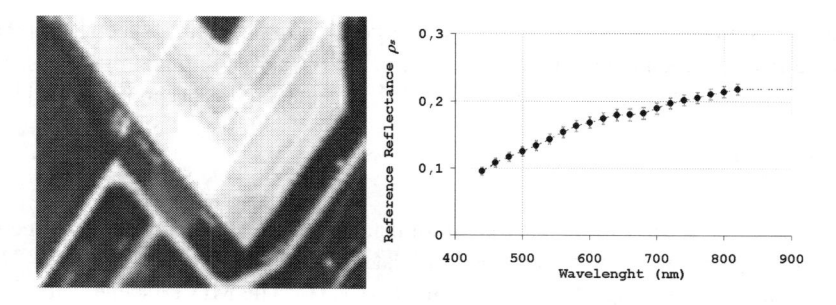

Figure 2: The study area selected on the runway of Venice airport in the coastal environment of the MIVIS image displayed in the figure 1. In the left side, zoom from the MIVIS image containing the selected square area composed by 112 pixel, being brighter than other ones; in the right side, the spectral reflectance of the area, made up of concrete material.

Table 2: Aerosol information retrieved from MIVIS data: aerosol optical thickness and the four aerosol types: dust-like, water soluble, oceanic and soot.

$\tau_a^M(\lambda = 550\text{ nm})$	% dust-like	% water soluble	% oceanic	% soot
0.142	19	32	38	11

forcing effects on the at-sensor radiance depends principally on the aerosol scattering of solar radiation. Hence, the parameters space of the inversion algorithm can be considered to be composed by aerosol optical properties only.

As concern the aerosol information retrieved by the inversion algorithm applied to the selected area of the MIVIS scene, the values of aerosol optical thickness and the four pure aerosol types (dust-like, water soluble, oceanic and soot) are displayed in the table 2. Unlike the aerosol optical thickness, the final percentage of the four aerosol types are very close to the initialization values and to the values expected in Venice lagoon (urban and maritime aerosol). Therefrom, $\tau_a^M(\lambda = 550\text{ nm})$ is the critical parameter to simulate the radiative effects of the aerosol scattering on the at-sensor radiance driving the inversion algorithm to the minimum of the cost function expressed by the equation 4.

During the MIVIS acquisition time, a sun photometer in at-ground station measured the aerosol optical thickness at the 550 nm wavelength. The station operated in the framework of the AERONET network described by Holben et al. [14]. The location, time of measurements and the values of aerosol optical thickness at wavelength 550 nm for the station considered here, are given in the table 3.

The aerosol optical thickness retrieved from MIVIS data, $\tau_a^M(\lambda = 550\text{ nm})$, is compared with ground-based AERONET measurements, $\tau_a^A(\lambda = 550\text{ nm})$. The correlation for $\tau_a^M(\lambda = 550\text{ nm})$ against $\tau_a^A(\lambda = 550\text{ nm})$ gives nearly a 1:1 relationship between retrieved from remote sensing and ground-based data, displayed in the table 2 and 3 respectively.

Table 3: The AERONET site.

Location	Position	Julian Day	Time	$\tau_a^A(\lambda = 550$ nm$)$
Venice	long = 12.5, lat = 45.3	234.49	11:46:13	0.138

The method for evaluating the goodness of airborne result is the percent deviation from the at-ground measurement by the following equation:

$$\Delta = \frac{\tau_a^M - \tau_a^A}{[\tau_a^M + \tau_a^A]/2} \qquad (5)$$

The low value, 3%, remarks the goodness of the results obtained from the first application of the inversion algorithm to the MIVIS data.

5 Conclusions

The first application of the inversion algorithm to retrieve aerosol optical thickness from MIVIS data has been presented in this work. The algorithm basically works by minimizing the cost function, expressed by equation 4, in the space defined by the most sensitive parameters. The retrieval relies on the knowledge of the underlying surface reflectance of the pixel which the algorithm has to be applied. The algorithm uses the 6S radiative transfer code to allow for lower computational cost maintaining the accuracy of the simulated data in the equation 1.

The preliminary results presented in this work highlight the potential of the optical MIVIS data in the assessment of aerosols optical properties related to the scattering of solar radiation. The algorithm has yielded information about the aerosol properties which helps to define, and providing understanding of, the atmospheric state during MIVIS airborne acquisition.

There are unsolved questions that have to be investigated further. In order to prove the robustness of the algorithm, increasing the MIVIS data where the method has to be applied on. The future plans also include to test the method in different aerosol content to analyse the performance of the inversion algorithm. Besides, the improvements of the algorithm are investigating about aerosol optical thickness without recourse to *a priori* information on the surface reflectance of the pixel observed, ρ_s of the equation 1.

References

[1] Houghton, J.T., Ding, Y., Griggs, D.J., Nouger, M., van der Linden, P.J., Dai, X., Maskell, K. & Johnson, C.A., *Climate Change 2001: The Scientific Basis*. Cambridge University Press: Berlin and New York, pp. 11–13, 2001.

[2] Kaufman, Y.J., *Theory and Applications of optical remote sensing*, Wiley, chapter The atmospheric effect on remote sensing and its correction, 1989.

[3] Charlson, R., Schwartz, S., Hales, J., Cess, R., Jr., J.C., Hansen, J. & Hofmann, D., Climate forcing by anthropogenic aerosol. *Science*, **255**, pp. 423–430, 1992.

[4] Kaufman, Y.J., Tanré, D. & Boucher, O., A satellite view of aerosols in the climate system. *Nature*, **419**, pp. 215–223, 2002.

[5] Penner, J.E., Dong, X.Q. & Chen, Y., Observational evidence of a change in radiative forcing due to the indirect aerosol effect. *Nature*, **427**, pp. 231–234, 2004.

[6] Kaufman, Y.J., Wald, A.E., Remer, L.A., Gao, B.C., Li, R.R. & Flynn, L., The MODIS 2.1-mum channel - correlation with visible reflectance for use in remote sensing of aerosol. *IEEE Transactions on Geoscience and Remote Sensing*, **35(5)**, pp. 1286–1298, 1997.

[7] Guanter, L., Alonso, L. & Moreno, J., A method for the surface reflectance retrieval from PROBA/CHRIS data over land: application to ESA SPARC campaigns. *IEEE Transactions on Geoscience and Remote Sensing*, **43(12)**, pp. 2908–2917, 2005.

[8] Vermote, E., Tanré, D., Deuzé, J., Herman, M. & Morcrette, J., Second simulation of the satellite signal in the solar spectrum: an overview. *IEEE Transactions on Geoscience and Remote Sensing*, **35(3)**, pp. 675–686, 1997.

[9] Press, W., Flannery, B., Teukolosky, S. & Vetterling, W., *Numerical Recipes*, volume 1. Cambridge University Press, 2nd edition, 1986.

[10] Bianchi, R., Marino, C. & Pignatti, S., Airborne hyperspectral remote sensing in Italy. *EUROPTO series recent advances in remote sensing and hypespectral remote sensing*, SPIE, volume 2318, pp. 29–37, 1994.

[11] Cavalli, R., Fusilli, L., Guidi, A., Marino, C., Pascucci, S., Pignatti, S., Casoni, L.V. & Vinci, M., The natural areas of Rome province detected by airborne remotely sensed data. *Annals of Geophysics*, **49(1)**, pp. 187–199, 2006.

[12] Bassani, C., Cavalli, R., Cavalcante, F., Cuomo, V., Palombo, A., Pascucci, S. & Pignatti, S., Deterioration status of asbestos-cement roofing sheets assessed by analyzing hypespectral data. *Remote Sensing of Environment*, **109**, pp. 361–378, 2007.

[13] A preliminary cloudless standard atmosphere for radiation computation. Technical Report WCP-112, WMO/TD-NO.24, International Association for Meteorology and Atmospheric Physics, Geneva, Switzerland, 1986.

[14] Holben, B.N., Eck, T.F., Slutsker, I., Tanré, D., Buis, J.P., Vermote, E., Reagan, J.A., Kaufman, Y.J., Nakajima, T., Lavenu, F., Jankowiak, I. & Smirnov, A., AERONET - a federate instrument network and data archive for aerosol characterization. *Remote Sensing of Environment*, **66**, pp. 1–16, 1998.

Section 8
Air pollution effects on ecosystems

Response of lichens to heavy metal and SO_2 pollution in Poland – an overview

K. Sawicka-Kapusta, M. Zakrzewska, G. Bydłoń & A. Sowińska
Environmental Monitoring Research Group,
Institute of Environmental Sciences, Jagiellonian University,
Kraków, Poland

Abstract

Epiphytic lichen Hypogymnia physodes (L) Nyl. from the natural environment and transplanted was used for ten years (1997–2006) to estimate air contamination in Poland. Lichens from natural environment were collected from the Base Station of the Integrated Nature Monitoring System and from Polish National Parks. Transplantation was carried out in Cracow city, small forest sites near Cracow conurbation and industrial areas in the Małopolska district. Hypogymnia physodes showed the changes in air pollution in Poland but also confirmed the presence of some areas still being contaminated. They are located near heavy industry, near large cities like Cracow and also in some areas located far from industrial sources.
Keywords: Poland, biomonitoring, air pollution, heavy metals, sulphur dioxide, lichens, Hypogymnia physodes.

1 Introduction

Biological monitoring can be defined as the measurement of the response of living organisms to changes in their environment [1, 2]. It is a very useful and sensitive method for estimating air and environment quality. This method has been successfully used for more than fifty years [3–5]. Lichens are very good bioindicators of air pollution as they accumulate contaminants as the function of their concentration in the air. This accumulation undergoes the passive way [6, 7]. Another advantage of lichens as an excellent bioindicator is that many species have a wide geographical distribution and are very common. Therefore they can be used at a local, regional and national scale [5, 8, 9]. Air pollution originates mainly from industrial manufacturing and energy production, coal and oil combustion, vehicular traffic and also small local sources. In Europe the main sources of air pollution have been increasing urbanisation and heavy traffic [10–12]. In the last ten years of 20th century Poland was one of the most polluted countries in Europe [13, 14]. The country was heavily affected by gaseous (SO_2, NO_X, CO, CO_2), particulate emissions (including heavy metals and polycyclic aromatic hydrocarbons) and other organic compounds. The situation has significantly improved in the last few years, but still high concentrations of these pollutants are being measured. Poland became a member of the European Union in 2004 and according to Environmental Protection Law showed adherence to standards of air quality. In 2005, Poland emitted 457 thousands tonnes of dust, 1222 thousand tonnes of sulphur dioxide, 811 thousand tonnes of nitrogen dioxide and 326 511 thousand tonnes of carbon dioxide. Heavy metal emissions also decreased when compared with the early nineties but in 2005 Poland emitted 1350 metric tonnes (t) of Zn, 536 t Pb, 46 t Cd, 20 t Hg, 54 t Cr, 237 t Ni and 356 t Cu [14]. At the present time the main sources of emissions are not heavy industry but large conurbations, heavy traffic and an abundance of local sources, located in small towns and villages.

The aim of this paper is to show the changes in air pollution in Poland in ten years using *Hypogymnia physodes* as a bioindicator.

2 Material and methods

Epiphytic lichen *Hypogymnia physodes* (L) Nyl. from the natural environment and transplanted were collected all over Poland over a ten year period (1997–2006). The investigated areas were located in different polluted parts of the country, around the heavy industry: steelworks, metal smelters, coal and metal mining, chemical factories, busy roads, small urban areas and Cracow conurbation. Samples were also taken from different forest areas, national parks, and Base Stations of the Integrated Nature Monitoring System for Poland. *Hypogymnia physodes* from unpolluted areas (Borecka Forest or Bory Tucholskie Forest) were transplanted to 7 small forest sites near Cracow agglomeration, 11 sites in Cracow city, and to 4 industrial sites in the Małopolska district. The transplants were exposed to a six month long period of winter and summer seasons. Unwashed lichen was digested in 4:1 nitric and perchloric acid. Concentration of heavy metals (Cd, Pb, Cu, Zn, Fe, Cr, Ni) were determined using IL 250 flame AAS [15], whereas sulphur content was determined using turbidimetric Butters-Chenry's method [16]. Simultaneously reference materials (SRM) were also analysed. Data are presented in $\mu g \cdot g^{-1}$ dry weight. Statistical analysis was carried out to determine the potential differences between the concentration of contaminants between sites and period of time.

3 *Hypogymnia physodes* the useful bioindicator of air pollution

Lichens are the symbiotic association between fungus and alga or cyanobacterium. Epiphytic lichens growing on the tree stems and branches use them only as a substrate. Instead lichens get nutrients from rainwater and deposited dust. They absorb substances for growth and survival through the exposed surface of the thallus [6, 7]. Lichens can accumulate trace metals (both essential and non essential) to levels far greater than their expected physiological needs [7, 17] Therefore they are used as sensitive bioindicators of heavy metals in the air [5, 8, 17]. Lichens have been shown to be highly sensitive to gaseous air pollution, particularly to sulphur dioxide. SO_2 caused high concentration of sulphur and acute injuries of lichen thalli [3, 4, 18, 19]. Lichens show the concentrations of metals and sulphur, also organic compounds and radionuclides as a function of atmospheric deposition amount. Thus it makes them widely used in monitoring of air pollution [4–7]. Lichens could be collected from natural environment or transplanted from clean area to the polluted sites [5, 15, 17, 21, 22].

During 1997 and 2006 several investigation using epiphytic lichens *Hypogymnia physodes*, collected from the natural environment or transplanted were carried out to estimate air pollution in different parts of Poland [12, 15, 23–27].

4 Investigation of air pollution in some areas in Poland

4.1 Monitoring of air pollution by heavy metals and SO_2 in the Base Stations of the Integrated Nature Monitoring System between 2001-2005

Heavy metals and sulphur concentration in *Hypogymnia physodes* collected from natural environment in the areas of seven Base Station of the Integrated Nature Monitoring System were determined [25]. In July 2001, 2003 and 2005 lichen samples were collected in Szymbark, Św. Krzyż, Pożary, Storkowo, Koniczynka, Puszcza Borecka and Wigry. The highest concentration of all heavy metals and sulphur was found in 2001 in Koniczynka Base Station situated 5 km north-east from Toruń agglomeration. Also high concentrations of cadmium, lead, zinc, iron and sulphur were noticed in lichens collected in Św. Krzyż and Szymbark Base Stations in 2001. The lowest level of investigated metals was found in Wigry and Storkowo Base Stations. Sulphur concentration ranged from 1200 $\mu g \cdot g^{-1}$ in Puszcza Borecka to 2889 $\mu g \cdot g^{-1}$ in Koniczynka Base Station. In the year 2003 the concentration of cadmium, lead, zinc, iron and sulphur increased in almost all Base Stations while copper remained at the same level. In 2005 a decrease of all investigated metals was noticed, with the exception of cadmium in Puszcza Borecka Base Station. In the case of sulphur, a higher concentration was noticed in Puszcza Borecka and Storkowo Base Station, lower in Pożary, Św. Krzyż and Szymbark (fig. 1). No changes were found in Wigry Base Station. The results of the following study confirmed air contamination by heavy metals and sulphur dioxide in three of the investigated Base Stations (Koniczynka, Św. Krzyż, Szymbark). Generally Base Stations located on the North of Poland were less contaminated by heavy metals (Cd, Pb, Cu, Zn, Fe) and SO_2 than Stations situated on the South. No statistical differences between years were found in all Base Stations. The sources of emissions which caused contamination are different in each of the investigated Base Stations [25–27].

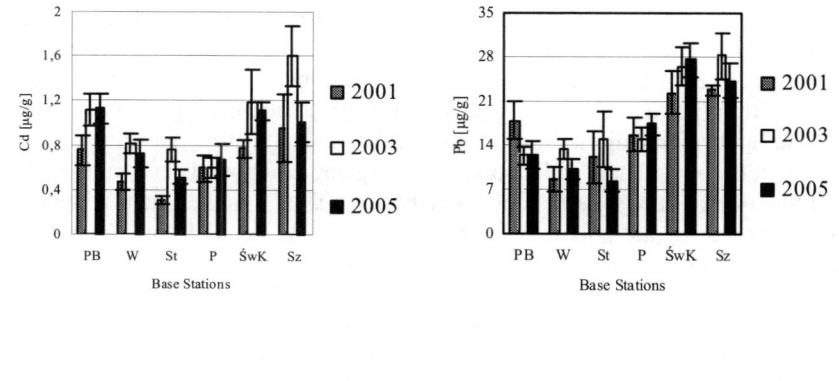

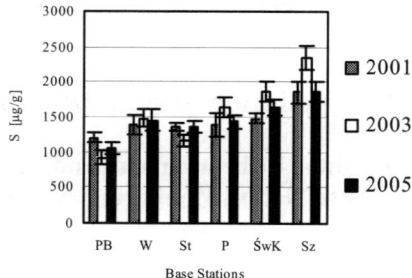

Figure 1: Lead, cadmium and sulfur concentrations ($\mu g \cdot g^{-1}$ d.w.) in the air of the Base Station of the Integrated Nature Monitoring System.

4.2 Changes in air pollution in Polish National Park

The first estimation of heavy metal contamination of the environment in Poland on a national scale, based on metal concentration in lichen *Hypogymnia physodes* were conducted in 1990 in 13 national parks [15]. The investigations were repeated in 1998 and 2005 in all Polish National Parks [23].

Samples of *Hypogymnia physodes* from natural environment were collected in July 1998 and 2003 from 23 Polish National Parks. The aim of this study was to estimate air pollution by heavy metals (Pb, Cd, Cu, Zn, Fe) (fig. 2) and sulphur dioxide and to compare the contamination over period of 5 years [22]. In 1998 and 2003 *Hypogymnia physodes* from Ojcowski National Park had the highest concentration of all heavy metals. In 1998 the lowest Cd and Pb concentration were found in Bory Tucholskie NP and in Woliński NP, respectively. The lowest Cu and Zn amount was observed in lichens from Białowieski NP while the concentration of Fe was lowest in Woliński NP. In 2003 the lowest content of Cd, Pb and Zn were determined in Drawieński NP while the concentration of Cu and Fe were found in Białowieski NP. The highest concentration of S in 1998 was detected in lichens from Świętokrzyski NP while in 2003 in lichens from Warta River Mouth NP. When comparing the concentration of heavy metals in lichens collected in 1998 and in 2003 it was found that the same twelve national parks were classified as relatively clean. They are: Woliński, Drawieński, Słowiński, Wigierski, Białowieski, Narwiański, Biebrzański, Bory Tucholskie NPs situated in the north of Poland and Kampinoski, Poleski, Roztoczański, Wielkopolski NPs located in the central part of the country. Warta River Mouth NP established in 2001 also belongs to relatively clean. Moderately polluted areas were represented by nine national parks (Świętokrzyski, Karkonoski, Góry Stołowe, Gorczański, Babiogórski Pieniński, Tatrzański, Bieszczadzki and Magurski NPs. Ojców NP situated in heavily industrialised region, remained heavily polluted. As far as SO_2, nine national parks were classified as clean; seven were recognized as moderately polluted, fifth as polluted and two were heavily polluted.

Sulphur dioxide concentration in the lichens from national parks has not decreased compared to 1998 as sulphur concentration is much higher than 5 years ago. Comparison of air pollution by heavy metals was not much lower as was expected. Cadmium concentration was even higher than 5 years ago [23]. Generally, national parks located in the southern part of Poland were more contaminated by heavy metals than those in the north of the country. Contamination by SO_2 is much more uniform across the country as high sulphur concentrations were noticed in national parks located in different parts of Poland. Air contamination in national parks originated from industrial sources, long distance transport but mainly from local sources located close to national parks border.

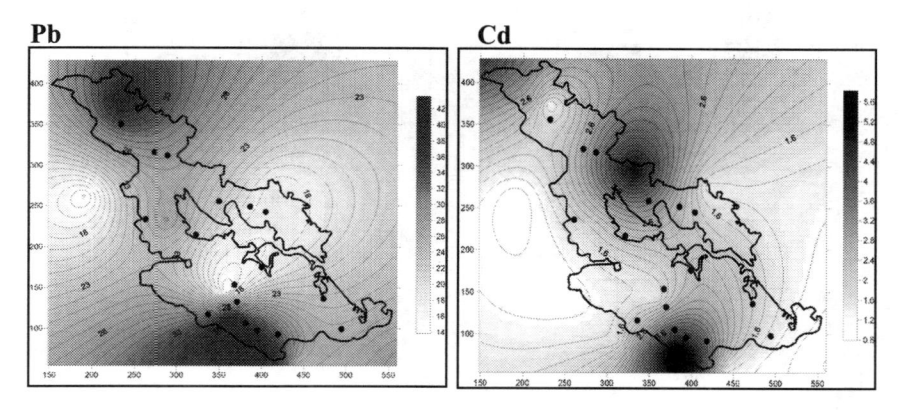

Figure 2: Lead and cadmium concentration ($\mu g \cdot g^{-1}$ d.w.) in the air of the Magurski National Park.

4.3 Lichens as bioindicators of urban pollution

Seven small forest sites located at different distance from Cracow conurbation and from busy road and along the prevailing wind direction were selected. Five sites were located along southern transect (Bonarka located 3 km from the town centre, Rajsko - 8 km, Mogilany - 14 km far from the city, Kornatka near Dobczyce reservoir - 26 km from Cracow and Węglówka - 35 km from agglomeration. Along the eastern transect only two sites: Koło (25 km from steelworks and 30 km from Cracow) and Ispina (30 and 35 km respectively) were located. *Hypogymnia physodes* from unpolluted area (Borecka Forest, north-eastern Poland) were transplanted to each of investigated sites for a six-month period. The transplants were exposed for two winter seasons 1998/1999 and 1999/2000 and for summer 2000. Observation for macroscopic injuries appearing on lichen thalli surface was performed monthly in each site. After transplantation concentrations of Cd, Pb, Cu, Zn, Fe and S in lichens were determined [12]. The highest metal concentration in *Hypogymnia physodes* was found in sites located close to Cracow conurbation. Generally concentrations of heavy metals and sulphur were higher after winter transplantations. The highest lead concentration (48 $\mu g \cdot g^{-1}$) was in Bonarka after winter 1999/2000 and in Bonarka and Rajsko (36 $\mu g \cdot g^{-1}$ on average) after winter 1998/1999 (fig. 3). The highest iron concentration (4200 $\mu g \cdot g^{-1}$) was found in Bonarka after winter 1998/1999 and in Koło and Ispina (2700 $\mu g \cdot g^{-1}$ on average) after in winter 1999/2000. High metal concentrations and acute injuries on lichen thalli observed during the winter transplantation were caused by high SO_2 emission from Cracow. High heavy metal and sulphur concentrations in *Hypogymnia physodes* confirm impact of Cracow conurbation (including steelworks emission) on small forest areas [12].

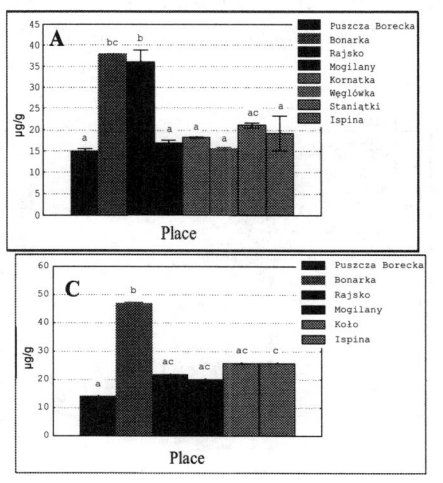

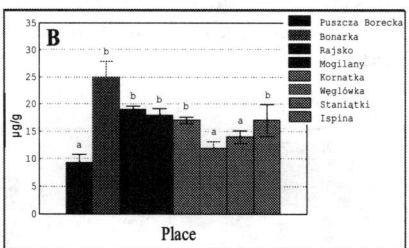

A – winter 1998/99
B – summer 1999
C – winter 1999/2000

a, b – different letters indicate statistical differences between investigated places

Figure 3: Lead concentration ($\mu g \cdot g^{-1}$ d.w.) in transplanted lichens *Hypogymnia physodes*.

4.4 *Hypogymnia physodes* identified metal pollution from industrial sources in Małopolska district

Lichen *Hypogymnia physodes* from Borecka Forest (unpolluted area) were transplanted in the middle of April 2002 for a six-month period to four forest sites located near industrial sources [27]. The following sites were chosen: 1. Bukowno situated in the immediate vicinity of the "Bolesław" Zn-Pb smelter, 2. Młoszowa located close to "Trzebinia" refinery and "Siersza" power plant, 3. Jankowice affected by emission from the chemical industry "Dwory" in Oświęcim and 4. Alwernia located close to the "Alwernia" chemical industry. After exposition, concentrations of heavy metals (Cd, Pb, Zn, Cr, Ni, Cu, Fe) and S in lichens were analysed. Statistical highest concentrations of Cd, Pb and Zn were detected in Bukowno as the result of zinc-lead smelter manufacturing. Extremely high Cr concentration in lichens was determined in Alwernia because of chromium compounds production. S concentration was rather uniform in all investigated sites with highest concentration in Młoszowa and Jankowice as the result of emission from power plant (based on coal) using coal and H_2SO_4 production in chemical industry [27].

Table 1: Average concentration (± SD) of elements in *Hypogymnia physodes* transplanted to the Cracow-Silesia industrial region.

Locations	Elements (μg^{-1} d.w.)			
	Cadmium	Lead	Copper	Zinc
Borecka Forest	0.54 ± 0.02 [a]	9.0 ± 0.4 [a]	3.7 ± 0.4 [a]	55 ± 7 [a]
Bukowno	7.70 ± 0.47 [d]	123.7 ± 20.0 [b]	10.8 ± 0.3 [d]	583 ± 56 [c]
Młoszowa	1.37 ± 0.05 [c]	11.2 ± 1.5 [a]	7.5 ± 0.4 [bc]	68 ± 6 [ab]
Jankowice	1.02 ± 0.12 [b]	11.3 ± 1.2 [a]	5.9 ± 0.3 [b]	55 ± 4 [a]
Alwernia	1.29 ± 0.10 [bc]	12.7 ± 3.1 [a]	9.9 ± 1.8 [cd]	77 ± 5 [b]

Locations	Elements (μg^{-1} d.w.)			
	Iron	Nickel	Chromium	Sulphur
Borecka Forest	350 ± 47 [a]	1.08 ± 0.19 [a]	0.22 ± 0.04 [a]	1237 ± 33 [a]
Bukowno	1306 ± 164 [c]	1.91 ± 0.12 [b]	0.79 ± 0.06 [b]	1842 ± 397 [b]
Młoszowa	720 ± 106 [b]	1.36 ± 0.19 [ab]	0.48 ± 0.06 [b]	1960 ± 109 [b]
Jankowice	597 ± 38 [b]	1.38 ± 0.06 [ab]	2.89 ± 0.54 [b]	1999 ± 242 [b]
Alwernia	1202 ± 368 [c]	9.18 ± 1.42 [c]	75.85 ± 4.62 [c]	1665 ± 109 [ab]

[a, b, c, d] – means within a column with different letters are statistically different in the element levels between studied sites at $P < 0.05$ for $N = 5$. Data from Białońska and Dayan [27].

5 Conclusions

1. Air contamination in Base Stations of the Integrated Nature Monitoring System was not improved during the 2001-2005 period
2. National parks located in the southern part of Poland were more contaminated by heave metals than those in the north of the country. SO_2 is much more uniform across the country as high sulphur concentrations were noticed in national parks located in different parts of Poland.
3. High metal and sulphur concentrations and acute injuries on *Hypogymnia physodes* thalli confirmed impact of Cracow conurbation (including steelworks emission) on small forest areas.
4. Transplanted *Hypogymnia physodes* is the excellent tool to identify metal emission from industrial sources.

References

[1] Burton M.A.S., *Biological Monitoring of Environmental Contaminants (Plants)*. MARC Rep. 32, Monitoring and Assessment Research Centre, King's College London, University of London, London, 1986.
[2] Martin, M.H. & Coughtrey, P.J., *Biological Monitoring of Heavy Metals Pollution - Land and Air*. Applied Science Publisher, London, 1982.
[3] Haksworth D.L.& Rose F., Qualitative scale for estimating sulphur dioxide air pollution in England and Wales using epiphytic lichens. *Nature* **227 (254)** pp. 145–148, 1970.
[4] Kranner I., Beckett R.P. & Varma A.K., (eds.) *Protocols in Lichenology. Culturing, Biochemistry, Ecophysiology and use in Biomonitoring*. Springer Lab Manual, Berlin, Heidelberg, New York, 2002.
[5] Conti, M.E. & Cecchetti, G., Biological monitoring: lichens as bioindicators of air pollution assessment - a review. *Environmental Pollution*, **114**, pp. 471–492, 2001.
[6] Tyler, G. Uptake, retention and toxicity of heavy metals in lichens. *Water, Air and Soil Pollution*, **47**, pp. 321–333, 1989.
[7] Puckett, K. J., Bryophytes and lichens as monitors as metal deposition. Lichens, Bryophytes and Air Quality. *Lichenologist*, **30**, pp. 231–267, 1988.
[8] Jeran, Z., Jacimovic R., Batič, F., Smodis B. & Wolterbeek H. Th. Atmospheric heavy metal pollution in Slovenia derived from results of epiphytic lichens. *Frasenius J Anal. Chem.*, **354**, pp. 681–687, 1996.
[9] Giordani P., Is the diversity of epiphytic lichens a reliable indicator of air pollution? A case study from Italy. *Environmental Pollution*, **146**, pp. 317–323, 2007.
[10] Cotrufo, M.F., De Santo, A.V., Alfani, A., Bartoli, G. & De Cristofaro, A., Effects of urban heavy metal pollution on organic matter decomposition in *Quercus ilex* L. woods. *Environmental Pollution*, **8**, pp. 81–87, 1995.
[11] Monaci, F., Moni, F., Lanciotti, E., Grechi, D. & Bargagli R., Biomonitoring of airborne metals in urban environments: new tracers of

vehicle emission, in place of lead. *Environmental Pollution*, **107**, pp. 321–327, 2000.
[12] Sawicka-Kapusta K., Gdula-Argasińska J., Zakrzewska M., Budka D. & Szpakowska K., The influence of Cracow urban pollution on small forest areas. *J.Phys. IV France* **107**, pp. 1197–1200, 2003.
[13] Nowicki, M., *Environment in Poland. Issues and Solutions*. Kluwer Academic Publishers. Dordrecht/Boston/London. Dordrecht, 1993.
[14] *Environment*. Central Statistical Office, Warsaw, Poland, 2007.
[15] Sawicka-Kapusta, K. & Rakowska, A., Heavy metal contamination in Polish national parks. *The Science of the Total Environment, Supplement, Part 1,* pp. 161–166, 1993.
[16] Nowosielski, O., *Metody oznaczania potrzeb nawożenia*. PWRiL. Warszawa, 1968.
[17] Jeran Z., Jacimovic R., Batic F. & Mavsar R., Lichens as integrating air pollution monitors. *Environmental Pollution*, **120**, pp. 107–113, 2002.
[18] Seaward M.R.D., Lichens and sulphur dioxide air pollution: field studies. *Environmental Reviews*, **1**, pp. 73–91, 1993.
[19] Takala K., Olkkonen H., Ikonen J., Jaaskelainen L & Puumalainen P., Total sulphur content of epiphytic and terricolous likens in Finland. *Ann. Botanici Fennici*, **2**, pp. 91–100, 1985.
[20] Pilegaard, K., Heavy metals in bulk precipitation and transplanted *Hypogymnia physodes* and *Dicranoweisia cirrata* in the vicinity of a Danish steelworks. *Water, Air and Soil Pollution*, **11**, pp. 77–91, 1979.
[21] Jeran, Z., Byrne, A.R. & Batič, F., Transplanted epiphytic lichens as biomonitors of air-contamination by natural radionuclides around the Žirovski Vhr uranium mine, Slovenia. *Lichenologist*, **27**, pp. 375–385, 1995.
[22] Van Dobben, H.F., Wolterbeek, H.T., Wamelink, G.W.W. & Ter Braak, C.J.F., Relationship between epiphytic lichens, trace elements and gaseous atmospheric pollutants. *Environmental Pollution*, **112 (2)**, pp. 163–169, 2001.
[23] Sawicka-Kapusta, K., Zakrzewska, M., Gdula-Argasińska, J. & Stochmal M., *Ocena narażenia środowiska obszarów chronionych. Zanieczyszczenie metalami i SO₂ parków narodowych*. Centrum Doskonałości Unii Europejskiej IBAES, Instytut Nauk o Środowisku, Uniwersytet Jagielloński. Kraków, 2005.
[24] Sawicka-Kapusta K., Zakrzewska M., Gdula-Argasińska J. & Bydłoń G., Air pollution in the base stations of the environmental integrated monitoring system in Poland. *Air Pollution XIII*, ed. C.A. Brebbia, WIT Transaction on Ecology and the Environment. WIT Press, Vol. 82, pp. 465–475, 2005.
[25] Sawicka-Kapusta K., Zakrzewska M. & Bydłoń G., Biological monitoring – the useful method for estimation air and environment quality. *Air Pollution XVI*, eds. C.A. Borrego &C.A. Brebbia, WIT Transaction on Ecology and the Environment. WIT Press, Vol 101, pp. 353–362, 2007.

[26] Sawicka-Kapusta K., Zakrzewska M. & Bydłoń G. Monitoring of air pollution by heavy metals and sulphur dioxide in the Base Stations of Integrated Nature Monitoring System using lichen *Hypogymnia physodes*. Biblioteka Monitoringu Środowiska, pp.217–226. Warszawa, 2007.

[27] Białońska D., Dayan F.E., Chemistry of the lichen Hypogymnia physodes transplanted to an industrial region. *Journal of Chemical Ecology*, **31**, pp. 2975–2991, 2005.

Forest ecosystem development after heavy deposition loads – case study Dübener Heide

C. Fürst, M. Abiy & F. Makeschin
*Institute for Soil Science and Site Ecology,
Dresden University of Technology, Germany*

Abstract

Forest ecosystems in the New Lander (Germany) were impacted for more than one century by industrial emissions. The deposition amount has decreased since the middle of the 1980s due to technological progress and closing of main emitters. In the research project ENFORCHANGE (www.enforchange.de), the impact of past industrial depositions on forest ecosystems is assessed in two model regions, and approaches how to integrate deposition residuals into forest management are developed. The here presented model region Dübener Heide is a ca. 300 km² large forest area in the industrial triangle Leipzig-Halle-Bitterfeld, which is one of the most polluted regions in the New Lander. A total deposition amount of 18 Mio t fly ash and of 12 Mio t SO_2 led to considerable changes of site properties, forest growth and health. The actual investigations in Dübener Heide revealed that the historical deposition impact still results in a spatial differentiation of forest growth conditions: nowadays, Dübener Heide can be divided into two parts with different impact level and intensity. Verifiable fly ash influence with high pH values and nutrient potential is limited to a zone of maximally 8–15 km distance to the former emitters, whereas SO_2 impacted the total 300 km² area, but its effects are no longer detectable. This spatial differentiation is relevant for tree species choice in the future: the heavily fly ash impacted sites are characterized by ample regeneration and growth of noble hardwood species and European beech, whereas the not measurably fly ash influenced sites are more or less suitable for Scots Pine and Oak. The prediction of the long-term development of the site potential and tree species suitability on heavily fly ash affected sites under different climate change scenarios are part of ongoing studies.

Keywords: forest ecosystem development, fly ash deposition, SO_2 deposition, forest growth and health, site potential, forest management planning.

1 Introduction – case study Dübener Heide and assessment of former deposition loads (ENFORCHANGE)

1.1 Deposition history in the model region Dübener Heide

For more than one century, forest ecosystems in the New Lander were heavily impacted by industrial depositions. In the industrial triangle Leipzig-Halle-Bitterfeld, one of the most polluted regions of the New Lander, this deposition originated from unfiltered brown-coal combustion and exhalations from chemical industry (Fürst et al. [5]). The extreme alteration of the natural conditions, which lasted until the early 1990s, is still impacting the site properties and vegetation dynamics and must be considered in actual forest management. Zooming into the region Leipzig-Halle-Bitterfeld and taking the Dübener Heide – the most important regional forest area – as an example, the historically documented deposition amounted from 1910 – 2000 to 18 Mio t fly ash and 12 Mio t SO_2. In the decade from 1961 – 1970, a fly ash deposition of up to 3–8 t / ha * a is reported by Lux [21, 24], Neumeister et al. [30], Nebe et al. [31] and Klose and Makeschin [13]. To demonstrate the extend of deposition impact on the forest soils: pH (KCl) values in the humus layer and upper mineral horizon of the regional forest soils (mainly poor sandy brown soils and podzols) increased in that period from originally 3–4 up to 7–9 and a base saturation of up to 100% is still detectable (Fritz and Makeschin [3]). From the 1980s on, the introduction of fly ash filters lead to a more or less acidic deposition regime (NO_x, SO_2 / SO_x). After 1989, a strong reduction of fly ash emission went along with raising atmospheric N deposition in a magnitude of 28 – 45 kg/ha*a and changed completely the regional deposition characteristics (Hüttl and Bellmann [11], Marquardt et al. [29], Gauger et al. [8]).

Lux [21, 24] and Lux and Stein [26] have shown that the Dübener Heide deposition situation is characterized by a wind direction and distance dependent gradient (Fig 1), starting in the eastern part of the forest mainly at the power plants and chemical industries clustered in Bitterfeld and its surroundings (Gräfenhainichen, Zschornewitz). The different deposition fractions SO_2 and fly ash, which contains "black" (tertiary) carbon, alkali / earth alkali metal salts, heavy metals and silicium compounds, were distributed along this gradient according to their aggregate state and particle size (Lux [22, 24], Niehus and Brüggemann [32], Magiera and Stryszcz [27], Stryszcz [35]).

Stein [34] and Lux [24] used a visual classification of forest decline for distinguishing up to five deposition zones along this gradient, where the differentiation between zone 1 a and 1 b (zones of highest intensity) was given up later on. Lux [21] and Lux and Pelz [25] proposed to take these deposition zones as basis for forest management planning. The deposition zones were defined on the basis of a sample plot supported evaluation system: visible crown damages (forest decline classes) in 150 plots in medium aged Scots pine stands were assessed on single tree level, then compiled for stand level and "regionalized" by subsequent spatial aggregation of comparable stands to the deposition zones. Each deposition zone was assumed to be homogenous

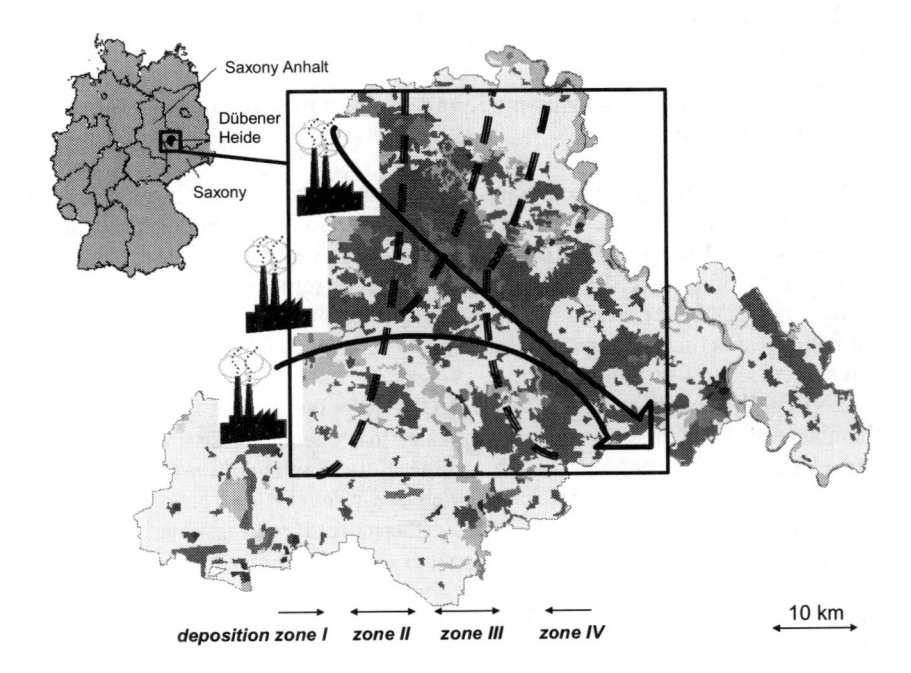

Figure 1: Schematic overview on the localization of Dübener Heide (black square) and the regional deposition gradient (black arrow), starting at the industrial sites in the East and following the dominant wind direction (map basis: CORINE LANDCOVER (CLC) 2000). The deposition zones are marked with scattered lines.

considering the deposition impact on forest growth and health, on specific risks and on possible silvicultural strategies and economic output. In deposition zone 1 e.g., Scots pine, the regionally dominating tree species, was heavily threatened by the alkaline fly ash deposition or dropped even totally out. In consequence, conversion efforts were concentrated to this zone and management intensity was reduced to deposition damage driven harvesting.

At the late 1980s, Herpel et al. [10] documented at heavily fly ash influenced sites a decrease of pH(KCl) of 0.4 units and base saturation decrease of 17% compared to the 1970s. This went along with incipient installations of fly ash filters at the main regional emitters. From 1988 to 2000, a further reduction of 0.7 pH-units was reported by Kurbel [18]. In the long run, a rapprochement of the site properties to the original regional characteristics is expected (Kopp [16], Kopp and Jochheim [17]).

As a result of deposition reduction, the health state of the forest and especially of Scots pine stands improved considerably since the 1990s. Actually, ample regeneration of noble hardwoods and European beech can be observed especially in the extremely fly ash influenced parts of the Dübener Heide. This however, might be a temporary phenomenon, whose sustainable development and ability to be integrated into regional silviculture must be evaluated beyond the background of the described re-acidification tendency.

1.2 ENFORCHANGE – assessment and evaluation of former deposition loads

"ENFORCHANGE" (Environment and Forests under Changing Conditions, www.enforchange.de) is a research project supported by the Federal Ministry of Education and Research (BMBF, Germany), which intends to assess the long-term effects of former depositions in two model regions in the New Lander, among them Dübener Heide. Based on this assessment, approaches are derived for better respecting this special situation and its expected impact on a number of forest services in forest management.

ENFORCHANGE started with the NULL-hypothesis, that the forest sites in the historically documented deposition zones in Dübener Heide are still different considering (a) their potentials such as nigh nutrient availability and (b) specific risks such as heavy metal release, which are relevant for forest management decisions (Makeschin and Fürst [28]). Furthermore, it was assumed that at least forest growth is still impacted by the spatial differentiation of the site properties along the former deposition gradient. Finally, ENFORCHANGE intended to model and regionalize ongoing ecosystem processes as basis for process-oriented forest management decisions.

Figure 2 resumes the ENFORCHANGE approach, how to come to an information pool providing spatially explicit time series data as basis for modelling and regionalization of ongoing ecosystem processes (Fürst et al. [5, 6]).

A number of 12 *key plots* was installed in the Dübener Heide along the historically documented deposition gradient. The key plots represent the major (terrestrial) soil type and stand type combinations in the region. They were preferably chosen at sites, where information from former deposition monitoring, forest health monitoring or growth and yield field trials could be involved. At the key plots, chemical and physical site properties are measured depth level-wise with focus on the humus layers and the upper mineral horizons and forest growth and yield characteristics are assessed.

The key plots were installed permanently for the total project duration, i.e. their geographical coordinates are documented, and geo-referenced to available GIS-information (site maps / geology, topography, etc.). Missing information, e.g. considering stand type development in a distinct deposition zone and on a distinct site type but in different age classes was collected at *satellite plots*, which are not permanently installed. Last but not least, field assessment of former fly ash deposition was carried out at the key plots and in a *regular sample grid* with two different grid densities (1*1 km² and 4*4 km²) as interface to the regionalization of the actually detectable deposition load. Here, ferrimagnetic susceptibility was used, which describes the amount of magnetizable iron-oxides, a distinctive component of fly ash from coal combustion (Fürst et al. [4]). Ferrimagnetic susceptibility was also measured depth-level wise with focus on the humus layers and the upper mineral horizons.

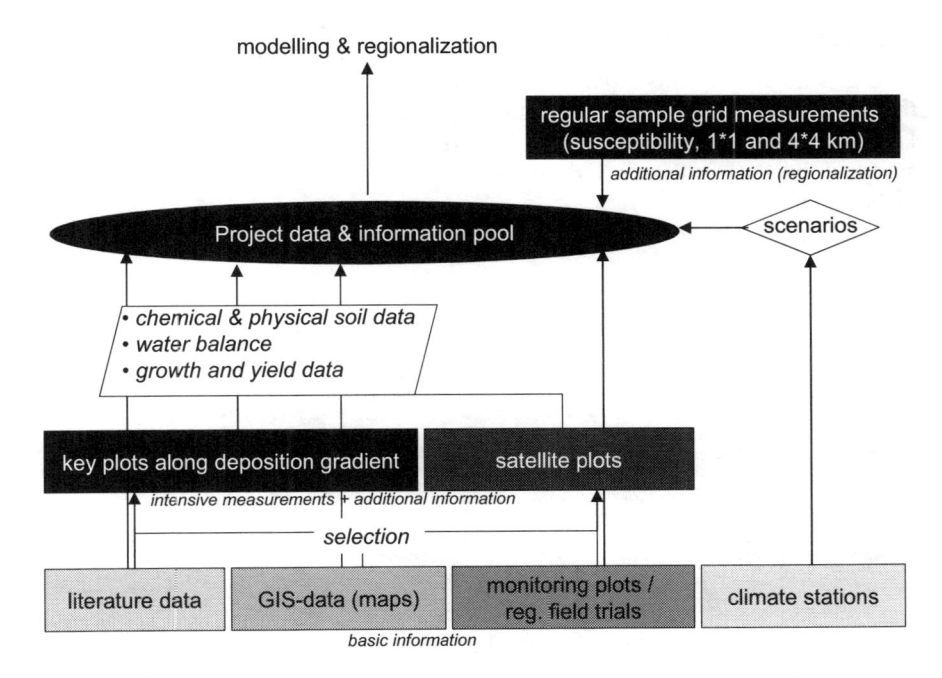

Figure 2: System of information bundling in ENFORCHANGE consisting of regionally available data, complementary information from own measurements and results from monitoring and regionalization.

Finally, information from *further regional monitoring and survey plots* (Level-I, Level-II monitoring, permanent soil monitoring sites, forest growth and yield field trials, climate stations), data from *literature analysis* and available *GIS-data* were integrated into the ENFORCHANGE information pool.

2 Long-term deposition impact on forest ecosystems – some first results

2.1 Deposition impact on the site potential

The analysis of chemical and hydrological site properties at the 12 key plots and the ferrimagnetic susceptibility based screening revealed that the differences along the historical deposition gradient still exist. They are mainly induced by former fly ash deposition. SO_2 deposition impact could not be detected anymore. The former deposition zones are still traceable by differences in the equipment with nutrients, especially base cations (Fig. 3a), by differences in physical humus properties, such as content of mineral matter in the humus layer (Klose et al. [14], Koch et al. [15]) and by different levels of ferrimagnetic susceptibility (Fig. 3b). Though, the borderlines of the former forest decline classification based spatial stratification and the actual spatial stratification according to chemical and physical characteristics are not completely identical.

Fig. 3a shows results from the multiple-regression based regionalization of the actual base saturation in the humus layer, Fig. 3b provides results from the kriging based regionalization of ferrimagnetic susceptibility (volume susceptibility) as indicator for the verifiability of fly ash deposition in the Dübener Heide.

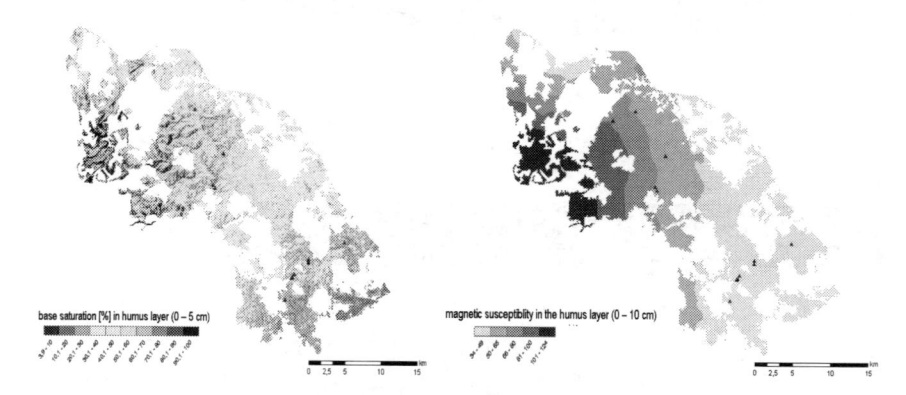

Figure 3: a (left): Regionalization of base saturation in the humus layer.
b (right): Regionalization of magnetic susceptibility in the humus layer.

Fig. 3a and b reveal major differences between the immediate vicinity to the former emitters in the eastern part of Dübener Heide and the western part, which is farthest from the emission sources. A differentiation of the part in between is possible, but the absolute values of the measured chemical and physical properties and their high variability do not support a clear separation into several deposition zones.

Actually, site potential differences, which are indeed relevant for differentiated forest management strategies, can only be ascertained for two zones: a "high influence zone" in up to 8–15 km distance to the former emitters and a "low influence zone" in more than 8–15 km distance. Fig. 4 (next page) introduces the results of a cluster analysis of the magnetic susceptibility values in the humus layers of the 12 key plots. The plots in a distance up to 8 and up to 15 km differentiate clearly from the rest. This finding is supported by similar cluster analysis results of further chemical and physical humus properties. The "high influence zone" is characterised by high base cation availability and base saturation in the humus layers, indicating a considerable nutrient pool far beyond from the natural level. The stock of extractable Ca in the humus layer and upper mineral soil until a depth of 30 cm e.g., reaches up to 4,000 kg / ha in the zone up to 8 km distance to the former emitters (Fritz et al. [3]). This is 10 to 20 times higher compared to the plots in a distance of 30 km, which are farthest to the former emitters and whose chemical properties represent more or less the original regional potential. Based on first tentative extrapolations, pH values might approximate to the original regional values in a time period of around 30–

50 yrs. Until now, it is however not foreseeable until when the high base cation potential can still be considered as silviculturally relevant factor. Considering the physical humus properties, a smaller fine pore volume going along with higher air capacity can be stated (Hartmann et al. [9]). In the "low influence zone", humus properties are much more dependent from the original site characteristics and the stand type.

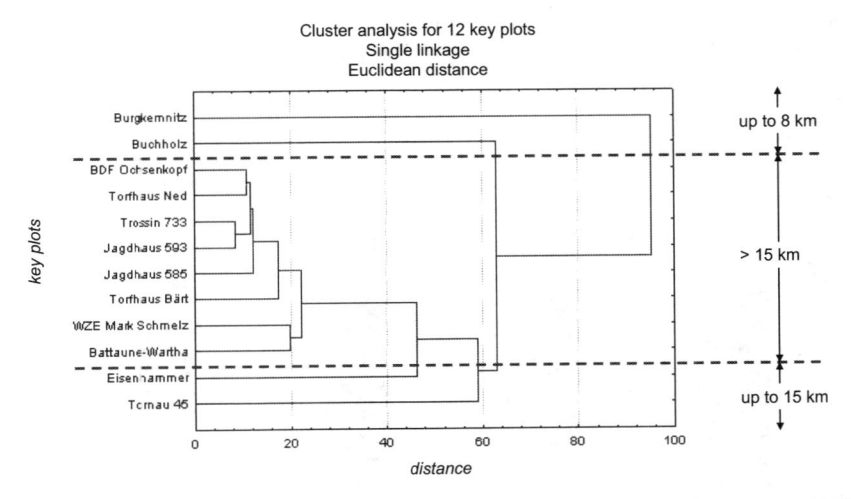

Figure 4:   Results from a cluster analysis of magnetic susceptibility (laboratory measurement, humus layer) at the 12 key plots.

One of the major challenges for future management of the forests in Dübener Heide is the change of the regional climate. The down-scaling of global climate change scenarios for Dübener Heide proved that a reduction of the mean annual precipitation of up to 100 mm and an increase of the mean annual temperature of up to 3.5 °C can be expected. Even worse, the water deficiency during the summer period is estimated to become aggravated, which affects especially the regionally dominating poor sandy soils (Bernhofer et al. [1], Franke and Köstner [2]). Beyond that background, results from effects of fly ash deposition on hydrological properties of the humus layers become important. Fly ash can not only be considered as multi-nutrient fertilizer (Fürst et al. [7]), but can also impact the properties of the regionally dominating moder–raw humus forms with their high hydrophobicity. In contrast to former findings (Thomasius et al. [36], Katzur et al. [12]), Hartmann et al. [9] revealed that fly ash reduces the water repellency and hydrophobicity of the humus layers in the "highly influenced zone" and increases the water conductivity. At the same time, the available water for plant growth, expressed by the field moisture capacity becomes smaller due to the fly ash caused decrease of the fine pore volume. Additionally, a tendencially decreased depth of the root zone due to high nutrient availability at the humus layer and upper mineral horizon at fly ash influenced sites might amplify the future risk of drought stress (Thomasius et al. [36], Koch et al. [15], Klose et al. [14]).

2.2 Deposition impact on forest growth

Forest health and consequently growth were extremely affected by the former depositions. Comparing different time strata, (a) the late 1960s until 1980, (b) the 1980s, (c) the time from 1940 until 1991 and (d) the mid of the 1970s until 1991, Lux [23] and Hüttl and Bellmann [11] proved the enormous impact of the industrial emissions on the forest development in Dübener Heide. Fig. 5 compares the reaction in radial increment of Scots pine for the four time strata at the historically documented four deposition zones.

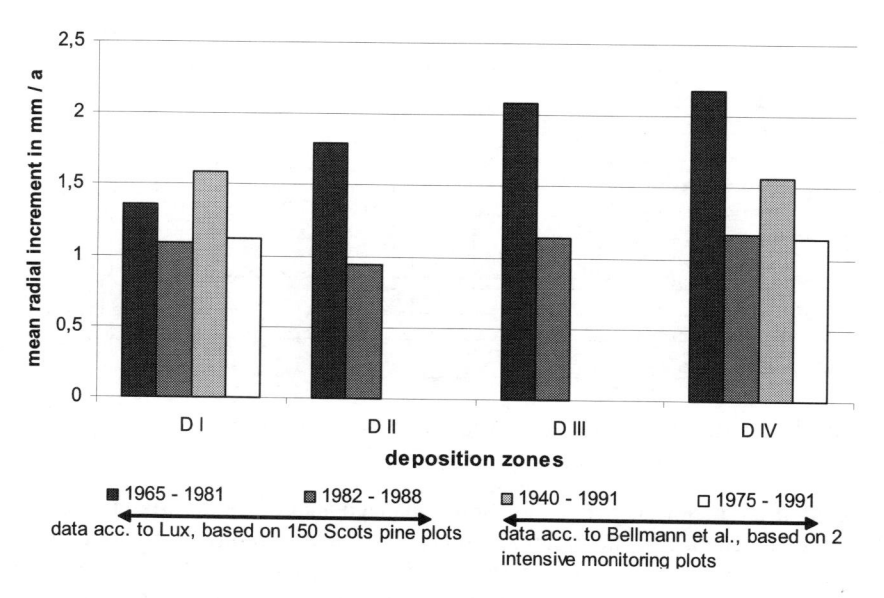

Figure 5: Reduction of the radial increment of Scots pine in four different time strata. In the 1980s, first fly ash filters were installed, whereas SO_2 emission was not yet stopped. Consequently, emission impact on forest health and growth became even worse (Lux [23], Hüttl and Bellmann [11]).

In tendency, radial increment was negatively impacted by the depositions at Dübener Heide. This resulted mainly from the extremely high SO_2 deposition: from 1965 to 1981, deposition showed the expected spatially differentiated impact on the mean radial increment, with decreasing intensity from deposition zone I (DI) to deposition zone IV (DIV). But afterwards, in the period 1982–1988, the spatial differentiation seemed to disappear. This period was characterized by beginning fly ash filtering, where at the same time, SO_2 deposition even increased. From the 1990s on, the last power plants were closed or were equipped with modern fly ash and SO_2 filtering techniques.

Comparing the radial increment tendencies between 1940–1991 and 1975–1991, where data were only available for zone I (DI) and zone IV (DIV), it can else be demonstrated that the influence of the deposition on differences in mean radial increment does not show the extreme spatial differentiation, which was assumed in the 1960s, the time, were the deposition zones were defined. The height growth tendencies followed comparable trends. This supports the impression that SO_2 deposition, which affected the forest over a wide area and not fly ash deposition with its more or less local importance, was the relevant agent. Of course, forests in the immediate vicinity of the former emitters reacted first and thus supported the stratification into four deposition zones at least in the first period of heavy deposition (Lux [22]).

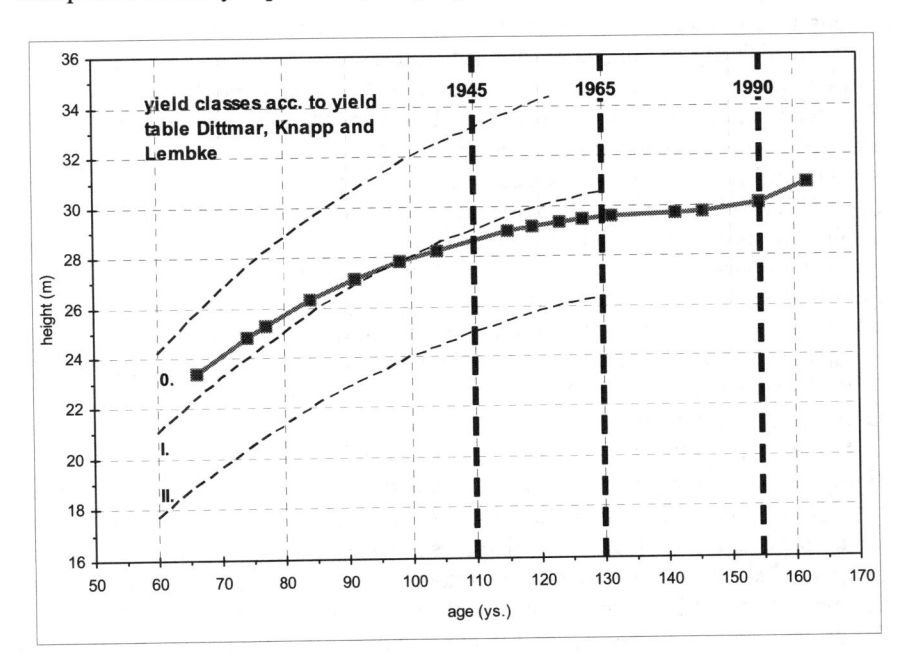

Figure 6: Trends of height growth development of a Scots pine stand (plot Tornau 45, 15 km distance to the emitters) before, during and after the deposition period. In the period from 1965–1990, a stagnation in height growth can be shown. After 1990, Scots pine restarted to grow in an age of even 155 yrs [data source: investigations of the former State Forest Research Centre Flechtingen, Saxony-Anhalt, 2005).

Investigations from intensive forest growth monitoring plots have proved that height growth of Scots pine recovered after the 1990s (Fig. 6).

Ongoing measurements at the ENFORCHANGE key plots show that nowadays, height and diameter growth of forests in Dübener Heide follow the general trend to be superior to the benchmark data in the regionally valid growth and yield tables (Pretzsch et al. [33]). This applies to all relevant stand types and especially for the regionally dominating Scots pine stands.

3 Conclusions and preview

After stopping the heavy depositions, the situation has been improved considerably for the regional forest ecosystem Dübener Heide. On the other hand, it should be highlightened that at least fly ash deposition effects can not only be considered as damaging factor. Fly ash deposition increased the available nutrient potential in the humus layers and the upper mineral horizons in the dominating poor sandy soils of Dübener Heide (Fürst et al. [5]). Furthermore, fly ash deposition tends to result in improved hydrological properties of the humus layers, a fact which gains in importance facing the problem of reduced water availability in the future. A visible consequence of the fly ash deposition caused improvement of the site potential is the ample noble hardwood and European beech regeneration, which can be observed in the zone of 8–maximally 15 km distance to the former emitters. Its potential to be integrated into silvicultural concepts must be discussed quite critically: considering the ongoing re-acidification of the fly ash influenced sites and the uncertainty how long the artificially increased nutrient potential is available for plant growth, the future regional suitability especially of noble hardwoods on sandy soils is doubtful. Furthermore, results of climate change modelling and regionalization suggest a severe decrease of regional precipitation, which amounts to almost 20–25% of the actual rainfall. This supports a turn back to drought resistant tree species such as Scots pine and Oak, which however are not able to benefit from the actually increased nutrient potential.

Some first results on the analysis of heavy metal loads in the regional forest sites as a result of fly ash deposition revealed total net values which exceed by far (up to 5-times) the thresholds given by national regulations for heavy metal values such as LABO [19]. Critical values however are more or less restricted to the immediate vicinity of the former emitters, where still high pH-values confine the mobility of endangering heavy metals and limit their possible discharge into the ground water. Prolonging the actual re-acidification tendency of the regional sites of 0.7 pH units within around 12–15 years after the closure of the former emitters and the additional acidification impact of regional N-deposition, a supposable potential of ground water quality impact can be expected in the next 50 ys. Conversion of the Scots pine and Oak dominated forests with European beech could be a countermeasure. Facing the problem of reduced water availability, this demands however for adapted conversion and transformation concepts with respect to the potential of different stand structures and tree species mixture types to reduce the evapotranspiration.

Therefore, model coupling approaches in ENFORCHANGE, linking forest growth (SILVA) with nutrient and water balance (BALANCE) and impact of forest structure and tree species composition on stand climate (HIRVAC) help to test and consider the above outlined multiple aspects in regional silvicultural planning. The future challenge will consist in using the still existing deposition driven site potentials under new climate conditions and to find strategies for responding on possible environmental risks.

References

[1] Bernhofer, C., Goldberg, V., Surke, M., Fischer, B., Meteorologie – Regionalisierung von Klima/Bestandesklimamodellierung, ENFORCHANGE – Wälder von heute für eine Umwelt von morgen, Statusbericht des Verbundes, Eigenverlag, 50–62, 2007.
[2] Franke, J., Köstner, B., Effects of recent climate trends on the distribution of potential natural vegetation in Central Germany, International journal of biometeorology **52(2)**, 139–147, 2006.
[3] Fritz, H., Makeschin, F., Chemische Eigenschaften flugaschebeeinflusster Böden der Dübener Heide, Arch. Naturschutz u. Landschaftsforsch. **46(3)**, 105–120, 2007.
[4] Fürst, C., Abiy, M., Makeschin, F., Regionalization of former fly ash deposition in forest systems in the industrial triangle Leipzig-Halle-Bitterfeld – a pre-test, Forestry, (submitted).
[5] Fürst, C., Lorz, C., Makeschin, F., Development of forest ecosystems after heavy deposition loads considering Dübener Heide as an example – challenges for a process-oriented forest management planning, SI "Meeting the challenges for process-oriented forest management" Forest Ecology and Management **248(1-2)**, 6–16, 2007.
[6] Fürst, C., Lorz, C., Abiy, M., Makeschin, F., Fly ash deposition in North-Eastern Germany and consequences for forest management, Contributions to Forest Sciences, Ulmer, 50–63, 2006 a.
[7] Fürst, C., Makeschin, F. Comparison of Wood Ash, Rock Powder, and Fly Ash – a review, Contributions to Forest Sciences, Ulmer, 63–81, 2006 b.
[8] Gauger, T., Anshelm, F., Schuster, H., Eirsman, J.W., Vermeulen, A.T., Draaijers, G.P.J., Bleeker, A., Nagel, H.-D., Mapping of ecosystem specific long-term trends in deposition loads and concentrations of air pollutants in Germany and their comparison with Critical Loads and Critical Levels, Part 1: Deposition Loads 1990-1999. Final Report 29942210 Umweltbundesamt, 207 p., 2002.
[9] Hartmann, P., Fleige, H., Horn, R., Flugascheeinfluss auf Böden in der Dübener Heide – Physikalische Eigenschaften, Hydrophobie und Wasserhaushalt, Arch. Naturschutz u. Landschaftsforsch. **46(3)**, 79–103, 2007.
[10] Herpel, J., Heinze, M., Fiedler, H.J., Veränderungen von Boden und Vegetation in Kiefernbeständen der Dübener Heide zwischen 1966 und 1990. Arch. Naturschutz u. Landschaftsforsch. **34(1)**, 17–41, 1995.
[11] Hüttl, R. F.; Bellmann, K. (eds.), Changes of Atmospheric Chemistry and Effects on Forest Ecosystems – A Roof Experiment Without Roof. Nutrients in Ecosystems 3, Springer Berlin, Heidelberg, New York, 384 p., 1999.
[12] Katzur, J., Strzyszcz, Z., Tölle, R., Liebner, F., Magnetisches Eisen als Tracer für die Bestimmung der Homogenität von Boden-Asche-Gemischen bei der Melioration extrem saurer Kippböden. Archiwum Ochrony Srodowiska **24**, 83–93, 1998.

[13] Klose, S., Makeschin, F., Chemical properties of forest soils along a fly ash deposition gradient in Eastern Germany. Europ. J. Forest Res. **123**, 3–12, 2004.
[14] Klose, S., Tölle, R., Bäucker, E., Makeschin; F., Stratigraphic Distribution of Lignite-Derived Atmospheric Deposits in Forest Soils of the upper Lusatian Region, East Germany. Water, Air and Soil Pollution **142**, 3–25, 2002.
[15] Koch, J., Klose, S., Makeschin, F., Stratigraphic and Spatial Differentiation of Chemical Properties in Long-term Fly Ash Influenced Forest Soils in the Dübener Heide Region, NE-Germany. Forstw. Cbl. **121**, 157–170, 2002.
[16] Kopp, D., Zusammenwirken von Standort und Vegetation bei der Erkundung des Zustandswandels von Waldnaturräumen in nordostdeutschen Tiefland. Arch. Naturschutz u. Landschaftsforsch. **42(1)**, 1–49, 2003.
[17] Kopp, D., Jochheim, H., Forstliche Boden- und Standortsdaten des Nordostdeutschen Tieflands als Datenbasis für die Landschaftsmodellierung. Kessel Remagen-Oberwinter, 207 p., 2002.
[18] Kurbel, R., Untersuchungen zur Mobilität und zum Gesamtgehalt von Schwermetallen mittels Säulenversuch und sequentieller Extraktion für zwei Standorte der Dübener Heide. Diploma thesis Dresden University of Technology, 65 p., 2002.
[19] LABO (Bund-Länder-Arbeitsgemeinschaft Bodenschutz), Hintergrundwerte für anorganische und organische Stoffe in Böden, 3. überarbeitete und ergänzte Auflage, Eigenverlag, 59 p., 2003.
[20] Lux, H., Ergebnisse von Zuwachsuntersuchungen (Bohrspananalysen) im Rauchschadensgebiet Dübener Heide, Arch. Forstwes. **11**, 1103–1121, 1964.
[21] Lux, H., Die großräumige Abgrenzung von Rauchschadenszonen im Einflussbereich des Industriegebietes um Bitterfeld, Wiss. Z. Techn. Univers. Dresden **14**, 433–442, 1965.
[22] Lux, H., Beitrag zur Trennung des Schadanteils gleichzeitig auf die Waldvegetation einwirkender Abgasquellen, Wiss. Z. Techn. Univers. Dresden **15**, 1533–1535, 1966.
[23] Lux, H., Zur Beeinflussung des Oberbodens von Kiefernbeständen durch basische Industriestäube. Wiss. Z. Techn. Univers. Dresden **23**, 915–920, 1974.
[24] Lux, H., Ausbreitung und Berechung (Immissionen). in Däßler, H.G (ed.).: Einfluss von Luftverunreinigungen auf die Vegetation: Ursachen – Wirkung – Gegenmaßnahmen, Fischer Jena, 26–33, 1976.
[25] Lux, H., Pelz, E., Schadzone und Schadstufe als Klassifizierungsbegriffe in rauchgeschädigten Waldgebieten. Forstwirtschaft **18**, 245–247, 1968.
[26] Lux, H., Stein, G., Die forstlichen Immissionsschadgebiete in Lee des Ballungsraumes Halle und Leipzig. Hercynia N.F., 352–354, 1977.
[27] Magiera, T., Strzyszcz, Z., Ferrimagnetic Minerals of Anthropogenic Origin in Soils of some Polish National Parks. Water, Air and Soil Pollution **124**, 37–48, 1999.

[28] Makeschin, F., Fürst, C., Influence of vectored changes of environmental factors (climate, site and human beings) on land-use systems – example forest land-use (ENFORCHANGE), Annals of Agrarian Science **5(2)**, ISSN 1512-1887, 86–91, 2007.
[29] Marquardt, W., Brüggemann, E., Auel, R., Herrmann, H., Möller, M., Trends of pollution in rain over East Germany caused by changing emissions, TELLUS 53 B, 529–545, 2001.
[30] Neumeister, H., Franke, C., Nagel, C., Peklo, G., Zierath, R., Peklo, P., Immissionsbedingte Stoffeinträge aus der Luft als geomorphologischer Faktor. Geoökodynamik **12**, 1–40, 1991.
[31] Nebe, W., Pommnitz, M., Jeschke, J., Aktueller Standortszustand und Waldumbau im nord-westsächsischen Tiefland. Forst und Holz **56(1)**, 3–8, 2001.
[32] Niehus, B., Brüggemann, L., Untersuchungen zur Deposition luftgetragener Stoffe in der Dübener Heide. Beitr. Forstwirtsch. u. Landsch. Ökol. **29(4)**, 160–163, 1995.
[33] Pretzsch, H., Rötzer, T., Moshammer, R., Waldwachstumsreaktionen und Systemanalyse, ENFORCHANGE – Wälder von heute für eine Umwelt von morgen, Statusbericht des Verbundes, Eigenverlag, 80–91, 2007.
[34] Stein, G., Der forstliche Zustandsvergleich – eine Diagnosemethode in rauchgeschädigten Waldgebieten. Wiss. Z. TU Dresden **14**, 1043–1049, 1965.
[35] Stryszcz, Z., Heavy Metal Contamination in Mountain Soils of Poland as a Result of Anthropogenic Pressure. Biology Bulletin **26**, 722–735, 1999.
[36] Thomasius, H.; Wünsche. M.; Selent, H., Bräunig, A., Wald- und Forstökosysteme auf Kippen des Braunkohlenbergbaus in Sachsen – ihre Entstehung, Dynamik und Bewirtschaftung – Schriftenreihe d. Sächs. Landesanst. f. Forsten **17**, 68 p., 1999.

Impact of biogenic volatile organic compound emissions on ozone formation in the Kinki region, Japan

A. Kondo, B. Hai, K. L. Shrestha, A. Kaga & Y. Inoue
Graduate School of Engineering Osaka University, Osaka, Japan

Abstract

The standard Biogenic Volatile Organic Compound (BVOC) emissions from ten Japanese plant species were measured by using a growth chamber where temperature and light intensity can be controlled. These species were selected due to their abundance in the estimated domain of the Kinki region. The BVOC emissions in Kinki region during July 2002 were estimated by revising the standard BVOC emissions from temperature and light intensity which were calculated by MM5. The two types of the ozone calculation were carried out by CMAQ. One was the calculation with BVOC emissions (BIO). Another was the calculation that assumes BVOC emissions to be zero (NOBIO). The maximum ozone concentrations of BIO reasonably reproduced the observed maximum concentrations in especially the fine days. The hourly differences of monthly average ozone concentrations between BIO and NOBIO had the maximum value of 6ppb at 2 p.m. The explicit difference appeared in urban area, though the place where the maximum of difference occurred changed. It was shown that the BVOC emitted from the forest area strongly affected the ozone generation in the urban area.

Keywords: biogenic volatile organic compound, ozone, MM5, CMAQ, growth chamber.

1 Introduction

The photochemical oxidant gives the damage to human and vegetations. In Japan the standard of the photochemical oxidant was regulated in 1970 and due to the useful countermeasure, the photochemical oxidant concentration had decreased until 1990. However, recently the photochemical oxidant concentration has been slowly increasing. The increase of the background ozone concentration due to the transboundary transport [1] is pointed out as one of the causes. From another view point, the temperature increase due to the global warming, the urban heat island and the increase of the ultraviolet rays [2] are also pointed out as one of the causes. It is well known that the biogenic volatile organic compound (BVOC) emissions increase accompanied with the temperature increase and that BVOC play the important role of the ozone generation. The BVOC emissions from the plants indigenous to Japan haven't enough been investigated, yet. In this study, the BVOC emissions from the dominant plants were obtained from the growth chamber experiment, the total emissions in the Kinki region were estimated, and the impacts of ozone generation due to the BVOV emissions were assessed by MM5/CMAQ.

2 Growth chamber experiment

2.1 Experimental procedure

Cryptomeria japonica, Chamaecyparis obtusa, Pinus densiflora, Quercus serrata, Quercus crispula, Fagus crenata, Quercus acutissima Carruthers, Quercus glauca and *Quercus myrsinaefolia*, the nine most abundant plants in the Kinki region, and *Oryza sativa* were selected. Emission measurements were performed by using a growth chamber that can manipulate temperature and light intensity. Plants were then adapted to the following experimental conditions: average temperature of Osaka and PAR was set at 1000 $\mu mol\ m^{-2}\ s^{-1}$ from 6 a.m. to 6 p.m. In order to collect air samples in a growth chamber, a 200 mg Tenax-TA adsorbent tube (Supelco, mesh 60/80) and a vacuum pump (GL Science SP208-1000Dual) with a flow rate of 100 ml min^{-1} were used. The trapped compounds into adsorbent tubes were thermally desorbed at 280°C by Thermal Desorber (Perkin Elmer ATD-50) connected to GC/MS (Shimadzu GC/MS-QP2010). Isoprene, α-pinene, β-pinene, myrcene, α-phellandrene, α-terpinene, p-cymene, limonene, γ-terpinene and terpinolene were analyzed [3].

2.2 Experimental results

The experiments for ten plant species conducted at several conditions, which were different from temperatures and light intensities. The BVOC emission from ten plant species at several conditions were converted to the standard condition (30°C, PAR: 1000 $\mu mol\ m^{-2}\ s^{-1}$) by using the Guenther equation [4] and Tingey equation [5]. Table 1 shows the monoterpenes emission at standard conditions. Table 2 shows the isoprene emission at standard conditions from six deciduous broadleaf trees, which are commonly found in Japan. A large amount of α-pinene and β-pinene was detected from *Cryptomeria japonica, Chamaecyparis obtusa* and *Pinus densiflora*. Previously, *Oryza sativa* has not been reported to emit BVOC, but five kinds of monoterpenes were detected. This detection could be very significant even though the amount is small because a large part of Japan is covered by paddy fields. A large amount of isoprene was detected from *Quercus serrata*.

Table 1: BVOC emissions from Coniferous trees at standard condition [30°C, PAR: 1000 μ mol m^{-2} s^{-1}].

compound	BVOC emissions [μg g$_{dw}^{-1}$h^{-1}]			
	Cryptomeria japonica	Chamaecyparis obtusa	Pinus densiflora	Oryza sativa
α-pinene	1.30	1.89	5.33	0.24
β-pinene	0.06	0.22	0.84	0.02
Myrcene	0.32	0.35	1.79	0.03
α-phellandrene	0.20	0.13	0.87	ND
α-terpinene	0.15	0.13	0.23	ND
p-cymene	0.10	0.28	0.14	0.03
Limonene	0.40	ND	0.82	0.08
γ-terpinene	0.21	0.49	ND	ND
Terpinolene	0.08	ND	0.27	ND
Total monoterpenes	2.81	3.48	10.28	0.40

Table 2: Isoprene emissions from broadleaf trees at standard condition [30°C, PAR: 1000 μ mol m^{-2} s^{-1}].

Plants name	Isoprene emission [μg g$_{dw}^{-1}$h^{-1}]
Quercus serrata	224.21
Quercus crispula	26.04
Fagus crenata	0.79
Quercus acutissima Carruther.	0.18
Quercus glauca	0.04
Quercus myrsinaefolia	0.03

3 Estimation of BVOC Emissions in the Kinki Region

3.1 Standard estimation

At first the emissions of isoprene and monoterpene at the standard conditions in the Kinki regions were estimated by using the forest database including dominant specie of plant, an area, an age, a biomass and so on, In the Kinki region, the deciduous biomass was 1.7 times larger than the broadleaf biomass but *Quercus serrata* emitted a large amount of isoprene. Therefore the emissions of isoprene were extremely larger than the emissions of monoterpene as shown in Fig.1 and Fig.2.

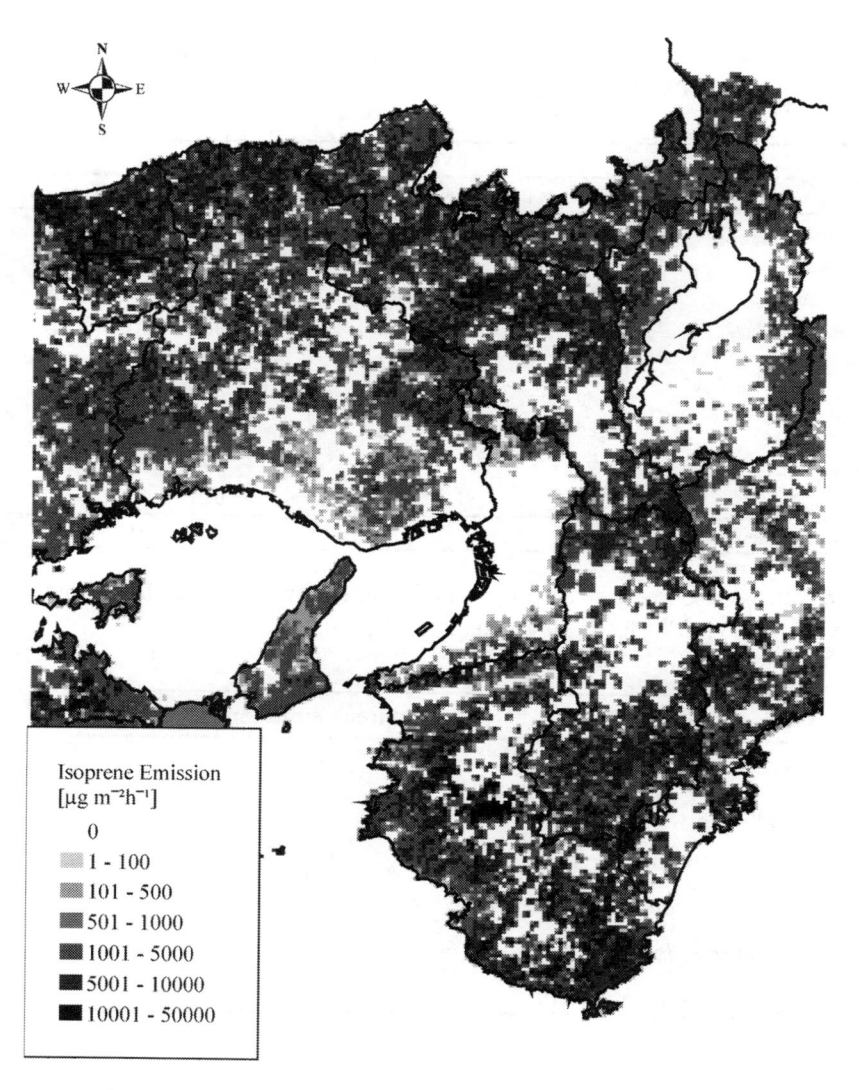

Figure 1: Standard emissions of isoprene in Kinki region.

3.2 Hourly emission

Emissions of isoprene and monoterpene vary according to the hourly variations of temperature and light intensity. The emissions on July 2002 were revised from the temperature and light intensity which were calculated by using MM5 (Meteorological Model version 5). Guenther equation for isoprene and Tingey equation for monoterpene were used to calculate the emissions variation due to temperature and light intensity. The equation of isoprene emissions E_{iso} are expressed by

$$E_{iso} = EF_{iso} \cdot C_T \cdot C_L \tag{1}$$

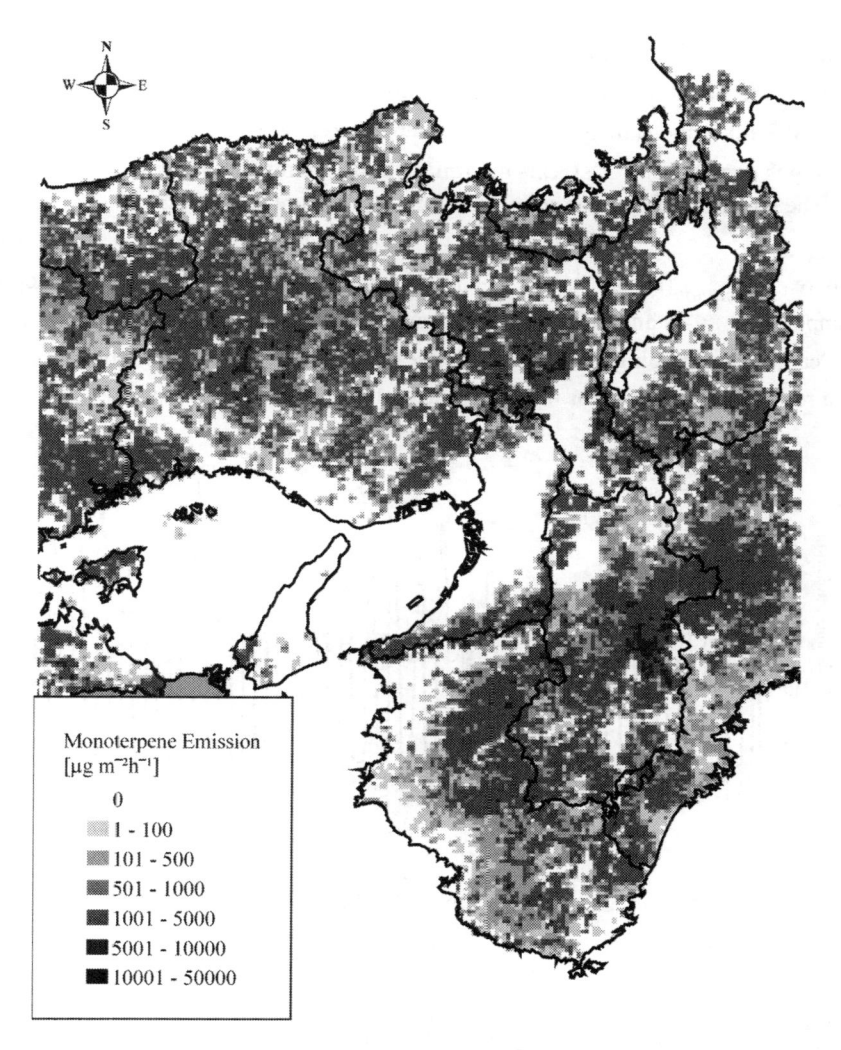

Figure 2: Standard emissions of monoterpene in Kinki region.

where EF_{iso} ($\mu g\ g_{dw}^{-1}\ h^{-1}$) is the isoprene emission rate at standard conditions. C_T is the correction factor due to temperature and C_L is the correction factor due to PAR. C_T and C_L are defined by

$$C_T = \frac{\exp\dfrac{C_{T1}\cdot(T-T_s)}{R\cdot T\cdot T_s}}{1+\exp\dfrac{C_{T2}\cdot(T-T_m)}{R\cdot T_s\cdot T}} \quad (2)$$

$$C_L = \frac{\alpha\cdot C_{L1}\cdot L}{\sqrt{1+\alpha^2\cdot L^2}} \quad (3)$$

where α (0.0027), C_{L1} (1.066), C_{T1} (95000 J/mol), C_{T2} (230000 J/mol) and T_m (314 K) are empirical coefficients, L (μmol m^{-2} s^{-1}) is the PAR flux, T_s (303 K) is the standard reference temperature, R (8.314 J K^{-1} mol^{-1}) is the ideal gas constant and T (K) is the foliar biomass temperature.

The equation of monoterpene emissions E_{mono} are expressed by

$$E_{mono} = EF_{mono} \cdot \exp(\beta \cdot (T - T_s)) \qquad (4)$$

where EF_{mono} (μg g$_{dw}^{-1}$ h^{-1}) is the monoterpene emission rate at standard temperature. β is an empirical coefficient ranging between 0.057 and 0.144 K^{-1}. β can vary according to chemical species and environmental conditions. 0.09^{-1} is a reasonable estimate for monoterpene emissions of most plants.

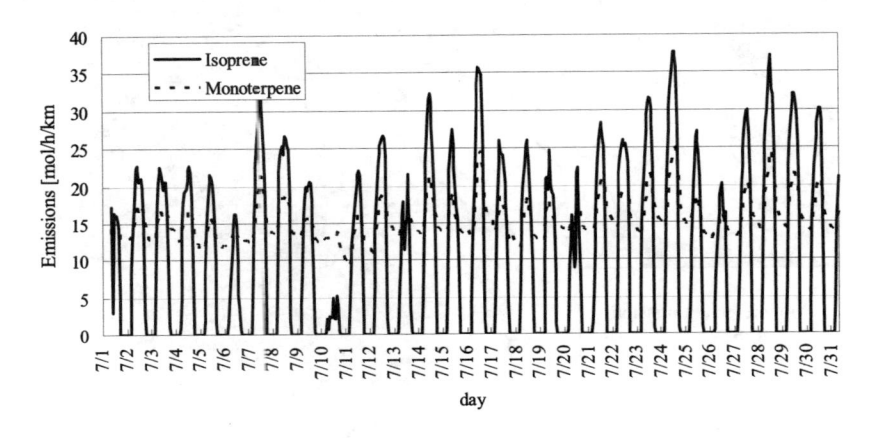

Figure 3: Hourly variations of isoprene and monoterpene.

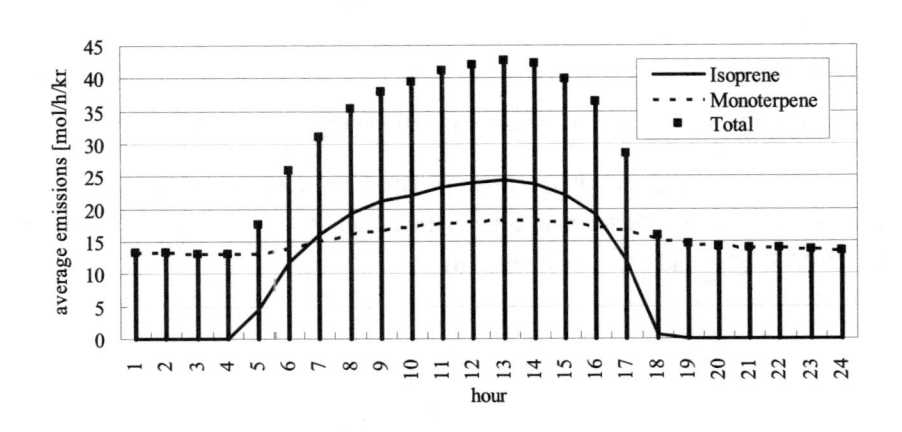

Figure 4: Hourly average emissions.

The hourly variations of isoprene and monoterpene on July 2002 are shown in Fig.3. As there were many days of cloudy or rainy in the first half of this month, the emissions were relatively low. The hourly average emissions are shown in Fig.4. In the daytime, isoprene and monoterpene emissions were almost same, but only monoterpene emitted in the night time.

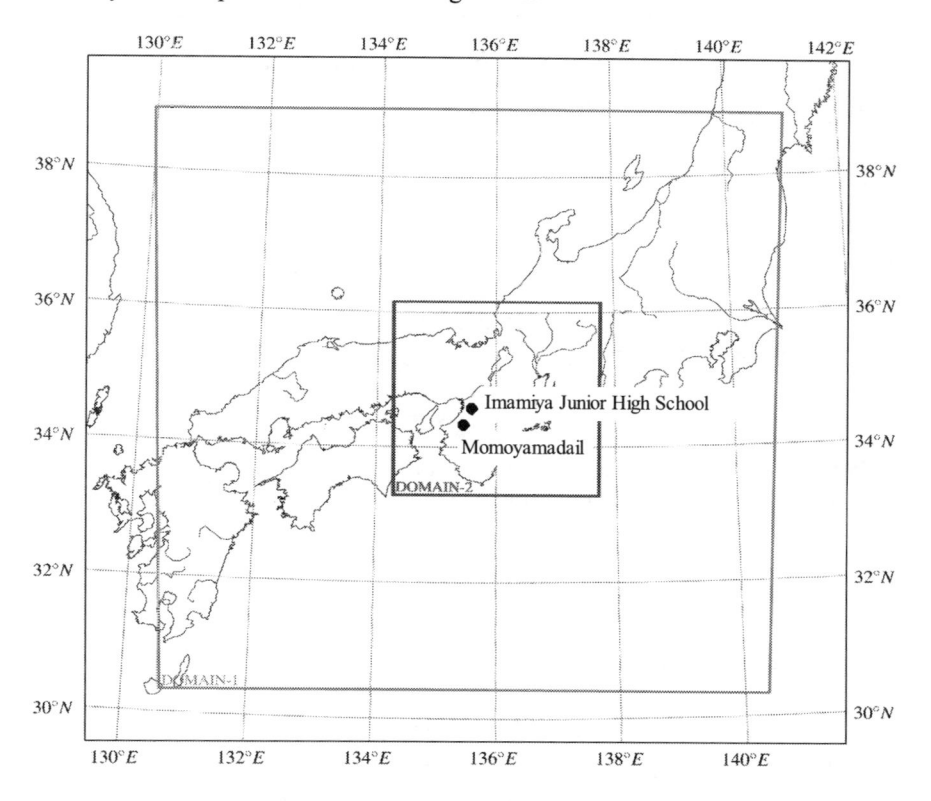

Figure 5:   Calculated domain.

4 Impact of BVOCs on ozone concentrations

Ozone concentrations were calculated by MM5/CMAQ (Community Multiscale Air Quality). The Grid Point Value–MesoScale Model (GPV–MSM) data from the Japan Meteorological Agency (JMA, http://www.jma.go.jp/jma/index.html) were assimilated as objective analysis data. The data are available for the Japan region and have a high horizontal resolution of 10 km × 10 km. The calculated region is shown in Fig.5. The domain-1 has the resolution of 9 km and includes almost whole of Japan. The domain-2 has the resolution of 3 km and includes all of the Kinki region. The two types of the calculation for one month of July 2002 were carried out. One is the calculation with BVOC emissions (BIO). Another is the calculation that assumes BVOC emissions to be zero (NOBIO).

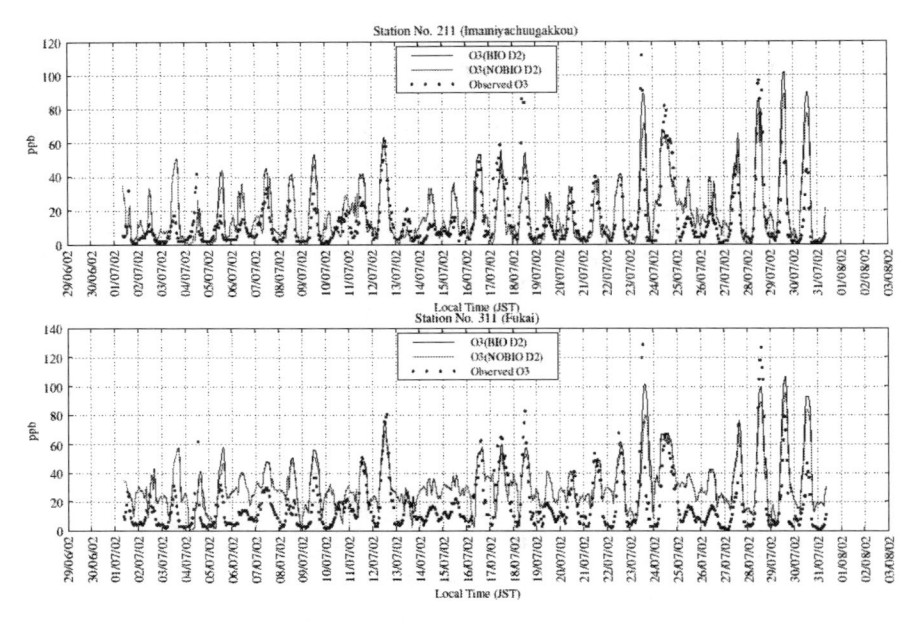

Figure 6: Calculated and observed ozone concentration at Imamiya Junior High School and Momoyamadai.

Table 3: Statistical results of BIO and NOBIO.

Metrics		BIO	NOBIO
Correlation Coefficient		0.41	0.37
Mean Bias	MB	-0.81	-6.03
Mean Normalized Bias	MNB	0.03	-0.06
Normalized Mean Bias	NMB	-0.01	-0.1
Normalized Mean Bias Factor	NMBF	-0.01	-0.11

4.1 Comparison with observations

The calculated ozone concentrations were compared at two observatories; Imamiya Junior High School and Momoyamadai, where the relatively high concentrations were observed. Imamiya Junior High School is located at the center of Osaka City and Momoyamadai is located at the southern part of Osaka Prefecture. The high ozone concentrations were observed on 23, 27, 28, 29, 30, and 31 July as shown in Fig.6. These days were clearly fine and the temperature was relatively high and the light intensity was strong. Consequently a lot of BVOC emitted from the forest area. The ozone concentrations of BIO became higher than NOBIO and reasonably reproduced the observed concentrations. For one example, the observed ozone concentration and the ozone concentration of BIO and NOBIO at Imamiya Junior High School on 23 July were 122 ppb, 110 ppb and 84 ppb, respectively. This means that the BVOC emissions make ozone concentration of 26 ppb increase. In the first half of this month when BVOC emissions were rather small due to the cloud days, the differences between the ozone concentration of BIO and NOBIO can't be seen.

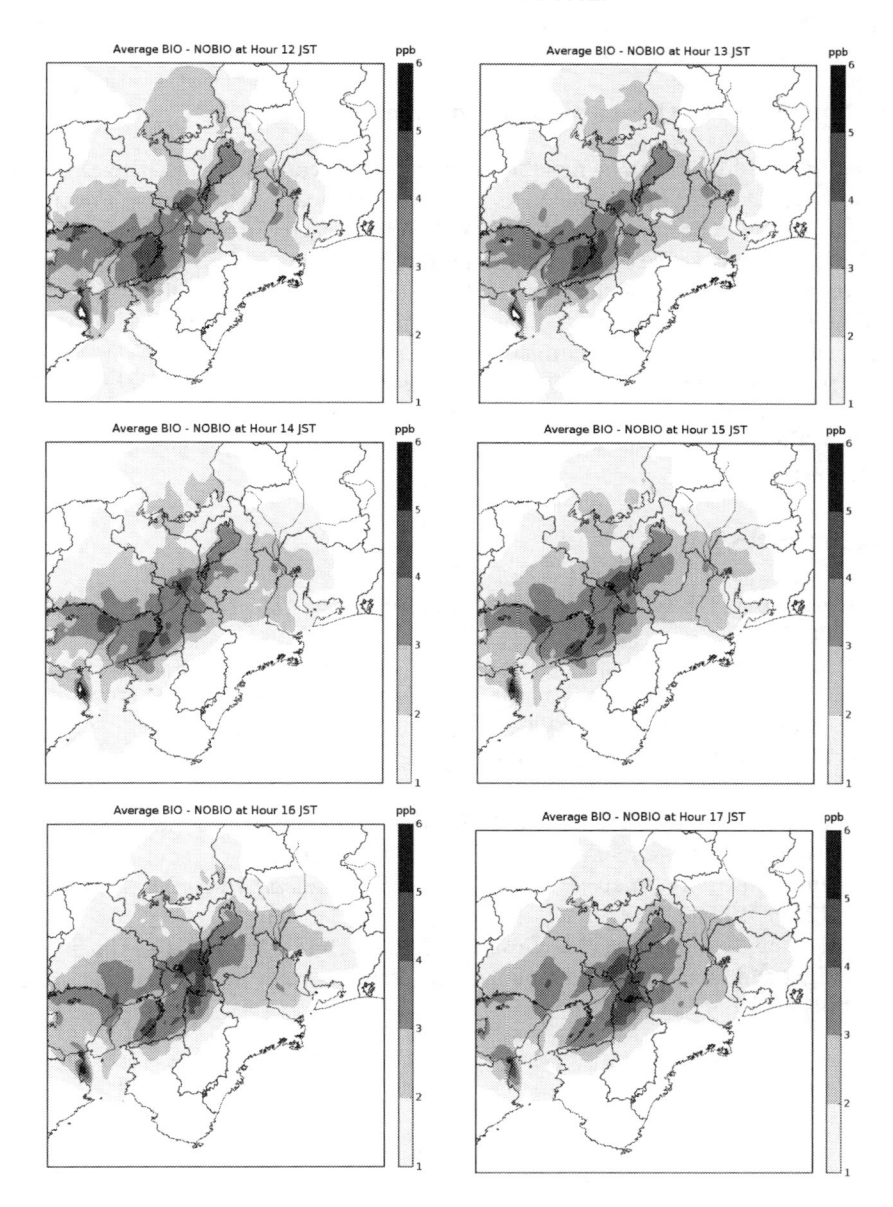

Figure 7: The hourly differences of monthly average ozone concentrations between BIO and NOBIO.

The statistical results are shown in Table 3. The correlation coefficient of BIO was improved to 0.41 from 0.31 ($p<0.05$) and the mean bias was also improved to -0.81 ppb from -6.05 ($p<0.01$). Mean normalized bias (MNB), normalized mean bias (NMB) and normalized mean bias factor (NMBF) which are the evaluation index for the ozone prediction by EPA were enhanced.

4.2 Impact of BVOCs on air pollutants concentrations

The hourly differences of monthly average ozone concentrations between BIO and NOBIO were investigated. As the photochemical reaction doesn't occur at night, ozone concentrations of BIO and NOBIO were almost the same. From sunrise, the difference of ozone concentrations appeared and it reached 6ppb at 2 p.m. when the temperature rose. The situation continued until 6 p.m. The differences of ozone concentrations at 12 a.m – 5 p.m. are shown in Fig.7. The difference appeared in the urban area, though the place where the maximum difference occurred changed. In all cases the maximum difference emerged in the border of Osaka City where a lot of NOx was emitted. It was shown that the BVOCs emitted from the forest area strongly affected the ozone generation in the urban area.

References

[1] Akimoto, H., Global air quality and pollution, *Science*, 302, 1716–1719, 2003.
[2] Wakamatsu, S., Ohara, T., and Uno, I., Recent trends in precursors in the Tokyo and Osaka areas, *Atmospheric Environment*, 30, 715–721, 1996.
[3] Hai Bao, Akira Kondo, Akikazu Kaga, and et al., Biogenic Volatile Organic Compound emission potential of forests and paddyfields in the Kinki region of Japan, Environmental Research, 106, 156–169, 2008.
[4] Guenther, A. B., Zimmerman, P. R., Harley, P. C., Manson, R. and Fall, R., Isoprene and Monoterpene Emission Rate Variability - Model Evaluations and Sensitivity Analyses -, *Journal of Geophysical Research*, 98, 12609–12617, 1993.
[5] Tingey, D., Manning, M., Grothaus, L., and Burns, W., (1980) Influence of light and temperature on monoterpene emission rates from slash pine, *Plant Physiol*, 65, 797–801, 1980.

Section 9
Policy studies

Potential contribution of local air quality management to environmental justice in England

I. Gegisian, M. Grey, J. W. S. Longhurst & J. G. Irwin
University of the West of England, Bristol, UK

Abstract

The United Kingdom has a well-developed system of local air quality management established to combat air quality problems at a local level. In England air quality management areas tend to cover areas of greater social deprivation, and action plans developed to address poor air quality provide an opportunity to take account of environmental justice. However, this is not explicitly required by Government guidance and evidence to date suggests that the opportunity has not been realised. Many measures in current air quality action plans are not specific to air quality. This holds out the potential for greater integration of air quality measures with other local strategies, most notably local transport plans, which take greater account of environmental justice issues. Although not primarily intended to improve air quality or address environmental inequality, the London Congestion Charge provides evidence of the potential of this type of measure.

Keywords: air pollution, deprivation, environmental justice, local air quality management.

1 Introduction

Environmental justice (EJ) may be defined as the equitable treatment of all people in the development and implementation of environmental policies, regardless of race, creed, income and social class. While the concept emerged in the United States of America in response to concerns regarding race [1], in the United Kingdom concern has focussed more generally on deprivation. Research by the Environment Agency [2] found a strong relationship between air quality and deprivation in England. Such relationships have added significance as deprived individuals may be more susceptible to the adverse health effects of air pollution due to their relative disadvantage or predisposing health conditions or behaviours [3]. One review has found that in five out of six cohort studies stronger associations were observed between mortality and exposure to air pollution among lower socioeconomic groups, although few of the results were statistically significant and the work revealed the difficulties in determining which characteristics of deprivation influenced the association [4].

The United Kingdom has a well-developed system of local air quality management established to combat air quality problems at a local level. The statutory framework was established in the Environment Act 1995 [5] which sets air quality objectives and standards for eight pollutants: benzene, 1,3 butadiene, carbon monoxide, nitrogen dioxide, lead, particles as PM_{10}, sulphur dioxide and ozone, although ozone is not included in local air quality management. If air quality objectives are, or are likely to be, exceeded, a local authority (LA) must define and declare an Air Quality Management Area (AQMA) and develop an Air Quality Action Plan (AQAP) to address the exceedances. The system and its implementation have been analysed in a number of publications eg Longhurst et al. [6]. In theory, the approach provides an opportunity to address issues of environmental justice when developing action plans. The research described here used a range of methods to address two questions. Firstly, it investigated equity in exposure to air pollution by using the presence or absence of an air quality management area as an indicator of exposure to pollution. Secondly, it examined the extent to which local authorities are taking account of social deprivation in developing their AQAPs.

2 Methodology

Qualitative and quantitative analytical methods were used together with GIS for visualisation and spatial analyses [7]. The practices of LAs in developing their AQAPs were assessed in three ways: an appraisal of the published action plan, a questionnaire to the authorities and a small number of case study interviews of officials. The appraisal used a checklist to collect information on the compilation of the AQAP and, in particular, any measures and mechanisms used to take account of EJ, or social issues in general. The questionnaire survey was undertaken to verify the appraisal, collect more up to date information on progress with the AQAP and assist in the selection of case studies. The case studies provided an opportunity to address themes emerging from the questionnaire in more detail and collect personal perspectives on the challenges of accommodating EJ issues within the local air quality management process.

The Index of Multiple Deprivation (IMD) is the English index for computing deprivation on a small area scale. The index consists of a number of parameters that express elements of deprivation and is divided into seven domains: income, employment, health deprivation and disability, education, skills and training, barriers and housing, crime and the living environment. Each of these domains is constructed of a number of indicators. A description of how these indices and domains are compiled and statistically manipulated has been published [8]. The indices were used as the measure of social deprivation in this study. Spatially they are based on the lower layer super output areas (SOA) from the UK census. These are areas with a minimum population of 1000 and a mean of about 1500 people based on postcode units.

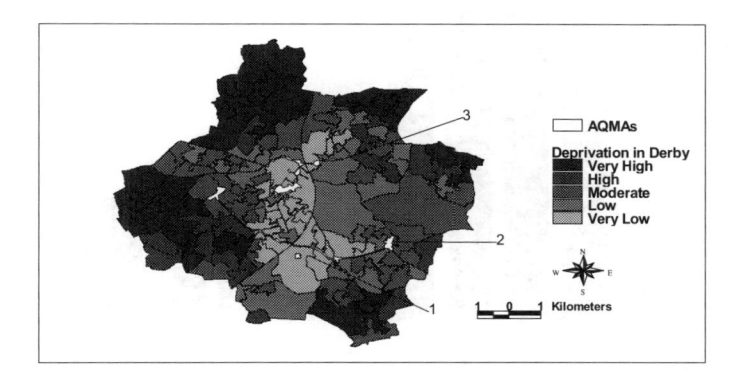

Figure 1:   Areas of deprivation and three AQMAs in Derby.

An example map showing deprivation by SOA with the AQMAs superimposed is shown in Figure 1.

3 Results

3.1 Deprivation results

An analysis of deprivation in AQMAs was conducted in order establish whether areas with elevated levels of pollution also suffer from high deprivation. Figure 2 considers the SOAs within AQMAs. The level of deprivation is predominately between the first and fourth deciles with over one third of the SOAs in the two most deprived deciles, compared with 10% in the two least deprived deciles. In Table 1 the distribution of deprivation in AQMAs is compared with the overall level of deprivation. The more deprived an SOA is, the more likely it is to be located in an AQMA.

The general conclusion can be drawn that AQMAs are more likely to have above average levels of deprivation rather than lower. Based on the average IMD score at district level, the North East and North West of England have the largest numbers of their districts in the most deprived deciles where 1 is the most deprived.

3.2 Taking account of deprivation

The second element of the research was to examine the extent to which LAs were taking account of deprivation in developing their action plans. The

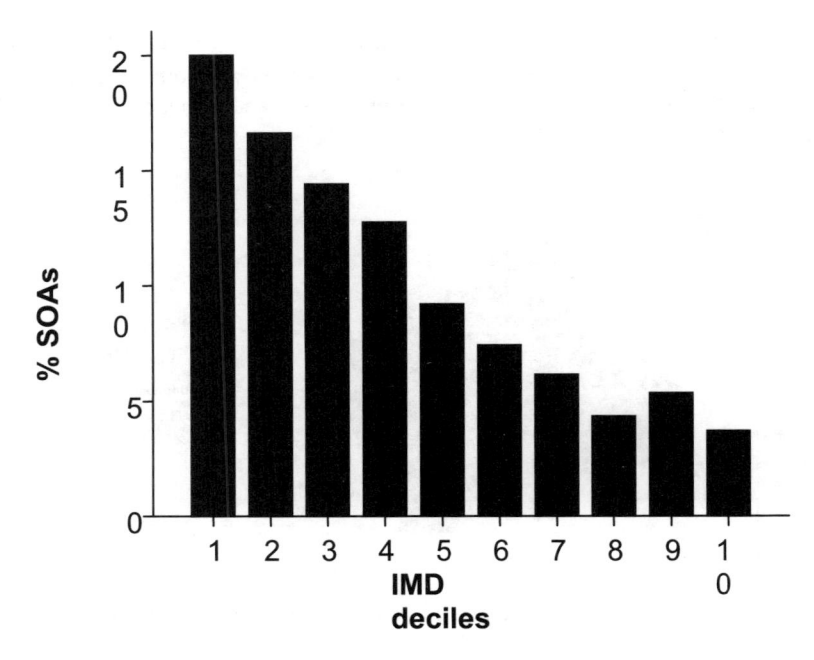

Figure 2: Distribution of deciles of deprivation in AQMAs (non whole boroughs) based on their associated SOAs (n = 2485).

Table 1: Distribution of deprivation in AQMAs compared to overall deprivation in Las.

Deciles of deprivation	No of SOAs in AQMAs	Total no of SOAs in associated LAs	%
1	497	1479	34
2	413	1274	32
3	358	1197	30
4	317	1115	28
5	229	1021	22
6	185	982	19
7	152	971	16
8	108	906	12
9	133	968	14
10	92	946	10

questionnaire survey which was sent to local authorities achieved a response rate of over 50% and identified 116 authorities that had produced AQAPs. Each plan was then examined to extract data on how it was prepared, the measures that were included and the extent to which EJ considerations had been taken into account through public participation or involvement of relevant bodies, as well as the direct use of social data, including deprivation.

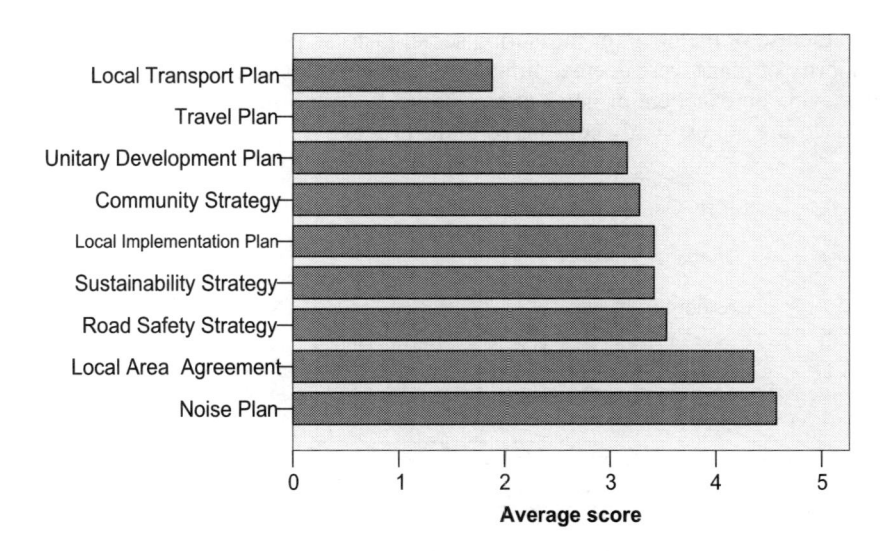

Figure 3: Internal strategies; the mean score for each option is shown where 1 is a highly considered strategy and 5 is a strategy that is not considered (n = 67).

The measures included in the plans were mainly transport related (Figure 3). The plans were also found to be flexible with consideration of educational and public awareness measures. These are the measures that most environmental health departments, responsible for LAQM, can actually implement. In line with government guidance these measures were prioritised according to their cost effectiveness, timetable and responsibility.

Although the integration of AQAPs with Local Transport Plans (LTP) is becoming widespread, about two-thirds of respondents found that this had made the process more complex. Results on internal collaboration indicated that, consistent with the strategies identified as important, transport and planning departments are most closely involved.

External organisations identified as relevant included County Councils, the Highways Agency and neighbouring authorities (Figure 4). In most cases the AQAPs included a list of consultees but did not provide information on their level of involvement. According to the appraisal, authorities also distributed their plan to many different organisations, the most common being the Primary Care Trusts or Health Authorities.

But the types of measures and the way they were reported in the AQAPs meant it was very difficult to assess their respective contribution. A classification of measures according to their origin and prospect of implementation revealed that the majority were not AQAP specific.

An EJ score was attributed to each plan [7]. This took account of evidence such as: specific statements relating to social impacts, consideration of relevant strategies, liaison with relevant local authority departments, public participation and the use of methods for measuring social factors. The EJ scores show that the majority of plans were deemed limited or failing with only 12 out of the 116 LAs achieving an excellent or good score (Figure 5). The results indicated that social impacts are not explicitly considered in the process.

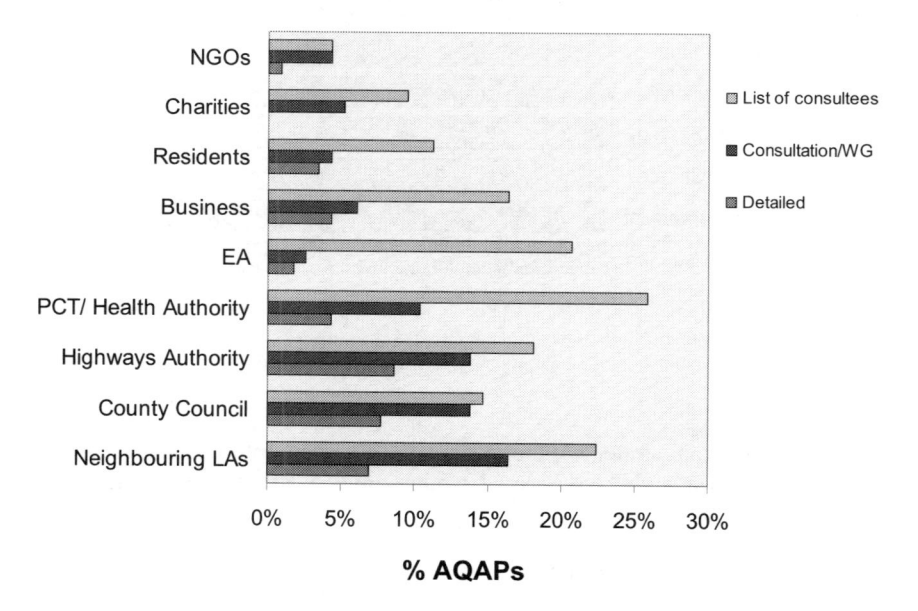

Figure 4: Involvement of external organisations (no of AQAPs = 116).

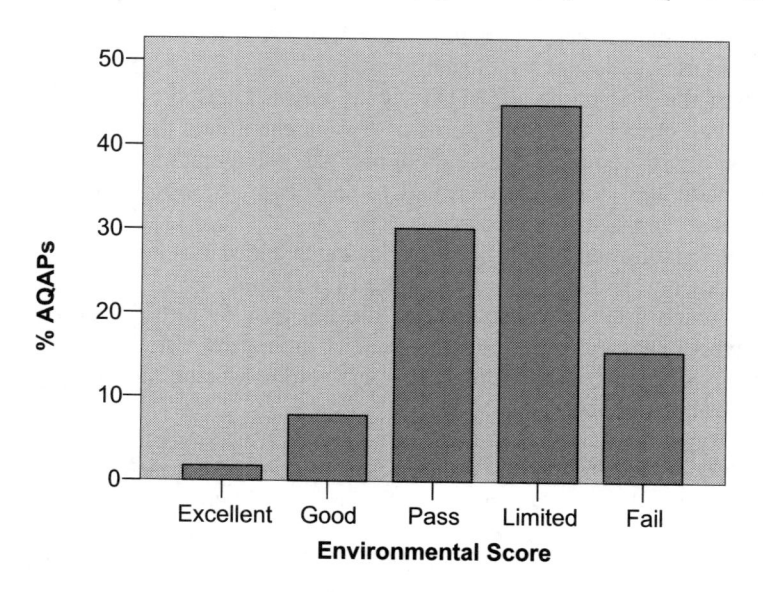

Figure 5: The Environmental Justice score for AQAPs (n = 116).

The case studies examined different types of local authorities in terms of their organisation, geographical region and deprivation levels. A number of overarching themes emerged: views on social impacts, incorporating social impacts into plans, LTPs, other relevant strategies and joint working. The first two refer to the views of interviewees on the relevance of social impacts and the possibility of incorporating them within LAQM. Interviewees were generally receptive to the idea of social impact and its significance but most had reservations about the possibility of incorporating them. Some stated that LAQM was not an appropriate process to take account of EJ.

The integration with LTPs was considered important as these plans tend to take account of the social needs of a local community. However, the success of integration was mixed. Some unitary authorities had successfully integrated their AQAP and LTP. But District Councils had found it more difficult to work with County Councils. 'Strategies' referred to other local authority policy documents which were considered relevant to EJ. These included Health Impact Assessments, Strategic Environmental Assessments, Sustainability Audits, Community Strategies and work undertaken as part of Local Strategic Partnerships.

Overall the case studies revealed that issues related to EJ are not usually taken into account in the AQAP process. But in some cases progress has been made through integration with LTPs which take greater account of local social issues.

4 Discussion

While the research revealed that AQMAs are often in areas of considerable deprivation, this is not specifically taken into account in developing AQAPs or prioritising measures within them. Local authorities follow government guidance and, in the main, do not take account of social data. However, the case study interviews revealed that local authority officers were aware of the level of social deprivation within their area and implicitly considered social impacts when assessing options. The prevailing view regarding the formal use of social data was negative; it was considered as an extra burden given the existing difficulties surrounding the production and implementation of AQAPs. This is not surprising as taking account of social impacts does not feature prominently in current Government guidance.

Public consultation was widely undertaken but it was less clear how the local knowledge gained was used in decision making. Communicating air pollution issues to the public and gathering their local knowledge can be problematic [9]. The types of measures and the way they were described in the AQAPs meant it was very difficult to assess their respective contributions to improving air quality. An analysis of the measures according to their origin and prospect of implementation revealed that their majority were not AQAP specific. This is not necessarily a negative finding; as if AQAPs are developed in conjunction with other local authority initiatives there may be greater potential to address EJ issues.

The morphology of urban space can have an impact on social deprivation. Space itself can be considered a factor in the geography of poverty and can contribute to persistent poverty [10]. Hence, where appropriate, AQAPs should be integrated with planning, neighbourhood renewal and regeneration initiatives. It would be beneficial to make use of whatever services a LA already has in place that could improve the inclusion of social impacts in decision making. These include community development, neighbourhood renewal and regeneration. A review of LA practices, policies and experiences with the use of social data across departments would help make an informed decision about using such data in AQAPs.

The principles of sustainable development coincide with EJ. At a global scale, local knowledge in conjunction with capital is believed to be a key to sustainable socioeconomic development [11]. In terms of LAQM and action planning the methods used by local authorities to gather this information have not been successful as the results of their efforts are not necessarily used. One problem may be a lack of understanding as to how best to use this information. There is also some scepticism surrounding its usefulness and relevance [12, 13]. Nevertheless it is important that public involvement is promoted. A better practice guide on LAQM consultation has been published providing an indicative guide for LAs [9].

An interesting example of the effect of traffic control measures on air quality and inequality is the London Congestion Charge. This was introduced in February 2003 to alleviate traffic congestion in central London. In a study to investigate the impacts on health and inequality Tonne et al. [14] used two modelled scenarios to isolate the changes in pollution due to traffic flow and speed, assuming meteorology was the same for both periods and no downward trend in emissions as the vehicle fleet was updated. Mortality impacts were estimated in terms of life years gained (LYG) by combining modelled changes in pollution concentration with pollution-mortality data. Population exposures to PM_{10} and NO_2 were analysed by quintile of socio-economic deprivation by overlaying modelled pollution concentrations on census areas of about 1500 population, using the Index of Multiple Deprivation. Modelled concentrations increased with deprivation although this relationship was less marked for PM_{10} than for NO_2. Within the congestion zone and adjacent wards LYG were 183 per 100,000 population compared with 26 across London as a whole. While the least deprived group experienced 0.02 ugm^{-3} and 0.01 ugm^{-3} decreases in NO_2 and PM_{10} respectively, the most deprived groups experienced 0.24 and 0.08 decreases. The authors conclude that the CCS had only a modest impact on concentrations of traffic-related pollutants and life expectancy but had also resulted in a slight reduction in inequalities in exposure to traffic-related pollution and mortality rates.

5 Conclusions

In England and Wales air quality management areas tend to cover areas of greater social deprivation. Action plans developed to address poor air quality in these areas do provide an opportunity to take account of environmental justice. But evidence suggests that, as this is not explicitly required by Government guidance, to date the opportunity has not been fully taken. It was found that many measures in AQAPs are not specific to air quality and this holds out the potential for integration of air quality measures with other local strategies, most notable local transport plans, but also strategies relating to planning. This may provide a better opportunity to take account of environmental justice while improving air quality. Although not intended principally to improve air quality or address environmental inequality, the London Congestion Charge provides initial evidence of the potential of this type of measure.

References

[1] Bullard, B., Dismantling environmental racism in the USA. *Local Environment*, **4**, pp. 5–19, 1999.
[2] Walker, G., Fairburn, J., Graham, S. & Gordon, M., *Environmental quality*, R&D technical report E2-067/1/TR. Environment Agency: Bristol, 2003.
[3] O'Neill, M. S., Jerrett, M., Kawachi, I., Levy, J. I., Cohen, A. J., Gouveia, N., Wilkinson, P., Fletcher, T., Cifuentes, L. & Schwartz, J., Health, Wealth and Air Pollution: Advancing Theory and Methods. *Environ Health Perspect*, **111(16)**, pp. 1861–1870, 2003.
[4] Laurent, O., Bard, D., Filleul, L. & Segala, C., Effect of socioeconomic status on the relationship between atmospheric pollution and mortality. *J Epidemiol Community Health*, **61(8)**, pp. 665–675, 2007.
[5] H M Government Environment Act 1995, Chapter 25. London: The Stationery Office, 1995.
[6] Longhurst, J. W. S., Irwin, J. G., Chatterton, T. J., Hayes, E. T., Leksmono, N. S. & Symons, J. K., A critical review of, and commentary on, the development of an effects based air quality management regime. *Atmos. Environ.*, in press.
[7] Gegisian, I., *Assessing the contribution of local air quality management to environmental justice in England and Wales*. PhD thesis, University of the West of England, 2007.
[8] ODPM, *The English Index of Deprivation 2004*. Wetherby: ODPM Publications, 2004.
[9] University of the West of England, Air Quality Review and Assessment Helpdesk. http://www.uwe.ac.uk/aqm/files/Steps_to_Better_Practice_Guidance_on_L AQM_Consultation.pdf
[10] Vaughan, L., Clark, D. L. C., Sahbaz, O & Hallay, M., Space and exclusion: does urban morphology play a part in social deprivation. *Area*, **37**, pp. 402–412, 2005.
[11] Corburn, J., Bringing local knowledge into environmental decision making – improving urban planning in communities at risk. *Journal of Planning Education and Research*, **22**, pp. 420–433, 2003.

[12] Petts, J. & Brooks, C., Expert conceptualisations of the role of lay knowledge in environmental decision making; challenges for deliberative democracy. *Environment and Planning A*, **38**, pp.1045–1059, 2006.
[13] Cinderby, S. & Forrester, J., Facilitating the local governance of air pollution using GIS for participation. *Applied Geography*, **25**, pp. 143–158, 2005.
[14] Tonne, C., Beevers, S., Armstrong, B.G., Kelly, F. & Wilkinson, P., Air pollution and mortality benefits of the London Congestion Charge: spatial and socioeconomic inequalities. *Occup. Environ. Med*, 2008. http://oem.bmj.com/cgi/content/zabstract/oem.2007.036533v1

Are environmental health officers and transport planners in English local authorities working together to achieve air quality objectives?

A. O. Olowoporoku, E. T. Hayes, N. S. Leksmono,
J. W. S. Longhurst & G. P. Parkhurst
Faculty of Environment and Technology,
University of the West of England, Bristol, UK

Abstract

Since 1997, Local Air Quality Management (LAQM) has been used as a process through which local authorities in England identify and manage specific air quality problems within their jurisdictions in order to achieve the air quality objectives (AQO). However, the limitation of this process is that of policy disconnect between diagnosis and solutions proffered within it. Over 90% of air quality 'hot-spots' identified through the LAQM are due to traffic-related sources. Hence, the air quality action plans prepared by the environmental health officers (EHO) are improperly calibrated as a policy instrument for tackling most of the problems discovered through the LAQM. The inclusion of air quality as one of the four shared priorities in the second round of the Local Transport Plan (LTP2) therefore implies that the EHO need to engage with the transport planners (TP) at the local level in order to address most of these problems i.e. traffic-related air pollution.

Since LAQM and LTP operate as two parallel frameworks with a separate agenda and timetable, adequate connectivity between both policy packages is thereby dependent on the type and level of inter-professional engagement between the departments and officials responsible for both policies at every level of government involved. This paper presents emerging issues from the questionnaire survey of EHO and TP in over 200 local authorities in 2007 as part of a three-year investigation into the effectiveness of achieving the AQO through the LTP in English local authorities. While there is wide support for the achievement of AQO through the LTP, the two groups identified differences in time-scale for delivering both policies, prioritisation of air quality within LTP, and unequal expectations as major factors affecting the integration. These factors indicate the existence of institutional complexities between parallel policy communities in ensuring integration.
Keywords: Local Air Quality Management (LAQM), Local Transport Plan (LTP), policy integration, environmental health officers, transport planners.

1 Introduction

Air quality and transport planning in England have a long history of targeted legislative and regulatory responses such as the Environment Act 1995 and Transport Act 2000, which established the Local Air Quality Management (LAQM) and Local Transport Planning (LTP) frameworks respectively. Although, both policies operate in separate institutional frameworks with different agendas and timetables, the contributions of transport-based sources to local air pollution has necessitated a more integrated approach. Part IV of the Environment Act 1995 provides the primary legislation for air quality management by requiring the Secretary of State to publish an air quality strategy (AQS). The strategy outlines methods and targets to be pursued by the UK Government and the Devolved Administrations of Scotland, Wales and Northern Ireland based on health effects standards and objectives for eight pollutants, seven of which are managed at the local scale through the LAQM regime [1]. The key aspect of this regime, as required in the legislation, is the *review* and *assessment* of local air quality against the seven pollutants. District councils, unitary and metropolitan authorities are therefore required by legislation to *review* air quality within their jurisdiction and *assess* whether the air quality standards and objectives are being achieved. In areas where these objectives may be compromised by the target date set in the NAQS, the local authorities are required to designate such as Air Quality Management Areas (AQMA) and put in place an Air Quality Action Plans (AQAP) to improve the local air quality [2].

The review and assessment process is designed to identify those areas where poor air quality coincides with public exposure [3]. Central Government has issued a series of descriptive guidance documents, accompanied by training sessions and helpdesks facilities to assist the local authorities in carrying out this process. The responsibility for managing the process at the local government level is usually undertaken by the environmental health department or their equivalent as a phased exercise which increases in depth and complexity consistent with the level of risk of failing to achieve the objectives [4]. (See Beattie *et al.* [5] and Longhurst *et al.* [6] for a full description of the Review and Assessment process.) Local authorities in England are currently undergoing the third round of the review and assessment process, resulting in the declaration of AQMA by over 205 local authorities in April 2007, accounting for 47% of local authorities in the UK [7]. Over 90% of AQMA declared, so far, were as a result of predicted exceedence in annual mean Nitrogen Dioxide (NO_2) and short-term PM_{10} objectives [8]. Both pollutants are largely due to traffic emissions from road transport sources, thereby undermining the power of LAQM, regardless of its intents and purposes, to remediate poor local air quality by itself. This creates an obvious limitation in the LAQM process in terms of policy disconnect between the diagnosis of the problem and the solutions proffered. While the review and assessment is effective in identifying the air pollution 'hot-spots' for subsequent declaration of an AQMA, the action planning is improperly calibrated as a policy instrument to the scale and nature of the discovered problem.

This is part of the necessity for the introduction of a parallel, but potentially more powerful, policy framework through which the traffic-related air pollution identified in the review and assessment can be properly addressed. In the Local Government White Paper, *Strong Local Leadership – Quality Public Services*, the government stated that it no longer require from the local authorities *"the production of a separate air quality management action plan where an air quality problem arises because of transport pollution. Instead, councils will be free to address this through their local transport plan"* [9]. In subsequent government guidance documents for LAQM, the integration of air quality action plans into the LTP was explicitly required for AQMA that have been declared due to traffic-related exceedences for roads that are under the jurisdiction of the local transport authority and falling within the scope of the LTP [10].

The current transport planning framework, LTP, was introduced in 1998 through the White Paper *"A New Deal for Transport: Better for everyone"* [11]. The subsequent legislation, Transport Act 2000 therefore require most local authorities in England (apart from Greater London) to produce a LTP as a form of financial bidding document submitted to the central government every five years, outlining comprehensive integrated transport strategies which will be implemented in this period. The LTP is aimed at ensuring certainty of funding for public transport initiatives to the local authorities, underpinned by a performance-based funding allocation system which is monitored and assessed by the government against a set of targets and objectives established in the LTP guidance documents [11]. The second round of the process which commenced in 2006 identified four shared priorities agreed between central government and the Local Government Association; congestion, accessibility, safety and air quality [12]. While the inclusion of air quality as a shared priority within the second round of LTP can be perceived as a required boost for achieving traffic-related air quality objectives within LAQM, there are potential limitations which may be due to the complex institutional arrangements in which both policy processes operate.

LAQM and LTP operate as two parallel frameworks with separate agendas and timetables, and are managed at the central government level by the Department for Environment, Food and Rural Affairs (Defra) and Department for Transport (DfT) respectively. At the local level, the responsibilities for LTP and LAQM are often in different departments within the local authority. Adequate connectivity between both policy packages is thereby dependent on the kind and level of inter-professional engagement between the departments and officials responsible for both policies at every level of government involved. Such engagement sometimes cuts across authorities due to variations in the institutional arrangements in which both policy processes operate (Fig. 1). English local governments operate either as a two-tier system where there is separation of functions between the upper-tier county council and the lower-tier district authorities, or in a single all-purpose system where a single unitary and metropolitan authority is responsible for all the local governments functions. However, some of the functions in an all-purpose system are often shared statutorily between metropolitan authorities (such as the Passenger Transport

Executives (PTE) that manages public transport on behalf of joint metropolitan authorities) or non-statutorily between unitary authorities with historical or geographical affiliation (such as in the case of unitary authorities having joint LTPs with neighbouring authorities as they share similar travel to work areas) [13]. The requirement for inter-professional collaboration between the transport planners (TP) responsible for the LTP and environmental health officials (EHO) responsible for the LAQM may be more complicated in a two-tier system where the air quality is traditionally managed by the lower tier authorities and the LTP by the county council (upper tier). In such arrangements there are possibilities for conflicts of priorities and resource allocation which are further compounded by the disparity in the time-scales of delivering both the LAQM and LTP framework.

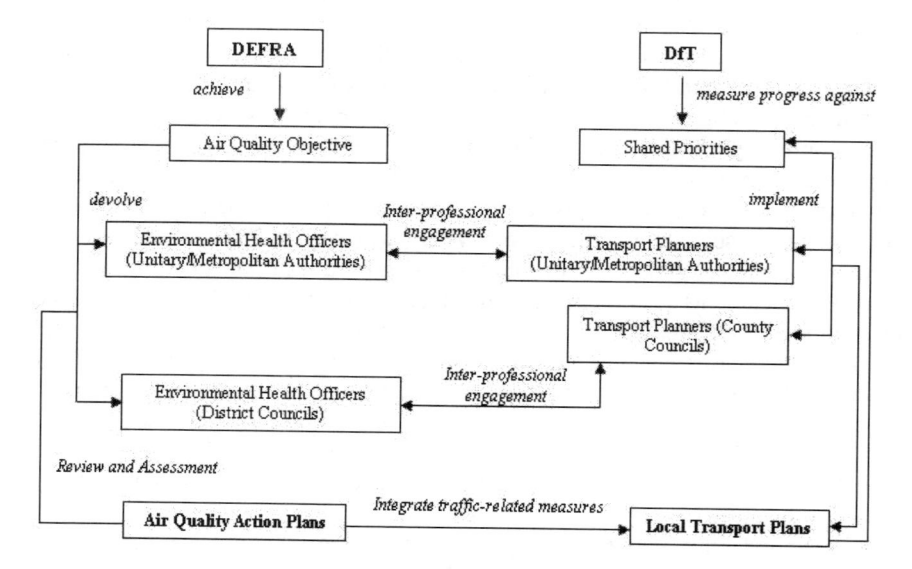

Figure 1: Institutional complexities involved in integrating air quality action plans into the local transport planning process in English local authorities.

Therefore, it can be argued that the success of achieving air quality objective (AQO) on the back of LTP is reliant on the capability of the process to overcome administrative complexities, conflicting timescales and other challenges of integration. This paper examines the perspectives and attitudinal approach of the primary stakeholders in the process (EHOs and TPs) towards the integration of both policies, and how this might influence the effectiveness of achieving the traffic-related AQO in the English local authorities. Evidence is presented from a web-based questionnaire survey of EHOs and TPs in over 200 local authorities between July and September 2007 as part of a three-year investigation into the effectiveness of integrating air quality into the LTP in England.

2 Methodology

Districts, unitaries and metropolitan authorities in England with current traffic-related AQMAs were identified from the Review and Assessment database held on behalf of Defra and the Devolved Administration at the University of the West of England, Bristol (UWE). The EHO from these authorities were selected for the survey due to the significance of transport-related issues in the preparation of their air quality action plans. The transport planning departments of local transport authorities such as counties and passenger transport authorities (PTA) or unitary authorities that included one or two authorities with traffic-related AQMA were selected due to the specific requirement for integrating air quality within their LTP. Specific EHO or TP whose responsibilities in the department included air quality management or LTP preparation respectively were selected and contacted for the survey. In some cases, the questionnaire was sent to a senior manager in the department with the expectation that it might be delegated it to the relevant officer.

The questionnaire was administered to 142 EHOs and 85 TPs through a web-link which further aids the retrieval and analysis of the questionnaire results. Responses were received from 70 EHOs and 41 TPs, an average of 49% response rate from both groups. The responses cut across the four major types of local authorities and all the regions in England (outside London). The questionnaire utilised the mixture of both the open-ended and close-ended questions. This is aimed at gathering as much detailed, yet structured information as possible within the limited available time and space. The data collected from both questionnaires was analysed using Statistical Package for Social Scientists (SPSS).

3 Results and discussion

By January 2005, at the time of preparing draft LTP2 documents, 138 local authorities in the UK had declared at least one AQMA, 90% of which were designated due to traffic emissions [14]. Due to the explicit requirement for the integration of action plans for traffic-related AQMA into the LTP2, it is expected that most of the AQMA declared by this time will be properly integrated into the LTP. However, 63% of the EHO surveyed said that their action plans were not integrated into the final LTP2, while 79% of the TP respondents acknowledged that the action plans were not effective for use during the preparation of the LTP.

These responses are indicative of some other factors which might influence the integration of both policies, three of which were identified in the survey.

3.1 Difference in timetable

First, an average of 60% from both groups recognised the disparity in the time-scales for achieving both the LAQM and LTP framework as a constraint to integration. 57% of EHO and 54% of TP respondents admitted that the action plans were not ready before the LTP submission deadline, thereby making it impossible for its integration into the final draft. In an attempt to overcome problems posed by this situation, the DfT recommended a new LTP progress reporting system starting from 2008, which will provide *'broader-based reviews, enabling authorities not only to assess their progress in meeting their objectives and targets during the first two years of [LTP2], but to consider any opportunities or threats to the effective delivery of the LTP2s in their remaining years'* [15]. This is intended to encourage local authorities who had missed the opportunity of integrating their actions plans into the final document at the time of the submission in 2006 to use the 2008 Progress Report as a means of so doing.

3.2 Degree of prioritising air quality within LTP

The second factor identified by both groups is the prioritisation of air quality within LTP. Although the Government through the DfT seeks the delivery of each of the shared priorities within the LTP, on the other hand, it provides opportunities for local authorities to decide the relative importance given to each priority in their area [12]. Table 1 presents the mean of Likert scale responses from the TP on the comparison of importance of the LTP priorities based on time, resources and fund allocation. Only 30% EHO disagreed with the statement that *"the transport planners do not put enough importance to achieving the air quality objectives within the LTP2"* this may be connected to the fact that other priorities such as safety, congestion and accessibility have longer histories of policy responses within the transport planning framework than air quality. Also, it can be argued that there are higher public opinion and concern over these priorities than air quality at the local level, thereby giving them political advantage over the latter. This is reflected in the way air quality is treated within the LTP. 42% of the EHO and 29% of the TP respondents thought that air quality is considered less important by the planners when preparing the LTP. However, in shire counties, metropolitan authorities or joint unitaries where an LTP covers more than one local authority, the risk of the air quality profile being reduced within the transport strategy was even greater if air quality problems arose in just one of the authorities. This view is echoed in these responses:

> *"Differing political priorities [is another reason why our local authority find it difficult to integrate action plans into LTP2]. It was politically uncomfortable for our county council to recognise the existence of one of our AQMAs and as a consequence, our action plan for this area was excluded from the LTP."*
> [EHO from a district authority in the East Midland region of England]
>
> *"Our AQMAs are very small and are stated by the county council as not 'even being on their list of problems areas."*
> [EHO from a district authority in the South East region of England]

Table 1: Mean of responses showing importance placed on priorities within LTP by the TP based on time, resources and fund allocation, where 1 = very high priority, 6 = very low priority (N = number of TP respondents).

PRIORITIES	N	MEAN (1-6)
Safety	41	1.46
Congestion	41	2.02
Accessibility	41	2.05
Other Local Priorities	39	2.33
Air Quality	41	2.98

The low relative importance of air quality within LTP raises a particular question on the implications of this on LAQM, given the broad support for the implementation of air quality through the transport planning (Fig. 2). This question leads to the third factor on the level of inter-professional working required by both groups for integration.

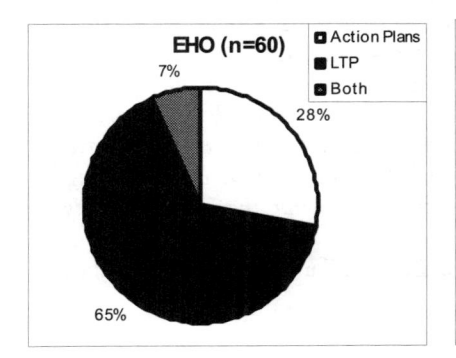

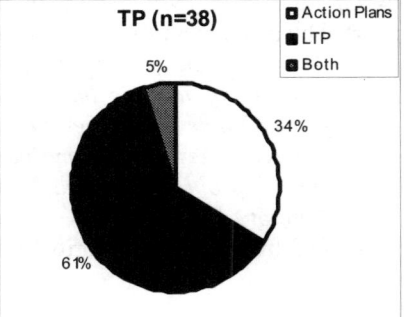

Figure 2: EHO and TP responses to question: In addressing transport-related AQ problems in your LA, which approach do you think is more effective?

3.3 Inter-professional engagement

Good communication and inter-professional engagement between the EHO and TP in promoting the air quality management profile within the LTP is crucial to the successful integration of both policies. However, the demonstration of such engagement in the LTP process is relatively weak as evidenced in the questionnaire survey. Although 23% of EHO and 12% of TP respondents agreed that the communication between both groups is very poor, there were disparities of opinion between both groups on the timing of consultation and the quality of communication necessary to integrate both policies. Despite the fact that over 90% of both groups surveyed agreed that the TP consulted with the EHO on

LTP, only 39% of EHO respondents thought that the consultation was early as compared to the 63% of TP respondents who considered it to be so. In addition, there were dissimilar perceptions on the quality of the communication between both groups, while 62% of TP respondents thought that the communication was sufficient to facilitate the integration, only 21% of the EHO respondents agreed with this.

Subsequently, lower percentage of EHO respondents (32%) thought that such communication is effective in comparison to 66% of TP respondents who considered it to be effective. It is therefore apparent that there are unequal expectations on the level of inter-professional engagement between both groups. As illustrated in Figure 3, there are sharp differences in the reflections of both groups regarding the attention given to the action plans by the TP during the implementation of LTP2. Such dichotomy is also reflected in their level of optimism towards addressing transport-related air quality problems through the LTP. Fewer EHO respondents (6%) thought that the integration has been successful in solving the problem, in comparison to 43% of TP respondents who think so. However, it can be argued that, rather than misconstruing the existing dichotomy between both groups as lack of support for air quality within LTP, it should be viewed as evidence of the existence of institutional complexities between parallel policy communities in ensuring integration. This argument is supported by a comment from one of the TP respondent:

> *"As with many joint ventures, successful operation requires that a measure of trust is built up between all involved. Two sets of professionals with different backgrounds and priorities will have different takes on the same subject. Progress is made by identifying shared problems and each contributing what they can towards solutions."*

[TP from unitary authority in the East Midlands region of England]

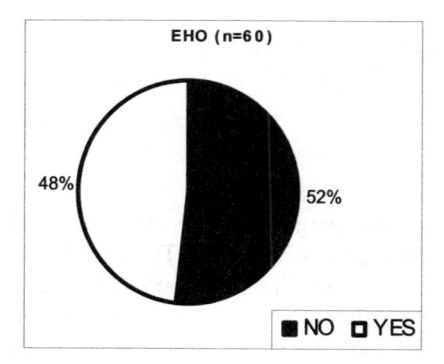

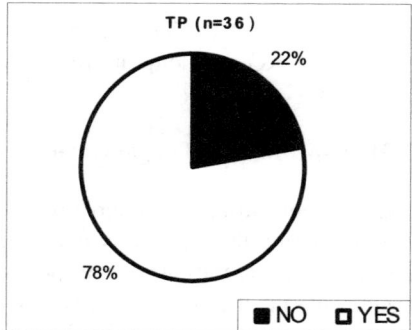

Figure 3: EHO and TP responses to question: Do you think the transport planners paid enough attention to the Action Plan during the implementation of the LTP2? (n= number of respondents).

4 Conclusions

While there is wide support for the achievement of traffic-related air quality objectives through the LTP, integrating both policies has been challenging due to timetable differences, differing interpretations by local authorities regarding the degree of prioritisation of air quality within LTP, and unequal expectations of both groups involved in the process. These factors are connected to the difficulties of facilitating wider collaboration and engagement between the two major stakeholders. Consequently, the existence of a collaborative platform, where the communication between the EHO and TP is promoted to a level acceptable and accessible to both groups, has positive potential for promoting the necessary integration.

Acknowledgements

The authors will like to thank Defra for allowing access to the review and assessment database, and the local authorities involved in this study for their participation.

References

[1] HM Government, Environment Act 1995. The Stationary Office: London, 1995.
[2] Department for Environment, Transport and the Regions, *The Air Quality Strategy for England, Scotland, Wales and Northern Ireland - Working Together for Clean Air*. The Stationary Office: London, 2000.
[3] Beattie, C.I., Longhurst, J.W.S. and Woodfield, N.K., Air Quality Management: evolution of policy and practice in the UK as exemplified by the experience of English local government. *Atmospheric Environment*, **35**, pp.1479–1490, 2001.
[4] Department for Environment, Food and Rural Affairs, National Assembly for Wales, Scottish Executive and Department of the Environment for Northern Ireland, *Part IV of the Environment Act 1995 Local Air Quality Management*. Technical Guidance LAQM, TG(03), Defra: London, 2003.
[5] Beattie, C.I., Chatterton, T.J., Hayes, E., Leksmono, N., Longhurst, J.W.S. and Woodfield, N.K., Air Quality Action Plans in the UK: an overview and evaluation of process and practice. *Proc. of the 14th Int. Conf. on Modelling, Monitoring and Management of Air Pollution*, eds. J.W.S. Longhurst & C.A. Brebbia, WIT Press: Southampton, pp. 503–512, 2006.
[6] Longhurst, J.W.S., Beattie, C.I., Chatterton, T.J., Hayes, E.T., Leksmono, N. S. and Woodfield, N. K., Local air quality management as a risk management process: Assessing, managing and remediation the risk of exceeding an air quality objective in Great Britain. *Environment International*, **32**, pp. 934–947, 2006.
[7] Hayes, E.T., Leksmono, N.S., Chatterton, T.J., Symons, J.K., Baldwin, S.T., and Longhurst, J.W.S., Co-management of carbon dioxide and local

air quality pollutants: identifying the 'win-win' actions. *Proc. Of the 14th IUAPPA World Congress,* Brisbane, Australia, 2007.
[8] Chatterton, T.J, Woodfield, N.K., Beattie, C.I., and Longhurst, J.W.S., Outcomes of the first round of local authority air quality Review and Assessments under the UK's Air Quality Strategy. *Journal of Environmental Monitoring*, **6,** pp. 849–853. 2004.
[9] Strong local leadership – Quality public services; UK. Department for Transport, Local Government and the Regions, 2001White Paper Online. http://www.communities.gov.uk/documents/localgovernment/pdf/143810
[10] Department for Environment, Food and Rural Affairs, National Assembly for Wales, Scottish Executive and Department of the Environment for Northern Ireland. *Part IV of the Environment Act 1995 Local Air Quality Management.* Policy guidance: Addendum LAQM, PGA (05), Defra: London, 2005.
[11] A New Deal for Transport; UK. Department of the Environment, Transport and the Regions, 1998 White Paper Online. www.dft.gov.uk/stellent/groups/dft_about/documents/pdf/dft_about_pdf_0 21588.pdf
[12] Department for Transport, *Full Guidance on Local Transport Plans: Second Edition.* DfT: London, 2004.
[13] Shepherd, S. P., Timms, P.M. and May, A.D., Modelling requirements for Local Transport Plans: An assessment of English experience. *Transport Policy*, **13,** pp. 307–317. 2006.
[14] Hassan, S. M., Evaluation of the local authority management process in Great Britain in its second round of Review and Assessment. PhD. Thesis University of the West of England, Bristol, Faculty of Applied Sciences, 2006.
[15] Department for Transport, *Strengthening local delivery: the draft Local Transport Bill.* HMSO: London, 2007.

A fuzzy MCDM framework for the environmental pollution potential of industries focusing on air pollution

R. K. Lad[1], R. A. Christian[1] & A. W. Deshpande[2]
[1]*Department of Civil Engineering,*
Sardar Vallabhabhai National Institute of Technology, Surat, India
[2]*Chair: Berkeley Initiative in Soft Computing (BISC),*
Special Interest Group (SIG),
Environment Management Systems (EMS),
Guest Faculty: University of California, Berkeley, California, USA

Abstract

The industrial policies of developing countries mainly focusing on the pursuit of economic growth with inadequate importance given to environmental pollution issues has resulted in rapid degradation of the natural environment. As the ambience has a dilution limit, industries that have been emitting air pollutants within the permissible pollution norms also contribute their share towards overall environmental degradation. Expressing permissible limits of pollution parameters on the dichotomous scale (Yes/No) needs a paradigm shift from crisp (Permissible OR Not Permissible) to fuzzy values (Permissible AND Not Permissible). A fresh look at the pollution control strategies is, therefore, necessary. An attempt has been made to address this problem and a new formalism of integrated effects of air pollutants is proposed for industries based on their air emissions. Vagueness in the perception of environmental experts for evaluating the techno-scientific parameters in linguistic terms for specific usage, coupled with imprecision in parametric data calls for the application of fuzzy modelling. In this study, importance is given to each air pollutant and a composite index is developed, which can reflect the air pollution potential of an industry. The study also reflects a case study of stringent environmental standards reflecting the rise in the pollution potential of industries. This can be linked to policy framing based on the principle of the polluter paying to control the pollution levels in the environment. The case study relates to the application of Fuzzy Multi Criteria Decision Making (FMCDM) for ranking the industries located in the State of Gujarat, India. The feasibility of the approach for ranking industries based on their air pollution potential is also discussed.
Keywords: Fuzzy Multi Criteria Decision Making, industrialization, linguistic variables, environmental issues, air pollution potential, fuzzy modelling, environmental experts' perception, stringent emission standards, fuzzy sets, ranking of industries.

1 Introduction

The rapid industrialization of India in the recent past has been the striking feature of the Indian economic development. The common indicators of economic welfare, such as national product and income have reflected the growth of the industry as a major indicator for the development of the nation. However, the other angle of industrialization has been the serious damage to the surrounding environment due to the wastes and pollutants generated from industries. The regulatory agencies set the norms to control the pollution, but as the ambience has a dilution limit, industries that have been emitting air pollutants within the permissible pollution norms also contribute their share towards overall environmental degradation. The situation is not very different from the case of the parameters relating to water, land and noise pollution. In this context, there is a need to see the pollution potential of industry on the basis of the permissible limit of pollutants. This study can be linked to the principle of "Polluter to Pay" on the basis of degree of certainty of pollution potential to reduce the pollution. This type of problem can be solved by using fuzzy logic. According to Hipel et al. [1], a decision problem is said to be complex and difficult, if there exist multiple criteria – both qualitative and quantitative in nature. Here a methodology is developed, employing the Fuzzy Multiple Criteria Decision Making (FMCDM) approach, for ranking of industries based on their air pollution potential. In this study separate weightage is given to the dust and gaseous criteria of air pollution. The ranking of industry can be done by jointly considering the water and air pollution potential indices [2].

2 Fuzzy Multi Criteria Decision Making (FMCDM) modelling

Many attempts have been made to study different methods of ranking alternatives and decision making for problems under fuzzy environment during the last few decades. For the evaluation of the modern concept of uncertainty readers can refer to the publication of a seminal paper by Lotfi A. Zadeh on fuzzy sets [3]. In his paper, Zadeh introduced a theory of objects – fuzzy sets – with boundaries that are not precise [4]. Bellman and Zadeh [5] used a concept of fuzzy goals and fuzzy constraints for fuzzy decision. Jain [6, 7] proposed a method of using the concept of membership level, whereas Baldwin and Guild [8] indicated that the above two methods suffer from some difficulties for comparing the alternatives and have disadvantages. Adamo introduced the α-preference rule using the concept of the α-level set. Chang indicated that the method proposed by Adamo [9] may lead to an inappropriate choice and went on to introduce the preference function concept of an alternative. A complete review of fuzzy numbers ranking methods was presented by Bortolan and Degani [10]. Hagemeister et al. [11] developed a methodology for hazard ranking of landfills using fuzzy composite programming, and presented a methodology to assess the environmental and public health hazard posed by an unregulated landfill when available data is imprecise, uncertain or subjective. Raj and Kumar [12] proposed the concept of a maximizing set and a minimizing set for ranking alternatives with fuzzy weights. Shen et al. [13] considered the characteristics of the construction business environment in China and identified the key parameters used in assessing contractors' competitiveness for awarding construction contracts in the market on a multi criteria basis. Seo et al. [14] developed a methodology with the help of a fuzzy decision making tool for the assessment of residential buildings, based on the acceptable level of environmental impact and socio-economic characteristics of the residential building. A more recent study by Singh and Tiong [15] highlights a fuzzy framework for contractor selection; this paper presents a systematic procedure based on fuzzy set theory to evaluate the capability of a contractor to deliver the project as per the owner's requirement. The approach developed for the ranking of industries based on their environmental pollution potential is somewhat analogous to the procedure suggested by Singh and Tiong [15]. The methodology discussed here has been successfully used for ranking of the same type of industries [2] considering the seven linguistic variables of Saaty [16].

2.1 Methodology

Figure 1 portrays an overview of the fuzzy decision framework to rank industries, which is self-explanatory. Identification of environmental experts is of prime importance. The importance weight for each of the criteria mentioned in Table 2 is developed by consulting environmental experts. To describe the level of performance on decision criteria Saaty [16] has proposed

```
┌─────────────────────────────────────────────────────────────────┐
│             Evaluation of pollution criteria                    │
└─────────────────────────────────────────────────────────────────┘
                              ▼
┌─────────────────────────────────────────────────────────────────┐
│             Define type(s) of fuzzy nos./fuzzy sets             │
└─────────────────────────────────────────────────────────────────┘
                              ▼
┌─────────────────────────────────────────────────────────────────┐
│       Define scale of preference and membership function        │
└─────────────────────────────────────────────────────────────────┘
                              ▼
┌─────────────────────────────────────────────────────────────────┐
│  Rating the preference of attribution on decision criteria (fuzzy value) │
└─────────────────────────────────────────────────────────────────┘
                              ▼
┌─────────────────────────────────────────────────────────────────┐
│ Fuzzy aggregation of scores, defuzzification of scores, x and normalization │
└─────────────────────────────────────────────────────────────────┘
                              ▼
┌─────────────────────────────────────────────────────────────────┐
│      Fuzzification and crisp score of industrial data           │
└─────────────────────────────────────────────────────────────────┘
                              ▼
┌─────────────────────────────────────────────────────────────────┐
│       Total score of dust and gaseous criteria                  │
└─────────────────────────────────────────────────────────────────┘
                              ▼
┌─────────────────────────────────────────────────────────────────┐
│       Weightage to dust and gaseous criteria                    │
└─────────────────────────────────────────────────────────────────┘
                              ▼
┌─────────────────────────────────────────────────────────────────┐
│   Overall score & ranking of different types of industries      │
└─────────────────────────────────────────────────────────────────┘
```

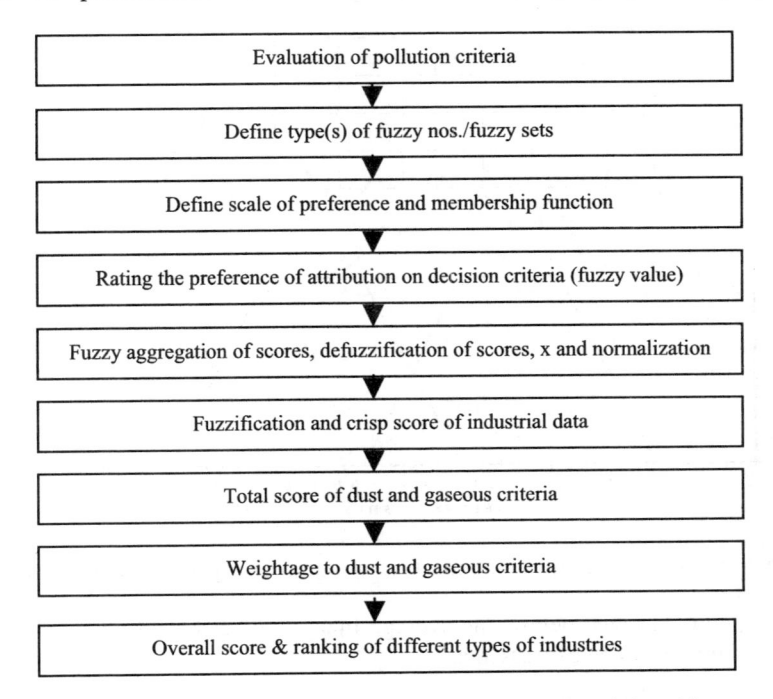

Figure 1:   Fuzzy decision framework for industrial ranking.

fuzzy numbers for seven linguistic variables. In this study, four fuzzy numbers are selected to describe the level of performance on decision criteria in the evaluation of pollution potential of industries. Four linguistic variables are used because it is convenient for an expert to distinguish subjectively between four alternatives. Table 1 shows the linguistic variables and fuzzy numbers used in this study. Figure 2 shows the graphical presentation of fuzzy numbers for the linguistics variables. The importance weight factors are computed for the sub criteria (parameters) of dust (SPM) and gases (SOx, NOx, Cl$_2$ and HCl). Table 2 shows experts' opinion for the sub criteria of air pollution.

Table 1: Linguistics variables and fuzzy numbers.

Linguistics Variables	Fuzzy Numbers
VI (Very Important)	(0.72,0.86,1.00,1.00)
I (Important)	(0.43,0.57,0.72,0.86)
A (Average)	(0.14,0.29,0.43,0.57)
NI (Not Important)	(0.00,0.00,0.14,0.29)

Table 2: Experts' opinion.

Sub Criteria	EE*1	EE2	EE3	EE4	EE5
Dust					
SPM, mg/Nm^3	I	I	VI	VI	VI
Gaseous					
SO$_2$, ppm	VI	VI	VI	VI	VI
NOx, ppm	I	VI	VI	A	VI
Cl$_2$, mg/Nm^3	NI	A	I	A	NI
HCl, mg/Nm^3	NI	NI	A	A	A

EE*- Environmental Expert

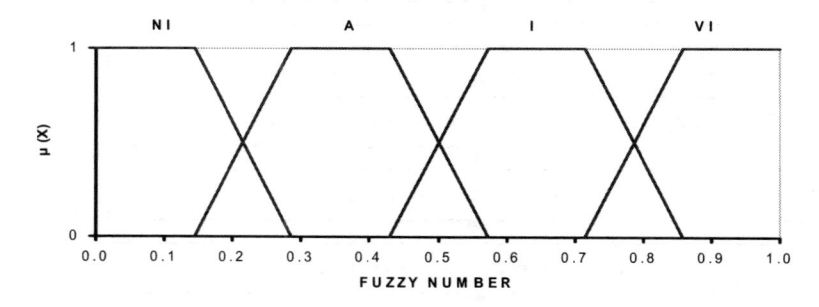

Figure 2: Graphical presentation of fuzzy numbers for linguistic variables.

Using eqn (1) given below, the average fuzzy number for all environmental experts' opinion can be expressed as

$$A_{ij}^k = \left(1/p\right) \otimes \left(a_{i1}^k \oplus a_{i2}^k \oplus ... \oplus a_{ip}^k\right) \text{ for } j=1,2,..,p \qquad (1)$$

where a_{ij}^k is a fuzzy number (weight) assigned to a parameter by environmental experts for the decision criterion C_k and p is the number of experts involved in the evaluation process. Using eqn (1) the matrix given above can be further simplified to calculate the average fuzzy number. The linguistic variables as assigned by the experts are converted to fuzzy numbers used in the above expression through Table 1 and Figure 2. Now, the defuzzified values for the sub criteria are obtained by using eqn (2).

$$E = (x_1 + x_2 + x_3 + x_4) / 4 \qquad (2)$$

For details about different types of fuzzy numbers, membership functions, aggregation and defuzzification methods, interested readers may refer to Zimmerman [17], Klir and Folger [18] and Kaufmann and Gupta [19].

The normalized weight for each sub criterion of dust and gases is obtained by dividing the scores of each sub criterion (C_{ij}) of dust and gases by the total of all sub criterions ($\sum C_{ij}$) of dust and gases respectively. The next step is to convert the parametric values of stack emissions to the fuzzy numbers (membership functions) based on the specified statutory norms.

Figure 3 shows the fuzzy set for not acceptable (membership function one) for gaseous parameter HCl. Similarly, fuzzy sets for other parameters of dust and gases can be developed.

The fuzzy decision matrix for sub criterion C_{15} (HCl) can be written as

$$X_{C_{15}} = \begin{bmatrix} \mu_1 \mu_2 \mu_3 \cdots \mu_n \\ (a_1, a_2, a_3, \ldots a_n,) \\ (b_1, b_2, b_3, \ldots b_n,) \\ (c_1, c_2, c_3, \ldots c_n,) \end{bmatrix} \begin{vmatrix} I_1 \\ I_2 \\ I_n \end{vmatrix}$$

where $a_1, a_2, a_3 \ldots a_n, b_1, b_2, b_3 \ldots b_n$ and $c_1, c_2, c_3 \ldots c_n$ are fuzzy values of HCl obtained from seasonal monitoring for Industry 1, Industry 2 and Industry n respectively.

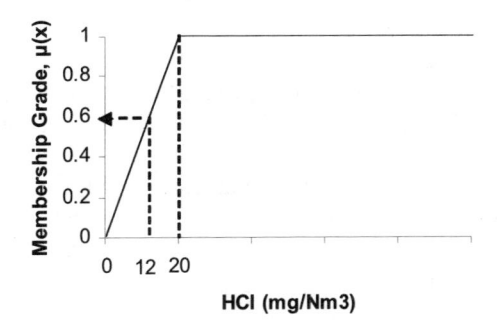

Figure 3:   Pollution parameter HCl: fuzzy set for not acceptable.

The crisp scores on the sub criterion C_{15} for each industry can be obtained using following equations.

Industry 1 = $(a_1 + a_2 + a_3 .. a_n)/n$,

Industry 2 = $(b_1 + b_2 + b_3 .. b_n)/n$ and

Industry n = $(c_1 + c_2 + c_3 .. c_n)/n$

Similarly, crisp scores can be computed for the other sub criteria of dust and gases. Using the simple additive weighing method (Hwang and Yoon [20]), the total scores (TS) for each industry for dust and gaseous criteria can be calculated by eqn (3).

$$TS = \sum (X_k \otimes W(C_k)) \text{ for } k = 1,2,3..n \quad (3)$$

where, $W(C_k)$ = weight or the importance value of the sub criterion k and

X_k = crisp score of the industry data against the sub criterion k.

Using pollution potential importance weight for both the criteria (dust and gaseous) such that their summation is equal to 1, an overall score (OS) for the industries can be calculated by eqn (4).

$$OS = \sum (TS_{ki} \otimes W(C_{ki})) \text{ for } k = 1,2,3..n \text{ and } i = 1,2,3...n \quad (4)$$

where TS_{ki} = total score of the industry i against the criterion k

$W(C_{ki})$ = weight or the importance value of the criterion k for industry i = $TS_{ki} / \sum TS_{ki}$

3 Case study

The case study relates to the available air emission characteristics from three chemical industries, three thermal power station units and three dying and printing textile industries located in Gujarat State, India. Table 3 shows the stack emissions for the above mentioned industries monitored for winter (M_1), summer (M_2) and the rainy season (M_3).

The parametric values of stack emissions are converted to the fuzzy numbers (membership functions) based on the specified statutory norms (see figure 3). For example, for 12 mg/Nm^3, the HCl normalized value is 0.6. Similarly, normalized values of stack emissions for all industries have been worked out. For the final total score, a unique membership value for each industry for different sub criteria can be obtained by using the simple average. Then the normalized weight for each sub criterion of dust and gases for different industries are calculated.

Using the simple additive weighing method (Hwang and Yoon [20]), the total score (TS) for each industry has been calculated and the same is as shown in Table 4. The matrix (Figure 4) is shown for the total score of the sub criteria of gases for chemical industries.

Table 3: Effluent characteristics of stack emissions.

Sub Criteria	GPCB[#] limit	Industry 1			Industry 2			Industry 3		
		M_1	M_2	M_3	M_1	M_2	M_3	M_1	M_2	M_3
Chemical Industries										
Air Pollution										
Dust										
SPM, mg/Nm^3	150	60.0	55.0	80.0	20.0	175	60.0	175	220	159
Gaseous										
SO_X, ppm	100	12.25	13.12	3.4	61.9	80.0	110	15.0	25.0	45.0
NO_X, ppm	50	21.5	30.0	36.0	58.0	49.0	55.0	14.0	35.0	36.0
Cl_2, mg/Nm^3	9	4.5	3.66	3.91	0.00	0.00	0.00	0.00	0.00	0.00
HCl, mg/Nm^3	20	0.00	0.00	0.00	0.00	0.00	0.00	0.00	0.00	0.00
Thermal Power Station Units										
Air Pollution										
Dust										
SPM, mg/Nm^3	150	0.00	7.00	3.8	3.6	3.1	2.4	0.00	0.00	0.00
Gaseous										
SO_X, ppm	100	3.61	6.22	12.25	12.23	9.21	6.1	4.80	4.20	6.1
NO_X, ppm	50	7.21	6.40	10.4	140	140	132	7.90	7.40	7.20
Dying and Printing Units										
Air Pollution										
Dust										
SPM, mg/Nm^3	150	149	128.3	132.1	117.3	96.1	101.7	148.5	134.06	135.9
Gaseous										
SO_X, ppm	100	51.48	40.80	41.5	55.05	42.82	43.62	58.21	52.74	55.25
NO_X, ppm	50	6.27	6.50	6.60	2.82	2.14	2.41	6.05	6.08	6.11

[#] *Gujarat Pollution Control Board*

$$\begin{bmatrix} \text{Sub Criteria} \\ SO_2 \\ NOx \\ Cl_2 \\ HCl \end{bmatrix} \begin{bmatrix} I_1 & I_2 & I_3 \\ 0.10 & 0.81 & 0.28 \\ 0.58 & 0.99 & 0.57 \\ 0.45 & 0.00 & 0.00 \\ 0.00 & 0.00 & 0.00 \end{bmatrix} \begin{bmatrix} W_{Ck} \\ 0.41 \\ 0.33 \\ 0.14 \\ 0.12 \end{bmatrix}$$

Figure 4: Matrix for total score for sub criteria of gaseous of chemical industries.

Table 4: Total score and summation for criteria of chemical industries.

Criteria	I1	I2	I3
Dust (C_1)	0.433	0.511	1.000
Gaseous (C_2)	0.298	0.660	0.305
Σ	0.731	1.171	1.305

$$\begin{bmatrix} \text{Criteria} \\ C1 \\ C2 \end{bmatrix} \begin{bmatrix} I1 & I2 & I3 \\ 0.433 & 0.511 & 1.000 \\ 0.298 & 0.660 & 0.305 \end{bmatrix} \begin{bmatrix} W_{Ck1} & W_{Ck2} & W_{Ck3} \\ 0.593 & 0.437 & 0.767 \\ 0.407 & 0.563 & 0.233 \end{bmatrix}$$

Figure 5: Matrix for overall score for dust and gaseous.

Table 5: Overall score and ranking for different types of industries.

Chemical Industries			Thermal Power Station Units			Dying and Printing Units		
0.378	0.595	0.838	0.097	0.483	0.095	0.758	0.580	0.771
7	4	1	8	6	9	3	5	2

Table 6: Overall score and ranking of chemical industries (sensitivity analysis).

GPCB limit	Criteria	I1	I2	I3
20 ppm	SO_2	0.444	0.645	0.849
100 ppm	SO_2	0.378	0.595	0.838

Using pollution potential importance weight for both the criteria (such that their summation is equal to 1) has been calculated. The matrix (Figure 5) is shown for an overall score of dust and gaseous criteria. From that an overall score (OS) for chemical industries has been calculated. Similarly, overall scores for thermal power station units and dying and printing units are calculated and the same are shown in Table 5.

3.1 Sensitivity analysis

In order to check the sensitivity of the model for the given sub criteria, it was proposed to operate the model with SO_2 emission norms of 20 and 100 ppm. The ranking was obtained for three chemical industries shown in Table 6.

3.2 Comments

As seen from the results (Table 6), the impact potential of an industry increases when the emission standards for the SO_2 are made more stringent from 100 to 20 ppm. Moreover, the same is also reflected on an overall score of the industry. However, the ranking of an industry does not change.

4 Discussions

From the results of different types of industries (Table 5), it can be inferred that the chemical industry ranks first in the list of nine industries with high pollution potential and the thermal power station unit ranks number nine in the list of these industries with minimum pollution potential. So in this way, it is possible to rank different types of industries on the basis of their pollution potential and it is possible to encourage industrial entrepreneurs to bring their industrial pollution potential to a minimum level by controlling pollution as an attempt to protect the environment.

5 Conclusion

The paper demonstrates the use of fuzzy modelling for the ranking of industries based on their air pollution potential with a case study. The sensitivity analysis reveals the pollution potential of the industry increases with stringent emission standards, however; it retains its raking. As the pollution levels in general are increasing, it is opined that the issue of pollution tax should be studied and considered by decision makers of developing countries to control the pollution levels in the environment.

References

[1] Hipel, K. W., Radford, K.J. & Fang, L., Multiple participant multiple criteria decision-making. *IEEE Trans. Syst. Man Cybern*, **23**, pp. 1184–1189, 1993.
[2] Lad, R. K., Desai, N. G., Christian, R. A. & Deshpande, A. W., Fuzzy modelling for environmental pollution potential ranking of industries. *International Journal for Environmental Progress, AIChE*, **27(1)**, pp. 84–90, 2008.
[3] Zadeh, L.A., Fuzzy Sets. *Information and Control.* **8(3)**, pp. 338–353, 1965.
[4] Klir, G.J &Yuan, B., *Fuzzy Sets and Fuzzy Logic-Theory and applications*, Prentice Hall of India Private Limited: New Delhi, pp.3, 2003.
[5] Bellman, R. E. & Zadeh, L. A., Decision-making in a fuzzy environment. *Management Science*, **17 (4)**, pp. B141–B164, 1970.
[6] Jain, R., Decision making in the presence of fuzzy variables. *IEEE Trans. Systems Man Cybern*, **6**, pp. 698–702, 1976.
[7] Jain, R., A procedure for multiple aspect decision making using fuzzy sets. *Int. J. Systems Sci.*, **8**, 1–7, 1977.
[8] Baldwin, J. F. & Guild, N. C. F, Comparison of fuzzy sets on the same decision space. *Fuzzy Sets and Systems*, **2**, pp. 213–231, 1979.
[9] Adamo, J. M., Fuzzy decision tree. *Fuzzy Sets and Systems*, **4**, pp. 207–219, 1980.
[10] Bortolan, G. & Degani, R., A review of some methods for ranking fuzzy subsets. *Fuzzy Sets and Systems.* **15**, pp. 1–19, 1985.

[11] Hagemeister, M.E., Jones, D.D. & Woldt, W.E., Hazard ranking of landfills using fuzzy composite programming. *Journal of Environmental Engineering, ASCE*, **122(4)**, pp. 248–258, 1996.
[12] Raj, A. P., & Kumar, N.D., Ranking alternatives with fuzzy weights using maximizing set and minimizing set. *Fuzzy Sets and Systems*, **105**, pp. 365–375, 1999.
[13] Shen, L. Y., Li, Q. M., Drew, D. & Shen, Q. P., Awarding construction contracts on multicriteria basis in China. *Journal of Construction Engineering and Management, ASCE*, **130 (3)**, pp. 385–393, 2004.
[14] Seo, S., Aramaki, T., Hwang, Y. & Hanaki, K., Fuzzy decision making tool for environmental sustainable buildings. *Journal of Construction Engineering and Management, ASCE*, **130 (3)**, pp. 415–423, 2004.
[15] Singh, D. & Tiong, R.L.K., A Fuzzy decision framework for contractor selection. *Journal of Construction Engineering and Management*, ASCE, **131(1)**, pp. 62–70, 2005.
[16] Saaty, T.L., A scaling method for priorities in hierarchical structures. *J. Math. Psychol.*, **15(1)**, pp. 234–281, 1977.
[17] Zimmerman, H.J, *Fuzzy Set Theory and its application*, Kluwer Nijhoff: Hingham Mass, 1985.
[18] Klir, G. J. & Folger, T. A., *Fuzzy sets, uncertainty and information*, Prentice-Hall: Englewood Cliffs, N. J., 1988.
[19] Kaufmann, & Gupta, M. M., *Fuzzy mathematical models in engineering and management science,* North-Holland, Amsterdam, The Netherlands, 1988.
[20] Hwang, C. L. & Yoon, K., *Multiple attribute decision making- methods and applications*, Springer: New York.

A Model Municipal By-Law for regulating wood burning appliances

A. Germain, F. Granger & A. Gosselin
Environment Canada, Environmental Protection Operations Division, Montreal, Canada

Abstract

This paper describes a Model Municipal By-Law, developed to support municipal or local governments that wish to control air pollution caused by the use of residential wood burning for heating purposes. Wood burning is the most important anthropogenic source of fine particulates ($PM_{2.5}$) in Canada. As a complement to a national regulation on new, cleaner burning wood burning appliances, initiatives were identified to address existing appliances. These initiatives include public outreach and a change-out program. As a result, a Model Municipal By-Law for regulating wood burning appliances was developed as an aid to local governments that want to regulate the use of residential wood burning appliances for residential use on their territory.

Keywords: residential wood combustion, emission, regulation, municipal by-law.

1 Introduction

In 2000, the Canadian Council of Ministers of the Environment developed a Canada-wide Standard for particulate matter less than or equal to 2.5 microns (also known as fine particulate or $PM_{2.5}$) as a result of the pollutant's adverse effects on human health [1]. It also indicated that measures where to be taken to reduce their emissions, including those from residential wood combustion for heating purpose [2].

This document is intended as an aid for municipalities where air quality problems due to residential wood burning are experienced and who therefore wish to put in place a municipal by-law for regulating woodburning appliances. The workshop summary of the Kelowna Residential Indoor Wood Burning

By-Law Workshop [3] served as a starting point for developing this document, as well as responses received by 17 out 26 Canadian and American jurisdictions invited to share their experience regarding the implementation and performance of their own by-laws.

2 Atmospheric emissions from residential wood combustion

There were approximately 3.6 million wood-burning appliances installed in Canada in 1997, distributed between fireplaces (almost 2 millions), wood stoves (1.3 millions) and central furnaces or boilers (0.3 million) [4]. Approximately 0.25 million (7%) of them were advanced technology appliances that comply with the United States Environmental Protection Agency's (EPA) *Standards of Performance for New Residential Wood Heaters* (40 CFR Part 60, Subpart AAA) [5] or the standard B415.1-00 of the Canadian Standard Association (CSA) [6]. We also refer to them as "certified" appliances or wood stoves.

Residential wood heating fulfils only 1% of Canada's energy demand yet it is responsible for 29% of the fine particulate matter ($PM_{2.5}$) and 48% of the polycyclic aromatic hydrocarbons (PAHs) emitted by all Canadian sources combined (excluding forest fires and dust from paved roads and unpaved roads). New-technology fireplaces and wood stoves that comply with the CSA's standard B415.1-00 or the Standards of Performance for New Residential Wood Heaters of the U.S. Environmental Protection Agency (U.S. EPA) have much lower levels of polluting emissions than non-certified appliances. The replacement, in whole or in part, of old wood-burning appliances with state-of-the-art units would reduce emissions of PM2.5, PAHs, volatile organic compounds (VOCs) and carbon monoxide (CO) by the home heating sector by 30 to 55% and lead to a 10% drop in emissions of greenhouse gases (GHG). Considering the lifespan of a wood stove, it could take another 20 to 40 years before those appliances currently in use are replaced.

Table 1 presents some of the Criteria Air Contaminants (CAC), PAH, D/F and GHG emissions data for different sectors of activity. For the purposes of comparison, the emissions of other sources of residential fuel combustion and total Canadian emissions (with and without open sources) are also presented. Fine particulate ($PM_{2.5}$), volatile organic compounds (VOCs) and carbon monoxide (CO) emission values were taken from Environment Canada's 2000 CAC Emissions Inventory [7]. The PAH emissions were taken primarily from the 1990 inventory prepared for the toxicity assessment of PAHs [8], except for residential wood combustion and the industrial sector. Emissions for residential wood combustion were calculated considering the amount of wood burned in Canada and emission factors derived from Fisher et al. [9] and from Valenti and Clayton [10]; for the industrial sector, PAH emissions from a 1990 inventory were updated to reflect emission reductions in the aluminium sector up to the year 2000. The dioxin and furan values come from Environment Canada's D/F Inventory [11] and the greenhouse gas values were taken from the Canadian GHG Inventory for 2000 [12].

Table 1: Air emissions of selected sectors of activity in Canada.

Selected sectors	PM$_{2.5}$ (kt)	VOC (kt)	CO (kt)	PAH (t)	D/F (g ET)	GHG (Mt eq. CO$_2$)
Year	2000	2000	2000	1990	1999	2000
Residential heating						
• natural gas and propane, oil, electricity, coal	3.6	2.3	14.0	32	7	42.7
• wood	101.3	147.4	662.0	1,381[a]	3	2.2
Total residential heating	104.9	149.7	676.0	1,413	10	44.9
Energy production	27.2	804.1	172.2	a.d.	5	195
Industrial sources	133.5[b]	190.8[b]	1,132.0[b]	610[a,b]	26[b]	119[b]
Transport	72.2	727.1	8,375.0	200	9	190
Commercial	3.1	6.5	8.1	a.d.	a.d.	31.9
Total emissions in Canada[c] without open sources	350	2,431	10,381	2,859	a.d.	a.d.
Total emissions in Canada[c]	963	2,604	11,282	4,945	163	726
Proportion from residential wood combustion (%)	29.0[d]	6.1[d]	6.4[d]	48.3[d]	2.0[d]	0.3[e]

Legend: PM$_{2.5}$: particles < 2.5 microns; CO: carbon monoxide; VOC: volatile organic compounds; PAH: polycyclic aromatic hydrocarbons; D/F: dioxins and furans; GHG: greenhouse gas.
a: Year 2000 data for residential wood combustion and industrial sector, reflecting improvement in aluminum sector.
b: Excluding energy sector industries.
c: Rounded off to closest unit.
d: Proportion (%) of Canada's total without open sources such as dust from roads and forest fires.
e: Proportion (%) of Canada's total greenhouse gas emissions.
f: NA: Not available.

3 Residential wood combustion control strategy in Canada

Environment Canada has worked on four inter-related elements to address the air emission from residential wood combustion is comprised of four inter-related elements. These four elements are:

1. Update of the Canadian Standard Association (CSA) B415-1.00 Standard on residential wood combustion appliances.
2. Integration of the CSA Standard in a regulation controlling the sale as well as the import and manufacture for sale of residential wood combustion appliances.
3. Public education and outreach.
4. Promote the change of existing old technology appliances for new certified appliances.

The first two elements address the new woodstoves and central-heating boilers coming into use. They will ensure that all new appliances sold in Canada are certified, cleaner burning appliances. However, once installed, a wood burning appliance will have an impact on air quality for a long period of time. The wood stove industry in Australia suggests that their working lifespan is 15 to 20 years [13] and an American source mentions that good-quality wood stoves can be expected to last 40 years [14]. In view of this slow turnover rate, other measures could be used to reduce emissions and their impact on air quality.

The two other elements of the strategy are meant to address existing, in-use appliances. The first one is a public education and outreach program. Its objective is to inform the general public on the impact of wood combustion on air quality and health, as well as the consumer's choice, technology and practices to reduce that impact.

The last two elements could help achieve a faster turnover rate of old technology stove to cleaner burning certified stoves by influencing the choice of the consumer in favour of the stoves using the best technology (certified stoves). This is appropriate when there are no regulations in place to limit the availability of conventional, uncertified appliances on the market. It could be more efficient if incentive reducing the price differential between advanced technology stoves and conventional stoves enough to shift consumer choice significantly was provided. However, it is not actually considered because of the elevated costs involved.

Another type of economic instrument, the change-out programs, is designed to accelerate the turnover rate by convincing the consumer to exchange its old technology stove for a certified one sooner than he would have without the incentive. These programs are best suited to local or regional scopes and are more efficient if coupled with a strong public education component.

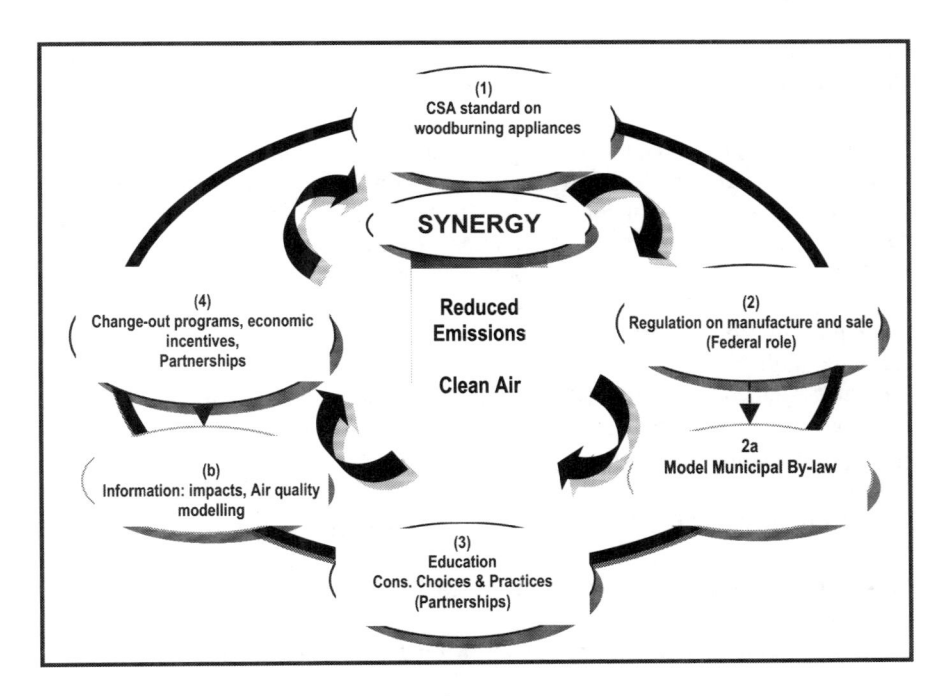

Figure 1.

4 Role of municipal governments in air quality management

The set of national-wide measures proposed in the strategy may not be sufficient to ensure that air quality problems associated with residential wood combustion are properly addressed everywhere in Canada. In certain areas, most notably where there is a dense concentration of wood burners and/or topographical or adverse weather conditions, a high concentration of fine particulates can result, and may indicate that wood burning may not be suitable in that area. In some cases, those conditions are only temporary; in other situations they are chronic and may require a controlled or more regulated approach to wood burning. In these instances, the most efficient approach may be through municipal governments. In that respect, the federal and provincial governments are best placed to control the entry of new appliances on the market, through regulating the sale, import and manufacture of appliances. The municipal or regional authorities are best placed to control the installation and use of residential wood combustion appliances.

5 Objective of the Model Municipal By-Law

However, many municipalities lack resources and expertise to design and implement efficient regulatory instruments to adequately manage air issues associated with residential wood combustion. The lack of adequate support to local governments who choose to regulate the use of wood as a residential energy source was indicated by stakeholders as an issue needing to be addressed. The model municipal by-law has been produced by Environment Canada in collaboration with representatives from the industry, municipalities, provincial and territorial governments, and environmental non-governmental organizations, under the Intergovernmental Working Group on Residential Wood Combustion (IGWGRWC). This document is a tool to be used by a municipality wishing to develop and put in place a municipal by-law to regulate wood burning appliances. It is intended to be used by municipalities that experience air quality problems because of residential wood combustion. This Model Municipal By-Law presents control strategies and options that may be adopted by municipal or regional authorities on residential wood burning. The appendix of the document contains elements of a model municipal by-law with the wording of control strategies and options contemplated to address reduction and control of particulate matter emissions.

6 Type strategies described in the Model By-Law

The control strategies presented in the Model By-Law can be categorized as follows.

- Strategies that specify limits on total emissions, measured either as a unit of production or as a reduction in emissions relative to a baseline. These include emission limits for wood burning appliances and mandatory curtailment strategies.
- Strategies that provide incentives or impose disincentives to limit emissions rather than making reductions compulsory. These include financial assistance strategies to encourage change-out of non-certified wood burning appliances and mitigation offset strategies.
- Strategies that do not yield quantifiable emission reductions, but still contribute to an area's overall attainment of air quality standards. These include public education and information strategies.

These strategies comprehensively address particulate matter pollution by establishing regulatory mechanisms, offering financial incentives and assistance and providing education and information. Municipalities can employ a mix of all three strategies. The document provides supporting information for the following control options.

- Restriction on Some Fuels. - The restrictions prohibit use specific fuels that promote adverse air quality conditions. These fuels include wet or unseasoned wood, garbage, plastics, treated wood, rubber products, waste oils, paint, solvents etc.
- Installation of wood burning Appliances. - This by-law would prohibit the installation of wood burning appliances that do not meet specific standards, for example Canadian Standard Association (CSA) standard or the US EPA standard.
- Non-certified Appliance Removal. – An example of such strategy is a provision that requires that, prior to the completion or consummation of a sale or transfer of any real property on or after a certain date, all

existing wood burning appliances that are not certified shall be replaced, removed or rendered permanently inoperable.
- No Burn Days. - The local authority may issue a declaration of an "Air Quality Advisory Period" through local communications media that would result in voluntary or mandatory curtailment of the use of non-certified or all wood burning appliances whenever conditions within the region are projected to cause ambient air quality concentrations of $PM_{2.5}$ that exceed a certain level (micrograms per cubic meter).
- Nuisance. - This strategy applies a nuisance by-law where woodburning appliance fires shall be maintained so as not to cause a nuisance for more than two minutes in succession except during a thirty-minute period following the starting or re-fuelling of the appliance.
- Opacity. - This strategy applies an opacity limit where, within the municipality, no person owning or operating a wood burning appliance shall at any time cause, allow or discharge emissions of an opacity greater than twenty (20) percent
- Outdoor Solid-fuel Combustion Appliances. – This strategy applies restriction to the installation and use of outdoor air or water heater. These types of appliances are not subjected to an enforceable standard as yet, and present a growing air quality issue in Canada.

For each control strategy, the following information is provided.
- Description. - A narrative of the strategy describing what and how the strategy may be accomplished.
- Target. - The particulate matter sources, type of development and emission goals that the strategy is intended to address.
- Advantages. - Various factors that support the strategy.
- Disadvantages. - Various factors that weigh against the strategy.
- Costs to private citizens. - The potential implementation costs of the strategy that will be directly passed on to the private citizens.
- Costs to the public sector. - The potential implementation costs to the municipal government and other public agencies. These costs may be considered indirect costs to the citizens.
- Emission savings. - Reductions from current levels of emissions and emission concentrations anticipated upon implementation of the strategy.
- Enforcement. - A discussion on how the strategy will be enforced to ensure compliance.
- Implementation guidelines. - Guidelines on how the strategy should be effectively implemented. This section often refers to the need for an education component.
- Who is doing it? - A non-exhaustive list of Canadian and United States jurisdictions that have implemented a similar control strategy

In addition to these control options supporting information is also provided on the emission offsets programs as well as awareness, education and communication strategies

7 Conclusion

At the moment, there are very few municipalities that have adopted, or even are planning to adopt a regulatory approach regarding residential wood combustion within their jurisdiction. The *Model Municipal By-Law for Regulating Woodburning Appliances* was published in 2006. It was initially distributed to more than 200 municipalities in Canada. Since then, at least one Canadian municipality has used the guide to prepare and adopt a by-law on residential wood combustion. However with increasing public awareness of the air quality and health issues resulting from the residential wood combustion, it is reasonable to think that public pressure will drive more local governments to consider this course of action.

References

[1] CCME, Canada-Wide Standards for Particulate Matter (PM) and Ozone, 2000, Online. http://www.ccme.ca/assets/pdf/pmozone_standard_e.pdf
[2] CCME, Joint Initial Actions to Reduce Pollutant Emissions that Contribute to Particulate Matter and Ground-level Ozone, 2000. Online. http://www.ccme.ca/assets/pdf/pmozone_joint_actions_e.pdf
[3] Workshop Summary, Residential Indoor Wood Burning Bylaw Workshop, 2002. Online. http://www.bvldamp.ca/pdf/rcsh/BurnBylawFinalSummary.pdf
[4] Niemi, D. 2002–2003. Personal communications. Head, Emissions Inventory Reporting and Outreach, Environment Canada, Pollution Data Branch.
[5] Subpart AAA—Standards of Performance for New Residential Wood Heaters; United States Environmental Protection Agency. Online. http://www.epa.gov/compliance/resources/policies/monitoring/caa/woodstoverule.pdf
[6] B415.1-00 Performance Testing of Solid-Fuel-Burning-Heating Appliances, CSA International, Toronto, 2000
[7] Criteria Air Contaminants Emission Summary, Environment Canada. Online. http://www.ec.gc.ca/pdb/cac/Emissions1990-2015/emissions1990-2015_e.cfm
[8] LGL – Lavalin-Girouard-Letendre, PAH Emissions into the Canadian Environment, 1990. Supporting Document No. 1 for the National Evaluation Report on PAHs. Report prepared by M. Fabbri-Forget for Environment Canada, Conservation and Protection, Quebec Region, Montreal, Quebec. 1993.
[9] Fisher, L.H., J.E. Houck, P.E. Tiegs, & J. McGaughey, Long-Term Performance of EPA-Certified Phase 2 Woodstoves, Klamath Falls and Portland, Oregon: 1998-1999. Report prepared for U.S. EPA, EPA contract 68-D7-001, WA 2-04, Washington, DC., 2000. Online. http://www.omni-test.com/Publications/Long-Term.pdf.

[10] Valenti, J.C. & R.K. Clayton, Project Summary Emissions from Outdoor Wood-Burning Residential Hot Water Furnaces. Report EPA/600/SR-98/017. National Risk Management Research Laboratory, Research and Development, U.S. EPA, Cincinnati, OH., 1998. Online. http://www.epa.gov/ttn/atw/burn/woodburn1.pdf

[11] Environment Canada, Inventory of Releases of PCDDs and PCDFs (updated February 2001). Report prepared for Environment Canada and the Federal/Provincial Working Group on Dioxins and Furans for the Federal-Provincial Advisory Committee for the Canadian Environmental Protection Act (CEPA-FPAC), Ottawa, Ontario 2001

[12] Olsen, K., P. Collas, P. Boileau, D. Blain, C. Ha, L. Henderson, C. Liang, S. McKibbon, & L. Morel-à-l'Huissier, *Canada's Greenhouse Gas Inventory 1990–1999*. Environment Canada, Greenhouse Gas Division, Ottawa, Ontario. 2002.

[13] Environment Australia, *Technical Report No. 4: Review of Literature on Residential Firewood Use, Wood-Smoke and Air Toxics*, 2002. Online. http://ea.gov.au/atmosphere/airtoxics/report4/exec-summary.html.

[14] Houck, J.E. & P.E. Tiegs, Residential Wood Combustion: $PM_{2.5}$ Emissions. Outline of presentation prepared for *WESTAR $PM_{2.5}$ Emission Inventory Workshop*, Reno, Nevada, July 22–23, 1998. Online. http://www.omni-test.com/Publications/westar.pdf

Author Index

Abiy M. 571
Adegoke J. O. 207
Akinyede J. O. 207
Al-Bassam E. 237
Al-Haider S. A. 189
Allegrini I. 521
Al-Madfai H. 49, 439, 483
Al-Salem S. M. 189
Alvarez I. D. 85
Alvarez J. T. 85
Angelino E. 197
Artíñano B. 115

Bacco D. 335
Baldwin S. T. 159
Baranka Gy. 77
Barberá J. 367, 375
Barero-Moreno J. M. 459
Barona A. 409
Barrera F. P. 13
Bartzis J. G. 459
Bassani C. 551
Berezin A. A. 533
Bessagnet B. 115
Bishop G. A. 247
Bonetta Sa. 543
Bonetta Si. 543
Borrego C. 511
Božnar M. Z. 39
Brizio E. 57
Brugarino T. 13
Bydłoń G. 561

Carraro E. 543
Cavalli R. M. 551
Chatterton T. J. 149, 207
Christian R. A. 617
Corona-Zambrano E. 419
Costa M. P. 197
Costabile F. 521
Cristina Martín M. 291
Cuenca A. 475

De Marco A. 125
Demetriou-Georgiou E. 459
Deshpande A. W. 617
Diaz de Argandona J. 409
Doval M. 367, 375

Eguía I. 291
El-Henshir A. K. 397
Elias A. 409
Everard M. 159
Ezcurra A. 409

Fagbeja M. A. 207
Falcon Y. I. 475
Farrugia P. S. 385
Fatsis A. 67
Fava G. 271
Ferreira J. 511
Forbes P. B. C. 345
Fürst C. 571

Geens A. J. 49, 439, 483
Gegisian I. 597
Gennaro M. C. 543
Genon G. 57
Germain A. 627
Gianotti V. 543
González E. 367, 375
González R. M. 3, 225
Gosetti F. 543
Gosselin A. 627
Granger F. 627
Granström K. M. 105
Grašič B. 39
Grey M. 597
Gridin V. V. 533
Gubanov A. F. 179
Guerrero F. 291

Hai B. 585
Hayes E. T. 149, 159, 607
Hellman S. J. 491

Herbarth O. 467
Herrera C. 475
Hillier D. 439
Hort C. 499

Ibarra-Berastegi G. 409
Inoue Y. 429, 585
Irwin J. G. 149, 597

Kaga A. 429, 585
Karim A. 29
Kgabi N. A. 323
Khan A. 237
Khiyami M. A. 447
Kondo A. 429, 585
Kotzias D. 459
Kulmala M. 323
Kuusik R. 311

Lad R. K. 617
Leksmono N. S. 149, 607
Leonardi C. 137
Letizia Ruello M. 271
Lindroos O. 491
Longhurst J. W. S. ....
.... 149, 159, 207, 597, 607
Lopes M. 511
Lougedo S. T. 85

Makeschin F. 571
Martinez E. 475
Marzal F. J. 367, 375
Massolo L. 467
Mercuriali M. 335
Merefield J. R. 159
Micallef A. 385
Michaelidou S. 459
Miñana A. 367
Miranda A. I. 511
Missia D. 459
Mkhatshwa G. V. 217
Mlakar P. 39
Monforti F. 125
Morant J. L. 3, 225
Müller A. 467
Nedre A. Y. 179

Nieblas-Ortiz E. C. 169
Nolan P. 29

Oddone M. 543
Olowoporoku A. O. 607
Ondarts M. 499

Palomino I. 115
Palukka T. 491
Panoutsopoulou A. 67
Parkhurst G. P. 607
Parra R. 95
Pérez J. L. 3, 225
Perez P. 21
Peroni E. 197
Piazza V. 13
Pienaar J. J. 323
Pietrogrande M. C. 335
Pignatelli T. 125
Pignato L. 13
Pignatti S. 551
Platel V. 499
Plaza J. 115
Popov V. 237
Priha E. 491
Pujadas M. 115

Qubian A. 29
Quintero-Nuñez M. 169

Racalbuto S. 125
Racero J. 291
Rantio T. 491
Rhoderick G. C. 357
Rohwer E. R. 345
Rojas-Caldelas R. 419

Sáenz J. 409
Sala C. 197
San José R. 3, 225
Saraga D. 459
Sawicka-Kapusta K. 561
Schellnhuber H.-J. 301
Scozia D. 543
Screpanti A. 125
Sharif T. A. 397

Sharratt B. S. 281	Trikkel A. 311
Shatilov R. A. 179	Trozzi C. 137
Shrestha K. L. 429, 585	Tuhkanen T. 491
Smit R. 255	
Snelson D. G. 49, 439, 483	Uibu M. 311
Sochard S. 499	
Sowińska A. 561	Vaccaro R. 137
Statharas J. 67	Valente J. 511
Stedman D. H. 247	Velts O. 311
Svirejeva-Hopkins A. 301	Vialetto G. 125
Symons J. K. 149	Vivanco M. G. 115
	Vlachakis N. 67
Tchepel O. 511	
Tinarelli G. 39	Zakrzewska M. 561
Tolis E. 459	Zani B. 521
Tovalin H. 467	Zanini G. 125
Treban M. M. 397	Zavala E. A. 475

WITPRESS

Coast–Valley Air Pollution

Edited by: **G. LATINI** and **G. PASSERINI**, Università Politecnica delle Marche, Italy

This book focuses on air pollution modeling in coast–valley environments. This includes the assessment of airborne pollutant emissions, transport, and depletion and the related chemistry. Modeling coast–valley environments is characterized by a high level of complexity due to the concurrent presence of complex orography and land-sea interface. As a first effect, both sea/land breezes and mountain/valley breezes are present and this book will show how they can be precisely modeled, including their interactions. Moreover, these areas are often densely inhabited and settled by industries and many other activities. To make things more complex, there is, often, a strongly uneven distribution of human activities, vegetation, roads and other emission sources.

Several other issues arise from the special wind regimes, especially in summer scenarios. As an example, breezes can trigger recirculation of pollutants while breeze fronts can generate pollution episodes. This book presents several techniques and case studies to help understand and model such special regimes. Also chemistry of airborne pollutants is very special in coast–valley environments. As an example photochemical smog behavior can be very unique (e.g. Coastal Ozone). This book will explain such peculiarities and will give advice for the modeling of such phenomena.

ISBN: 978-1-84564-098-9   2008
apx 300pp
apx £85.00/US$170.00/€127.50

Air Pollution XV

Edited by: **C.A. BREBBIA**, Wessex Institute of Technology, UK and **C.A. BORREGO**, University of Aveiro, Portugal

This book contains the edited proceedings of the Fifteenth Annual International Conference on the Modelling, Monitoring and Management of Air Pollution. With pollution becoming a serious worldwide problem, it is imperative that risks to human health be eliminated. This series of volumes is aimed at the development of computational and experimental techniques to achieve a better understanding of air pollution problems and seek their solution.

This volume encompasses a wide range of topics such as: Air Pollution Modelling; Air Quality Management; Urban Air Management; Transport Emissions; Emissions Inventory; Comparison of Model and Experimental Results; Monitoring and Laboratory studies; Global and Regional Studies; Aerosols and Particles; Climate Change and Air Pollution; Atmospheric Chemistry; Indoor Pollution; Environmental Health Effects; Remote Sensing.

WIT Transactions on Ecology and the Environment, Vol 101
ISBN: 978-1-84564-067-5   2007   624pp
£195.00/US$355.00/€292.50

WITPress
Ashurst Lodge, Ashurst, Southampton, SO40 7AA, UK.
Tel: 44 (0) 238 029 3223
Fax: 44 (0) 238 029 2853
E-Mail: witpress@witpress.com

WITPRESS

Regional and Local Aspects of Air Quality Management

Edited by: **D.M. ELSOM**, Oxford Brookes University, UK and **J.W.S. LONGHURST**, University of the West of England, UK

The resolution of local and regional air pollution problems requires the development of an appropriate scientific and decision-making framework within which effective air quality management may be undertaken.

Drawn from nine countries around the world – Argentina, Australia, Colombia, India, Iran, Italy, Mexico, the United Kingdom and United States – this collection of case studies describes the development and implementation of selected aspects of local or regional management frameworks and/or measures adopted in the pursuit of achieving and sustaining acceptable air quality.

Partial Contents: Air Quality Management in Australia; A Critical Evaluation of the Local Air Quality Management Framework in Great Britain – Is It a Transferable Process? Monitoring and Modelling Air Quality in Mendoza, Argentina; Management of Motor Vehicle Emissions in the United States; Sectoral Analysis of Air Pollution Control in Delhi; Air Quality Management in the Greater Tehran Metropolitan Area.

Series: Advances in Air Pollution, Vol 12
ISBN: 1-85312-952-6   2004   336pp
£124.00/US$198.00/€186.00

Modelling Urban Vehicle Emissions

M. KHARE, Indian Institute of Technology, India and **P. SHARMA**, Indraprastha University, India

"...a worthwhile source to dip into when establishing an emissions inventory or setting up an Air Quality Management District. ...it probably will (and should) appear on the lists of more specialised libraries."

LOCAL TRANSPORT TODAY

Comprehensive and well-organised, this unique book presents various air quality modelling techniques, previously scattered throughout the literature, together with their applications. It also provides a step-by-step guide to using these models, followed by case studies to illustrate the points discussed.

Partial Contents: Urban Air Quality Management and Modelling; Air Pollution Due to Vehicular Exhaust Emissions – A Review;Development and Application of Vehicular Exhaust Models.

Series: Advances in Transport, Vol 9
ISBN: 1-85312-897-X   2002   232pp
£79.00/US$123.00/€118.50

We are now able to supply you with details of new WIT Press titles via
E-Mail. To subscribe to this free service, or for information on any of our titles, please contact the Marketing Department, WIT Press, Ashurst Lodge, Ashurst, Southampton, SO40 7AA, UK
Tel: +44 (0) 238 029 3223
Fax: +44 (0) 238 029 2853
E-mail: marketing@witpress.com

WITPRESS

Mesoscale Atmospheric Dispersion

Edited by: **Z. BOYBEYI**, *Science Applications International Corporation, Virginia, USA*

This book presents a collection of invited review papers by leading scientists in the field covering this complex phenomenon.

Partial Contents: Basic Aspects of Mesoscale Atmospheric Dispersion; Real Time Modeling and Emergency Response Forecast; Urban Parameterizations for Mesoscale Meteorological Models.

Series: Advances in Air Pollution, Vol 9
ISBN: 1-85312-732-9   2001   448pp
£160.00/US$262.00/€240.00

methods.

Contributions in this volume address current developments in the science and application of air pollution modelling: Air Quality Management; Climate Change; Atmospheric Chemistry; Urban and Rural Air Pollution Issues; Emission Inventories; Transport and Air Pollution; Comparison of Model and Experimental Results; Aerosols and Particles; Monitoring and Laboratory Studies; and Indoor Pollution.

WIT Transactions on Ecology and the Environment, Vol 86
ISBN: 1-84564-165-5   2006   824pp
£260.00/US$470.00/€390.00

Air Pollution XIV

Edited by: **J.W.S. LONGHURST**, *University of the West of England, UK* and **C.A. BREBBIA**, *Wessex Institute of Technology, UK*

This volume contains papers accepted for the Fourteenth International Conference on Air Pollution. Pollution is widespread throughout the world, and elimination of the risks to human health is of the utmost importance. Air Pollution XIV examines the development of experimental as well as computational techniques to achieve a better understanding of air pollution problems and seek their solutions.

This volume contains papers featuring case studies, particularly those discussing the evaluation of proposed emission techniques and strategies, as well as papers of a more theoretical nature dealing with advanced mathematical and computational

WIT eLibrary

Home of the Transactions of the Wessex Institute, the WIT electronic-library provides the international scientific community with immediate and permanent access to individual papers presented at WIT conferences. Visitors to the WIT eLibrary can freely browse and search abstracts of all papers in the collection before progressing to download their full text.

Visit the WIT eLibrary at
http://library.witpress.com

All prices correct at time of going to press but subject to change.
WIT Press books are available through your bookseller or direct from the publisher.

WITPRESS

Air Quality Management

Edited by: **J.W.S. LONGHURST**, University of the West of England, UK, **D.M. ELSOM**, Oxford Brookes University, UK and **H. POWER**, Wessex Institute of Technology, UK

This book evaluates and reviews the development and application of the air quality management process from a European, North American and Australian perspective. The approaches described provide a critical assessment of practice as well as important pointers for the future.

Partial Contents: Air Quality Management in the United Kingdom – Development of the National Air Quality Strategy; Operation of the Air Pollution Warning System in Cracow During Pollution Episodes; Air Pollution Management in Australia – The Example of Newcastle, NSW; Air Quality Management in the Bulkley Valley of Central British Columbia, Canada.

Series: *Advances in Air Pollution, Vol 7*
ISBN: 1-85312-528-8   2000   312pp
£98.00/US$149.00/€147.00

WIT Press is a major publisher of engineering research. The company prides itself on producing books by leading researchers and scientists at the cutting edge of their specialities, thus enabling readers to remain at the forefront of scientific developments. Our list presently includes monographs, edited volumes, books on disk, and software in areas such as: Acoustics, Advanced Computing, Architecture and Structures, Biomedicine, Boundary Elements, Earthquake Engineering, Environmental Engineering, Fluid Mechanics, Fracture Mechanics, Heat Transfer, Marine and Offshore Engineering and Transport Engineering.

Air Pollution XIII

Edited by: **C.A. BREBBIA**, Wessex Institute of Technology, UK

Air Pollution continues to be a major cause of concern all over the world and is a problem that requires urgent attention. The contamination of our atmosphere affects the quality of life and has serious consequences for human health and climatic change. As the energy demands of the world's population continue to increase at an accelerating rate, air pollution increases and the problem is becoming more difficult to solve.

This situation, which is in danger of becoming out of control, has resulted in a widespread public awareness of the degrees of air pollution. This has led to demands to find ways of stopping further deterioration of air quality and to start implementing remedial initiatives.

Air Pollution XIII presents some of the latest developments in this field, bringing together recent results and state-of-the-art contributions from researchers around the world. It contains the papers presented at the Thirteenth International Conference on Modelling, Monitoring and Management of Air Pollution. The aim of the Conference was to develop a better understanding of the problem and new tools for managing air quality.

WIT Transactions on Ecology and the Environment, Vol 82
ISBN: 1-84564-014-4   2005   696pp
£245.00/US$392.00/€367.50